Lecture Notes in Networks and Systems

Volume 449

The series "Lecture Notes in Networks and Systems" publishes the latest developments in Networks and Systems—quickly, informally and with high quality. Original research reported in proceedings and post-proceedings represents the core of LNNS.

Volumes published in LNNS embrace all aspects and subfields of, as well as new challenges in, Networks and Systems.

The series contains proceedings and edited volumes in systems and networks, spanning the areas of Cyber-Physical Systems, Autonomous Systems, Sensor Networks, Control Systems, Energy Systems, Automotive Systems, Biological Systems, Vehicular Networking and Connected Vehicles, Aerospace Systems, Automation, Manufacturing, Smart Grids, Nonlinear Systems, Power Systems, Robotics, Social Systems, Economic Systems and other. Of particular value to both the contributors and the readership are the short publication timeframe and the world-wide distribution and exposure which enable both a wide and rapid dissemination of research output.

The series covers the theory, applications, and perspectives on the state of the art and future developments relevant to systems and networks, decision making, control, complex processes and related areas, as embedded in the fields of interdisciplinary and applied sciences, engineering, computer science, physics, economics, social, and life sciences, as well as the paradigms and methodologies behind them.

Indexed by SCOPUS, INSPEC, WTI Frankfurt eG, zbMATH, SCImago.

All books published in the series are submitted for consideration in Web of Science.

For proposals from Asia please contact Aninda Bose (aninda.bose@springer.com).

More information about this series at https://link.springer.com/bookseries/15179

Leonard Barolli · Farookh Hussain ·
Tomoya Enokido
Editors

Advanced Information Networking and Applications

Proceedings of the 36th International Conference on Advanced Information Networking and Applications (AINA-2022), Volume 1

Editors
Leonard Barolli
Department of Information
and Communication Engineering
Fukuoka Institute of Technology
Fukuoka, Japan

Farookh Hussain
University of Technology Sydney
Sydney, NSW, Australia

Tomoya Enokido
Faculty of Bussiness Administration
Rissho University
Tokyo, Japan

ISSN 2367-3370 ISSN 2367-3389 (electronic)
Lecture Notes in Networks and Systems
ISBN 978-3-030-99583-6 ISBN 978-3-030-99584-3 (eBook)
https://doi.org/10.1007/978-3-030-99584-3

This Springer imprint is published by the registered company Springer Nature Switzerland AG
The registered company address is: Gewerbestrasse 11, 6330 Cham, Switzerland

Welcome Message from AINA-2022 Organizers

Welcome to the 36th International Conference on Advanced Information Networking and Applications (AINA-2022). On behalf of AINA-2022 Organizing Committee, we would like to express to all participants our cordial welcome and high respect.

AINA is an international forum, where scientists and researchers from academia and industry working in various scientific and technical areas of networking and distributed computing systems can demonstrate new ideas and solutions in distributed computing systems. AINA was born in Asia, but it is now an international conference with high quality thanks to the great help and cooperation of many international friendly volunteers. AINA is a very open society and is always welcoming international volunteers from any country and any area in the world.

AINA International Conference is a forum for sharing ideas and research work in the emerging areas of information networking and their applications. The area of advanced networking has grown very rapidly, and the applications have experienced an explosive growth especially in the area of pervasive and mobile applications, wireless sensor networks, wireless ad-hoc networks, vehicular networks, multimedia computing and social networking, semantic collaborative systems, as well as grid, P2P, IoT, big data, and cloud computing. This advanced networking revolution is transforming the way people live, work, and interact with each other and is impacting the way business, education, entertainment, and health care are operating. The papers included in the proceedings cover theory, design, and application of computer networks, distributed computing, and information systems.

Each year AINA receives a lot of paper submissions from all around the world. It has maintained high-quality accepted papers and is aspiring to be one of the main international conferences on the information networking in the world.

We are very proud and honored to have two distinguished keynote talks by Prof. Mario A. R. Dantas, University of Juiz de Fora, Minas Gerais, Brazil, and Prof. Isaac Woungang, Ryerson University, Toronto, Ontario, Canada, who will present their recent work and will give new insights and ideas to the conference participants.

An international conference of this size requires the support and help of many people. A lot of people have helped and worked hard to produce a successful AINA-2022 technical program and conference proceedings. First, we would like to thank all authors for submitting their papers, the session chairs, and distinguished keynote speakers. We are indebted to program track co-chairs, program committee members and reviewers, who carried out the most difficult work of carefully evaluating the submitted papers.

We would like to thank AINA-2022 General Co-chairs, PC Co-chairs, and Workshops Co-chairs for their great efforts to make AINA-2022 a very successful event. We have special thanks to Finance Chair and Web Administrator Co-chairs.

We do hope that you will enjoy the conference proceedings and readings.

Organization

AINA-2022 Organizing Committee

Honorary Chair

Makoto Takizawa	Hosei University, Japan

General Co-chairs

Farookh Hussain	University of Technology Sydney, Australia
Tomoya Enokido	Rissho University, Japan
Isaac Woungang	Ryerson University, Canada

Program Committee Co-chairs

Omar Hussain	University of New South Wales, Australia
Flora Amato	University of Naples "Federico II," Italy
Marek Ogiela	AGH University of Science and Technology, Poland

Workshops Co-chairs

Beniamino Di Martino	University of Campania "Luigi Vanvitelli," Italy
Omid Ameri Sianaki	Victoria University, Australia
Kin Fun Li	University of Victoria, Canada

International Journals Special Issues Co-chairs

Fatos Xhafa	Technical University of Catalonia, Spain
David Taniar	Monash University, Australia

Award Co-chairs

Arjan Durresi	Indiana University Purdue University in Indianapolis (IUPUI), USA
Fang-Yie Leu	Tunghai University, Taiwan

Publicity Co-chairs

Markus Aleksy	ABB AG, Germany
Lidia Ogiela	AGH University of Science and Technology, Poland
Hsing-Chung Chen	Asia University, Taiwan

International Liaison Co-chairs

Nadeem Javaid	COMSATS University Islamabad, Pakistan
Wenny Rahayu	La Trobe University, Australia

Local Arrangement Co-chairs

Rania Alhazmi	University of Technology Sydney, Australia
Huda Alsobhi	University of Technology Sydney, Australia
Ebtesam Almansour	University of Technology Sydney, Australia

Finance Chair

Makoto Ikeda	Fukuoka Institute of Technology, Japan

Web Co-chairs

Phudit Ampririt	Fukuoka Institute of Technology, Japan
Kevin Bylykbashi	Fukuoka Institute of Technology, Japan
Ermioni Qafzezi	Fukuoka Institute of Technology, Japan

Steering Committee Chair

Leonard Barolli	Fukuoka Institute of Technology, Japan

Tracks and Program Committee Members

1. Network Protocols and Applications

Track Co-chairs

Makoto Ikeda	Fukuoka Institute of Technology, Japan
Sanjay Kumar Dhurandher	Netaji Subhas University of Technology, New Delhi, India
Bhed Bahadur Bista	Iwate Prefectural University, Japan

TPC Members

Admir Barolli	Aleksander Moisiu University of Durres, Albania
Elis Kulla	Okayama University of Science, Japan
Keita Matsuo	Fukuoka Institute of Technology, Japan
Shinji Sakamoto	Kanazawa Institute of Technology, Japan
Akio Koyama	Yamagata University, Japan
Evjola Spaho	Polytechnic University of Tirana, Albania
Jiahong Wang	Iwate Prefectural University, Japan
Shigetomo Kimura	University of Tsukuba, Japan
Chotipat Pornavalai	King Mongkut's Institute of Technology Ladkrabang, Thailand
Danda B. Rawat	Howard University, USA
Amita Malik	Deenbandhu Chhotu Ram University of Science and Technology, India
R. K. Pateriya	Maulana Azad National Institute of Technology, India
Vinesh Kumar	University of Delhi, India
Petros Nicopolitidis	Aristotle University of Thessaloniki, Greece
Satya Jyoti Borah	North Eastern Regional Institute of Science and Technology, India

2. Next-Generation Wireless Networks

Track Co-chairs

Christos J. Bouras	University of Patras, Greece
Tales Heimfarth	Universidade Federal de Lavras, Brazil
Leonardo Mostarda	University of Camerino, Italy

TPC Members

Fadi Al-Turjman	Near East University, Nicosia, Cyprus
Alfredo Navarra	University of Perugia, Italy
Purav Shah	Middlesex University London, UK
Enver Ever	Middle East Technical University, Northern Cyprus Campus, Cyprus
Rosario Culmone	University of Camerino, Camerino, Italy
Antonio Alfredo F. Loureiro	Federal University of Minas Gerais, Brazil
Holger Karl	University of Paderborn, Germany
Daniel Ludovico Guidoni	Federal University of São João Del-Rei, Brazil
João Paulo Carvalho Lustosa da Costa	Hamm-Lippstadt University of Applied Sciences, Germany
Jorge Sá Silva	University of Coimbra, Portugal

Apostolos Gkamas	University Ecclesiastical Academy of Vella, Ioannina, Greece
Zoubir Mammeri	University Paul Sabatier, France
Eirini Eleni Tsiropoulou	University of New Mexico, USA
Raouf Hamzaoui	De Montfort University, UK
Miroslav Voznak	University of Ostrava, Czech Republic
Kevin Bylykbashi	Fukuoka Institute of Technology, Japan

3. Multimedia Systems and Applications

Track Co-chairs

Markus Aleksy	ABB Corporate Research Center, Germany
Francesco Orciuoli	University of Salerno, Italy
Tomoyuki Ishida	Fukuoka Institute of Technology, Japan

TPC Members

Tetsuro Ogi	Keio University, Japan
Yasuo Ebara	Osaka Electro-Communication University, Japan
Hideo Miyachi	Tokyo City University, Japan
Kaoru Sugita	Fukuoka Institute of Technology, Japan
Akio Doi	Iwate Prefectural University, Japan
Hadil Abukwaik	ABB Corporate Research Center, Germany
Monique Duengen	Robert Bosch GmbH, Germany
Thomas Preuss	Brandenburg University of Applied Sciences, Germany
Peter M. Rost	NOKIA Bell Labs, Germany
Lukasz Wisniewski	inIT, Germany
Angelo Gaeta	University of Salerno, Italy
Graziano Fuccio	University of Salerno, Italy
Giuseppe Fenza	University of Salerno, Italy
Maria Cristina	University of Salerno, Italy
Alberto Volpe	University of Salerno, Italy

4. Pervasive and Ubiquitous Computing

Track Co-chairs

Chih-Lin Hu	National Central University, Taiwan
Vamsi Paruchuri	University of Central Arkansas, USA
Winston Seah	Victoria University of Wellington, New Zealand

TPC Members

Hong Va Leong	Hong Kong Polytechnic University, Hong Kong
Ling-Jyh Chen	Academia Sinica, Taiwan
Jiun-Yu Tu	Southern Taiwan University of Science and Technology, Taiwan
Jiun-Long Huang	National Chiao Tung University, Taiwan
Thitinan Tantidham	Mahidol University, Thailand
Tanapat Anusas-amornkul	King Mongkut's University of Technology North Bangkok, Thailand
Xin-Mao Huang	Aletheia University, Taiwan
Hui Lin	Tamkang University, Taiwan
Eugen Dedu	Universite de Franche-Comte, France
Peng Huang	Sichuan Agricultural University, China
Wuyungerile Li	Inner Mongolia University, China
Adrian Pekar	Budapest University of Technology and Economics, Hungary
Jyoti Sahni	Victoria University of Technology, New Zealand
Normalia Samian	Universiti Putra Malaysia, Malaysia
Sriram Chellappan	University of South Florida, USA
Yu Sun	University of Central Arkansas, USA
Qiang Duan	Penn State University, USA
Han-Chieh Wei	Dallas Baptist University, USA

5. Web-Based and E-Learning Systems

Track Co-chairs

Santi Caballe	Open University of Catalonia, Spain
Kin Fun Li	University of Victoria, Canada
Nobuo Funabiki	Okayama University, Japan

TPC Members

Jordi Conesa	Open University of Catalonia, Spain
Joan Casas	Open University of Catalonia, Spain
David Gañán	Open University of Catalonia, Spain
Nicola Capuano	University of Basilicata, Italy
Antonio Sarasa	Complutense University of Madrid, Spain
Chih-Peng Fan	National Chung Hsing University, Taiwan
Nobuya Ishihara	Okayama University, Japan
Sho Yamamoto	Kindai University, Japan
Khin Khin Zaw	Yangon Technical University, Myanmar
Kaoru Fujioka	Fukuoka Women's University, Japan
Kosuke Takano	Kanagawa Institute of Technology, Japan
Shengrui Wang	University of Sherbrooke, Canada
Darshika Perera	University of Colorado at Colorado Spring, USA
Carson Leung	University of Manitoba, Canada

6. Distributed and Parallel Computing

Track Co-chairs

Naohiro Hayashibara	Kyoto Sangyo University, Japan
Minoru Uehara	Toyo University, Japan
Tomoya Enokido	Rissho University, Japan

TPC Members

Eric Pardede	La Trobe University, Australia
Lidia Ogiela	AGH University of Science and Technology, Poland
Evjola Spaho	Polytechnic University of Tirana, Albania
Akio Koyama	Yamagata University, Japan
Omar Hussain	University of New South Wales, Australia
Hideharu Amano	Keio University, Japan
Ryuji Shioya	Toyo University, Japan
Ji Zhang	The University of Southern Queensland
Lucian Prodan	Universitatea Politehnica Timisoara, Romania
Ragib Hasan	The University of Alabama at Birmingham, USA
Young-Hoon Park	Sookmyung Women's University, Korea
Dilawaer Duolikun	Cognizant Technology Solutions, Hungary
Shigenari Nakamura	Tokyo Metropolitan Industrial Technology Research Institute, Japan

7. Data Mining, Big Data Analytics and Social Networks

Track Co-chairs

Omid Ameri Sianaki — Victoria University, Australia
Alex Thomo — University of Victoria, Canada
Flora Amato — University of Naples "Frederico II," Italy

TPC Members

Eric Pardede — La Trobe University, Australia
Alireza Amrollahi — Macquarie University, Australia
Javad Rezazadeh — University Technology Sydney, Australia
Farshid Hajati — Victoria University, Australia
Mehregan Mahdavi — Sydney International School of Technology and Commerce, Australia
Ji Zhang — University of Southern Queensland, Australia
Salimur Choudhury — Lakehead University, Canada
Xiaofeng Ding — Huazhong University of Science and Technology, China
Ronaldo dos Santos Mello — Universidade Federal de Santa Catarina, Brazil
Irena Holubova — Charles University, Czech Republic
Lucian Prodan — Universitatea Politehnica Timisoara, Romania
Alex Tomy — La Trobe University, Australia
Dhomas Hatta Fudholi — Universitas Islam Indonesia, Indonesia
Saqib Ali — Sultan Qaboos University, Oman
Ahmad Alqarni — Al Baha University, Saudi Arabia
Alessandra Amato — University of Naples "Frederico II," Italy
Luigi Coppolino — Parthenope University, Italy
Giovanni Cozzolino — University of Naples "Frederico II," Italy
Giovanni Mazzeo — Parthenope University, Italy
Francesco Mercaldo — Italian National Research Council, Italy
Francesco Moscato — University of Salerno, Italy
Vincenzo Moscato — University of Naples "Frederico II," Italy
Francesco Piccialli — University of Naples "Frederico II," Italy

8. Internet of Things and Cyber-Physical Systems

Track Co-chairs

Euripides G. M. Petrakis — Technical University of Crete (TUC), Greece
Tomoki Yoshihisa — Osaka University, Japan
Mario Dantas — Federal University of Juiz de Fora (UFJF), Brazil

TPC Members

Akihiro Fujimoto	Wakayama University, Japan
Akimitsu Kanzaki	Shimane University, Japan
Kawakami Tomoya	University of Fukui, Japan
Lei Shu	University of Lincoln, UK
Naoyuki Morimoto	Mie University, Japan
Yusuke Gotoh	Okayama University, Japan
Vasilis Samolada	Technical University of Crete (TUC), Greece
Konstantinos Tsakos	Technical University of Crete (TUC), Greece
Aimilios Tzavaras	Technical University of Crete (TUC), Greece
Spanakis Manolis	Foundation for Research and Technology Hellas (FORTH), Greece
Katerina Doka	National Technical University of Athens (NTUA), Greece
Giorgos Vasiliadis	Foundation for Research and Technology Hellas (FORTH), Greece
Stefan Covaci	Technische Universität Berlin, Berlin (TUB), Germany
Stelios Sotiriadis	University of London, UK
Stefano Chessa	University of Pisa, Italy
Jean-Francois Méhaut	Université Grenoble Alpes, France
Michael Bauer	University of Western Ontario, Canada

9. Intelligent Computing and Machine Learning

Track Co-chairs

Takahiro Uchiya	Nagoya Institute of Technology, Japan
Omar Hussain	UNSW, Australia
Nadeem Javaid	COMSATS University Islamabad, Pakistan

TPC Members

Morteza Saberi	University of Technology Sydney, Australia
Abderrahmane Leshob	University of Quebec in Montreal, Canada
Adil Hammadi	Curtin University, Australia
Naeem Janjua	Edith Cowan University, Australia
Sazia Parvin	Melbourne Polytechnic, Australia
Kazuto Sasai	Ibaraki University, Japan
Shigeru Fujita	Chiba Institute of Technology, Japan
Yuki Kaeri	Mejiro University, Japan
Zahoor Ali Khan	HCT, UAE
Muhammad Imran	King Saud University, Saudi Arabia

Ashfaq Ahmad	The University of Newcastle, Australia
Syed Hassan Ahmad	JMA Wireless, USA
Safdar Hussain Bouk	Daegu Gyeongbuk Institute of Science and Technology, Korea
Jolanta Mizera-Pietraszko	Military University of Land Forces, Poland

10. Cloud and Services Computing

Track Co-chairs

Asm Kayes	La Trobe University, Australia
Salvatore Venticinque	University of Campania "Luigi Vanvitelli," Italy
Baojiang Cui	Beijing University of Posts and Telecommunications, China

TPC Members

Shahriar Badsha	University of Nevada, USA
Abdur Rahman Bin Shahid	Concord University, USA
Iqbal H. Sarker	Chittagong University of Engineering and Technology, Bangladesh
Jabed Morshed Chowdhury	La Trobe University, Australia
Alex Ng	La Trobe University, Australia
Indika Kumara	Jheronimus Academy of Data Science, Netherlands
Tarique Anwar	Macquarie University and CSIRO's Data61, Australia
Giancarlo Fortino	University of Calabria, Italy
Massimiliano Rak	University of Campania "Luigi Vanvitelli," Italy
Jason J. Jung	Chung-Ang University, Korea
Dimosthenis Kyriazis	University of Piraeus, Greece
Geir Horn	University of Oslo, Norway
Gang Wang	Nankai University, China
Shaozhang Niu	Beijing University of Posts and Telecommunications, China
Jianxin Wang	Beijing Forestry University, China
Jie Cheng	Shandong University, China
Shaoyin Cheng	University of Science And Technology of China, China

11. Security, Privacy and Trust Computing

Track Co-chairs

Hiroaki Kikuchi	Meiji University, Japan
Xu An Wang	Engineering University of PAP, China
Lidia Ogiela	AGH University of Science and Technology, Poland

TPC Members

Takamichi Saito	Meiji University, Japan
Kouichi Sakurai	Kyushu University, Japan
Kazumasa Omote	Univesity of Tsukuba, Japan
Shou-Hsuan Stephen Huang	University of Houston, USA
Masakatsu Nishigaki	Shizuoka University, Japan
Mingwu Zhang	Hubei University of Technology, China
Caiquan Xiong	Hubei University of Technology, China
Wei Ren	China University of Geosciences, China
Peng Li	Nanjing University of Posts and Telecommunications, China
Guangquan Xu	Tianjing University, China
Urszula Ogiela	AGH University of Science and Technology, Poland
Hoon Ko	Chosun University, Korea
Goreti Marreiros	Institute of Engineering of Polytechnic of Porto, Portugal
Chang Choi	Gachon University, Korea
Libor Měsíček	J.E. Purkyně University, Czech Republic

12. Software-Defined Networking and Network Virtualization

Track Co-chairs

Flavio de Oliveira Silva	Federal University of Uberlândia, Brazil
Ashutosh Bhatia	Birla Institute of Technology and Science, Pilani, India
Alaa Allakany	Kyushu University, Japan

TPC Members

Rui Luís Andrade Aguiar	Universidade de Aveiro (UA), Portugal
Ivan Vidal	Universidad Carlos III de Madrid, Spain
Eduardo Coelho Cerqueira	Federal University of Pará (UFPA), Brazil

Christos Tranoris	University of Patras (UoP), Greece
Juliano Araújo Wickboldt	Federal University of Rio Grande do Sul (UFRGS), Brazil
Yaokai Feng	Kyushu University, Japan
Chengming Li	Chinese Academy of Science (CAS), China
Othman Othman	An-Najah National University (ANNU), Palestine
Nor-masri Bin-sahri	University Technology of MARA, Malaysia
Sanouphab Phomkeona	National University of Laos, Laos
Haribabu K.	BITS Pilani, India
Shekhavat, Virendra	BITS Pilani, India
Makoto Ikeda	Fukuoka Institute of Technology, Japan
Farookh Hussain	University of Technology Sydney, Australia
Keita Matsuo	Fukuoka Institute of Technology, Japan

AINA-2022 Reviewers

Abderrahmane Leshob
Abdullah Al-khatib
Adil Hammadi
Admir Barolli
Adrian Pekar
Ahmad Alqarni
Aimilios Tzavaras
Akihiro Fujihara
Akihiro Fujimoto
Akimitsu Kanzaki
Akio Doi
Akira Sakuraba
Alaa Allakany
Alex Ng
Alex Thomo
Alfredo Cuzzocrea
Alfredo Navarra
Amita Malik
Angelo Gaeta
Anne Kayem
Antonio Esposito
Antonio Loureiro
Apostolos Gkamas
Arcangelo Castiglione
Arjan Durresi
Ashutosh Bhatia
Asm Kayes
Baojiang Cui
Beniamino Di Martino
Bhed Bista
Caiquan Xiong
Carson Leung
Chang Choi
Christos Bouras
Christos Tranoris
Danda Rawat
David Taniar
Dimitris Apostolou
Dimosthenis Kyriazis
Eirini Eleni Tsiropoulou
Elis Kulla
Enver Ever
Eric Pardede
Ernst Gran
Eugen Dedu
Evjola Spaho
Farookh Hussain
Fatos Xhafa
Feilong Tang
Feroz Zahid
Flavio Silva
Flora Amato
Francesco Orciuoli
Francesco Palmieri

Funabiki Nobuo
Gang Wang
Goreti Marreiros
Guangquan Xu
Hideharu Amano
Hiroaki Kikuchi
Hiroshi Maeda
Hsing-Chung Chen
Indika Kumara
Irena Holubova
Isaac Woungang
Jana Nowaková
Javad Rezazadeh
Ji Zhang
Jianxin Wang
Jolanta Mizera-Pietraszko
Jordi Conesa
Jorge Sá Silva
Kazunori Uchida
Kazuto Sasai
Keita Matsuo
Kevin Bylykbashi
Kin Fun Li
Kiyotaka Fujisaki
Koki Watanabe
Konstantinos Tsakos
Kosuke Takano
Kouichi Sakurai
Leonard Barolli
Leonardo Mostarda
Libor Mesicek
Lidia Ogiela
Lucian Prodan
Luigi Coppolino
Makoto Ikeda
Makoto Takizawa
Marek Ogiela
Mario Dantas
Markus Aleksy
Masakatsu Nishigaki
Masaki Kohana
Mingwu Zhang
Minoru Uehara
Miralda Cuka
Mirang Park
Miroslav Voznak
Nadeem Javaid
Naeem Janjua
Naohiro Hayashibara
Nobuo Funabiki
Norimasa Nakashima
Omar Hussain
Omid Ameri Sianaki
Othman Othman
Øyvind Ytrehus
Paresh Saxena
Pavel Kromer
Philip Moore
Pornavalai Chotipat
Purav Shah
Quentin Jacquemart
Ragib Hasan
Ricardo Rodríguez Jorge
Rosario Culmone
Rui Aguiar
Ryuji Shioya
Safdar Hussain Bouk
Salimur Choudhury
Salvatore Venticinque
Sanjay Dhurandher
Santi Caballé
Satya Borah
Sazia Parvin
Shahriar Badsha
Shigenari Nakamura
Shigeru Fujita
Shigetomo Kimura
Shinji Sakamoto
Somnath Mazumdar
Sriram Chellappan
Stefan Covaci
Stefano Chessa
Takahiro Uchiya
Takamichi Saito
Tarique Anwar
Tetsuro Ogi
Tetsuya Oda
Tetsuya Shigeyasu

Thomas Dreibholz
Tomoki Yoshihisa
Tomoya Enokido
Tomoya Kawakami
Tomoyuki Ishida
Urszula Ogiela
Vamsi Paruchuri
Vinesh Kumar
Wang Xu An
Wei Ren
Wenny Rahayu
Winston Seah Isaac Woungang
Xiaofeng Ding
Yaokai Feng
Yoshitaka Shibata
Yuki Kaeri
Yusuke Gotoh
Zahoor Khan

AINA-2022 Keynote Talks

Data Intensive Scalable Computing in Edge/Fog/Cloud Environments

Mario A. R. Dantas

University of Juiz de Fora, Minas Gerais, Brazil

Abstract. In this talk are presented and discussed some aspects related to the adoption of data intensive scalable computing (DISC) paradigm considering the new adoption trend of edge/fog/cloud environments. These contemporaneous scenarios are very relevant for all organizations in a world where billion of IoT and IIoT devices are being connected, and an unprecedent amount of digital data is generated. Therefore, they require special processing and storage.

Resource Management in 5G Cloudified Infrastructure: Design Issues and Challenges

Isaac Woungang

Ryerson University, Toronto, Canada

Abstract. 5G and Beyond (B5G) networks will be featured by a closer collaboration between mobile network operators (MNOs) and cloud service providers (CSPs) to meet the communication and computational requirements of modern mobile applications and services in a mobile cloud computing (MCC) environment. In this talk, we enlighten the marriage between the heterogeneous wireless networks (HetNets) and the multiple clouds (termed as InterCloud) for a better resource management in B5G networks. First, we start with an overview of the building blocks of HetNet and InterCloud, and then we describe the resource managers in both domains. Second, the key design criteria and challenges related to interoperation between the InterCloud and HetNet are described. Third, the state-of the-art security-aware resource allocation mechanisms for a multi-cloud orchestration over a B5G networks are enlighten.

Contents

An Approach for Mitigating Disruptions on Resources' Consumption Cycles

Zakaria Maamar[1(✉)], Fatma Masmoudi[2], and Ejub Kajan[3]

[1] Zayed University, Dubai, United Arab Emirates
zakaria.maamar@zu.ac.ae
[2] Prince Sattam Bin Abdulaziz University, Alkharj, Kingdom of Saudi Arabia
[3] State University of Novi Pazar, Novi pazar, Serbia

Abstract. This paper examines the impact of disruptions on consumption cycles of resources. Such a cycle consists of states and transitions that depict how a resource is prepared, consumed, locked, unlocked, and withdrawn. It happens that events like last-minute upgrades and urgent fixes arise disrupting the resource's ongoing consumption. Disruption leads to suspending an ongoing consumption to accommodate these events according to 3 scenarios referred to, in this paper, as co-existence, taking turns, and co-existence/taking turns. To verify the correctness of the resources' consumption cycles with respect to each scenario, Petri Nets (PN) are developed linking this verification to properties like liveness and deadlock freeness.

1 Introduction

The democratization of the Internet, through an explosive penetration rate, compounded with an increasing number of on-the-move users are putting a lot of pressure on available computation, storage, and communication *resources* despite multiple technical advances like virtualization and load balancing. This pressure becomes severe when events like urgent upgrades to counter attacks and unexpected demands to execute last-minute requests *disrupt* the ongoing consumption plans of resources.

Simply put, disruption means suspending ongoing operations, initiating operations linked to the disruption, and, finally, resuming the suspended operations with the "hope" of not being subject to penalties by regulatory authorities nor raising concerns among users, for example. How to handle sudden changes with minimal impact on committed resources, and, how to make resources ready for such changes are 2 questions that we addressed in the past from a *consumption* and *transactional* perspectives [8]. On the one hand, the *consumption* perspective uses 3 properties (*limited*, *limited-but-renewable*, and *non-shareable*, [7]) to capture resources' characteristics. For instance, some resources are limited like storage while others are (temporarily) *non-shareable* like bandwidth. On the other hand, the *transactional* perspective uses 3 properties (*pivot*, *retriable*, and *compensatable*, [6]) to ensure that consumption demands of resources would remain compliant with these resources' consumption properties, should disruptions arise.

Although disruptions are sometimes beyond the control of organizations even when they are beneficial like upgrading security measures [11], suspending and resuming

L. Barolli et al. (Eds.): AINA 2022, LNNS 449, pp. 1–12, 2022.
https://doi.org/10.1007/978-3-030-99584-3_1

ongoing operations would most probably lead to "unpleasant" situations. Some business opportunities could be missed, some users' demands could be delayed, and some deadlines could be violated. In this paper, we examine how suspension/resumption could be reduced and ideally avoided by allowing both regular operations and disruptive operations to proceed, should resources allow the co-existence of these operations. Indeed, these latter could concurrently consume resources or take turns to consume resources. Another option is to mix concurrent and take-turn consumption. Each situation would require special arrangements to avoid excessive and/or deadlock consumption of resources, for example. We resort to *Petri Nets* (*PN*, [9]) to verify the correct definitions of the 3 consumption scenarios, co-existence, take-turns, and co-existence/take-turns. *PN*s have been used successfully to model, control, and analyze discrete event dynamic systems that are characterized by: concurrency or parallelism; asynchronous processes; distributed, deterministic and/or stochastic, deadlocks, conflicts, and event driven-processes [2].

We apply our approach for mitigating disruptions on resources' consumption cycles to the *cloud*. In [8], we treated *AWS EC2* as a resource and set it as *limited* to 2 min consumption time-period and *pivot* to tolerate either the success or the failure of its consumption. During the experiments, we let the regular consumption happen for 30 s before it becomes suspended in order to make room for the disruptive consumption. The completion of the disruptive consumption lasts for 50 s allowing the regular consumption to resume within the time limit of 2 min. Would the remaining 120 s have been enough for the regular consumption to complete, then this consumption would succeed. Otherwise, it would fail. The question that we raise in this paper is the following: *should not we accommodate both regular and disruptive consumption instead of suspension and resumption?* 3 scenarios are identified: co-existence, taking turns, and co-existence/taking turns. Due to lack of space, the last scenario is not analyzed. Section 2 presents some related works. Section 3 briefly discusses our previous work on resource consumption. Section 4 examines how disruptions could be mitigated. *PN*-based verification of the mitigation is detailed in Sect. 5. Section 6 concludes the paper and presents future work.

2 Related Work

According to Baker et al., resources abstract some entities, whether physical or logical, that could be discovered, composed, and consumed so, that, certain business goals are achieved [1]. Our current work on resource consumption is at the crossroads of many disciplines that examine, among other things, the impact of disruptions on this consumption. A simple definition of how to manage resources is:

> the process by which businesses manage their various resources effectively. Those resources can be intangible - people and time - and tangible - equipment, materials, and finances. It involves planning so that the right resources are assigned to the right tasks. Managing resources involves schedules and budgets for people, projects, equipment, and supplies[1].

[1] www.shopify.com/encyclopedia/resource-management

Although the work of Cades et al. is not related to *ICT* [3], their findings suggest many insights into the impact of disruptions on performing tasks. Some disruptions are positive when organizations' resources need to be protected from attacks, for example. Contrarily, some are negative when organizations' resources become shared with others, for example. The severity of disruptions would depend on their types as well as the types of tasks that these disruptions will target.

In [4], Fernández el al. present a framework to model and simulate the supply process monitoring for disruptive events detection and prediction. These events produce negative effects and hence, affected schedules should be fixed. The authors developed a Web service that organizations could use to develop discrete event-based simulation models for monitoring processes so these organizations can evaluate their readiness to detect and anticipate disruptive events. By analogy with Fernández el al.'s work, our disruptive events have, depending on their nature, either negative or positive effects on resources and enforcing these effects happen through transactional properties. In addition, our consumption properties nicely fit into Fernández el al.'s framework when resources like trucks would be defined as *non-shareable* and transport licences as *limited-but-renewable*. In [10], Puvvadi et al. propose a solution for handling flow disruptions that pump failures caused in a virtualized environment. The solution decreases the degradation of the requests serving latency during the affected servers migration. The off-the-rack migration of affected virtual machines is ensured by estimating the residual cooling capacity detected in the failed cooling system. The migration also adjusts the CPU clock frequencies through a proposed adaptive algorithm. The implementation proved the efficiency of this approach, however it treats all disruptions in the same way, while each one has its own type. Finally, in [12], Zaman et al. note that although many techniques address resource-constrained project scheduling problems, practical issues like resource unavailability and developing recovery plans are overlooked. The authors suggest a proactive and reactive scheduling technique. The former determines the make-span with focus on maximizing the floating resources that could be used to handle any future disruption effectively. The latter minimizes the revised make-span and recovery cost when disruptions occur. Our scenarios are in line with the reactive technique since disruptions already occurred and thus, need to be accommodated.

3 Resource Consumption in Brief

Figure 1 summarizes the work we carried out in the past with extensive details given in [8]. For the sake of consistency, *planned operations* form a resource's *regular* consumption cycle and *unplanned operations* form a resource's *disruptive* consumption cycle. In this figure, disruption and completion are events that trigger suspending and resuming planned operations, respectively.

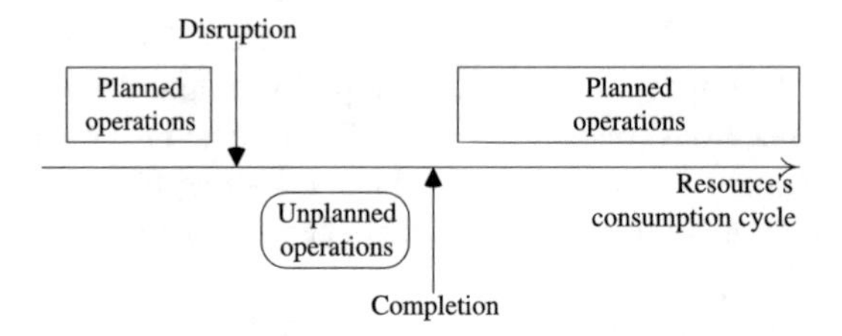

Fig. 1. Impact of disruption on a resource's consumption cycle

3.1 Consumption Properties

With respect to our past work on social coordination of business processes [7], we label resources with consumption properties: *unlimited* (*ul*), *shareable* (*s*), *limited* (*l*), *limited-but-renewable* (*lr*), and *non-shareable* (*ns*). Unless stated a resource is by default *unlimited* and/or *shareable*. Briefly, *limited* means that the consumption of a resource is restricted to a particular capacity and/or time period. *Limited-but-renewable* means that the consumption of a resource continues to happen since the (initial) agreed-upon capacity has been increased and/or the (initial) agreed-upon time period has been extended. And, *non-shareable* means that the concurrent consumption of a resource must be coordinated (e.g., one at a time).

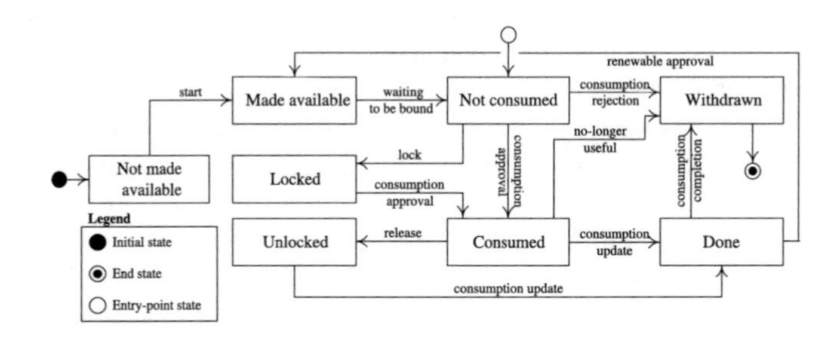

Fig. 2. Resource's consumption cycle as a state diagram (adopted from [7])

Figure 2 is a state diagram of a resource's *consumption cycle* (*cc*) per property. On the one hand, the states (s_i) in this diagram are: *not-made-available* (the resource is neither created nor produced, yet), *made-available* (the resource is either created or produced), *not-consumed* (the resource waits to be bound by a consumer), *locked* (the resource is reserved for a consumer in preparation for its consumption), *unlocked* (the resource is released by a consumer after its consumption), *consumed* (the resource is bound by a consumer due to the consumer's ongoing performance), *withdrawn* (the resource ceases to exit after unbinding all consumers that were consuming this resource), and, finally, *done* (the resource is unbound by a consumer after completing the consumption). In this list of states, *not-made-available* is the initial state, *withdrawn* is the final state, and *not-consumed* is an entry state avoiding to go through the initial state again. On the other hand, the transitions ($trans_j$) connecting a resource's states together include *start, waiting-to-be-bound, consumption-approval, consumption-update, lock, release, consumption-rejection, consumption-completion, renewable-approval*, and *no-longer-useful*. Readers are referred to [8] for the complete list of consumption cycles.

3.2 Disruption Handling

We consider 2 types of disruptive events that could impact the consumption of resources: making room to accommodate unplanned consumption requests and making room to service resources (e.g., upgrade and fixing). We use *l* and *lr* properties to illustrate how these events impact the ongoing consumption of a resource.

On the one hand, an *EC2 instance* is assigned to a consumer for consumption for a limited time period. However, the consumption of this instance is suspended due to another consumer's pressing demand that constitutes the disruptive event. Resuming the suspended consumption cycle would, only, happen outside this time period, which is not possible confirming the suspension of this cycle.

On the other hand, an *Amazon RDS database-instance* is assigned to a consumer for consumption for a limited time period that can be renewed if necessary. However, this consumption is *suddenly* suspended during this time period to increase the storage size after receiving an unexpected huge amount of new data. After the increase that constitutes the disruptive event, the ongoing resource consumption is allowed to complete processing the initial data but the renewal of the resource to process the extra data clashes with other consumers' consumption requests, which results into suspending the consumption cycle. To examine the impact of disruptive (*d*) events (*e*) on the completion of consumption cycles, we revisit Fig. 2 that is the comprehensive state diagram of a resource's consumption cycle. The results of this revision are 1 new state, *suspended*, and 2 new transitions, *suspension* and *conditional resumption* (Fig. 3); in all figures, dashed lines correspond to disruptions. These transitions connect *consumed* and *suspended* states together signaling that an ongoing regular consumption-cycle would be put on-hold and then, **probably** resumed, should the resumption conditions[2] become satisfied with the ultimate goal of reaching the *withdrawn* (final) state.

Definition 1. An event is *disruptive* when the impacts of its occurrence during the consumption cycle of a resource *r* would suspend this resource's regular consumption cycle $r.cc_{cp}$ forcing it to transition from *consumed* state to *suspended* state. In conjunction with this transition, a resource becomes associated with a disruptive consumption cycle $r.cdc_{cp}$ (on top of the regular consumption cycle $r.cc_{cp}$). Resuming $r.cc_{cp}$ is subject to satisfying some conditions; i.e., $suspended \overset{conditional-resumption}{\longrightarrow} consumed$.

[2] Resumption conditions vary according to the resource's type and consumption property. Conditions could be availability for *l* resources and access rights for *ns* resources.

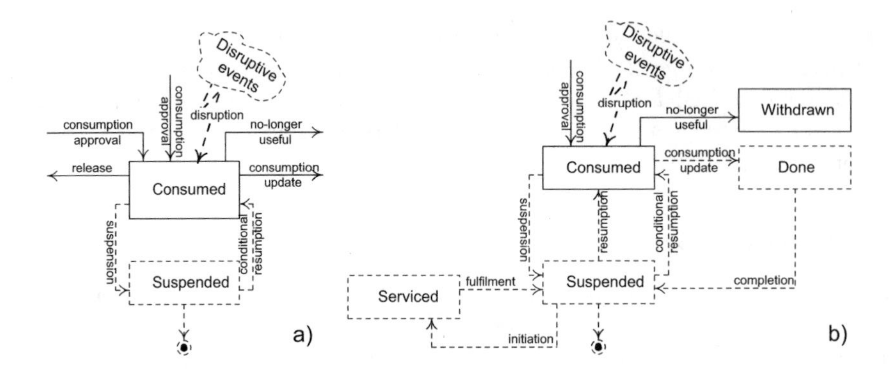

Fig. 3. First/Second revisions of a resource's state diagram to capture disruptions

The states and transitions in a disruptive consumption cycle depend on the event's trigger: make room for extra consumption or make room for service. To capture both triggers, we review again the resource's state diagram included in Fig. 3. As a result, 1 new state, *serviced*, along with 4 new transitions, *resumption*, *initiation*, *fulfilment*, and *completion*, are created to form 2 separate disruptive consumption cycles, one per trigger (Fig. 3 with focus on dashed lines).

When a disruptive event occurs, it causes both the suspension of the regular resource consumption cycle ($r.cc_{cp}.S_{active} = \{...suspended\}$) and the initiation of the disruptive consumption cycle $r.cdc_{cp}$. Once this cycle ends ($r.cdc_{cp}.S_{active} = \{suspended, serviced, suspended\}$ for a disruption of type "service" and $r.cdc_{cp}.S_{active} = \{suspended, consumed, done, suspended\}$ for a disruption of type "extra cycle"), $r.cc_{cp}$ resumes as follows: *conditional resumption* **holds**, $r.cc_{cp}$ completes successfully: $suspended \xrightarrow{conditional-resumption} consumed \xrightarrow{no-longer-useful} withdrawn$; and *conditional resumption* **does not hold**, $r.cc_{cp}$ fails to complete and hence, remains in *suspended*.

4 Scenarios to Mitigate Disruptions

We identify 3 scenarios for mitigating the impact of disruptions on resource consumption cycles: planned/unplanned operations co-existence, planned/unplanned operations taking turns, and planned/unplanned operations co-existence/taking turns. Due to lack of space, the last scenario is not discussed.

Scenario 1: Co-existence. Figure 4 illustrates scenario 1 that consists of allowing (some) planned operations and unplanned operations to co-exist; i.e., both operations concurrently consume resources. After a first (and unique) suspension of the regular consumption cycle $r.cc_{cp}$ to make room for the disruptive consumption cycle $r.cdc_{cp}$, the regular consumption cycle resumes without waiting for the disruptive consumption cycle to complete like shown in Fig. 1. Allowing this co-existence to happen requires that the resource accommodates both cycles, which is most probably dependent on this resource's nature (e.g., hardware *versus* software) and consumption property as per the discussions below.

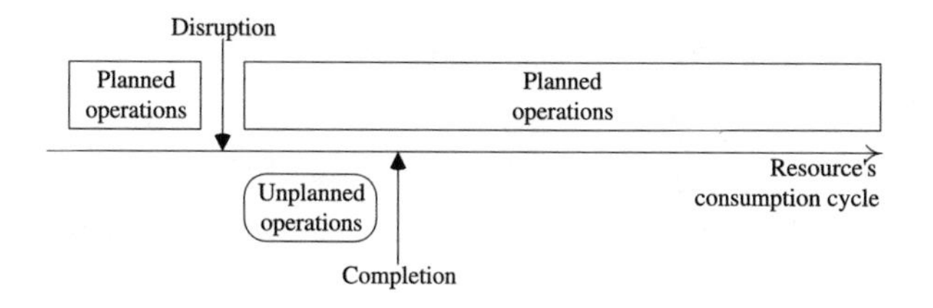

Fig. 4. Scenario 1 where planned and unplanned operations co-exist

Figure 5 captures the impact of scenario 1 on a disrupted resource's state diagram. This impact means adjusting the second revision of Fig. 3 (b) where *conditional resumption* transition from *suspended* to *consumed* states is renamed into *immediate resumption* transition, *completion* transition from *done* to *suspended* states is dropped, *consumed* state becomes common to both cycles (regular/disruptive consumption, plain/dashed lines), *suspended* state is no longer a potential final state, and *service update* transition is added from *suspended* to *done* states.

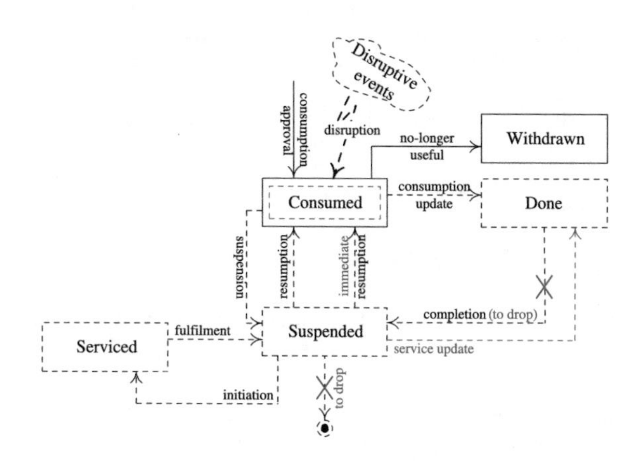

Fig. 5. Adjustment of a resource's state diagram because of scenario 1

As stated above, accommodating both regular consumption cycle and disruptive consumption cycle by a resource would depend on this resource's consumption property. We adopt *l* (*limited*) property for scenario 1, *lr* (*limited-but-renewable*) for scenario 2, and *ns* (*non-shareable*) for scenario 2.

A *l* resource has the following regular and disruptive consumption cycles (Fig. 6):

- $r.cc_{l_1}$: *not-made-available* $\xrightarrow{start}$ *made available* $\xrightarrow{waiting-to-be-bound}$ *not-consumed* $\xrightarrow{consumption\ rejection}$ *withdrawn*.
- $r.cc_{l_2}$: *not-made-available* $\xrightarrow{start}$ *made available* $\xrightarrow{waiting-to-be-bound}$ *not-consumed* $\xrightarrow{consumption-approval}$ *consumed* $\xrightarrow{consumption-update}$ *done* $\xrightarrow{consumption-completion}$ *withdrawn*.

– $r.cc_{l_3}$: $\textit{not-made-available} \xrightarrow{start} \textit{made available} \xrightarrow{waiting-to-be-bound} \textit{not-consumed} \xrightarrow{consumption-approval}$ *consumed*(disruption occurs so either $r.cdc_{l_1}$ or $r.cdc_{l_2}$ is triggered: $\xrightarrow{suspension}$) *consumed*[3] $\xrightarrow{consumption-update} \textit{done} \xrightarrow{consumption-completion} \textit{withdrawn}$. In compliance with Fig. 4, a part of $r.cc_{l_3}$ and either $r.cdc_{l_1}$ or $r.cdc_{l_2}$ concurrently progress.
  - $r.cc_{l_3}/r.cdc_{l_1}$: $\textit{suspended} \xrightarrow{resumption/immediate-resumption\ of\ r.cc_{l_3}} \textit{consumed} \xrightarrow{consumption-update} \textit{done} \xrightarrow{consumption-completion} \textit{withdrawn}$.
  - $r.cc_{l_3}/r.cdc_{l_2}$: $\textit{suspended} \xrightarrow{initiation/immediate-resumption\ of\ r.cc_{l_3}} \textit{serviced} \xrightarrow{fulfilment} \textit{suspended} \xrightarrow{service-update} \textit{done} \xrightarrow{service-completion} \textit{withdrawn}$.

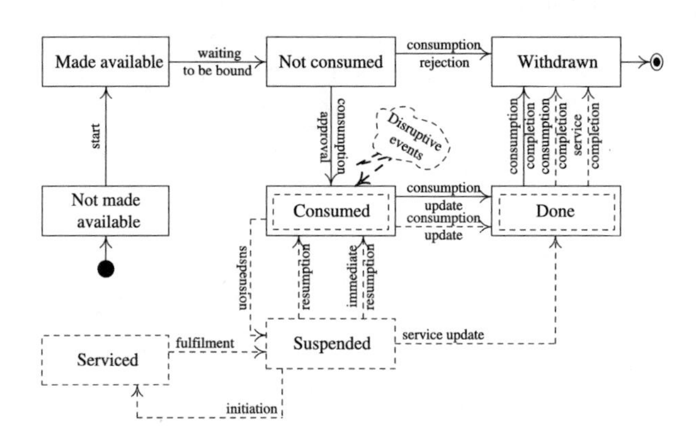

Fig. 6. *Limited* resource's state diagram associated with scenario 1

Scenario 2: Taking Turns. Figure 7 illustrates scenario 2 that consists of allowing planned operations and unplanned operations to take turns during the consumption of resources. After a first suspension (that will become recurrent) of the regular consumption cycle $r.cc_{cp}$ to make room for the disruptive consumption cycle $r.cdc_{cp}$, the planned operations associated with the regular consumption cycle and the unplanned operations associated with the disruptive consumption cycle need to synchronize[4] their resource consumption so, that, each would have specific duration and/or quantity that would be necessary for its completions. Contrarily to scenario 1 where the resource was required to accommodate both cycles, which is dependent on this resource's nature and consumption property, this requirement is dropped in scenario 2. However, we assume in scenario 2 that the resource would be "enough" to accommodate the consumption of both planned operations and unplanned operations.

[3] Thanks to immediate-resumption transition.

[4] Consumption synchronization does not fall into the scope of this paper.

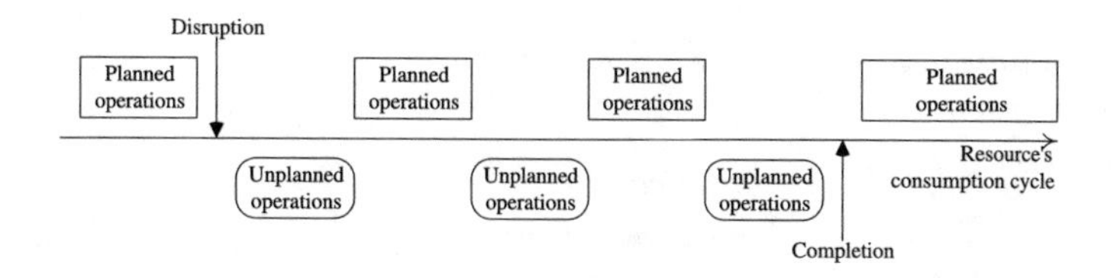

Fig. 7. Scenario 2 where planned and unplanned operations take turns

Figure 8 captures the impact of scenario 2 on a disrupted resource's state diagram. This impact means adjusting the second revision of Fig. 3 (b) where *conditional resumption* transition from *suspended* to *consumed* states is renamed into *R-recurrent resumption* (*R* for Regular) transition, *completion* transition from *done* to *suspended* states is dropped, *consumed* state becomes common to both cycles (regular/disruptive consumption, plain/dashed lines), *resumption* transition is renamed into *D-recurrent resumption* (*D* for disruption) transition, *suspension* transition from *consumed* to *suspended* states is renamed into *recurrent suspension* transition, and *suspended* state is no longer a potential final state. We use *recurrent* to reflect the repetitiveness of consuming resources by planned and unplanned operations.

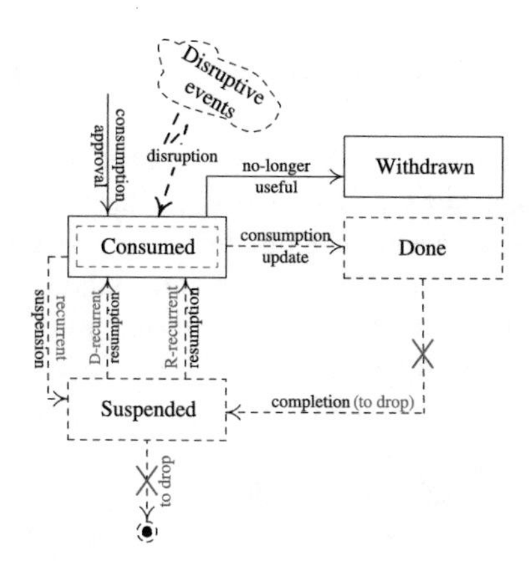

Fig. 8. Adjustment of a resource's state diagram because of scenario 2Adjustment of a resource's state diagram because of scenario 2

By analogy with scenario 1, we illustrate how planned and unplanned operations take turns with respect to a resource's consumption property. This time we consider *lr* property. A *lr* resource has the following regular and disruptive consumption cycles (Fig. 9):

– $r.cc_{lr_1}$: $\textit{not-made-available} \xrightarrow{start} \textit{made available} \xrightarrow{waiting-to-be-bound} \textit{not-consumed} \xrightarrow{consumption-approval} \textit{consumed} \xrightarrow{consumption-update} \textit{done} \xrightarrow{consumption-completion} \textit{withdrawn}.$

- $r.cc_{lr_2}$: *not-made-available* $\overset{start}{\longrightarrow}$ *made available* $\overset{waiting-to-be-bound}{\longrightarrow}$ *not-consumed* $\overset{consumption-approval}{\longrightarrow}$ *consumed* $\overset{consumption-update}{\longrightarrow}$ *done* $\overset{renewable-approval}{\longrightarrow}$ *made available* $\overset{waiting-to-be-bound}{\longrightarrow}$ *not-consumed* $\overset{consumption-approval}{\longrightarrow}$ *consumed* $\overset{consumption-update}{\longrightarrow}$ *done*either $\overset{renewable-approval}{\longrightarrow}$ *made available ...*[5] or $\overset{consumption-completion}{\longrightarrow}$ *withdrawn*[6].
- $r.cc_{lr_3}$: *not-made-available* $\overset{start}{\longrightarrow}$ *made available* $\overset{waiting-to-be-bound}{\longrightarrow}$ *not-consumed* $\overset{consumption-approval}{\longrightarrow}$ *consumed* $\overset{recurrent-suspension}{\longrightarrow}$

  At this stage the disruption[7] occurs, which means either $r.cdc_{lr1}$ or $r.cdc_{lr2}$ will be triggered.

  - $r.cc_{lr_3}/r.cdc_{lr1}$: *suspended* $\overset{D-recurrent-resumption}{\longrightarrow}$ *consumed* $\overset{consumption-update}{\longrightarrow}$ *done* $\overset{consumption-completion}{\longrightarrow}$ *withdrawn*[8].
  - $r.cc_{lr_3}/r.cdc_{lr2}$: $r.cc_{lr_3}/r.cdc_{lr2.1}$ $\overset{consumption-update}{\longrightarrow}$ *done* $\overset{consumption-completion}{\longrightarrow}$ *withdrawn*, where $r.cc_{lr_3}/r.cdc_{lr2.1}$ is *suspended* $\overset{D-recurrent-resumption}{\longrightarrow}$ *consumed* $\overset{recurrent-suspension}{\longrightarrow}$ *suspended* $\overset{R-recurrent-resumption}{\longrightarrow}$ *consumed* $\overset{recurrent-suspension}{\longrightarrow}$. $r.cc_{lr_3}/r.cdc_{lr2.1}$ is a sub-cycle of $r.cc_{lr_3}/r.cdc_{lr2}$ and remains in the loop until the disruption is over (i.e., *consumed* $\overset{consumption-update}{\longrightarrow}$) and the consumption is resumed.

5 PN-based Verification of Consumption Cycles

PN is a graphical modelling language based on 3 concepts: state modeling that consists of an initial marking, places that could hold tokens, and states that show the distribution of tokens [9]. It also relies on event modeling that consists of directed arcs that connect places and execution that is is ensured via transition enabling.

To ensure the correctness of the resource consumption cycles, we adopted *PN*s along with *Real Time Studio (RT-Studio)* [5]. *RT-Studio* is an integrated environment for editing, simulating, and automating the verification of real-time systems modeled as networks of *PN*s.

Let p, d, and r, be 3 global variables that correspond to *time period*, *disruption*, and *resumption*, respectively. We also consider a *time interval* consisting of an *inferior bound* (i) and a *superior bound* (s) in order to approve or reject resource consumption as well as verify whether there is sufficient time after handling disruptions. Then, we proceed with verifying the correctness of the planned and unplanned operations co-existence during resource consumption. We note that *consumption-approval* transition is triggered because there is enough time period to consume the resource ($i < p < s$). In addition, we note that *suspension* transition is triggered since d is set to 1 (i.e., d is a binary digit stating whether there is disruption (1) or not (0)). Afterwards, $r.cdc_{l_1}$

[5] Another resource consumption starts after the renewal.

[6] End of consumption cycle.

[7] 1st suspension that will be followed by other "artificial" suspensions to allow either the planned operations or the unplanned operations to resume resource consumption.

[8] No consumption renewal is requested by the planned operations.

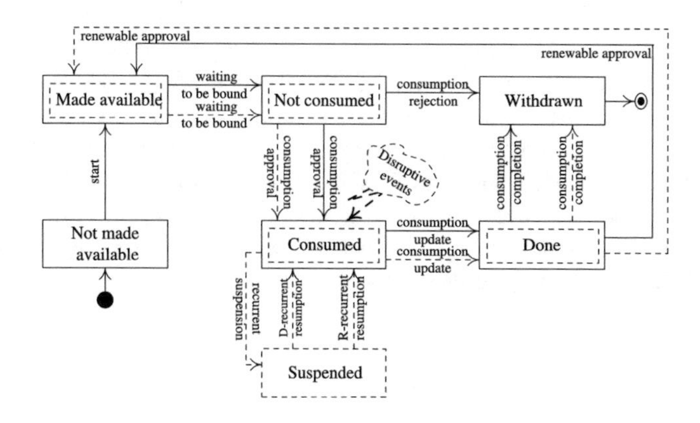

Fig. 9. *Limited-but-renewable* resource's state diagram associated with scenario 2

goes through *consumed*, *done*, and *withdrawn* places. This comes from the fact that r is set to 1 (i.e., r is a binary digit stating whether there is resumption (1) or not (0)). An online demo is available at https://youtu.be/-O3vrHdToqk. We targeted 4 properties that scenario 1's *PN* should satisfy as follows:

Fairness and *liveness* describing how a collection of transitions make progress during execution. Typically, *fairness* is deemed necessary and critical when checking a system's *liveness*. First, *fairness* requires that every enabled marking gets to execute frequently. This property is covered since some states occur over and over such as *consumed* that is executed twice during the normal and disruptive cycles. Second, for any consumption cycle cc whose marking s can be reached from an initial state, then cc is said to be *alive* denoting absence of deadlock.

Reachability assesses if a marking M_n is reachable from the initial marking M_0 via a sequence of transitions that transforms M_0 into M_n. Indeed, each enabled transition triggered the move of a token from a marking to another.

Boundedness states that the number of tokens in a place of the *PN* does not exceed a finite number k for any marking reachable from the initial marking M_0. For instance, we can verify that lr resource's is bounded in a way that each place in a consumption cycle holds a limited number of tokens at any time.

The performed simulation illustrated the co-existence of both a resource's regular consumption cycle and disruptive consumption cycle. The resource's *PN* tracked its consumption by enabling the necessary nodes and firing the necessary transitions. The verification of properties also showed the pure nature of the *PN* in terms of fairness, liveness, reacheability, and boundedness.

6 Conclusion

This paper presented an approach for mitigating the impact of disruptions on resources' consumption cycles and verifying the correctness of these cycles so, that, situations like deadlocks are prevented ahead of time. A consumption cycle consists of states and transitions illustrating how a resource is prepared, consumed, locked/unlocked, withdrawn, and probably disrupted. Because of disruptions that suspend ongoing resource consumption, the cycles were specialized into regular and disruptive shedding light on 3 scenarios that accommodate operations associated with the completion of each cycle. We referred to these scenarios as co-existence, taking turns, and co-existence/taking turns. We resorted to *PN*s to ensure the correctness of these scenarios with focus on the co-existence scenario. In term of future work, we would like to verify the third scenario, adopt other properties such as *shareable*, and finally add concurrency between transitions by using Colored *PN*s.

References

1. Baker, T., Ugljanin, E., Faci, N., Sellami, M., Maamar, Z., Kajan, E.: Everything as a resource: foundations and illustration through Internet-of-Things. Comput. Ind. **94** (2018)
2. Bong Wan, C.: Petri Net approaches for modeling, controlling, and validating flexible manufacturing systems. Ph.D thesis, Iowa State University (1994)
3. Cades, D.M., Werner, N., Trafton, J.G., Boehm-Davis, D.A., Monk, C.A.: Dealing with interruptions can be complex, but does interruption complexity matter: a mental resources approach to quantifying disruptions. Proc. Hum. Fact. Ergon. Soc. **1**, 09 (2008)
4. Fernández, E., Bogado, V., Salomone, E., Chiotti, O.: Framework for modelling and simulating the supply process monitoring to detect and predict disruptive events. Comput. Ind. **80**, 30–42 (2016)
5. Hadjidj, R., Boucheneb, H.: RT-studio: a tool for modular design and analysis of realtime systems using interpreted time Petri Nets. In: Joint Proceedings of PNSE 2013 and ModBE 2013, Milan, Italy (2013)
6. Little, M.: Transactions and web services. Commun. ACM **46**(10), 49–54 (2003)
7. Maamar, Z., Faci, N., Sakr, S., Boukhebouze, M., Barnawi, A.: Network-based social coordination of business processes. Inf. Syst. **58**, 56–74 (2016)
8. Maamar, Z., Sellami, M., Masmoudi, F.: A transactional approach to enforce resource availabilities: application to the cloud. In: Cherfi, S., Perini, A., Nurcan, S. (eds.) RCIS 2021. LNBIP, vol. 415, pp. 249–264. Springer, Cham (2021). https://doi.org/10.1007/978-3-030-75018-3_16
9. Peterson, J.L.: Petri Nets. ACM Comput. Surv. **9**(3), 223–252 (1977)
10. Puvvadi, U., Desu, A., Stachecki, T., Ghose, K., Sammakia, B.: An adaptive approach for dealing with flow disruption in virtualized water-cooled data centers. In: Proceedings of CLOUD 2019, Milan, Italy (2019)
11. Ristov, S., Fahringer, T., Peer, D., Pham, T.P., Gusev, M., Mas-Machuca, C.: Resilient techniques against disruptions of volatile cloud resources. In: Rak, J., Hutchison, D. (eds.) Guide to Disaster-Resilient Communication Networks. Computer Communications and Networks, pp. 379–400. Springer, Cham (2020). https://doi.org/10.1007/978-3-030-44685-7_15
12. Zaman, F., Elsayed, S.M., Sarker, R.A., Essam, D.: Resource constrained project scheduling with dynamic disruption recovery. IEEE Access **8**, 144866–144879 (2020)

Text Detection and Recognition Using Augmented Reality and Deep Learning

Imene Ouali(✉), Mohamed Ben Halima, and Ali Wali

REGIM: REsearch Groups on Intelligent Machines,
National Engineering School of Sfax (ENIS), University of Sfax, Sfax, Tunisia
imene3ouali@gmail.com, {mohamed.benhlima,ali.wali}@isims.usf.tn

Abstract. In recent years, the detection and recognition of text in natural images has become a very attractive and important subject for researchers. Many applications were developed for text detection and recognition and the majority of them are based on deep learning (DL) and augmented reality (AR). In this article, we propose a perfect solution based on both deep learning and augmented reality in order to make the text reading process more efficient, clear and safer. The system purpose is to help visually impaired people read a text from natural images. First of all, the user has to hover his smartphone's camera over the image of the text present in his environment. Then, the system executes the detection and recognition module using the DL model. Finally, the system displays the associated graphical data augmented on the identified text on the screen of the smartphone using the AR method. AR method is used to improve the visualization of the detected and recognized word so that the user can read that text more efficiently. This mobile application has the highest-level visual features to improve the reading process of the detected and recognized text. To validate the system performance, the application is tested on a group of people who answer a questionnaire that reflects their experience with our proposed approach. In addition, user study test is performed to test user friendliness and satisfaction.

Keywords: Text detection · Text recognition · Natural image · Augmented Reality · Deep Learning

1 Introduction

Recent developments in information and communication technologies allow people to obtain interesting information to improve their lives. Technological advances make the smartphone accessible to everyone. The smartphone is supposed to support different types of multimedia, which makes the reading experience more interesting. Hence, several researchers were motivated to develop a mobile system that uses this concept. Smartphones are common in this field with both iOS and Android devices active all over the world. As it was the case with other information and communication technologies, text detection and recognition are suitable for mobile technology and applications. The aim of the design and development of the project is to produce a mobile application to

L. Barolli et al. (Eds.): AINA 2022, LNNS 449, pp. 13–23, 2022.
https://doi.org/10.1007/978-3-030-99584-3_2

help the visually impaired or people with difficulty reading the text, to obtain clearer and improved textual information and to offer more interactivity to users. The objective of our proposed approach is to produce an application to guide visually impaired people or people who have difficulty reading a text to facilitate navigation and understand their environment. The application has a simple and easy to use graphical interface. This Android application uses both deep learning and augmented reality to offer more interactivity to users. When augmented reality technology is applied to it, the application adds some objects (displays non-real, digital processed text) to the real world. First of all, the user needs to download and install the app. The user logs into the application, and hover over the camera on the image. The first step is to automatically open the camera to capture the image of text that the user wants to read it. Subsequently, a detection module is applied to this captured image to detect the text which is in the captured image. then, a text recognition module is applied to this detected text image. These two modules use the DL. Finally, the associated recognized text is displayed on the user's smartphone screen using AR. The user can interact with the viewed content. Each time the user wants to read a word or a sentence, he has to hover over this word again and he will receive the information necessary to understand this text. The information is provided in 3D image form. Thus, the user will have clear and readable text. This application will allow people to better understand the world around them. Several methods based on augmented reality and deep learning are studied in the state-of-the-art section The remainder of this paper is organized as follows: Sect. 2 defines several related works. Section 3, present the proposed approach. Section 4 present the result and the discussion. Finally, Sect. 5 is reserved for conclusion and future work.

2 Related Work

In this section, we present some related works dealing with the problem of detection and recognition of text from natural images. There are several works that attempt to find solutions using deep learning and other methods aim to find solutions using augmented reality.

2.1 Text detection and recognition based on Deep learning method

In this sub-section, a comparative study between the different methods and systems for text detection and recognition from natural images using deep learning are presented in the Table 1.

Table 1. Comparison between Deep learning based text detection and recognition method.

Ref	Year	Méthode	Orientation	Language	Experimental results of benchmarks
[16]	2020	FC2RN	Multi-oriented scene text	Multi-language	MSRA-TD500, ICDAR2017-RCTW, IC-DAR2015, and COCO-Text
[6]	2020	Encoder-decoder structure	Irregular shape texts	Multi-language texts	MSRA-TD500, ICDAR2015, ICDAR2017-MLT and CTW1500
[22]	2019	CNN+ RNN	Irregular shape texts	Multi-language texts	CTW1500, TotalText, ICDAR2013, IC-DAR2015 and MSRATD500
[3]	2019	CRAFT	complicated scene text images	Multi-language	ICDAR2013, ICDAR2015, ICDAR2017, MSRA-TD500, TotalText, CTW-1500
[9]	2019	CNN	Horizantal	Latin	ICDAR 2013, Oxford
[7]	2019	FCN	Irregular shape texts	multilingual	ICDAR 2015 and ICDAR 2017
[5]	2019	Mask R-CNN	Multi-oriented and curved text	Multi-language	ICDAR-2015, ICDAR-2017 MLT, SCUT-CTW1500
[23]	2019	DSRN	Multi-oriented scene text	Multi-language	ICDAR2015 and MSRA-TD500
[24]	2019	VGG16-based network	Irregular scene texts	Multi-language	TotalText and SCUT-CTW1500, ICDAR 2015 and MSRA-TD500
[1]	2018	CNN	Cursive text	Arabic and Urdu	Own database
[15]	2018	FCN	Irregular scene texts	Multi-language	ICDAR 2003, ICDAR 2011 et SVT
[8]	2018	Mask TextSpotter	Irregular shape texts	Latin	ICDAR2013, ICDAR2015 and Total-Text
[21]	2018	CNN+ FCN	Irregular scene texts	Multi-language	ICDAR2011, ICDAR2013 and Street View text

Furthermore, Xugong et al. [16] aim to improve multi-oriented detection while keeping the pipeline simple. A fully convolutional corner refinement network (FC2RN) is provided for precise multi-oriented text detection. An initial corner prediction and a refined corner prediction are obtained in a single pass. With a new quadrilateral RoI convolution operation for multi-oriented scene text, the initial quadrilateral prediction is encoded into feature maps which can be used to predict the shift between the initial prediction and ground truth as well as to produce a refined confidence score. Besides, Xi Liu et al. [6] propose a new approach based on a fully convolutional network with an encoder-decoder structure. In the encoder part, VGG-16 is used as a backbone and the last two layers are converted from fully connected layers to convolutional layers. In the decoder part, the authors propose an accurate segmentation-based detector, equipped with contextual attention and a repulsive text border. In addition, Xiaobing Wang et al. [22] proposes a new method of detecting text regions of arbitrary shapes in two steps: text proposition using Convolutional Neural Network (CNN) and proposition refinement using Recurrent Neural Network (RNN). For text proposal, a text region proposal network (Text-RPN) is used to generate text proposals from an input image. Meanwhile, the CNN feature maps of the input image are obtained here, which can be used in the following. Then, the text proposals are checked and refined via a refinement network. Moreover, Youngmin et al. [3] proposed the CRAFT model as a novel text detector.

Such model have the ability to detect individual characters even with lack of information about character level annotations. The authors offer a new fully convolutional network architecture based on VGG-16. The proposed method provides the character region score and the affinity score effectively detect the text box. Besides, Kulsoom et al. [9] proposed a new technique of text detection and a text character classification based on CNN model. In addition, Antonio et al. [7] proposed an efficient fully convolutional neural network named OctShuffleMLT, with fewer layers and parameters, which can accurately detect multilingual scene text. Furthermore, Zhida et al. [5] present a new Mask R-CNN based text detection approach. Such model can detect multi-oriented and curved text from natural scene images. Furthermore, the authors propose the Pyramid Attention Network (PAN) as a new backbone of Mask R-CNN To improve the Mask R-CNN functionality for text detection tasks. Moreover, Yuxin et al. [23] propose an end-to-end architecture called DSRN (Deep Scale Relationship Network) to map multiscale convolution characteristics to a scale invariant space to achieve uniform activation of multisize text instances. First, they are developing a scale transfer module to transfer multi-scale feature maps to a unified dimension. Due to feature heterogeneity, simply concatenating feature maps with multiscale information would limit detection performance. For that reason, they propose a scale relation module to aggregate multi-scale information by bidirectional convolution operations. Finally, to further reduce instances detected in error, a new callback loss is proposed to force the network to be more concerned with instances of text detected in error by increasing the weight of misclassified examples. Besides, Yongchao et al. [24] introduced a new text detector named TextField to detect jagged scene texts. In addition, Asghar et al. [1] studied the difficult problem of recognizing cursive text in natural scene images. In particular, they focus on recognizing single characters in Urdu in natural scenes that could not be processed by traditional optical character recognition (OCR) techniques developed for documents scanned in Arabic and Urdu. They also present a dataset of Urdu characters segmented from images of signs, street scenes, store scenes, and banner ads that contain Urdu text. A convolutional neural network (CNN) is applied as a classifier to provide high accuracy for natural scene text detection and recognition. A manually segmented character dataset was developed and deep learning based data augmentation techniques were applied to increase the size of the dataset. Furthermore, Siyang et al. [15] proposed a new technique for the precise segmentation of text lines in an image. The algorithm takes a cropped image containing a word. It first performs coarse segmentation using fully convolutional network (FCN). Then, The segmentation is refined using a fully connected Conditional Random Field (CRF) with a new kernel definition that includes line width information. Besides, Pengyuan Lyu et al. [8] present a Mask TextSpotter model based on a simple and smooth end-to-end learning procedure. Such model a semantic segmentation based on deep neural networks trained from end to end in order to perform a high text detection and recognition accuracy. This technique is inspired from Mask R-CNN. It handles instances of irregularly shaped text (ex. curved text). Moreover, Youbao et al. [21] proposed a new method of scene text detection. The proposed method involves superpixel-based stroke function transformation (SSFT) and deep learning-based region classification (DLRC). SSFT is developed for candidate character region extraction (CCR). It partitions an input image into multiple regions via superpixel-based clustering, removing most of the regions based on predefined criteria satisfied by characters, and refines the remaining regions to obtain CCRs by calculating a stroke width map. Character regions are identified from CCRs using DLRC, in which several hand-made low-level features, namely color, texture, and geometric features, and some high-level features based on a Deep convolutional neural networks (CNNs) are first extracted from the regions. Then, these features are merged using two fully connected networks (FCNs) for classification of the regions. In the DLRC step, the deep feature extraction CNN and feature fusion FCNs are formed together. Then, the extracted character regions are merged to form candidate text regions, from which the final scene texts are detected.

2.2 Text Detection and Recognition Based on Augmented Reality Method

In this sub-section, a comparative study of different methods and systems for text detection and recognition from natural images using augmented reality are presented in the Table 2.

Table 2. Comparison between Augmented reality based text detection and recognition method.

Ref	Year	Engine	Beneficiary	Lanague
[12]	2021	Vuforia	Visually impaired people	Arabic language
[4]	2020	Vuforia	Tourists	English language
[11]	2020	Vuforia	Visually impaired people	English language
[13]	2019	ARx2	Tourists	Arabic language, English language
[2]	2018	Vuforia	Tourists	Aceh language, Indonesian language
[18]	2018	Tesseract OCR	Tourists	Arabic language
[20]	2018	Vuforia	Tourists	Banjar language, Indonesian language
[14]	2017	OCR	Tourists	Thai language, Malay language

Furthermore, Imene et al. [12] developed a mobile application based AR for Detection and Recognition text for Visually impaired people. This method detect English texte using smart device features such as a camera. Besides, the collected data will be extracted and recognized through a tool based on Vuforia AR engine. In addition, Prithvi Bhatt et al. [4] developed a mobile application with augmented reality, geared towards the tourism sector. This application gives users the opportunity to discover places through an experience mixing reality with digital content and knowledge of tourism resources. It chooses the content to view and provides viewing options in 2D and 3D modes when the content is filtered appropriately. Moreover, Imene et al. [11] developed a mobile application based on AR in order to detect and recognize Arabic letters with diacritical marks in real-time. This architecture shows great potential for using AR engine to detect Arabic words with diacritics within orientation, writing style and complex background. In addition, it improves the visualization by reading the detected Arabic words with diacritics through a created dataset. This application aims to improve the user's experience and simplicity for partially sighted and visually impaired people.

Besides, Henda et al. [13] developed an AR mobile application for real-time Arabic text translation. The user simply hovers the device's camera over the desired text and it will be translated automatically and quickly. This application consists of three main components which are: text detection, text extraction and text translation. In addition, Zalfie Ardian et al. [2] developed a mobile application called ARgot. ARgot is a real-time translation application from Aceh to Indonesian language. This application allows the detection of text available within the company. This research also aims to see how augmented reality technology can be applied in text without markers and have different fonts. Moreover, Abdul Khader et al. [18] developed a mobile application based on augmented reality. The application targets the community by offering a reliable service in the form of text translation. It extracts the text from the image captured using a mobile camera and translates it into the required target language. It helps people understand banners, signs, etc., and allows people to easily navigate from one place to another. The Tesseract Optical Character Recognition (OCR) library for the Arabic language is used in this work to improve the quality and accuracy of extraction, recognition and translation of Arabic text. In addition, Aulia et al. [20] developed a mobile application for translating from the Banjar language into Indonesian called BandoAR. This application can work in real-time without requiring any input media such as text which needs to be typed using a good quality camera and has auto focus features on the camera. This augmented reality application can be applied in text without markers and has different font sizes, backgrounds and distances. Moreover, Muhammad Pu et al. [14] have developed a mobile AR-based translator named ARThaiMalay to translate the menu of printed Thai food into Malay. This application consists of three main components: translation based on optical character recognition (OCR) technology, dictionary development using SQLite database and viewing database data local.

3 Proposed Approach

In this section, we present a new architecture for text detection and recognition. This architecture helps the visually impaired or people who have difficulty reading a text. This application works on mobile phones. Therefore, a smartphone camera is needed to capture an image of the text and then store and process it. A local database is used to map the text of the captured image to match the text of the image in the database. The job is to identify the textual data and display that text in a 3D image. To obtain this result, it is wrong to follow a process of the approach that we propose (see Fig. 1). This process consists of 2 main parts. The first is text detection and recognition using DL and the second is text visualization using AR. The two modules have been detailed in the following two subsections. And as we know that mobile applications based on augmented reality combine both the real world and the digital world. In other words, we add digital objects to the real world to solve some problems. The use of augmented reality technology for innovation in more and more widespread in people lifecycle. Currently, the portable mobile application model to improve the use of technology has been created using the innovation of augmented reality. Unity 3D software with Vuforia Engine is used for the implementation of augmented reality technology used for the development of the proposed mobile application. This application aims to give interactivity to the user when the item is selected so that more data can be displayed.

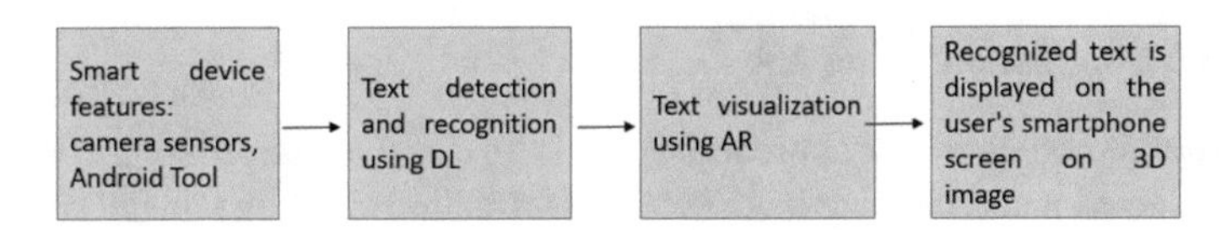

Fig. 1. Proposed architectures.

3.1 Text Detection and Recognition Using DL

For the first module of our approach which is the detection and recognition of text using DL. The second step is that the user hovers over the image of the text they want to read it using the smartphone's camera. We have defined a personalized word list for the database. The second step is the detection of the text. This step focuses the camera on the content of the image so that the text can be easily detected. After text detection, the information is retrieved from the database by matching the words. The third step is text recognition. The words are compared against a predefined word list. Thus, the process identifies whether the target text exists in the database or not. A search is performed, once the target text is identified, the corresponding information is extracted from the database and returned to the system. VGG-16 is used as training architecture in our proposed approach. VGG16 is a convolutional neural network (CNN) architecture. The default input size for the VGG16 model is 224×224 pixels with 3 channels for the RGB image. It has 3×3 filter convolution layers with a stride 1 and a maxpool 2×2 filter layer of stride 2. It follows this arrangement of convolution and maximum pool layers consistently throughout the architecture. Ultimately it has 2 FC (fully connected layers) followed by a softmax for the output. The 16 in VGG16 refers to 16 diapers that have weights. This network is a fairly large network and it has about 138 million parameters (see Fig. 2).

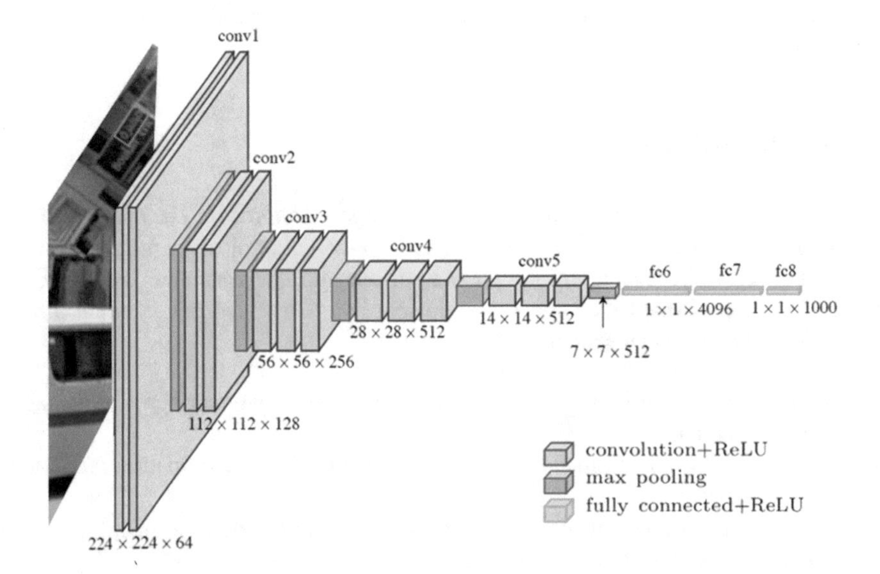

Fig. 2. Architecture VGG-16 [19]

3.2 Text Visualization Using AR

For the second module of our approach which is Text visualization using AR. Once the image is detected and recognized, additional information about the image will be available to the user. The technologies used to develop this module are Unity and Vuforia. Unity is a cross-platform game engine. It was developed and published by Unity Technologies. This engine is used to create three-dimensional and two-dimensional virtual and augmented reality games. Vuforia uses computer vision for image identification and tracking. Using these features allow us to add and create virtual objects to real world objects. The virtual object appears to be part of the real-world scene. In our app, the user will get a clearer and larger text image directly on their smartphone screen. The purpose of using the Vuforia engine is to have a good visualization of the word (Fig. 3). In the example shown in this figure, the poster image of the bakery name will be detected and recognized. Subsequently, the text will be displayed as a 3D image.

(a)

(b)

Fig. 3. Image after using DL for text detection and recognition (a), image after using AR for text visualization (b)

4 Result and Discussion

In this section, we assess the contribution of our proposed approach. The purpose of this evaluation is to test the potential of our DL and AR-based application proposal for the general public. When evaluating the features of our app, the participants commented positively and confirmed their usefulness in real life. We tested our system with 18 participants divided into 4 groups.

- Group 1: This group is made up of 5 visually impaired women participants, aged between 20 and 45.
- Group 2: This group is made up of 6 visually impaired participating men, aged between 22 and 50.
- Group 3: This group is made up of 4 participating women having visual difficulties, aged between 27 and 42.

- Group 4: This group is made up of 3 participating men having visual difficulties, aged between 25 and 40.

These participants are invited to complete a short questionnaire which consists of 4 questions.

- Question 1: Is the application difficult to use and requires training?
- Question 2: Are you satisfied with the results of this application?
- Question 3: Is this app useful?
- Question 4: Is this application expensive and requires a lot of hardware?

Each participant must answer the questionnaires with a score, it gives the note 1 if it spreads the question with yes and gives 0 in the reverse case. We add the notes of each group and we have displayed as a single note. The results are displayed in Table 3.

Table 3. The test results of our proposed method.

	Question 1	Question 2	Question 3	Question 4
Group 1	1	4	5	0
Group 2	0	5	4	0
Group 3	0	5	4	0
Group 4	0	5	5	1

Based on the analysis performed on the test (Table 3), the results can be concluded that our proposed system does not require any training and does not require expensive hardware, just a smartphone. Everyone owns smartphones these days. Any age of the person can use this app. Difficulties and problems with writing such as size, colors, etc. are solved with our proposed system. The visualization module supports the detection phase and the recognition of the text. Our real-time Android based smartphone-based system. A detection and recognition method based on ontology such as [10] or internet of things and big data such as [17] are proposed as a Future work.

5 Conclusion and Future Work

In this paper, an augmented reality mobile application is developed. It gives users the ability to read text through improved and better experience, mixing augmented reality with digital content. Our project, based on the Android platform, once the text is recognized, it provides visualization in 3D modes. The proposed system of interactive virtual assistance with augmented reality is implemented in which a user can pass a camera of the smartphone over a text image and obtain augmented textual information through a 3D image in explanation of the displayed text. The system helps the user learn new concepts with the aid of graphical assistance. Since the app can be implemented on any smart phone, the user can use it as per their convenience. In addition, no additional maintenance is required for this system, making it an economical solution. The interactivity aspect like the presentation of the 3D image allows the user to understand the concept from all angles. The system can be further expanded and used for different age groups, but also to help users view text faster. It offers an interesting way to understand their environment.

References

1. Ali, A., Pickering, M., Shafi, K.: rdu natural scene character recognition using convolutional neural networks. In: 2018 IEEE 2nd International Workshop on Arabic and Derived Script Analysis and Recognition (ASAR), pp. 29–34. IEEE (2018)
2. Ardian, Z., Santoso, P.I., Hantono, B.S.: Argot: text-based detection systems in real time using augmented reality for media translator aceh-indonesia with android-based smartphones. J. Phys. Conf. Ser. **1019**, 012074 (2018)
3. Baek,Y., Lee, B., Han, D., Yun, S., Lee, H.: Character region awareness for text detection. In: Proceedings of the IEEE/CVF Conference on Computer Vision and Pattern Recognition, pp. 9365–9374 (2019)
4. Bhatt, P., Panchal, K., Patel, H., Rote, U.: Tourism application using augmented reality. Available at SSRN 3568709 (2020)
5. Huang, Z., Zhong, Z., Sun, L., Huo, Q.: Mask R-CNN with pyramid attention network for scene text detection. In: 2019 IEEE Winter Conference on Applications of Computer Vision (WACV), pp. 764–772. IEEE (2019)
6. Liu, X., Zhou, G., Zhang, R., Wei, X.: An accurate segmentation-based scene text detector with context attention and repulsive text border. In: Proceedings of the IEEE/CVF Conference on Computer Vision and Pattern Recognition Workshops, pp. 550–551 (2020)
7. Lundgren, A., Castro, D., Lima, E., Bezerra, B.: OctShuffleMLT: a compact octave based neural network for end-to-end multilingual text detection and recognition. In: 2019 International Conference on Document Analysis and Recognition Workshops (ICDARW), vol. 4, pp. 37–42. IEEE (2019)
8. Lyu, P., Liao, M., Yao, C., Wu, W., Bai, X.: Mask textspotter: an end-to-end trainable neural network for spotting text with arbitrary shapes. In: Proceedings of the European Conference on Computer Vision (ECCV), pp. 67–83 (2018)
9. Mansoor, K., Olson, C.F.: Recognizing text with a CNN. In: 2019 International Conference on Image and Vision Computing New Zealand (IVCNZ), pp. 1–6. IEEE (2019)
10. Ouali, I., Ghozzi, F., Taktak, R., Sassi, M.S.H.: Ontology alignment using stable matching. Procedia Comput. Sci. **159**, 746–755 (2019)
11. Ouali, I., Sassi, M.S.H., Halima, M.B., Ali, W.: A new architecture based AR for detection and recognition of objects and text to enhance navigation of visually impaired people. Procedia Comput. Sci. **176**, 602–611 (2020)
12. Ouali, I., Hadj Sassi, M.S., Ben Halima, M., Wali, A.: Architecture for real-time visualizing arabic words with diacritics using augmented reality for visually impaired people. In: Barolli, L., Woungang, I., Enokido, T. (eds.) AINA 2021. LNNS, vol. 225, pp. 285–296. Springer, Cham (2021). https://doi.org/10.1007/978-3-030-75100-5_25
13. Ouertani, H.C., Tatwany, L.: Augmented reality based mobile application for real-time arabic language translation. Commun. Sci. Technol. **4**(1), 30–37 (2019)
14. Pu, M., Majid, N., Idrus, B.: Framework based on mobile augmented reality for translating food menu in Thai language to Malay language. Int. J. Adv. Sci. Engl. Inf. Technol. **7**, 153–159 (2017)

15. Qin, S., Ren, P., Kim, S., Manduchi, R.: Robust and accurate text stroke segmentation. In: 2018 IEEE Winter Conference on Applications of Computer Vision (WACV), pp. 242–250. IEEE (2018)
16. Qin, X., Zhou, Y., Guo, Y., Wu, D., Wang, W.: Fc2rn: a fully convolutional corner refinement network for accurate multi-oriented scene text detection. arXiv preprint arXiv:2007.05113 (2020)
17. Sassi, M.S.H., Jedidi, F.G., Fourati, L.C.: A new architecture for cognitive internet of things and big data. Procedia Comput. Sci. **159**, 534–543 (2019)
18. Saudagar, A.K.J., Mohammad, H.: Augmented reality mobile application for arabic text extraction, recognition and translation. J. Stat. Manag. Syst. **21**(4), 617–629 (2018)
19. Simonyan, K., Zisserman, A.: Very deep convolutional networks for large-scale image recognition. arXiv preprint arXiv:1409.1556 (2014)
20. Syahidi, A.A., Tolle, H., Supianto, A.A., Arai, K.: Bandoar: real-time text based detection system using augmented reality for media translator Banjar language to Indonesian with smartphone. In: 2018 IEEE 5th International Conference on Engineering Technologies and Applied Sciences (ICETAS), pp. 1–6. IEEE (2018)
21. Tang, Y., Wu, X.: Scene text detection using superpixel-based stroke feature transform and deep learning based region classification. IEEE Trans. Multimedia **20**(9), 2276–2288 (2018)
22. Wang, X., Jiang, Y., Luo, Z., Liu, C.-L., Choi, H., Kim, S.: Arbitrary shape scene text detection with adaptive text region representation. In: Proceedings of the IEEE/CVF Conference on Computer Vision and Pattern Recognition, pp. 6449–6458 (2019)
23. Wang, Y., Xie, H., Fu, Z., Zhang, Y.: DSRN: a deep scale relationship network for scene text detection. In: IJCAI, pp. 947–953 (2019)
24. Xu, Y., Wang, Y., Zhou, W., Wang, Y., Yang, Z., Bai, X.: Textfield: learning a deep direction field for irregular scene text detection. IEEE Trans. Image Process. **28**(11), 5566–5579 (2019)

An Energy Consumption Model of Servers to Make Virtual Machines Migrate

Dilawaer Duolikun[1(✉)], Tomoya Enokido[2], Leonard Barolli[3], and Makoto Takizawa[1]

[1] RCCMS, Hosei University, Tokyo, Japan
makoto.takizawa@computer.org
[2] Faculty of Business Administration, Rissho University, Tokyo, Japan
eno@ris.ac.jp
[3] Department of Information and Communications Engineering, Fukuoka Institute of Technology, Fukuoka, Japan
barolli@fit.ac.jp

Abstract. It is critical to reduce the electric energy consumption of information systems to realize green societies. In this paper, we discuss the migration approach to reducing the energy consumption of servers by taking advantage of live migration technologies of virtual machines. We propose a VM (Virtual machine Migration) algorithm to make virtual machines migrate from a host server to a guest server so that the total energy consumption of the servers can be reduced. In the evaluation, we show the total energy consumption of servers in a cluster can be reduced in the VM algorithm compared with other algorithms.

Keywords: Server selection algorithm · Migration of virtual machines · Green computing systems · Estimation algorithm

1 Introduction

Scalable information systems like the IoT (Internet of Things) [33,34] are composed of millions of servers, clients, and devices and accordingly consume huge amount of energy [2,29]. Especially, the electric energy consumption of clusters of servers [3–6,18] have to be reduced to decrease the carbon dioxide emission [29]. In this paper, we discuss the migration approach to reducing the energy consumption of clusters by taking advantage of live migration technologies [1] of virtual machines [1,3–7,29]. Energy-aware algorithms [8–10,14–17,26,28] are proposed to select a virtual machine on a host server to perform an application process. In addition, virtual machines migrate from a host server to a guest server to reduce the energy consumption of the servers in the migration approach [8,10,18–22,27]. Algorithms to make a system more reliable by replicating processes on virtual machines are also proposed [11,12,20].

L. Barolli et al. (Eds.): AINA 2022, LNNS 449, pp. 24–36, 2022.
https://doi.org/10.1007/978-3-030-99584-3_3

Power consumption and computation models of servers [3–6,14,17] and fog nodes [32–34] are proposed. By using the models, the execution time and energy consumption of a server to perform processes are obtained by simulating the computation steps of the processes as discussed in the SM (SiMulation) algorithm [14–17]. However, it is difficult to *a priori* obtain the computation residue of each active process and it takes time to do the simulation. In the SP (SimPle) algorithm [18,19,30,31], only the number of active processes on each server is used to simply estimate the energy consumption. The MI (Monotonically Increasing) [23] and MMI (Modified MI) [24,25] algorithms are also proposed to more precisely estimate the energy consumption of a server.

Suppose $n_t(\geq 0)$ processes $p_{t1}, \ldots, p_{t,n_t}$ are active, i.e. performed on a server s_t at current time. We assume an average value aT of the minimum execution time of each process is *a priori* known, which is the execution time of a process without any other process on a fastest server in a cluster. In order to estimate the execution time of the active processes, we assume the total computation residue RS_t of the processes is $aT{\cdot}n_t/2$. The computation residue RP_{ti} of each active process p_{ti} on a server s_t is modeled to be $c(1 + (n_t - i)a)$ where $a \geq 0$ and c is such a constant that $RS_t = \sum_{i=1}^{n_t} RP_{ti}$ given RS_t and n_t. We propose a model to estimate the total energy to be consumed by servers among which virtual machines migrate. We also propose a VM (Virtual machine Migration) algorithm to find a virtual machine on a host server and a guest server to which the virtual machine migrates to reduce the total energy consumption of the servers. In the evaluation, we show the energy consumption of servers can be reduced in the VM algorithm compared with the non-migration RB (round robin) algorithm.

In Sect. 2, we present the computation and power consumption models of a server. In Sect. 3, we propose the energy estimation algorithm. In Sect. 4, we propose the VM algorithm. In Sect. 5, we evaluate the VM algorithm.

2 Power Consumption and Computation Models of a Server

A cluster S is composed of servers $s_1, \ldots, s_m (m \geq 1)$. Each server s_t is equipped with $np_t(\geq 1)$ homogeneous CPUs. Each CPU supports $nc_t(\geq 1)$ homogeneous cores and each core supports $tn_t(\geq 1)$ homogeneous threads. The server s_t totally supports nc_t $(= nc_t{\cdot}pc_t)$ cores and nt_t $(= nc_t{\cdot}tn_t)$ threads. In this paper, a *process* stands for an application process which uses CPU resources [4]. A process is *active* if and only if (iff) the process is being performed on a thread. Time is modeled to be a discrete sequence of time units [tu]. $SP_t(\tau)$ is a set of active processes on a server s_t at time τ. A server is *active* if at least one process is active, otherwise *idle*.

We consider how much electric energy a server consumes to perform processes at time τ. The power consumption $NE_t(n_t)$ [W] of a server s_t with n_t $(= |SP_t(\tau)|)$ active processes is given in the MLPCM (Multi-Level Power Consumption) model [14,16–18]. An *idle* server s_t just consumes the minimum power $minE_t$ [W], where $n_t = 0$. Each time a CPU, core, and thread are activated, the power consumption NE_t of a server s_t increases by bE_t, cE_t, and tE_t [W], respectively. A server s_t consumes the maximum power $maxE_t$ if every thread is active, i.e. $n_t \geq nt_t$.

$$NE_t(n_t) = \begin{cases} minE_t & \text{if } n_t = 0. \\ minE_t + n_t{\cdot}(bE_t + cE_t + tE_t) & \text{if } 1 \leq n_t \leq np_t. \\ minE_t + np_t{\cdot}bE_t + n_t{\cdot}(cE_t + tE_t) & \text{if } np_t < n_t \leq nc_t. \\ minE_t + np_t{\cdot}bE_t + nc_t{\cdot}cE_t + n_t{\cdot}tE_t & \text{if } nc_t < n_t < nt_t. \\ maxE_t(= minE_t + np_t{\cdot}bE_t + nc_t{\cdot}cE_t + nt_t{\cdot}tE_t) & \\ & \text{if } n_t \geq nt_t. \end{cases} \quad (1)$$

The power $E_t(\tau)$ [W] consumed by a server s_t to perform n_t $(= |SP_t(\tau)|)$ processes at time τ is $NE_t(n_t)$. Energy consumed by a server s_t from time st [time unit] to et is $\sum_{\tau=st}^{et} NE_t(|SP_t(\tau)|)$ [W tu].

The execution time of each process depends on how many processes are active on a thread. The *minimum execution time* $minT_{ti}$ [tu] shows the execution time of a process p_i on a server s_t where only the process p_i is active on a thread without any other process. Let $minT_i$ be a minimum one of $minT_{1i}, \ldots, minT_{mi}$ of the servers $s_1, \ldots, s_m$ in the cluster S. The amount of computation of each process p_i is defined to be $minT_i$ [3–6]. $minT_i/minT_{ti} = minT_j/minT_{tj}$ for any pair of processes p_i and p_j. The *thread computation* rate TCR_t of a server s_t is $minT_i/minT_{ti} (\leq 1)$ for any process p_i, which shows the computation speed of a server s_t. If only one process p_i is active on a thread of a server s_t, the process p_i is performed at rate TCR_t. If n_t processes are active on a thread, each of the processes is performed at rate TCR_t/n_t. On a server s_t with n_t active processes, each process is performed at rate $NPR_t(n_t)$ in the MLC (Multi-Level Computation) model [14,16,17] where $NPR_t(n_t) = TCR_t$ for $0 < n_t \leq nt_t$, $nt_t{\cdot}TCR_t/n_t$ for $n_t > nt_t$. The *server computation* rate $NSR_t(n_t)$ $(\leq nt_t{\cdot}TCR_t)$ of a server s_t is $NPR_t(n_t){\cdot}n_t$ for $n_t > 0$.

We present a model of the energy consumption of a server s_t and the execution time of each process. C_t, T_t, and E_t denote a set of active processes, active time and energy consumption of each server s_t, respectively. RP_i and T_i show the computation residue and execution time of each process p_i, respectively. τ shows time. At each time τ, if a process p_i starts on a server s_t, RP_i is $minT_i$. T_i and E_t are incremented by 1 if $|C_t| > 0$ and $NE_t(|C_t|)$, respectively. RP_i of each process p_i in the set C_t is decremented by the process computation rate $NPR_t(n_t)$. Then, if $RP_i \leq 0$, p_i terminates and is removed from the set C_t.

[Computation model of processes on a server s_t]

1. Initially, $E_t = 0$; $C_t = \phi$; $T_t = 0$; $\tau = 1$;
2. **while** ()
   (a) **for** each process p_i which starts on a server s_t at time τ,
   $C_t = C_t \cup \{p_i\}$; $RP_i = minT_i$; $T_i = 0$;
   (b) $n_t = | C_t |$; $E_t = E_t + NE_t(n_t)$; **if** $n_t > 0$, $T_t = T_t + 1$;
   (c) **for** each process p_i in C_t,
   $T_i = T_i + 1$; $RP_i = RP_i - NPR_t(n_t)$; **if** $RP_i \leq 0$, $C_t = C_t - \{p_i\}$;

(d) $\tau = \tau + 1$;

Processes issued by clients are performed on a virtual machine vm_k of a server s_t. Let $VP_k(\tau)$ be a set of active processes on a virtual machine vm_k at time τ. A virtual machine vm_k is *active* iff $|VP_k(\tau)| > 0$, otherwise *idle*. A virtual machine vm_k is *smaller* than vm_h iff $|VP_k(\tau)| < |VP_h(\tau)|$. A virtual machine vm_k on a host server s_h can migrate to a guest server s_g in a live manner [1]. Here, active processes on vm_k do not terminate but are just suspended during the migration.

3 An Estimation Model of Energy Consumption

Suppose n_t processes $p_{t1}, \ldots, p_{t,n_t}$ are active on a server s_t at curent time τ, i.e. $SP_t(\tau) = \{p_{t1}, \ldots, p_{t,n_t}\}$. In this paper, we assume the computation residue RP_{ti} of each active process p_{ti} is $c(1 + a(n_t - i))$ where $a \geq 0$ at time τ, $RP_{t,i-1} = RP_{ti} + a$ and $RP_{t1} > \ldots > RP_{t,n_t}$. The total computation residue RS_t of the server s_t is $RP_{t,n_t} + \ldots + RP_{t1} = c[1 + (1+a) + (1+2a) + \ldots + (1+(n_t-1)a)] = c[2 + a(n_t - 1)]n_t/2$. c is given as a function $RF_t(RS_t, n_t)$ where $RF_t(x, y) = 2x/[(2 + a(y-1))y]$. In this paper, RS_t is assumed to be $aT(n_t/2)$. The residue RS_t and the number n_t of active processes are given for each server s_t. By using RS_t, n_t, and a, c is obtained as $RF_t(RS_t, n_t)$.

We present a model on how processes in $SP_t(\tau)$ are performed on a server s_t. A variable SP initially denotes a set $SP_t(\tau)$ of active processes on a server s_t at current time τ. An active process whose computation residue is the smallest is referred to as *top* in SP. Now, p_{t,n_t} is a top process since the residue RP_{t,n_t} is the smallest. First, every process p_{ti} in SP is performed at the computation rate $NPR_t(n_t)$ and the residue RP_{ti} of each process p_{ti} is decremented by $RP_{t,n_t}(= c = RF_t(RS_t, n_t))(i = 1, \ldots, n_t)$. Then, the top process p_{t,n_t} terminates at time $T_{t1}(RS_t, n_t)(= RF_t(RS_t, n_t)/NPR_t(n_t)) + \tau$. If $n_t \geq nt_t$, $T_{t,n_t}(RS_t, n_t) = RF_t(RS_t, n_t)/(nt_t{\cdot}TCR_t)$, otherwise $RF_t(RS_t, n_t)/TCR_t$. Totally the computation $RF_t(RS_t, n_t){\cdot}n_t$ of the n_t processes $p_{t,n_t}, \ldots, p_{t1}$ is performed for $T_{t,n_t}(RS_t, n_t)$ [tu]. Hence, the server s_t consumes the energy $E_{t,n_t}(RS_t, n_t) = NE_t(n_t){\cdot}T_{t,n_t}(RS_t, n_t)$ [W tu]. Here, $(n_t - 1)$ processes $p_{t,n_t-1}, \ldots, p_{t1}$ are active where p_{t,n_t-1} is the top process. The top process p_{t,n_t-1} has the residue $a{\cdot}RF_t(RS_t, n_t)$ and terminates time $T_{t,n_t-1}(RS_t, n_t)$ $(= a{\cdot}RF_t(RS_t, n_t)/NPR_t(n_t - 1))$ [tu] after the top process p_{t,n_t} terminates. The computation $a{\cdot}RF_t(RS_t, n_t)(n_t - 1)$ of the $(n_t - 1)$ processes is performed. The server s_t consumes the energy $E_{t,n_t-1}(RS_t, n_t) = NE_t(n_t - 1){\cdot}T_{t,n_t-1}(RS_t, n_t)$ [W tu]. Thus, the execution time $T_{ti}(RS_t, n_t)$ is $a{\cdot}RF_t(RS_t, n_t)/NPR_t(i)$ and the energy consumption $E_{ti}(RS_t, n_t)$ is $NE_t(i){\cdot}T_{ti}(RS_t, n_t)$ where p_{ti} is a *top* process, i.e. $(n_t - i)$ processes $p_{t,n_t}, \ldots, p_{t,i+1}$ are terminated while i processes $p_{ti}, \ldots, p_{t1}$ are active $(i = n_t, \ldots, 1)$. In the MLC model, the computation rate $NPR_t(i)$ is $nt_t{\cdot}TCR_t/i$ if $i \geq nt_t$, otherwise TCR_t. We introduce a function $D_t(x, y) = RF_t(x, y)/(nt_t{\cdot}TCR_t)$ for $x = n_t$, $a{\cdot}RF_t(x, y)/(nt_t{\cdot}TCR_t)$ for $x < n_t$. Here, the execution time $T_{ti}(RS_t, n_t)$ is $D_t(RS_t, n_t){\cdot}i$ if $i \geq nt_t$, otherwise $D_t(RS_t, n_t){\cdot}nt_t$. The execution time $T_{t,n}(RS_t, n_t)$ since the process $p_{t,n+1}$

terminates until the top process $p_{t,n}$ terminates is given by the following function $TN_t(RS_t, n_t, n)$:

$$TN_t(x, y, z) = \begin{cases} D_t(x, y) \cdot z & \text{if } x \geq y \text{ and } z \geq y. \\ D_t(x, y) \cdot nt_t & \text{otherwise.} \end{cases} \tag{2}$$

$TN_t(RS_t, n_t, n-1)/TN_t(RS_t, n_t, n) = (n-1)/n$ for $n \geq nt_t$ and $n < n_t$, $a(n-1)/n$ for $n = n_t$. This means the execution time of each top process p_{tn} monotonically decreases as n decreases since the number of processes on each thread decreases. On the other hand, the execution time of each top process p_{tn} is the same, i.e. $TN_t(RS_t, n_t, n-1)/TN_t(RS_t, n_t, n) = 1$ for $n < nt_t$ since at most one process is performed on each thread.

The execution time $ET_t(RS_t, n_t, n)$ is $\sum_{i=n}^{n_t} TN_t(RS_t, n_t, i)$ to terminate $(n_t - n + 1)$ processes $p_{t,n_t}, \ldots, p_{t,n+1}, p_{tn}$ $(n \leq n_t)$ in the set $SP_t(\tau)$. Here, the other $(n-1)$ processes $p_{t,n-1}, \ldots, p_{t1}$ are still active. First, suppose $n_t > nt_t$. The execution time $ET_t(RS_t, n_t, n)$ is $D_t(RS_t, n_t)[n_t + a(n_t - 1) + \ldots + a \cdot n] = D_t(RS_t, n_t)[a(n - n^2 - n_t + n_t^2) + 2n_t]/2$ for $n \geq nt_t$. If $n < nt_t$ and $n_t \geq nt_t$, $ET_t(RS_t, n_t, n)$ is $ET_t(RS_t, n_t, nt_t) + D_t(RS_t, n_t) nt_t (nt_t - n)a = D_t(RS_t, n_t)[(nt_t - nt_t^2 + n_t - n_t^2)a/2 + nt_t(nt_t - 1)] = D_t(RS_t, n_t)(nt_t^2 - nt_t - n_t + n_t^2)a/2$. If $n_t \leq nt_t$, at most one process is performed on each thread. Hence, $ET_t(RS_t, n_t, n)$ is $D_t(RS_t, n_t)(1 + a(n_t - n))nt_t$.

$$ET_t(r, n_t, n) = \begin{cases} D_t(r, n_t)[a(-n^2 + n + n_t^2 - n_t)/2 + n_t] \\ \qquad\qquad \text{if } n \geq n_t \text{ and } n_t > nt_t. \\ D_t(r, n_t)[(a(n_t^2 - n_t + nt_t^2 - nt_t) + 2n_t)/2 + \\ \qquad a(nt_t(nt_t - n))] \quad \text{if } n < n_t \text{ and } n_t > nt_t. \\ D_t(r, n_t)[1 + a(n_t - n)]nt_t \quad \text{if } n_t \leq nt_t. \end{cases} \tag{3}$$

The total execution time $TET_t(RS_t, n_t)$ to perform all the n_t processes in the set $SP_t(\tau)$ is $ET_t(RS_t, n_t, 1) = D_t(RS_t, n_t)[a(n_t^2 - n_t + nt_t^2 - nt_t) + 2n_t]/2$ if $n_t > nt_t$, else $D_t(RS_t, n_t)[1 + a(n_t - 1)]n_t$.

A server s_t consumes the total energy $TES_t(RS_t, n_t)$ to perform n_t processes $p_{t1}, \ldots, p_{t,n_t}$ in the set $SP_t(\tau)$. $TES_t(r, n_t)$ is $\sum_{n=1}^{n_t}(NE_t(n) \cdot TN_t(r, n_t, n))$.

By using the equation $tm = ET_t(RS_t, n_t, n)$ (3), the number n of active processes to be performed at time $\tau + tm$ [tu] is $\lceil EN_t(RS_t, n_t, tm) \rceil$:

$$EN_t(r, n_t, tm) = \begin{cases} (1/2 + \sqrt{1/4 + (n_t^2 + n_t) + (2/a)(n_t - tm/D_t(r, n_t))}) \\ \qquad\qquad \text{if } n_t > nt_t \text{ and } tm \leq ET_t(RS_t, n_t, nt_t). \\ nt_t + (ET_t(RS_t, n_t, nt_t) - tm)/(a \cdot nt_t \cdot D_t(RS_t, n_t)) \\ \qquad\qquad \text{if } n_t > nt_t \text{ and } tm > ET_t(RS_t, n_t, nt_t). \\ (n_t + 1) - tm/(a \cdot nt_t \cdot D_t(r, n_t)) \qquad \text{if } n_t \leq nt_t. \end{cases} \tag{4}$$

We consider the computation residue $TRS_t(RS_t, n_t, tm)$ of the server s_t at time $\tau + tm$ [tu]. First, the number np_1 of processes active at time $\tau + tm$ is $\lceil EN_t(RS_t, n_t, tm) \rceil$. Let t_1 and t_2 be a pair of the execution time

$ET_t(RS_t, n_t, np_1)$ and $ET_t(RS_t, n_t, np_1 + 1)$, respectively. Here, $t_1 > t_2$. The amount A of computation performed by time $\tau + t_2$ is $RF_t(RS_t, n_t)[1 + (1+a) + \ldots + (1 + a(n_t - 1))] = RF_t(RS_t, n_t)[n_t + a(n_t + np_1 - 1)(n_t - np_1)/2]$. From time t_2 to tm, the amount B of computation is $a \cdot RF_t(RS_t, n_t) \cdot np_1 \cdot (tm - t_2)/(t_1 - t_2)$. The computation residue $TRS_t(RS_t, n_t, tm)$ is $RS_t - (A + B)$.

The energy $TEE_t(RS_t, n_t, tm)$ to be consumed by the server s_t for tm [tu] from current time τ is given as follows, where $np_1 = \lceil EN_t(RS_t, n_t, tm) \rceil$:

$$TEE_t(r, n_t, tm) = \begin{cases} \sum_{n=np_1+1}^{n_t} (NE_t(n) \cdot TN_t(r, n_t, n)) + \\ \qquad NE_t(np_1) \cdot TN_t(r, n_t, np_1)(tm - t_2)/(t_1 - t_2). \end{cases} \tag{5}$$

4 Migration of Virtual Machines

4.1 Energy Consumption for Migration of a Virtual Machine

We discuss how to estimate the total energy to be consumed by a pair of a host server s_h and a guest server s_g where a virtual machine vm_k migrates. Suppose n_h and n_g processes are active, whose total computation residues are RS_h and RS_g, on the servers s_h and s_g, respectively. We also suppose nv_k processes of the computation residue RV_k are active on the virtual machine vm_k. RS_h, RS_g, and RV_k are assumed to be $aT \cdot n_h/2$, $aT \cdot n_g/2$, and $aT \cdot nv_k/2$, respectively, where aT is the average minimum execution time of processes.

First, we consider case no virtual machine migrates from the server s_h to s_g. Here, the servers s_h and s_g consume the energy $NE_h = TES_h(RS_h, n_h)$ and $NE_g = TES_g(RS_g, n_g)$ [W tu] for $NT_h = TET_h(RS_h, n_h)$ and $NT_g = TET_g(RS_g, n_g)$ [tu] to perform every current active process, respectively. Even an idle server s_t consumes the power $minE_t$ as presented in the MLPC model (1). For example, even if every process terminates on the server s_h, the server s_h consumes the minimum power $mimT_h$ until the other server s_g gets idle. Let t_h and t_g be a pair of the execution time of the servers s_h and s_g, respectively. In order to take into consideration the energy consumption of an idle server, we introduce the following function [15,25]:

$$mE_{hg}(t_h, t_g) = \begin{cases} minE_h \cdot (t_g - t_h) & \text{if } t_h \leq t_g. \\ minE_g \cdot (t_h - t_g) & \text{otherwise.} \end{cases} \tag{6}$$

The total energy consumption NEE_{hg} of the servers s_h and s_g from time τ to $\tau + \max(NT_h, NT_g)$ is $NE_h + NE_g + mE_{hg}(NT_h, NT_g)$ where no virtual machine migrates.

Next, we consider another case the virtual machine vm_k starts migrating at time τ and restarts on the guest server s_g at time $\tau + tm$. First, suppose some processes are active on the guest server s_g at time $\tau + tm$. Until the virtual machine vm_k restarts, the n_g processes in the set $SP_g(\tau)$ are performed on the guest server s_g. Hence, the guest server s_g consumes the energy

$NEE_g = TEE_g(RS_g, n_g, tm)$ from time τ to $\tau + tm$. Some processes terminate until time $\tau + tm$ and $nn_g(\leq n_g)$ $(= EN_h(RS_g, n_g, tm))$ processes are still active as discussed in the estimation model. Hence, the total computation residue of the server s_g is reduced to $NRS_g = TRS_g(RS_g, n_g, tm)$. Now, nv_k processes on the virtual machine vm_k restart on the guest server s_g. Totally $(nn_g + nv_k)$ processes are active on the guest server s_g and the total computation residue of the server s_g is $NRS_g + RV_k$. The execution time MTT_g of the $(nn_g + nv_g)$ processes is $TET_g(NRS_g + RV_k, nn_g + nv_k)$ and the energy consumption MEE_g is $TES_g(NRS_g + RV_k, nn_g + nv_k)$. The total execution time MT_g of the guest server s_g is $tm + MTT_g$. The guest server s_g totally consumes the energy $ME_g = NEE_g + MEE_g$ from time τ to $\tau + MT_g$.

Next, suppose no process is active at time $\tau + tm$, i.e. every process in the set $SP_g(\tau)$ terminates at $\tau + NT_g$ before $\tau + tm$, i.e. $tm > NT_g$. Here, only the nv_k processes on the virtual machine vm_k are active. Hence, the total execution time MT_g is $tm + TET_g(RV_k, nv_k)$. The server s_g just consumes the energy $minE_g$ from time NT_g to tm since no process is active. The guest server s_g consumes the energy $ME_g = NE_g + (tm - NT_g) \cdot minE_g + TES_g(RV_k, nv_k)$ from time τ to $\tau + MT_g$.

Next, we consider the host server s_h which the virtual machine vm_k leaves for the guest server s_g. At time τ, n_h processes are active on the host server s_h. Since nv_k processes on the virtual machine vm_k leave the host server s_h, $(n_h - nv_k)$ processes are active on the server s_h at time τ. Hence, the computation residue of the server s_h is $RS_h - RV_k$. The execution time MT_h of the $(n_h - nv_k)$ processes is $TET_h(RS_h - RV_k, n_h - nv_k)$ and the energy consumption ME_h is $TES_h(RS_h - RV_k, n_h - nv_k)$.

The total energy consumption $MEE_{hg:k}$ of the servers s_h and s_g from time τ to $\tau + \max(MT_h, MT_g)$ is $ME_h + ME_g + mE_{hg}(MT_h, MT_g)$ where a virtual machine vm_k migrates from s_h to s_g.

If $NEE_{hg} > MEE_{hg:k}$, the total energy to be consumed by the host and guest servers s_h and s_g can be reduced by the migration of the virtual machine vm_k from s_h to s_g.

4.2 Migration Algorithm

Suppose n_t processes are active on each server s_t whose total computation residue is RS_t in a cluster. If a client issues a process p_i, a server s_h hosting at least one virtual machine is first selected. Since the process p_i is performed in addition to the n_h processes, totally $n_h + 1$ processes are acttive on the server s_h. The total computation residue RS_h is incremented by the residue $minT_i$ of the process p_i, i.e. $RS_h + minT_i$. Hence, the energy to be consumed by the host server s_h is $TEE_h(RS_h + minT_i, n_h + 1)$ as discussed in the preceding section. A server s_h whose $TEE_h(RS_h + minT_i, n_h + 1)$ is the smallest is selected as a host server. Then, a smallest virtual machine vm_k is selected on the host server s_h and the process p_i is performed on vm_k.

In this paper, each server s_h is checked every $mgint$ time units [tu] in the following migration algorithm. If the condition "$NEE_{hg} > MEE_{hg:k}$" is satisfied, a virtual machine vm_k on a host server s_h migrates to a guest server s_g. In order to reduce the number of migrations of virtual machines, a host server whose virtual machine migrates to a guest server is not taken as a guest server of another host server. A guest server to which a virtual machine migrates is also not selected to be a host server. This means each server is selected to be either a host or guest so that virtual machines do not migrate to and from the same server. On the other hand, more than one virtual machine on the host server s_h can migrate to the guest server s_g.

[VM (Virtual machine Migration) algorithm]

1. S = a set of all the servers; $AS(\subseteq S)$ = a set of active servers;
2. **for** every active server s_t in AS,
   n_t = the number of active processes; $RS_t = aT \cdot n_t/2$;
3. **while** $(AS \neq \phi)$ **do**
   (a) Select an active server s_h in AS whose energy $TES_h(RS_h, n_h)$ is the largest; $S = S - \{s_h\}$; /* s_h is not selected as a guest server */
   (b) Select a smallest active virtual machine vm_k on the host server s_h;
   (c) Select a server s_g in S where $NEE_{hg} > MEE_{hg:k}$ and $MEE_{hg:k}$ is the smallest;
   (d) If s_g is found, the virtual machine vm_k migrates from s_h to s_g;
   If s_g in AS, $AS = AS - \{s_g\}$; /* s_g is not selected as a host server */
   Then, if there is no active virtual machine on s_h, $AS = AS - \{s_h\}$;
   (e) If s_g is not found, $AS = AS - \{s_h\}$;

5 Evaluation

We evaluate the VM algorithm in terms of the total energy consumption TE [W tu] and execution time TT [tu] of servers, and the average execution time AT [tu] of processes compared with the non-migration RR (Round Robin) algorithm. Suppose a new process p_i is issued by a client. On each server s_t, n_t processes are active whose total computation residue is $RS_t = aT \cdot n_t/2$. In the RB algorithm, a host server s_h is selected in the round-robin way and no virtual machine migrates. Then, a smallest virtual machine on the host server s_h is selected to perform the process. Thus, processes are uniformly issued to each server, i.e. n_t and n_u are similar for each pair of servers s_t and s_u in the RB algorithm. In the VM algorithm, a host server s_h is first selected where the total energy consumption $TES_h(RS_h + minT_i, n_h + 1)$ is the smallest. Then, a smallest virtual machine on s_h is selected as the RB algorithm. Furthermore, a virtual machine vm_k on a server s_h migrates to a guest server s_g if the migration condition on s_h, s_g, and vm_k is satisfied. Every four time units [tu], the migration condition is checked for every server, i.e. $mgint = 4$. In this paper, the migration time tm is one [tu]. If a virtual machine vm_k migrates to a guest server s_g at time τ, processes on vm_k are suspended and then restarts on s_g at time $\tau + tm$. The constant a is 1.0 in the VM algorithm.

There are eight servers $s_1, \ldots, s_8$ ($m = 8$) whose energy and performance parameters are shown in Table 1. Each server initially hosts twenty virtual machines. Totally n processes $p_1, \ldots, p_n$ are issued from time 1 to $xtime$, where $xtime$ is 1,000 [tu]. For each process p_i, the starting time $stime_i$ is randomly taken from 1 to $xtime$ [tu]. The minimum execution time $minT_i$ of each process p_i is randomly taken from 1 to 20 [tu] so that the average value aT of $minT_1, \ldots, minT_n$ is 10 [tu]. A process configuration $PF(n)$ of n processes $p_1, \ldots, p_n$ is a set $\{\langle i, minT_i, stime_i \rangle \mid i = 1, \ldots, n\}$ of n tuples. Each tuple $\langle i, minT_i, stime_i \rangle$ denotes a process p_i. Given the number n of processes, $n/50$ process configurations are randomly generated. Then, for each process configuration $PF(n)$, the VM and RB algorithms are executed. Then, the total energy consumption TE and total execution time TT of the servers, and the average execution time AT of n processes are obtained. The average values of TE, TT, and AT obtained for $n/2$ process configurations $PF(n)$ are calculated for each number n of processes.

Table 1. Parameters of servers

sid	*TCR*	*nb*	*nc*	*nt*	*minE*[W]	*bE*[W]	*cE*[W]	*tE*[W]	*maxE*[W]
1,5	1	1	8	16	250	25	12	4	435
2,6	0.9	1	8	16	200	20	10	2	332
3,7	0,7	1	4	8	180	15	8	1	235
4,8	0.5	1	2	4	100	10	6	1	126

Figure 1 shows the total energy consumption TE [k W tu] of the servers $s_1, \ldots, s_8$ for number n of processes. As shown in Fig. 1, the total energy consumption TE of the VM algorithm is about 50 to 70 [%] smaller than the RB algorithm. In the RB algorithm, TE linearly increases for the number n of processes. In the VM algorithm, TE is around 2,000 [k W tu] for $n < 6,000$ and linearly increases for $n \geq 6,000$.

Figure 2 shows the total execution time, i.e. active time TT [tu] of the servers for the number n of the processes. TT of the VM algorithm is 20 to 40 [%] smaller than the RB algorithm. TT of the RB algorithm linearly increases as n increases.

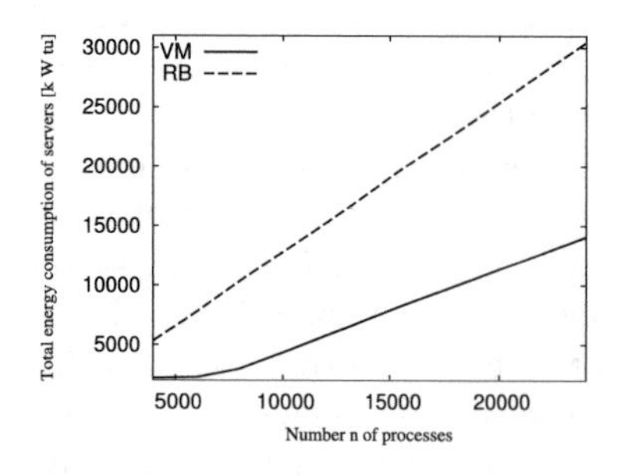

Fig. 1. Total energy consumption TE of servers.

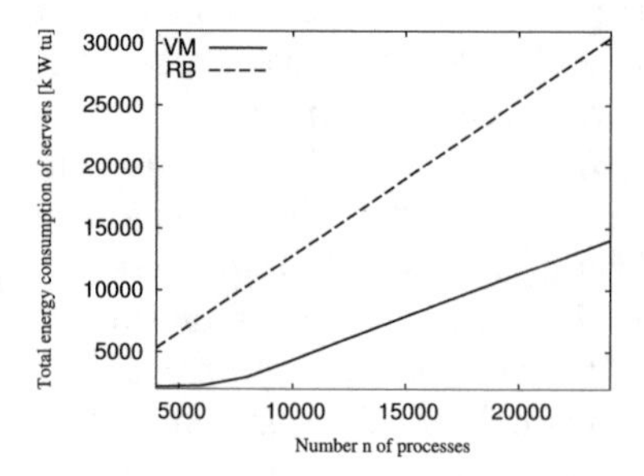

Fig. 2. Total execution time TT of servers.

On the other hand, TT of the VM algorithm is about 8,200 [tu] for $n < 6,000$ and linearly increases for $n \geq 6,000$.

Figure 3 shows the average execution time AT of n processes. AT of the VM algorithm is 20 to 40 [%] smaller than the RB algorithm. AT of the VM algorithm linearly increases as n increases for $n > 6,000$ and linearly increases for $n \geq 6,000$ as shown in Fig. 3.

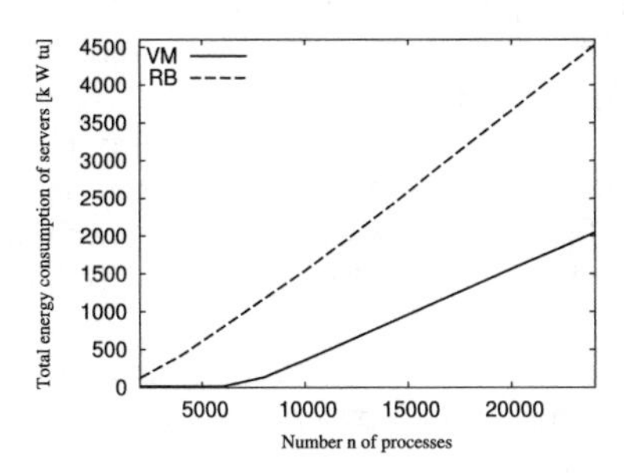

Fig. 3. Average execution time AT of processes.

6 Concluding Remarks

It is critical to reduce the energy consumption of servers to realize green societies. In this paper, we first proposed the algorithm to estimate the energy consumption of a host and guest servers where a virtual machine migrates. Then, we proposed the VM algorithm where virtual machines on host servers migrate to guest servers so that the total energy consumption of the servers can be reduced. In the evaluation, we showed the total energy consumption of servers in the VM algorithm is about 20–40 [%] than the non-migration RB algorithm.

References

1. KVM: Main Page - KVM (Kernel Based Virtual Machine) (2015). http://www.linux-kvm.org/page/Main_Page
2. Dayarathna, M., Wen, Y., Fan, R.: Data center energy consumption modelling: a survey. IEEE Commun. Surv. Tutorials **18**(1), 732–787 (2016)
3. Enokido, T., Aikebaier, A., Takizawa, M.: Process allocation algorithms for saving power consumption in peer-to-peer systems. IEEE Trans. Ind. Electron. **58**(6), 2097–2105 (2011)
4. Enokido, T., Aikebaier, A., Takizawa, M.: A model for reducing power consumption in peer-to-peer systems. IEEE Syst. J. **4**(2), 221–229 (2010)
5. Enokido, T., Aikebaier, A., Takizawa, M.: An extended simple power consumption model for selecting a server to perform computation type processes in digital ecosystems. IEEE Trans. Ind. Inform. **10**(2), 1627–1636 (2014)

6. Enokido, T., Takizawa, M.: Integrated power consumption model for distributed systems. IEEE Trans. Ind. Electron. **60**(2), 824–836 (2013)
7. Enokido, T., Takizawa, M.: An energy-efficient load balancing algorithm to perform computation type application processes for virtual machine. In: Proceedings of the 18th International Conference on Network-Based Information Systems (NBiS-2016), pp. 32–39 (2015)
8. Enokido, T., Takizawa, M.: Power consumption and computation models of virtual machines to perform computation type application processes. In: Proceedings of the 9th International Conference on Complex, Intelligent, and Software Intensive Systems (CISIS-2015), pp. 126–133 (2015)
9. Enokido, T., Duolikun, D., Takizawa, M.: An energy efficient load balancing algorithm based on the active time of cores. In: Proceedings of the 12th International Conference on Broad-Band Wireless Computing, Communication and Applications (BWCCA-2017), pp. 185–196 (2017). https://doi.org/10.1007/978-3-319-69811-3_16
10. Enokido, T., Duolikun, D., Takizawa, M.: The energy consumption laxity-based algorithm to perform computation processes in virtual machine environments. Int. J. Grid Util. Comput. **10**(5), 545–555 (2019)
11. Enokido, T., Duolikun, D., Takizawa, M.: The improved redundant active time-based (IRATB) algorithm for process replication. In: Proceedings of the 35th International Conference on Advanced Information Networking and Applications (AINA-2021), pp. 172–180 (2021). https://doi.org/10.1007/978-3-030-75100-5_16
12. Enokido, T., Duolikun, D., Takizawa, M.: The redundant active time-based algorithm with forcing meaningless replica to terminate. In: Proceedings of the 15th International Conference on Complex, Intelligent, and Software Intensive Systems (CISIS-2021), pp. 216–213 (2021)
13. Enokido, T., Duolikun, D., Takizawa, M.: The improved redundant active time-based algorithm with forcing termination of meaningless replicas in virtual machine environments. In: Proceedings of the 24th International Conference on Network-Based Systems (NBiS-2021), pp. 50–58 (2021)
14. Kataoka, H., Duolikun, D., Enokido, T., Takizawa, M.: Energy-efficient virtualisation of threads in a server cluster. In: Proceedings of the 10th International Conference on Broadband and Wireless Computing, Communication and Applications (BWCCA-2015), pp. 288–295 (2015)
15. Kataoka, H., Duolikun, D., Sawada, A., Enokido, T., Takizawa, M.: Energy-aware server selection algorithms in a scalable cluster. In: Proceedings of IEEE the 30th International Conference on Advanced Information Networking and Applications (AINA-2016), pp. 565–572 (2016)
16. Kataoka, H., Sawada, A., Dilawaer, D., Enokido, T., Takizawa, M.: Multi-level power consumption and computation models and energy-efficient server selection algorithms in a scalable cluster. In: Proceedings of the 19th International Conference on Network-Based Information Systems (NBiS-2016), pp. 210–217 (2016)
17. Kataoka, H., Nakamura, S., Duolikun, D., Enokido, T., Takizawa, M.: Multi-level power consumption model and energy-aware server selection algorithm. Int. J. Grid Util. Comput. **8**(3), 201–210 (2017)
18. Duolikun, D., Enokido, T., Takizawa, M.: Static and dynamic group migration algorithms of virtual machines to reduce energy consumption of a server cluster. In: Transactions on Computational Collective Intelligence, vol. XXXIII, pp. 144–166 (2019)

19. Duolikun, D., Enokido, T., Takizawa, M.: Simple algorithms for selecting an energy-efficient server in a cluster of servers. Int. J. Commun. Netw. Distrib. Syst. **21**(1), 1–25 (2018)
20. Duolikun, D., Enokido, T., Hsu, H.H., Takizawa, M.: Asynchronous migration of process replicas in a cluster. In: Proceedings of IEEE the 29th International Conference on Advanced Information Networking and Applications (AINA-2015), pp. 271–279 (2015)
21. Duolikun, D., Watanabe, R., Enokido, T., Takizawa, M.; An eco migration algorithm of virtual machines in a server cluster. In: Proceedings of IEEE the 32nd International Confernce on Advanced Information Networking and Applications (AINA-2015), pp. 189–196 (2018)
22. Duolikun, D., Enokido, T., Takizawa, M.: Energy-efficient group migration of virtual machines in a cluster. In: Proceedings of the 33rd International Conference on Advanced Information Networking and Applications (AINA-2019), pp. 145–155 (2019). https://doi.org/10.1007/978-3-030-15032-7_12
23. Duolikun, D., Enokido, T., Barolli, L., Takizawa, M.: A monotonically increasing (MI) algorithm to estimate energy consumption and execution time of processes on a server. In: Barolli, L., Chen, H.-C., Enokido, T. (eds.) NBiS 2021. LNNS, vol. 313, pp. 1–12. Springer, Cham (2022). https://doi.org/10.1007/978-3-030-84913-9_1
24. Duolikun, D., Enokido, T., Barolli, L., Takizawa, M.: An energy-efficient algorithm to make virtual machines migrate in a server cluster. In: Proceeding of the 21st International Conference on Broadband and Wireless Computing, Communication and Applications (BWCCA-2021), pp. 25–26 (2021)
25. Duolikun, D., Enokido, T., Barolli, L., Takizawa, M.: An energy-efficient algorithm to make virtual machines migrate in a server cluster, accepted at In: Proceedings of the 10th International Conference on Emerging Internet, Data and Web Technologies (EIDWT-2022) (2022)
26. Inoue, T., Aikebaier, A., Enokido, T., Takizawa, M.: Algorithms for selecting energy-efficient storage servers in storage and computation oriented applications. In: Proceeding of IEEE the 26th International Conference on Advanced Information Networking and Applications (AINA-2016), pp. 920–927 (2016)
27. Noaki, N., Saitto, T., Duolikun, D., Enokido, T., Takizawa, M.: An energy-efficient algorithm for virtual machines to migrate considering migration time. In: Proceedings of the 15th International Conference on Broadband and Wireless Computing, Communication and Applications (BWCCA-2020), pp. 341–354 (2020). https://doi.org/10.1007/978-3-030-61108-8_34
28. Noguchi, K., Saito, T., Duolikun, D., Enokido, T., Takizawa, T.: An algorithm to select a server to minimize the total energy consumption of a cluster. In: Proceedings of the 15th International Conference on P2P, Parallel, Grid, Cloud and Internet Computing (3PGiC-2020), pp. 18–28 (2020). https://doi.org/10.1007/978-3-030-61105-7_3
29. Natural Resources Defense Council (NRDS): Data center efficiency assessment - scaling up energy efficiency across the data center industry: Evaluating key drivers and barriers. http://www.nrdc.org/energy/files/data-center-efficiency-assessment-IP.pdf (2014)
30. Watanabe, R., Duolikun, D., Enokido, T., Takizawa, M.: An eco model of process migration with virtual machines. In: Proceedings of the 19th International Conference on Network-Based Information Systems (NBiS-2016), pp. 292–297 (2016)

31. Watanabe, R., Duolikun, D., Takizawa, M.: Simple estimation and energy-aware migration models of virtual machines in a server cluster. Concurr. Comput. Pract. Exper. **30**(21), e4771 (2018)
32. Oma, R., Nakamura, S., Duolikun, D., Enokido, T., Takizawa, M.: An energy-efficient model for fog computing in the Internet of Things (IoT). Internet Tings **1–2**, 14–26 (2018)
33. Oma, R., Nakamura, S., Enokido, T., Takizawa, M.: A tree-based model of energy-efficient fog computing systems in IoT. In: Proceedings of the 12th International Conference on Complex, Intelligent, and Software Intensive Systems (CISIS-2018), pp. 991–1001 (2018). https://doi.org/10.1007/978-3-319-93659-8_92
34. Oma, R., Nakamura, S., Duolikun, D., Enokido, T., Takizawa, M.: A fault-tolerant tree-based fog computing model. Int. J. Web Grid Serv. **15**(3), 219–239 (2019)

Development of a Blockchain-Based Ad Listing Application

Hamza Salem(✉), Manuel Mazzara, Hadi Saleh, Rami Husami, and Siham Maher Hattab

Innopolis University, Republic of Tatarstan, Russia
{h.salem,h.saleh,m.husami}@innopolis.university, m.mazzara@innopolis.ru

Abstract. In today's time, ad-listing websites work as the main leads generator for businesses. Such websites are managed by a central authority called website admins. Because of the centralization, admins can change and remove some business information and manipulate the reviews and ratings to increase the benefits for other businesses. So, in this paper, we have developed a smart contract for an ad-listing website using Solidity on the Ethereal blockchain. Thus, Implementing a decentralized app architecture that uses Blockchain Technology. We have developed a decentralized web application that works on the Ethereal Ropsten test network and uses Infura platform as a virtual node. Virtual nodes provide the connection between our NodeJS app and the Blockchain, eliminating the need for running an Ethereal full node. By using such architecture, users don't need to sign in with a wallet extension or any Web3 component. From a user's perspective, the website is like any web app using a database. But, we had the advantage of utilizing the Blockchain test network for free as a data warehouse. Also, we calim that this app guarantees a more honest user experience because of the transparent nature of Blockchains.

Keywords: Decentralized application · Smart contract · Blockchain · Ethereal · Solidity

1 Introduction

Currently, ad listing websites work as a centralized client-server system. It takes a lot of time and human labor for approval. Due to centralization, users have to place their trust in that server. Possibly, the server data may get damaged or misused for someone's profit. So, we need to utilize a modern scheme to solve the aforementioned limitations. One candidate that stands out is a smart contract decentralized application scheme [1].

When the ads listing process got first introduced to the real world, it took a middle man to verify that the ads are legit and have the correct information. According to authors [1], for an application to be considered a decentralized application it must meet the following three criteria:

L. Barolli et al. (Eds.): AINA 2022, LNNS 449, pp. 37–45, 2022.
https://doi.org/10.1007/978-3-030-99584-3_4

1. The application must be completely open-source, anyone can access the code. It must operate autonomously without the interference of third parties. All its data and records must be cryptographically stored in a public, decentralized blockchain.
2. The application must mint tokens according to a basic algorithm or set of criteria and it should distribute some or all at the beginning of its operation.
3. The application will be enhanced pursuant to changes and market feedback proposed by the majority consensus of its users.

In this case, an application that is built on the Ethereal network is a decentralized application. It satisfies all the mentioned conditions thanks to the Etherum's design model. We have implemented a smart contract using Solidity as a programming language and Ethereal as an underlying Network. Regarding the interface, we have created our UI using HTML, JavaScript, and Web3.js. Web3.js is a famous collection of JavaScript libraries that facilitate querying data from the Blockchain and writing to it. Further, as Blockchain Apps require wallet browser plugins to interact with the their interface, most of the current implementations lose several users. This happens because users may not have these plugins, not care enough to install plugins, or not prefere to interact with a complex architecture. To solve this problem, we propose a decentralized app architecture using NodeJS to view data from Ethereal directly without any plugins. MetaMask is by far the most famous Ethereal wallet, it is a plugin that provides an interface to connect to Ethereal. But, it can be daunting for some users to use it. For our design to be complete, we need to secure the data any central authority modification. Our solution proposes an open-source smart contract that allows users to interact with Ads and see business information as written by the owner. Thus, Our application does not have a single point of failure like conventional database systems.

2 Terminology and Literature Review

Here we list some important terms mentioned in this paper:

Ethereal: Ethereal is a blockchain platform with its own cryptocurrency (ETH) and programming language (Solidity). As a blockchain network, Ethereal is a decentralized public ledger for verifying and recording transactions. Users can create, publish, monetize, and use applications on the platform. They Ether currency serves two purposes. First, an incentive for users to use, hold, and invest. Second, a price unit to pay for computations and complex logic.

Etheruem Virtual Machine: The EVM serves as a computing architecture that executes specific programs. It is a shared computer between Ethereal's clients. It holds Ethereal's internal state and records all state changes. For example, if users exchange tokens or interact with smart contracts, the EVM takes care of reflecting these changes.

Solidity: Solidity is a Contract Oriented Language, used for writing smart contracts which can be deployed to the EVM. It follows an object-oriented approach and supports features like inheritance, complex data types, and many others.

A growing body of literature has investigated the process of creating Dapps demonstrating the advantages and disadvantages of using Blockchains in web applications. Various approaches have been proposed to create Dapps using different Blockchains. Most of these approaches have a unique architecture based on different industries.

In [2] the authors have presented an implementation of a smart contract for the Logistics industry. They have shown an example of how Blockchains can be utilized within Logistics as it enables clients to send and track products through a web interface connected to Ethereal. The main weakness in their study is that they have focused on the smart contract part, and they need to dive more into the depth of how the user will interact with their to sign and send transactions [8].

A sub-case from the previous use case in logistics, the authors have demonstrated a decentralized application for the traceability process of a pharmaceutical industry [3]. Still, the authors did not investigate other blockchains which allow for higher transaction throughput and lower latency than Ethereal. Ethereal Blockchain for sure is a good choice for data security. Still, the issue of fees and transaction latency needs to be solved too. For instant transactions, Ethereal could not be the best choice to implement these Dapps. A well-known criticism of building Dapps is the transaction time and the idea of pre-requirement for using it through a Metamask plugin on the browser just to retrieve information. According to dapp.review [4], several directions have been leading the trend of building Dapps. The most popular use case was found to be in finance and exchange applications, see Fig. 1. On the other hand, the architecture for these Dapps can be different based on the category. The main purpose of creating this paper was to highlight how architecture makes Dapps more friendly for Users and to facilitate the relationship between regular Apps and Dapps through architecture [5].

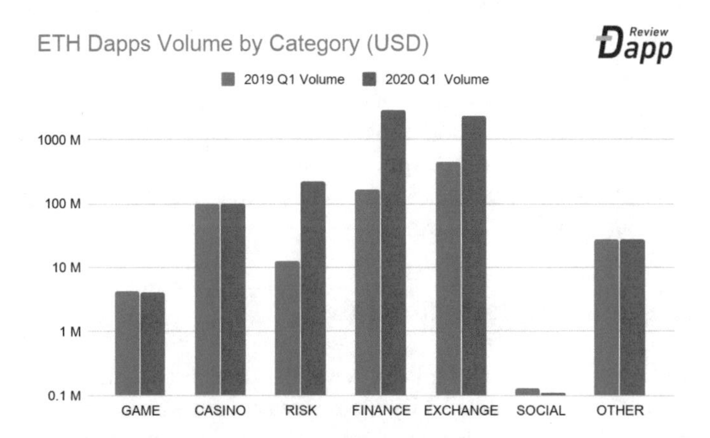

Fig. 1. Ethereal dapps volume by category USD [4].

3 Methodology

This experimental design was implemented as shown in Fig. 2 to solve the issue that most of the Dapps have right now. This is achieved by using a NodeJS server to fetch data; regular users do not need to install the Metamask plugin to see data coming from the Blockchain [6]. Furthermore, our approach uses NodeJS not just for avoiding the injected web3 component. Also, it is a way to fix data retrieval in Blockchain Dapps [7]. For example, some of the utility code or functions such as pagination and data cleaning should be modified based on changes in requirements. This could be a problem for smart contracts because a smart contract can not be modified. The principle of using middleware as a channel to reform content is necessary to avoid huge loading and to assure flexibility of data retrieval.

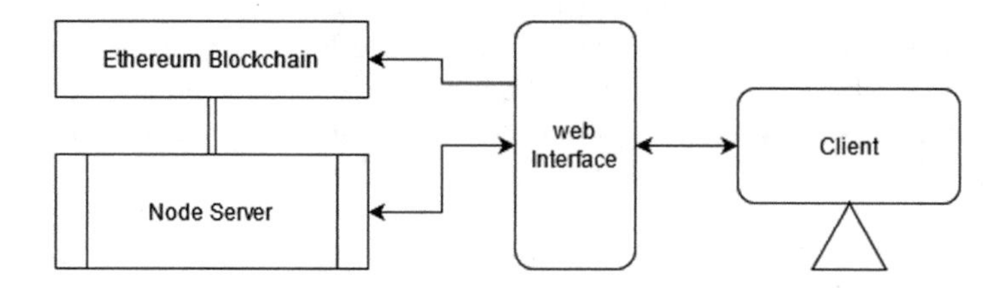

Fig. 2. Dapp Architecture using NodeJS server

As shown in Fig. 2, a client can access our Dapp through a web interface. There are two options for them to communicate with it. Firstly, retrieving data using a NodeJS server to solve the two issues that were mentioned above. Secondly, sending data using a web interface supported by Web3 integration. A user can use a plugin like Metamask to sign transactions and sending information to be saved in the Blockchain.

Subsequently, such architecture is needed for Dapps that aim to provide a better experience for mainstream users. For example, any user with knowledge in integrated web3 techniques and Blockchain technology knows how to connect a Metamask account with any Dapp. On the other hand, the users with zero knowledge about such technologies can not see data as data coming from the NodeJS server without any integration.

4 Design and Implementation

To implement the above methodology, we started with the smart contract provided below in Code snippet 1. By using remix as an online IDE, we have deployed the contract on the Ropsten test network and interacted with it using web3.js and Node JS server to retrieve the data [9,10].

```
pragma experimental ABIEncoderV2;
contract adsContract {
    event Creation(address from, string vin);

    event Transfer(address from, address to, string vin);

    struct ads {
        string title;
        string description;
        string phone;
        string mobile;
        string img;
        string location_text;
        string map_link;
        string face;
        string inst;
        string vin;
        address owner;
    }

    mapping(string => ads) adss;
    string[] public collesctads;
    address public owner = msg.sender;
    uint256 public creationTime = now;

    constructor() public {
    owner = msg.sender}

    modifier onlyBy(address _account) {
        require(msg.sender == _account, "Sender not authorized.");
        _;
    }

    function changeOwner(address _newOwner) public onlyBy(owner) {
        owner = _newOwner;
    }

    /**
     * Creates a track record of a new Ads.
     * Transaction will fail (and burn gas!) if the Ads already exists.
     */
    function createAds(
        string vin,
        string title,
        string description,
```

```
        string phone,
        string mobile,
        string img,
        string location_text,
        string map_link,
        string face,
        string inst
    ) public {
        assert(adss[vin].owner == 0x0);
        adss[vin].vin = vin;
        adss[vin].owner = msg.sender;
        adss[vin].title = title;
        adss[vin].description = description;
        adss[vin].phone = phone;
        adss[vin].mobile = mobile;
        adss[vin].img = img;
        adss[vin].location_text = location_text;
        adss[vin].map_link = map_link;
        adss[vin].face = face;
        adss[vin].inst = inst;
        collesctads.push(vin);
        emit Creation(msg.sender, vin);
    }

    function transferOwnership(string vin, address _owner) public {
        ads storage transferObject = adss[vin];
        assert(transferObject.owner == msg.sender);
        transferObject.owner = _owner;
        emit Transfer(msg.sender, _owner, vin);
    }

    /**
     * Returns the current data of the given Ads
     */
    function getAds(string vin) public view returns (ads memory) {
        return adss[vin];
    }

    function getAllAds() public view returns (ads[] memory) {
        ads[] memory addss = new ads[](collesctads.length);
        for (uint256 i = 0; i < collesctads.length; i++) {
            addss[i] = adss[collesctads[i]];
        }
        return addss;
    }
}
\caption{\textbf{Code Snippet 1}:Ads Listing Smart Contract}
```

In the code above, a struct is used as a data type to save the object on the Blockchain. This struct includes most of the fields in any regular ad. additionally, the contract owner was saved using msg.sender object that represents the account address that is deploying the contract. the following table shows all functions and their usage in the contract (Table 1).

Table 1. Contracts Functions and their usage

Function name	Usage
createAds()	Save the ad object in the array of structs
getAds()	Retrieve ad object using the id (vin)
getAllAds()	Retrieve the array of ads in the blockchain
changeOwner()	transfer the ownership for the main contract
transferOwnership()	transfer the ownership of an ad in the Blockchain

Because of using Node Server, we were able to organize the array of ads and implement the paging functionality in the server without connecting directly to the Blockchain. As shown in Code Snippet 2, a JavaScript function is written in node js to retrieve the ads as pages in the Dapp. The reason for selecting Node server was to avoid the inconvenience of loading all data on the client-side using web3. Most of the Dapps work with front-end web3 components. Because of that, a significant loading time will be saved for clients when the server fetches data without waiting to retrieve all objects.

```
app.post('/getAllads', function(req, res) {
    var json = [];
    var page=req.body.page;
    AdsContract.methods.getAdss().call({}, function(error, result) {
        console.log(result);
        for (i in result) {
            json.push({
                vin: i,
                title: result[i][0],
                description: result[i][1],
                phone: result[i][2],
                mobile: result[i][3],
                img: result[i][4],
                location_text: result[i][5],
            });
        }
        res.send(json.slice(-5 *page));
    });
});
\caption{\textbf{Code Snippet 2}: Node JS Post-request to retrieve JSON array}
```

5 Conclusion and Future Work

There is always room for misuse when a central authority controls a system. To eliminate the need for intermediaries in ad listing platforms, we use Blockchain technology to remove the need for such controlling entities. By using this decentralized application, users ensure that their data is being treated with no manipulation. The smart contract used in this application guarantees trust, credibility, and immutability. Also, by using Node Js as a server to fetch data; our Dapp does not need any web3 validation on the client-side. Users do not need to install a plugin like Metamask to see data. In addition, by using an Ethereal test network, the Dapp architecture does not need any database to handle sorting data. In future work, we also plan to include additional features to this application like:

1. A voting function that allows users to up-vote or down-vote an ad. This feature also relies on Blockchain and smart contracts so voters can be sure that their votes are unaltered.
2. A feature that allows users to remove sensitive content. This is to empower the community and guarantee a clean user experience. This feature can be built above the above-mentioned voting mechanism. If any ad surpasses a certain threshold of down-votes, users can vote to remove this ad.
3. Building a fully-functional interface for this dapp. The client-side components are still in progress.
4. Deploying the contract to different Blockchains, EVM-based (Polygon) and non EVM-based (NEAR and Solana). These Blockchains provide lower fees and higher throughput.

References

1. Johnston, D., et al.: The General Theory of Decentralized Applications, DApps (2014)
2. Andreou, A.S., Christodoulou, P., Christodoulou, K.: A decentralized application for logistics: using blockchain in real-world applications. The Cyprus Review (2018)
3. Chiacchio, F., et al.: A decentralized application for the traceability process in the pharma industry. Procedia Manuf. **42**, 362–369 (2020)
4. Dappreview (2019). https://analytics.dapp.review/
5. Taş, R., Tanrıöver, Ö.Ö. : Building a decentralized application on the Ethereum blockchain. In: 2019 3rd International Symposium on Multidisciplinary Studies and Innovative Technologies (ISMSIT). IEEE (2019)
6. Teixeira, P.: Professional Node. js: Building Javascript Based Scalable Software. Wiley, Hoboken (2012)
7. Lee, W.-M.: Using the web3. js APIs, Beginning Ethereum Smart Contracts Programming, pp. 169–198. Apress, Berkeley (2019)
8. Lee, W.-M.: Using the Metamask Chrome Extension. Beginning Ethereum Smart Contracts Programming, pp. 93–126. Apress, Berkeley (2019)

9. Latif, R.M.A., et al.: A remix IDE: smart contract-based framework for the healthcare sector by using Blockchain technology. Multimedia Tools Appl. 1–24 (2020)
10. Hu, Y.-C., et al.: Hierarchical interactions between ethereum smart contracts across testnets. In: Proceedings of the 1st Workshop on Cryptocurrencies and Blockchains for Distributed Systems (2018)

Sequential Three-Way Decisions for Reducing Uncertainty in Dropout Prediction for Online Courses

Carlo Blundo, Giuseppe Fenza, Graziano Fuccio, Vincenzo Loia, and Francesco Orciuoli(✉)

University of Salerno, Fisciano, Italy
{cblundo,gfenza,gfuccio,loia,forciuoli}@unisa.it

Abstract. Massive Open Online Courses (MOOCs) allow accessing qualitative online educational resources for huge amounts of online students. In this context, the dropout phenomenon is known as a nasty problem faced by several existing studies proposing methods and techniques to make predictions on students who are at risk of dropping out. Although the majority of such studies adopt traditional classification algorithms based on supervised methods, the present work proposes a sequential approach based on Three-Way Decisions and Neighborhood Rough Sets. The underlying idea is to exploit weekly data in order to classify, with high levels of precision, students who are likely going towards dropout or not. In cases of uncertainty, the classification decision is deferred to the next week, when new data is available. Such an approach has the advantage to preserve resources and avoiding wasting them with students erroneously classified at risk of dropout. The sequential application of the approach makes the recall increase as new data is gathered.

1 Introduction

Nowadays, more and more students and practitioners are taking advantage of Massive Open Online Courses (MOOCs), also offered by Universities [4]. The most important idea underlying MOOCs is to reach a wider audience and remove barriers to high-quality education offered by institutions. The COVID-19 pandemic has increased the attention towards MOOCs and, in general, online courses. In fact, during the year 2020, such courses have seen exponential growth in terms of participants [5]. MOOCs and, in general, online courses suffer high dropout rates (students who don't complete the course) [1]. Researchers in several study areas are facing this problem and, consequently, numerous solutions to mitigate the dropout problem can be recognized from the specialized literature, in the fields of machine learning, data science, and so on. For example, the authors of [2] trained several models to predict the dropout in the first week of the course by using two features: time spent on the learning materials and the number of accesses to the web pages of the online course. Moreover, the authors of [20] use a deep learning model to predict dropouts and provide an explanation of their possible causes. Such explanation allows the didactic staff to set and provide personalized interventions for the students who are classified to be at risk of dropout. In [19]

L. Barolli et al. (Eds.): AINA 2022, LNNS 449, pp. 46–55, 2022.
https://doi.org/10.1007/978-3-030-99584-3_5

the authors focus on feature engineering and provide a deep learning model to automatically extract features from raw data (coming from the activity log files). Furthermore, an unsupervised approach is proposed in [9]. In such a work, the authors use the K-Means method to divide students into two clusters: active students and low completion students. Then a quantitative analysis on the students' activity log is accomplished in order to extract rules capable of automatically identifying inactive students. The objective of the previously described works is mainly to achieve good levels of classification precision. Although some existing approaches consider the temporal aspects during the construction of the classification models and, in particular, during the feature engineering phase, their methods do not natively and systematically include such temporal aspects in their definition. In the present paper, a novel approach based on sequential three-way decisions [23] is introduced to face the dropout problem. The primary goal of the approach is to detect students who are likely to be heading toward dropping out. The underlying idea is that the classification (dropout or not) should lead to a specific course of action only when the classification decision is supported by enough information. In the contrary case, the decision is postponed to allow gathering more information about the students' behaviour. In such a way, *time* and *uncertainty* are explicitly considered by the proposed approach and the advantage is to achieve good levels of precision at each iteration and an increasing level of recall as the iterations go on. In other terms, the approach provides the didactic staff with the capability to reduce the time to make a classification decision when there is sufficient information available and to postpone the decision when there is insufficient information, i.e., when the evaluated uncertainty is too high to proceed.

This paper is organized as follows. Section 2 describes the proposed approach and the techniques supporting it; Sect. 3 reports the details related to a case study including the description of the executed experiments and the discussion on the evaluation results. Lastly, Sect. 4 provides conclusions and some ideas about future works.

2 Proposed Approach

As previously explained, the proposed approach deals with two main factors, which are related to the two main original aspects of the work: *uncertainty* and *time*. Three-Way Decisions (3WD) theory [22] is employed to handle uncertainty to be evaluated in order to make or defer a classification decision. Moreover, the sequential application of 3WD, i.e., Sequential Three-Way Decisions (S3WD), is adopted to systematically handle the time periods over which the students' behaviour manifest themselves. Handling time is fundamental because an uncertain situation, conducting to a deferment of the classification decision at a given time period, could evolve in a clearer one in the next time period.

2.1 Main Workflows

The proposed approach is composed of two main workflows: training and execution. The first one concerns the extraction of the rules, from a training dataset, to weekly classify students in positive, negative, and boundary regions. The second one is executed to apply the extracted rules, every week, to new incoming data in order to assign each student to one of the three regions.

The training workflow (see Fig. 1) starts with the data preparation task, where the training dataset is cleaned, normalized, and processed to create a decision table that is suitable input for the 3WD process. Each row of such table represents a student enrolled on a given course (from now on, for the sake of simplicity, the objects in the Universe reported by the decision tables will be considered simply students). The table columns are divided into two sets. The first set contains the conditional attributes, i.e., the features characterizing the students' behaviour during the executed learning activities. In particular, for the first iteration (executed at the end of the first week of learning activities), the values of the conditional attributes aggregate data related to the first week of learning activities. The second set contains a decision attribute with boolean values stating if a specific student made a dropout (value 1) or not (value 0). The construction of the decision table is repeated for each iteration because the Universe, at a given iteration, will include only rows for which there was no decision in the previous iterations and, moreover, its features will change their values as more data arrive. Let us return to the description of the first iteration. Once the decision table is ready, the Neighbourhood Rough Set [18] analysis is performed. The output of such analysis is the construction of a set of information granules, namely neighbours, which are used to partition the Universe into the three regions, i.e., to classify the students along with three classes. Now, it is possible to extract the classification rules [10] for the first-week data. In particular, rules to classify students to positive (at risk of dropout) and negative (not at risk of dropout) regions are generated.

The described process is repeated, at the end of the second week of learning activities, by considering a new decision table including only students (objects) belonging to the boundary region obtained as a result of the first iteration, the same set of conditional features and the same decision feature. The values for such table are obtained through the aggregation of the first and second-week data coming from the activity log of the MOOC platform. A new tri-partitioning is obtained and a new set of rules can be extracted for the positive and the negative regions. Such rules can be used for the second-week decisions. The process is repeated also for the next weeks.

Let us briefly explain the execution workflow. The result of the training workflow is represented by $N-1$ sets of classification rules, one for each week of the course except the last one because has no sense to generate a classification set of rules to predict students' dropout in the last week. These rules have to be applied to new incoming data. Assume that a new course started on the MOOC platform. Therefore, after the first week of activities, it is possible to prepare data by considering, for each student, the same conditional features adopted for the training phase. The rules for week 1 are applied to the aforementioned data in order to assign the students to the positive region or the negative one. The didactic staff can now apply suitable actions to such students according to the region to which they belong and the overall strategy. At the end of the second week, once new data is collected for students belonging to the boundary region and aggregated to the data coming from the first week, the rules for week 2 can be applied to such aggregated data, providing a new classification and the chance for the

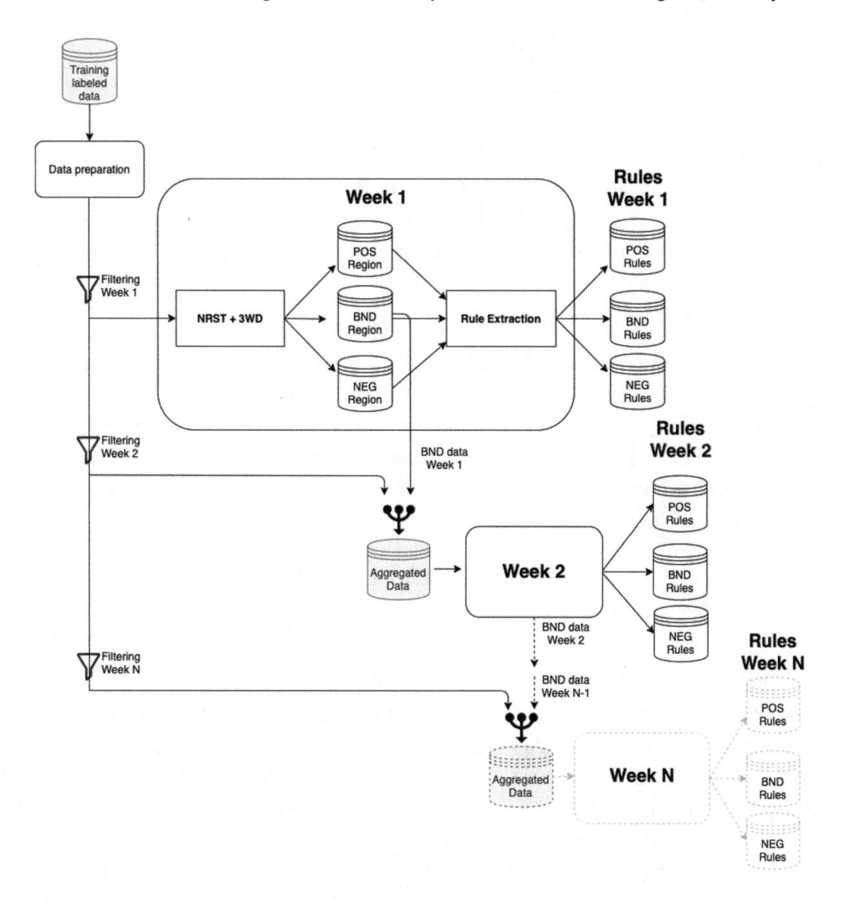

Fig. 1. Training workflow

didactic staff to better understand the dropout risk of those students whose behaviour was uncertain after the first week. This process can be repeated for all the $N-1$ sets of classification rules.

2.2 Neighborhood Rough Set and Three-Way Decisions

Rough Set Theory (RST) [15] has become a significant data analysis tool to deal with uncertainty in several heterogeneous application domains. The idea underlying RST is that in the case of vague data it makes sense to study approximations of sets (concepts) rather than the crisp sets. In particular, RST introduces two approximation operators: lower approximation and upper approximation.

Let U be the universe of the discourse, A the set of features characterizing the universe, $X \subset U$ a concept to be approximated, and let $B \subset A$, upper and lower approximations of X in the feature space B [13] are defined as follows:

$$\underline{B}(X) = \{x \in U | [x]_B \subseteq X\},$$
$$\overline{B}(X) = \{x \in U | [x]_B \cap X \neq \emptyset\}, \tag{1}$$

where $[x]_B$ is the equivalence class of $x \in X$ with respect to the indiscernibility relation based on the subset of attributes B. In the case of numerical data, it is possible to consider an extension of RST, namely Neighborhood Rough Set Theory (NRST) [18]. In NRST, the equivalence classes $[x]_B$ are replaced by the neighbours [8]:

$$\delta(x) = \{y \in U | \Delta_B(x,y) \leq \delta\}. \tag{2}$$

Distance function Δ is can be defined as, for instance, *Euclidean*, *Manhattan* or *Chebyshev* distances. Moreover, δ is a suitable threshold [25].

Therefore, let $\mathscr{K}_{\mathscr{N}} = (U, N)$ be a neighborhood approximation space, where U is the universe and $N = \{\delta(x) \mid x \in U\}$ is a neighborhood relation on U defined as the family of neighborhood granules (neighbors). Hence, the approximations of concept $X \subset U$, in terms of N, is defined as follows:

$$\underline{N}_{(\alpha,\beta)}(X) = \{x \in U | Pr(X|\delta(x)) \geq \alpha\},$$
$$\overline{N}_{(\alpha,\beta)}(X) = \{x \in U | Pr(X|\delta(x)) > \beta\}, \tag{3}$$

where β and α are two thresholds [24] and $Pr(X|\delta(x)) = \frac{|X \cap \delta(x)|}{|\delta(x)|}$.

Another important tool for dealing with uncertain data and supporting data-driven decision-making is the Three-Way Decisions Theory [17] (3WD). Such theory is used to support decision-makers by performing a ternary classification. More in detail, 3WD analysis partitions the Universe of discourse into the acceptance (positive), rejection (negative) and non-commitment (boundary) regions in a way that decision-makers can apply different action strategies for the objects of the Universe based on the specific region they belong.

The three regions are defined as follows:

$$POS_{(\alpha,\beta)} = \underline{N}_{(\alpha,\beta)}(X),$$
$$BND_{(\alpha,\beta)} = \overline{N}_{(\alpha,\beta)}(X) - \underline{N}_{(\alpha,\beta)}(X), \tag{4}$$
$$NEG_{(\alpha,\beta)} = U - \overline{N}_{(\alpha,\beta)}(X)$$

It is noteworthy that the selection of suitable values for thresholds β and α is crucial to set the dimensions of the three regions. In the rest of the paper, U is the set of students and A is the set of the features that describes a student's learning activities.

In the present work, 3WD based on NRST is adopted to determine the following regions: i) students who are likely to go towards a dropout (positive region), ii) students who are likely to regularly finish the course (negative region), and iii) students manifesting uncertain behaviour (boundary region).

2.3 Time-Related Aspects of the Approach

One of the main characteristics of the proposed approach is the application of the Sequential Three-Way Decisions [23] where the 3WD is iteratively applied. Therefore the Sequential Three-Way Decisions algorithm defined in [9] has been contextualised to the case of dropout predictions and re-formulated by considering 3WD based on NRST (see Algorithm 1).

Algorithm 1. Sequential 3WD based on Neighborhood Rough Sets

Require: a set of objects U and a couple of thresholds $\{\alpha, \beta\}$

$POS = \emptyset$
$NEG = \emptyset$
$i = 0$
$U_0 = U$
while $U_i \neq \emptyset$ and $i < n$ **do**
  $POS_{(\alpha,\beta)}(U_i) = \{x \in U_i | Pr(X|\delta(Des_{w_{i+1}})(x))) \geq \alpha\}$
  $NEG_{(\alpha,\beta)}(U_i) = \{x \in U_i | Pr(X|\delta(Des_{w_{i+1}})(x)))) \leq \beta\}$
  $BND_{(\alpha,\beta)}(U_i) = \{x \in U_i | \beta < Pr(X|\delta(Des_{w_{i+1}})(x)))) < \alpha\}$
  $POS = POS \cup POS_{\alpha,\beta}(U_i)$
  $NEG = NEG \cup NEG_{\alpha,\beta}(U_i)$
  $U_{i+1} = BND_{(\alpha,\beta)}(U_i)$
  $i = i + 1$
end while

Assume that U is the entire set of objects. The object $u \in U$ represents a student enrolled on a given MOOC course. Such student u is described by means of a set of features related to her interaction with the course content and activities. In particular, $Des_{w_i}(u)$ is the description of u along with the feature set (conditional attributes) of the decision table during the first i weeks of study. More in detail, $Des_{w_1}(u)$ represents the activities of u during the first week and $Des_{w_2}(u)$ represents the activities of u during the first two weeks of the course and so on. Let $\delta(Des_{w_i}(u))$ the same neighbourhood relation defined in the Eq. (2) taking into account only the description of u obtained for the first i weeks. Such relation is used to process positive, boundary and negative regions with respect to U_i and the concept modelling dropout. Once the boundary region has been obtained, the objects belonging to it are used to create U_{i+1} to support the next iteration. Hence, iteration $i+1$ will consider U_{i+1} but the objects $u \in U_{i+1}$ will be described by using data coming from weeks $1, 2, \ldots, i, i+1$. Note that in Algorithm 1 the formulas used to calculate positive, negative and boundary regions are obtained by simply integrating Eqs. (2), (3), and (4). Lastly, as explained before, classification rules for positive and negative regions can be mined to build a time-based prediction model.

3 Case Study

In order to evaluate the proposed approach, the dataset KDD CUP 2015 [7] has been exploited. The experiment, realized by using Python 3.6 [14], Pandas [12], Numpy [11], Jupyter [6], and its results are described in the following sections.

3.1 Dataset Description and Preparation

The dataset KDD CUP 2015 was built by gathering and organizing data, coming from the MOOC platform XuetanX [21], in different *CSV* files [26]. In particular, these files were joined to obtain a single table reporting all the interactions of students with the Web resources of the courses they are enrolled in. The table is supervised in the sense that for each interaction (row) there is a value informing if such interaction belongs to a student who abandoned the course or not.

The obtained table has been processed in order to consider the students' behaviour in fixed time slots [3]. In other terms, the finer information about all the interaction events is aggregated for the first week, the first two weeks and the first three weeks. More in detail, the features obtained through the aggregation process have been also analysed by using the Lasso method in order to obtain the most relevant ones: *session duration* that reports the total time the student was connected to the course site, *activity days* representing the number of consecutive days of student's activities, *average session* that reports the average study session time, *average time between sessions* reporting the average time between two study sessions. The values for the aforementioned features have been calculated with respect to a reference time slot.

Furthermore, because of unbalanced class (dropout and no-dropout), the dataset has been undersampled, taking 20k rows of no-dropout and 20k rows of dropout students, to improve the performances and reliability of the process and values have been scaled (min-max strategy) before starting the Sequential 3WD process.

3.2 Experimentation and Evaluation

Starting from the final dataset described in the previous section, it was possible to execute the training workflow described in Sect. 2.1. More in detail, the first iteration of 3WD has been applied to a decision table obtained by filtering the aforementioned dataset with respect to the first week.

After the completion of the first iteration of Sequential 3WD (Algorithm 1), the three regions (for the first week) were available for the application of the rule mining [16] algorithm. In this experimentation, a rule mining algorithm based on the discernibility matrix was used. In order to execute such an algorithm, the numerical fields of the decision table were first discretized. This algorithm has been applied to obtain the first set of classification rules. Some interesting mined rules for the first-course week are described in Table 1. For instance, students having low levels of session duration and very low levels of activity days are classified in the positive region. Moreover, students having high levels of session duration and high levels of activity days are classified in the negative region.

Once the first iteration was completed, an updated decision table has been constructed for supporting the second iteration. In particular, the decision table for the second iteration has been constructed with the aggregated (first and second week) data of students belonging to the boundary region of the previous iteration. The second iteration of NRST and 3WD is now applied on the new decision table and, again, for the positive and negative regions, the classification rules have been extracted. A total of three iterations have been executed.

Table 1. First week rules (VL = Very Low, L = Low, M = Medium, H = High, VH = Very High)

Rule	Antecedent	Consequent
1	$Session\ Duration = L \wedge Activity\ Days = VL$	*POS*
2	$Discussion = M \wedge Activity\ Days = VL \wedge Problem = L$	*POS*
3	$Session\ Duration = H \wedge Activity\ Days = VH$	*NEG*
4	$Activity\ Days = VH \wedge Average\ Session = VH \wedge Discussion = H$	*NEG*

Table 2. Results for each week

Metric	Week 1	Week 1 + 2	Week 1 + 2 + 3
Precision	91%	92%	92%
Recall	47%	62%	73%
F1-score	62%	74%	81%

With respect to the parameters adopted for the training phase, the thresholds β (0.33) and α (0.63) for the 3WD and δ (0.005) for the *Manhattan* distance function (used to generate neighbours) are selected by using the grid search approach.

Table 2 reports precision, recall and F1-score for each iteration of the Sequential 3WD algorithm.

The results shed light on the high precision in predicting dropout (and no-dropout) students for all the iterations. Moreover, another interesting outcome is the growing trend of recall, meaning that also the correct prediction rate grows week-by-week when more information is added to the universe.

4 Final Remarks

This study proposes an approach to mitigate the dropout phenomenon week-by-week through a sequential decision support system. The evaluation results confirm the initial idea underlying the proposed approach. In particular, at the end of each week, the constructed classifier is able to achieve high precision levels while recognizing students who are going towards a dropout or to regularly finish the course. Moreover, the advantage of the approach emerges clear from such results. In fact, in the presence of uncertainty, the approach suggests the decision-maker defer the decision and wait for more information about students involved in such uncertainty. Once more information is available in the following weeks, one can reconsider the situation for these students. This aspect is demonstrated by the recall values increasing along the three weeks. In further works, we will try to evaluate the performance of the rule mining algorithm to improve such a task in order to empower the whole approach.

Acknowledgement. Thanks to Tiziana Coppola for the initial discussion on the used dataset.

References

1. Adams, A., Liyanagunawardena, T., Williams, S.: MOOCs: a systematic study of the published literature 2008–2012. Int. Rev. Res. Open Dist. Learn. **14**, 202–227 (2013)
2. Alamri, A., et al.: Predicting MOOCs dropout using only two easily obtainable features from the first week's activities. In: Coy, A., Hayashi, Y., Chang, M. (eds.) ITS 2019. LNCS, vol. 11528, pp. 163–173. Springer, Cham (2019). https://doi.org/10.1007/978-3-030-22244-4_20
3. Blundo, C., Fenza, G., Fuccio, G., Loia, V., Orciuoli, F.: A time-driven FCA-based approach for identifying students' dropout in MOOCs. Int. J. Intell. Syst. **37**(4), 2683–2705 (2021)
4. Deng, R., Benckendorff, P., Gannaway, D.: Progress and new directions for teaching and learning in MOOCs. Comput. Educ. **129**, 48–60 (2019)
5. Impey, C., Formanek, M.: MOOCs and 100 days of COVID: enrollment surges in massive open online astronomy classes during the coronavirus pandemic. Soc. Sci. Humanit. Open **4**(1), 100177. ISSN 2590-2911 (2021)
6. Jupyter. https://jupyter.org/
7. Knowledge Discovery and Data Mining. MOOC dataset from KDD cup 2015 (2015). http://data-mining.philippe-fournier-viger.com/the-kddcup-2015-dataset-download-link/
8. Kumar, S.U., Inbarani, H.H.: A novel neighborhood rough set based classification approach for medical diagnosis. Procedia Comput. Sci. **47**, 351–359. ISSN 1877-0509 (2015)
9. Liu, T.-Y., Li, X.: Finding out reasons for low completion in MOOC environment: an explicable approach using hybrid data mining methods. DEStech Trans. Soc. Sci. Educ. Hum. Sci. 376–384 (2017)
10. Menardi, G., Torelli, N.: Training and assessing classification rules with imbalanced data. Data Min. Knowl. Discov. **28**, 92–122 (2014)
11. Numpy. https://numpy.org/
12. Pandas. https://pandas.pydata.org/
13. Pawlak, Z.: Rough set approach to knowledge-based decision support. Eur. J. Oper. Res. **99**(1), 48–57 (1997)
14. Python 3. https://www.python.org/
15. Qian, Y., et al.: Local rough set: a solution to rough data analysis in big data. Int. J. Approximate Reasoning **97**, 38–63. ISSN 0888-613X (2018)
16. Shao, M.-W., Leung, Y., Wu, W.-Z.: Rule acquisition and complexity reduction in formal decision contexts. Int. J. Approximate Reasoning **55**(1, Part 2), 259–274. ISSN 0888-613X (2014). Special issue on Decision-Theoretic Rough Sets
17. Sun, B., Chen, X., Zhang, L., Ma, W.: Three-way decision making approach to conflict analysis and resolution using probabilistic rough set over two universes. Inf. Sci. **507**, 809–822 (2019)
18. Wang, Q., Qian, Y., Liang, X., Guo, Q., Liang, J.: Local neighborhood rough set. Knowl. Based Syst. **153**, 53–64 (2018)
19. Wang, W., Yu, H., Miao, C.: Deep model for dropout prediction in MOOCs. In: ICCSE 2017: Proceedings of the 2nd International Conference on Crowd Science and Engineering, pp. 26–32 (2017)
20. Xing, W., Du, D.: Dropout prediction in MOOCs: using deep learning for personalized intervention. J. Educ. Comput. Res. **57**, 073563311875701 (2018)
21. XuetanX. https://www.xuetangx.com/
22. Yao, Y.: Three-way decisions with probabilistic rough sets. Inf. Sci. **180**(3), 341–353. ISSN 0020-0255 (2010)
23. Yao, Y.: Granular computing and sequential three-way decisions. In: Lingras, P., Wolski, M., Cornelis, C., Mitra, S., Wasilewski, P. (eds.) RSKT 2013. LNCS (LNAI), vol. 8171, pp. 16–27. Springer, Heidelberg (2013). https://doi.org/10.1007/978-3-642-41299-8_3

24. Yao, Y., Deng, X.: Sequential three-way decisions with probabilistic rough sets. In: Proceedings of the 10th IEEE International Conference on Cognitive Informatics and Cognitive Computing, ICCI*CC 2011, pp. 120–125 (2011)
25. Zhang, J., Li, T., Ruan, D., Liu, D.: Neighborhood rough sets for dynamic data mining. Int. J. Intell. Syst. **27**, 317–342 (2012)
26. Zhang, T., Yuan, B.: Visualizing MOOC user behaviors: a case study on XuetangX. In: Yin, H., et al. (eds.) IDEAL 2016. LNCS, vol. 9937, pp. 89–98. Springer, Cham (2016). https://doi.org/10.1007/978-3-319-46257-8_10

Enhanced Autonomous Driving Through Improved 3D Objects Detection

Razvan Bocu[1(✉)] and Maksim Iavich[2]

[1] Transilvania University of Brasov, Brasov, Romania
razvan.bocu@unitbv.ro
[2] Caucasus University, Tbilisi, Georgia
miavich@cu.edu.ge

Abstract. The detection of 3D objects is fundamental in the field of autonomous driving. The involved computations consist of sets of 3D bounding boxes that determine specific significant objects. The existing scientific contributions generally do not report applied studies that assess the reliability of 3D objects detection considering various weather conditions, including adverse scenarios like heavy rain and thick fog, and also other relevant problematic use cases. This paper presents an applied research process that describes the core of a 3D objects detection system, which considers two particular road topologies, a roundabout and a T-junction. The experimental data is collected through a partnership with several car manufacturers.

1 Introduction

The autonomous driving systems rely on the relevant software modules, which determine an accurate overview regarding the driving environment. The accuracy of the computations that are performed by these components is fundamental for the autonomous vehicles, which drive through environments that are featured by a certain topological complexity. Thus, the inaccurate detection of environmental objects may cause tragic incidents, which are determined by the improper detection and classification of the encountered objects and patterns.

The efficient and accurate detection of 3D objects is essential for any reliable autonomous driving process, which relates to the proper estimation of the 3D bounding boxes that define the respective object's position and orientation in space. Furthermore, the category to which the object belongs is also determined. The 3D objects are generally detected through the consideration of machine learning techniques Chen (2017), Ku (2018). Additionally, it is relevant to note that the practical experiments consider real-world or synthetic data Roberts (2011), Arnold (2019). Thus, KITTI Geiger (2012) is one of the datasets that store image data, which has been collected by frontal cameras. It also includes metadata that define the 3D boxes annotations. The accurate and efficient classification of the 3D objects is also determined by the sufficiently rich texture images. The ranging laser-based cameras (Lidar) Beltran (2018), Li (2016), and the light detection are able to offer data that can be used for the computation processes that determine the detected objects' orientation and position in space Chen (2019), Ma

L. Barolli et al. (Eds.): AINA 2022, LNNS 449, pp. 56–66, 2022.
https://doi.org/10.1007/978-3-030-99584-3_6

(2019), Hurl (2020). Thus, existing 3D objects' detection processes consider the consolidated data that is obtained from several data acquisition sensors Ghamisi (2019), Yin (2018). This approach improves the accuracy and performance of the 3D objects detection processes.

The combined data that is collected by the data acquisition sensors can be affected by particular functional issues. Thus, the practical experience demonstrates that certain issues are of particular interest, such as restricted perception horizon that is determined by a limited field of view, occlusion, and also the inability to properly process distant regions that are naturally determined by a low-point density Simony (2018). Therefore, the collection of the data in a cooperative manner, which uses several data acquisition sensors, represents a viable solution to this problem Schlosser (2016), Simonyan (2015). The existing scientific papers in the field report limited contributions that regard practical aspects, such as maneuver coordination, lane selection, and autonomous crossings of intersections. This paper describes an approach that implies the utilization of data that is cooperatively gathered from several sensors Du (2018), LinT-Y(2017). This determines a significantly improved 3D objects detection process, which benefits from an enhanced perception of the horizon, that is also affected by less noisy data.

The cooperative detection of 3D objects may consider two particular algorithmic modalities: late fusion (LF) and early fusion (EF) Castanedo (2013). These are labelled in consideration of the chronological sequence, as the process of data fusion may take place after or before the proper 3D objects detection phase Zhou (2018), Yang (2019). Thus, the late fusion regards each data collection instance in an independent fashion, and the determined 3D boxes are fused into an aggregated 3D object Ren (2015), Dosovitskiy (2017). The early fusion processes the collected data and combines it before the actual detection phase.

The rest of this paper is structured considering the following parts. Thus, the following section presents the reported 3D objects detection model. Furthermore, the experimental data is described. Additionally, the algorithmic model's training process and the real-world field performance assessment are described. Following this, the outcomes of the practical evaluation process are presented, with an emphasis on some essential theoretical and practical aspects, which also concern the description of an extension of the system's field trial relative to problematic use case scenarios and road topologies. The last section concludes the paper.

2 Presentation of the Proposed Model

The described model determines an improvement and algorithmic optimization of the Frustum PointNet (F-PointNet) fusion based model Qi (2018), which is designed to process data that is collected from n sensors. Each sensor offers depth sensing capabilities that include laser-based Lidar, and also functions that compute depth data. Moreover, each sensor possesses a data processing unit (processor). We describe a centralized data collection model for the detection of 3D objects. This suggests that the individual sensors send the acquired data to the central processing components through wired or wireless data links. Additionally, it is important to assert that the sensors are properly calibrated, which implies that their accurate position in space and orientation are transmitted to the central data processing components. The initial F-PointNet fusion based approach usually processes the image data that is generated by front view cameras. This limitation determines the difficulties that this approach has concerning the proper detection of 3D objects in particular environmental conditions, as an example at night time. Although F-PointNet is considered as one of the reliable fusion based approaches, the mentioned drawback, which affects a significant proportion of the real-world use case scenarios, make it unacceptable for our automotive industry collaborators. Consequently, we have extended and improved the plain F-PointNet fusion based approach, in such a way that the model that we present is capable to precisely detect 3D objects in virtually any environmental conditions. The following sections present and qualitatively assess the proposed model.

The central data processing components are featured with sufficient computational power, which allows them to process and aggregate the data that is provided by the individual sensors. Moreover, the central data processing components perform data synchronization functions and transmit the obtained data to passing autonomous vehicles through a radio communication system. This suggests that autonomous vehicles possess the required basic hardware components, which allow them to receive the radio signals from the central data processing components. It is also relevant to state that the autonomous vehicles are featured with the necessary computational components, which allow them to locally process received data, which includes data chunks that pertain to environmental perception and trajectory control. The central data processing components are not tasked with the transmission of actual trajectory control signals to the enrolled autonomous vehicles. The mentioned vehicles use the data that is sent to them by the central data processing components, which allows them to conduct local data processing operations that support them in order to make their own trajectory control decisions. The central data processing components have the role to offload data processing operations that are computationally expensive from the actual vehicles, and thus allow them to make more precise and time efficient decisions.

2.1 Description of the Objects Detection Component

This component essentially determines the functional features of the designed architectural model considering three layers. Thus, it is defined by a feature learning network, several convolutional middle layers, and a Localized Proposal Network (LPN).

The feature learning network transforms the 3D point cloud data into a representation that is determined by a constant size, which can be efficiently processed by the convolutional layers of the neural network. It is important to state that the plain Voxelnet model uses lasers-based reflection intensity channels, as well as the 3D spatial coordinates of the detected point, let us refer to them as (a, b, c). The 3D objects detection model that is described in this paper considers the spatial coordinates in an exclusive manner. This allows for the data processing stage to efficiently generate the points that can be processed by the central data processing components in a computationally efficient way. The input point cloud is divided into voxels of equal size, let us refer to them as (vox_a, vox_b, vox_c). Here, the components determine the width, length and height, respectively. In the case of each voxel, a set of t points is generated in order to build a proper feature vector. Moreover, let us state that T represents the threshold regarding the maximum number of points that pertain to each individual voxel. If t is greater than T, then a sample of T random points is produced. This has the role to decrease the computational load. It also improves the balance concerning the distribution of points between different voxels. The coordinates of these points are sent over to a chain of Voxel Feature Encoding (VFE) layers. Each particular VFE layer is made of fully connected layers, which are determined by local aggregations, and max-pooling operations Shi (2019). The output of this neural network consists of a 4D tensor that is indexed considering the following features of the voxel: dimension, height, length, and width.

The middle layers that are part of the convolutional neural network are especially important. Essentially, they add three additional stages to the data processing pipeline relative to the 4D voxel tensor, which has been presented. These supplementary stages include spatial data that relate to the nearby voxels. Thus, it is immediate to note that they have the role to add the mandatory three-dimensional context to the considered features map.

The Localized Proposal Network (LPN) receives the tensor that is obtained. The LPN network is structured relative to three stages of convolutional layers, which are part of the neural network. They are subsequently processed using three additional transposed convolutional layers. Furthermore, a high resolution features map is obtained. This features map is used in order to generate two output branches. The algorithmic model considers a confidence score, which designates the probability for a 3D object to be present in a particular analyzed scene. It also considers a regression map, which determines the position, orientation, and size of the processed bounding box. The interested reader may find out additional interesting information considering alternate similar approaches in Knorr (2013).

3 The Experimental Dataset

The experimental dataset includes data that were collected by the sensors, which are statically positioned at the side of the road. The sensors' placement is proper in order to define the two basic scenarios, the roundabout and the T-junction. The sensors are able to acquire depth and RGB image data at a resolution of 640×480 pixels, while the horizontal field of view is supplied at a 90 degree angle. Ten sensors are placed in the area of the T-junction. They are deployed on vertical masts with a height of 4.3 m. The balanced collection of data is determined by the field deployment of the sensors, which ensures that five of them monitor the incoming direction, while the other five point to the T-junction's opposite direction. Additionally, the roundabout scenario uses twelve sensors, which are installed on masts with a height of 6.1 m. Relative to the roundabout sensors, it can be asserted that six of them monitor the incoming road lanes, while the other six sensors monitor the outgoing road lanes. The precise physical placement of the sensors, in the case of both scenarios, was iteratively optimized through an empirical onsite process.

The experimental dataset is structured considering four distinct sections. Thus, two data sections pertain to the T-junction scenario, while the other two sustain the system's work in the case of the roundabout scenario. Each data section includes 24,000 image samples, which are used for the training process, and 1000 test image samples. The image sample, as a compound entity, is defined as the overall set of RGB and depth images, which are gathered by all the deployed sensors at a precise moment in time. Additionally, each image sample also provides data that pertains to the detected 3D objects' spatial position and orientation, dimension and category.

The experimental dataset includes entity patterns that define four main categories: pedestrians, cyclists, motorcyclists, and vehicles. These categories are specified in the dataset with the relative weights 0.2, 0.2, 0.2, and 0.4, respectively. This has the role to assign to the actual vehicles a proper weight, which properly model the real-world situation. The data curation phase has the role to ensure that each 3D object is represented using eight image samples. This approach enhances the structural diversity of the 3D objects that are provided by the experimental dataset. It also allows for more spatial positions and orientations to be stored. Naturally, the actual movement of the autonomous vehicles considers the standard road traffic regulations.

The detection areas are defined by a rectangle, which has the length and width of 100 by 50 m, relative to the T-junction scenario. The detection areas that define the roundabout scenario relate to squares with sides of 90 m. The data acquisition sensors cover an area with a size of 4300 m^2 square metres in the case of the T-junction scenarios, and 12,200 m^2 relative to the roundabout scenarios. It is important to note that the algorithmic core of the described 3D objects detection model considers the mathematical model of a laser-based Lidar sensor, which is reported in Castanedo (2013). Moreover, further theoretical and practical aspects may be studied in papers Ortiz (2013), Feng (2019).

4 The Training Process

The experimental dataset is processed considering a training process that is defined by the following steps. Thus, a certain instance of the 3D objects detection model is trained using the data that is collected by several sensors, in the context of the algorithmic process that has been presented. The training process uses a Stochastic Gradient Descent (SGD) optimisation during 90 epochs. The learning rate is 10^{-4}, while the momentum is 0.8. Furthermore, a loss function is defined, which adds a penalty to the regression relative to the position, size and yaw angle.

Let us suppose that the voxel size is defined as (vox_a, vox_b, vox_c). Consequently, the anchor stride that goes along the dimensions X and Y, in the case of the T-junction, is set to (0.3,0.3,0.5) m and 0.5 m, respectively. The usage of precisely the same hyperparameters for the roundabout is not appropriate, considering that the covered area is approximately three times larger. This would produce computational problems that are determined by the impossibility to store all the features maps in the graphics processing unit's (GPU) memory. Therefore, the spatial coverage of the axis X and Y is reduced, in the case of the roundabout, through a voxel size of (0.4,0.4,0.4) m, and an anchor stride of 0.8 m.

The algorithmic core that is concerned with the 3D objects detection is mostly designed to extract vehicles from the image samples that are processed. The remaining categories, which pertain to pedestrians, cyclists, and motorcyclists, prevent the statistical phenomenon of overfitting. Thus, they allow for the model that is trained to learn the necessary distinct features that define the vehicles.

It is also relevant to note that the proposed algorithmic core is designed to apply rotations to the determined bounding boxes. The angle of these rotations is selected, considering a randomized model, from the interval $[-26, 26]$. The rotation is enforced in order to determine the value of the rotation angle in an as general as possible way. Furthermore, the rotation also prevents the model from reaching a state of overfitness.

5 Presentation of the Performance Assessment Process

The actual real-world performance of the described 3D objects detection approach is assessed through the consideration of two road scenarios (topologies), the T-junction and the roundabout. Moreover, the variation in the number of data acquisition sensors is evaluated relative to the accuracy of the 3D objects detection process itself.

5.1 Performance Assessment Metrics

The 3D objects detection system is evaluated through the consideration of four performance metrics. Thus, these are intersection relative to union (IRTU), recall, precision, and the communication cost, which is calculated considering the average volume of data that is transferred between a sensor and the central data processing components relative to each image sample. The communication cost is measured in kilobits. Let us recall that the concept of image sampling has already been described in a previous section.

The intersection relative to union determines the spatial similarity of a pair of bounding boxes. Thus, one of them is normally selected from the set of determined bounding boxes, while the other one is chosen from the ground-truth set. This is computed considering the following formula: $IRTU(B_{gt}, B_e) = \frac{volume(B_{gt} \cap B_e)}{volume(B_{gt} \cup B_e)}$. Here, B_{gt} and B_e designate the ground-truth and estimated bounding boxes, respectively. The set of estimated bounding boxes includes all the positive entities, which are the bounding boxes that are generated by the detection system considering a confidence score that is greater than a certain threshold, let us call it Trs. It is relevant to state that the metric IRTU also relates to the location, size, and the yaw angle of both bounding boxes, which defines the orientation. The value of this metric is 0, if the respective bounding boxes do not intersect, while it is 1 if the two bounding boxes are identical in terms of their size, orientation, and location. The value of the Trs has been optimally determined considering an iterative experimental calibration process. It was determined that a value of 0.75 implies an optimal balance between the quality of the 3D objects detection process output, and the required computational resources.

The precision is computed as a ratio between the number of estimated bounding boxes that are matched, considering the already mentioned definition, and the total number of bounding boxes that are contained in the estimated set. Moreover, the recall is calculated as the ratio between the number of estimated bounding boxes that are matched relative to the total number of bounding boxes that are included in the ground-truth set. It is immediate to observe that precision and recall essentially represent functions of Trs. The arithmetic and computational relationship between precision and recall is analyzed in some existing scientific contributions. Thus, further details may be consulted in paper Everingham (2010). Moreover, let us analyze the quantitative performance data that are presented in Table 1, which demonstrate that the system scales well considering the relatively complex use case scenario that we considered.

Table 1. Comparative performance analysis between late and early fusion schemes.

Scheme and topology	Communication cost (Kilobits)	Computation time (ms)
LF T-junction	0.39	217
EF T-junction	471	296
LF roundabout	0.17	139
EF roundabout	541	228

5.2 Investigation Concerning the Number and Placement of Sensors

This particular phase of the detection system's field evaluation relates to the assessment of the influence that the number, spatial position and orientation of the sensors have on the detection of the 3D objects. This real-world assessment relates to the early and late fusion schemes. The same algorithmic core is used in order to detect the actual 3D objects. In Table 2, the detection performance is determined considering the actual accuracy of the identified 3D objects. The accuracy is calculated as the number of accurate 3D object detections relative to the total number of 3D objects that are part of the experimental dataset.

Table 2. Comparative performance analysis considering various numbers of the active sensors.

No. sensors	T-Junction EF	T-Junction LF	Roundabout EF	Roundabout LF
1	0.212	0.186	0.196	0.174
2	0.243	0.205	0.207	0.198
3	0.324	0.308	0.305	0.289
4	0.439	0.413	0.424	0.402
5	0.547	0.532	0.536	0.514
6	0.657	0.635	0.638	0.618
7	0.789	0.768	0.778	0.769
8	0.858	0.837	0.849	0.828
9	0.957	0.946	0.948	0.939
10	0.992	0.989	0.978	0.968
11	–	–	0.989	0.985
12	–	–	0.994	0.992

The values of the accuracy, which are displayed in Table 2, prove that the values of this metric increase in a directly proportional dependence on the number of active data collection sensors. It is relevant to note that, in the case of the T-junction scenario, the optimal accuracy of the detection is obtained with ten active data collection sensors. Consequently, the balanced detection accuracy in the case of the roundabout scenario is obtained with twelve data collection sensors. The experimental process that was conducted demonstrate that the optimal level of the accuracy is greater or equal than 0.98, which ensures that the 3D objects are precisely detected. Consequently, the autonomous driving process flows in an optimal way. Furthermore, a larger number of data collection sensors are necessary, in the case when the detection accuracy level should be maintained over a larger area. It is important to note that the monitored area is $4300\,m^2$ in the case of the T-junction topology, while the roundabout covers a surface of $12{,}200\,m^2$. Furthermore, it can be noticed that the early fusion scheme generates a superior level of the accuracy regarding all the assessed use cases, as the early fusion processing routines consider more data throughout the preprocessing stage, as compared to the late fusion detection model.

Additionally, the experimental evaluation process relates to the spatial position and orientation of the data collection sensors. This is especially significant for the T-junction topology, which uses groups of three data collection sensors that are installed in order to monitor specific sub-sections of the entire detection zone. The next paragraph comments on the outcomes of the experimental evaluation that were obtained during the evaluation of the impact that the spatial diversity has on the accuracy of the 3D objects detection.

The experimental process proves that the problematic of the data collection sensors' spatial diversity Xu (2019) is relevant. The prevention or, at least, minimization of the multihop data collection links that are established from the sensors to the central data processing components is mandatory in order to preserve the discussed level of the 3D objects detection accuracy. It was determined that, in the case of the T-junction topology, a group of two sensors performs better than the individual most efficient data collection sensor by 57%, while the cluster of three sensors, which gather data from specific sub-sections of the entire area, generate an improvement of 96%, as compared to the most efficient data collection sensor. Moreover, the same performance improvements are 46% and 82%, respectively, if the roundabout topology is considered. The magnitude of the improvement is computed relative to the base value of the accuracy, as it is generated by a specific data collection sensor in the case of the roundabout and T-junction topologies. This sufficiently proves that the clusters of sensors should offer an optimal overlap between their members. As a consequence, this additionally proves that the early fusion scheme may further prevent the occurrence of incorrectly detected 3D objects. Thus, the clusters of sensors, which include members that are properly overlapped, determine an increase of the 3D objects detection accuracy.

5.3 Remarks Regarding Additional Road Traffic Scenarios

The reported system was deployed to another national road, apart from the already presented one. The supplementary experimental setup considers both road topologies, the T-junction and the roundabout. The experimental process evaluated the deployed system over a period of one month. The system has been assessed considering several weather conditions, which include thick fog and heavy rain. Additionally, the number of the installed sensors has been successively decreased. The outcomes of the experimental process prove that the system is not significantly affected by the adverse weather conditions. Moreover, it may be noticed that the system offers an acceptable level regarding the detection performance, if less sensors are installed. The results of the initial experimental evaluation stages show that the system can be regarded as a sufficiently economical solution for the field deployment of an efficient autonomous driving system. It may be readily observed that six data collection sensors are sufficient in order to accurately monitor the area of the T-junction proper, while eight sensors are sufficient in order to ensure an efficient coverage of the T-junction zone. The supplementary data collection sensors, if they are present, may be utilized in order to conduct the pre-calibration of the system before the arrival of the autonomous vehicle in the monitored zones.

6 Conclusions and Future Work

This paper presents a complex study that proposes an enhanced integrated approach, which may improve the functional accuracy, reliability and safety of autonomously driven vehicles. It is essential to note that the system that is presented uses early and late fusion data processing and detection schemes. The system employs data collection sensors, which are installed on masts at the edge of the road. The acquired data are aggregated and transmitted to the central data processing components. The determined bounding boxes are sent to the vehicles that drive in the monitored area. The described system's validity and effectiveness are evaluated considering two relatively problematic road topologies, a T-junction and a roundabout. The results of the experimental process prove that a greater number of data collection sensors is required in order to generate image samples that can be processed during the 3D objects detection process. The reported experiments prove that the presented system is capable to substantially decrease the rate of falsely detected 3D objects, while also enhancing the accuracy of the detection process. It can be asserted that the system accurately processes the image samples, and it consequently determines the 3D objects in virtually all of the considered use cases. Furthermore, it is relevant to note that the actual hardware architecture of the deployed system uses existing and easily available technologies and data transmission protocols. This architectural approach minimizes the implementation costs, and it also improves the system's economical efficiency. The data that is shared by the central data processing components to each autonomous vehicle ensures that even the cars that are not featured with the most recent technologies may consider a safe and efficient autonomous driving experience. The system's 3D objects detection accuracy is evaluated relative to the reference F-PointNet model. The results demonstrate that the improved detection core and architecture of the presented integrated system allows for the system to detect the 3D objects in a more accurate and more efficient manner than similar existing contributions. The field research deployment that produced the experimental data that was considered, which has been built and installed with the support of our industry partners, already supports the reasearch and development projects that are conducted by our automotive industry collaborators, which aim to enhance the currently available autonomous driving approaches.

Several aspects of the described system will be improved. Thus, the algorithmic design and implementation of the early and late fusion models will be enhanced, with an emphasis on the early fusion schemes, which exhibit the most efficient computational behaviour in practice. Additionally, the field placement of the data acquisition sensors should be improved through successive empirical trials. The goal is to obtain the greatest possible coverage with the smallest number of data acquisition sensors, while maintaining the high levels of detection accuracy that are reported in this paper. Furthermore, the relevant research requirements that are issued by our automotive industry partners will be considered during the upcoming development process of the integrated 3D objects detection system.

References

Geiger, A., Lenz, P., Urtasun, R.: Are We ready for autonomous driving? The KITTI vision benchmark suite. In: Proceedings of the 2012 IEEE Conference on Computer Vision and Pattern Recognition (CVPR), Providence, RI, USA, 16–21 June 2012, vol. 2012, pp. 3354–3361 (2012)

Chen, X., Ma, H., Wan, J., Li, B., Xia, T.: Multi-view 3D object detection network for autonomous driving. In: Proceedings of the 2017 IEEE Conference on Computer Vision and Pattern Recognition (CVPR), Honolulu, HI, USA, 21–26 July 2017

Ku, J., Mozifian, M., Lee, J., Harakeh, A., Waslander, S.: Joint 3D proposal generation and object detection from view aggregation. In: Proceedings of the 2018 IEEE/RSJ International Conference on Intelligent Robots and Systems (IROS), Madrid, Spain, 1–5 October 2018

Qi, C.R., Liu, W., Wu, C., Su, H., Guibas, L.J.: Frustum PointNets for 3D object detection from RGB-D data. In: Proceedings of the 2018 IEEE Conference on Computer Vision and Pattern Recognition (CVPR), Salt Lake City, UT, USA, 18–23 June 2018

Ortiz, L.E., Cabrera, E.V., Gonçalves, L.M.: Depth data error modeling of the zed 3D vision sensor from stereolabs. ELCVIA Electron. Lett. Comput. Vis. Image Anal. **17**, 1–15 (2013)

Roberts, R., Sinha, S.N., Szeliski, R., Steedly, D.: Structure from motion for scenes with large duplicate structures. In: Proceedings of the CVPR 2011, Colorado Springs, CO, USA, 20–25 June 2011, pp. 3137–3144 (2011)

Arnold, E., Al-Jarrah, O.Y., Dianati, M., Fallah, S., Oxtoby, D., Mouzakitis, A.: A survey on 3D object detection methods for autonomous driving applications. IEEE Trans. Intell. Transp. Syst. **20**, 3782–3795 (2019)

Beltran, J., Guindel, C., Moreno, F.M., Cruzado, D., Garcia, F., De La Escalera, A.: BirdNet: a 3D object detection framework from Lidar information. In: Proceedings of the 2018 21st International Conference on Intelligent Transportation Systems (ITSC), Maui, HI, USA, 4–7 November 2018, pp. 3517–3523 (2018)

Li, B., Zhang, T., Xia, T.: Vehicle detection from 3D lidar using fully convolutional network. In: Proceedings of Robotics: Science and Systems, Ann Arbor, MI, USA, 18–22 June 2016

Simony, M., Milzy, S., Amendey, K., Gross, H.-M.: Complex-YOLO: an Euler-region-proposal for real-time 3D object detection on point clouds. In: Proceedings of the European Conference on Computer Vision (ECCV) Workshops, 8–14 September 2018

Castanedo, F.: A review of data fusion techniques. Sci. World J. **2013**, 142–149 (2013)

Feng, D., et al.: Deep multi-modal object detection and semantic segmentation for autonomous driving: datasets, methods, and challenges. arXiv 2019 arXiv:1902.07830

Zhou, Y., Tuzel, O.: Voxelnet: end-to-end learning for point cloud based 3D object detection. In: Proceedings of the IEEE Conference on Computer Vision and Pattern Recognition (CVPR), Salt Lake City, UT, USA, 18–23 June 2018

Shi, S., Wang, X., Li, H.: PointRCNN: 3D object proposal generation and detection from point cloud. In: Proceedings of the IEEE Conference on Computer Vision and Pattern Recognition (CVPR), Long Beach, CA, USA, 15–20 June 2019
Yang, S., Sun, Y., Liu, S., Shen, X., Jia, J:. STD: Sparse-to-Dense 3D object detector for point cloud. In: Proceedings of the IEEE International Conference on Computer Vision (ICCV), Seoul, Korea, 27 October–2 November 2019, pp. 1951–1960 (2019)
Qi, C.R.; Su, H.; Mo, K.; Guibas, L.J. PointNet: Deep learning on point sets for 3D classification and segmentation. In: Proceedings of the 2017 IEEE Conference on Computer Vision and Pattern Recognition (CVPR), Honolulu, HI, USA, 21–26 July 2017, pp. 652–660 (2017)
Chen, Q., Tang, S., Yang, Q., Fu, S.: Cooper: cooperative perception for connected autonomous vehicles based on 3D point clouds. In: Proceedings of the 39th IEEE International Conference on Distributed Computing Systems (ICDCS), Dallas, TX, USA, 7–10 July 2019
Chen, Q., Ma, X., Tang, S., Guo, J., Yang, Q., Fu, S.: F-Cooper: feature based cooperative perception for autonomous vehicle edge computing system using 3D point clouds. In: Proceedings of the IEEE/ACM Symposium on Edge Computing (SEC), 7–9 November 2019
Hurl, B., Kohen, R., Czarnecki, K., Waslander, S.: TruPercept: trust modelling for autonomous vehicle cooperative perception from synthetic data. In: Proceedings of the 2020 IEEE Intelligent Vehicles Symposium (IV), Las Vegas, NV, USA, 19 October–13 November 2020
Ghamisi, P.: Multisource and multitemporal data fusion in remote sensing: a comprehensive review of the state of the art. IEEE Geosci. Remote. Sens. Mag. **7**, 6–39 (2019)
Yin, L., Wang, X., Ni, Y., Zhou, K., Zhang, J.: Extrinsic parameters calibration method of cameras with non-overlapping fields of view in airborne remote sensing. Remote Sens. **10**, 1298 (2018)
Yue, R., Xu, H., Wu, J., Sun, R., Yuan, C.: Data registration with ground points for roadside Lidar sensors. Remote Sens. **11**, 1354 (2019)
Knorr, M., Niehsen, W., Stiller, C.: Online extrinsic multi-camera calibration using ground plane induced homographies. In: Proceedings of the 2013 IEEE Intelligent Vehicles Symposium (IV), Gold Coast, Australia, 23–26 June 2013, pp. 236–241 (2013)
Ren, S., He, K., Girshick, R., Sun, J.: Faster R-CNN: towards realtime object detection with region proposal networks. In: Proceedings of the Advances in Neural Information Processing Systems, vol. 7–12, pp. 91–99 (2015)
Dosovitskiy, A., Ros, G., Codevilla, F., Lopez, A., Coltun, V.: CARLA: an open urban driving simulator. In: Proceedings of the 1st Annual Conference on Robot Learning, 13–15 November 2017, pp. 1–16 (2017)
Everingham, M., Van Gool, L., Williams, C.K., Winn, J., Zisserman, A.: The Pascal visual object classes (VOC) challenge. Int. Comput. Vis. **88**, 303–338 (2010)
Xu, S., Liu, H., Gao, F., Wang, Z.: Compressive sensing based radio tomographic imaging with spatial diversity. Sensors **19**, 439 (2019)
Schlosser, J., Chow, C.K., Kira, Z.: Fusing Lidar and images for pedestrian detection using convolutional neural networks. In: Proceedings of the 2016 IEEE International Conference on Robotics and Automation (ICRA), Stockholm, Sweden, 16–21 May 2016, pp. 2198–2205 (2016)
Simonyan, K., Zisserman, A.: Very deep convolutional networks for large-scale image recognition. arXiv 2015 arXiv:1409.1556
Du, X., Ang, M.H., Karaman, S., Rus, D.: A general pipeline for 3D detection of vehicles. In: Proceedings of the 2018 IEEE International Conference on Robotics and Automation, Brisbane, Australia, 21–25 May 2018, pp. 3194–3200 (2018)
Lin, T.-Y., Dollar, P., Girshick, R., He, K., Hariharan, B., Belongie, S.: Feature pyramid networks for object detection. In: Proceedings of the IEEE Conference on Computer Vision and Pattern Recognition, Honolulu, HI, USA, vol. 21–26, pp. 2117–2125 (2017)

Performance Analysis of Wake-Up Radio Based Protocols Considering Non-ideal Transmission Channel

Mayssa Ghribi[1(✉)] and Aref Meddeb[2]

[1] NOCCS Laboratory, National Engineering School of Tunis, University of Tunis-El Manar, Tunis, Tunisia
mayssa.ghribi@enit.utm.tn

[2] NOCCS Laboratory, National Engineering School of Sousse, University of Sousse, Sousse, Tunisia
aref.meddeb@eniso.u-sousse.tn

Abstract. Since its emergence, Wake-up radio technology has solved idle listening and overhearing problems of the duty-cycled approach in wireless sensor networks (WSN). Using collision avoidance mechanisms has furthered the success of WuR by improving the overall performance of wireless communications. These latter are characterized by transmission errors in real environments. The effect of non-ideal transmission channel on the performance of WuR has not yet been investigated. In this paper, a discrete-time Markov chain (DTMC) model that takes into account transmission errors is provided. This model is used to evaluate and compare the performance of two WuR based protocols, with and without collision avoidance, referred to as WuR-CA and WuR-NCA, respectively. We demonstrate that WuR-CA provides better performance than WuR-NCA in terms of reliability and throughput, at the cost of higher delay and power consumption, in both ideal and no-ideal channel conditions.

1 Introduction

In recent years, Wake-up Radio (WuR) approach has been extensively used in IoT networks. WuR gained success over traditional duty-cycled approaches by suppressing idle listening and overhearing [5,7,8]. WuR consists in using two radios instead of one; the Main Radio (MR) and the Wake-up Radio itself, which are respectively used for data and Wake-up Call (WuC). The MR is initially powered off and the WuR is kept active in order to sense the shared communication medium [3]. Upon detecting a WuC by WuR, the MR is powered ON and the data transmission is started. After finishing data transmission, MR is turned off while WuR remains active. Several studies were proposed in literature to prove the effectiveness of WuR [10]. For example, authors in [12] used analytical formulas to compare the performance of a WuR-based MAC (SCM-WuR) with two duty-cycled protocols (a synchronous MAC (S-MAC) and an asynchronous MAC X-MAC) in terms of energy consumption. Results show that SCM-WuR provides the best energy efficiency in unsaturated conditions. In the case of saturated condition, SCM-WuR outperforms X-MAC, but does not systematically save more energy than

L. Barolli et al. (Eds.): AINA 2022, LNNS 449, pp. 67–78, 2022.
https://doi.org/10.1007/978-3-030-99584-3_7

S-MAC. In [7], authors illustrate the benefits of WuR by comparing SCM-WuR with a transmitter initiated MAC based on duty-cycled approach (B-MAC), a receiver initiated MAC (RI-MAC), X-MAC, and IEEE 802.15.4 using simulation. Further, in [13], a comparison of a WuR based protocol with ContikiMAC was presented. Results demonstrate that WuR outperforms duty-cycled protocol in terms of latency, PDR and power consumption when using single-hop networks. However, for large and multi-hop networks, the power consumption of WuR is higher than ContikiMAC especially in the presence of collisions. Latency exhibited with WuR also increases, but it is lower than that incurred using ContikiMAC.

All proposed evaluations of WuR only consider an ideal transmission channel (i.e., the channel is considered to be error-free and only collisions can occur in the network). However, real environments consider communications over a Rayleigh fading channel (ideal and non-ideal transmission channel), which is characterized by the presence of transmission errors [1].

In this paper, we propose a new analytical model using discrete-time Markov chain (DTMC) for a WuR protocol using collision avoidance (WuR-CA) [11], referred to as CSMA-WuR in [2]. The new proposed model takes into account transmission errors. As far as we know, this is the first work that uses DTMC in order to model WuR considering non-ideal transmission channel. We derive mathematical expressions of metrics i.e., reliability, throughput, latency, and power consumption, used to analyze and compare the performances of WuR-CA and a WuR-based protocol that does not implement collision avoidance (WuR-NCA).

The remainder of this paper is structured as follows. Section 2 gives an overview of the evaluated protocols. In Sect. 3, we introduce the analytical WuR-CA model. The metrics analytical expressions are provided in Sect. 4. Numerical results are discussed in Sect. 5 and Sect. 6 concludes the paper.

2 The Principle of Studied WuR-Based MAC

In this section, we introduce the operation principle of WuR-CA and WuR-NCA. Like the unslotted CSMA-CA [6], WuR-CA uses collision avoidance before sending WuCs while these latter are sent without using any collision avoidance mechanism in the case of WuR-NCA.

2.1 WuR-CA

With WuR-CA, each node configures two counters, NB and BE. The former is set to zero and the latter is initialized to the lowest value of the backoff "BO" exponent "macMinBe". The sender performs CCA after arbitrarily picking a BO period in $[0, 2^{BE} - 1]$. The value of BE is set to $\min(BE + 1, macMaxBe)$ and NB is incremented if the channel is busy. Otherwise, the sender starts WuC transmission. When the desired node receives a WuC, it powers ON its MR and data transmission can start. This process is finished when the sender receives an Ack. Thus, both the receiver and the sender power-off their MR and keep their WuRx active in order to detect incoming WuCs.

2.2 WuR-NCA

WuR-NCA, also termed SCM-WuR in [7], adopts the transmitter initiated mode and does not use collision avoidance before sending WuCs. The sender initiates the data transmission process by first transmitting a WuC to the desired node. When this latter receives the WuC, its MR is turned on. Data is then sent to the receiver and an ACK is sent back to the sender at the end of the process. MR of both communicating nodes are then turned off and the WuRx are kept active to permanently sense the channel.

3 DTMC Modeling

Further, Fig. 1 illustrates the DTMC utilized for modeling the behavior of WuR-CA, which adopts an approach similar to the one proposed in [9]. Notations used to elaborate the DTMC are provided in Table 1. In our model, we assume that retransmission is not permitted.

When a node does not have data to send, it remains in state *SL*. Upon the generation of a data packet with a probability *n*, the node moves to the *ON* state, which consists of turning on the MR. Then the node begins the first BO stage. All BO stages are depicted by the state (j, i), $j \in [0, m]$, $i \in [0, W_0)$ in the Markov chain.

If the channel is busy after CCA, the node transits from state $(j, 1)$ to the next BO stage $(j, 2)$. The same process is repeated until the node gains access to the channel or reaches the maximum number of *BO* stages. Thus, the packet will be discarded. The failure and successful transmission states are illustrated using states from $(-2, 0)$ to $(-2, L_f - 1)$ and from $(-1, 0)$ to $(-1, L_s - 1)$, respectively.

The steady state probability of each state (j, i) can be written as:

$$p_{j,i} = \frac{W_j - i}{W_j} p_{j,0} = \frac{W_j - i}{W_j} \alpha^j p_{0,0} = \frac{W_j - i}{W_j} \alpha^j nP_{sl}, \tag{1}$$

Table 1. DTMC notations

Notation	Description
n	Probability of the availability of a new packet in the queue
N	Number of nodes
α	Probability of finding channel occupied after CCA
σ	The duration of a time slot
P_f	Probability of failure transmission
m	Max number of BO stages
P_{sl}	Probability that there is no packet to transmit
P_{on}	Probability of powering ON the MR
L_s	Successful packet size in terms of Backoff
L_f	Failed packet size in terms of Backoff
$p_{j,i}$	Probability states of WuR-CA

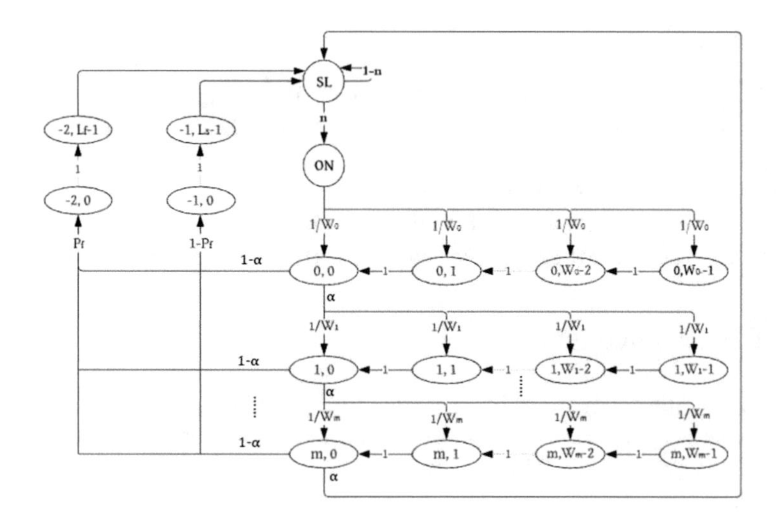

Fig. 1. The Markov chain modeling for WuR-CA protocol.

The successful state probability is:

$$p_{-2,i} = \left(1 - P_f\right)\left(1 - \alpha^{m+1}\right) nP_{sl}. \tag{2}$$

The failure state probability can be expressed as:

$$p_{-1,i} = P_f\left(1 - \alpha^{m+1}\right) nP_{sl}. \tag{3}$$

The probability τ that a node performs CCA after choosing a random BO slot time can be obtained as.

$$\tau = \sum_{j=0}^{m} p_{j,\,0}\,. \tag{4}$$

The probability P_f indicates a failed packet transmission due to either collision (P_{col}) or transmission errors (P_e), and is given by:

$$P_f = P_{col} + P_e = 1 - (1-\tau)^{N-1} + 1 - (1 - BER)^l \tag{5}$$

where l is the packet size and BER is the bit error rate.

The probability α of sensing a busy channel is expressed as:

$$\alpha = P_f(1-\alpha)\left((1-P_f)L_s + P_f L_f\right) = \frac{P_f\left((1-P_f)L_s + P_f L_f\right)}{1 + P_f\left((1-P_f)L_s + P_f L_f\right)} \tag{6}$$

The normalization property of the DTMC is:

$$P_{sl} + P_{on} + \sum_{j=0}^{m}\sum_{j=0}^{W_j-1} p_{j,i} + \sum_{i=0}^{L_s-1} p_{-2,i} + \sum_{i=0}^{L_f} p_{-1,i} = 1\,. \tag{7}$$

4 Performance Analysis

We now use the DTMC model proposed in Sect. 3 to derive the metrics used for performance analysis of WuR-CA. Note that based on the model proposed in [2], we can derive the performance metrics related to WuR-NCA. Recall that WuR-NCA does not perform BO or CCA prior WuC transmission.

4.1 Reliability

Reliability is defined as the ratio of packets successfully received by the sink to the total number of generated packets. P_{fcer} expresses the probability of packet transmission failure due to collision and channel error and $P_{fc} = \alpha^{m+1}$ expresses the probability of failing to access the channel within (m+1) backoff stages.

- **WuR-CA**

$$R_{CA} = 1 - P_{fcer} - P_{fc} = 1 - ((P_f\left(1 - \alpha^{m+1}\right)) + \alpha^{m+1}), \tag{8}$$

- **WuR-NCA**

Since WuR-NCA does not perform BO or CCA prior to WuC transmission, its reliability can be express as:

$$R_{NCA} = 1 - P_{fNCA} = 1 - (P_c + P_e) = 1 - ((1 - e^{-(N-1)\lambda T_s\left(1+e^{T_s\lambda}\right)}) + P_e). \tag{9}$$

4.2 Throughput

The throughput S as defined in [4] is the fraction of time the channel is utilized to successfully send data and is expressed as:

$$S = \frac{P_t P_s (1-P_e) L_p}{(1-P_t)\sigma + P_t(1-P_s)T_f + P_t P_s(1-P_e)T_s + P_t P_s P_e T_f} \tag{10}$$

where,

- $P_t = 1 - (1-\tau)^N$ is the probability that the channel is busy.
- $P_s = \frac{N\tau(1-\tau)^{N-1}}{P_{tr}}$ is the probability of successful transmission.
- T_s and T_f are the time needed to successfully transmit the packet and the duration of failed packet transmission due either collision or channel errors.

4.3 Delay

The average delay of a successful packet transmission includes the time required to access the channel and to send the packet to the sink. In the case of WuR-CA, we have:

$$D_{CA} = T_{succ} + T_{bo} \sum_{j=0}^{m} \frac{(1-\alpha)\alpha^j}{1-\alpha^{m+1}} [(j+1) T_{cca} + \sum_{k=0}^{j} D], \tag{11}$$

where $D = \sum_{i=0}^{W_j-1} \frac{i}{W_j}$, $T_{succ} = T_{WuC} + T_{on} + T_{data} + SIFS + T_{ack}$ is the total duration needed to successfully transit the data packet, T_{bo} is the duration of one BO period, and T_{cca} is the amount of time for sensing. In the case of WuR-NCA, the average delay can be written as:

$$D_{NCA} = P_f T_f + (1-P_f) T_s. \tag{12}$$

4.4 Power Consumption

The Total power consumption is defined as the sum of power consumed by a node during sleep, MR switching, CCA, idle listening, successful transmission, and failed transmission states. In the case of WuR-CA, this can be expressed as:

$$\rho_{CA} = \rho_s P_{sl} + \rho_{sw} P_{on} + \rho_{cca} \sum_{j=0}^{m} p_{j,0} + \rho_i \sum_{j=0}^{m} \sum_{i=1}^{W_j-1} p_{j,i} + \rho_{succ}^{tx} \sum_{i=0}^{L_s-1} p_{-2,i} + \rho_f^{tx} \sum_{i=0}^{L_f-1} b_{-1,i}, \tag{13}$$

where ρ_s, ρ_{sw}, ρ_{cca}, ρ_i, ρ_{succ}^{tx}, and ρ_f^{tx} represent the power consumed at sleep, MR switching, CCA, idle listening, successful transmission, and failed transmission. In the case of WuR-NCA, the power consumed by a given node can be expressed as:

$$\rho_{NCA} = \rho_f^{tx} P_f + \rho_{succ}^{tx} \left(1 - P_f\right). \tag{14}$$

Performance metrics used for performance evaluation over an ideal transmission channel are obtained by setting the value of BER to zero in Eq. (5).

5 Results and Discussion

In this section, we present and discuss numerical results obtained using Matlab. We analyze the performance of WuR-CA in comparison, with WuR-NCA which doesn't perform neither BO nor CCA before WuC transmission.

We consider a network of 15 nodes with three different scenarios. In the first scenario, we study the effect of varying the number of nodes N and WuC duration on the performance metrics under both error-free and error-prone channel conditions. In the second and third scenarios, we vary the arrival rate λ and the data packet size L_p, respectively, in ideal and non-ideal transmission channel conditions (i.e., when BER = 0 and BER = 10^{-3}). In the forth scenario, we vary the BER and fix the arrival rate to $\lambda = 1$ packets/s and packet size to $L_p = 30$ octets The values of ρ_s, ρ_{sw}, ρ_{cca}, ρ_i, ρ_{data}^{tx}, ρ_{WuC}^{tx}, ρ_{data}^{rx} and ρ_{WuC}^{rx} are set to 20 µa, 2.7 µA, 20.28 mA, 3.5 µA, 17.4 mA, 152 mA,

18.8 mA and 8 μA. The values of macMinBE, macMaxBE, data rate and m are equals to 3, 8, 250 kbps and 5. Finally, we set the values of $T_{WuC}, T_{on}, SIFS, T_{ack}, T_{bo}, T_{cca}$, MAC and Phy header lengths to be 6 ms, 1.79 ms, 192 μs, 160 μs, 320 μs, 1.92 ms, 16 octets and 6 octets, respectively.

5.1 Variation in Number of Nodes

We first study the impact of the number of nodes and WuC duration (i.e., WuC length) on the performance of both protocols under error-free and error-prone channel conditions.

5.1.1 Error-Free Channel Conditions

Figure 2 illustrates the effect of number of nodes *N* on reliability, throughput, delay, and power consumption under error-free channel conditions. As shown in this figure, WuR-CA outperforms WuR-NCA in terms of reliability and throughput regardless of the value of WuC duration. This is because it adopts collision avoidance mechanism prior to WuC transmission. Furthermore, to achieve a better reliability when *N* increases, WuR-CA needs to perform additional BO and CCA, which generates a longer average delay and a higher power consumption as shown in Fig. 2(c) and 2(d). The average delay of WuR-NCA remains constant when *N* increases and with the same WuC duration. The reason is that WuR-NCA needs the same duration for a failed or successful transmission. It is also noticeable from Fig. 2(d) that both variation in *N* and WuC duration do not affect the power consumption of WuR-CA. In fact, when the channel is error-free, WuR-CA almost provides the same reliability (always higher than 96%) in all cases of WuC duration, and it needs to perform the same number of CCA therefore it requires the same amount of energy.

5.1.2 Error-Prone Channel Conditions

Figure 3 illustrates the effect of the number of nodes *N* on reliability, throughput, delay, and power consumption under error-prone channel conditions. It is clearly observable that the performance of both protocols significantly decreases when the channel is error-prone. WuR-CA still outperforms WuR-NCA in terms of reliability and throughput at the cost of a higher delay. It also consumes less energy than WuR-NCA mainly when the transmission fails. Figure 3(d) also shows that the power consumption of WuR-CA decreases with the WuC duration. The reason is that the probability of packet rejection after CCA due to collision increases, therefore the power consumption of WuR-CA decreases since it consumes less energy with a failed packet transmission. This figure also illustrates the benefits of using collision avoidance. In fact, we can clearly remark from Fig. 3(a) that WuR-NCA totally loses its effectiveness when the number of nodes is equal or higher than 25 and the WuC duration is equal to 12.5 ms. Hence, it is strongly recommended using a shorter WuC duration (i.e., a shorter WuC size and a higher data rate) to achieve a better performance when the network is dense, and the channel is error-prone.

5.2 Effect of the Traffic Load

Figure 4 depicts the effect of varying the traffic load (λ). The performance provided by WuR-CA is always better than that offered by WuR-NCA in both ideal and non-ideal channel conditions and with an increased density. Figure 4(a) shows that the reliability of WuR-CA decreases slightly when we increase λ. In fact, the reliability of WuR-CA is kept above 99% in non-fading network for $\lambda = 5$ packets/s. The gap between the reliability of WuR-CA when BER = 0 and BER = 10^{-3} is important, despite the use of CCA and BO. This is because in the presence of fading, the probability of collisions increases and yields a higher rate of transmission failures.

Figure 4(b) shows that the throughput of WuR-CA exhibits a linear behavior as function of λ in fading and non-fading conditions. However, the throughput of WuR-NCA increases when BER = 0 and λ is equal to or below 4. When BER = 10^{-3}, the throughput of WuR-NCA slightly increases for low traffic loads and λ below 3. In all other cases, the throughput decreases with the traffic load.

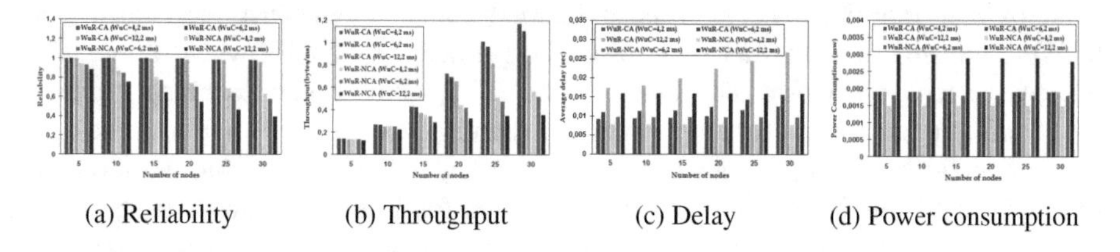

(a) Reliability (b) Throughput (c) Delay (d) Power consumption

Fig. 2. Effect of N and WuC duration on reliability, throughput, delay and power consumption under error-free channel.

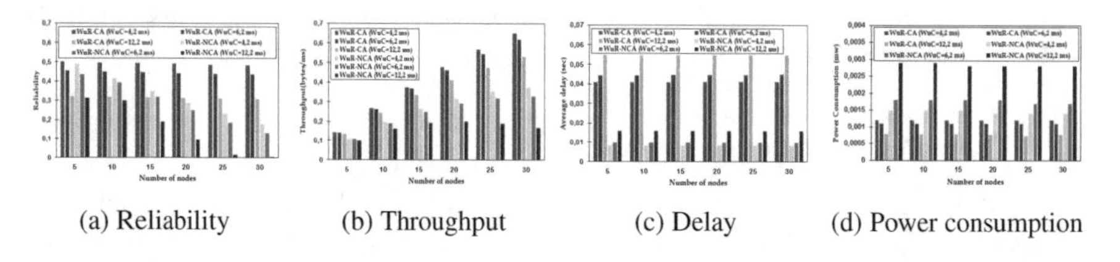

(a) Reliability (b) Throughput (c) Delay (d) Power consumption

Fig. 3. Effect of N and WuC duration on reliability, throughput, delay and power consumption under error-prone channel.

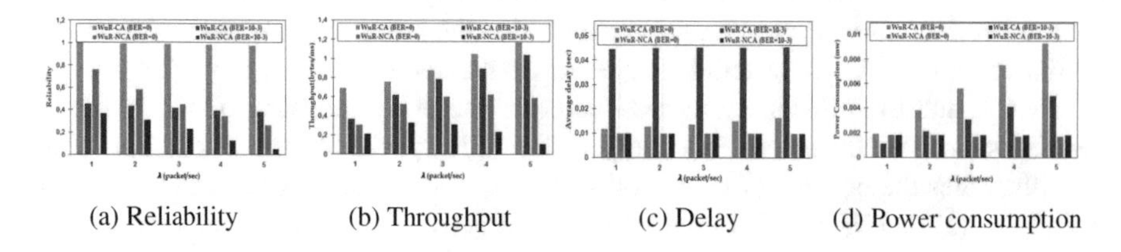

(a) Reliability (b) Throughput (c) Delay (d) Power consumption

Fig. 4. Effect of traffic load on reliability, throughput, delay and power consumption.

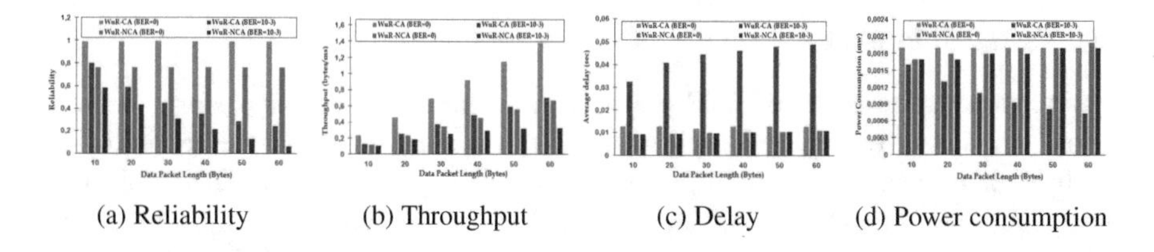

(a) Reliability (b) Throughput (c) Delay (d) Power consumption

Fig. 5. Effect of data packet length on reliability, throughput, delay and power consumption.

As shown in Fig. 4(c) and 4(d), the average delay and power consumption of WuR-NCA are constant, even for high traffic loads and in presence of fading. In fact, WuR-NCA needs the same duration and consumes the same amount of energy for both successful and failed packet transmission, and does not use any collision avoidance mechanism. Therefore, it provides a smaller delay than that offered by WuR-CA. Increasing λ yields a higher average delay with WuR-CA, in both ideal and non-ideal channel conditions, due to the need for additional BO stages caused by transmission failure. This increases power consumption of WuR-CA.

5.3 Variation in Data Packet Length

Figure 5 depicts the effect of data packet length (L_p). It is clearly observable from Fig. 5(a) that when BER = 0, L_p doesn't affect the reliability of WuR-CA, which is practically constant and nearly equal to 1. The reliability of WuR-NCA exhibits the same behavior as WuR-CA in ideal channel conditions. However, in the presence of fading, the reliability of both protocols decreases significantly. This is because the probability of transmission errors increases linearly with L_p.

As shown in Fig. 5(b), the throughput of both protocols increases with L_p. Moreover, the throughput provided in ideal channel conditions is higher than that offered in the presence of fading. This figure also highlights the benefits of using collision avoidance. In fact, due to the use of CCA plus BO, WuR-CA provides a throughput two times higher than that offered by WuR-NCA when BER = 0 and $L_p = 60$ octets.

Figure 5(c) shows that in the absence of fading, the delay with WuR-CA is constant as far as the reliability is maintained constant (Fig. 5(a)). In presence of fading, the delay of WuR-CA increases with L_p because additional time is needed to transmit a large packet. The delay of WuR-NCA (in ideal and non-ideal channel) slightly increases with L_p. This can be explained by the fact that the duration of data packet mainly depends on L_p.

As shown in Fig. 5(d), the power consumption of WuR-NCA increases slightly with L_p in both channel conditions. WuR-CA consumes the same amount of energy in absence of fading, with an increased L_p. Since the reliability of WuR-CA is constant, the same amount of energy is consumed. However, in the presence of fading, the power consumption of WuR-CA decreases as reliability decreases. As we mentioned earlier, L_p has a major impact on the probability of transmission errors which causes earlier data packet discard. This requires less BO stages and less power consumption.

5.4 Variation in BER

Figure 6 depicts the effect of BER on the various metrics used for performance evaluation. As shown in Fig. 6(a) and 6(b), when we increase BER, the reliability and the throughput decrease for both protocols. WuR-CA provides a better reliability and throughput than WuR-NCA. This is because WuC is sent without performing CCA or BO in the case of WuR-NCA, which increases the possibility of collisions. We can also notice that the average delay increases significantly with WuR-CA when we increase BER. This can be explained by the fact that more backoff stages are required by WuR-CA in the presence of many errors. On the other hand, WuR-NCA maintains constant delay and power consumption, as illustrated in Fig. 6(c) and 6(d), respectively. This is due to the fact that WuR-NCA consumes the same amount of energy in the case of successful or failed transmissions. However, the power consumption of WuR-CA decreases notably with BER. In fact, WuR-CA consumes less energy in the case of failed transmissions compared to successful transmissions.

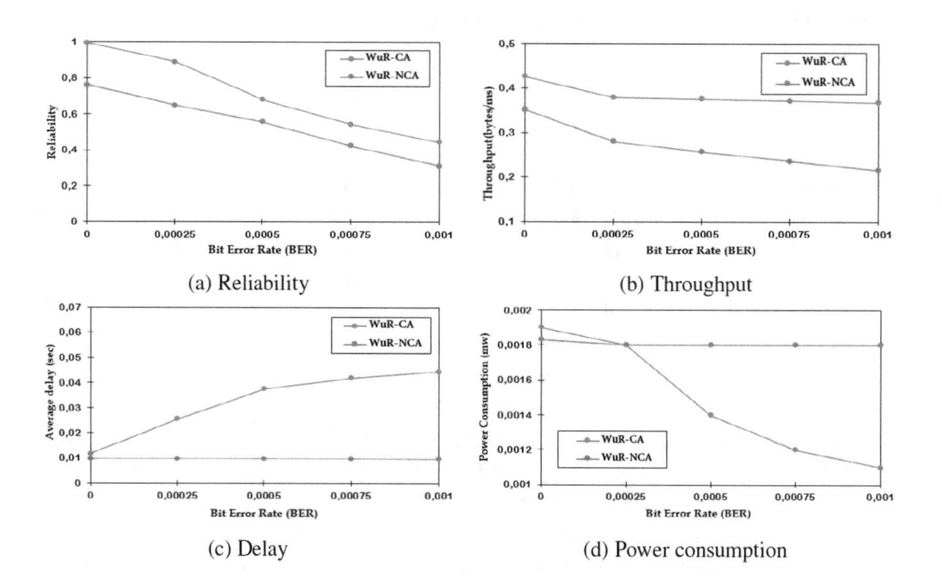

Fig. 6. Effect of BER on reliability, throughput, delay and power consumption.

6 Conclusion

In this paper, we studied the performance of WuR under fading channel by developing a DTMC model that takes into account transmission errors. We have illustrated the benefits of using collision avoidance prior to WuC transmission. A comparison of two WuR based protocols, with and without Collision Avoidance (WuR-CA and WuR-NCA), was conducted by considering ideal and non-ideal transmission channel. Numerical results show that when the number of nodes increases and the channel is error-free, WuR-CA provides a better performance than WuR-NCA. This latter loses its effectiveness with an error prone channel and a dense network. In fact, reducing the duration of WuC is the best alternative to obtain a better performance with WuR-NCA because a shorter duration is needed to send a WuC, which leads to a smaller channel occupancy probability and therefore, a smaller probability of collision. Furthermore, when increasing BER, WuR-CA outperforms WuR-NCA in terms of reliability, throughput, and power consumption at the cost of a higher delay. Variation in arrival rates has no impact on the reliability of WuR-CA when BER = 0, but it affects its delay and power consumption because WuR-CA needs to perform more BO stages to achieve high reliability. Increasing data packet size only affects the throughput of WuR-CA when BER = 0 and it impacts all metrics for WuR-NCA when BER = 10^{-3}. Power consumption and delay of WuR-NCA are only affected when increasing L_p because the probability of transmission errors mainly depends on data packet size. As a future work, we will focus in studying the applicability of Quality of Service (QoS) mechanisms and service differentiation in WuR-based networks.

References

1. Alkama, L., Bouallouche-Medjkoune, L., Bachiri, L.: Modeling and performance evaluation of the IEEE 802.15.4K CSMA/CA with priority channel access mechanism under fading channel. Wireless Pers. Commun. **115**(1), 527–556 (2020). https://doi.org/10.1007/s11277-020-07584-9
2. Ghose, D., Li, F.Y., Pla, V.: MAC protocols for wake-up radio: principles, modeling and performance analysis. IEEE Trans. Industr. Inf. **14**(5), 2294–2306 (2018)
3. Spenza, D., Magno, M., Basagni, S., Benini, L., Paoli, M., Petrioli, C.: Beyond duty cycling: wake-up radio with selective awakenings for long-lived wireless sensing systems. In: 2015 IEEE Conference on Computer Communications (INFOCOM), pp. 522–530 (2015)
4. Daneshgaran, F., Laddomada, M., Mesiti, F., Mondin, M.: Unsaturated throughput analysis of IEEE 802.11 in presence of non ideal transmission channel and capture effects. IEEE Trans. Wirel. Commun. **7**(4), 1276–1286 (2008)
5. Demirkol, I., Ersoy, C., Onur, E.: Wake-up receivers for wireless sensor networks: benefits and challenges. IEEE Wirel. Commun. **16**(4), 88–96 (2009).
6. IEEE Standard for Low-Rate Wireless Networks. In: IEEE Std 802.15.4-2015 (Revision of IEEE Std 802.15.4-2011), pp. 1–709, 22 April 2016
7. Oller, J., Demirkol, I., Casademont, J., Paradells, J., Gamm, G.U., Reindl, L.: Has time come to switch from duty-cycled MAC protocols to wake-up radio for wireless sensor networks? IEEE/ACM Trans. Networking **24**(2), 674–687 (2016)
8. Kozłowski, A., Sosnowski, J.: Energy efficiency trade-off between duty-cycling and wake-up radio techniques in IoT networks. Wireless Pers. Commun. **107**(4), 1951–1971 (2019). https://doi.org/10.1007/s11277-019-06368-0
9. Kiran, M.P.R.S., Rajalakshmi, P.: Performance analysis of CSMA/CA and PCA for time critical industrial IoT applications. IEEE Trans. Industr. Inf. **14**(5), 2281–2293 (2018)
10. Ghribi, M., Meddeb, A.: Survey and taxonomy of MAC, routing and cross layer protocols using wake-up radio. J. Network Comput. Appl. **149**, 102465 (2020)
11. Ghribi, M., Meddeb, A.: Performance evaluation of collision avoidance techniques using wake-up radio in WSNs. In: 2020 International Conference on Software, Telecommunications and Computer Networks (SoftCOM), pp. 1–6 (2020)

12. Zhang, M., Ghose, D., Li, F.Y.: Does wake-up radio always consume lower energy than duty-cycled protocols?. In: 2017 IEEE 86th Vehicular Technology Conference (VTC-Fall), pp. 1–5 (2017)
13. Sampayo, S.L., Montavont, J., Prégaldiny, F., Noël, T.: Is Wake-up radio the ultimate solution to the latency-energy tradeoff in multi-hop wireless sensor networks?. In: 2018 14th International Conference on Wireless and Mobile Computing, Networking and Communications (WiMob), pp. 1–8 (2018)

CaWuQoS-MAC: Collision Avoidance and QoS Based MAC Protocol for Wake-Up Radio Enabled IoT Networks

Mayssa Ghribi[1](✉) and Aref Meddeb[2]

[1] NOCCS Laboratory, National Engineering School of Tunis, University of Tunis-El Manar, Tunis, Tunisia
mayssa.ghribi@enit.utm.tn

[2] NOCCS Laboratory, National Engineering School of Sousse, University of Sousse, Sousse, Tunisia
aref.meddeb@eniso.u-sousse.tn

Abstract. IoT applications have different requirements in terms of quality of service (QoS). For real-time applications, data must be processed as fast as possible. In contrast, some applications are delay-tolerant and support long data delivery delay. Recently, Wake-up radio (WuR) technology has gained great attention by eliminating idle listening and overhearing. However, so far QoS has not been a concern in WuR-based MAC protocols. In this paper, a Collision Avoidance QoS- and WuR-based MAC protocol (CaWuQoS-MAC) is proposed in order to support delay requirements for real-time IoT applications. CaWuQoS-MAC offers the possibility to specify the delay limit within which the packet needs to be delivered and after which is considered as expired. We develop a discrete-time Markov chain model to evaluate the performance of the proposed protocol. CaWuQoS-MAC is compared with another WuR protocol that does not perform Collision Avoidance prior to Wake-up Call (WuC) transmission in terms of reliability, delay, and power consumption under both error-free and error-prone channel conditions. Numerical results show that CaWuQoS-MAC supports the desired QoS requirements when the channel is error-free under both light and heavy traffic loads and with a high network density.

1 Introduction

Internet of Things (IoT) concerns all objects connected to the Internet. IoT applications are varied and cover many areas such as industry, healthcare, home automation, etc. [1]. Some IoT applications are time sensitive [16] and require a bounded delay guarantees [7] to process their data in real time. Others are less demanding in terms of time and tolerate delays. This variety of IoT application requirements necessitates the use of QoS mechanisms [15].

WuR is one of several technologies that have promoted the success of IoT by eliminating idle listening and overhearing in wireless communications [9,11].

L. Barolli et al. (Eds.): AINA 2022, LNNS 449, pp. 79–90, 2022.
https://doi.org/10.1007/978-3-030-99584-3_8

In WuR-based networks, a node uses two radios instead of one: the main radio (MR), which is used for data transmission and the WuR, which is utilized for Wake-up Call (WuC) transmission. In such network, data transmission process is initialized by sending a WuC to the intended node, which powers on its MR upon receiving this WuC. The MR of both communicating nodes are powered off after finishing data transmission, while WuRx remain active in order to sense the medium permanently [5]. Many researchers investigated energy efficiency, collision avoidance, and protocol design using WuR [13]. To the best of our knowledge, so far QoS has not yet been studied in the WuR context.

In this paper, we propose a MAC protocol, called CaWuQoS-MAC, that supports both collision avoidance prior WuC transmission and QoS. In fact, CaWuQoS-MAC enables Clear Channel Assessment (CCA) and backoff (BO) before the transmission of WuCs. Furthermore, in order to control delays incurred by collision avoidance mechanisms [14] and to guarantee QoS in terms of end to end delay, CaWuQoS-MAC offers the possibility to specify the delay limit within which the packet needs to be sent before it is considered as expired. Moreover, we developed a Discrete Time Markov Chain (DTMC) model that takes both collision and transmission errors into account in order to evaluate the performance of CaWuQoS-MAC. We then derive mathematical expressions of the various metrics including reliability, delay, and power consumption. In our study, two different traffic classes with different delay limits were considered in the performance evaluation.

The remainder of this paper is structured as follows. In Sect. 2, we give an insight on related work. In Sect. 3, we introduce the operation principle of the proposed protocol and the analytical model used to analyze CaWuQoS-MAC. Performance metrics are provided in Sect. 4. Numerical results and discussion are illustrated in Sect. 5, before the paper is concluded in Sect. 6.

2 Related Work

Only a few researchers considered collision avoidance between WuCs [13]. Authors in [3] propose three MAC protocols (CCA-WuR, CSMA-WuR, and ADP-WuR) in order to improve the performance of WuR under both heavy and light traffic loads by enabling collision avoidance before WuC transmission. CCA-WuR activates clear channel assessment to verify the availability of the channel before the transmission of WuCs and offers the possibility to set the maximum number of CCA attempts. CSMA-WuR uses a collision avoidance mechanism similar to the unslotted CSMA proposed in [10]. In fact, it performs both CCA and BO procedure prior WuC transmission.

ADP-WuR [3] adapts its operation according to traffic conditions. It first uses CCA-WuR for two attempts. If it gains access to the channel, it proceeds to data transmission. Otherwise, it utilizes CSMA-WuR at the third attempt. CCA and BO are performed until the maximum number of attempts is reached. Results show that CCA is a good alternative when the traffic is light. CSMA-WuR and ADP-WuR provide a good performance in terms of WuC loss probability at the cost of a higher delay and energy consumption when the traffic is heavy [3].

Another WuR protocol is proposed in [4] to avoid collision between WuCs by using BO procedure and CCA with a fixed-size contention window (CW). If the channel is available after performing CCA, a node randomly chooses a BO period in the range [0, CW] and it then sends the WuC. Otherwise, the CCA procedure is repeated until the maximum number of attempts is attained. Bo-WuR was evaluated using DTMC under saturated traffic conditions. Results indicate that BO-WuR has better performance in terms of throughput and collision probability, delay and energy efficiency [4].

3 Protocol Design and DTMC Modeling of CaWuQoS-MAC

In this section, we describe the protocol design and introduce the Markov chain used to model and evaluate CaWuQoS-MAC protocol.

3.1 Protocol Design

The main purpose of the CaWuQoS-MAC protocol is to avoid collision among WuCs, while respecting the delay constraints imposed by IoT applications. CaWuQoS-MAC uses a collision avoidance mechanism similar to the one used by the Prioritized Contention Access proposed in IEEE 802.15.4-2015 [10]. CaWuQoS-MAC specifies a parameter termed "TolDelay" as the maximum delay accepted for packet transmission. Data transmission process is initialized by setting the BE (Backoff Exponent) counter to $\max(1, macMaxBe{-}1)$ and TolDelay to zero. The sender node randomly chooses a BO period (TB) in [0, $2^{BE} - 1$]. CCA is then performed to verify the availability of the channel. If this latter is busy, the elapsed time (delay) is incremented. Otherwise, if TB is equal to zero, the node gains access to the channel and starts the transmission of the WuC. It then waits for the MR to be turned ON before it starts transmission. TB is decremented by one when it is not equal to zero. If the measured delay is higher than $TolDelay$, the transmission is failed and the packet is rejected.

3.2 DTMC Modeling

We use the DTMC presented in Fig. 1 and inspired by the mechanism adopts in [12]. A node remains in the SL state until it has a packet to send. Thus, it transits to ON state, which means that the node is turning on its MR. Then, it moves to the state $(i, 1)$ after choosing a random BO period. All backoff waiting time is represented by the state (i, j) in the DTMC. If the channel is occupied after CCA, the node transits from state $(i, 1)$ to $(i, 2)$. The same process is repeated until the node gains access to the channel or surpasses the delay limit d, which generates the packet rejection. Successful and failed transmission attempts are represented by states $(-2, 0)$ and $(-1, 0)$, respectively. Symbols L_s and L_f represent the successful and failed packet size in terms of backoff periods. In this paper, we do not implement acknowledgements and retransmissions because they both increase energy consumption and yield high probability of collision [2]. Hence, we rely on the application or a reliable transport protocol to deal with transmission errors.

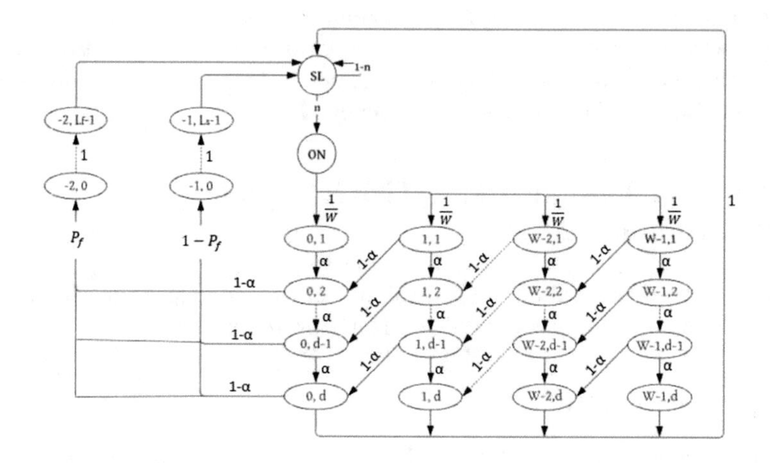

Fig. 1. The Markov chain modeling for CaWuQoS-MAC protocol.

Based on the Markov chain presented in Fig. 1, we can formulate the transition probabilities of CaWuQoS-MAC as follows.

- Transition probability from the last BO stage to the success state:

$$P(-2, 0 \mid 0, j) = (1 - \alpha)(1 - P_f) \quad \forall\, j \in [0,\ d) \tag{1}$$

- Transition probability from the last BO stage to the failure state:

$$P(-1, 0 \mid 0, j) = (1 - \alpha)\, P_f \quad \forall\, j \in [0,\ d) \tag{2}$$

- The probability of increasing delay in the same BO period due to channel occupancy:

$$P(i, j \mid i,\ j - 1) = \alpha \quad \forall\ \ j\ \in (1, d],\ i \in [0,\ W) \tag{3}$$

The steady probabilities related to the Markov chain expressed as follows.

- The steady state probability of each state (i, k) can be presented as.

$$b_{i,j} = \sum_{k=0}^{\min(W-1-i,\ j-1)} C_{j-1}^{k}\, (1 - \alpha)^{k} \alpha^{j-k-1} b_{i,1}, \tag{4}$$

Where,

$$b_{i,1} = \frac{nP_{sl}}{W} \quad \forall\, i \in [0, W - 1] \tag{5}$$

- The successful state probability is:

$$b_{-2,0} = (1 - P_f)\,(1 - \alpha) \sum_{j=1}^{d-1} b_{0,j} \quad . \tag{6}$$

- The failure state probability is:

$$b_{-1,0} = P_f\,(1 - \alpha) \sum_{j=1}^{d-1} b_{0,\ j} \quad . \tag{7}$$

The normalization property of the proposed Markov chain can be expressed as:

$$P_{sl} + P_{on} + \sum_{i=0}^{W-1} \sum_{j=1}^{d} b_{i,j} + \sum_{j=0}^{L_s-1} b_{-2,j} + \sum_{j=0}^{L_f-1} b_{-1,j} \ = 1 \ . \tag{8}$$

The probability τ that a node attempts CCA is derived from (9).

$$\tau = \sum_{j=1}^{d-1} b_{0,j} \tag{9}$$

P_f gives the probability of failed transmission due either collision (P_c) or transmission errors (P_e), and it is given by:

$$P_f = \ P_c + \ P_e \ = 1 - (1 - \tau)^{N-1} + 1 - (1 - BER)^l, \tag{10}$$

where l is the packet size and BER is the bit error rate.

The probability of channel occupancy α can be expressed as:

$$\alpha = P_f\,(1 - \alpha)\,(L_s\,(1 - P_f) + L_f P_f) = \frac{P_f\,(L_s\,(1 - P_f) + L_f P_f)}{1 + P_f\,(L_s\,(1 - P_f) + L_f P_f)} \tag{11}$$

4 Performance Analysis

In this Section, we derive mathematical expressions of the performance metrics based on the DTMC given in Fig. 1.

4.1 Reliability

Reliability is defined as the percentage of data packets that are successfully delivered to the sink among the total number of generated packets.

$$R = 1 - P_d - P_f(1 - P_d) \tag{12}$$

where P_d indicates the probability of packet rejection (discard) due to delay limit surpassing, and it can be written as:

$$P_d = \frac{\sum_{i=0}^{w-1} b_{i,d}}{nP_{sl}} = \frac{\sum_{i=0}^{w-1} \sum_{k=0}^{\min(w-1-i,\ d-1)} C_{d-1}^{k} (1 - \alpha)^k \alpha^{d-1-k}}{w} \tag{13}$$

To obtain the reliability over an error-free channel, we must set the value of BER to zero.

4.2 Delay

The delay of a successful data packet transmission is defined as the time required to successfully deliver the packet to the sink, including the time needed to access the channel. We have:

$$D = T_s + T_{bo} \frac{Path_s}{\sum_{j=1}^{d-1} b_{0,j}} \quad (14)$$

where

- $T_s = L_s T_{bo}$ is the duration of a successful packet transmission.
- T_{bo} gives the duration of a BO period.
- L_s indicates the successful packet size in terms of backoff periods.
- $Path_s$ indicates all paths in the Markov chain that conduct to successful transmission state from the "sl" state.

4.3 Power Consumption

The power consumption ρ is obtained by the summation of the power consumed in each state. The power consumed by a node can be expressed as:

$$\rho = \rho_s P_{sl} + \rho_{sw} P_{on} + \rho_{cca} \sum_{i=0}^{W-1} \sum_{j=1}^{d} b_{i,j} + \rho_s^{tx} \sum_{j=0}^{L_s-1} b_{-2,j} + \rho_f^{tx} \sum_{j=0}^{L_f-1} b_{-1,j} \quad (15)$$

where ρ_s, ρ_{sw}, ρ_{cca}, ρ_i, ρ_{succ}^{tx} and ρ_f^{tx} indicates the power consumed at sleep, main radio switching, CCA, idle listening, successful, and failed transmission, respectively.

5 Results and Discussion

We compare the performances of our protocol CaWuQoS-MAC and SCM-WuR [9]. Simulation parameters are specified in Table 1. We consider the following scenarios:(1) We firstly vary the number of nodes from 5 to 30 with a pitch of 5 while utilizing two data packet sizes (35 bytes and 50 bytes) under both error-free and error-prone channel conditions. (2) We vary the BER for a fixed λ (1 packet/s). (3) We vary the arrival rates λ for $BER = 0$ and $BER = 0.0005$. Two service classes are considered for CaWuQoS-MAC, with a delay limit of 15 ms and 20 ms [8].

Table 1. Simulation parameters [3,17]

	Parameter	Value
Common	Supply power	3.3 V
Main radio	Bit-rate	250 kbps
	Tx current	17.4 mA
	Rx current	18.8 mA
	Idle current	20 μa
	SIFS	192 μs
	L_P	35 bytes
	MAC header length	16 bytes
	Phy header length	6 bytes
Wake-up radio	TX current	152 mA
	Rx current	8 μA
	Idle state Current	3.5 μA
	Switching Current	2.7μA
	T_{CCA}	128 μs
	T_{ON}	1.79 ms
	T_{BO}	320 μs
	T_{WuC}	6 ms
	macMinBE	3
	macMaxBE	8
	macMaxBackoffs	5

5.1 Performance Analysis with Variable N

We first evaluate the effect of the number of nodes under error-free and error-prone channel conditions.

5.1.1 Error-Free Channel Conditions

Figure 2 depicts the effect of number of nodes N on reliability, delay, and power consumption under error-free channel conditions. As shown in Fig. 2(a), when the number of nodes is below 15, CaWuQoS-MAC has the same reliability regardless of the value of d and payload size. Otherwise, the reliability of CaWuQoS-MAC decreases when N increases or d decreases. This is due to the fact that with a dense network, the probability of channel occupancy increases and leads to a higher probability of collision. For the same reason, the reliability of SCM-WuR decreases with N. CaWuCoS-MAC outperforms SCM-WuR in all cases because as it performs collision avoidance prior WuC transmission.

As shown in Fig. 2(b), the delay of CaWuQoS-MAC is constant and almost equal to the delay limit, with the same value of d, even if N increases. This can be explained by the fact that CaWuQoS-MAC continues to perform the

CCA procedure until the maximum delay limit is reached. It is also noticeable according to Fig. 2(b) that when the reliability of CaWuQoS-MAC is higher than 0.9, its power consumption remains constant, while it decreases below this reliability value.

Both delay and power consumption of SCM-WuR, as illustrated in Fig. 2(b) and Fig. 2(c), are constant and do not depend on N. This is because SCM-WuR consumes the same amount of energy and needs the same duration for both failed and successful data transmissions.

5.1.2 Error-Prone Channel Conditions

Figure 3 plots the effect of number of nodes N on reliability, delay, and power consumption under error-prone channel. It is clearly remarkable from Fig. 3(a) that the reliability of CaWuQoS-MAC decreases with an increased N and reduced d. This can be explained by the fact that with a higher value of d, a node has more chances to gain access to the channel because it performs more CCA. The reliability of SCM-WuR also decreases with N. This is because there is no collision avoidance, which increases the probability of collision when the network is dense. Moreover, the reliability of the two protocols decreases with a larger payload size L_p. This is because the likelihood of transmission errors increases with L_p, yielding a higher probability of failed transmissions.

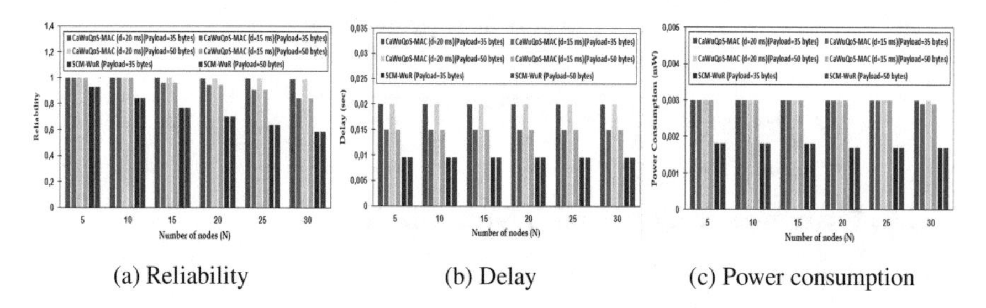

(a) Reliability (b) Delay (c) Power consumption

Fig. 2. Effect of number of nodes on reliability, delay, and power consumption under error-free channel.

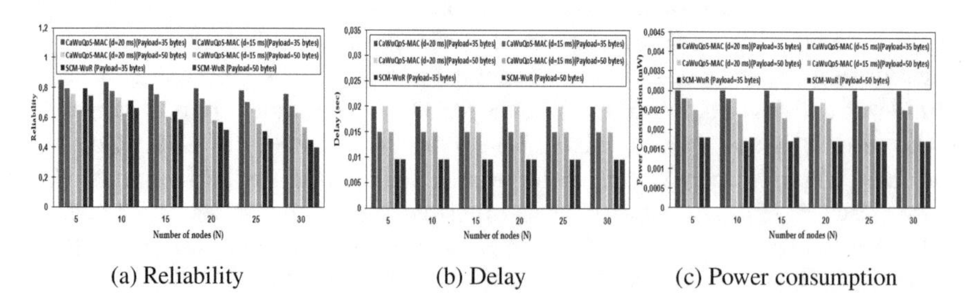

(a) Reliability (b) Delay (c) Power consumption

Fig. 3. Effect of number of nodes on reliability, delay and power consumption under error-prone channel.

As illustrated in Fig. 3(a), the delay of the two protocols exhibits the same behavior as in error-free channel conditions for all values of d and L_p. As we mentioned earlier, CaWuQoS-MAC continues to perform CCA until it attains the delay limit, while SCM-WuR does not perform CCA or BO. Thus, SCM-WuR does not require additional delay to successfully deliver data packets to the sink, compared with a failed transmission attempt.

As shown in Fig. 3(c), with CaWuQoS-MAC, the higher the reliability, the higher the power consumption. In fact, CaWuQoS-MAC performs an increased number of CCA and BO for large values of d. Further, the power consumption of CaWuQoS-MAC decreases with the payload when d is small because of the high probability of packet discard. Furthermore, the power consumption of SCM-WuR with a large payload is slightly high because the probability of transmission errors mainly depends on L_p.

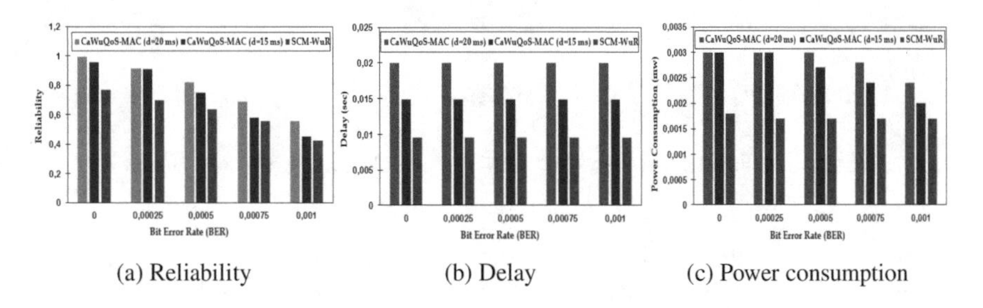

(a) Reliability (b) Delay (c) Power consumption

Fig. 4. Effect of BER on reliability, delay, and power consumption.

5.2 Performance Analysis vs. BER

Figure 4(a) depicts the reliability as function of BER. This figure shows that CaWuQoS-MAC outperforms SCM-WuR for both $d = 15\,\text{ms}$ and $d = 20\,\text{ms}$ due to collision avoidance. However, this reliability decreases for both protocols with BER. In fact, under error-prone channel conditions, the collision probability increases in the network and causes a raised failure transmission probability. We can also notice that CaWuQoS-MAC provides better reliability for an increased value of d.

Figure 4(b) illustrates the variation in delay vs BER. We can notice that the delay of CaWuQoS-MAC with $d = 20\,\text{ms}$ is the highest compared to with $d = 15\,\text{ms}$. In fact, for $d = 20\,\text{ms}$, CaWuQoS-MAC performs more CCA in order to gain access to the channel. It is also remarkable that the delay of the three protocols remains constant regardless of BER (almost equal to d in the case of CaWuQoS-MAC. This can be explained, in the case of CaWuQoS-MAC, by the fact that it performs CCA until the maximum limit d to discard the packet is reached. SCM-WuR provides the shortest delay because it does not perform any collision avoidance.

As shown in Fig. 4(c), SCM-WuR exhibits a constant power consumption when increasing BER, because it consumes the same amount of energy in the case of successful or failed transmissions. The power consumption of CaWuQoS-MAC decreases with BER for both values of the delay limit d. In fact, the probability of transmission failure increases when the channel is error-prone leading to less power consumption because the power consumed when the transmission fails is lower than that consumed with successful transmissions.

5.3 Performance Analysis with Variable Traffic Load

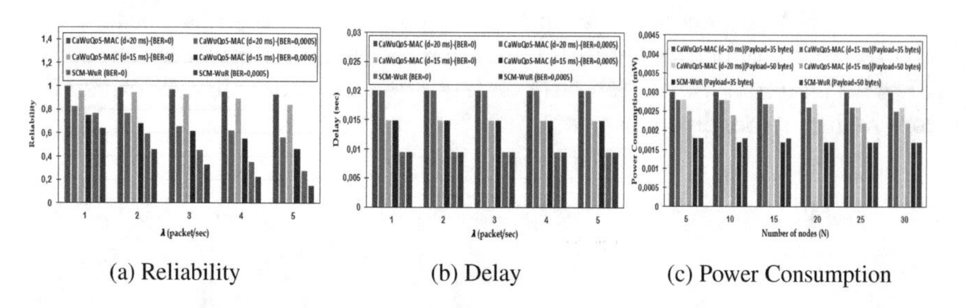

(a) Reliability (b) Delay (c) Power Consumption

Fig. 5. Reliability, delay, and power consumption with variable λ.

Figure 5 plots the effect of λ on the performance metrics. As shown in Fig. 5(a), the reliability of all protocols exhibits the same behavior and decreases with λ because the channel occupancy increases for higher traffic loads. When $d = 20$ ms, CaWuQoS-MAC exhibits the best reliability for both error-free and error-prone channel conditions. This is because the probability of packet discard caused by exceeding the delay limit d deceases for a large delay tolerance, yielding a better reliability. The reliability of SCM-WuR decreases significantly when the channel is error-prone and λ is equal to or higher than 3 packets/s. In contrast, CaWuQoS-MAC provides a higher reliability (higher than 0.5) when the channel is error-prone, $\lambda = 3$ packets/s, and $d = 15$ ms.

Figure 5(b) shows that the delay incurred by SCM-WuR is lower than that provided by CaWuQoS-MAC. This is because the latter performs CCA and BO, which increases the delay. The delay of SCM-WuR is constant even for high traffic loads and in presence of fading because it does not utilize collision avoidance. In fact, the same duration is required for successful and failed transmission attempts.

Figure 5(c) shows that the power consumption of CaWuQoS-MAC increases when increasing the traffic load. This is because the busy channel probability increases with λ. Thus, CaWuQoS-MAC needs to perform more CCA in order to successfully transmit a packet. The power consumed by CaWuQoS-MAC when the channel is error-prone is lower than that consumed when the channel is error-free for both values of d. This is because the probability of transmission failures due both collisions and transmission errors is much higher when the channel is error-prone, yielding earlier packet discard reducing the number of packets that need to be sent.

6 Conclusion

In this paper, we proposed CaWuQoS-MAC, a QoS- and WuR-based MAC protocol that ensures Collision Avoidance among WuCs and supports delay constraints for IoT applications. A major CaWuQoS-MAC feature is that it ensures a maximum tolerable delay for packet transmissions. To evaluate the performance of CaWuQoS-MAC, we derived a DTMC based model. We compared CaWuQoS-MAC to a non-collision avoidance WuR-based MAC protocol i.e., SCM-WuR, under both error-free and error-prone channel conditions.

Numerical results show that CaWuQoS-MAC provides better performance in terms of reliability while respecting the delay limit, at the cost of a slightly higher power consumption when the channel is error-free, and under both light and heavy traffic loads and high network density. Variation in number of nodes has no impact on the delay and power consumption, but it slightly affects the reliability of both protocols when the channel is error-free. Under error-prone channel conditions, the performance of CaWuQoS-MAC strongly depnds on the configured delay limit because it needs more time to perform additional CCA and BO. The data packet size also impacts the reliability and the power consumption of CaWuQoS-MAC when the channel is error-prone. In general, SCM-WuR loses its effectiveness when the traffic is heavy and under both ideal and non-ideal transmission channel. As an additional QoS feature, we aim to address the issue of service differentiation in IoT networks based of WuR MAC protocols.

References

1. Meddeb, A.: Internet of things standards: who stands out from the crowd? IEEE Commun. Mag. **54**(7), 40–47 (2016)
2. Buratti, C., Verdone, R.: Performance analysis of IEEE 802.15.4 non beacon-enabled mode. IEEE Trans. Veh. Technol. **58**(7), 3480–3493 (2009)
3. Ghose, D., Li, F.Y., Pla, V.: MAC protocols for wake-up radio: principles, modeling and performance analysis. IEEE Trans. Industr. Inf. **14**(5), 2294–2306 (2018)
4. Ghose, D., Li, F.Y.: Enabling Backoff for SCM Wake-Up Radio: protocol and modeling. IEEE Commun. Lett. **21**(5), 1031–1034 (2017). https://doi.org/10.1109/LCOMM.2017.2653779
5. Spenza, D., Magno, M., Basagni, S., Benini, L., Paoli, M., Petrioli, C.: Beyond duty cycling: Wake-up radio with selective awakenings for long-lived wireless sensing systems. In: 2015 IEEE Conference on Computer Communications (INFOCOM), pp. 522–530 (2015)
6. Ait Aoudia, F., Gautier, M., Berder, O.: OPWUM: opportunistic MAC protocol leveraging wake-up receivers in WSNs. J. Sens. (2016). Hindawi Publishing Corporation
7. Liebeherr, J., Wrege, D.E., Ferrari, D.: Exact admission control for networks with a bounded delay service. IEEE/ACM Trans. Networking **4**(6), 885–901 (1996)
8. Guck, J.W., Reisslein, M., Kellerer, W.: Function split between delay-constrained routing and resource allocation for centrally managed QoS in industrial networks. IEEE Trans. Industr. Inf. **12**(6), 2050–2061 (2016)

9. Oller, J., Demirkol, I., Casademont, J., Paradells, J., Gamm, G.U., Reindl, L.: Has time come to switch from duty-cycled MAC protocols to wake-up radio for wireless sensor networks? IEEE/ACM Trans. Networking **24**(2), 674–687 (2016)
10. IEEE standard for low-rate wireless networks. In: IEEE Std 802.15.4-2015 (Revision of IEEE Std 802.15.4-2011), pp. 1–709, 22 April 2016
11. Kozłowski, A., Sosnowski, J.: Energy efficiency trade-off between duty-cycling and wake-up radio techniques in IoT networks. Wireless Pers. Commun. **107**(4), 1951–1971 (2019). https://doi.org/10.1007/s11277-019-06368-0
12. Kiran, M.P.R.S., Rajalakshmi, P.: Performance analysis of CSMA/CA and PCA for time critical industrial IoT applications. IEEE Trans. Industr. Inf. **14**(5), 2281–2293 (2018)
13. Ghribi, M., Meddeb, A.: Survey and taxonomy of MAC, routing and cross layer protocols using wake-up radio. J. Network Comput. Appl. **149**, 102465 (2020)
14. Ghribi, M., Meddeb, A.: Performance evaluation of collision avoidance techniques using wake-up radio in WSNs. In: 2020 International Conference on Software, Telecommunications and Computer Networks (SoftCOM), pp. 1–6 (2020)
15. Nandi S., Yadav A. (2011) Adaptation of MAC Layer for QoS in WSN. Wyld, D.C., Wozniak, M., Chaki, N., Meghanathan, N., Nagamalai, D. (eds.): NeCoM/WeST/WiMoN -2011. CCIS, vol. 197. Springer, Heidelberg (2011). https://doi.org/10.1007/978-3-642-22543-7
16. Watteyne, T., Tuset-Peiro, P., Vilajosana, X., Pollin, S., Krishnamachari, B.: Teaching communication technologies and standards for the industrial IoT? Use 6TiSCH! IEEE Commun. Mag. **55**(5), 132–137 (2017)
17. AMS : AS3933 3D Low Frequency Wakeup Receiver. AMS Datasheet (2015)

A Dynamic ID Assignment Approach for Modular Robots

Joseph Assaker[2], Abdallah Makhoul[1(✉)], Julien Bourgeois[1], Benoît Piranda[1], and Jacques Demerjian[2]

[1] FEMTO-ST Institute, Univ. Bourgogne Franche-Comté, CNRS, 1 cours Leprince-Ringuet, 25200 Montbéliard, France
{abdallah.makhoul,julien.bourgeois,benoit.piranda}@femto-st.fr
[2] LaRRIS, Faculty of Sciences, Lebanese University, Fanar, Lebanon
jacques.demerjian@ul.edu.lb

Abstract. In this paper, we present a distributed Id assignment algorithm for modular robots. Our proposed solution supports both the removal and the addition of particles in the system, while maintaining particular characteristics in the logical tree, allowing for fast and efficient inter-module communications. The key goal here is to maintain easily calculated routes between any two particles in the system, with the minimal overhead possible. The idea of "holes" or free IDs is introduced in the system by three main alterations to our previous unique id assignment algorithm [2]. The first being the modification of the unique ID assignment phase. The second being the handling of particles removal from the system. And the third being the handling and initiation of a newly added particle anywhere in the system.

1 Introduction

Modular robots are autonomous systems composed of a large number of modules/particles that change their morphology/shape according to the surrounding environment and the application requirements [1,12]. Self-reconfiguration by rearranging connections between modules is the main task of modular robots. This task and others [9,11] rely on message transmission between particles directly connected or connected through other particles. Therefore, it is important to have some kind of identification for each module that remains constant throughout all of the modifications the system may go through. To accomplish this goal one might suggest at the manufacturing level and whenever a module is manufactured to directly assign a global unique ID to it (similar to a MAC on most electronics). Another simple technique that comes to mind is a random ID assignment [10]. However, such approaches are highly inefficient and limiting, as it would impose very long IDs which is not suitable for small and energy constraint modules. If having a system composed of 100 modules for example, why would one use long globally unique IDs between all existing modules whereas IDs composed of 7 bits (total of 128 unique IDs) are enough. A similar related problem is ID Assignment in Sensor Networks. In this context, Several schemes have been proposed to assign locally unique identifiers for sensor nodes [4,5,7,13]. However, theses

L. Barolli et al. (Eds.): AINA 2022, LNNS 449, pp. 91–104, 2022.
https://doi.org/10.1007/978-3-030-99584-3_9

approaches can not be applied to modular robots. Indeed, there are significant differences between the two systems, notably that the topology of modular robots will evolve continuously as the robot changes its morphology. The communications in modular robots are generally done only with the adjacent neighbour modules [3], and with the absence of a sink base station).

In our previous work [2], we presented a distributed algorithm that assigns unique IDs to modules. It is composed of three phases. The first one consists in discovering the whole system while building a logical tree. The second one finds the total size of particles in the system needed, and the third one is dedicated to the unique ID assignment. After the execution of this algorithm, the final system tree distributing the unique IDs could be compared to a B-Tree; where in order to reach the module with ID x, we need to take the route passing by the module with ID less or equal to x and its sibling with ID greater than x. This would render message passing between modules anywhere on the system very efficient. In this work, we assumed that all modules remain static and do not change their positions until the end of the algorithm. However, modular robots are by nature 'modular', meaning that the modules' positions will vary throughout the operation the system. This would result in the shuffling of IDs in the system, thus losing the B-Tree-like property.

In this paper, we propose an extension to our previous algorithm that relaxes this assumption to accommodate a dynamic topology where modules join and leave at any time during the execution of the algorithm or after its termination. This new version of the algorithm allows the system to handle various modules position changes with the goal of maintaining the order of the logical tree, by introducing holes or free IDs in the system. The goal here is to have available free spots for newly added modules to fit in, without breaking the order of the logical tree. We also aim at addressing this obstacle, that is of post-communication and routing between any two modules in the system after the termination of the ID assignment algorithm. Maintenance of the B-Tree-like property is one of the proposed solutions.

The remainder of this paper is organized as follows. In Sect. 2, we recall the main steps of our previous algorithm and present the main idea of the new version proposed in this paper. In Sect. 3, we develop the distributed dynamic ID assignment algorithm. Section 4 presents some discussions. In Sect. 5 we present several simulation results. Finally, Sect. 6 concludes with a brief description of future work.

2 Background on Unique ID Assignment

In our previous work [2], we proposed a three phases distributed ID assignment algorithm. In the first phase, the goal is to discover the whole system of modules while along the way building a tree structure rooted at the leader module. The second phase is dedicated to collect the system size (i.e., number of modules in the system) with the final goal of reporting the total size from the whole system to the leader module. Having the system size in hand, the leader module can calculate the least amount of bits needed in order to code global unique IDs for every module in the system. In the third phase, and after building the tree structure that logically connects modules and calculating the least amount of bits needed, the final step of unique ID assignment is launched from the leader to the whole system. In this algorithm we used 5 types of message: 1 (Explore neighbours for potential children); 2 (Confirm that the explored node is a child); 3 (Decline that the explored node is a child); 4 (Report the node's sub-tree size to its parent); 5 (Distribute the global unique IDs to children) [2].

One of the most useful benefits derived from this ID distribution method, is that the unique IDs are allocated in such a way throughout the logical tree that they resemble a B-tree structure. In the sense that, given any node in the system with ID i, all of its children will have IDs x, with $i < x < j$, where j is the ID of the node's direct sibling (or its parent's sibling if it is the last child node) [2]. This is particularly useful in the context of inter-module communication. First, given the constructed logical tree, each module would reach the root node relatively quickly and efficiently. In the other hand, if the root node wanted to reach any other node in the system, and without having this B-tree-like characteristic in the IDs distribution, a broadcast would have been required in order to ensure that the message reach its intended destination. However, having this ordered characteristic allows the root module to find any other module in the system following the same, relatively short, path from the module in question to the root, by abiding straightforward binary decisions while searching for the module in question. This can be further extended to any communication between any two modules in the system. Short communication paths between any two modules can be easily determined. Albeit this being a powerful feature to add to the system, its practicality, as is, falls short due to the variable morphology of the modular robots systems. In other words, if a single module changes its position while maintaining the same unique ID, it will break the ordered structure rules and render the implementation of efficient inter-module communication unfeasible, unless we impose very strict rules on the system by preventing it from changing any module position without a re-execution of the algorithm. Thus, the necessity of maintaining this characteristic throughout various morphology changes arises. Here, we propose the extended version of the unique ID assignment algorithm, which supports both the removal and addition of modules in the system after the initial execution of the algorithm, while maintaining the characteristics of an ordered tree structure. In order to be able to achieve such a goal, the IDs distribution phase will have to be revamped by introducing a new concept during this phase: the intended addition of unused IDs, or holes, in the IDs logical tree of the system. These holes will consist of "free" IDs, added in various nodes in the logical tree of the system, which will serve as available IDs for whenever a new module is connected to the system, via an existing module. The idea here is to have available IDs ready to be assigned to new modules, without breaking the order of the logical tree structure, and with the minimal amount of alterations possible made in the system. For instance, by adding just 1 bit to the ID length, we can double the space of IDs in the system, and allocate all unused IDs as free IDs throughout the system. The following will describe the three main, altered or newly added, functionalities of the ID Assignment algorithm. Starting with the alteration of the IDs distribution phase by introducing the idea of free IDs, followed by the newly added functionalities of module addition and module removal handling, while maintaining the logical tree characteristics with the minimum cost possible.

The main new idea added to the algorithm, which will allow us to handle various modules position changes with the goal of maintaining the order of the logical tree, is the introduction of holes or free IDs in the system. The goal here is to have available free spots for newly added modules to fit in, without breaking the order of the logical tree. This concept will be incorporated in the addition of free IDs at various levels of the tree. For example, considering a module with ID 10 and a reserved free ID, 11, having children with IDs going from 12 till 20, with its direct sibling having the ID 21. If a newly added module connects to this module, it will be able to take ID 11, without breaking the order of the logical tree. i.e., all of its children would still have IDs x with $10 < x < 21$. The initial distribution, and later various re-distributions, of free IDs will be handled by modifying or upgrading our IDs distribution phase from the original algorithm. Furthermore, the management of those free IDs will be handled by both the module removal and module addition sections of the algorithm.

3 Dynamic ID Assignment Algorithm

It is worth mentioning that whenever a module receives a type 5 message (cf Sect. 2), now containing both its unique ID and the number of free IDs available for its sub-tree, it is crucial for the algorithm to allocate the unique ID as the node's ID, and the directly following IDs as free IDs. For example, let's consider a type 5 message with the unique ID 10, and F free IDs. If the module determines that it should store 1 free ID at its level, this ID have to be ID 11. That is because, if the module gives ID 11 to its first child and stores for example ID 15 at its level, it will create an ambiguity in the system while searching for the module with ID 15. In other words, while searching for ID 15 and arriving at the module with ID 10, the algorithm would assume that all IDs in the sub-tree of its first child should be greater than 11. So the algorithm would not be able to determine whether to search for a potential ID 15 in the sub-tree of the first child, or the sub-tree of the second child, or if the module with ID 15 doesn't exist. Whereas if the module reserved ID 11 as a free ID, while providing its first child with ID 12, the algorithm can directly determine that the module with ID 11 is missing, as it is neither the module with ID 10, nor its first child with ID 12.

Keeping the first two phases of the original algorithm intact, and after the leader module calculates the least amount of bits needed to cover all IDs in the system, it adds 1 bit to this length, creating at least N free IDs for a system composed of N modules. The leader module would then proceed by assigning the ID 0 to itself, calculating the number of free IDs it should store at its level, and distribute unique IDs and free IDs to its children via type 5 messages. A given module determines the number of free IDs, M, to store following this formula: $M = \lfloor (message.n_free_ids/subtree_size) \rfloor$, which assures the balanced distribution of free IDs throughout the system, with more weight being given to lower level modules in the logical tree. This basically states that, if all the modules in a module's sub-tree can store at least Y free IDs, the module itself will also store Y free IDs.

Next, each child's share of the available free IDs is calculated with a basic percentage model, where each child will receive free IDs relative to its sub-tree size. The distribution of IDs to a module's children should respect the following rues: (i) send the module's ID + the number of the module's free IDs + 1 to the first child, (ii) send the module's ID + its number of free IDs + 1 + $\sum_{n=1}^{i-1}(S_n + F_n)$ to the ith child, where S_n is the sub-tree size of the nth child, and F_n the number of free IDs allocated to the nth child. The algorithm's part concerning this modification of the unique ID assignment phase is presented in Algorithm 1. In this algorithm, the variable n_free_ids contains the total number of free IDs in the module's sub-tree. The variable all_ids contains the total size of the ID space allocated to its sub-tree size (i.e., the sum of both the number of allocated unique IDs and the number of available free IDs). Finally, the variable $free\_ids$ is a set which contains an ordered list of available free IDs specifically at this module. The reason for this variable to be a set, and not an integer for example, is because the set of available free IDs in a given module at a given time t could be a non-contiguous set of IDs.

An illustrative example of the execution of this modified phase is presented in Fig. 1. In the example, we consider that the execution of phase 1 and 2 of the algorithm have finished (c.f. Sect. 2). Let's assume that the number of free IDs is equal to the number of unique IDs in the system (i.e., each module would store exactly 1 free ID). Supposing that the module to the left of the leader has a sub-tree size of 2, the leader would send the ID $0+1+1=2$ (i.e., its own ID + the number of free IDs it will store + 1) to it, while sending the ID $0+1+1+(2+2)=6$ to the module below it (following the formula defined above). This reserves all IDs from 2 till 5 to the sub-tree of the module to the left of the leader. The same principle applies supposing that the module below the leader has a sub-tree size of 7 and the module to the right of the leader has a sub-tree size of 87, the ID $0+1+1+[(2+2)+(7+7)+(87+87)]=194$ will be sent to the fourth and last child of the leader (the one above it). Numbers in the upper right corner of each module represent free IDs in this module.

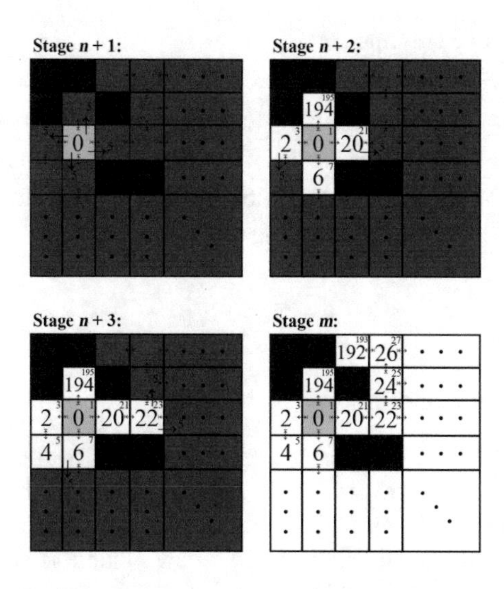

Fig. 1. Illustrative example of the first three stages of the unique IDs + free IDs assignment phase execution and the final stage, m, of the system.

3.1 Module Removal Handling

In this section we present how our approach deals with module removal. We distinguish two cases, the first one when the module leaves completely the system and the second one when it moves from a position to another one. Whenever a module decides to leave the system, whether to change its location in the system or to turn off completely, it must firstly notify its parent of that decision. This notification will be carried out via a type 6 message (cf. Table 1). Type 6 messages will contain extra information concerning the leaving module's free IDs. Given the scheme through which the IDs and free IDs are distributed (a module's set of free IDs contain IDs directly after the module's ID), a simple integer could be sent in order to denote the set of free IDs available in the leaving module. When a module receives a type 6 message, from one of its children, it will remove this child from its children list, update its sub-tree size and number of free ids and add the newly collected set of IDs, including both the child's ID and free IDs set, to its own set of free IDs. Doing so, the algorithm would have successfully "recycled" the leaving module's IDs, ready to be given away to any new module that comes back to this location in the system. One extra step required in this process is to send out type 7 messages (cf. Table 1), from parent to parent up to the root, in order to notify these modules about the modification that occurred in their sub-trees, updating their respective sub-tree sizes and number of free IDs. Algorithm 2 presents the steps concerning the handling of module removal and free ids recycling.

Table 1. Messages' role description

Type	Role Description
6	Notifying parent that module is leaving
7	Notifying ancestors, up to the root, that a module left
8	Notifying neighbor that it is now parent to the module
9	Notifying ancestors, up to the root, that a module joined
10	Delegating ID re-assignment to parent module

3.2 Module Addition Handling

Many schemes could be adopted for adding one module to the system. Whenever a newly added module connects to an already existing module in the system, it is the role of the latter to provide the former with a unique ID, and either with or without a set of free IDs depending on the situation. When a new module is added to the system, it would send a type 8 message (cf. Table 1) to any of its neighbors, which it now considers as its parent (note that this is a situation where the child chooses its parent, and not the inverse). A module receiving a type 8 message would firstly add the origin of this message to its set of children, with a sub-tree size of 1. Next, it would study its options as to how to instantiate this newly added module by providing it with an appropriate

Algorithm 1: Dynamic ID Assignment Algorithm

```
switch message.type do
  case 5
    n_free_ids ← message.n_free_ids ; all_ids ← subtree_size + n_free_ids
    M ← floor(message.n_free_ids/subtree_size) ; free_ids ← ∅
    for K ← 1toM do
      free_ids ← free_ids ∪ {message.id + k}
    unique_id ← message.id ; next_id ← message.id + M + 1
    for each child in children do
      percentage ← child.subtree_size/subtree_size ; child_n_free_ids ← in(free_ids *
percentage) ; send type 5 message to child
      next_id+ = child.subtree_size + child_free_ids
procedure CHECK( )
  if children.size = 0 OR received all children subtree_size then
    if leader = false then . . .
    else
      calculate least necessary bits ; calculate n_free_ids
      all_ids ← subtree_size + n_free_ids ; unique_id ← 0
      M ← floor(n_free_ids/subtree_size) ; free_ids ← ∅
      for K ← 1 to M do
        free_ids ← free_ids ∪ {k}
      next_id ← message.id + M + 1
      for each child in children do
        percentage ← child.subtree_size/subtree_size
        child_n_free_ids ← int(n_free_ids * percentage)
        send type 5 message to child
        next_id+ = child.subtree_size + child_n_free_ids
end procedure
```

unique ID. Multiple scenarios could occur at this stage. If the module has a non-empty set of free IDs, it will directly assign to the newly added module a unique ID, with the possibility of also providing it with free IDs. If the set of free IDs available in the module is a contiguous set, the module would split this set in half, and provide the second half to the new module. This action will be carried out with a type 5 message, where the first element of this set would be treated as the unique ID of the new module, and all other IDs as free IDs in the new module. For instance, if the module's set of free IDs is equal to $\{10,11,12,13,14\}$, ID 12 will be assigned as the new module's unique ID and the set $\{13,14\}$ as its free IDs set. Whereas if the set of free IDs available in the module is a non-contiguous set, the module would provide the right-most contiguous subset of its free IDs set to the new module, with the same method as in the prior scenario (i.e., first element as the unique ID, and the rest as free IDs). For example, considering the free IDs set of $\{10,11,12,15,16\}$, ID 15 will be assigned as the new module's unique ID and the set [16] as its free IDs set. On the other hand, and if the module does not have any available free ID, a re-distribution of IDs will be carried out. Two cases present themselves: if \n_free_ids" is 0 or greater than 0. If the number of free ids in the module's sub-tree is greater than 0, this means that a free ID is available somewhere

Algorithm 2: Module removal handling

```
n_free_ids ← 0
all_ids ← 0
free_ids ← ∅
if received message then
  switch message.type do
    ...
    case 6
      children ← children − {< message.origin, 1 >} ; subtree_size− = 1 ; n_free_ids+ = 1
      for K ← 0 to M do
        free_ids ← free_ids ∪ {message.id + k}
      free_ids ← SORT(free_ids)
      if leader = false then
        send type 7 message to parent
    case 7
      child ← find message.origin in children
      child.subtree_size− = 1
      subtree_size− = 1
      n_free_ids+ = 1
      if leader = false then
        send type 7 message to parent
procedure NODE_LEAVE( )
  send type 6 message to parent
end procedure
```

in its sub-tree. A re-distribution of unique IDs and free IDs will be performed from this module, down to all modules in its sub-tree. This will be carried out by having the module send out a type 5 message to itself. This means, it will reconsider its options of free IDs and how to re-distribute IDs to its children, given the new modification (i.e., addition of a new module) in its sub-tree. Finally, and if the number of free ids in the module's sub-tree is 0, the module would have to seek out free IDs from a higher level module in the logical tree of the system. Basically, the module's role now would be to search for an appropriate temporary root module from which a re-distribution phase will be instantiated from. Consequently, the module would send out a type 10 message (cf. Table 1) to its parent, which basically transfers the task of re-distribution to it. Whenever a module receives a type 10 message, and after updating its sub-tree size and its child's sub-tree size, it would check its options. If the module's \"n_free_ids" is 0, it would send out to its turn a type 10 message to its parent. If \"n_free_ids" is greater than 0 (i.e., there are holes in its sub-tree that can be utilized), a re-distribution will be performed from this module, down to all modules in its sub-tree. This also will be carried out by having the module send out a type 5 message to itself, which will lead to all modules in its sub-tree to accordingly receive type 5 messages and update their IDs and sets of free IDs. One extra step required in this process is to send out type 9 messages (cf. Table 1), from parent to parent up to the root, in order to notify these modules about the modification that occurred in their sub-trees, updating their respective sub-tree sizes and number of free IDs. The steps concerning the handling of modules addition to the system is presented in Algorithm 3. It should be noted that in the algorithm, four functions are assumed to be available. Function "*length*", which takes as input a given set and returns its length. Function "*sort*", which takes as input a given set and returns the sorted version of this set. Function "*is_contiguous*", which takes as input a given sorted set returns a Boolean whether this set is composed of a contiguous block of items or not. And function "*get_last_subset*", which takes as input a given sorted set and returns the right-most contiguous subset from this set.

4 Discussion

To better understand the benefits provided as a result of our approach, Fig. 2 is presented, in order to showcase a small scale example of the execution of the various sections of this algorithm. We assume having an initial system size of 13 modules, with a total ID space of 26 IDs resulting in an initial 13 unique IDs and 13 free IDs. At stage k, the initial execution of the algorithm has already been terminated with success, assigning to each module in the system an appropriate unique ID and free IDs. In addition, a new module is connected to the system in the bottom right corner of the system, right next to the module with ID 8. This new module, and in order to instantiate itself in the system, sends out a message of type 8 to the module with ID 8 and now considers it, its parent. In stage k + 1, it is now up to the module with ID 8 to instantiate the new module added to the system. Given that this module has an available free ID, 9, it sends out this ID to the new module via a type 5 message, and the new module is now assigned the ID 9. Also, type 9 messages are forwarded to the modules with IDs 6, 4 and 0, in order to notify them about the modification which occurred in their sub-trees, however these messages are omitted in the illustration. Going to stage k + 2, two main activities can be observed in the system. Firstly, the module on the upper left of the system, with ID 22, left the system. Before its departure, it sends out a message of type 6 to its parent, notifying it about its departure and providing it with its own unique ID and set of free IDs to be stored in it. The parent, with ID 20, received this request, removes the module with ID 22 from its children set, and adds the IDs 21 and 22 to its own set of IDs, being now [21, 22, 23]. Also, type 7 messages are forwarded to the modules with IDs 18 and 0, in order to notify them about the modification which occurred in their sub-trees, however these messages are omitted in the illustration. Secondly, yet another new module is added to the system, below the module with ID 8. As in the previous case, the new module would send out a message of type 8 to the module with ID 8 and considers it now its parent. Finally, and in stage k + 3, it is now the responsibility of the module with ID 8 to instantiate the new module. However, the module with ID 8 does not have any free IDs left stored in it. Furthermore, *n_free_ids*" of the module with ID 8 is equal to 0 (i.e., there are no free IDs within its own sub-tree). Thus, this module would have to delegate the task of ID re-assignment to its parent, via a message of type 10. Upon receiving the type 10 message, the module with ID 6 determines that free IDs are available within its sub-tree, and can be utilized in order to redistribute the IDs evenly to its children and descendants. It does so by sending a type 5 message to itself, launching a re-assignment task for its own sub-tree. This eventually results in an equal redistribution of IDs, with the new module being now assigned the unique ID 9.

Algorithm 3: Module addition handling

switch *message.type* **do**
  case 8
    $children \leftarrow children \cup < message.origin, 1 >$; $subtree\_size+=1$
    $len \leftarrow LENGTH(free\_ids)$
    if $len > 0$ **then**
      $contiguous \leftarrow IS\_CONTIGUOUS(free\_ids)$
      if $contiguous = true$ **then**
        $id\_index \leftarrow len/2$; $id \leftarrow free\_id[id\_index]$; $child\_n\_free\_ids \leftarrow len - id\_index - 1$
      else
        $subset \leftarrow GET\_LAST\_SUBSET(free\_ids)$; $id \leftarrow subset[0]$
        $child\_n\_free\_free\_ids \leftarrow LENGTH(subset) - 1$
      for $K \leftarrow 0$ *to* $child\_n\_free\_ids$ **do**
        $free\_ids \leftarrow free\_ids - \{id + k\}$
      $n\_free\_ids -= 1$
      if $leader = false$ **then**
        `send type 9 message to parent`
      `send type 5 message to message.origin`
    else
      if $n\_free\_ids > 0$ **then**
        if $leader = false$ **then**
          `send type 9 message to parent`
        `send type 5 message to itself`
      else
        `send type 10 message to parent`
  case 9
    $child \leftarrow$ `find message.origin in children` ; $child.subtree\_size+=1$
    $subtree\_size+=1$; $n\_free\_ids -= 1$
    if $leader = false$ **then**
      `send type 9 message to parent`
  case 10
    $child \leftarrow$ `find message.origin in children` ; $child.subtree\_size+=1$
    $subtree\_size+=1$
    if n_free_ids **then**
      if $leader = false$ **then**
        `send type 9 message to parent`
      `send type 5 message to itself`
    else
      `send type 10 message to parent`
procedure NODE_ADDED( )
  $parent \leftarrow neighbor$; $subtree\_size \leftarrow 1$; `send type 8 message to neighbors`
end procedure

As showcased above, the overhead cost of maintaining the ordered logical tree of IDs is exceptionally low. This is a particularly attractive option when the alternatives are the re-execution of the whole algorithm, or the last phase of the algorithm, or even using

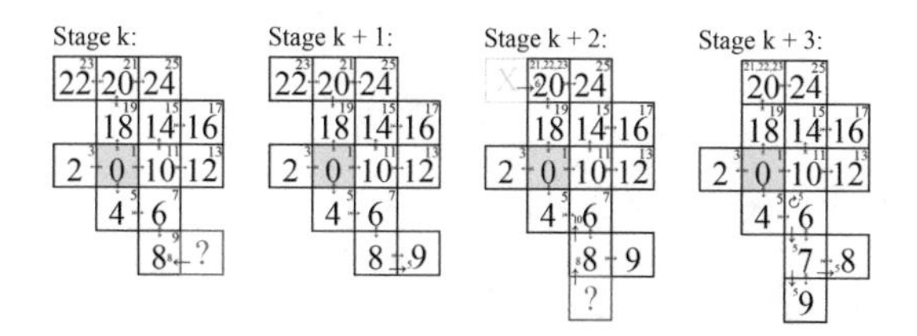

Fig. 2. example of the execution of various algorithm tasks related to the management of free IDs in the a given modular robots system.

broadcasting as a mean of message exchange between any two modules of the system. In the example above, in the case of the addition of the first new module, only a handful of messages were exchanged in order to assign to the new module a unique ID. In the second case of module addition, we can see that only 4 modules, out of 13, handled the assignment of a unique ID to the new module while maintaining the order of the IDs logical tree. The benefits of this algorithm only grow with larger system sizes. In a system composed of thousands of modules, no modules may need to be altered in order to accommodate to a new addition to the system, or just a dozen modules may be affected by this module addition, instead of the whole system. Furthermore, one can start understanding the wide range of possible configuration for this algorithm. For example, if many modules are expected to move on the leaf levels of the system, a free IDs distribution scheme giving more weight to lower level nodes could be implemented. If modules are predicted to move from the top area to the bottom area of a given system, more free IDs could be reserved in the latter area. Moreover, the number of free IDs in the system could be easily altered by choosing how many bits to add to the ID length, basically defining the space of available IDs. One could think of this configuration as a trade-off between IDs length and overhead costs when moving modules around. A system with a high morphology variance and low number of messages transmitted may benefit greatly from a bigger IDs space. A system with few modules changing their position in the system and a high rate of message transmission may benefit more from a smaller IDs space.

5 Experimental results

In order to evaluate our proposed algorithm, we conducted various simulations. We implemented the distributed approach in VisibleSim [8] which is a simulator supporting large-scales ensembles and different modular robot systems including Blinky Blocks used in our simulations. In fact, Blinky Blocks are centimeter-size blocks placed in a cubic lattice able to communicate and coordinate together through serial links on the block faces (neighbor-to-neighbor communication model).

To evaluate our proposed algorithm and test disposal of free IDs throughout the system, we carried out two different scenarios. The first scenario consists of adding modules to already existing shapes. The idea is to add modules on different shapes consisting of hundreds of modules. The second one is a reconfiguration scenario consisting of transforming a chain shaped system into an 'S' letter like shape.

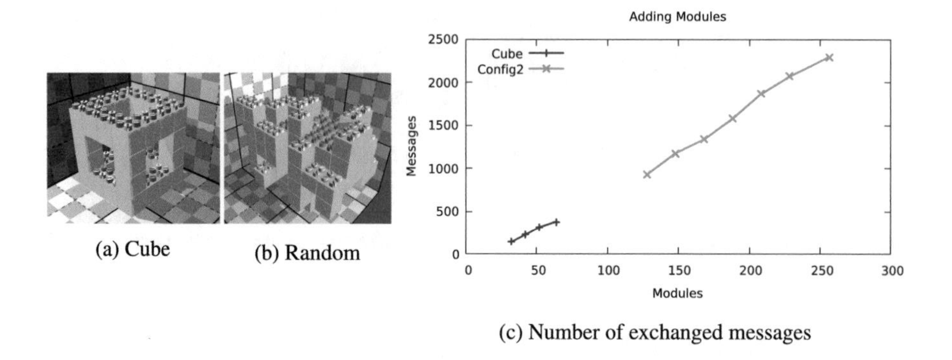

(a) Cube (b) Random (c) Number of exchanged messages

Fig. 3. Modular robots configurations and results

First Scenario: Adding New Modules to Existing Shapes

In this scenario, we carried out several additions of modules on two different configurations. The first is a cube shaped one presenting very regular volumes with symmetries (cf. Fig. 3a) and the second one representing a large set of randomly assembled blocks resulting in a variable number of neighbours disposed at each module (cf. Fig. 3b). For the first configuration (cube shape), We added several modules to the system until we reached double the initial number of modules. We considered a system composed of 32 blocks and at each stage of the simulation, we added 10 new modules, until we reached the total number of 64 modules. The second example illustrated in Fig. 3b shows a set of blocks assembled randomly with some irregularities. This kind of shapes is resulting in a variable number of neighbours. In this case, we considered a network of 128 modules and at each stage of the simulation, we added 20 new modules, until we reached the total number of 256 modules.

In the two cases and for the different scenarios, it has been clearly shown that our approach ensures the assignment of new IDs to all newly added modules. Furthermore, after the execution of our algorithm the tree has been updated and the leader knows all the routes to reach these new modules. However, the cost of this assignment is in the number of messages sent in the system. Therefore, we evaluated the number of messages exchanged while varying the number of added modules. In Fig. 3c, we showed the results of the total number of messages exchanged during the execution of our approach. As aforementioned, we studied the number of messages exchanged while adding each time 10 or 20 blocks respectively to the cube and random configurations. The messages in the system are used to assign id for each block and distribute free spots for each one as explained in previous sections. As well, they are used for inter-module communication and notification purposes whenever a module joins/leaves the system. The results show that the number of messages linearly increases with the number of added modules. In order to be able to assign new ids, in the proposed approach the modular robot has to exchange an additional number of messages. On the other hand, it is more cost-effective than the case where the whole process has to be started all over again and from the beginning. Furthermore, our approach ensures easy access to specific identified nodes, which can be very useful for fault detection issue.

Second Scenario: Self-reconfiguration
In these series of simulations, we considered the free ids assignment in a self reconfiguration scenario where the number of modules is fixed and just the positions and neighbors of some modules change. We considered an initial chain shape of the modular robot to be reconfigured into the letter 'S' like shape. The self-reconfiguration algorithm proposed in [6] is used in this simulation. As a matter of fact, the idea is having modules moving from one position to another by disconnecting from the module's interface and joining another one having a free spot ready to fit in. The results showed that our approach can handle the reconfiguration from initial to final shape by involving both the removal and the addition of modules in the system after the initial execution of the algorithm. Furthermore, the order of the logical tree is maintained due to the free Ids disposal at each block resulting in a minimum number of modifications to be made in the system as discussed earlier in previous sections. We also evaluated the additional number of messages exchanged during this reconfiguration task. This increase is about 30% more messages compared to the messages sent just for the reconfiguration process.

6 Conclusion and Future Work

In this paper, we presented an extended version of our approach concerning the unique ID assignment of modular robots [2]. This new version was proposed, rendering the feasibility of fast and efficient inter-module communications in modular robots system much more likely. This was possible due to the implementation of a reconfiguration adapting algorithm, that handles the distribution and handling of holes or free IDs throughout the system. The three main tasks of this extended algorithm were thoroughly explained. The discussion and obtained results showed clearly the benefits of the extension of the algorithm, while providing insights as to how to define the IDs length in a particular modular robots system. In a future work, we plan on studying the effects of having multiple leaders distributed evenly throughout the system on the time and energy complexity of the algorithm.

Acknowledgments. This work has been supported by the EIPHI Graduate School (contract ANR- 17-EURE-0002).

References

1. Alattas, R.J., Patel, S., Sobh, T.M.: Evolutionary modular robotics: survey and analysis. J. Intell. Robot. Syst. **95**(3–4), 815–828 (2019)
2. Assaker, J., Makhoul, A., Bourgeois, J., Demerjian, J.: A unique identifier assignment method for distributed modular robots. In: 2020 IEEE/RSJ International Conference on Intelligent Robots and Systems (IROS), pp. 3304–3311 (2020)
3. Brunete, A., et al.: Current trends in reconfigurable modular robots design. Int. J. Adv. Robot. Syst. **14**(3), 1728 (2017)
4. Haque, M.R., Naznin, M., Shahriyar, R.: Distributed low overhead id in a wireless sensor network. In: Proceedings of the 17th International Conference on Distributed Computing and Networking. ICDCN 2016, pp. 12:1–12:4 (2016)

5. Lin, J., Liu, Y., Ni, L.M.: SIDA: self-organized ID assignment in wireless sensor networks. In: IEEE International Conference on Mobile Adhoc and Sensor Systems, pp. 1–8 (2007)
6. Barolli, L., Woungang, I., Enokido, T. (eds.): AINA 2021. LNNS, vol. 225. Springer, Cham (2021). https://doi.org/10.1007/978-3-030-75100-5
7. Petroccia, R.: A distributed ID assignment and topology discovery protocol for underwater acoustic networks. In: Third Underwater Communications and Networking Conference, pp. 1–5 (2016)
8. Piranda, B., Fekete, S., Richa, A., Römer, K., Scheideler, C.: Your simulator for programmable matter. In: Algorithmic Foundations of Programmable Matter, Visiblesim (2016)
9. Pratissoli, F., Reina, A., Kaszubowski Lopes, Y., Sabattini, L., Groß, R.: A soft-bodied modular reconfigurable robotic system composed of interconnected Kilobots. In: MRS (2019)
10. Smith, J.R.: Distributing identity symmetry breaking distributed access protocols. IEEE Robot. Autom. Mag. **6**(1), 49–56 (1999)
11. Thalamy, P., Piranda, B., Bourgeois, J.: Distributed self-reconfiguration using a deterministic autonomous scaffolding structure. In: AAMAS 2019, pp. 140–148 (2019)
12. Yao, M., Belke, C.H., Cui, H., Paik, J.: A reconfiguration strategy for modular robots using origami folding. Int. J. Robot. Res. **38**(1), 73–89 (2019)
13. Zhou, H., Mutka, M.W., Ni, L.M.: Reactive ID assignment for sensor networks. In: IEEE International Conference on Mobile Adhoc and Sensor Systems Conference, p. 6. IEEE (2005)

Open-Source Publish-Subscribe Systems: A Comparative Study

Apostolos Lazidis[1], Euripides G. M. Petrakis[1](✉), Spyridon Chouliaras[2], and Stelios Sotiriadis[2]

[1] School of Electrical and Computer Engineering, Technical University of Crete (TUC), Chania, Greece
alazidis@isc.tuc.gr, petrakis@intelligence.tuc.gr
[2] Department of Computer Science, Birkbeck, University of London, London, UK
{s.chouliaras,s.sotiriadis}@bbk.ac.uk

Abstract. Publish-Subscribe systems are designed to facilitate communication between services and applications. A typical system comprises the publisher, the subscriber, and the broker but, may also feature message queues, log files or databases for storing messages, clusters or federations of brokers and, apply message delivery policies, communication protocols, context manager for linked data, security services, and a streaming API. Not all these features are supported by all systems or, others may be optional. Therefore, there is no common ground for the comparison of Publish-Subscribe systems. The evaluation is about seven popular open-source and state-of-the-art systems namely, Apache Kafka, RabbitMQ, Orion-LD, Scorpio, Stellio, Pushpin, and Faye. All systems are evaluated and compared in terms of functionality and performance under real-case scenarios.

1 Introduction

Publish-subscribe systems handle a wealth of information that is transmitted around the world via the Internet [7]. In a real-case scenario, an academic conference takes place to discuss advances in various fields such as medicine and computer science. In this scenario, each presentation involves several speakers that publish information to a group of listeners. The listeners can also subscribe to presentations on various topics that take place in different rooms. Internet applications should follow this communication pattern. There should be a central service (i.e. a broker) where applications send information (i.e. presentations) and, this service forwards information to listeners who subscribed to specific topics. Systems that support this type of communication are referred to as *Publish-Subscribe*.

In Publish - Subscribe communication, the *publisher* is the service that publishes messages (e.g. events) and the *subscriber* is the one that receives them. These two services operate independently of each other: the publishers do not need to be aware of the existence of subscribers. However, the subscribers may be aware of the publishers who are is sending them the messages. The *broker* has a central role in the communication and is responsible for collecting the published messages and for sending

L. Barolli et al. (Eds.): AINA 2022, LNNS 449, pp. 105–115, 2022.
https://doi.org/10.1007/978-3-030-99584-3_10

them to the subscribers. This decoupling of roles is a feature that differentiates the Publish-Subscribe communication pattern from traditional messaging methods based on the request-response model.

Publish-Subscribe systems must scale up well to handle large message loads (i.e. more brokers must be added as needed). A Publish-Subscribe system may feature message queues [4,8,18], log files [11] or databases [9,11,12] for storing messages. Some implementations feature also a replication manager [9,11,18] (i.e. to ensure that no message is lost, copies of messages are stored in multiple brokers; if one broker fails, it will be replaced by another), a context manager [9] for context data such as messages and subscriptions, a federation service [9,11,18] (i.e. data can be moved from one broker to another), security services (e.g. data encryption, user authentication, and authorization [11,12,18], streaming API [11] for filtering messages in real-time, etc. A Publish-Subscribe system defines a message format (e.g. byte array [11], JSON [4,8], NGISI-LD [6,9,12]), a communication protocol (e.g. TCP [11] or HTTP [6,9,12], Webscockets [4,8]) etc. The subscribers receive messages from brokers using either a push-based (*push policy*) or a pull-based approach (*pull policy*). In the push-based approach, the broker checks the communication and forwards the messages to the subscribers as soon as they arrive. In the pull-based approach, the subscriber polls the broker for new messages. The format (i.e. message structure) most systems choose to exchange messages is JSON, which, however, tends to be replaced by JSON-LD [17].

There are several implementations of Publish-Subscribe systems that meet different demands. Most systems ensure reliable and secure communication, others are designed for faster communication, others for reliability (i.e. ensure that no message is lost), others for handling large message loads without affecting performance. In addition, some features are optional (i.e. can be activated by the user). Each system brings a different set of features onboard an application and, the decision on which solution is the best fit for a particular use case is not always easy. Each solution must be evaluated concerning the requirements of an application (e.g. whether security or scalability is more important than error tolerance).

Many popular solutions are open source, while others are fully proprietary and have been incorporated into commercial (cloud) platforms. The latter is known to be highly scalable, fault-tolerant, and secure, can be ported to external (e.g. stream processing) platforms, and, most important, all provide good performance guarantees. Choosing one or another solution is not a real dilemma as all of them support the most essential features. The present study focuses on seven popular highly performance open-source systems most of which have not been studied before or compared against state-of-the-art systems such as Apache Kafka and RabbitMQ. The most critical factors of their design and functionality and their impact on performance are identified, analyzed, and measured. For reasons of a fair comparison, a minimum set of common features is applied to all systems. Performance is the most important factor and reveals how fast the communication takes place and how it is affected by scaling the message load. Previous studies did not consider more than two or three systems [13,20].

Systems design principles and functionality are discussed in Sect. 2 along with results from previous surveys in the field. All systems are evaluated in terms of functionality and performance in Sect. 3 and Sect. 4 respectively. Lessons learned and issues for future research are discussed in Sect. 5.

2 Related Work and Background

Esposito [7] dictates the satisfaction of the properties *System agreement*, *System validity* and, *System integrity* to guarantee reliability. System agreement suggests that, if a non-defective subscriber receives a message, then all non-defective subscribers will receive the message as well. System validity requires that, if a non-defective publisher sends a new message, then at least one of the non-defective subscribers will receive it. Finally, system integrity requires that each non-defective subscriber must receive each message at most once. These properties are not sufficient to make a Publish-Subscribe model reliable. Often, real-time applications impose strict control on transmission and require transmitting messages within a time frame, otherwise, messages may be lost. Some researchers considered *timeliness* property to satisfy that all subscribers receive messages within a time limit [5].

Publish-Subscribe systems comprise a publisher, a subscriber, and a broker connected by channels. These system components often behave in an undesirable way due to various errors including node crashes, churn, network anomalies, and link crashes [3]. These errors need to be addressed to implement a reliable Publish-Subscribe system. Two definitions related to the reliability emerged referred to as *resiliency* and *reconfiguration* [16]. The former refers to the ability of a system to guarantee the properties of the agreement, validity, and integrity (in the event of network errors) while the latter refers to the ability of a system to handle errors so as not to affect system connectivity.

Performance is an important factor to consider in Publish-Subscribe systems [2]. As the workload scales up, system reliability is being challenged and often performance is being affected. Oh, Kim and Fox [14] address the problem of performance, propose a cost model for in a real-time domain, and compare the Publish-Subscribe model with other communication models such as request-reply and polling models. Esposito [7] addresses the issue of reliability, that is the capability of the system to tolerate faults in the network or computing nodes and to support reliable event notification. Xiao, Zhang, and Chen [19] apply a tree structure of topic clusters that saves memory and improves the real-time performance and reliability of message forwarding. Khoury et al. [10] address the problem of privacy of the communication that typically appears when information about the published content and the subscribers' interests are exposed at the broker.

In this landscape, there are no standard benchmarks (i.e. message loads, metrics, and tools) for measuring the performance of Publish-Subscribe systems. The Open Benchmarking Framework[1] (OMB) provides benchmarking suites (i.e. workloads and metrics) tailored to Apache Kafka, Apache RocketMQ[2] and Apache Pulsar[3]. For reasons of a fair comparison, both OMB and the systems need to be fine-tuned on the same platform but this is done for AlibabaCloud or Amazon Web Services (AWS) cloud platforms only.

[1] http://openmessaging.cloud/docs/benchmarks/.

[2] https://rocketmq.apache.org.

[3] https://pulsar.incubator.apache.org.

Throughput and *latency* are two common performance metrics of performance. Latency is a tangible metric that measures the time it takes a system to deliver messages to consumers (end-to-end latency). Throughput measures the number of messages produced or consumed per second and depends on how efficiently a system utilizes the hardware (e.g. the disks, RAM and CPU). The highest throughput for a system requires a high degree of parameterization of system and hardware. This requires that the user be familiar with the peculiarities of both but, this is not easy. On the other hand, the systems are not equivalent in terms of parameterization (the same parameterization does not apply to all systems). Compared to all other systems, Apache Kafka offers the highest degree of parameterization. For example, the user of Apache Kafka can define many partitions for a topic and sent several copies of each message. However, RabbitMQ (the same as many other systems) does not support partitions. To match Kafka's setup, a single direct exchange (equivalent to a topic) can be linked to queues in the RAM (which are defined as durable).

Various studies on the Web compare the performance of specific systems. Yigal [20] compare Apache Kafka [11] with Redis[4] (a cache database that can be used as Publis-Subscribe system) for message processing in log aggregation applications. Nikhil and Chandar [13] present a benchmarking on Apache Kafka, Apache Pulsar[5] and RabbitMQ [18] focusing on system throughput and system latency.

3 Functional Evaluation and Comparison

In the following, all systems are compared in terms of functionality (i.e. features and operations). Table 1 summarizes the features of all systems. FIWARE[6] is an open-source initiative and software platform that supports smart context data management in several application domains (e.g. smart cities, smart energy) based on EU standards[7]. In 2019, the European Telecommunications Standards Institute (ETSI) proposed NGSI-LD [1], a new data exchange protocol based on Linked Data (LD)[8]. NGSI-LD provides an API for publishing, querying, and subscribing to context information. Entities are the key components of NGSI-LD. An entity can be the description of a concept, a subscription to another entity, or a property of an entity. An NGSI-LD document is a valid JSON-LD document (but not the other way around). Listing 1 is an example of the NGSI-LD declaration of a room entity with a temperature value (the temperature is a property in the room entity) [15]. The *@context* label defines a hyperlink that points to the ontology that describes the entity. There are no serialization and de-serialization libraries for this information model. This is a disadvantage for application developers who must be aware of NGSI-LD syntax to use Fiware systems for routing messages.

[4] https://redis.io.
[5] https://pulsar.apache.org.
[6] https://www.fiware.org.
[7] https://ec.europa.eu/cefdigital/wiki/display/CEFDIGITAL/EU+Standards.
[8] https://www.w3.org/standards/semanticweb/data.

Listing 1. NGSI-LD model example

```
1   {
2    ''id'':''urn:ngsi-ld:Room:01'',
3    ''type'':''Room'',
4    ''temperature'':{
5     ''type'':''Property'',
6     ''value'':17,
7     ''observedBy'':{
8      ''type'':''Relationship'',
9      ''object'':''urn:ngsi-ld:Sensor:01''
10    }
11   },
12   ''@context'':[''https://uri.etsi.org/ngsi-ld/v1/ngsi-ld
        -core-context.jsonld'']
13  }
```

Orion-LD [9], Scorpio [12] and Stellio [6] are popular Publish-Subscribe (referred to as *Context - Broker*) solutions conformant with the NGSI-LD standard. Stellio and Scorpio were developed in parallel aiming to enrich Apache Kafka with REST NGSI-LD interfaces. They not only differ from Apache Kafka in message format but, also offer additional functionality such as create, read, update and delete (CRUD) operations on entities. An information provider sends information to a Scorpio or Stellio entity. A user or service can subscribe to an entity and get informed on possible changes of that entity. These two systems comprise more services than Apache Kafka such as databases for storing context information (entities, subscriptions, users). Scorpio features a PostgreSQL database, Stellio a Neo4j database data, and the TimescaleDB and PostGIS of PostgreSQL. Orion-LD implements custom push and pulls message delivery mechanisms using MongoDB.

All Fiware systems adopt the NGSI-LD information model for context data. RabbitMQ, Scorpio, Stellio, and Apache Kafka are error-tolerant and this makes them useful in cases where messages should not be lost (e.g. in bank transactions). For applications that process messages in real-time, Apache Kafka, Scorpio, and Stellio are the best solutions. Orion-LD, Pushpin, and Faye are easy to learn although the related documentation is not extensive. In terms of documentation, Apache Kafka and RabbitMQ are the most complete. However, both these systems, because of their extensive functionality and complexity are the most difficult to learn. Stellio and Scorpio are easy to learn. However, if a developer needs to write software extensions for these systems, she/he must be familiar with Apache Kafka. Pushpin and Faye are less complete, but in terms of learning difficulty, they are the easiest to learn. They are recommended mostly to less demanding applications where performance or fault-tolerance are less mandatory (e.g. chat applications, collaborative text editors).

Apache Kafka implements a custom binary protocol over TCP. Most systems support communication over HTTP protected by TLS/SSL. RabbitMQ supports a variety of protocols (i.e. AMQP, MQTT, STOMP, and HTTP). Communication in Pushpin takes place via HTTP, HTTP long-polling, HTTP streaming, and Websockets. while communication in Faye takes place via HTTP, HTTP long-polling, and Websockets. Kafka, Scorpio, and Stellio can connect to other external systems thanks to the Connector API of Kafka. Apache Kafka applies a pull-based approach for message delivery (i.e. the subscriber queries the broker for new messages), while Pushpin and Faye adopt the

Table 1. Publish-Subscribe systems comparison.

Feature	*Apache Kafka*	*Orion-LD*	*Scorpio*	*Stellio*	*RabbitMQ*	*Pushpin*	*Faye*
Open-Source	√	√	√	√	√	√	√
Language	Scala, Java	C, C++	Java	Kotlin	Erlang	C++	Node.js, Ruby
Learning Difficulty	5/5	2/5	2/5	2/5	4/5	1/5	1/5
Format	Byte array	NGSI-LD	NGSI-LD	NGSI-LD	Byte array	JSON	JSON
Stream Processing	-	Stream API	Stream API	Stream API	-	-	-
Broker Clustering	√	√	√	√	√	-	Using Redis
Broker Federation	√	√	√	√	√	-	-
Fault Tolerance	√	-	√	√	√	-	-
Message Queries	√	√	√	√	-	-	-
Message Updates	√	√	√	√	-	-	-
Message Retention	√	√	√	√	√	-	-
Message Replication	√	√	√	√	√	-	-
Acknowledgment of Receipt	√	√	√	√	√	√	√
Message Delivery	Pull	Push, Pull	Push, Pull	Push, Pull	Push, Pull	Push	Push
Message Storage	Disk partitions	MongoDB	PostgreSQL	Neo4j, TimescDB, PostGIS	Queue (disk, RAM)	Queue (RAM)	Queue (RAM)
Connection Encryption	TLS/SSL	SSL	No Info	No Info	TLS/SSL	TLS/SSL	TLS/SSL
Authentication, Authorization	OAuth2.0, Kerberos, Authorization lists	CORS	LDAP, SASL, OAuth2.0	OAuth2.0	LDAP, SASL, OAuth2.0	CORS, JWT	CORS
Protocol	Custom TCP	HTTP	HTTP	HTTP	AMQP, STOMP, MQTT, HTTP	HTTP (Long Polling, Streaming), Websockets	HTTP (Long-Polling), WebSockets

push-based approach (i.e. the messages are forwarded to the subscribers as soon as they are received). All other systems support both approaches. The publishers at Orion-LD, Scorpio, and Stellio can update stored messages and so does Apache Kafka using *Log Compaction* (i.e. for messages with the same key in the partitions of a topic are updated).

All Fiware systems, the same as Apache Kafka, support a federation of brokers and broker clustering. Faye can also operate a cluster of brokers to achieve high availability. Messages in Scorpio, Stellio, Apache Kafka, and RabbitMQ keep multiple copies in partitions and queues respectively. Scorpio and Stellio also have this feature because they use Apache Kafka. RabbitMQ works similarly to Apache Kafka. If a broker fails, then the communication will be continued by another broker. Broker federation is not similar to clustering and refers to the ability of a broker to send messages directly to another broker automatically. This feature is supported by Orion-LD, Scorpio, Stellio, and RabbitMQ.

In Orion-LD, Scorpio, and Stellio the messages are stored on the disk. In Apache Kafka and RabbitMQ the messages are kept for a certain period or until the partitions and queues reach a certain size. In Apache Kafka and Fiware systems, the messages can be read more than once. In RabbitMQ, Pushpin, and Faye, once read, the messages are deleted. In Pushpin and Faye, the messages are sent in message queues and, if there is no subscriber to receive them, they are deleted. In regards to security, all systems support at least basic user identification and authorization mechanisms.

4 Performance Evaluation and Comparison

Apache Kafka is the industry's messaging workhorse and has been designed to support the parallel processing of message streams. Various experiments confirmed that Apache Kafka is very fast even under extreme message loads. An important factor is the high degree of parameterization of Apache Kafka to perfect its performance. Experiments by Nikhil and Chandar [13] revealed that Apache Kafka (with 100 partitions) delivered the highest throughput (message load) compared to RabbitMQ and Apache Pulsar while delivering messages very fast even for very high throughput (i.e. 5ms latency for 200 MB/s throughput). RabbitMQ can achieve lower latency but only for much lower throughput (e.g. 1ms latency for 30 MB/s throughput). The purpose of the following experiment is not the reproduce the results of previous studies, but rather to show which solution is best for a small-scale application with low throughput (i.e. small message loads).

A service at the front-end accepts user requests and forwards them to the Publish-Subscribe system by issuing a POST request. The messages are send as NGSI-LD entities. Systems that support NGSI-LD (i.e. Orion-LD, Scorpio, Stellio) process the *@context* field accordingly. All other systems (i.e. Kafka, RabbitMQ, Pushpin, Faye) handle NGSI-LD as ordinary JSON (i.e. the *@context* field is an ordinary JSON attribute). Latency is defined as the time lapsed from the time of issuing a POST request until the messages are written to the broker. For reasons of fairness of the comparison, all systems must support permanent message storage. RabbitMQ, Pushpin and Faye are connected to a MongoDB where the entity is published. Before an entity is published, the service checks if the entity was previously posted by another user. If not, it is published in the MongoDB for Orion-LD or, to a topic (e.g. a disk partition) in the case of Apache Kafka, Scorpio and Stellio. Then, a subscriber can retrieve the entity from MongoDB by issuing a GET operation.

Apache Kafka will search a topic to determine whether the message is there already. When the publisher publishes an entity, is awaiting acknowledgment of receipt that the message was published successfully. This can be asynchronous (i.e. the publisher is blocked until the confirmation is received) or an asynchronous operation (i.e. the publisher is not blocked and can send more POST requests before a confirmation is received). In addition, each time a message is read, its offset is published so that, the reader stores the position of the last message read (i.e. in case of failures the subscriber can continue reading messages from the last position read). Synchronous offset commitment and synchronous acknowledgment of receipt are two factors that affect speed. For Apache Kafka or RabbitMQ, the mode of operation is set to synchronous and, asynchronous for Scorpio and Stellio. This will allow us to study the impact of the synchronous policy on performance.

4.1 Experimental Results

Apache Bench[9] has been used to evaluate system performance in terms. The user specifies the number of requests as well as the number of requests that will be served simultaneously (to evaluate how performance is affected as the number of users escalates). Each Apache Bench command contains the IP address where the messages are sent, the number of requests, the data that will be sent to the publisher (e.g. NGSI-LD entities), and the number of requests to be processed simultaneously (simulating the effect of many message producers). The performance is measured in two scenarios: (a) the number of requests is 1,000 with a concurrency of 200 and, (b) the number of requests is 10,000 with a concurrency of 2,000. In the case of concurrent requests (e.g. users issuing requests simultaneously), Apache Bench opens 200 or 2,000 TCP connections respectively and attempts to send requests at the same time. All systems in this study are hosted in Google Cloud Platform[10] (GCP). A Virtual Machine (VM) has been deployed in GCP with 4 CPU cores, 16 GB Memory, and 80 GB HDD running Ubuntu 18.04 LTS. A Docker environment in this VM deploys each Publish-Subscribe system (including databases) in separate containers.

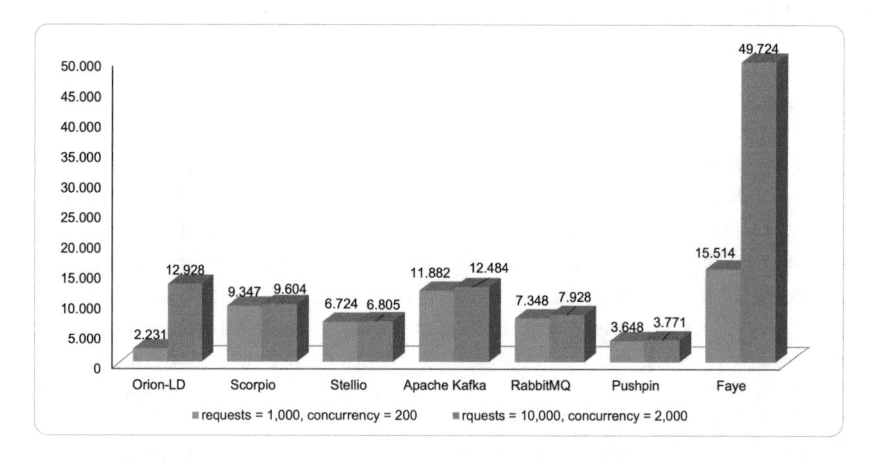

Fig. 1. Latency (ms) for publishing 1,000 and 10,000 entities with concurrency 200 and 2,000 respectively.

Figure 1 reports the time for publishing 1,000 and 10,000 message entities to the broker. For systems applying a pull policy (i.e. Apache Kafka, Orion-LD, Scorpio, and Stelio), this corresponds to latency, since the messages remain in the broker until they are read by a subscriber. For systems that apply a push policy (i.e. RabbitMQ, Faye, and Pushpin) this measurement has no practical use and is only for comparison (i.e. the messages do not remain on the broker but, are automatically forwarded to the subscribers). Figure 1 represents the relative performance of all systems.

[9] https://httpd.apache.org/docs/2.4/programs/ab.html.

[10] https://cloud.google.com/.

For 1,000 requests (concurrency = 200), Orion-LD outperforms all other systems. Orion-LD does not store messages in queues or disk partitions. Messages are saved on a MongoDB database and feature a *connection pool* to connect. These links to MongoDB are reusable and Orion-LD does not need to open a new connection upon each request. The size of the connection pool is set to 10 (by default). For 10,000 requests (concurrency = 2,000), the response time rises to 12,928 ms, indicating that Orion-LD does not scale up very well. The reason for this poor performance (compared to all other systems except Faye) is, again, the connection pool. For many simultaneous requests, the connections are busy and many requests wait for a connection to become available. Another reason is that Orion-LD processes only one request at a time (mutual exclusion or *mutex* policy).

Scorpio and Stellio perform about the same since their architecture is very similar. For 1,000 requests, both are slower than Orion-LD. Before publishing a message, they search all messages (i.e. one after the other) in a topic to determine whether the message has been published before. Orion-LD does the same, but searching MongoDB is much faster. In Scorpio, entities are published in a queue. Checking whether an entity was created before, the messages in a topic are read one by one. In Stellio, a new topic is created for each new entity with the name of the entity. This operation is faster in Stellio than it is in Scorpio. For 10,000 requests (again) both Scorpio and Stellio exhibit the same good performance (as for 1,000 requests) thanks to Apache Kafka (i.e. topic partitions are read in parallel).

What makes Apache Kafka fast is that it uses page cache (i.e. disk cache) to write and read messages. When a publisher sends a message to a partition it is kept in the page cache until the system decides to flush it on the disk. The data transfer time from disk to page cache is very fast thanks to *zero-copy*[11] technique. However, for 1,000 requests, Apache Kafka is slower than both Scorpio and (although both systems incorporate Kafka). This is probably because of the synchronous mechanism that is applied. If set to asynchronous, the performance of Apache Kafka would be at least as good as that of Scorpio and Stellio (in both, Apache Kafka is set to asynchronous). As expected, for 10,000 requests, Apache Kafka maintains the same good performance (i.e. Apache Kafka is known to scale up very well under stress).

Pushpin is very fast. This is because it does not apply any confirmation of receipt mechanism. The message queues in Pushpin are stored in a memory cache while, those of Scorpio, Stellio, and Apache Kafka are stored on partitions. Pushpin (similar to all non-Fiware systems) does not parse the NGSI-LD message according to the ontology in the hyperlink of the *@context* field. For 10,000 requests Pushpin maintains the same good performance, making it suitable for applications running a large load scale.

In Apache Kafka and RabbitMQ, the acknowledgment of receipt is set to synchronous. By setting this mechanism to asynchronous or, by configuring the system to sent no acknowledgments at all, both systems can run much faster. In RabbitMQ, before a message is sent to a queue, it is initially sent to an *exchange agent*, which is responsible for routing the messages to the appropriate queue (i.e. this introduces a delay). In Pushpin and Faye messages are sent directly to queues. RabbitMQ adopts a push policy for delivering messages to subscribers. This is not necessarily fast espe-

[11] https://developer.ibm.com/articles/j-zerocopy/.

cially when the message production rate is faster than the consumption rate, and harms performance when the queues are overloaded with messages. In that case, the broker controls the transmission rate as messages wait in a queue until they are transmitted. Pushpin applies a push delivery policy as well but is more lightweight as a system overall. Faye is the slowest system in all cases and supports the least functionality. It is not intended to compete with other systems but rather to provide a lightweight and easy-to-use solution to everyday applications with no particular performance demands (e.g. chat). It adopts a push policy and the broker controls the transmission rate to the subscribers.

5 Conclusions and Future Work

This study presents a comparison of popular open-source Publish-Subscribe systems. The comparison in terms of functionality reveals that Publish-Subscribe systems like Scorpio and Stellio have more features than Orion-LD, Apache Kafka, RabbitMQ, Pushpin, and Faye. However, Scorpio is the most complete system overall. Pushpin and Faye offer the least functionality. Pushpin and Faye are not fault-tolerant, they cannot connect to other external systems, the data in the queues cannot be updated and even worst, the messages are lost if there are no subscribers to receive them. Scorpio and Stellio, the same as Apache Kafka, are high error-tolerant, support live-stream processing operations (i.e. can be linked to external systems), and provide advanced user identification and authorization mechanisms. Scorpio and Stellio have enriched Apache Kafka with REST NGSI-LD interfaces and allow updates and queries on message queues or databases. NGSI-LD format is not supported by RabbitMQ and Apache Kafka. Almost all systems are fast and can meet the speed requirements of modern fast-paced applications (e.g. news, Internet of Things, services communication in Software Oriented Architectures).

References

1. Bees, D., Frost, L., Bauer, M., Fisher, M., Li, W.: NGSI-LD API: for Context Information Management (2019). https://www.etsi.org/images/files/ETSIWhitePapers/etsi_wp31_NGSI_API.pdf. (ETSI White Paper No. 31, 1st edition)
2. Bellavista, P., Corradi, A., Reale, A.: Quality of service in wide scale publish-subscribe systems. IEEE Commun. Surv. Tutorials **16**(3), 1591–1616 (2014). https://ieeexplore.ieee.org/document/6803100
3. Chen, C., Vitenberg, R., Jacobsen, H.: OMen: overlay mending for topic-based publish/subscribe systems under churn. In: ACM International Conference on Distributed and Event-based Systems (DEBS 2016), pp. 105–116 (2016). https://doi.org/10.1145/2933267.2933305
4. Coglan, J.: Faye: Simple Pub/Sub Messaging for the Web (2021). https://faye.jcoglan.com
5. Corsaro, A., Querzoni, L., Scipioni, S., Tucci-Piergiovanni, S., Virgillito, A.: Quality of service in publish/subscribe middleware. In: Baldoni, R., Cortese, G. (eds.) Global Data Management, pp. 79–97. IOS Press (2006). https://ebooks.iospress.nl/volumearticle/22318
6. EGM: FIWARE STELLIO Context Broker (2021). https://stellio.readthedocs.io/en/latest/index.html

7. Esposito, C.: A tutorial on reliability in publish/subscribe services. In: ACM International Conference on Distributed Event-Based Systems (DEBS 2012), pp. 399–406 (2012). https://dl.acm.org/doi/10.1145/2335484.2335537
8. Fanout: Pushpin: Add push to your API (2021). https://pushpin.org
9. FIWARE: ORION-LD Linked Data Context Broker (2021). https://fiware-academy.readthedocs.io/en/latest/core/orion-ld/index.html
10. Khoury, J., Lauer, G., Pal, P., Thapa, B., Loyall, J.: Efficient private publish-subscribe systems. In: IEEE International Symposium on Object/Component/Service-Oriented Real-Time Distributed Computing, pp. 64–71 (2014). https://ieeexplore.ieee.org/document/6899132
11. Narkhede, N., Shapira, G., Palino, T.: Kafka The Definitive Guide, Real Time Data and Stream Processing at Scale. O'Reilly Media, Newton (2017). https://kafka.apache.org/
12. NECTI: ScorpioBroker Documentation (2020). https://scorpio.readthedocs.io/_/downloads/en/latest/pdf/
13. Nikhil, A., Chandar, V.: Benchmarking Apache Kafka, Apache Pulsar, and RabbitMQ: Which is the Fastest? (2020). https://www.confluent.io/blog/kafka-fastest-messaging-system/
14. Oh, S., h. Kim, J., Fox, G.C.: Real-time performance analysis for publish/subscribe systems. Future Gener. Comput. Syst. (FGCS) **26**(3), 318–323 (2010). https://dl.acm.org/doi/10.1016/j.future.2009.09.001
15. Privat, G.: Guidelines for Modelling with NGSI-LD (2021). https://www.researchgate.net/publication/349928709_Guidelines_for_Modelling_with_NGSI-LD_ETSI_White_Paper. ETSI White Paper
16. Sivaharan, T., Blair, G., Coulson, G.: Green: a configurable and re-configurable publish-subscribe middleware for pervasive computing. In: On the Move to Meaningful Internet Systems 2005: CoopIS, DOA, and ODBASE, pp. 732–749. Agia Napa, Cyprus (2005). https://link.springer.com/chapter/10.1007/11575771_46#citeas
17. Sporny, M., Longley, D., Kellogg, G., Lanthaler, M., Champin, P., Lindström, N.: JSON-LD 1.1: A JSON-based Serialization for Linked Data (2021). https://w3c.github.io/json-ld-syntax/. W3C Editor's Draft
18. VMware: RabbitMQ: Messaging that Works (2021). https://www.rabbitmq.com
19. Xiao, D., Zhang, Y., Chen, J.: A real-time and reliable forwarding scheme of publish/subscribe system. In: International Conference on Measuring Technology and Mechatronics Automation, pp. 51–54 (2014). https://ieeexplore.ieee.org/document/6802634
20. Yigal, A.: Kafka vs. Redis: Log Aggregation Capabilities and Performance (2016). https://logz.io/blog/kafka-vs-redis/

Conceptual Foundations of Code Rationalization Through a Case Study in Haskell

Razvan Bocu(✉) and Dorin Bocu

Department of Mathematics and Computer Science, Transilvania University of Brasov, 500036 Brasov, Romania
{razvan.bocu,d.bocu}@unitbv.ro

Abstract. The syntagm "code rationalization" denotes, in this paper, the approach by which the programmer chooses from the multitude of known alternatives that transpose the algorithmic solution of a problem into the target language of a software system, maximizing the benefits that the user pursues through the utilization of the software system. This paper emphasizes the importance of streamlining the code considering a case study in Haskell.

1 Introduction

The Haskell language is a stylish representative of the functional paradigm. The literature has presented over time its many advantages. Although the language did not receive the attention of the industry, the academic environment and the enthusiasm of the fans kept the Haskell project in the attention of a large number of curious, enterprising and contributors to the huge potential of the functional paradigm. This paper aims to present Haskell templates for solving problems of general interest. In fact, it is not the problems that are the target of the approach, but the algorithmic implications that the newcomer to Haskell has to face. The Haskell language does not claim to rethink problem-solving algorithms, but rather to choose the optimal template for implementing these algorithms in Haskell. More precisely, as is natural, there are differences between the implementation templates of an algorithm in a procedural language, such as Java, C++, or C#, and Haskell. The presentation of such templates can help to shorten the time needed to familiarize newcomers with the style of coding that is specific to Haskell. What are the reasons why the Haskell language is worth the effort to adapt with certain recipes for the implementation, at first sight unintuitive, of some algorithms encountered in many other languages? Are they the amazing compactness of the code, its memory usage efficiency, its code readability, or remarkable theoretical foundations? These aspects will be considered in this paper, which is structured considering the following sections. The following section presents the most relevant existing contributions. Moreover, the third section discusses specific aspects related to the theoretical and technological fundamentals of code rationalization. Following, significant considerations concerning the generic code rationalization, and Haskell code rationalization are included. Furthermore, the case study is presented and analyzed. The case study that is presented may be considered as a model concerning the functional code rationalization in Haskell. It also describes the

L. Barolli et al. (Eds.): AINA 2022, LNNS 449, pp. 116–128, 2022.
https://doi.org/10.1007/978-3-030-99584-3_11

beneficial effects of code rationalization in connection to some of the code's quality factors. Furthermore, the study illustrates the important role that conceptualization plays, together with the precise knowledge regarding the functional language's potential for conceptualization. The last section concludes the paper.

2 Relevant Existing Contributions

The methodical study of functional code rationalization, particularly considering the Haskell programming language [1–5], is a sparsely approached topic in the relevant scientific literature [6–10]. Thus, the authors of contribution [11] discuss on a contribution regarding Liquid Haskell, which improves Haskell with refinement types. Although this approach determines the optimization of functional code specification, it does not value the code compactness and runtime efficiency, as this paper does. The authors of [12] present an interesting extension of the plain Haskell programming language, which is represented by the embedded domain specific language Haski. This is used in order to implement Internet of Things (IoT) networked structures. The refinement of Haskell into a functional programming variant that is able to sustain the reliable operation of resources constrained IoT devices demonstrates the functional paradigm's potential to support various types of extensions and performance improvements. This is one of the factors that allowed for the research effort that is reported in this paper to specify the relevant functional code rationalization approach. Furthermore, it is possible to write highly idiomatic code and easy-to-write session types, as it is demonstrated by the authors of paper [13]. The emblematic concept of functional pearl is illustrated through a case study that describes the representation of correct-by-construction matrices [14]. This constitutes another proof concerning the possibility to write elegant and computationally efficient code in a purely functional style. The versatility and relevance of the functional approach for standard development scenarios is illustrated in paper [15], which introduces a possible programming environment that may be used to develop full-stack applications in a functional way. Additionally, functional programming is mentioned or effectively presented as a potential or fully effective solution for various academic or industrial real-world scenarios [16–20]. The relevant scientific literature demonstrates that there are only a few contributions that report scientific efforts, which aim to improve the design and implementation of functional code. This is the explanation for the relative brevity of this literature review section. Furthemore, it can be stated that this paper describes a code rationalization approach that is, to the best of our knowledge, singular in the scope of similar scientific works.

3 The Theoretical and Technological Fundamentals of Code Rationalization

The world of programming languages is constantly expanding. Perhaps this is also why programming in any language, at present bears, in many respects, the imprint of the theoretical and practical experience gained by each programmer. The methodologies of many software companies, aware of the disadvantages of such an approach, come with improvements, aiming to comply with predominantly formal standards, which concern code writing. These standards are of real interest in terms of project management. Concerning the software system solution engineering, other approaches are needed that could contribute to the optimization of the result of the programming activity, in general. Nevertheless, the programming activity is only a stage of the complex process, which realizes a software system. The quality of the programming activity is even strongly influenced by other activities, such as analysis, design, testing, which represent only the essential processes from a technical point of view. This paper discusses on the idea of code rationalization, and it takes into account the essential factors that can influence the outcome of the rationalization effort. The syntagm code rationalization encompasses aspects that are discussed in the related scientific literature, and pertain to requirements that are specific to a certain phase during the development of the software system, such as:

- **Compliance with the fundamental principles of the software systems engineering;**
- **Elaboration and systematic utilization of design patterns;**
- **Elaboration and systematic utilization of implementational patterns;**
- **The utilization of the optimal programming paradigm in order to solve a certain type of problem** (object orientation, aspect orientation, component orientation, services orientation, etc.).

3.1 Compliance with the Fundamental Principles of Software Systems Engineering

This is an extensive topic, due to the fact that its substance is born at the confluence of two fundamental concerns, namely the engineering of the solution of a software system and the management of the engineering process of the solution of a software system. The scientific contributions in the field have researched these two major sub-themes of software engineering. A series of books written by influential authors in the field of software engineering [21–23] tried to emphasize, together with the invariants of the software system development process, the importance of good practice, which shows openness to the way in which the theoretical knowledge acquired in software engineering and IT project management is applied in a world that is severely affected by the spiral of changes, considering financial rather than philosophical reasons.

Even when software engineering specialists show methodological courtesy, there are many cases in which the appeal to the methodical abstraction of the solution and the process of obtaining the solution is bypassed. This is a kind of ignorance, from a pragmatic perspective of the saying "festina lente", according to the current stage of development of the human civilization, a stage in which the time allocated to deep reflection is reduced. There are many recommendations that we find beautifully lined up in the literature or on the websites of numerous software engineering experts, who share with enthusiasm and good faith their theoretical and practical knowledge.

3.2 Elaboration and Systematic Utilization of Design Patterns

Software engineers are systematically concerned with hoarding the practical experience they gain over time. Design templates are one of the important ways that software developers can choose in order to make quality software systems. Relative to the contribution that is reported in [24], each design template is associated with a problem that can occur countless times in software projects. It describes the essential components of the solution to that problem, so that the solution can be used countless times. In each context, the template is subject to specific adjustments. The design patterns [26] represent a world in which the well-trained person can make quality reuse of design experience, while developing the necessary horizon for the correct assessment of the quality of software artifacts, at different levels of abstraction.

3.3 Elaboration and Systematic Utilization of Implementational Patterns

Implementational templates are the natural product of specialists for whom the solution of a problem is a systematic effort of abstraction. The contribution that is reported in paper [25] suggests that the development of implementational templates is an approach that capitalizes on the fundamental ideas of software engineering in general and design templates. The main target is represented by the code reuse, which is related to obtaining advantages that come bundled with the programming language in which the implementational template is made. Without insisting on the formal definition of the concept of implementational template, the author of the paper [25] is rather concerned with exemplifying the benchmarks that a programmer who specifies an implementation template must consider. Guided by his obvious pragmatism, Kent Back has the merit of associating the process of specifying an implementation template with valuable ideas, including from a theoretical perspective, such as the crystallization of a certain level of excellence in code writing.

3.4 Consideration of the Optimal Programming Paradigms for a Certain Type of Problem

The choice of the language in which certain components of a project are coded presupposes, among others, the exact knowledge of the advantages and disadvantages offered by the paradigm that the language promotes in a sustainable way. For example, if we have a lot of other reasons why we would be interested in the functional paradigm, then we have to choose one of the programming languages that rigorously respects the requirements of functional programming. Naturally, once we have opted for a paradigm, the search continues using other criteria. For example, if once we have chosen the functional paradigm as we want pure functional code, then the Haskell language can be an inspired choice.

4 Generic Code Rationalization

4.1 Problem Statement

It may already have been intuited that the idea of code rationalization, considering relatively isolated perspectives, is already a consistent presence in the world of programmers

[21, 24, 25]. This is also confirmed and augmented by the latest versions of established programming environments, which already incorporate support for the use of architectural templates, design templates, partial automation of testing or implementation. Nevertheless, the theoretical reasons, although usually well-founded, are not always a priority at the level of the so-called good practices. The systematic following of the exhortations of these theoretical reasons could become attractive from a practical point of view if we elaborate software development scenarios in which the rationalization of the code written in a certain language is assimilated to a process of searching for the state of symmetry of the code. This is the state in which the requirements assumed by developers in the process of realizing a software system are harmonized. The random walk in the programming activity favors the appearance of some contradictions between the assumed requirements. Furthermore, significant loopholes are also possible in meeting these requirements. Therefore, in order to systematically monitor these contradictions and to eliminate leaks, this paper proposes a framework that underlies the code rationalization.

Thus, let us consider a software artefact, which may be the object of several requirements. Let us note with ***C*** such a requirement. The identification of a requirement becomes useful from a methodological perspective if:

- **It is associated with a metric, so that the requirement is quantifiable;**
- **It has at least one associated constraint, let us note it with *C**.**

The specification of a metric for the requirement ***C*** involves, without excessive formalism, the identification of two fundamental components of the metric:

- **The list or the domain of valid values that are associated to requirement *C*;**
- **The method to calculate the current values, which are associated to requirement *C*.**

Concerning the constraint ***C****, it is specified through a type, and the limit value of the requirement, let us refer to it as LV.

We consider that there are two fundamental types of constraints:

- Constraints of the type **lower threshold (LT);**
- Constraints of the type **upper threshold (UT).**

Thus, a constraint of the type UT is satisfied if the current value that is associated to the requirement is less or equal to LV. Furthermore, a constraint of the type LT is satisified if the current value that is associated to the requirement is greater or equal to LV.

Obviously, **the limit value LV is established by the developer in such a way that the software artefact meets the expectations of its potential user**.

We consider that such an approach regarding the concept of requirement mandates the consideration of the generic code rationalization processes. Additionally, an automated framework that may support the actual code rationalization processes can be specified. The systematic monitoring of the progresses concerning the code rationalization turns from a conscience-related fact into a task that is considered during the planning of a software project.

4.2 Code Rationalization as a Process to Seek the Optimal State of the Software Systems

The assimilation of a program written in a certain language with a system implies that the choice for the optimal circumstantial version (OCV) of the program is an approach whose success can be made more efficient if the programmer capitalizes on the contribution of code rationalization in specifying it. The following paragraphs discuss about the usefulness of the concept of symmetry relative to the programming activity.

In order to ensure a more precise presentation, let us state that if $\mathbf{C}_1^*, \mathbf{C}_2^*, \ldots, \mathbf{C}_n^*$ **are the constraints that are considered during the implementation of a programme, then we can state that the programme is in an OCV state if the decisions that determine the code writing ensure the fulfillment of all the constraints at an appropriate circumstantial level.**

By contrast, a programme that does not fulfill at least a constraint is specified in a non-optimal state.

Beyond the many recommendations that the literature makes to developers in general and, consequently, to programmers, in this paper we focus on the following three issues, considered in many papers as a kind of resistance structure of software systems:

- The search for the optimal abstraction scheme of the problem's data (OASD);
- The search for the optimal algorithmic scheme (OAS), which values the potential of OASD;
- The search for the optimal implementation scheme (OIS).

The Search for the Optimal Abstraction Scheme of the Problem Data

As there are not many programs that have nothing to do with the world of data, it goes without saying that the realization of a program invariably involves the effort of finding the optimal scheme for the abstraction of the data.

The first type of abstraction of a program's data is the conceptual approach by which data from external format (associated with certain circumstances) are progressively transformed into an internal format.

The second type of abstraction of a program's data is the conceptual approach by which data from internal format are returned to external format, which is associated with certain circumstances. The challenge in this type of abstraction is to find the minimum core of redundancies with which to obtain a maximum impact on the external user of the data. Practice clearly shows that this type of abstraction must, at the same time, control the impact that the excessive specification of the external data format can have on the associated algorithmic schemes.

The Search for the Optimal Algorithmic Scheme to Capitalize the Potential of OASD

There are many requirements we can have when critically evaluating algorithmic schemes to capitalize on the potential of the types of data we work with. Here is a list of essential requirements:

- The execution time;
- The internal memory requirements, which is tightly connected to the processed data's structure;
- The index for favoring some optimal subsequent processings;
- The coding simplicity.

If we consider only these four requirements, then we understand that choosing the optimal algorithmic scheme for processing a type of data (in certain circumstances) is an activity in which the effort of abstraction (in another plane of course) is intense, often necessarily assorted with exemplary bursts of creativity, which determine basic pieces of real problem-solving templates of a certain type. If requirements, such as those exemplified above, are associated with constraints, then it is clear that the need to meet a constraint may lead us to rethink the data on which the algorithmic scheme operates. The naturally iterative searches continue until the constraint is met.

The Search for the Optimal Implementation Scheme
The translation of OASD and OAS into a certain programming language is partially dependent on the level of the programmer's expertise relative to the language. The choice of the optimal variant for implementing OASD and OAS decisions is a search process in which the acquired expertise must be related to the state in which the code is from OCV point of view.

5 Case Study

5.1 The Considered Problem

The illustration of the content and usefulness of a code streamlining approach in Haskell considers the implementation of the optimal Haskell code to determine how often words appear in a list of words. This code could be of interest in a project dedicated to the analysis of texts in order to identify some patterns in the use of words. At the same time, we specify that, for the function that generates the list of frequencies with which the words appear in the randomly generated text, there is a requirement that its execution time respect the constraint of not exceeding 2 s at a test volume of 2,000,000 words. Thus, given a text, let us make the following assumptions:

1. The text we intend to analyze is stored in a file;
2. The code that solves all tokenization issues is assumed to exist. Using this code, we can generate a list that contains all the words in the text.
3. We will assume that sorting the list of words, if necessary, is solved. This is a well-founded assumption, since in the <Prelude> module, which is delivered and accessible with the installation of the Haskell platform, there are several predefined solutions, including the <sort> function.

In general, but also in this case, algorithmic considerations can contribute to a positive approach to coding activity. The specifications required in accordance with the requirements of the case study are:

1. The list of words is sorted;
2. Consequently, the occurrences of a certain word are grouped.
3. Considering this basic logical structure, the first iteration of the algorithm may consist of the following steps:

   a) The first element of the list is fetched;
   b) Considering this first element, the number of occurrences in the list is determined;
   c) The occurrences of the first element are deleted from the initial list;
   d) The iteration resumes at step (a) if the remaining list is not empty.

5.2 Example of Code Rationalization in Three Haskell Iterations

Coding by a beginner in Haskell would involve the use of three functions, through the composition of which, the problem of determining the frequency with which each word appears in the given list can be solved. As you can see below, these three functions could be:

- genFIt_1 – in order to determine the list of frequencies;
- contapIt_1 – in order to count the number of occurrences of a word in the list, making the assumption that the occurrences are grouped;
- delap – in order to delete the occurrences of a word in the list, which is the word on the first position in the list.

```
-- Iteration 1
data Token=Token String|Err
     deriving (Read,Show,Eq,Ord)
genFIt_1::[Token]->[(Token,Int)]
genFIt_1 []=[]
genFIt_1 lt=((head lt),(contapIt_1 (head lt)(tail lt))):
            (genFIt_1 (delap (head lt) lt))
-- The function determines the number of consecutive
-- occurrences of a certain Token
contapIt_1::Token->[Token]->Int
contapIt_1 _ []=1
contapIt_1 tok (tc:rl)  |(tok==tc)=1+(contapIt_1 tok rl)
                        |otherwise=1
-- The function deletes the consecutive occurrences
-- of a certain Token
delap::Token->[Token]->[Token]
delap _ []=[]
delap tok (tc:rl) |(tok==tc)=(delap tok rl)
                  |otherwise=(tc:rl)
```

The above code illustrates part of the virtues of Haskell language regarding its code rationalization capabilities. Thus:

1. The type Token, which is declared with the reserved word <data>, in addition to representing the programmer's choice as to the name of the type constructor, the value constructor and the way a word is represented in internal format, requires the compiler to generate instances of the classes of types Read, Show, Eq, Ord, which would allow their use in the context of the type Token;
2. The function genFIt_1, according to the definition of its type, has as input a list of Token data and as output a list for which each element is a tuple, the first element of the tuple being a Token data and the second being the frequency with which the Token appears in the list. This way of defining the type of the genFIt_1 function will put its mark on the successive schemes for specifying the behavior of the genFIt_1 function;
3. The problem with this way of implementing the genFIt_1 function is that it is based on the contapIt_1 function, which for each word, traverses twice the group to which it belongs, once when counting the occurrences of a word, and once again when deleting the occurrences of the word. This becomes problematic in the case of test lists with many elements in terms of the necessary execution time.

Table 1. The data that is generated by the function genFIt_1

Tested function	Number of elements in the test list	Execution time in seconds	LV	Type of constraint
genFIt_1	2,000,000 words	5.78 s	2 s	UT

As it can be observed in Table 1, considering a test with 2,000,000 elements, the execution of the genFIt_1 function takes about 11.72 s. It is a reality that a demanding programmer should not be accustomed to. Many queries can be categorized as inadmissible in terms of response times if they were based on a function such as genFIt_1. Given the value of LV, the contribution of the genFIt_1 function to the symmetry of the code of an application that incorporates it, even if it is one-dimensional, is unacceptable.

```
-- Iteration 2
genFIt_2::[Token]->[(Token,Int)]
genFIt_2 []=[]
genFIt_2 lt=((head lt),ftok):(genFIt_2 rl)
              where (ftok,rl)=contapIt_2 (head lt) lt

contapIt_2::Token->[Token]->(Int,[Token])
contapIt_2 _ []=(0,[])
contapIt_2 tok (tc:rl)=if (tok==tc) then ((1+fi),rli)
                                    else (0,(tc:rl))
                            where (fi,rli)=contapIt_2 tok rl
```

Starting from the observation that the source of excessive processor time consumption is the contapIt_1 function, the natural solution seemed to us to be an amendment to the signature of the contapIt_1 function, obviously having implications for its implementation. The change caused the execution time of the genFIt_2 function to decrease to approximately 0.80 s, according to Table 2, which means that the constraint for which LV was set to 2 s is properly satisfied. **It should also be noted that:**

- The functional purity of Haskell code is preserved;
- The functions composition is elegantly combined with recursion, and with the lists work;
- The usage of the clause where is particularly useful for the compactness of the function contapIt_2 code.

Table 2. The data that is generated by the function genFIt_2

Tested function	Number of elements in the test list	Execution time in seconds	LV	Type of constraint
genFIt_2	2,000,000 words	0.80 s	2 s	UT

```
-- Iteration 3
  genFIO::[Token]->IO [(Token,Int)]

  genFIO lnr=do
  {
    if (lnr==[])   then return []
                   else do
       {
         (nrap,rl)<-return (contapIt_2 (head lnr) lnr);
         lsi<-genFIO rl;
         return (((head lnr),nrap):lsi)
       }
  }
```

The genFIO function is indispensable when the methodical observance of the monadic style of code writing is desired. It is interesting to see that the logic of using the monadic style brings advantages both in terms of the necessary ingredients and, somewhat surprisingly, in terms of execution time, according to the data in Table 3.

Table 3. The data that is generated by the function genFIO

Tested function	Number of elements in the test list	Execution time in seconds	LV	Type of constraint
genFIO	2,000,000 words	0.86 s	2 s	UT

5.3 Final Remarks Related to the Presented Use Case

The approach presented is intended to be a concrete proof of the fact that the programming activity has two fundamental dimensions:

- The circumstantial dimension;
- The speculative dimension.

The Circumstantial Dimension of the Programming Approach
We described a case study in order to draw attention to the close connection of the products of the human mind with the well-being of the environment to which man himself refers. **The metaphorical well-being we are talking about is the result of a search process, which can be even tree-like, at the end of which we can live the satisfaction of finding a shorter, less invasive, less polluting way of relating to the world we belong to.** The related search process led us to the conclusion that a possible optimal solution to generate the list of frequencies with which words appear in a text is, according to taste or needs, the functions genFIt_2 or genFIO. The difference between them is measured in tenths of a second in the case of a list of two million items. This level of performance should be satisfactory at the moment. The complete Haskell code may be downloaded for testing purposes from the following Git repository: https://github.com/Brasoveanul/HaskellCodeRationalization. The presented code occasionally uses lists of integers instead of lists of words, as test data.

The Speculative Dimension of the Programming Approach
A genuine programmer knows that the odds for success during a research process are higher if the speculative module of thinking is properly used. The immersion into the concrete without being endowed with an adequate speculative armor, which effectively combines thought templates with bursts of creativity, does not guarantee the success of a research approach. **Even if humanity is going through a period in which the respect for speculative thinking is declining, the opinion of the authors of this paper is that without the contribution of speculative thinking humanity, as a species, can not evolve.**

6 Conclusions

The consideration of the conceptual framework that is presented and the way it is used in the case study implies that the next possible step is concerned with the creation of a framework, which would be useful in order to monitor the software artefacts' state considering the OCV perspective.

References

1. Bird, R.: Pearls of Functional Algorithm Design. Cambridge University Press, Cambridge (2010)
2. Bocu, D.: Conceptual și aplicativ în programarea logică și funcțională, MATRIX ROM, București (2020)
3. Thompson, S.: The Craft of Functional Programming. Boston MA USA, Addison-Wessley (1999)
4. en.wikibooks.org. https://en.wikibooks.org/wiki/Haskell. Accessed 14 June 2021
5. Zilberstein, N.: Eliminating bugs with dependent Haskell (experience report). In: Proceedings of the 13th ACM SIGPLAN International Symposium on Haskell, USA, 7 August 2020 (2020)
6. Jantschi, L.: Detecting extreme values with order statistics in samples from continuous distributions. Mathematics **8**(216), 2020 (2020)
7. Breitner, J., et al.: Ready, Set, Verify! Applying hs-to-coq to real-world Haskell code. J. Funct. Program. **31**, e5 (2021). https://doi.org/10.1017/S0956796820000283
8. Figueroa, I., Leger, P., Fukuda, H.: Which monads Haskell developers use: an exploratory study. Sci. Comput. Program. **201**, 102523 (2021)
9. Jantschi, L.: A test detecting the outliers for continuous distributions based on the cumulative distribution function of the data being tested. Symmetry **11**(835), 2019 (2019)
10. Egi, S., Nishiwaki, Y.: Functional Programming in Pattern-Match-Oriented Programming Style. arXiv 2020, arXiv:2002.06176
11. Liu, Y., Parker, J., Redmond, P., Kuper, L., Hicks, M., Vazou, N.: Verifying replicated data types with typeclass refinements in Liquid Haskell. Proc. ACM Program. Lang. **4**, 1–30 (2020)
12. Valliappan, N., Krook, R., Russo, A., Claessen, K.: Towards secure IoT programming in Haskell. In: Proceedings of the 13th ACM SIGPLAN International Symposium on Haskell, USA, 7 August 2020, pp. 136–150 (2020)
13. Kokke, W., Dardha, O.: Deadlock-free session types in linear Haskell. arXiv 2021, arXiv:2103.14481
14. Santos, A., Oliveira, J.N.: Type your matrices for great good: a haskell library of typed matrices and applications (functional pearl). In: Proceedings of the 13th ACM SIGPLAN International Symposium on Haskell, USA, 7 August 2020, pp. 54–66 (2020)
15. Melo, C.A.R., Liu, P., Ying, R.: A platform for full-stack functional programming. In: Proceedings of the 2020 IEEE International Symposium on Circuits and Systems (ISCAS), Seville, Spain, 12–14 October 2020, pp. 1–5 (2020)
16. Nandi, C., Wilcox, J.R., Panchekha, P., Blau, T., Grossman, D., Tatlock, Z.: Functional programming for compiling and decompiling computer-aided design. Proc. ACM Program. Lang. **2**, 1–31 (2018)
17. Ponce, L.M., Lezzi, D., Badia, R.M., Guedes, D.: DDF library: enabling functional programming in a task-based model. J. Parallel Distrib. Comput. **151**, 112–124 (2021)
18. Rubio, F., de la Encina, A., Rabanal, P., Rodriguez, I.: A parallel swarm library based on functional programming. In: Proceedings of the International Work-Conference on Artificial Neural Networks, Cadiz, Spain, 14–16 June 2017, pp. 3–15 (2017)
19. Gamari, B., Dietz, L.: Alligator collector: a latency-optimized garbage collector for functional programming languages. In: Proceedings of the 2020 ACM SIGPLAN International Symposium on Memory Management, London, UK, 16 June 2020, pp. 87–99 (2020)
20. Chifamba, T.D., Motara, Y.M.: Gamification of functional programming. In: Proceedings of the 2020 2nd International Multidisciplinary Information Technology and Engineering Conference (IMITEC), Kimberley, South Africa, 25–27 November 2020, pp. 1–8 (2020)
21. Bertrand, M.: Object-Oriented Software Construction Prentice-Hall International Series in Computer Science. Prentice-Hall, Upper Saddle River (1997)

22. Bjorner, D.: Software Engineering 1. Abstraction and Modelling. Springer, Berlin (2006)
23. Bjorner, D.: Software Engineering 2. Specification of Systems and Languages. Springer, Berlin (2006)
24. Blaha, M.: Patterns of Data Modeling. CRC Press, Boca Raton (2010)
25. Beck, K.: Implementation Patterns. Addison-Wesley, Boston (2008). ISBN: 978-0321413093
26. Gamma, E., Vlissides, J., Helm, R., Johnson, R.: Design Patterns: Elements of Reusable Object Oriented. Addison-Wesley, Boston (1994)
27. Hutton, G.: Programming in Haskell. Cambridge University Press, Cambridge (2007)

Energy-Efficient Concurrency Control by Omitting Meaningless Write Methods in Object-Based Systems

Tomoya Enokido[1(✉)], Dilawaer Duolikun[2], and Makoto Takizawa[3]

[1] Faculty of Business Administration, Rissho University, 4-2-16, Osaki, Shinagawa-ku, Tokyo 141-8602, Japan
eno@ris.ac.jp

[2] Department of Advanced Sciences, Faculty of Science and Engineering, Hosei University, 3-7-2, Kajino-cho, Koganei-shi, Tokyo 184-8584, Japan

[3] Research Center for Computing and Multimedia Studies, Hosei University, 3-7-2, Kajino-cho, Koganei-shi, Tokyo 184-8584, Japan
makoto.takizawa@computer.org

Abstract. In object-based systems, applications are composed of multiple objects and transactions created on clients issue methods to manipulate the objects. Multiple conflicting transactions have to be serialize to keep all the objects mutually consistent. If the more number of transactions are concurrently performed in a system, the throughput of a system decreases since the overhead to serialize conflicting transactions increases. In addition, a large amount of electric energy is consumed on servers which hold objects since the larger number of methods are performed on the servers to manipulate objects. In this paper, the EE2PL (Energy-Efficient Two-Phase Locking) protocol is newly proposed to reduce not only the total electric energy consumption of servers but also the execution time of each transaction by omitting meaningless write methods on each object. We show the total electric energy consumption of servers and the average execution time of each transaction can be reduced in the EE2PL protocol than the 2PL (Two-Phase Locking) protocol in the evaluation.

Keywords: Two-phase locking (2PL) protocol · Energy-efficient two-phase locking (EE2PL) protocol · Concurrency control · Transactions · Object-based systems

1 Introduction

An object [1,2] is a unit of computation resource like a database system [3]. Each object is an encapsulation of data and methods to manipulate the data in the object. In object-based systems [1,2,4], objects are distributed on multiple servers and each application is composed of multiple objects. A transaction [3,5] is an atomic sequence of methods to manipulate objects. Each transaction

L. Barolli et al. (Eds.): AINA 2022, LNNS 449, pp. 129–139, 2022.
https://doi.org/10.1007/978-3-030-99584-3_12

created on a client issues methods supported by each target object to utilize an application service. Here, conflicting methods issued by multiple transactions have to be serialized [2,3,5] to keep all the objects mutually consistent. In our previous studies, the *Object-base Group* (*OG*) protocol [6] was proposed to not only serialize multiple conflicting transactions but also order only messages which are required to be ordered at an application level based on the conflicting relation among methods. The *Role Ordering* (*RO*) [7] scheduler was proposed so that multiple conflicting transactions are serializable in the significant dominant relation of roles [8]. The *Legal Information Flow* (*LIF*) [9] scheduler was proposed to not only prevent illegal information flow but also serialize conflicting transactions based on the role concept [8]. The *Purpose-Marking* (*PM*) protocol [10] based on the *purpose* [10] concept was proposed to not only prevent illegal information flow but also serialize conflicting transactions. The *Two-Phase Locking* (*2PL*) protocol [5,11] is widely used to serialize multiple conflicting transactions in object-based systems. In the 2PL protocol, each object is locked by a transaction before manipulating the object. By using the 2PL protocol, multiple conflicting transactions can be serialized and objects can be kept mutually consistent. On the other hand, every object has to be locked for every method issued by each transaction. As a result, the throughput of a system decreases since the overhead to lock objects increases if the more number of transactions are concurrently performed in the system. In addition, every method issued by each transaction is surely performed on each target object in the 2PL protocol. As a result, the more number of transactions are issued in a system, the larger electric energy is consumed in servers. Hence, an energy-efficient concurrency control mechanism is required to not only reduce the total electric energy consumption of servers but also increase the throughput of a system as discussed in Green computing [12–14].

In this paper, *meaningless write methods* which are not required to be performed on each object is defined based on the precedent relation among transactions and semantics of methods. Next, the *Energy-Efficient Two-Phase Locking* (*EE2PL*) protocol is newly proposed to reduce the total electric energy consumption of servers by omitting meaningless write methods on each object. We evaluate the EE2PL protocol in terms of the total electric energy consumption of servers and the average execution time of each transaction compared with the 2PL protocol. The evaluation results show the total electric energy consumption of servers and the average execution time of each transaction in the EE2PL protocol can be more reduced than the 2PL protocol.

In Sect. 2, we present the 2PL protocol, data access model, and power consumption model of a server. In Sect. 3, we propose the EE2PL protocol. In Sect. 4, we evaluate the EE2PL protocol compared with the 2PL protocol.

2 System Model

2.1 Object-Based Systems

A system is composed of a set S of multiple servers s_1, ..., s_n ($n \geq 1$) and clients interconnected in reliable networks. An object o [1] is an unit of computation resource like a file. Each object o is an encapsulation of data and methods to manipulate data in the object o. Let O be a set of objects o_1, ..., o_m ($m \geq 1$) in the system. Each object o_h is allocated to a server s_t in the server cluster S. Methods are classified into *read* (r) and *write* (w) methods in this paper. Write methods are furthermore classified into *full* write (w^f) and *partial* write (w^p) methods, i.e. $w \in \{w^f, w^p\}$. A full write method fully writes a whole data in an object o_h. A partial write method writes a part of data in an object o_h. Suppose a file object F supports *insert*, *delete*, *modify*, and *read* methods. Insert and delete methods are partial write methods. A modify method is a full write method. Let $op(o_h)$ be a state obtained by performing a method op ($\in \{r, w\}$) on an object o_h. Let $op_1 \circ op_2(o_h)$ be a state obtained by performing a method op_1 after another method op_2 on the object o_h. A pair of methods op_1 and op_2 on an object o_h are *compatible* if and only if (iff) $op_1 \circ op_2(o_h) = op_2 \circ op_1(o_h)$. Otherwise, a method op_1 *conflicts* with another method op_2. In this paper, we assume conflicting relations among methods are given as shown in Table 1.

Table 1. Conflicting relation among methods.

		read (r)	write (w)	
			full (w^f)	partial (w^p)
read (r)		compatible	conflict	conflict
write (w)	full (w^f)	conflict	conflict	conflict
	partial (w^p)	conflict	conflict	conflict

2.2 Two-Phase Locking (2PL) Protocol

A *transaction* is an atomic sequence of methods [5]. A transaction T^i issues read (r) and write (w) methods to manipulate objects. Let $\mathbf{T}$ be a set $\{T^1$, ..., $T^k\}$ ($k \geq 1$) of transactions issued in a system. Multiple conflicting transactions are required to be *serializable* [3,5] to keep all the objects mutually consistent. Let H be a schedule of transactions in $\mathbf{T}$. A transaction T^i *precedes* another transaction T^j ($T^i \rightarrow_H T^j$) in a schedule H iff a method op^i issued by the transaction T^i is performed before a method op^j issued by the transaction T^j and the method op^i conflicts with the method op^j. A schedule H is serializable iff the precedent relation $\rightarrow_H$ is acyclic [5].

The *Two-Phase Locking* (*2PL*) protocol [2,3,5,11] is widely used to serialize multiple conflicting transactions. In this paper, multiple conflicting transactions are serialized based on the 2PL protocol. Let $\mu(op)$ be a *lock mode* of a method op

($\in \{r, w\}$). In this paper, $\mu(w)$ shows a lock mode of a full/partial write method w. $\mu(r)$ indicates a lock mode of a read method r. If op^1 is compatible with op^2 on an object o_h, the lock mode $\mu(op^1)$ is compatible with $\mu(op^2)$. Otherwise, a lock mode $\mu(op^1)$ conflicts with another lock mode $\mu(op^2)$. In the 2PL protocol, a transaction T^i locks each object o_h by the following procedure [2,3,5,11]:

1. If an object o_h can be locked by a lock mode $\mu(op)$, the object o_h is manipulated by the method op.
2. When the transaction T^i commits or aborts, the lock on the object o_h is released. Once a transaction T^i releases a lock on an object, the transaction T^i does not lock any object until the transaction T^i commits or aborts.

2.3 Data Access Model

Methods which are being performed and already terminate are *current* and *previous* at time τ, respectively. Let $RP_t(\tau)$ and $WP_t(\tau)$ be sets of current *read* (r) and *write* (w) methods on a server s_t at time τ, respectively. A notation $P_t(\tau)$ shows a set of current read and write methods on a server s_t at time τ, i.e. $P_t(\tau)$ $= RP_t(\tau) \cup WP_t(\tau)$. Let $r_t^i(o_h)$ and $w_t^i(o_h)$ be methods issued by a transaction T^i to read and write data in an object o_h on a server s_t, respectively. Each read method $r_t^i(o_h)$ in a set $RP_t(\tau)$ reads data in an object o_h at rate $RR_t^i(\tau)$ [Byte/sec (B/sec)] at time τ. Each write method $w_t^i(o_h)$ in a set $WP_t(\tau)$ writes data in an object o_h at rate $WR_t^i(\tau)$ [B/sec] at time τ. Let $maxRR_t$ and $maxWR_t$ be the maximum read and write rates [B/sec] of read and write methods on a server s_t, respectively. The read rate $RR_t^i(\tau)$ ($\leq maxRR_t$) and write rate $WR_t^i(\tau)$ ($\leq maxWR_t$) are $dr_t(\tau) \cdot maxRR_t$ and $dw_t(\tau) \cdot maxWR_t$, respectively. Here, $dr_t(\tau)$ and $dw_t(\tau)$ are degradation ratios. $1/(|RP_t(\tau)| + rw_t \cdot |WP_t(\tau)|)$ and $1/(wr_t \cdot |RP_t(\tau)| + |WP_t(\tau)|)$, respectively, where $0 \leq rw_t \leq 1$ and $0 \leq wr_t \leq 1$. $0 \leq dr_t(\tau) \leq 1$ and $0 \leq dw_t(\tau) \leq 1$.

The *read laxity* $rl_t^i(\tau)$ [B] and *write laxity* $wl_t^i(\tau)$ [B] of methods $r_t^i(o_h)$ and $w_t^i(o_h)$ show how much amount of data are to be read and written in an object o_h by the methods $r_t^i(o_h)$ and $w_t^i(o_h)$ at time τ, respectively. Suppose that methods $r_t^i(o_h)$ and $w_t^i(o_h)$ start on a server s_t at time st_t^i, respectively. At time st_t^i, the read laxity $rl_t^i(\tau) = rb_h$ [B] where rb_h is the size of data in an object o_h. The write laxity $wl_t^i(\tau) = wb_h$ [B] where wb_h is the size of data to be written in an object o_h. The read laxity $rl_t^i(\tau)$ and write laxity $wl_t^i(\tau)$ at time τ are rb_h - $\Sigma_{\tau=st_t^i}^{\tau} RR_t^i(\tau)$ and wb_h - $\Sigma_{\tau=st_t^i}^{\tau} WR_t^i(\tau)$, respectively.

2.4 Power Consumption Model of a Server

In our previous studies, the *power consumption model for a storage server* (*PCS* model) [15] to perform storage and computation processes are proposed. Let $E_t(\tau)$ be the electric power [W] of a server s_t at time τ. $maxE_t$ and $minE_t$ show the maximum and minimum electric power [W] of the server s_t, respectively. In this paper, we assume only read and write methods are performed on a server s_t. According to the PCS model [15], the electric power $E_t(\tau)$ [W] of a server s_t to perform multiple read and write methods at time τ is given as follows:

$$E_t(\tau) = \begin{cases} WE_t & \text{if } |WP_t(\tau)| \geq 1 \text{ and } |RP_t(\tau)| = 0. \\ WRE_t(\alpha) & \text{if } |WP_t(\tau)| \geq 1 \text{ and } |RP_t(\tau)| \geq 1. \\ RE_t & \text{if } |WP_t(\tau)| = 0 \text{ and } |RP_t(\tau)| \geq 1. \\ minE_t & \text{if } |WP_t(\tau)| = |RP_t(\tau)| = 0. \end{cases} \tag{1}$$

A server s_t consumes the minimum electric power $minE_t$ [W] if no method is performed on the server s_t, i.e. the electric power in the idle state of the server s_t. The server s_t consumes the electric power RE_t [W] if at least one r method is performed on the server s_t. The server s_t consumes the electric power WE_t [W] if at least one w method is performed on the server s_t. The server s_t consumes the electric power $WRE_t(\alpha)$ [W] $= \alpha \cdot RE_t + (1 - \alpha) \cdot WE_t$ [W] where $\alpha = |RP_t(\tau)|/(|RP_t(\tau)| + |WP_t(\tau)|)$ if both at least one r method and at least one w method are concurrently performed. Here, $minE_t \leq RE_t \leq WRE_t(\alpha) \leq WE_t \leq maxE_t$. The total electric energy $TEE_t(\tau_1, \tau_2)$ [J] of a server s_t from time τ_1 to τ_2 is $\Sigma_{\tau=\tau1}^{\tau2} E_t(\tau)$. The processing electric power $PEP_t(\tau)$ [W] of a server s_t at time τ is $E_t(\tau)$ - $minE_t$. The total processing electric energy $TPEE_t(\tau_1, \tau_2)$ of a server s_t from time τ_1 to τ_2 is given as $TPEE_t(\tau_1, \tau_2) = \Sigma_{\tau=\tau1}^{\tau2} PEP_t(\tau)$.

3 Energy-Efficient Two-Phase Locking (EE2PL) Protocol

3.1 Meaningless Write Methods

A method op^1 *precedes* op^2 in a schedule H ($op^1 \rightarrow_H op^2$) iff 1) the method op^1 is issued before op^2 by a same transaction T^i, 2) the method op^1 issued by a transaction T^i conflicts with the method op^2 issued by a transaction T^j and $T^i \rightarrow_H T^j$, or 3) $op^1 \rightarrow_H op^3 \rightarrow_H op^2$ for some method op^3. Let H_h be a *local schedule* of methods which are performed on an object o_h in a schedule H. A method op^1 *locally precedes* another method op^2 in a local schedule H_h ($op^1 \rightarrow_{H_h} op^2$) iff $op^1 \rightarrow_H op^2$.

Suppose a partial write method $w^p(o_h)$ locally precedes another full write method $w^f(o_h)$ in a local schedule H_h ($w^p(o_h) \rightarrow_{H_h} w^f(o_h)$) on an object o_h. Here, the partial write method $w^p(o_h)$ is not required to be performed on the object o_h if the full write method $w^f(o_h)$ is surely performed on the object o_h just after the partial write method $w^p(o_h)$, i.e. the full write method $w^f(o_h)$ can *absorb* the partial write method $w^p(o_h)$.

[Definition] A full write method op^1 *absorbs* another partial or full write method op^2 in a local subschedule H_h on an object o_h iff (if and only if) one of the following conditions is hold:

1. $op^2 \rightarrow_{H_h} op^1$ and there is no read method op' such that $op^2 \rightarrow_{H_h} op' \rightarrow_{H_h} op^1$.
2. op^1 absorbs op'' and op'' absorbs op^2 for some method op''.

[Definition] A write method op is *meaningless* iff the write method op is absorbed by another write method op' in the local subschedule H_h of an object o_h.

3.2 Energy-Efficient 2PL (EE2PL) Protocol

In this paper, the *Energy-Efficient Two-Phase Locking* (*EE2PL*) protocol is newly proposed to reduce not only the total electric energy consumption of a server cluster but also the average execution time of each transaction by omitting meaningless write methods on each object.

Suppose an object o_h allocated to a server s_t is locked by a transaction T^i with a lock mode $\mu(w)$ and a partial write method $w_t^{pi}(o_h)$ is issued by the transaction T^i as shown in Fig. 1. In the 2PL protocol, the partial write method $w_t^{pi}(o_h)$ is performed on the object o_h as soon as the object o_h receives the partial write method $w_t^{pi}(o_h)$. On the other hand, the object o_h sends a termination notification of the partial method $w_t^{pi}(o_h)$ to the transaction T^i as soon as the object o_h receives the partial write method $w_t^{pi}(o_h)$ but the partial write method $w_t^{pi}(o_h)$ is not performed until the next method is performed on the object o_h in the EE2PL protocol. This means that the partial write method $w_t^{pi}(o_h)$ is delayed until the next method is performed on the object o_h. Suppose the object o_h is locked by another transaction T^j with a lock mode $\mu(w)$ after the transaction T^i commits and a full write methods $w_t^{fj}(o_h)$ is issued by the transaction T^j. Here, the partial write method $w_t^{pi}(o_h)$ issued by the transaction T^i is meaningless since the full write method $w_t^{fj}(o_h)$ issued by the transaction T^j absorbs the partial write method $w_t^{pi}(o_h)$ on the object o_h. Hence, the full write method $w_t^{fj}(o_h)$ can be performed on the object o_h without performing the partial write method $w_t^{pi}(o_h)$. This means that the meaningless write method $w_t^{pi}(o_h)$ can be omitted on the object o_h.

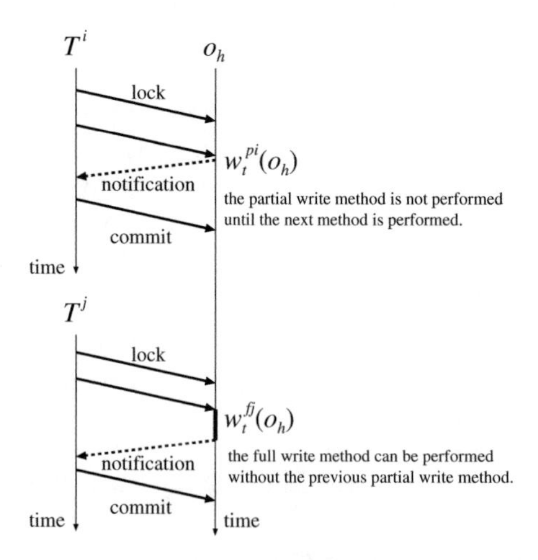

Fig. 1. Omission of a meaningless write method.

Suppose an object o_h allocated to a server s_t is locked by a transaction T^j with lock mode $\mu(r)$ after a transaction T^i commits and a read methods $r_t^j(o_h)$ is issued by the transaction T^j as shown in Fig. 2. Here, the partial write method $w_t^{pi}(o_h)$ issued by the transaction T^i has to be performed before the read method $r_t^j(o_h)$ is performed since the read method $r_t^j(o_h)$ has to read data written by the partial write method $w_t^{pi}(o_h)$.

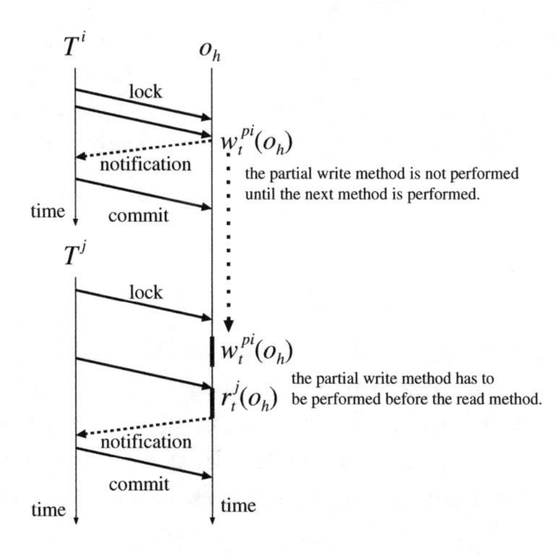

Fig. 2. Execution of a write method.

Let $o_h.Dw$ be a write method $w_t^i(o_h)$ issued by a transaction T^i to write data of an object o_h in a server s_t, which is waiting for the next method *op* to be performed on the object o_h. Suppose a transaction T^i issues a method *op* to an object o_h. In the EE2PL protocol, the method *op* is performed on the object o_h by the following **EE2PL** procedure:

EE2PL($op(o_h)$) {
  if $op(o_h) = r$, { /* *op* is a read method. */
    if $o_h.Dw = \phi$, perform($op(o_h)$);
    else { /* $o_h.Dw \neq \phi$ */
      perform($o_h.Dw$);
      $o_h.Dw = \phi$;
      perform($op(o_h)$);
    }
  }
  else { /* *op* is a write method. */
    if $o_h.Dw = \phi$, $o_h.Dw = op(o_h)$;
    else { /* $o_h.Dw \neq \phi$ */
      if $op(o_h)$ absorbs $o_h.Dw$, $o_h.Dw = op(o_h)$; /* $o_h.Dw$ is omitted. */
      else {
        perform($o_h.Dw$);

```
                $o_h.Dw = op(o_h)$;
            }
        }
    }
}
```

In the EE2PL protocol, the total electric energy consumption of a server cluster can be more reduced than the 2PL protocol since the number of write methods performed on each object can be reduced in the EE2PL protocol. In addition, the execution time of each transaction can be more reduced in the EE2PL protocol than the 2PL protocol since each transaction can commit or abort without waiting for performing meaningless write methods.

4 Evaluation

4.1 Environment

We evaluate the EE2PL protocol in terms of the average execution time of each transaction and the total processing electric energy of a server cluster S compared with the 2PL protocol. A homogeneous server cluster S is composed of ten homogeneous servers s_1, ..., s_{10} ($n = 10$). In the server cluster S, every server s_t ($t = 1$, ..., 10) follows the same data access model and power consumption model as shown in Table 2. Parameters of each server s_t are given based on the experimentations [15]. There are fifty objects o_1, ..., o_{50} in a system. The size of data in each object o_h is randomly selected between 50 and 100 [MByte]. Each object o_h supports *read* (r), *full write* (w^f), and *partial write* (w^p) methods. Each object is randomly allocated to a server in the server cluster S.

Table 2. Homogeneous cluster S ($t = 1$, ..., 10)

server s_t	$maxRR_t$	$maxWR_t$	rw_t	wr_t	$minE_t$	WE_t	RE_t
s_t	80 [MB/sec]	45 [MB/sec]	0.5	0.5	39 [W]	53 [W]	43 [W]

The number m ($0 \leq m \leq 5{,}000$) of transactions are issues to manipulate objects. Each transaction issues three methods randomly selected from one-hundred fifty methods on the fifty objects. The total amount of data of an object o_h is fully written by each full write (w^f) method. On the other hand, a half size of data of an object o_h is written and read by each partial write (w^p) and read (r) methods, respectively. The starting time of each transaction T^i is randomly selected in a unit of one second between 1 and 360 [sec].

4.2 Total Processing Electric Energy Consumption

Figure 3 shows the total processing electric energy consumption [KJ] of the server cluster S to perform the number m of transactions in the 2PL and EE2PL protocols. For $0 \leq m \leq 5{,}000$, the total processing electric energy consumption of the server cluster S can be more reduced in the EE2PL protocol than the 2PL protocol. In the EE2PL protocol, meaningless write methods are omitted on each object. As a result, the total processing electric energy consumption of the server cluster S can be more reduced in the EE2PL protocol than the 2PL protocol.

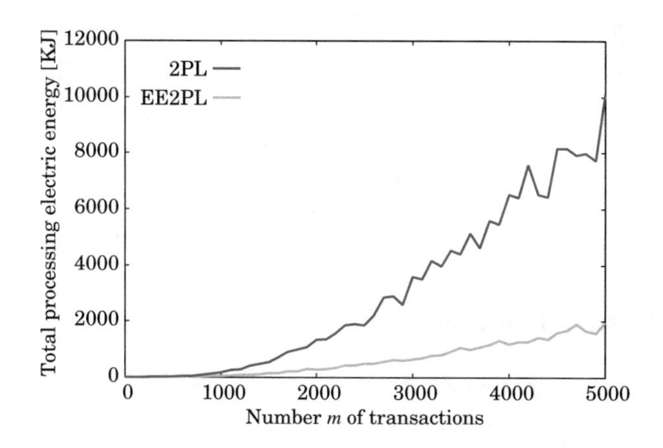

Fig. 3. Total processing electric energy consumption [KJ] of a server cluster S.

4.3 Average Execution Time of Each Transaction

Figure 4 shows the average execution time [sec] of the m transactions in the 2PL and EE2PL protocols. In the 2PL and EE2PL protocols, the average execution time increases as the total number m of transactions increases since more number of transactions are concurrently performed. For $0 < m \leq 5{,}000$, the average execution time of each transaction can be more reduced in the EE2PL protocol than the 2PL protocol. In the EE2PL protocol, each transaction can commit without waiting for performing meaningless write methods. Hence, the average execution time of each transaction can be more reduced in the EE2PL protocol than the 2PL protocol.

Following the evaluation, the total processing electric energy consumption of a server cluster and the average execution time of each transaction in the EE2PL protocol can be more reduced than the 2PL protocol, respectively. Hence, the EE2PL protocol is more useful than the 2PL protocol.

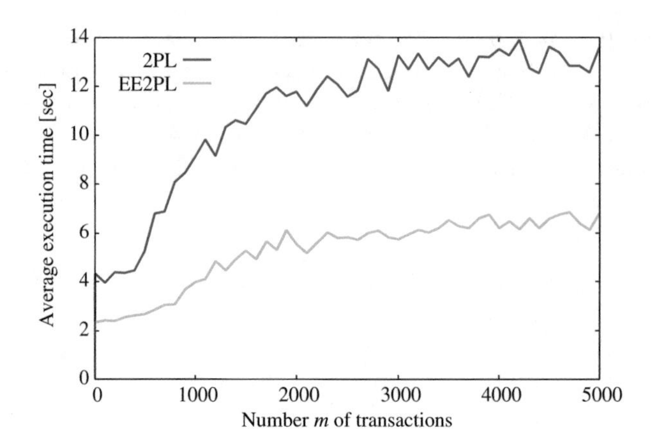

Fig. 4. Average execution time [sec] of each transaction.

5 Concluding Remarks

In this paper, we newly proposed the EE2PL (Energy-Efficient Two-Phase Locking) protocol to reduce not only the total processing electric energy consumption of a server cluster but also the average execution time of each transaction by omitting meaningless write methods. We evaluated the EE2PL protocol compared with the 2PL protocol. In the evaluation, we showed the total processing electric energy consumption of a server cluster and the average execution time of each transaction can be more reduced in the EE2PL protocol than the 2PL protocol. Following the evaluation, the EE2PL protocol is more useful than the 2PL protocol.

References

1. Object Management Group Inc.: Common object request broker architecture (CORBA) specification, version 3.3, part 1 – interfaces (2012). http://www.omg.org/spec/CORBA/3.3/Interfaces/PDF
2. Tanaka, K., Hasegawa, K., Takizawa, M.: Quorum-based replication in object-based systems. J. Inf. Sci. Eng. **16**(3), 317–331 (2000)
3. Gray, J.N.: Notes on data base operating systems. In: Bayer, R., Graham, R.M., Seegmüller, G. (eds.) Operating Systems. LNCS, vol. 60, pp. 393–481. Springer, Heidelberg (1978). https://doi.org/10.1007/3-540-08755-9_9
4. Enokido, T., Duolikun, D., Takizawa, M.: Energy-efficient quorum selection algorithm for distributed object-based systems. In: Barolli, L., Terzo, O. (eds.) CISIS 2017. AISC, vol. 611, pp. 31–42. Springer, Cham (2018). https://doi.org/10.1007/978-3-319-61566-0_4
5. Bernstein, P.A., Hadzilacos, V., Goodman, N.: Concurrency Control and Recovery in Database Systems. Addison-Wesley, Boston (1987)
6. Enokido, T., Higaki, H., Takizawa, M.: Object-based ordered delivery of messages in object-based systems. In: Proceedings of the 1999 International Conference on Parallel Processing (ICPP-1999), pp. 380–387 (1999)

7. Enokido, T., Takizawa, M.: Role-based concurrency control for distributed systems. In: Proceedings of the 20th IEEE International Conference on Advanced Information Networking and Applications - Volume 1 (AINA-2006), pp. 407–412 (2006)
8. Sandhu, R.S., Coyne, E.J., Feinsteink, H.L., Youman, C.E.: Role-based access control models. IEEE Comput. **29**(2), 38–47 (1996)
9. Enokido, T., Barolli, V., Takizawa, M.: A legal information flow (LIF) scheduler based on role-based access control model. Comput. Stand. Interfaces **31**(5), 906–912 (2009)
10. Enokido, T., Takizawa, M.: Purpose-based information flow control for cyber engineering. IEEE Trans. Ind. Electron. **58**(6), 2216–2225 (2011)
11. Garcia-Molina, H., Barbara, D.: How to assign votes in a distributed system. J. ACM **32**(4), 814–860 (1985)
12. Natural Resources Defense Council (NRDS): Data center efficiency assessment - scaling up energy efficiency across the data center lndustry: Evaluating key drivers and barriers (2014). http://www.nrdc.org/energy/files/data-center-efficiency-assessment-IP.pdf
13. Enokido, T., Duolikun, D., Takizawa, M.: The improved redundant active time-based (IRATB) algorithm for process replication. In: Barolli, L., Woungang, I., Enokido, T. (eds.) AINA 2021. LNNS, vol. 225, pp. 172–180. Springer, Cham (2021). https://doi.org/10.1007/978-3-030-75100-5_16
14. Enokido, T., Duolikun, D., Takizawa, M.: The redundant active time-based algorithm with forcing meaningless replica to terminate. In: Barolli, L., Yim, K., Enokido, T. (eds.) CISIS 2021. LNNS, vol. 278, pp. 206–213. Springer, Cham (2021). https://doi.org/10.1007/978-3-030-79725-6_20
15. Sawada, A., Kataoka, H., Duolikun, D., Enokido, T., Takizawa, M.: Energy-aware clusters of servers for storage and computation applications. In: Proceedings of the 30th IEEE International Conference on Advanced Information Networking and Applications (AINA-2016), pp. 400–407 (2016)

A Multi-agent Model to Support Privacy Preserving Co-owned Image Sharing on Social Media

Farzad N. Motlagh(✉), Anne V. D. M. Kayem, and Christoph Meinel

Hasso-Plattner-Institute, University of Potsdam, Potsdam, Germany
{farzad.motlagh,anne.kayem,christoph.meinel}@hpi.de

Abstract. Privacy on online social networks is a concern, specifically in relation to sharing co-owned images. Co-owned images raise a privacy conundrum, in that enforcing privacy policies impacts negatively on performance and usability. In most existing work, this issue is addressed by requiring users from the same or overlapping friendship networks to contribute privacy opinions vis-a-vis posting the co-owned image. This poses two issues: (1) privacy posting decisions cannot be made, resulting in delay; and (2) ineffective user opinion computation necessitates large amounts of image distortion, resulting in low levels of user satisfaction. In this paper, we present a multi-agent system in which an opinion formulation algorithm computes offline user opinions based on user personality and behaviour information. Our results indicate that posting decisions, for images in which all co-owners are offline, take 3.81 s on average.

1 Introduction

Social media offers a feature-rich platform for co-owned image sharing [1,2]. Co-owned images are ones that may have been taken by an individual, but include images of other individuals. In certain cases, however, users involved may not be aware that the picture was taken and would typically not have been consulted before the image is shared publicly. For users wanting to keep their information private, this raises a privacy a conflict [3,4].

For instance, consider a scenario in which Alice wishes to upload co-owned graduation party images to her social media account. Using her privacy settings, she controls the exposure of the uploaded co-owned images so that the images are visible only to her close friends. However, one of her friends, say Bob, would have preferred to keep his presence at the graduation party secret from his employer, say John, who also happens to be in Alice's friendship network. By sharing the image without Bob's consent, Alice inadvertently puts Bob's privacy vis-a-vis his employer at risk.

Problem Statement. Most existing approaches [1,5–12] consider mainly cases in which the users involved are online and can actively provide privacy opinions vis-a-vis posting the co-owned image. What has received less attention, is the case in which

L. Barolli et al. (Eds.): AINA 2022, LNNS 449, pp. 140–151, 2022.
https://doi.org/10.1007/978-3-030-99584-3_13

the users involved may be offline for extended periods. This, however, poses two issues in terms of privacy preservation and usability, namely: (1) When users are offline, a privacy decision regarding whether or not to post the image cannot be reached, thereby resulting in delay. (2) Ineffective user opinion computation necessitates large amounts of image distortion, resulting in a low level of user satisfaction.

Contributions. In this paper, we present a multi-agent model to address both problems by handling opinions from both offline and online users. Our multi-agent model maps each user profile to an agent. An **uploader agent** acting on behalf of the user wishing to post the co-owned image, contacts the **user agents** acting on behalf of the other users who appear in the image, to reach a privacy agreement with respect to posting the image. Users who are online submit their opinions directly to the **uploader agent**. For users who are offline and/or users who fail to provide an opinion before a pre-defined response time threshold, opinions must be computed to determine the associated users' privacy preferences with respect to the co-owned image. To formulate the offline user opinions, we employ an opinion computation algorithm that infers the user's privacy preferences based on personality profile and behaviour history information. Finally, based on the user privacy opinions, the **uploader agent** invokes a **filtering agent** to blur out (enforce privacy by concealing) the images belonging to the users who declined to have their image displayed. Our empirical results demonstrate that we can reach a privacy-preserving decision regarding posting a co-owned image in a time-efficient manner, even when most of the users involved are offline.

The rest of the paper is organized as follows. Section 2 presents the related work. Our proposed agent-based model is presented in Sect. 3. In Sect. 4, we discuss results from our empirical model. We offer conclusions and suggestions for future work in Sect. 5.

2 Related Work

Work on the problem of posting co-owned images in a privacy preserving manner is initiated by Squicciarini et al. [5]. The Squicciarini et al. approach was based on game theory and employed a collaborative private box based on inference to handle privacy concerns regarding posting co-owned images on platforms such as Facebook. However, a key drawback to this approach is that it lacks an efficient technique to obtain the opinions of the individuals appeared in the image, which may put the privacy of these individuals at risk.

Some approaches [6,7,13–15] consider the problem of co-owned image sharing as one of the multi-user privacy conflicts. For instance, Hu et al. [7] used decision and sensitivity voting for conflict resolution. To do this, each image was assigned a sensitivity level, and users were given a decision value as a vote, based on the perceived user sensitivity level in relation to the image. The final decision was made through a collaborative decision making process based on the number of total votes. However, the Hu et al. [7] approach does not handle time restricted scenarios where user satisfaction is conditional on posting speed.

Other solutions consider friends of friends from an adversarial stance [8, 15–18]. In addition to the Hu et al. [7] multiparty access control concept, Suvitha [8] applied a flexible sensitivity level and a majority consent to share an online content on an OSN. The Suvitha's [8] approach, however, does not offer a time-efficient approach for dealing with delays in posting, when the sensitivity model cannot reach a privacy consensus.

Joseph [9] proposed a model to compute privacy risk and information loss to address multi-user privacy conflict. Using multiparty access approach, the algorithm separates publishers' sharing groups to mark them with trust and distrust labels. However, when there is a large group of individuals listed in the conflicting group, this strategy limits content uploaders.

Ali et al. [1] proposed a cryptographic technique to handle the data uploading problem raised by the Joseph solution. Each data owner creates a secret share for the data co-owner. Viewers can access published data if they get a certain threshold of secret shares. Similarly, some cryptographic models such as non-interactive public key exchange [19], saleable group key management [20], consensus encryption algorithm [21] have been focused on content sharing conflicts on OSNs. However, since the majority of viewers are unable to obtain secret shares, the posting delay increases.

To support collaborative privacy management, Ulusoy [10] used a tax-based approach to limit co-owner uploads. Users are expected to pay more tax if they collaborate more in the decision-making process. Nevertheless, this approach restricts former users actively engaging in the voting process and so puts privacy at risk as reduces usability.

Other approaches that build on the extended Squicciarini et al. [5] approach focus on user behavior [11, 22–25]. A game theory algorithm was proposed by Du et al. [11] to influence clients' interactions and to encourage participation in configuring privacy settings. However, data publishers' opinions are not always taken into account which poses a privacy risk. Fuzzy group decision-making models [12, 26–28] have been used to support data publishing based on consensus. In this vein, Akkuzu et al. [12] proposed using dynamic trust values for weighting co-owner opinions. If co-owners are concerned about the potential security aspects of the co-owned online content and do not want the content to be shared, but the owner agrees, then Akkuzu et al. [12] approach reduces the trust value in the owner side. But this approach requires the co-owners to share a lot of private data, which publishers considered undesirable.

To tackle the issue of publishing private data, Mosca et al. [29] proposed ELVIRA algorithm as an agent-based collaborative sharing approach. ELVIRA supports users in enforcing privacy preferences with respect to co-owned images. While ELVIRA addresses the issues of consensus, one issue remains: dealing with offline users, belonging to overlapping friendship networks, whose privacy opinion is uncertain. In the following section, we present our approach to addressing both issues.

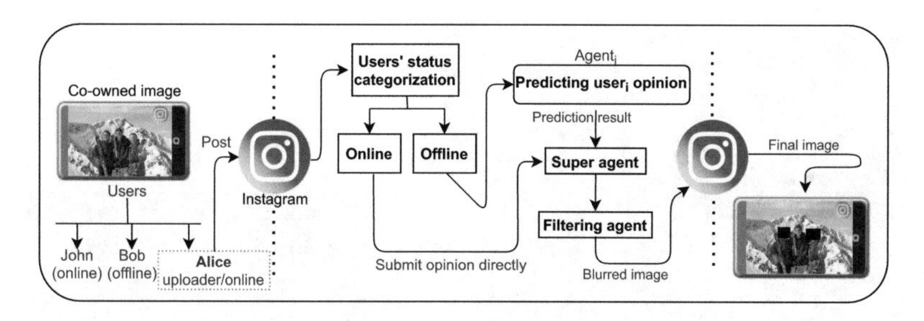

Fig. 1. Agent supported image transformation

3 Co-owned Image Sharing

We assume that images contain user tags that describe each user on an online social network. Moreover, tags are associated with a user account. So, a link between a tag and a user refers to an active (valid) account on the social network platform. We also assume that a user who wishes to post/share an image owns the image, in that he/she appears (has a tag) within the image. In the agent framework, the profile of this user is bound to a super agent called an **uploader agent**. Likewise, all tags belonging to the other users within the image to be posted are linked to user agents. In order to post an image, the uploader image must reach a consensus with these user agents. For instance, in Fig. 1, Alice wishes to post an image containing tags associated with John and Bob. In this case, Alice's profile is linked to an *uploader agent*, while John and Bob's profiles are bound to **user agents**.

3.1 Multi-agent Model

In our multi-agent model an *uploader agent* (agent bound to the profile of the user initiating the posting request) begins by broadcasting a message to the other user agents associated with the profiles of the users who appear in the co-owned image. As shown in Fig. 2(a), to determine which user agents to invoke, the uploader agent analyses the co-owned image to identify the pinned tags within the image and binds these tags to the associated user agents on a per-user profile basis.

As shown in Fig. 2(b), the *uploader agent* then broadcasts a message to all the user agents involved in the co-owned image requesting opinions on whether or not the associated users agree to have their image posted. Users who are online and actively using the social network platform, submit responses directly to the *uploader agent* during a pre-defined time window. If all the users are online and the *uploader agent* receives responses before a pre-defined delay threshold, a decision can be made on how to transform the image for posting. When users are offline, or online but fail to provide an opinion before the delay threshold expires, each corresponding *user agent* must determine what opinion its associated user is likely to have vis-a-vis the image to be posted.

Fig. 2. Agent activation (a) Binding tags to associated user agents [34]; (b) Uploader agent interaction with offline users' agents

In order to decide, each *user agent* uses its **opinion computation algorithm** to formulate an opinion on behalf of the user. The opinion, once obtained, is submitted to the *uploader agent*. We now explain how the *user agent* computes an opinion on behalf of a user.

3.1.1 Opinion Computation

Each user agent maintains a case-base of previous opinions that the associated user submitted with respect to co-owned images. We use α to denote the minimum threshold of opinions in the case-base. This occurs when a user is new or has not previously participated in a co-owned image posting decision. The maximum number of opinions is denoted by F and F is such that $\alpha \leq F$.

Based on the opinions in the case-base, the *user agent* computes a Mean Score Opinion (MSO) using Eq. (1) as follows:

$$MSO = \frac{\sum_{i=n}^{i=n-m} R_i}{m+1} \tag{1}$$

where n is the last opinion registered in the case-base, m is a list of opinions chosen from a case-base such that $n \geq m \geq \alpha$, and R_i is ith user opinion.

Once the MSO has been computed, the *user agent* must evaluate the computed score to determine if the value obtained is a satisfactory representation of the user's real opinion. If the case-base is empty - that is the user is new or has not previously participated in a co-owned image sharing scheme, then m $\leq \alpha$. In this case, the *user agent* submits the value of α to the *uploader agent* to indicate that the user's decision is "Disagree" and so his/her image should be blurred.

We now consider the case in which a user's opinion is unclear. That is, the uploader agent has not received a firm "Disagree" or "Agree" message from a user agent. In this case, the uploader agent qualifies the user opinion as being within an *Uncertain Range*. We formulate the *Uncertain Range* mathematically as follows:

We formulate the uncertain range mathematically as follows:

$$\frac{Z}{2} - \beta \leq UR \leq \frac{Z}{2} + \beta \tag{2}$$

where Z is the in range maximum value of submitting an opinion, and (β) as the maximum distance from $Z/2$.

When the MSO $\leq UR$, the *user agent* submits the MSO value to the *uploader agent* to indicate that the user has "Disagreed" to his/her image being posted. If the MSO $\geq UR$, the *user agent* submits the MSO value to the *uploader agent* to show that the user has "Agreed" to have his/her image posted.

We now consider what happens if a *user agent* submits an opinion but the *uploader agent* rejects it because it is an indecisive opinion, that is, it falls within the uncertain range. The *uploader agent* uses personality profile data to train a machine learning model to obtain a personality score (*PS*) that respects Eq. (2). Our machine learning model employs the Random Forest and Support Vector Regression algorithms, as examples of regression-based algorithms that are useful in supporting continuous scoring schemes. Based on the results of the personality computation, the user's personality is then mapped onto a personality score scale (ranging from one to six) similar to that used in [30]. The *PS* is compared against Eq. (2), and if the $PS \leq UR$, *user agent* concludes that the decision is "Disagree" and if the $PS \geq UR$, then the decision is "Agree". If a clear opinion has not been computed, the *uploader agent* considers that the user opinion is "Disagree". Figure 3(a) provides a visualisation of the agent opinion submission process. Once the user opinions have been computed, the **uploader agent** activates the **filtering agent**.

3.1.2 Filtering and Blurring

The filtering agent uses the python OpenCV library and the enhanced Haar-cascade based face detection algorithm [31], which is a functional object detection technique. Furthermore, to blur out images in OpenCV, the filtering agent uses a Gaussian smoothing kernel to have a bell-curve around the center pixel [32] instead of using a black box to blur the faces.

The faces of users who decline to share their images online are blended with the Gaussian blurred segments (see Fig. 3(b)), and then inserted on an independent layer over the face positions, as shown in Fig. 3(c).

Algorithm (1) summarizes the operation of the multi-agent model in terms of reaching a decision in a time efficient manner on transforming an image for utility (minimal blurring) and privacy (adhering to user privacy opinions). There are two procedures in Algorithm (1). The first procedure computes and checks offline users' opinions; the second procedure blurs faces of disagreed users and returns the modified image to the uploader agent.

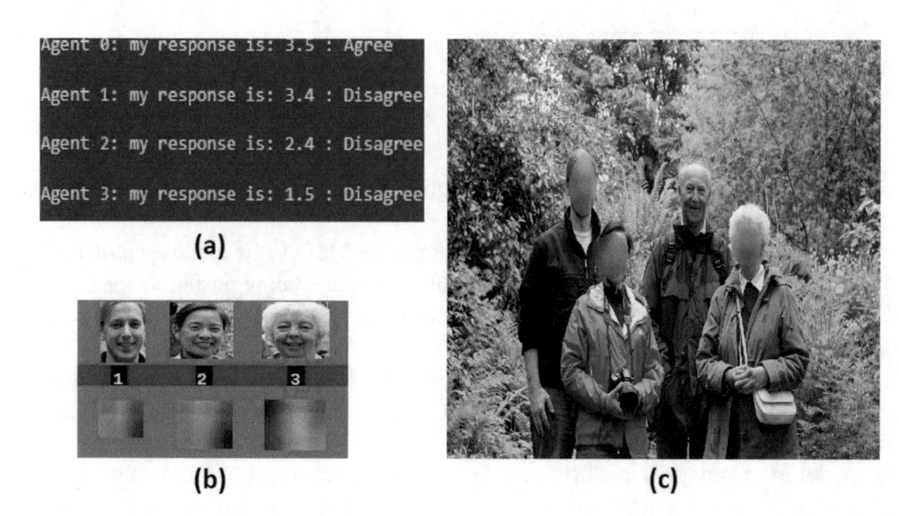

Fig. 3. Applying opinions on the co-owned image (a) Agents' participation on behalf of offline users; (b) Detected faces [34]; (c) Blurred faces

4 Experimental Setup and Results

Our multi-agent model was implemented on a machine with an Intel CPU 2.3 GHz core i7, 16 GB of RAM, and run over a Windows 10 operating system using Google Colab. For the personality score scheme, we used the Big Five Personality Test (BFPT) dataset [30] containing 1,015,342 questionnaire answers. In our experiments, we only used the first 300K records and the first ten items ('EXT1' to 'EXT10') as users' recent online opinions about sharing images on social networks and we attempted to predict 'EXT9' feature in the dataset. Furthermore, we used Mesa project [33] to construct a multi-agent system since it provides foundations for running agent-based models operating on python.

4.1 Results

Table 1 shows our inputs into multi-agent model based on BFPT dataset [30]. Briefly, in Table 1, α denotes the least number of opinions to activate an agent, Z is the highest opinion value in the dataset, and F is the number of opinions submitted directly by $user_i$. Because the neutral opinion on the BFPT dataset equals three, we consider *beta* to be 0.5 as the maximum distance in the uncertain range.

Algorithm 1. Opinion Computation Algorithm (OCA)

```
1: Input ← a co-owned image
2: Output → a filtered image considering users opinions
3: procedure OCA (CO-OWNERS, IMAGE)
4:     UA ← activate a super agent
5:     for all co-owners do UA → request user(i) to submit opinion end for
6:     WAIT (Time_window); % receiving co-owners opinions
7:     %Status check:
8:     for all co-owners do
9:         if UA did not receive opinion from user_i then
10:        status ← offline else status ← online end if
11:    end for
12:    for all offline co-owners do
13:        activate user_i agent; agent_i checks database_i;
14:        top:
15:        if database_i is not empty then
16:            compute_i ← compute user_i opinion; % run opinion computation algorithm
17:            if opinion_i located within uncertain range then goto top end if
18:            FilteringAgent ← collected opinions;
19:        end if
20:    end for
21: end procedure
22: procedure FILTERINGAGENT (Face_Positions, OPINIONS, IMAGE)
23:    faces ← Face_Positions;
24:    if opinion_i == disagree then Blur (face_i); % Gaussian filter end if
25:    Return filtered image
26: end procedure
```

Table 1. Inputs into multi-agent model based on BFPT dataset

	α	β	Z	UR	F	Disapproval	Approval	Neutral
Values	3	0.5	6	2.5–3.5	10	0	6	3

Figure 4(a) shows the number of agents whose opinions fall in UR (infected agents) during the opinion formulation process. For 30 co-owned images, we put six agents to the test to compute users' opinions. Based on the results, a maximum of two agents were infected agents. Accordingly, Fig. 4(b) depicts the effectiveness of the multi-agent model in terms of time complexity. Bonding users to agents and considering infected agents take about 3.81 s to compute offline users' opinions.

We use a variety of opinion computation algorithms to compute user opinions, including MSO, Random Forest (RF), and Support Vector Regression (SVR). Furthermore, We split the BFPT data into two sets: 80% for training and 20% for testing opinions accuracy in comparison to Personality Score. Figure 5 shows the performance of opinion computation accuracy for middle-sized dataset (1K to 11K user previous opinions in BFPT dataset). On average, the RF algorithm computed the users opinions with 57.82% accuracy, while the results for SVR and MSO were 61.71 and 22.79%,

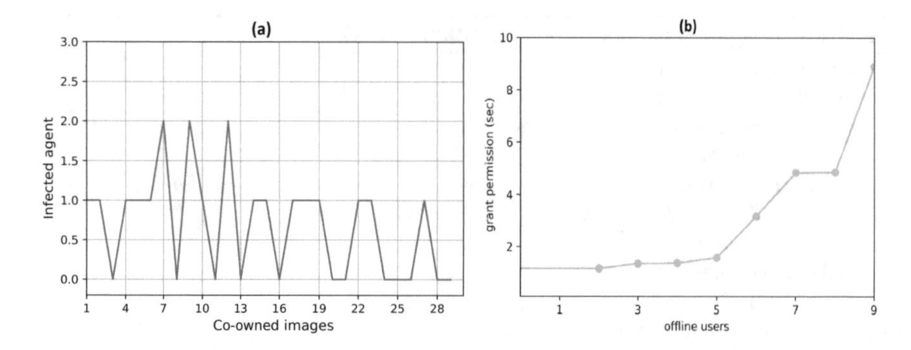

Fig. 4. (a) the number of agents in the UR; (b) effectiveness of the multi-agent model in terms of time complexity considering infected agents.

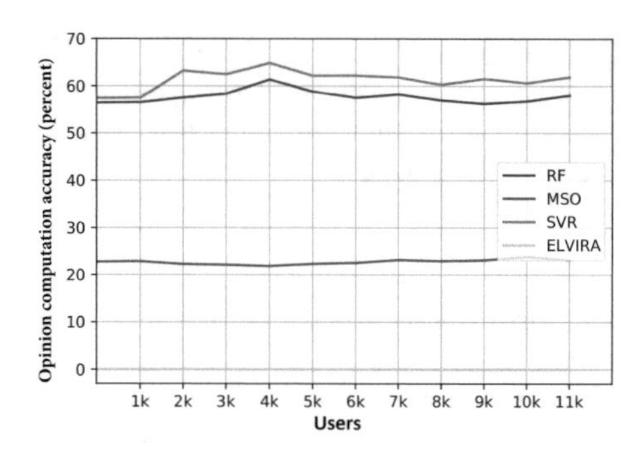

Fig. 5. User opinion computation accuracy for middle-sized test dataset with $11k$ users

respectively. Moreover, the opinion computation results for large-sized dataset with 300K users were 58.73, 24.02, 63.82 and zero percent for RF, MSO, SVR and ELVIRA [29] algorithms, respectively. Generally, SVR provides the best-performing model in terms of computing users' opinions based on their recent opinions.

Moreover, Fig. 6(a) delineates the performance of our agent-based model in the direction of blurring accuracy for multiple offline users. The results achieved from 60 co-owned images present that the SVR and RF perform more effectively in terms of face blurring accuracy than the MSO algorithm. Based on the results, however, the RF algorithm is suitable for co-owned images with at least four offline users. On the other hand, the SVR algorithm performs well when the co-owned images contain between one to three offline users. Figure 6(b) indicates the training time required to compute users' opinions. According to the results, the training time for SVR algorithm surges when faced with online behaviours of more than 3k user profiles. Based on the results, we can conclude that the training time for the SVR algorithm can be reduced by training a middle-sized dataset since the SVR algorithm performance for 4k users is close to the large-sized dataset (300K).

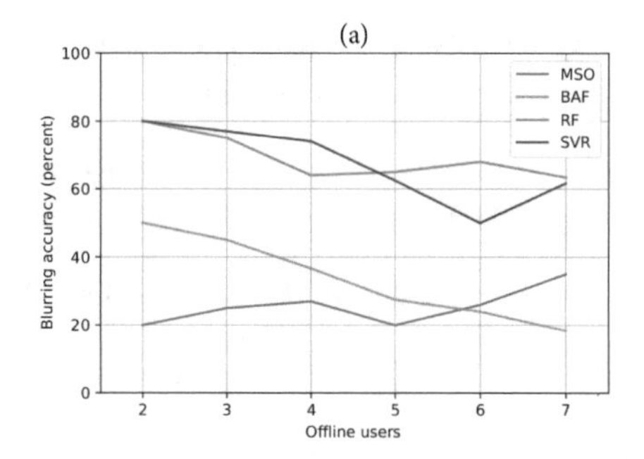

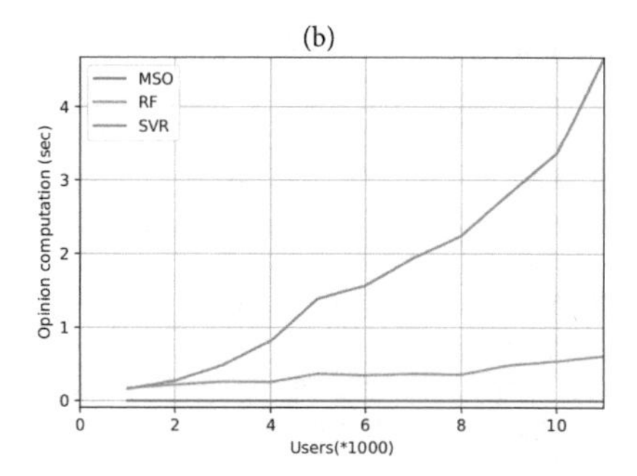

Fig. 6. (a) Face blurring accuracy for multiple offline users based on various opinion computation algorithms (BAF: Blur all Faces, SVR: Support Vector Regression, MSO: Mean Score Opinion, RF: Random Forest); (b) Training time to formulate users' opinions

5 Conclusion

Enforcing privacy on co-owned images raises a conflict in terms of usability and privacy when some or most of the users involved are offline. Two key issues arise: (1) when users are not online, a privacy decision regarding whether or not to post the image cannot be reached, thereby resulting in delay; and (2) ineffective user opinion computation necessitates large amounts of image distortion, resulting in low levels of user satisfaction. We employed a multi-agent model to address both issues in a time-efficient manner. To formulate offline user opinions, we employ an opinion computation algorithm that infers the users' privacy preferences based on the users' historical online usage behaviours. Our empirical results demonstrate that by supporting predictions with machine learning algorithms, agents can reliably reach a consensus on posting the image in 3.81 s on average without negatively impacting performance, which is beneficial in terms of usability.

References

1. Ali, S., Rauf, A., Islam, N., Farman, H.: A framework for secure and privacy protected collaborative contents sharing using public OSN. Cluster Comput. **22**(3), 7275–7286 (2017). https://doi.org/10.1007/s10586-017-1236-2
2. Severo, M., Feredj, A., Romele, A.: Soft data and public policy: can social media offer alternatives to official statistics in urban policymaking? J. Policy Internet **8**(3), 354–372 (2016)
3. Xu, L., Jiang, C., He, N., Han, Z., Benslimane, A.: Trust-based collaborative privacy management in online social networks. J. IEEE Trans. Inf. Forensics Secur. **14**(1), 48–60 (2018)
4. Giovannetti, E., Hamoudia, M.: Understanding the different pre and post peak adoption drivers in the process of mobile social networking diffusion, pp. 1–4. International Telecommunications Society (ITS), Seoul (2018)
5. Squicciarini, A.C., Shehab, M., Paci, F.: Collective privacy management in social networks. In: Proceedings of the 18th International Conference on World Wide Web, pp. 521–530 (2009)

6. Wishart, R., Corapi, D., Marinovic, S., Sloman, M.: Collaborative privacy policy authoring in a social networking context. In: IEEE International Symposium on Policies for Distributed Systems and Networks, pp. 1–8 (2010)
7. Hu, H., Ahn, G.: Multiparty authorization framework for data sharing in online social networks. In: IFIP Annual Conference on Data and Applications Security and Privacy, pp. 29–43 (2011)
8. Suvitha, D.: Mechanisms of multiparty access control in online social network. J. Recent Dev. Eng. Technol. **2**(3), 8–13 (2014)
9. Joseph, N.S.: Collaborative data sharing in online social network resolving privacy risk and sharing loss. IOSR J. Comput. Eng. **16**(5), 55–61 (2014)
10. Ulusoy, O.: Collaborative privacy management in online social networks. In: Proceedings of the 17th International Conference on Autonomous Agents and MultiAgent Systems, pp. 1788–1790 (2018)
11. Du, J., Jiang, C., Chen, K.C., Ren, Y., Poor, H.V.: Community-structured evolutionary game for privacy protection in social networks. J. IEEE Trans. Inf. Forensics Secur. **13**(3), 574–589 (2018)
12. Akkuzu, G., Aziz, B., Adda, M.: Towards consensus-based group decision making for co-owned data sharing in online social networks. J. IEEE Access **8**, 91311–91325 (2020)
13. Kumar, A., Bezawada, R., Rishika, R., Janakiraman, R., Kannan, P.K.: From social to sale: the effects of firm-generated content in social media on customer behavior. J. Market. **80**(1), 7–25 (2016)
14. Thomas, K., Grier, C., Nicol, D.M.: unFriendly: multi-party privacy risks in social networks. In: Atallah, M.J., Hopper, N.J. (eds.) PETS 2010. LNCS, vol. 6205, pp. 236–252. Springer, Heidelberg (2010). https://doi.org/10.1007/978-3-642-14527-8_14
15. Baden, R., Bender, A., Spring, N., Bhattacharjee, B., Starin, D.: Persona: an online social network with user-defined privacy. In: Proceedings of the ACM SIGCOMM 2009 Conference on Data Communication, pp. 135–146 (2009)
16. Dürr, M., Maier, M., Dorfmeister, F.: Vegas–a secure and privacy-preserving peer-to-peer online social network. In: International Conference on Privacy, Security, Risk and Trust and 2012 International Conference on Social Computing, pp. 868–874 (2012)
17. Jahid, S., Nilizadeh, S., Mittal, P., Borisov, N., Kapadia, A.: Vegas–a secure and privacy-preserving peer-to-peer online social network. In: IEEE International Conference on Pervasive Computing and Communications Workshops, pp. 326–332 (2012)
18. Schwittmann, L., Boelmann, C., Wander, M., Weis, T.: SoNet–privacy and replication in federated online social networks. In: IEEE 33rd International Conference on Distributed Computing Systems Workshops, pp. 51–57 (2013)
19. Lv, X., Mu, Y., Li, H.: Non-interactive key establishment for bundle security protocol of space DTNs. J. IEEE Trans. Inf. Forensics Secur. **9**(1), 5–13 (2013)
20. Ali, S., et al.: SGKMP: a scalable group key management protocol. J. Sustain. Cities Soc. **39**, 37–42 (2018)
21. Srilakshmi, P., Aaratee, S., Subbalakshmi, S.: Privacy my decision: control of photo sharing on online social networks. Int. J. Comput. Sci. Inf. Technol. (IJCSIT) **7**(2), 780–782 (2016)
22. Liu, F., Pan, L., Yao, L.H.: Evolutionary Game Based Analysis for User Privacy Protection Behaviors in Social Networks. In: IEEE Third International Conference on Data Science in Cyberspace (DSC), pp. 274–279 (2018)
23. Tosh, D., Sengupta, S., Kamhoua, C., Kwiat, K., Martin, A.: An evolutionary game-theoretic framework for cyber-threat information sharing. In: IEEE International Conference on Communications (ICC), pp. 7341–7346 (2015)
24. Chen, J., Kiremire, A.R., Brust, M.R., Phoha, V.V.: Modeling online social network users' profile attribute disclosure behavior from a game theoretic perspective. J. Comput. Commun. **49**, 18–32 (2014)

25. Squicciarini, A.C., Griffin, C.: An informed model of personal information release in social networking sites. In: International Conference on Privacy, Security, Risk and Trust and 2012 International Conference on Social Computing (ICC), pp. 636–645 (2012)
26. Capuano, N., Chiclana, F., Fujita, H., Herrera-Viedma, E., Loia, V.: Fuzzy group decision making with incomplete information guided by social influence. J. IEEE Trans. Fuzzy Syst. **26**(3), 1704–1718 (2017)
27. Martinez-Cruz, C., Porcel, C., Bernabé-Moreno, J., Herrera-Viedma, E.: A model to represent users trust in recommender systems using ontologies and fuzzy linguistic modeling. J. Inf. Sci. **311**, 102–118 (2015)
28. Akkuzu, G., Aziz, B., Adda, M.O.: Fuzzy logic decision based collaborative privacy management framework for online social networks. In: 3rd International Workshop on FORmal Methods for Security Engineering, pp. 674–684 (2019)
29. Mosca, F., Such, J.: ELVIRA: an explainable agent for value and utility-driven multiuser privacy. In: International Conference on Autonomous Agents and Multiagent Systems (AAMAS), pp. 916–924 (2021)
30. Tunguz, B.: Big Five Personality Test Dataset (2018). https://www.kaggle.com/tunguz/big-five-personality-test/version/1. Accessed 16 Sept 2021
31. Gangopadhyay, I., Chatterjee, A., Das, I.: Face detection and expression recognition using Haar cascade classifier and Fisherface algorithm. J. Recent Trends Sig. Image Process. **922**, 1–11 (2019)
32. Getreuer, P.: A survey of Gaussian convolution algorithms. J. Image Process. On Line **3**, 286–310 (2013)
33. Masad, D., Kazil, J.: Effective substances. In: 14th PYTHON in Science Conference, pp. 153–160 (2015)
34. Gallagher, A., Chen, T.: Understanding groups of images of people. In: IEEE Conference on Computer Vision and Pattern Recognition, pp. 256–263 (2009)

Efficient Restoration of Structural Controllability Under Malicious Edge Attacks for Complex Networks

Bader Alwasel(✉)

Department of Applied Natural Sciences, Applied College, Qassim University, Buraydah, Saudi Arabia
bwasel@qu.edu.sa

Abstract. In recent years, protecting large-scale complex networks, such as electric power grids and their monitoring systems, has become a critical part of cutting-edge research. Adversaries often attempt to intercept and hack these systems by interfering with intermediate nodes or preventing controllers and physical systems from receiving actuator signals and industrial sensor measurements. The present work concentrates primarily on the structural controllability of such systems, as this topic has recently received significant interest through the equivalent problem of the power dominating set within the field of electrical power and network control. Nonetheless, it is well established that these problems are NP-hard with low approximation. Developing strategies that can restore a network's performance is critical given their importance for a wide range of networks, particularly power networks. Thus, this paper focuses on regaining the structural controllability for Erdős-Rényi random directed networks following attacks on edges. Furthermore, these strategies will be evaluated using Matlab simulations to examine their applicability in real-world contexts.

1 Introduction

Two structural characteristics of dynamic systems, observability and controllability, have been well established in the design and maintenance of networked systems under control. However, the necessary increased focus on large complex systems and networks as the most appropriate context for such notions has itself piqued academics' attention [1, 2]. To ensure that a linear time-invariant (LTI) model's networks are controlled by external inputs and that the state of each network vertex can be forced from an arbitrary configuration in a finite number of steps, Kalman [3] established state controllability and observability as necessary properties for LTI systems. As a result, these linear network models offer a solid foundation for network controllability research, and this focuses on the linear time-invariant system with external inputs that is represented using the differential equations:

$$\dot{x}(t) = \mathbf{A}x(t) + \mathbf{B}u(t); \quad x(t_0) = x_0 \tag{1}$$

where $x(t) \in R^n$ is the state vector of the network at time t; $u(t) \in R^m$ is the input vector that contains the control inputs at time t; and $\boldsymbol{A}$ is an $n \times n$ adjacency matrix that represents

L. Barolli et al. (Eds.): AINA 2022, LNNS 449, pp. 152–166, 2022.
https://doi.org/10.1007/978-3-030-99584-3_14

the network topology based on identifying the linking topology between the state nodes. The input matrix is thus $\boldsymbol{B} \in R^{n \times m}$, where $(m \leq n)$, elucidates the set of nodes that are controlled by a time-dependent input vector $u(t) = (u_1(t), \ldots, u_m(t))$ which forces the system to move towards the requisite state in a finite number of steps. According to Kalman's rank criterion, therefore, the system given by Eq. (1) is completely controllable if and only if the rank matrix used is:

$$\text{rank}\ [\boldsymbol{B}, \boldsymbol{AB}, \boldsymbol{A}^2\boldsymbol{B}, \ldots, \boldsymbol{A}^{n-1}\boldsymbol{B}] = n \tag{2}$$

Despite the fact that the controllability rank condition illustrated in Eq. (2) provides a complete, thorough, and detailed framework for the design and study of LTI systems, computing Kalman's rank criterion is excessively expensive, particularly for large complex networks such as power networks or similarly large control systems, as the number of possible input combinations increases exponentially with the number of nodes to $(2^N - 1)$ in the worst case. To manage this potential difficulty while still designing a completely controllable system, Lin [1] thus proposed the idea of structural controllability, which is a powerful notion that can be utilised to develop graph-theoretical interpretations of Kalman's algebraic criteria. Such interpretation enables the identification of the necessary and sufficient requirements for identifying the specific driver nodes (N_D) capable of controlling a given system with a certain structure (topology) as defined by Liu et al. [2]. They also demonstrated that the entire system, represented as $\boldsymbol{AB}$, can be depicted using a directed graph to determine a minimum set of driver nodes required to control the given LTI-dynamic digraph so that $\boldsymbol{G}(\boldsymbol{A}, \boldsymbol{B}) = (V, E)$ where $V = V_A \cup V_B$ is the vertex set and $E = E_A \cup E_B$ is the edge set.

To obtain the minimal driver nodes for V_B from G($\boldsymbol{A}$, $\boldsymbol{B}$) = (V, E) based on a given $G(V, E)$, several approaches have been employed to locate driver nodes; the most-studied strategy being that proposed by Liu et al., the Maximum Matching approach [2]. However, the current paper focuses on an alternative technique based on the power dominating set (PDS) problem, which provides an equivalent framework for studying structural controllability via the identification of minimal driver nodes (N_D). This was proposed by Haynes et al. [4] as an extension to the well-known dominating set (DS) problem as a means for assessing the structure of electric power networks and monitoring their systems with as few measurement devices as feasible. One approach to monitoring these variables is thus to install as few real-time phase measurement units (PMUs) as feasible at selected locations in the electric power systems, as the expensive nature of these devices makes reducing the quantity of PMUs extremely desirable if this can be done without reducing efficacy. However, the challenge of identifying the minimum set of measuring devices is exactly equivalent to the domination problem introduced by Haynes et al. [4], which was already a well- known problem in combinatorial optimisation.

Definition 1 (Power Dominating Set Problem [4]): Given an undirected graph $G = (V, E)$, find a minimum-size set $P \subseteq V$ such that all vertices in V are observed by the vertices in P.

Haynes et al. established two important observation rules (**OR1**, **OR2**) that enable the extraction of the minimum set of driver nodes capable of controlling a network via power domination in graphs; these were subsequently simplified by Kneis et al. [5]. PDS can be formulated as a set of conditions that must hold for a set of driver nodes:

[**OR1**] A vertex in N_D observes itself and all of its neighbours.
[**OR2**] If an observed vertex $_v$ of degree $d \geq 2$ is adjacent to $d - 1$ observed vertices, then any remaining unobserved neighbour becomes observed as well.

Certain specific classes of graphs have been studied using an algorithmic approach to computed the number of power dominating sets [6, 7]. Several recent researchers have also examined the resilience and controllability of various random graph classes where some edges or nodes are removed [8–18].

The remaining sections of the paper are organised as follows. Section 2 defines the basic assumptions and conditions for constructing the input network used to restore structural controllability after the removal of edges as well as threat scenarios. Section 3 describes the four recovery strategies for restoring the controllability of directed networks. Section 4 presents the network and adversarial model to illustrate the main results of this work together with the practical validation of the recovery strategies. Finally, Sect. 5 concludes the paper.

2 Preliminary Assumptions for Analysis

The network model is based on a directed ER random graph, which is frequently used as a benchmark to evaluate the topology and connectivity of a wide range of complex networks and allows for the analysis of network dynamics. The restoration strategies proposed are thus analysed based on multi-round edge scenarios focused on power edges (denoted here as $\mathbf{TS_i}$), as extensively studied in previous work [16] and detailed below. In Fig. 1, the driver nodes (i.e. PDS) are represented by green circles that observe both themselves and the observed nodes, which are likewise represented as grey circles, using observation rules **D1** and **D2**. Meanwhile, the unobserved nodes that have been attacked by removing their edges are marked by white circles.

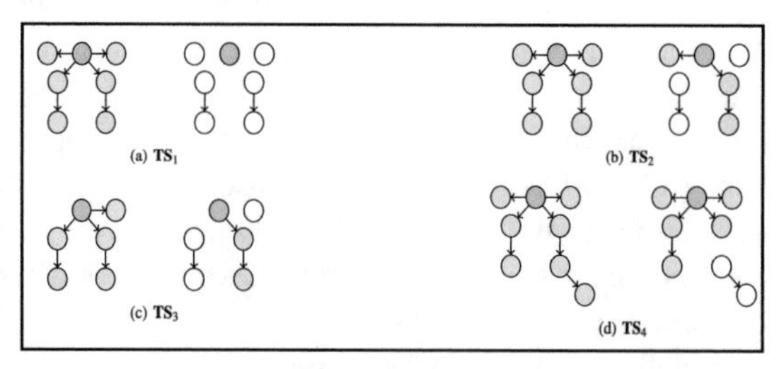

Fig. 1. An illustration of various network edge removal patterns that target power edges in a controlling network before and after an attack.

$\mathbf{TS_1}$: By iteratively removing all of the node's edges, an attacker targets the node in a PDS with the greatest out-degree.
$\mathbf{TS_2}$: Repeatedly attempts to undermine structural controllability by deleting a few (but not all) edges from a vertex in a PDS with the largest out-degree.

TS$_3$: Removes some (but not all) edges from a vertex within a PDS with the smallest out-degree continuously.
TS$_4$: Removes a maximum of one edge every attack round from vertices that are not within a PDS.

ASM-1: For the network model, the Erdős-Rényi random graph class ER (n, p) is used to generate a sparse random digraph $G(V, E)$ as presented in this study, which typically has about the same number of edges as vertices. The selected ER (n, p) model thus has two parameters, the number of nodes in the graph, n, and the edge probability, p, which together allow for the construction of a graph where, for each pair of unique vertices, v and w, p is the probability that the edge (v, w) exists independently of all other edges.
ASM-2: For the resulting instances of ER (n, p), only unweighted directed graphs $G(V, E)$ based on an arbitrary set of nodes V and a set of edges E are examined; this thus does not include self-loops, parallel edges or isolated vertices (i.e. a vertex that has no edges, denoted here as V_{isolated}, but does allow for cases with two edges with opposite directions on the same two end vertices (called antiparallel edges).
MTHD-1: To determine the minimum set of driver nodes for $G(V, E)$, an approach based on the PDS problem is used, as discussed in more depth in [13].
MTHD-2: The adversary model, which assumes that attackers with prior knowledge of structural control of the network and its driver nodes are capable of performing malicious operations such as edge removals, is applied by building an adversarial model based on the threat scenarios presented above, as further examined in [16].

3 Structural Controllability Restoration

For the purpose of simulating structural controllability restoration strategies in the presence of perturbations, it is assumed that malicious attackers remove power edges from nodes in a given ER directed network as in the threat (**TS$_i$**) noted above. To address this further, two distinct types of targeted nodes are identified:

N-1: The driver node, v, which is a member of the control node set N_D.
N-2: Node u, a member of the set of controlled nodes (i.e. a dependent node) controlled by a driver node.

Each proposed strategy requires a given ER directed graph to be structured into a tree-like structure based on the structural controllability properties as described in Lin's Structural Controllability Theorem [1] through the computation of PDS, achieved using the algorithm proposed in [13].

3.1 STG-1: Strategy Using Backup Instances

When used for attack scenarios **TS$_1$**, **TS$_2$**, **TS$_3$** and **TS$_4$**, this approach recovers the damaged network's controllability by restoring the power edges to the nodes attacked in a directed network. The strategy relies on the use of a backup list of edges, which thus requires initial pre-computation prior to any attack, for the control relationships between the nodes in a given network. For each unobserved node, the class of node (**N-1** or **N-2**) the vulnerable node belongs to must be verified. If the compromised node is part of N_D (i.e. **N-1**), the control edges that lie between the attacked driver node and its dependent nodes in a given backup list must be checked. However, if the compromised node is part of **N-2**, the control edges between the attacked node and its dependent nodes must be checked. As a result, the necessary edge restorations between the attacked node and its unobserved nodes must then be established, and a verification process must be completed to ensure that none of the new restoration edges violates the two observation rules (**D1**, **D2**).

3.2 STG-2: Strategy Using Edge Allocation

This strategy uses an edge allocation approach that reroutes edges from a vertex within the driver nodes to the unobserved nodes. The approach thus enables the establishment of a new edge between the best candidate in N_D and the unobserved node such that the following properties are satisfied:

- The number of edges allocated equals the number of removed edges, implying that the overall number of edges in a given network remains constant.
- Where a driver node is selected from the set of N_D, this is capable of observing an unobserved node through a new control link that fulfils the first observation rule **D1**, which requires observation of both the driver node and its dependent nodes.
- The second observation rule **D2** is not violated, such that each unobserved node thus has at most one incoming edge pointed at it.

The following steps demonstrate the constructive approach:

Step-1: Search for unobserved nodes U.
Step-2: Identify the candidates in N_D that can cover each vertex included in U through a new edge.
Step-3: Establish a new edge connecting a node in N_D to a node in U.
Step-4: Repeat **Step 3** until candidates in N_D bserve all nodes in U.

At the end of this algorithm, the number of driver nodes in a given network will remain unchanged, except in the case of threat scenario $\mathbf{TS_1}$, in which all edges of an attacked driver node are destroyed entirely due to the nature of the attack; in that case only, new edges must be allocated to the compromised nodes to allow them to become observed by the current driver nodes in the network.

3.3 STG-3: Strategy Using Diameter-Based Relinking

As in **STG-2**, this strategy applies an edge allocation mechanism; however, in this case, it involves the establishment of a new link between each candidate node with an out-degree equal to zero and an unobserved node such that the candidate does not belong to the driver nodes. The strategy uses network diameter to demonstrate changes in the network topology and its structural control under exposure to various attacks. The advantage of this use of graph diameter is that it may be used to reduce the cost of restoring network controllability based on calculating the minimum diameter among all the maximum distances between a driver node and its dependent nodes prior to establishing a new control edge. The steps of the algorithm then proceed as follows: Using the breadth-first search approach, the minimum diameter between a node in the driver nodes and the head node of each dependency path (P_i) should be computed, where a dependency path from a vertex x to a vertex y refers to path P with all edges in P are directed from x to y such that it starts from a vertex x, without incoming edges (denoted as the tail or initial vertex of the path), and ends at a vertex y, without out-going edges (denoted as the head or terminal vertex of the path). The following step is to search for the tails of all unobserved dependency paths (P_j) and select the best candidates from among head nodes within the minimum diameter of P_i. Now a new edge must be established which connects each candidate head of P_i to each tail of P_j. Finally, the process of establishing a new edge should be repeated until no tails of P_j remain. At the end of the algorithm, each restoration of a control edge must adhere to the following rules:

- The new restoration edges must guarantee controllability while abiding by the structural control properties specified by the two observation rules (**D1**, **D2**).
- The number of remaining driver nodes is reduced by one in the case of $\mathbf{TS_1}$, due to the nature of the attack and the method used in this strategy, which restores control by relinking dependency paths, unlike the other strategies that involve the driver node in the edge recovery process.

3.4 STG-4: Driver Node Injection Strategy

The strategy works by imposing a driver node at each unobserved node with an in-degree of zero; this typically increases the number of driver nodes required to control the whole network, which is often undesirable given the exorbitantly high cost of controllers. However, it remains applicable to directed networks for which the cost of establishing an edge is greater than the cost of imposing a time-variant controller at a node. The following steps should thus be followed to inject external controllers into the disconnected components, where the deletion of edges in a dependency path results in the emergence of new disconnected components, denoted by (**DCC**):

Step-1: Search for all unobserved dependency paths (P_j) after an attack.
Step-2: Compute the size of **DCC** required to control these paths.
Step-3: Inject a driver node into each tail node with an in- degree zero for each **DCC**.
Step-4: If there exists an isolated compromised node (an attacked node that does not belong to P_j), inject a driver node into it.
Step-5: Repeat **Steps 3** and **4** until all **DCC** are controlled by the set of driver nodes.

While inserting driver nodes into the unobserved dependency paths, the following constraints must be taken into consideration, however:

- The injection of N_D into **DCC** must comply with the observation rule (**D1**).

- The number of driver nodes required to control all unobserved dependency paths must be equal to the size of the **DCC**.

When the algorithm halts, the total number of the original edges will be decreased because of the nature of the strategy, which injects driver nodes into each disconnected component rather than recovering edges during restoration. Additionally, the set of driver nodes will be increased in size in comparison to the number of initial driver nodes prior to the attack, with the size of the increase contingent upon the level of damage, based on the strategy's nature, which involves injecting a driver node into each detached component to initiate the restoration process.

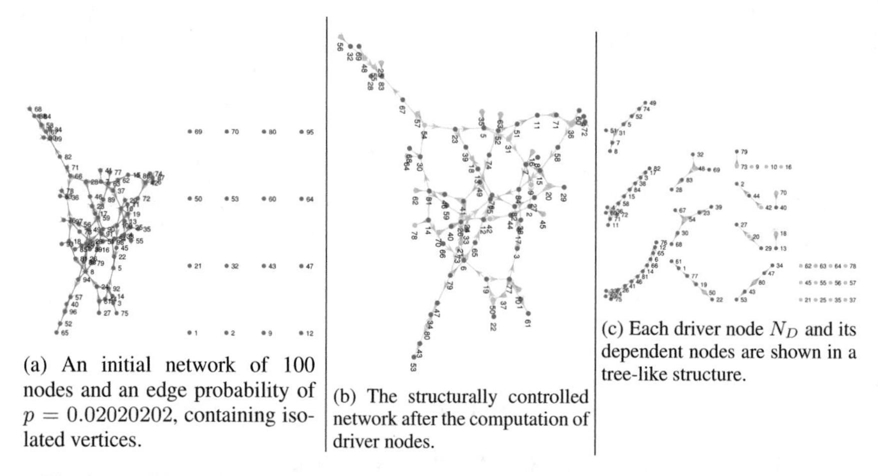

(a) An initial network of 100 nodes and an edge probability of $p = 0.02020202$, containing isolated vertices.

(b) The structurally controlled network after the computation of driver nodes.

(c) Each driver node N_D and its dependent nodes are shown in a tree-like structure.

Fig. 2. An illustration of the structural controllability of an network via driver nodes.

4 Experimental Results and Discussion

This section presents the main results of this work, including the practical validation of the recovery strategies proposed as well as an empirical analysis of simulated experiments. Several simulation examples, including graphical figures, are also illustrated to accomplish this and validate the theoretical results of this paper.

4.1 Network and Attacks Models

According to **ASM-1** and **ASM-2**, the network model is constructed using a directed ER random graph, which produces networks of different sizes, including small ($\leq$100), medium ($\leq$800), and large ($\geq$1300) networks with 100, 800, and 1300 nodes respectively, and with low connectivity probabilities to simulate sparse real-world networks. The algorithm developed in the previous work [13] is used to compute the set of PDS (the set of driver nodes) for directed ER networks of various sizes, as shown in Fig. 2, where a minimum set of N_D is marked in green and the dependent nodes controlled by N_D

are highlighted in blue. To demonstrate how the suggested recovery strategies restore structural controllability of ER directed networks, we use Matlab simulations[1] to follow an adversarial model that is based on threat scenarios $\mathbf{TS_i}$ (as described in Sect. 2) that target the power edges of ER directed networks [16]. The table summarizes the computation of the set of driver nodes for each given network, taking into account that these acquired driver nodes are not unique since the structural control edges between nodes are constructed independently throughout network generation. The computational results in Table 1 indicate that there is a direct relationship between the size of the original networks and their driver nodes. When the number of nodes in the networks increases, the minimum number of N_D required to control the networks also thus increases. This is because of the networks' low connectivity probabilities, which result in a reduction in overall network edges. Additionally, this reduction may produce some isolated vertices due to the lack of edges incident on them.

Table 1. The simulation results of computing the set of N_D (or PDS) for different directed ER network sizes.

N	p	E	PDS	$V_{connected}$	$V_{isolated}$
100	0.02020202	100	57	27	16
800	0.00250313	800	238	457	105
1300	0.00153965	1300	410	717	173

4.2 Analysis and Validation

The four recovery strategies (**STG-i**) were developed using Matlab[2] and applied to the compromised ER directed networks ($n \in 100, 800, 1300$) to restore their structural controllability and validate the results obtained based on:

- Ascertaining that the structural control properties (**D1** and **D2**) are not violated after recovery.
- Ensuring that the entire network is structurally observable by calculating the network's observation degree (**D1**) after recovery.
- Wherever feasible, guaranteeing that the number of power dominating sets (driver nodes) and their control edges remain constant after recovery.

The correctness of the approaches can be proved by the execution of the experimental simulation.

Precondition: There exists a structural perturbation caused by a potential threat of types $\mathbf{TS_{1,2,3,4}}$ at a time t.

[1] The full code is available in [16].

[2] The full code is available upon request from the author.

Postcondition: Restored structural controllability by the implementation of **STG-1,2,3,4**, validating the structural control properties (**D1** and **D2**) are fulfilled as defined in Sect. 1.

Consider the graphs in Fig. 3, 4, 5, 6, and their edge restorations. After recovery, check that all of the degree constraints at nodes are met. Two states are defined for each node in a given ER directed graph as the number of incoming edges $s^{-}(v)$ and out-going edges $s^{+}(v)$ between a vertex in a driver node set and its dependent nodes, or between the dependent nodes themselves. There are thus two distinct types of nodes that must be distinguished during the recovery process:

Case 1: If a vertex $u \in \mathbf{N}{-}1$, then the condition **D1** should be satisfied: For each driver node (u), the state $\mathrm{s}^{-}(u)$ must have a value of 0, whereas the state $\mathrm{s}^{+}(u)$ may have a value equal or greater than zero (i.e. $\mathrm{s}^{+}(u) \geq 0$), such that the following properties holds:

- $\forall\, u \in G : \mathrm{s}^{-}(u) = 0$, and
- $\forall\, u \in G : \mathrm{s}^{+}(u) \geq 0$

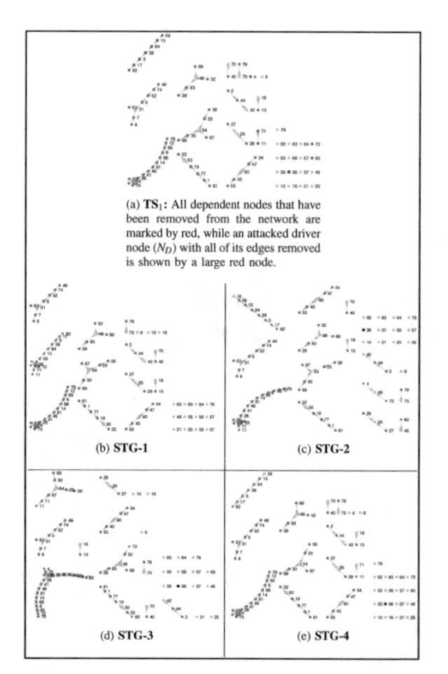

(a) TS$_1$: All dependent nodes that have been removed from the network are marked by red, while an attacked driver node (N_D) with all of its edges removed is shown by a large red node.
(b) STG-1
(c) STG-2
(d) STG-3
(e) STG-4

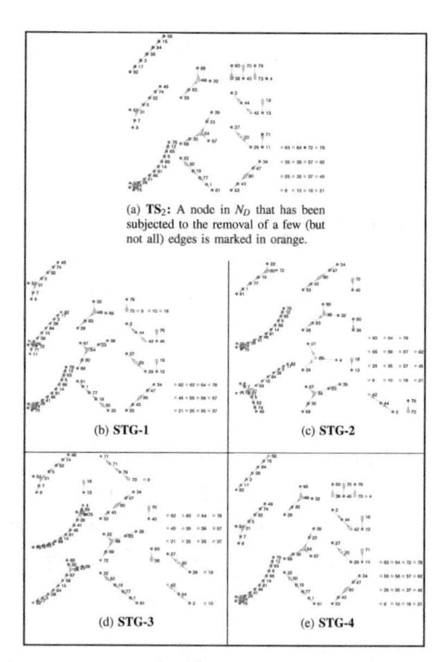

(a) TS$_2$: A node in N_D that has been subjected to the removal of a few (but not all) edges is marked in orange.
(b) STG-1
(c) STG-2
(d) STG-3
(e) STG-4

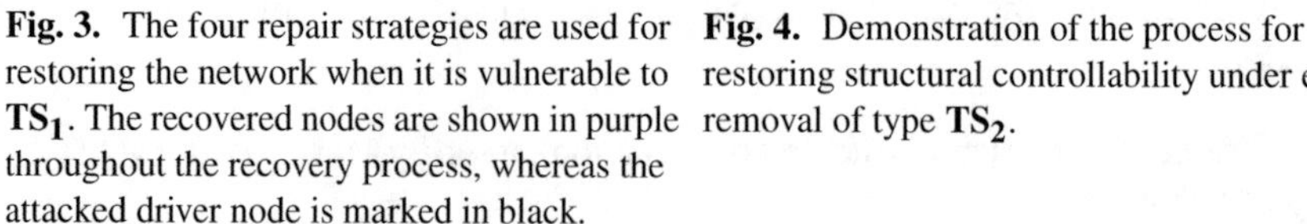
Fig. 3. The four repair strategies are used for restoring the network when it is vulnerable to $\mathbf{TS_1}$. The recovered nodes are shown in purple throughout the recovery process, whereas the attacked driver node is marked in black.

Fig. 4. Demonstration of the process for restoring structural controllability under edge removal of type $\mathbf{TS_2}$.

By the degree constraints at u, the results of the graphical simulations for **STG-1,2,3,4** when a network is vulnerable to threats of type **TS1,2,3,4** show that each driver

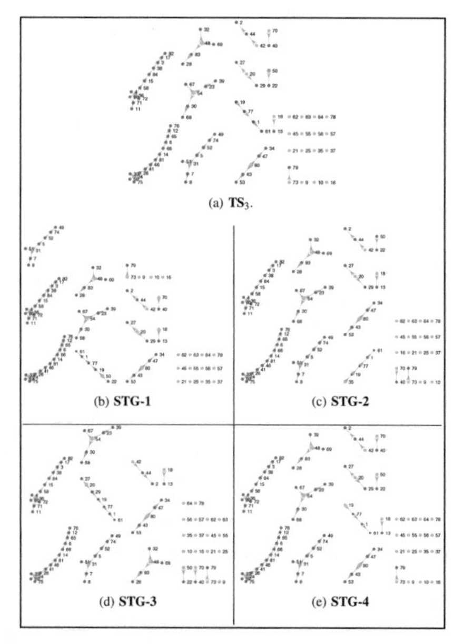

Fig. 5. Illustration of **STG-i** for regaining structural controllability after edge removal of type $\mathbf{TS_3}$.

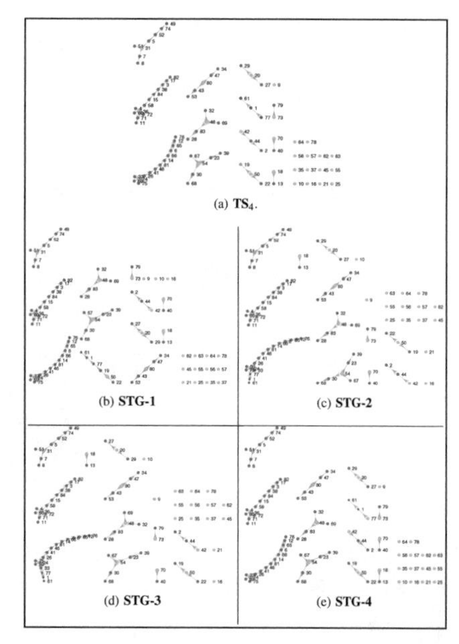

Fig. 6. The model **STG-i** when the structural controllability of the network is targeted by edge removal of type $\mathbf{TS_4}$.

node has an in-degree equal to zero, and all out-going edges (out-degree) are incident on dependent nodes from the driver node. However, it is not always the case that each driver node has dependent nodes, meaning that $s^{+}(u) = 0$, but each driver node must have no incoming edge incidents on it, such that $s^{-}(u) = 0$; otherwise, the condition (**D1**) is violated because a driver node u has in-degree ≥ 1. Note that the verification process for the degree constraints is restricted to the driver nodes after recovery, rather than the whole graph, resulting in a reduction in computation time.

Case 2: If a vertex $v \in \mathbf{N-2}$, then the condition **D2** should be met. According to the observation rules by Haynes et al., the difference between propagation nodes **D2** and domination nodes **D1** is that each node in the propagation nodes (the dependent nodes) must have exactly one incoming edge incident on it from either a driver node or another dependent node, and/or at most one out-going edge incident to dependent nodes, such that the following property holds: $\forall v \in G : s^{-}(v) = 1 \Rightarrow s^{+}(v) \leq 1$.

By induction, if an affected dependent node does not fulfil constraint **D2**, the unobserved node becomes part of the N_D to ensure that at least **D1** is satisfied; otherwise, it will be challenging to construct the tree-like structure at the end of the recovery. As a consequence of **Cases 1** and **2**, the postcondition is true. The recovery approaches **STG-1,2,3,4** terminate therefore when the structural control properties (**D1** and **D2**) are fulfilled.

As the dual problem related to controllability, the simulation findings can also be used to analyse the structural observability of such ER directed networks. One important

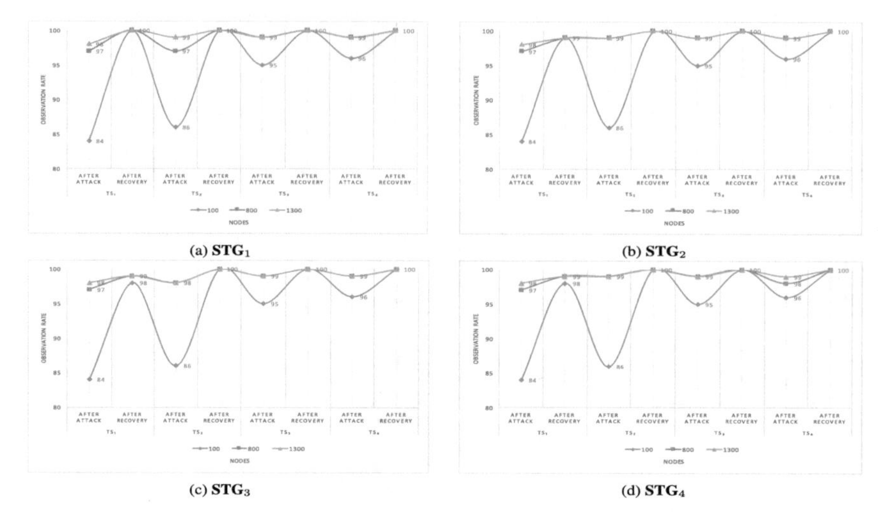

(a) $\mathbf{STG}_1$ (b) $\mathbf{STG}_2$ (c) $\mathbf{STG}_3$ (d) $\mathbf{STG}_4$

Fig. 7. The rate of observation before and after recovery is measured for each restoration strategy using the computation of **D1**.

indication of network deterioration when the control network is subjected to random edge removal is to measure the rate of observation before and after an attack. This indicator is also useful for verifying whether the whole network is structurally observable after recovery by computing the network's observation degree (**D1**). Figure 7 presents the rate of observation for the networks that are vulnerable to various edge removal threats of type $\mathbf{TS}_{1,2,3,4}$. For any threat scenario, **STG-1** is very efficient in restoring the structural observation of the compromised networks, with the observation percentage reaching 100% after recovery without affecting the network topology. Additionally, **STG-2,3,4** are still feasible and effective for $\mathbf{TS}_2$, $\mathbf{TS}_3$ and $\mathbf{TS}_4$, where the networks are similarly capable of regaining complete control after recovery (100% of observability) without modifying their structures. However, when the networks are exposed to $\mathbf{TS}_1$, the rate of observability reduces slightly after recovery in the case of **STG-2,3,4**. This is due to the nature of the implementation process for these restoration strategies, which exclude the attacked driver nodes in addressing the restoration process.

Due to the high computational complexity associated with computing PDS, it is worth noting that the primary goal of the recovery process is not to restore controllability as a whole via full re-computation if the PDS properties are only partially compromised. Rather, the goal is to tackle the damaged parts through re-computing the controlling nodes and their dependent nodes when their edges have been attacked without changing the size of N_D or the total number of edges used to establish control in a network, if possible. Figure 8 indicates that there is a variation in the set of driver nodes in each repair strategy. In particular, **STG-4** is the worst-case recovery strategy, since the number of N_D required to control the whole network increases throughout the recovery process, which is undesirable given the high cost of controllers. This increase in N_D size is related to the nature of **STG-4**, which is based on injecting a driver node into each disconnected component to re-establish structural controllability. While this is not the optimal approach, it is relevant to directed networks when the cost of establishing an edge exceeds the

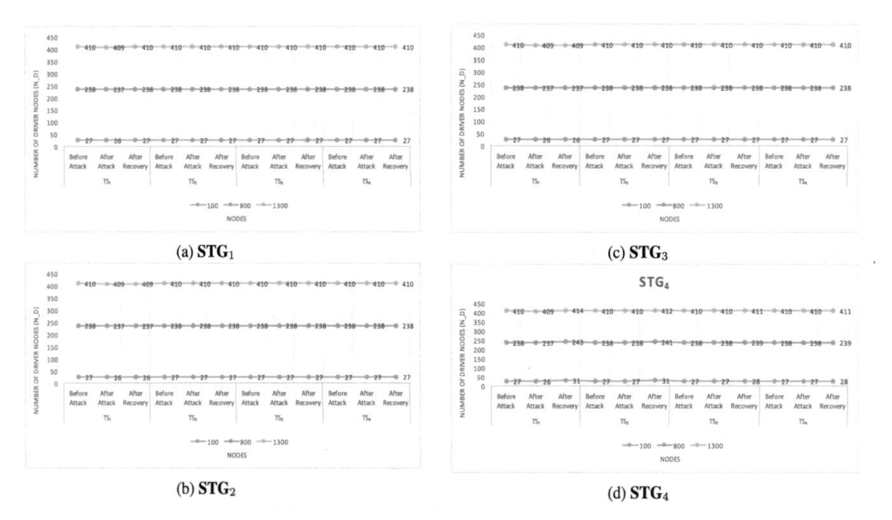

(a) $\mathbf{STG_1}$

(b) $\mathbf{STG_2}$

(c) $\mathbf{STG_3}$

(d) $\mathbf{STG_4}$

Fig. 8. The total number of a PDS (driver nodes) capable of controlling the network before and after the implementation of each restoration strategy.

cost of imposing a time-variant controller at each node. Conversely, when compromised nodes are restored via **STG-1**, the number of N_D remains unchanged, which is an efficient solution for restoring the PDS problem following attacks, as the strategy relies on repairing control relationships (edges) between driver nodes and their dependent nodes using a backup list of driver nodes stored on an external backup infrastructure before an attack. On the other hand, the set of N_D remains constant during the execution of **STG-2** and **STG-3** after attacks ($\mathbf{TS_2}$, $\mathbf{TS_3}$ and $\mathbf{TS_4}$), which both strategies relink the head of P_i to the tail of P_j without changing the number of N_D. However, the exception is threat scenario $\mathbf{TS_1}$ in which all edges of an attacked driver node are destroyed, thereby reducing the total number of N_D in a network. Therefore, the repair strategies **STG-2** and **STG-3** isolate the set of attacked driver nodes from the restoration process, affecting the number of N_D.

Preserving the number of edges is just as critical as maintaining the set of N_D after a restoration. Therefore, the suggested strategies seek to re-establish structural controllability while maintaining the size of network's edges under any threat scenario. The total number of edges before and after recovery is calculated for each repair strategy when different network sizes undergo topological changes in the form of edge removal, as shown in Table 2. This table indicates that the overall number of edges remains constant after the recovery process in the majority of instances, particularly in the case of **STG-1**, **STG-2**, and **STG-3**. While in **STG-4**, the total number of edges varies because of the nature of this strategy, which imposes a driver node on each unobserved dependency path rather than restoring edges, resulting in an increase in the overall number of driver nodes in a network.

Table 2. The number of power links in each recovery strategy following topological changes.

Recovery Strategy	Attack Scenarios		Nodes		
			100	800	1300
STG-1	TS_1	After Attack	95	794	1295
		After Recovery	100	800	1300
	TS_2	After Attack	96	795	1299
		After Recovery	100	800	1300
	TS_3	After Attack	99	799	1299
		After Recovery	100	800	1300
	TS_4	After Attack	99	799	1299
		After Recovery	100	800	1300
STG-2	TS_1	After Attack	95	794	1295
		After Recovery	100	800	1300
	TS_2	After Attack	96	799	1298
		After Recovery	100	800	1300
	TS_3	After Attack	99	799	1299
		After Recovery	100	800	1300
	TS_4	After Attack	99	799	1299
		After Recovery	100	800	1300
STG-3	TS_1	After Attack	95	794	1295
		After Recovery	100	800	1300
	TS_2	After Attack	96	796	1297
		After Recovery	100	800	1300
	TS_3	After Attack	99	799	1299
		After Recovery	100	900	1300
	TS_4	After Attack	99	799	1299
		After Recovery	100	800	1300
STG-4	TS_1	After Attack	95	794	1295
		After Recovery	95	794	1295
	TS_2	After Attack	96	797	1298
		After Recovery	96	797	1298
	TS_3	After Attack	99	799	1299
		After Recovery	99	799	1299
	TS_4	After Attack	99	799	1299
		After Recovery	99	799	1299

4.3 Time Complexity Analysis

This section discusses the computational complexity associated with the restoration strategies **STG-1**, **STG-2**, **STG-3** and **STG-4**. For simplicity, the notation $|V| = n$, $|E| = e$ and $|N_D| = n_d$ are used to represent the vertices and edges of a network, respectively, and the set of driver nodes. Let $U|n_d| = A$ and $U|n(e) - n_d(e)| = b$ denote the set of unobserved driver nodes and their out-going edges, respectively. Meanwhile unobserved dependent nodes and their edges are denoted by $U|n - n_d| = B$ and $U|n(e) - n_d(e)| = b$, respectively. The tail vertices are represented by t(v), whereas the head vertices zero are denoted by $h(u)$.

STG-1: It is required to traversing the unobserved nodes twice $k = 2$ at most, with a cost of $O\ (k.(A + B) + k.(a + b))$. The computational cost of the whole algorithm is therefore $O\ ((n + e) + (k.(A + B) + k.(a + b))) = 0(n + e)$ time.
STG-2: It is required to search for all driver nodes n_d in a network and randomly select the best candidate; this part of the algorithm can be computed in $O\ (n_d)$ time. As a result, the whole algorithm runs in $O\ (2(n + e) + n_d) = O\ (2n)$ time.
STG-3: The algorithm needs to process whole nodes to compute the diameter between a driver node and its head vertices $h(u)$ (its child nodes); the overhead of this part is thus $O\ (n + e)$. The second part of this algorithm involves the establishment of an edge with a cost of $O\ (h(u) + t(v))$. The total running time of this algorithm is therefore $((3n + 3e) + (h(u) + t(v))) = O(3n)$ time.
STG-4: The algorithm should search for tail nodes that are not yet controlled with a cost of $O\ (n + e)$. Following this, each unobserved tail node can be injected by a driver node

with a running time of $O(t(v))$. As a result, the total complexity of this strategy is $O((2n + 2e) + t(v)) = O(2n)$ time.

5 Conclusion and Future Work

The main contribution of this work is the development of four repair strategies based on the PDS formulation for directed network controllability following topological changes in the form of multi-round edge attacks. These four restoration strategies can effectively restore structural control of the Erdős-Rényi directed networks, where the simulation results showed that the observability rate for the majority of repair strategies has been retrieved after attacks. Besides restoring network controllability, one of the primary objectives is to address the problem of network controllability at a low computational cost while maintaining the network topology and its structural control when exposed to different attacks. This may be accomplished by establishing a restoration link instead of imposing a time-variant controller on a node in order to preserve a fixed number of driver nodes. Additionally, the simulation findings showed that in the majority of cases, especially for **STG-1**, **STG-2**, and **STG-3**, the total number of edges remains unchanged after the recovery process. In contrast, **STG-4** can be an inefficient recovery strategy with respect to **STG-1,2,3,4** due to the change in network topology induced by increasing the number of driver nodes during the restoration. The results also highlighted that the time complexity of structural control recovery can be effectively executed in polynomial time via various strategies; however, **STG-1** and **STG-2** have significantly higher computational efficiency than the other strategies, which either change the topological structure of networks throughout the recovery process, as **STG-4** does, or use high-cost computing, as **STG-3**. Future work will expand the analysis to investigate the potential of regaining structural control after single edge attacks, as discussed in [15], on a variety of directed networks and comparable control topologies, primarily random Erdős-Rényi, small-world (Watts-Strogatz) and scale-free (Barabási-Albert) graphs.

References

1. Lin, C.-T.: Structural controllability. IEEE Trans. Autom. Control **19**(3), 201–208 (1974)
2. Liu, Y.-Y., Slotine, J.-J., Barabási, A.-L.: Controllability of complex networks. Nature **473**(7346), 167–173 (2011)
3. Kalman, R.E.: Mathematical description of linear dynamical systems. J. Soc. Ind. Appl. Math. Series Control **1**(3), 152–192 (1963)
4. Haynes, T.W., Hedetniemi, S.M., Hedetniemi, S.T., Henning, M.A.: Domination in graphs applied to electric power networks. SIAM J. Discrete Mathematics **15**(4), 519–529 (2002)
5. Kneis, J., Mölle, D., Richter, S., Rossmanith, P.: Parameterized power domination complexity. Inf. Process. Lett. **98**(4), 145–149 (2006)
6. Zhao, M., Shan, E., Kang, L.: Power domination in the generalized petersen graphs. Discussiones Mathematicae: Graph Theory **40**(3), 695–712 (2020)
7. Hedetniemi, S.T., McRae, A.A., Mohan, R.: Algorithms and complexity of power domination in graphs. In: Haynes, T.W., Hedetniemi, S.T., Henning, M.A. (eds.) Structures of Domination in Graphs. DM, vol. 66, pp. 461–484. Springer, Cham (2021). https://doi.org/10.1007/978-3-030-58892-2_15

8. Sudakov, B., Vu, V.H.: Local resilience of graphs. Random Struct. Algorithms **33**(4), 409433 (2008)
9. Alwasel, B., Wolthusen, S.D.: Recovering structural controllability on Erdős-Rényi graphs in the presence of compromised nodes. In: Rome, E., Theocharidou, M., Wolthusen, S. (eds.) CRITIS 2015. LNCS, vol. 9578, pp. 105–119. Springer, Cham (2016). https://doi.org/10.1007/978-3-319-33331-1_9
10. Lou, Y., Wang, L., Chen, G.: Toward stronger robustness of network controllability: a snapback network model. IEEE Trans. Circ. Syst. I: Regular Papers **65**(9), 2983–2991 (2018)
11. Li, X., Zhang, Z., Liu, J., Gai, K.: A new complex network robustness attack algorithm. In: Proceedings of the ACM international Symposium on Blockchain and Secure Critical Infrastructure, pp. 13–17 (2019)
12. Chen, G., Lou, Y., Wang, L.: A comparative study on controllability robustness of complex networks. IEEE Trans. Circ. Syst. II: Express Briefs **66**(5), 828–832 (2019)
13. Alwasel, B.: Recovery of structural controllability into critical infrastructures under malicious attacks. Int. J. Adv. Comput. Sci. Appl. **11**(4), 723–728 (2020)
14. Lou, Y., He, Y., Wang, L., Chen, G.: Predicting network controllability robustness: a convolutional neural network approach. IEEE Trans. Cybern. (2020)
15. Alwasel, B.: Robustness analysis of structural controllability for directed networks against single edge attacks. Int. J. Innovative Technol. Expl. Eng. **9**(8), 944–951 (2020)
16. Alwasel, B.: Simulating robustness of structural controllability for directed networks under multi-round edge strategies. IEEE Access **9**, 84688–84699 (2021)
17. Ramos, G., Silvestre, D., Silvestre, C.: The robust minimal controllability and observability problem. Int. J. Robust Nonlinear Control **31**(10), 5033–5044 (2021)
18. Zhao, J., Sun, S., Shen, H., Xia, C.: An effective network repair strategy against both random and malicious edge attacks. In: 40th Chinese Control Conference, pp. 8628–8633, IEEE (2021)

Resource Authorization Methods for Edge Computing

Ryu Watanabe[1(✉)], Ayumu Kubota[1], and Jun Kurihara[2]

[1] KDDI Research, Inc., 2–1–15 Ohara, Fujimino, Saitama 356–8502, Japan
{ry-watanabe,kubota}@kddi-research.jp
[2] Graduate School of Information Science, University of Hyogo, 7–1–28 Minatojima-Minamimachi, Chuo, Kobe, Hyogo 650–0047, Japan
kurihara@ieee.org

Abstract. To realize the distribution of processing load and prompt response, the concept of edge computing is drawing attention. Under the edge computing environment, server processing is carried out on an edge node located near various devices with a communication module instead of on a central server. These edge nodes can be provided by another entity other than service providers such as network operators. The edge nodes have fewer computing resources than a central server has. Therefore, appropriate dynamic resource management is required to avoid resource exhaustion. For this purpose, authorization techniques, e.g., OAuth, can be applied. In this paper, we consider applying the OAuth protocol for a privilege delegation on edge computing. Firstly, we clarify the authentication flows differ depending on the relationship of edge computing players (edge provider, user, service provider). We then describe the unique problems of resource authorization on edge computing.

1 Introduction

With the diversification of electronic devices and the evolution of communication modules, a wide variety of devices are connected to the Internet, and the amount of exchanged data is increasing. In addition, there are new applications, such as connected cars, that require immediate response in computational processing over the network. From these perspectives, "edge computing" is attracting attention as a means of distributing the processing load and ensuring immediate responsiveness, where computational processing, which is traditionally performed on a central server hosted by a service provider, is performed on "edge nodes" located in the vicinity of user devices on the network. Unlike the central server, edge nodes can be generally provided by not service providers but network operators. Furthermore, as in the case of a central server, the computational resources of each edge node are shared among multiple users, but the scale of sharing resources is usually limited compared to that of a central server. Therefore, it is important to avoid the exhaustion of computational resources by dynamically allocating privilege to use computational resources to appropriate

L. Barolli et al. (Eds.): AINA 2022, LNNS 449, pp. 167–179, 2022.
https://doi.org/10.1007/978-3-030-99584-3_15

edge nodes so that only authorized users or services can execute computational operations. To achieve this, we can apply the "authorization" method, in which the owner of a computation resource delegates the privilege to execute a computation to a user or a service.

The OAuth [5,8] is a typical authorization method on the Internet. With OAuth, authorization is performed by passing a "token" describing the privilege that the owner of the resource wishes to delegate to the authorization recipient. Therefore, in this paper, the authors denote the application of this OAuth authorization flow to the authorization of computational resources in edge computing. In particular, we target the case where a network operator is the provider of the edge node, and the subscribers of its network service are the users of the computation service. We clarify that the authorization flow that should be applied depends on the interrelationships among edge computing participants (network operators, users, and service providers). In addition, we will discuss the dynamic allocation of computing resources, low-latency response, and other considerations for edge computing authorization.

2 Related Work

Soni et al.'s survey on edge computing security [11] identifies Distributed Denial of Service (DDoS) attacks, side-channel attacks, malware infections, as well as authentication and authorization methods as issues that have a significant impact on edge nodes, users, and user terminals. Regarding authentication and authorization methods in edge computing, there have been many studies on "access from edge nodes to central servers". For example, Grande et al. [3] proposed a method based on OAuth [5] to allow edge nodes to take over the authorization process on behalf of poor ability IoT devices. In addition, Echeverría et al. [1] proposed an authentication and authorization method based on the ACE protocol [7] for IoT devices in environments with limited network resources, such as battlefields. However, the authorization of the "use of edge node resources by users or services" in edge computing has been rarely considered.

Moreover, the GSMA [4], and ETSI [2] consider edge computing based on mobile networks, with handover between edge nodes and roaming between network operators. For their authentication and authorization of edge nodes, the use of NEF (Network Exposure Function), i.e., network APIs provided by operators, is suggested, but the specific authorization flow is not mentioned.

A lot of research has already been done on the management (registration and discovery) of computational resources of nodes in edge computing (ex., [6,10,13]). For example, Murturi et al. [9] give a method that enables user terminals to discover nearby computational resources by managing edge node metadata on multiple neighboring edge nodes in a distributed manner. However, the computational resource authorization on an edge node always requires "searching for edge nodes near the user that have available resources. Therefore, the authorization of use of resources of edge nodes needs to be integrated with resource management methods.

Table 1. Teaminology of OAuth

Team	Description
Protected resource	Information resource on a resource server
Resource server	A server that maintains protected resources and provides protected resources upon request
Client	A service or application that accesses resource servers
Resource owner	The entity that owns protected resources. Also known as end users. Resource owner authorizes access to its protected resources
Authorization server	A server that authenticates the resource owner, issues access token, and manages authorization flow. It has an authorization endpoint that performs the actual authorization process and a token endpoint that issues and manages access tokens
Access token	A token for an authenticated request, presented by the client to the resource server
Scope	Scope of access privileges to be delegated (authorization scope)

"Authorization" on the Internet means the delegation of privilege (or authority) in service to another entity. OAuth [8] is one of the most widely used authorization methods. The current version of OAuth is 2.0 (OAuth2.0) [5]. In this paper, OAuth refers to OAuth2.0 unless otherwise specified. Definitions of terms in OAuth are given in Table 1. The Authorization code flow, which is a typical OAuth authorization flow, is shown in Fig. 1. In OAuth 2.0, other authorization flows, such as the implicit flow, have been defined, but the basic flow is this authorization code flow. In the authorization flow of OAuth, only the resource owner can delegate the access privileges of the protected resources to a federated service application called the client. Under OAuth flow, access privilege to protected resources can include not only the read attribute but also the execute attribute. At first, the resource owner and the authorization server (authorization endpoint) consent to the authentication and the authorization in response to the client's request (Fig. 1: (1)–(5)). The client obtains an access token from the authorization server using a short-lived authorization code issued by the authorization endpoint (Fig. 1: (6)–(8)). Thereafter, the client presents this access token to the resource server as "proof of authorization" to access the protected resource (Fig. 1: (9)–(10)). The scope of the delegated privileges is described in the access token in a way that cannot be falsified. Therefore, the client cannot access information or execute functions without the permission of the resource owner, and can only exercise the specified scope of privileges. In edge computing, the OAuth authorization flow using access tokens can be applied for access to computational resources of edge nodes [14, Sect. III-D].

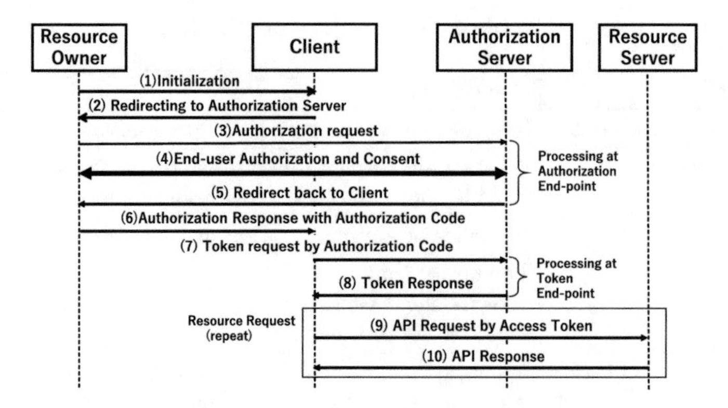

Fig. 1. Authorization flow on OAuth2.0 [5, Sect. 4.1])

Table 2. Terms definition in this paper

Term	Definition
Edge Provider (EP)	The owner of edge nodes
Edge Node (EN)	A node that provides computational resources and performs computation
Service Provider (SP)	An entity that provides a service by utilizing computational resources on edge nodes
Computational resource	Resources held by an edge node for the performing a computation (CPU, memory, electrical power, and so on.)
Instance	A computation performed using computational resources on an edge node for a service (or part of a function) provided by an SP
User	A entity who uses a service provided by a service provider

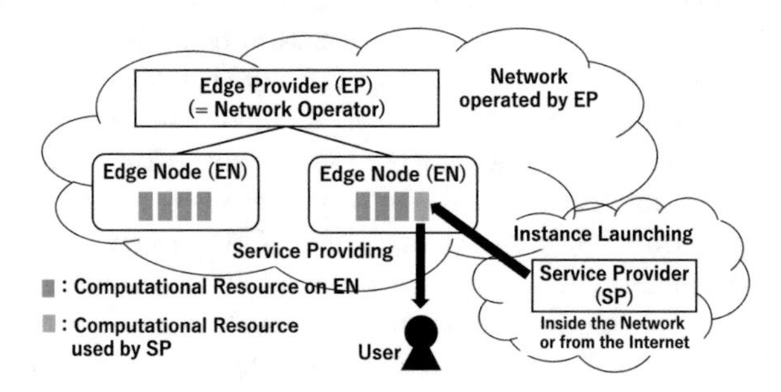

Fig. 2. Edge computing model

3 Requirements for the Authorization on Edge Computing Models

3.1 Edge Computing Models

Table 2 shows the definitions of the edge computing terms used in this paper, and Fig. 2 shows the edge computing model assumed in this paper. The definition of the participants (entities) on the model assumed in this paper is given below.

- An **Edge Provider** (EP) manages multiple **Edge Nodes** (EN) in the network.
- Each EN has **computational resources** that are consumed as they perform computations in edge computing.
- Computational resources are defined as the unified amount of CPU spec and free time, free memory, and available electrical power.
- **Service Provider** (SP) provides services accompanied by edge computing, which is "computation performed using EN's computational resources."
- The aforementioned computation for providing a service for an SP running on an EN is called an **instance** of that service or an **instance** of part of the service function.
- **User** refers to the user of the service provided by an SP as described above. When it indicates the users and subscribers of the network, explicitly state them.
- The EP is assumed to be the operator of the network in which the EN is deployed, i.e., the network operator. In light of the current situation, we note that, in many cases, the network operator is a mobile network operator.
- In addition, we assume that the users shall be a subscriber of the network operator in question.
- The communication between a user and an EN is considered to occur in the communication network provided by the network operator.

In summary, our model assumes that the network operator (=EP) provides access to ENs and computational resources on them to the subscribers (=users) of its network and the services they use. In our model, an instance of edge computing is started only after "authorization," which delegates the privilege of using computational resources, is performed. Considering the authorization in this model, we expect that the appropriate authorization flows for edge resource usage of edge nodes (ENs) will differ based on the respective interrelationships among EP, SP, and user. We will clarify these flows in the following sections.

3.2 Requirement for Applying the OAuth Flow to Resource Authorization on Edge Computing

In this section, we describe the general prerequisites for applying the authorization flow of OAuth2.0 to "authorization for the use of computational resources" in edge computing. First, the Edge Provider (EP) or Edge Node (EN) defined in the previous section corresponds to the resource server in OAuth2.0 in terms of the entity that holds the resources. Therefore, the computation resources of EN correspond to the protected resources in OAuth2.0. The authorization of this protected resource is done by the entity that is the resource owner delegating its usage privileges to the entity that is the client. At this time, the resource owner itself specifies the scope and target of the privilege to be delegated and requests authorization from the authorization server. In general, for edge computing, it is necessary to select an EN that satisfies both of the following requirements for an instance execution.

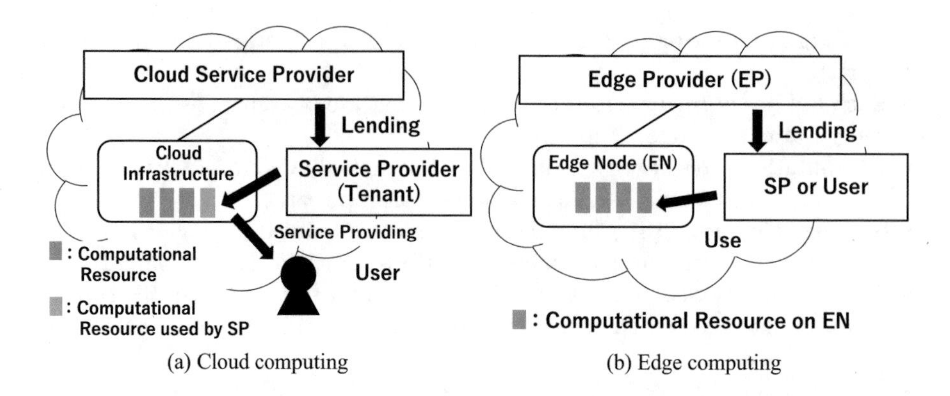

Fig. 3. Concept of cloud computing and Edge computing

- The EN must be in the vicinity of the user.
- The EN must have enough free computational resources for the computations the user wants to perform.

Based on these considerations, authorization of computational resources in edge computing requires setting up an authorization scope that is limited to the selected EN and computational resources (in terms of time and quantity) on it. If the authorization scope is unlimited, it is easy to make the computation resources consumed by the DoS (Denial of Service) attack. Therefore, the following requirements need to meet for the application of OAuth authorization flow to the resource authorization in edge computing.

1. Scope generation by pre-selection of appropriate EN(s) with nearby and free resources.
2. Delegation of resource usage privilege that is limited only to the temporal and quantitative computing resources of the relevant EN(s).

However, these requirements need "real-time resource management" of all ENs in the communication network operated by a network operator. Therefore, in edge computing, where immediate response is desired, low latency is also required for authorization flow with computational resource management, discovery, and scheduling.

4 Participant Relationships in Edge Computing Authorization

4.1 Concept of Privilege Delegation Model

First, for comparative purposes, we show the concept of "cloud computing," which is the original concept of edge computing in Fig. 3-(a). In cloud computing, a service provider is a "tenant" and is provided a computing infrastructure from

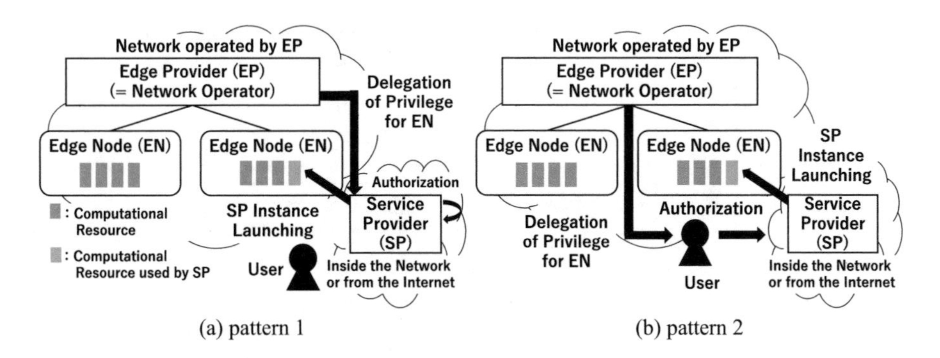

Fig. 4. Patterns in Edge computing authorization

a cloud service provider. The service provider uses the computing infrastructure to provide services to end users. Reflecting the relationship in cloud computing mentioned above, edge computing in this paper is considered as the provision of the EN and its computational resources owned by the EP to another participant. (Figure 3-(b)). We note that the resource server and the authorization server in the OAuth2.0 authorization flow correspond to the EN that possesses the computing resource and the EP that is its actual owner.

4.2 Pattern Classification by Participant Relationship

Pattern 1: First, we consider the case where the EP and SP have a relationship similar to cloud computing, i.e., a contractual arrangement where the SP, as a "tenant," provides services to users using the computational resources on the EN provided by the EP. This pattern in which **SPs are delegated the privilege to use computational resources is called "pattern 1"** (Fig. 4(a)). Service provision in pattern 1 is performed by delegating the SP's own privilege to the SP's individual service, so the SP itself corresponds to both the resource owner and client in OAuth authorization flow.

Pattern 2: Next, consider the other situation. This pattern shows the form where the network operator, which is the EP, delegates the privilege to use computational resource on the EN(s) to its subscribers under contract with it. The pattern in which **users are delegated the privilege to use the computational resources is called "pattern 2"** (Fig. 4(b)). The service provision in pattern 2 can be regarded as the delegation of the user's privilege to use the computing resources to the SP. Therefore, the user and the SP correspond to the resource owner and the client in the OAuth authorization flow, respectively.

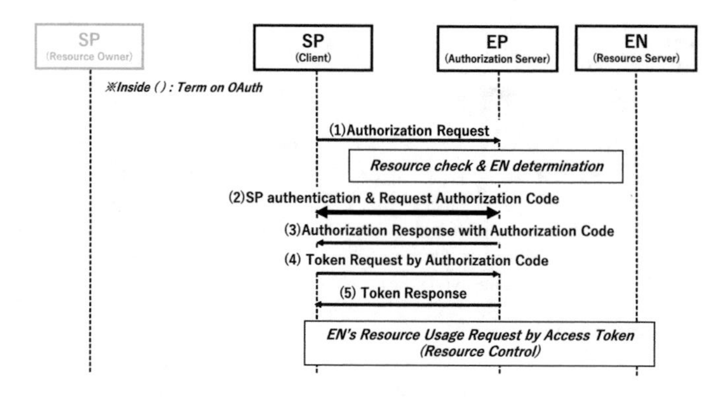

Fig. 5. Authorization flow in pattern 1

4.3 Use Case of Each Pattern

The pattern to be applied changes according to the use case. For example, if an SP provides a service to users in the vicinity of an EN on a "regularly," then pattern 1 is applied to reduce complexity. On the other hand, in case of the service such as a virtual desktop service that is provided to a specific user, who wants and requests prompt service response from the EN in the vicinity of the user, pattern 2 is applied for it.

4.4 Authorization Flow on Each Pattern

Pattern 1: The authorization flow in pattern 1 is shown in Fig. 5. In pattern 1, the usage privilege of ENs is delegated to an SP, i.e., the SP itself reuses them. Therefore, the usage privilege delegated to the SP itself is transferred (re-delegated) to the SP's service. The resource owner and the client on OAuth flow correspond to the SP, or the SP and the service of the SP, respectively. As defined in the previous section, authorization in edge computing requires the process of "management, discovery, and scheduling of computational resources." Therefore, in this flow, the EP that receives an authorization request from the SP determines the EN in the vicinity of the location specified by the SP (i.e., the location of the user to whom the service is provided), reserves, cancels, releases, and confirms the release of the computational resources of the EN in question.

Pattern 2: The authorization flow in pattern 2 is shown in Fig. 6. In this flow, the usage privilege of ENs is delegated to a user. Therefore, this usage privilege delegated to the user is re-delegated to the SP (service). For this reason, the resource owner on OAuth flow corresponds to the user, and the client on OAuth flow corresponds to the SP or the service of the SP, respectively. As in pattern 1, this flow also requires the process of management, discovery, and scheduling of computational resources on ENs by the EP. However, unlike pattern 1, the EP can know the user's location at all times or in advance, because the EP is a network operator and the user can make authorization requests to the EP.

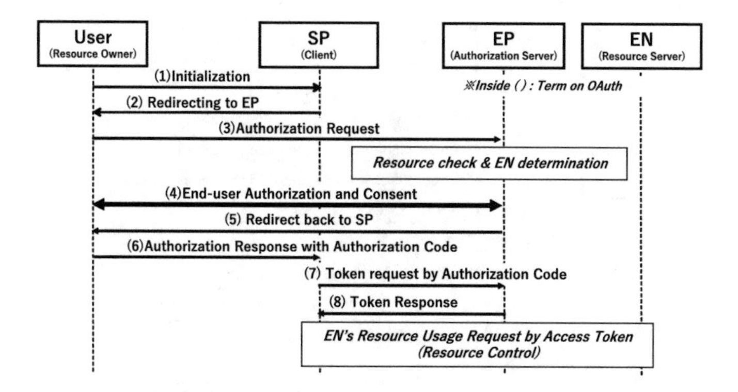

Fig. 6. Authorization flow in pattern 2

Table 3. Processing on EP for each pattern

	Discovery of EN near the user	Discovery and scheduling of EN on single network
Pattern 1	✓	✓
Pattern 2	–	✓

5 Discussion for Authorization Flow

For each authorization flow, we consider the reduction of the processing load at the EP, which is a consideration specific to edge computing. First, we summarize the processing required on the EP for resource discovery and scheduling for each pattern in Table 3.

5.1 Identifying User Location

From Table 3, pattern 2 allows the EP to identify the user's location immediately or in advance. Therefore, the discovery of ENs in the vicinity of the user is easy. On the other hand, in pattern 1, the SP, not the user, has to notify the EP of the user's location. If the user privacy is taken into consideration, the SP should not know the user's exact location, but some kind of user information like the user ID for the network operator, to the EP. Therefore, the EP needs to search for ENs without prior knowledge of the user's location in the network. It cannot immediately provide an "appropriate EN in the vicinity of the user" that meets the requirements, which may result in a heavy overhead.

5.1.1 Identifying User Location with Two-Step Authorization Method

One of the solutions to the user location problem, we can use the following two-step authorization method shown in Fig. 7.

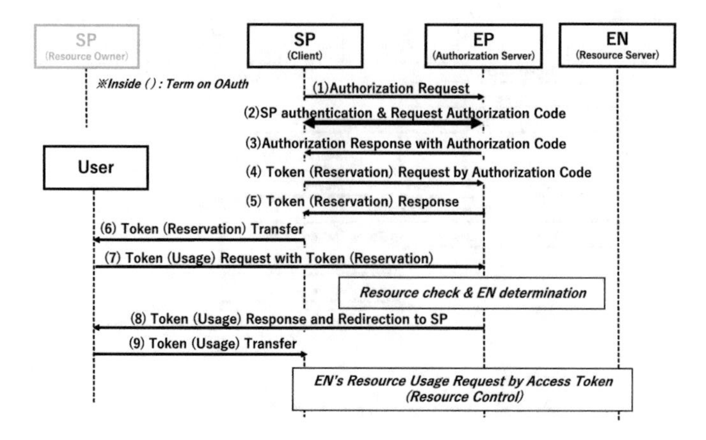

Fig. 7. Authorization flow in pattern 1 with two-step verification

1. In the authorization flow, the SP shall obtain an "access token" of the '**right for the resource reservation**' as the scope.
2. The SP sends this access token (reservation) to the user who want to use the service in step (6).
3. The user sends this access token (reservation) to the EP's "resource discovery and reservation endpoint" and requests resource allocation and authorization.
4. The EP issues a new "access token" of the '**right for the computational resource usage**' as the scope and sends it to the user.
5. The user sends the access token (usage) to the SP in step (9).
6. The SP sends the access token for computational resource usage to the EN and requests the use of the resource.

Using the method mentioned above, the EP receives the request for resource allocation from the user, who is explicitly a subscriber of its network, not the SP. Therefore, it is expected that the EP can correctly discover and reserve ENs near the user as in pattern 2.

5.1.2 Identifying User Location Using DNS Support

Another solution to the user location problem is to use the enhanced support of DNS query [12]. There are cases where DNS can be used as a means of finding nearby EN. This is a special use of DNS, which uses a mechanism to return information on EN in the vicinity of the user in response to the user's DNS query according to the user's location. Actually, for SPs, it is not important to know the user's exact location. Instead of the user's exact location, they should know the EN in the vicinity of the user. As mentioned before, in edge computing, the EP who manages the ENs is mostly the network operator. Therefore, it can be assumed that they also provide a DNS server to the subscribers. In other words, since both the ENs and the DNS server are operated by the same entity, it is possible to add the information of the EN in the vicinity of a user as a DNS response and return it.

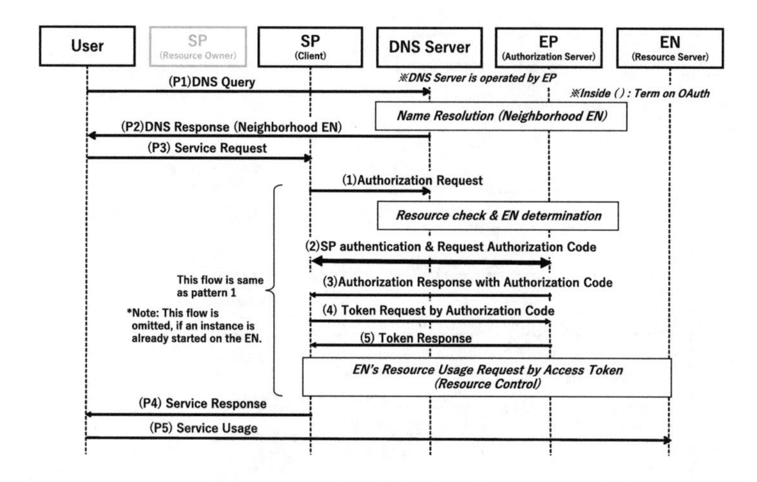

Fig. 8. Authorization flow in pattern 1 using DNS support

The flows of using DNS support to perform EN lookups in the neighborhood are showed below. In these cases, we consider the use of DNS in terms of the use of the service provided by the SP. Therefore, the flows start with the service use request by the user and finish with the service use access to the service on the EN in the vicinity of the user by the user.

Pattern 1 with DNS: The Fig. 8 shows the flow when EN lookup by DNS is applied to pattern 1. Pattern 1 is a case where the SP has been delegated the privilege to use ENs by EP. Before the user requests the SP to use the service, the user uses DNS to get the information of nearby ENs. If the SP is already providing a service on the EN in question, it will respond to the user with the service response. If the service is not being provided, the SP will request the EP to authorize the use of the resources of the EN in question, and after the service instance is started, the SP will respond with information about the service use.

Pattern 2 with DNS Support: Figure 9 shows the flow when EN lookup is provided by DNS in pattern 2. Pattern 2 is a case where the user has been delegated the privilege of using EN. The user first requests the SP to use the service, and the SP returns the URL of the service as a response. When resolving the name of the URL, the DNS also performs a search for nearby ENs and returns the response. Then, the user authorizes to use the EN for the SP, and the SP executes the instance construction on the EN. Note that in the case of pattern 2, resource authorization will always occur because the user's privilege is used to make an instance for the user.

5.2 Comparison with Original Flow

As a simple evaluation, a comparison with the original OAuth2.0 flow is shown in Table 4. The proposed method requires dynamic determination of the target ENs, which increases the number of messages. For the appropriate resource

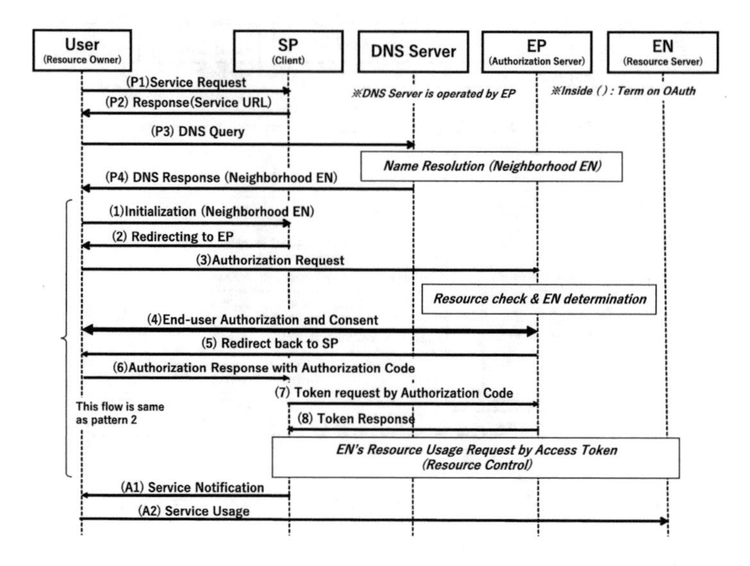

Fig. 9. Authorization flow in pattern 2 using DNS support

Table 4. Comparison with OAuth2.0 flow

Protocol	Sequence	Specifying resources	Resource management
OAuth2.0	Simple	Static (pre assign)	Resource server
Proposal	Need additional message	Dynamic (depends on user location)	Collaboration among EP and ENs

management, the EP and ENs need to work together. In addition, as mentioned above, the protocol is heavier if it also needs to work with the DNS server. It is necessary to quantitatively evaluate how much additional load is added.

6 Conclusion

In this paper, the authors have classified authorization flows according to the interrelationships among each participant in edge computing. In addition, we have proposed the solution to the notifying user location problem. For this purpose, we have applied two-step authorization and the DNS-based neighbor EN exploration method to the authorization flows. As future work, we denote a detailed design and evaluation of a method for minimizing the overhead on the authorization flow in each pattern.

Acknowledgement. This work was supported in part by JSPS KAKENHI Grant Number JP20K23329 and JP21H03442, and University of Hyogo Special Grant for Young Researchers.

References

1. Echeverría, S., Lewis, G.A., Klinedinst, D., Seitz, L.: Authentication and authorization for IoT devices in disadvantaged environments. In: Proceedings IEEE WF-IoT 2019, pp. 368–373 (2019)
2. ETSI: Multi-access edge computing (MEC): MEC 5G integration. ETSI GR MEC 031 V2.1.1: https://www.etsi.org/deliver/etsi_gr/MEC/001_099/031/02.01.01_60/gr_MEC031v020101p.pdf (2020)
3. Grande, E., Beltrán, M.: Edge-centric delegation of authorization for constrained devices in the Internet of Things. Comput. Commun. **160**, 464–474 (2020)
4. GSMA: Operator platform concept, phase 1: Edge cloud computing (2020). https://www.gsma.com/futurenetworks/resources/operator-platform-concept-whitepaper/
5. Hardt, D.: The OAuth 2.0 authorization framework. RFC6749 (2012): https://datatracker.ietf.org/doc/html/rfc6749
6. Hong, C.H., Varghese, B.: Resource management in fog/edge computing: a survey on architectures, infrastructure, and algorithms using edge computing. ACM Comput. Surv. **52**(5), 1–37 (2019)
7. IETF ACE Working Group: https://tools.ietf.org/wg/ace/
8. IETF OAuth Working Group: https://tools.ietf.org/wg/oauth/
9. Murturi, I., Avasalcai, C., Tsigkanos, C., Dustdar, S.: Edge-to-edge resource discovery using metadata replication. In: Proceedings of the IEEE ICFEC 2019, pp. 1–6 (2019)
10. Pradhan, M., Poltronieri, F., Tortonesi, M.: Dynamic resource discovery and management for edge computing based on SPF for HADR operations. In: Proceedings of the IEEE ICMCIS 2019, pp. 1–6 (2019)
11. Soni, N., Malekian, R., Thakur, A.: Edge computing in transportation: security issues and challenges. arXiv:2012.11206 (2020)
12. Suzuki, M., et al.: Enhanced DNS support towards distributed MEC environment. ETSI White Paper No. 39 (2020)
13. Toczé, K., Nadjm-Tehrani, S.: A taxonomy for management and optimization of multiple resources in edge computing. Wirel. Commun. Mob. Comput. **2018**, 23 (2018). https://doi.org/10.1155/2018/7476201. Article ID 7476201
14. Xiao, Y., Jia, Y., Liu, C., Cheng, X., Yu, J., Lv, W.: Edge computing security: state of the art and challenges. Proc. IEEE **107**(8), 1608–1631 (2019). https://doi.org/10.1109/JPROC.2019.2918437

Impact of Self C Parameter on SVM-based Classification of Encrypted Multimedia Peer-to-Peer Traffic

Vanice Canuto Cunha[1,2], Damien Magoni[3], Pedro R. M. Inácio[2], and Mario M. Freire[2](✉)

[1] Universidade Federal de Mato Grosso, Cuiabá, Brazil
vanice@ic.ufmt.br
[2] Instituto de Telecomunicações, Universidade da Beira Interior, Covilhã, Portugal
{inacio,mario}@di.ubi.pt
[3] LaBRI-CNRS, Université de Bordeaux, Talence, France
damien.magoni@u-bordeaux.fr

Abstract. Home users are increasingly acquiring, at lower prices, electronic devices such as video cameras, portable audio players, smartphones, and video game devices, which are all interconnected through the Internet. This increase in digital equipment ownership induces a massive production and sharing of multimedia content between these users. The supervised learning machine method Support Vector Machine (SVM) is vastly used in classification. It is capable of recognizing patterns of samples of predefined classes and supports multi-class classification. The purpose of this article is to explore the classification of multimedia P2P traffic using SVMs. To obtain relevant results, it is necessary to properly adjust the so-called Self C parameter. Our results show that SVM with linear kernel leads to the best classification results of P2P video with an F-Measure of 99% for C parameter ranging from 10 to 70 and to the best classification results of P2P file-sharing with an F-Measure of 98% for C parameter ranging from 30 to 70. We also compare these results with the ones obtained with Kolmogorov-Smirnov (KS) tests and Chi-square tests. It is shown that SVM with linear kernel leads to a better classification performance than KS and chi-square tests, which reached an F-Measure of 67% and 70% for P2P file-sharing and P2P video, respectively, for KS test, and reached an F-Measure of 85% for both P2P file-sharing and P2P video for chi-square test. Therefore, SVM with linear kernel and suitable values for the Self C parameter can be a good choice for identifying encrypted multimedia P2P traffic on the Internet.

This work was financed by CAPES (Brazilian Federal Agency for Support and Evaluation of Graduate Education) within the Ministry of Education of Brazil under a scholarship supported by the International Cooperation Program CAPES/COFECUB - Project 9090-13-4/2013 at the University of Beira Interior. This work is also funded by FCT/MCTES through national funds and, when applicable, co-funded by EU funds under the project UIDB/50008/2020 and by FCT/COMPETE/FEDER under the project SECURIoTESIGN with reference number POCI-01-0145-FEDER-030657, and by operation Centro-01–0145-FEDER-000019 - C4 - Centro de Competêencias em Cloud Computing, co-funded by the European Regional Development Fund (ERDF) through the Programa Operacional Regional do Centro (Centro 2020), in the scope of the Sistema de Apoio á Investigação Cientfica e Tecnológica - Programas Integrados de IC&DT.

L. Barolli et al. (Eds.): AINA 2022, LNNS 449, pp. 180–193, 2022.
https://doi.org/10.1007/978-3-030-99584-3_16

Keywords: Chi-square test · Kolmogorov-smirnov test · P2P Traffic · Support vector machine · SVM

1 Introduction

According to the 2020 report from Sandvine [1], 80% of the current Internet traffic is generated by three key application classes: video, gaming, and social sharing. Among these applications, video corresponds to the largest traffic volume. More specifically, video streaming grew its overall traffic share during lockdown, which included accelerated video releases to streaming, binge-watching multiple seasons of TV shows, search for entertainment and information on what is happening in the world and video traffic from social networks like TikTok. Among video streaming applications, we pay a particular attention in this paper to peer-to-peer (P2P) video streaming. According to the global application total traffic share in 2020 reported by Sandvine [1], BitTorrent is the fourth most used application/platform after YouTube, NetFlix and HTTP-based streaming.

For P2P media streaming, users can take advantage of their aggregated upload bandwidth capacity for efficiently distributing video content among themselves. However, P2P traffic, including BitTorrent traffic, is difficult to detect, prioritize or mitigate, namely inside organizations, specially when protocol obfuscation techniques are used.

Streaming sessions among peers can last for long periods, which can interfere with the available network bandwidth in organizations required to perform critical network-based enterprise tasks. For this reason, Internet Service Providers (ISPs) and network administrators in organizations consider the identification and classification this type of traffic as an important matter, enabling to appropriately managing resource allocation and planning future network growth [2,3].

On the other hand, nowadays P2P traffic is often encrypted and has varying packet lengths. It is important to classify encrypted multimedia P2P traffic to properly manage the network's resources. In that context, recognizing the different types of apps that use the network's resources and classify them is a pre-requirement that contributes for an advanced management of the network, such as providing quality of service (QoS) and price, besides identifying anomalies.

P2P multimedia applications can affect the performance of servers, services or critical applications of organizations or tasks dependent on the network. In this situation, a network administrator may need to impose limitations on P2P traffic, by limiting the transmission rate, differentiating services or even blocking those connections, to ensure a good performance of the internal applications, and/or to enforce rules to regulate the use of P2P systems.

The purpose of this article is to investigate the impact of both adjusting the Self C parameter and selecting a particular SVM kernel for specifically classifying multimedia P2P traffic.

2 Related Work

Recently, many studies have been carried out to classify traffic with the help of the SVMs [4–18]. Some of them have optimized the kernel settings and SVM parameters to improve the classification results, such as in [6, 18]. Self C is one of the parameters of SVM, also denominated as C Penalty, corresponding to the degree of punishment and causing implications on the experimental results. It is important to properly adjust this parameter, as it will directly affect the network traffic classification effectiveness. This parameter is responsible for the optimization of the SVM, avoiding an incorrect classification, being thus a regularization parameter [19].

Several works addressed the classification of Internet traffic using SVM, as we show concisely in Table 1. However, to our knowledge, the current literature is lacking a study presenting the impact of the adjustment of specific SVM parameters for the classification of multimedia P2P traffic. Therefore, this article addresses this issue.

Table 1. Summary of the main points on traffic classification using SVM addressed in articles found in the literature. In the Performance column: Precision - P, Recall - R, Accuracy - A, F-Measure - FM.

Work	Method	Real-time operation	Detection of encrypted traffic	Performance(%)
Mavroforakis *et al.* [7]	SVM	No	No	A: –
Yuan *et al.* [9]	SVM	No	Yes	A: 81.75 and 95.98
Aggarwal *et al.* [8]	SVM and Naïve Bayes	Yes	Yes	A: 88.88
Aamir *et al.* [16]	SVM + KNN + RF	No	No	A: 95–96.66
Rezvani *et al.* [4]	Fuzzy + SVM	No	No	A: 99.44
Tang *et al.* [5]	SVM + Wavelet (WL)	No	No	A:–
Akinyelu *et al.* [10]	SVM	No	No	A: –
Sankaranarayanan *et al.* [12]	SVM	No	Yes	A: -
Han *et al.* [13]	Entropy + SVM	No	No	–
Luo *et al.* [14]	SVM and Genetic Algorithm	Yes	No	A: 100; FM: 61–66.67
Budiman *et al.* [18]	SVM	No	No	A:–
Şentaş *et al.* [15]	SVM	Yes	No	A:–
Raikar *et al.* [17]	SVM, NB, Nearest Centroid	Yes	No	A: 91–96

3 Methodology

3.1 Classification Method

SVM takes ground on the static learning theory, which aims to provide requirements to pick a classifier that has a good performance. SVM is a supervised learning machine consisting of training and test phases for the available data groups.

It is capable of recognizing sample patterns of pre-defined classes, of supporting multiclass learning, and of implementing the *one-against-one* approach. In this approach, for k classes, $k(k-1)/2$ classifiers are built. Depending on the number of classes, each classifier is trained as if there were two classes only: the intended one and all others. For the implementation of this work, the *one-against-one* approach was used [20].

The choice of the Kernel function is vital in the learning process and classification with the SVM. This choice can have a meaningful role in the results. For Zhongsheng *et al.* in [21], when we use SVM, and properly choose the kernel functions, better results are reached.

As an example, in the training phase the SVM uses techniques to divide data that are not divided with the linear kernel function use. To determine the separation hyperplane, the *smooth* margin technique allows an error margin of the classification. In SVM's training phase, there is a parameter set by the user that specifies the allowed *smoothness* of this margin.

Some parameters of the SVM method for classification are defined by the user, including the Self C parameter. The C parameter is responsible for the optimization of the SVM, preventing the classification from being done incorrectly. Self C is the main parameter in the SVM, this parameter is responsible for the tolerance and the level of acceptance of error in the classification [22]. The application of SVM for traffic identification requires fine-tuning the algorithm and the adjustment of its parameters for the classification of multiclass traffic. A trade-off must found between the efficiency and the Accuracy of the detection. The proposed method is also applicable to encrypted network traffic.

One of the problems encountered in configuring the classification with the SVM method was the selection of the kernel and its parameter values.

In this work, we explored the usage of four different kernels for SVM: the linear kernel, the sigmoid kernel, the Radial Basis Function (RBF) kernel, and the polynomial (degree = 3) kernel. The higher the C value, the higher the probability to get all training points classified correctly [23]. The main settings for the SVM algorithm are the kernel employed and the error or cost penalty parameter C, which is beneficial in network traffic classification problems as shown in [24]. With respect to the cost variable, we tried several values in the interval $[0.1; \quad 70.0]$. In most implementations of an SVM technique (e.g., in Python), the Self C parameter comes with a default value of 1.0. Table 2 shows the parameters used for the classification.

Figure 1 shows the architecture of the classifier adopted to perform the classification. Raw data were pre-processed, extracting the distribution of the size of the packets by flows, forming a new database. This new base served as input for the SVM method, where 30% of the base sample was used for the training set, generating the training models and the other 70% of the sample was used for the test set. For classification, SVM uses the models generated in the training set together with the test set. After these procedures, we obtained the exit from the classification.

The experiments were executed on a desktop computer running Ubuntu 14.04.5 Operating System and equipped with a 64-bit Intel core i7, 2.93 GHz, 6 GiB of system memory.

For classification with SVM, the sklearn module[1] provided by the scikit-learn python library [25] was used and applied to our data set. The classification was divided into 3 steps, as follows:

- Step 1 - Data treatment - Generation of the new database: For this step, a script *GeraBaseSVMNew.py* was created using the python language, whose objective is

[1] https://scikit-learn.org/stable/modules/generated/sklearn.svm.SVC.html.

Table 2. SVM parameters used for optimizing encrypted multimedia traffic detection.

Parameter	Value
Self C	[0.1; 70.0]
Kernel	'linear', 'sigmoid', 'RBF', 'poly'
Degree	3
Gamma	Auto deprecated
Coef 0	0.0
Shrinking	True
Probability	False
Tol	0.001
Cache size	200
Class weight	None
Verbose	False
Max iter	–1
Decision function shape	ovr
Random state	None

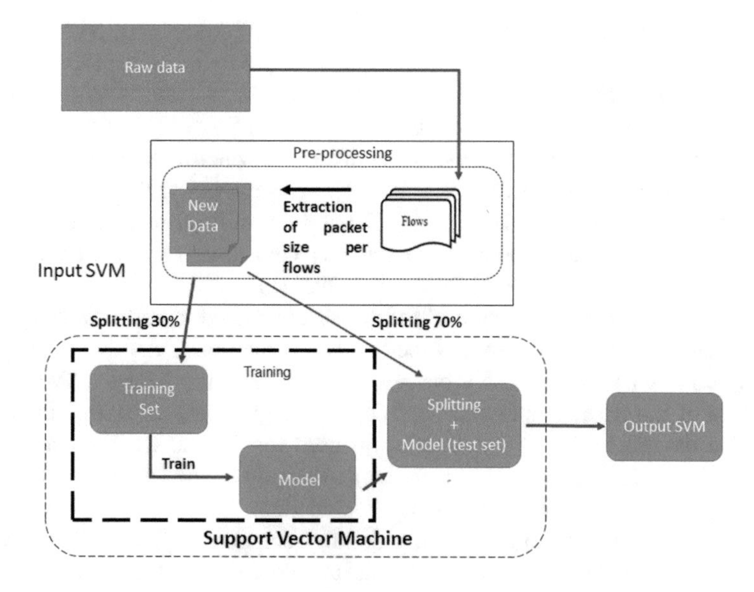

Fig. 1. Architecture of the classifier.

to convert the raw database into a new database, which was used as input in SVM. First, we treat the flows using the tuple [source ip, destination ip, packet size] we extract from the streams the distribution of the relative frequencies of the size of the packets per stream, forming a new database.

We create buckets to calculate the distribution of the relative frequency. The features, were the buckets, where each row, has 100 columns, considered a feature. The conversion of the raw database into a new database of relative frequency was necessary to improve computational performance.
Mapping the classes - The classes were defined based on the IP of each application, for each collective file, formed a Target database with the protocols.

- Step 2 - Training and test phase - The database generated by the script in step 1, was used to generate the models (training phase) and test. To perform the tests, the models were created using the script *SVM_Multiclass.py* [26] also implemented in python, in addition, this script was used to classify and return the classification reports [27].
- Step 3 - Data validation and performance evaluation.

3.2 Dataset and Classification Features

In this research work, we use a dataset which was also described in a previously published work [28]. The data set contains approximately 25 GB of network traffic traces generated by different Internet applications and services, captured using the `tcpdump` tool and stored on disk. Since the flows were previously stored in a database, all tests carried out in this work have used offline classification only.

The data stored and generated by machines dedicated to a specific traffic, allowed us by construction to obtain the ground truth for the classes.

To accomplish step 1, it is necessary to calculate or update the cumulative probability distribution of the size of the packets per each type of flow, so that we can later obtain the values of the relative frequencies by type of flow, as shown in the Table 3. With the amount of data obtained, the calculation of the distribution function was performed as follows:

- 100 buckets were defined for counting the occurrences of packet sizes.
- In each bucket, the number of observed packets having a size falling within the bounds of the bucket will be counted (Observed Frequency f_i).
- Once the observed frequencies are obtained, the relative frequencies are calculated by Eq. (1).

$$fr_i = \frac{f_i}{n}, \tag{1}$$

where n represents the total number of transmissions observed in each "Traffic Class" or "Application/Protocol"; Table 3 shows the distribution of flows. The classes considered for the traffic analysis are commonly used on the Internet, and are briefly presented in Table 4.

Table 3. Definition of the buckets for the distribution of packet sizes.

Bucket	Packet size bounds	Frequency	Relative Frequency
1	0–15	f_1	fr_1
2	16–31	f_2	fr_2
3	32–47	f_3	fr_3
.	.	.	.
.	.	.	.
.	.	.	.
100	1584–1600	$f100$	$fr100$

Table 4. Analyzed traffic flows.

Application/Protocol	Traffic Class	Number of flows
Bittorrent	P2P file-sharing	961
Edonkey	P2P file-sharing	961
Gaming runscape	P2P Video	418
Gaming war of legends	P2P Video	418
Ppstream	P2P Video	419
Sopcast	P2P Video	419
Tvu	P2P Video	418
Http, web browsing, telnet	Others[a]	179

[a]The other classes are those that are not mapped.

4 Evaluation

4.1 Classification Results

After obtaining the results (output) provided by the classifier, the results were validated through the ground truth and evaluated using the confusion matrix, the Recall, Precision and F-measure metrics as defined in [29].

The features used as entrance to our classification were relative frequencies and accumulated frequencies. The results obtained through SVM were compared to the Kolmogorov-Smirnov(KS) and Chi-squared tests [30]. KS was used with the aim to select the distribution that best represents the applications (flows). On the other hand, Chi-squared test [31] was used to compare the relative frequency distribution to the relative frequency of a distribution previously selected that represents a traffic or application class. KS is defined by [30]:

$$D = MAX_x \mid F_{1,n}(x) - F_{2,n'}(x) \mid , \tag{2}$$

where $F_{1,n}$ and $F_{2,n'}$ are the accumulated distributions that were compared and for each variable n, n' were determined, that represents the observation numbers.

Chi-squared is defined by [31]:

$$X^2 = \sum_{i=1}^{k} \frac{(x_i - E_i)^2}{E_i}, \tag{3}$$

where x_i and E_i $(0 \leq i \leq k)$ are respectively the observed and expected frequencies, and $k \in \mathbb{N}$ represents the number of buckets.

The resulting classifications using the SVM classifier with the linear, RBF, sigmoid and polynomial kernels are shown in Figs. 2, 3 and 4. The amount of *support* was 2091 for the P2P video class and 1922 for the P2P file-sharing. The support is the number of occurrences of the class specified in the data set. In the case of our article, it corresponds to the number of items in the class (flows).

We observe that SVM can classify multimedia traffic and that we can optimize the results by adjusting the C parameter, specifically for P2P multimedia traffic. The results demonstrate that there is an impact of the parameter self C on the classification. The factor of that impact for the values self $C = [0.1; 70.0]$, are shown in Figs. 2, 3 and 4 for each SVM kernel.

The results obtained in the linear kernel with C = (0.1, 0.5) were below the values obtained with the default parameter, corresponding to 91% of Precision for the P2P video class with the self C = (0.1) and 89% of Precision for the P2P class file-sharing. For both the P2P video and P2P file-Sharing classes, we obtained the best results with the linear kernel from self C = (30.0), when the classifier reached its highest classification level for both classes, reaching 99% of Precision, 100% Recall, and 98% F-Measure, as shown in Figs. 2, 3 and 4.

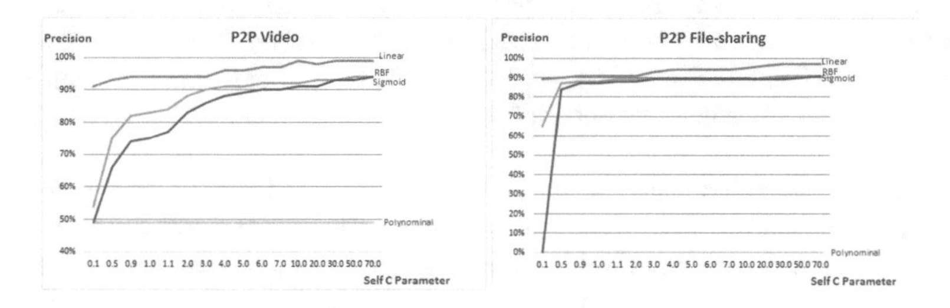

Fig. 2. Precision, as a function of Self C parameter, of SVM-based classification for P2P video and P2P file sharing traffic.

The results obtained with the RBF kernel showed a significant impact when compared to values of self C lower than the default and values greater than the default, mainly for the P2P file-sharing class. For this class, the impact was a 74% improvement in the performance of the F-measure with self C = (30.0), as shown in Fig. 4.

The calculation of F-Measure was important to evaluate the efficiency of the classification, since it represents the value of the harmonic mean between the values found for Recall and Precision. For the P2P video and P2P file-sharing classes, Precision was higher than Recall, indicating that the methodology has greater ability to reduce false positive samples (type II error), than false negative samples (type I error).

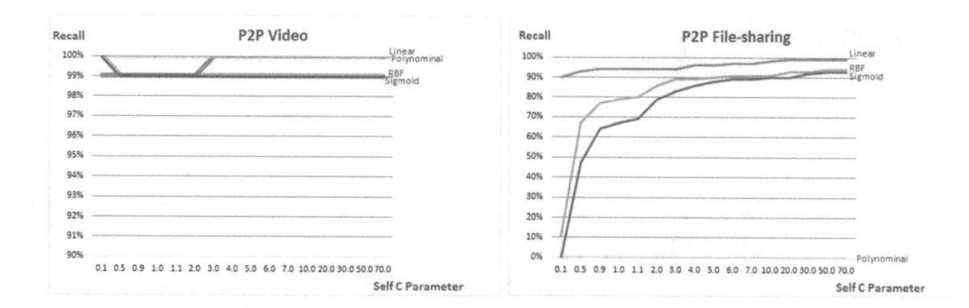

Fig. 3. Recall, as a function of Self C parameter, of SVM-based classification for P2P video and P2P file sharing traffic.

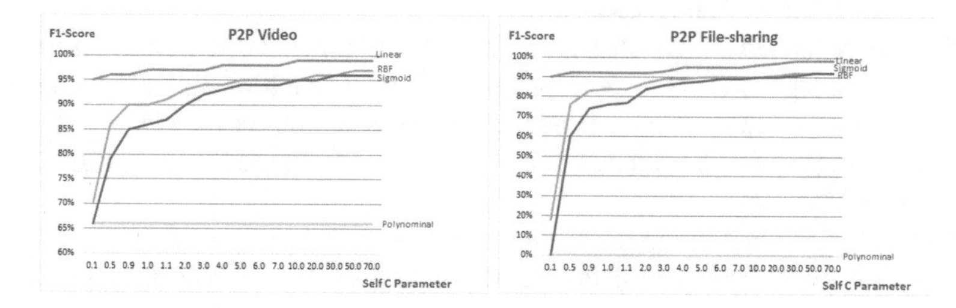

Fig. 4. F1-Score, as a function of Self C parameter, of SVM-based classification for P2P video and P2P file sharing traffic.

Analyzing the impact of self C, in the classification with the sigmoid kernel, we can see that the biggest impact was on the performance of the P2P file-sharing class. With the self C = (0.1) we have a performance so low that it reached 0% of Precision, Recall and F-measure. For self C = (70.0), we reached the highest performance point for the P2P file-sharing class where we obtained 91% of Precision, 93% of Recall and 92% of F-measure.

For the classification with the polynomial kernel, for both the P2P video class and the P2P file-sharing class, there was no impact. The performance for both classes remained the same for all tested self C values. We can conclude that given the analysis of Figs. 2, 3 and 4 and for our test scenario, the self C in the polynomial kernel did not have any impact on the classification performance.

For results with C = 50, it can be seen by the analysis that the P2P video and P2P file-sharing classes achieved 99% and 97% of Precision with the linear kernel and 94% and 91% of Precision, respectively, with the RBF, showing an excellent performance to discriminate how many instances are correctly classified in these classes. However, the linear kernel exhibited higher Precision results for P2P video and a slightly better one for P2P file-sharing. The P2P video classes obtained 100% of Precision with the linear and RBF kernels, and 99% of Recall, which means that both are able with high performance to identify how many of this class are encounters across the number of elements of that class. For the P2P file-sharing class, the linear kernel presented a better result for the Recall, although the result obtained for the RBF kernel is also considered. The method using the polynomial kernel obtained 49% of Precision for the P2P video class. It could not classify the data set with the relative frequencies used in this article. The results achieved for the P2P file sharing class were very low or close to 0, for all values of C in [0.1; 70.0].

Table 5 presents a comparison among classification results obtained with SVM with linear and RBF kernels, KS, and Chi-square tests. In the classification with the KS statistical method, we obtained a Precision of 84% for P2P file-sharing and 100% for P2P Video. For P2P file-sharing and P2P video, we obtained a Recall of 56%. The F-Measure values were 67% for P2P file-sharing and 70% for P2P video. This means that the classification with the statistical method KS had a lower average performance when compared to the classification with the linear kernel associated with a C parameter in the range of 30 to 70, and with the RBF kernel associated with a C parameter in the range of 50 to 70.

Table 5. Summary of results - comparative table of the results obtained with SVM-Linear and RBF in the best range of C parameter, KS and Chi-Square.

Performance	Methods							
	Linear kernel (C = [30–70])		RBF kernel (C = [50–70])		KS		Chi-Square	
	P2P file-sharing	P2P Video	P2P file-sharing	P2P Video	P2P file-sharing	P2P Video	P2P file-sharing	P2P Video
Precision	97%	99%	91%	94%	84%	100%	91%	100%
Recall	99%	100%	94%	99%	56%	56%	80%	74%
F-Measure	98%	99%	92%	97%	67%	70%	85%	85%

In the classification with the Chi-square statistical method, we obtained a Precision of 91% and a Recall of 80% for P2P file-sharing, and a Precision of 100%, and a Recall of 74% for the P2P video. The F-Measure values achieved 85% for P2P file-sharing and P2P video. This means that the Chi-square achieved performance average better than KS, with 15% higher for P2P video and 18% higher for P2P file-sharing. Although these values are better than compared to KS, the statistical method chi-square was low to the mean performance when compared to linear kernel and RBF kernel with the adjusted C parameter. In linear kernel with the parameter C in the range of 30 to 70, we obtained 15% more than the performance average when compared to chi-square. On RBF kernel with C parameter in the range of 50–70, we obtained 7% more than Chi-square for P2P file-sharing and 12% for P2P video.

Our results have shown that the linear kernel leads to the best classification results of P2P video with an F-Measure of 99%, which is achieved for C parameter ranging from 10 to 70. The linear kernel also leads to the best classification results of P2P file-sharing with an F-Measure of 98%, which is achieved for values of C parameter between 30 to 70.

4.2 Computational Performance

We evaluate the computational performance by measuring CPU consumption (in %) and memory consumption (in MB) during the execution time needed to classify the database using psrecord[2]. Figures 5 and 6 show the computational performance of the linear, RBF, sigmoid, and polynomial kernels which presented the most significant results in the classification. During our tests, we have seen that the memory consumption was more significant when compared to the CPU consumption.

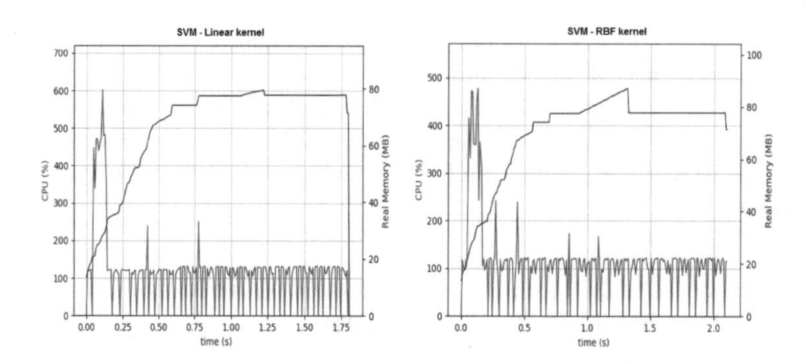

Fig. 5. Computational resource usage in terms of CPU (%) and memory (MB) of linear and RBF kernels.

Analyzing the results, it can be seen that the memory is released by the process at the end of the execution of the linear kernel.

Note that the shortest execution time among the four kernels was obtained for SVM with the linear kernel, with an execution time of 1.75 s.

This does not happen in the RBF kernel at the end of the execution, as we can see in the graph that the process does not release the memory. The execution time of the linear kernel is relatively shorter when compared to the RBF kernel. The CPU usage (in %) is almost the same for both cases.

The computational performance of the sigmoid kernel is lower when compared to the polynomial kernel. The sigmoid kernel has an execution time which is 2 s shorter than the polynomial kernel. However, it has a longer execution time when compared to the linear and RBF kernels. As with the linear kernel, the memory is released as soon as the process is released. CPU consumption is about the same for both kernels. These results show that the linear kernel, in addition to showing better classification results, correctly identifies flows that belong to the class and correctly identifies flows that do not belong to the class and has a lower computational cost.

The computational costs of the classification using the KS and Chi-square methods were higher compared to the linear, RBF, poly, and sigmoid kernels. Memory consumption exceeded 600 MB for both, and execution time achieved 3000 s for the KS method and almost 400 s for the Chi-square method. These execution times were considered high when compared to the ones of the SVM kernels.

[2] https://pypi.org/project/psrecord.

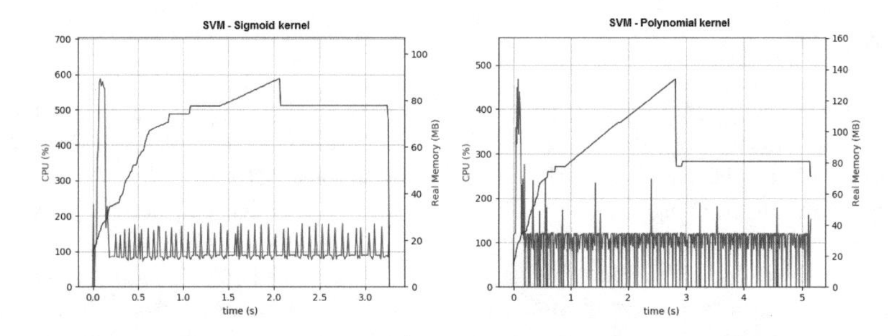

Fig. 6. Computational resource usage in terms of CPU (%) and memory (MB) of sigmoid and polynomial kernels.

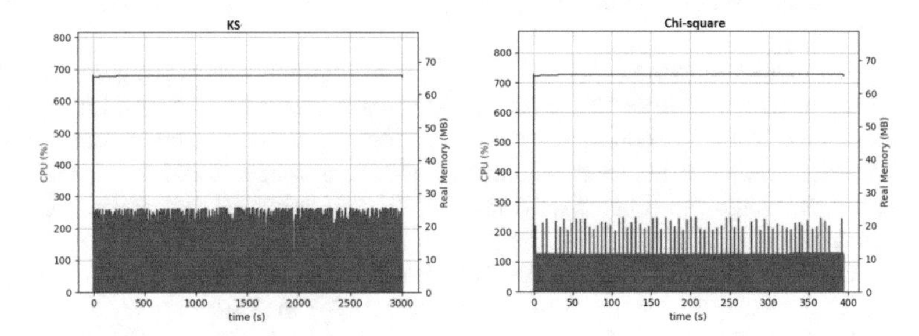

Fig. 7. Computational resource usage in terms of CPU (%) and memory (MB) of KS and Chi-square.

5 Conclusion

SVM classification has shown significantly better results for the linear kernel, RBF, and Sigmoid, when compared to the Polynomial kernel for the data set presented in this paper. These results can be attributed to the fact that SVM considers properties of the multimedia P2P traffic flow, such as the distribution of packets per flow, an important characteristic to differentiate it from the other protocols and classes found in internet traffic. With the adjustment of the self C parameter, SVM has demonstrated a high discrimination capacity for P2P protocols. The more data for the training are entered, the better the classification will be. When we increase the value of self C, we notice that the Precision and Recall values also increase. We can conclude that increasing the values in parameter C reduces type I and II errors and improves the ability to identify flows. The computational cost for the execution of the SVM method was presented taking into account the use of both CPU and memory during the classification. We have observed that over time, the CPU usage remained the same, while the memory usage increased. Our results show that SVM can indeed be a good choice for identifying multimedia P2P traffic on the internet. In comparison with the statistical methods KS and Chi-square, the linear kernel has shown the best F-measure performance for both P2P file-sharing and P2P video results. For future work, we intend to implement new classifiers for the internet traffic based on statistical methods such as distances and divergences and compare them with the ones investigated in this article.

References

1. Sandvine, The global internet phenomena report covid-19 spotlight, 7 May 2020. https://www.sandvine.com/covid-internet-spotlight-report?hsCtaTracking=69c3275d-0a47-4def-b46d-506266477a50%7Cac52173f-34c1-42df-8469-a091e7219e7a
2. Yang, J., Yuan, L., Dong, C., Cheng, G., Ansari, N., Kato, N.: On characterizing peer-to-peer streaming traffic. IEEE J. Select. Areas Commun. **31**(9), 175–188 (2013)
3. Pal, K., Govil, M.C., Ahmed, M., Chawla, T.: A survey on adaptive multimedia streaming. In: Recent Trends in Communication Networks, pp. 185–202, IntechOpen (2019)
4. Rezvani, S., Wang, X., Pourpanah, F.: Intuitionistic fuzzy twin support vector machines. IEEE Trans. Fuzzy Syst. **27**(11), 2140–2151 (2019)
5. Tang, J., Chen, X., Hu, Z., Zong, F., Han, C., Li, L.: Traffic flow prediction based on combination of support vector machine and data denoising schemes. Phys. Stat. Mech. Appl. **534**, 120642 (2019)
6. Syarif, I., Prugel-Bennett, A., Wills, G.: SVM parameter optimization using grid search and genetic algorithm to improve classification performance. Telkomnika **14**(4), 1502 (2016)
7. Mavroforakis, M.E., Theodoridis, S.: A geometric approach to support vector machine (SVM) classification. IEEE Trans. Neural Netw. **17**(3), 671–682 (2006)
8. Aggarwal, R., Singh, N.: A new hybrid approach for network traffic classification using svm and naïve bayes algorithm. Int. J. Comput. Sci. Mobile Comput **6**, 168–174 (2017)
9. Yuan, R., Li, Z., Guan, X., Xu, L.: An SVM-based machine learning method for accurate internet traffic classification. Inf. Syst. Front. **12**(2), 149–156 (2010)
10. Akinyelu, A.A., Ezugwu, A.E.: Nature inspired instance selection techniques for support vector machine speed optimization. IEEE Access **7**, 154581–154599 (2019)
11. Xiao, J.: SVM and KNN ensemble learning for traffic incident detection. Phys. Stat. Mech. Appl. **517**, 29–35 (2019)
12. Sankaranarayanan, S., Mookherji, S.: SVM-based traffic data classification for secured IoT-based road signaling system. Int. J. Intell. Inf. Technol. (IJIIT) **15**(1), 22–50 (2019)
13. Han, W., Xue, J., Yan, H.: Detecting anomalous traffic in the controlled network based on cross entropy and support vector machine. IET Inf. Secur. **13**(2), 109–116 (2019)
14. Luo, C., et al.: Short-term traffic flow prediction based on least square support vector machine with hybrid optimization algorithm. Neural Process. Lett. **50**(3), 2305–2322 (2019)
15. Şentaş, A., et al.: Performance evaluation of support vector machine and convolutional neural network algorithms in real-time vehicle type and color classification. Evol. Intell. **13**(1), 83–91 (2018). https://doi.org/10.1007/s12065-018-0167-z
16. Aamir, M., Zaidi, S.M.A.: Clustering based semi-supervised machine learning for DDOS attack classification. J. King Saud Univ. Comput. Inf. Sci (2019)
17. Raikar, M.M., Meena, S., Mulla, M.M., Shetti, N.S., Karanandi, M.: Data traffic classification in software defined networks (SDN) using supervised-learning. Procedia Comput. Sci. **171**, 2750–2759 (2020)
18. Budiman, F.: SVM-RBF parameters testing optimization using cross validation and grid search to improve multiclass classification. Sci. Vis. **11**(1), 80–90 (2019)
19. Singla, M., Shukla, K.K.: Robust statistics-based support vector machine and its variants: a survey. Neural Comput. Appl. **32**(15), 11173–11194 (2019). https://doi.org/10.1007/s00521-019-04627-6
20. Marwala, T.: Support vector machines. In: Handbook of Machine Learning, Wold Scientific, pp. 97–112 (2018)
21. Zhu, Y., Zheng, Y.: Traffic identification and traffic analysis based on support vector machine. Neural Comput. Appl. **32**(7), 1903–1911 (2019). https://doi.org/10.1007/s00521-019-04493-2

22. Fan, Z., Liu, R.: Investigation of machine learning based network traffic classification. In : 2017 International Symposium on Wireless Communication Systems (ISWCS), pp. 1–6, IEEE (2017)
23. Duan, K.-B., Keerthi, S.S.: Which is the best multiclass SVM method? an empirical study. In: International workshop on multiple classifier systems, pp. 278–285, Springer (2005)
24. Velasco-Mata, J., Fidalgo, E., González-Castro, V., Alegre, E., Blanco-Medina, P.: Botnet detection on TCP traffic using supervised machine learning. In: International Conference on Hybrid Artificial Intelligence Systems, pp. 444–455, Springer (2019)
25. Pedregosa, F., et al.: Scikit-learn: machine learning in python. J. Mach. Learn. Res. **12**, 2825–2830 (2011)
26. Vanice-ufmt. https://github.com/Vanice-ufmt/Codigo, 30 October 2020
27. Reports, C.: 28 April 2020. https://scikit-learn.org/stable/modules/generated/sklearn.metrics.classification_report.html
28. Cunha, V.C., Zavala, A.A., Inácio, P.R., Magoni, D., Freire, M.M.: Classification of encrypted internet traffic using kullback-leibler divergence and euclidean distance. In: International Conference on Advanced Information Networking and Applications, pp. 883–897, Springer (2020)
29. Han, J., Pei, J., Kamber, M.: Data Mining: Concepts and Techniques. Elsevier (2011)
30. Neto, M., Gomes, J.V., Freire, M.M., Inácio, P.R.: Real-time traffic classification based on statistical tests for matching signatures with packet length distributions. In: 2013 19th IEEE Workshop on Local & Metropolitan Area Networks (LANMAN), pp. 1–6. IEEE (2013)
31. Pandis, N.: The chi-square test. Am. J. Orthod. Dentofac. Orthop. **150**(5), 898–899 (2016)

Machine Learning-Based Communication Collision Prediction and Avoidance for Mobile Networks

Khaled Abid(✉), Hicham Lakhlef, and Abdelmadjid Bouabdallah

Sorbonne University, University of Technology of Compiègne,
57 Avenue de Landshut, 60200 Compiègne, France
{khaled.abid,hicham.Lakhlef,bouabdal}@hds.utc.fr

Abstract. Communication collision between devices represents a critical issue in wireless networks. It can cause network disruption, packet loss, communication delay and energy wastage. In the case of mobile networks, collision occurs frequently because of nodes' mobility. Existing works rely on resolving collision after it happens or use GPS coordinates to predict neighborhood status in the future, and both methods cause collision. Those solutions cause packet loss and/or energy wastage.

In this paper, we introduce a GPS-free Machine Learning-based models to predict and avoid collision in mobile networks and resolve it before it occurs. Our models can be implemented in almost every mobile device, since they use only neighboring information to predict collision on top of that a GPS module is not required. The results demonstrate a promising potential of using Artificial Intelligence via Machine Learning modeling as a novel approach to avoid communication collision by using only neighboring information. Our models can achieve at least 70% accuracy for traffic in vehicular network in two-way highway, four-way intersection and four-way roundabout. Furthermore, they can avoid 80% of collisions in two-way highway and four-way intersection, as well as 65% in four-way roundabout. Which means we can decrease packet loss caused by collision by at least 65%, hence improve energy consumption.

1 Introduction

In recent years, Internet of Things "IoT" has gained unprecedented popularity, reaching billions of users [1] with a lot of them connected to the network wirelessly. Nevertheless, wireless communication encounters the problem of limited communication resources. Precisely, two or more nodes must not transmit data simultaneously when they are in each other's vicinity, since they cause conflict and collision [2]. A collision occurs when a node receives messages from two or more neighbors at the same time. Moreover, a conflict occurs when two neighbors send a message to each other at the same time. Researchers have proposed several solutions [3–5] to deliver efficient communication in wireless networks. Among them, contention-based protocols [6], where nodes compete to access the communication channel, prove to give poor Quality of Service "QoS" for dense networks and heavy loads. On the other side, contention-free

L. Barolli et al. (Eds.): AINA 2022, LNNS 449, pp. 194–204, 2022.
https://doi.org/10.1007/978-3-030-99584-3_17

protocols [7] such as Time Division Multiple Access (TDMA) [8] are better suited for wireless networks, especially under peak load conditions. Furthermore, they provide guaranteed end-to-end delay performance in multi-hop wireless communication.

TDMA consists of a contention-free channel access mechanism in which TDMA time-frames are divided into time-slots. In order to ensure collision-free transmission, a time-slot assignment algorithm should guarantee that the same time-slot is not assigned to nodes in 2-hop neighborhood. For mobile networks, even if the time-slot assignment algorithm provides a collision-free schedule, mobile nodes induce time-slot collision because of their movement into each other's vicinity. That collision causes packet loss, communication delay and disruption within the network, as well as energy loss. Therefore, generating an adaptive conflict-free schedule in a TDMA-based MAC protocol for networks where nodes are in movement is a challenging problem. Most of the existing works have developed MAC protocols that resolve collision after it happens, which may cause packet loss and instantaneous network disruption, as well as energy wastage. Since most collisions occur due to nodes' movement, the collision rate can be reduced if each node can predict collisions in its neighborhood before it occurs and advertise collided nodes to resolve it.

Some researchers have proposed GPS-based protocols to avoid collisions. However, in the case of heterogeneous networks, it is not obvious that every node is equipped with GPS module. As well, GPS module consumes a lot of energy, so many users can switch it off for energy saving. Heterogeneity of devices and existing obstacles in the field (e.g. buildings, tunnel) makes it difficult for classical approaches (which use GPS) to predict collisions either for the fact that GPS coordinates are not available or because of the existence of precision leak.

Machine learning (ML), as a part of artificial intelligence (AI), has been successfully applied in a variety of communication fields to improve the QoS and the efficiency of resource management in the network. In this context, it is not trivial for wireless devices to correctly recognize collision without position coordinates information. That's why, developing reliable ML model to classify collision between nodes is very important to improve the QoS and to save energy.

In this paper, we propose a novel methodology for collision avoidance in TDMA-based mobile networks. Every node uses the proposed method to predict and avoid collision before it happens by using exchanged neighboring information. We employ different ML classifiers to model the neighboring relationship between nodes by using extracted features from neighboring-related information for the purpose of avoiding collisions. For classification, we exploit a large dataset in order to evaluate the performance of our models and check whether a collision will occur in the future or not. We tested the developed algorithms on vehicular networks. Our results show that it is possible to obtain good performance, reaching an accuracy of more than 70% and avoid more than 65% of collisions for traffic in two-way highway, four-way intersection and four-way roundabout. Moreover, our solution is based solely on the historic neighboring interactions between nodes, allowing the system to predict collision independently of GPS coordinates. Indeed, our system does not need any feature coming from GPS module which makes it robust and energetically efficient. The proposed solution was proved to improve communication delay, Packet Loss Ratio (PLR) as well as energy consumption in the network.

2 Related Work

Researchers have conducted many works to improve the network communication performance. In [9], the authors have proposed a collision discovery mechanism which is not based on topology information exchange between unmanned aerial vehicles. In this mechanism, each time-slot is divided into three sub-slots, where each one of them have a different function: broadcasters broadcast in sub-slot 1, obtain feedback about the existence of collision in sub-slot 2 and obtain a report of the time-slot occupancy in sub-slot 3. This mechanism has a bandwidth efficiency issue, as it necessitates time for collision discovery for each time-frame. Add to that, the collision between nodes engenders high delay and packet loss.

In [10], the authors have proposed a procedure to detect and resolve slot collision in Mobile Ad-hoc Networks, "MANETs". In the proposed work, every node allocates a control and a data slot. Through control information, each node knows about occupied slots of its 1-hop neighbors. If a node finds out that two neighboring nodes are using the same data slot (collision), it informs them to resolve it. The chosen method engenders high communication delay and packet loss, as the two neighboring nodes cannot change their chosen slot until they receive the collision alert.

To eliminate collision in the network, some works have proposed prediction-based collision avoidance protocols. In [11], the authors have proposed Prediction-Based TDMA MAC protocol "PTMAC" for Vehicular Ad-hoc Networks "VANETs" designed for two-way traffic and four-way intersections. In this protocol, most of the encountered collisions can be predicted and, potentially, eliminated before they really happen in both two-way traffic and four-way intersections, regardless of the traffic loads on different road segments. Using exchanged location, speed, moving direction and slot occupation information, intermediate vehicles between 3-hop neighbors can predict potential collision and inform the closest one to change its occupied slot.

In [12], the authors have proposed a collision avoidance method based on vector-based mobility model in TDMA-based VANET. The proposed protocol aims to avoid access and merge collisions by predicting the mobility of nearby vehicles which use exchanged control information. Every node exploits exchanged information (direction, longitude, latitude, speed and acceleration) from more than 2-hop neighbors to recognize the existence of possible collision in the future by using a vector-based mobility model.

The proposed protocols are applied to VANET networks in which vehicles have a known mobility model, which is not the case for mobile networks in general where node's mobility models are variant, such as in smart cities, smart agriculture and military applications. Furthermore, these protocols necessitate the implementation of a GPS module in every node to extract position-related information, which is not feasible in many applications. Moreover, a GPS module engenders high deployment cost and a significant energy consumption for communication with satellites.

In this paper, we propose a GPS-free ML-based protocol that exploits node's previous interactions with neighbors to predict possible collision between 3-hop neighbors via an intermediate node. This method is cost-effective as it increases the network performance and saves energy.

3 Proposed Solution

In this section, we explain our proposed solution by detailing the system architecture, data extraction and model implementation.

3.1 System Architecture

Considering a network composed of mobile nodes which need to communicate important information such as security-related information between each others. The wireless medium is shared between the nodes using the TDMA channel access method. Time is divided into time-frames, and every time-frame is composed of N time-slots. In the network entry, every node selects one free time-slot for broadcasting and keeps listening in other time-slots. We adopt, in this paper, the range propagation loss model for radio propagation, so messages can only be received by nodes in the range of m meters. One node can face collision either with primary neighbors, since two adjacent nodes cannot transmit and receive in the same time, or with secondary neighbors, since one node cannot receive packets from two or more adjacent nodes in the same time.

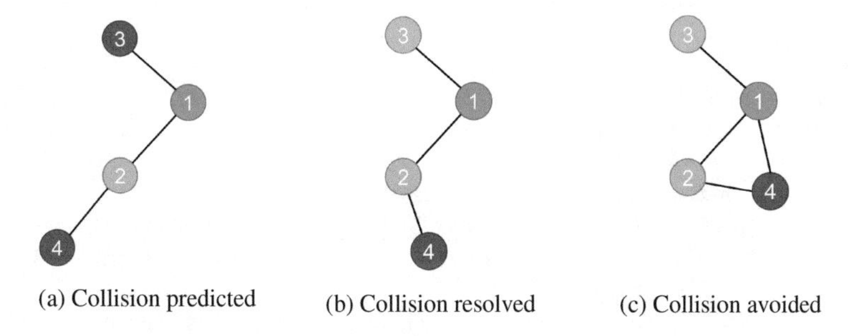

(a) Collision predicted (b) Collision resolved (c) Collision avoided

Fig. 1. Collision prediction & avoidance: Node 4 is on movement. In Fig. (1a) node 1 process exchanged control information, predicts a collision and sends an alert to node 3. In Fig. (1b), node 3 receives the alert from node 1 and resolves collision. In Fig. (1c), when node 4 becomes a 2-hop neighbor of node 3 (as predicted by node 1), the collision is already resolved.

Our goal is to identify possible collision in the near future between 3-hop neighbors via an intermediate node using only the history of interactions of every node in the 2-hop neighborhood. In other words, if a 1-hop and a 2-hop neighbor of an intermediate node share the same time-slot, a collision will occur if the 2-hop neighbor becomes a 1-hop one. Therefore, predicting the neighborhood status will guide us to collision prediction. For every 3-hop neighbor sharing the same time-slot in the network, we want from an intermediate node to classify them if they will be 2-hop neighbors in the future or not. We consider that during one time-frame, 3-hop neighbors cannot be 1-hop neighbors as node's speed does not exceed a predefined limit. In the case of a probable collision, the intermediate node alerts the conflicting node to resolve the collision before it happens by switching to a free time-slot from the time-slot occupancy table "SOT".

An example is shown in Fig. 1. In this figure, colors represent the allotted time-slots. Node 2 broadcasts the control information about itself and its 1-hop neighbors (node 4). Since node 4 and 3 occupy the same time-slot (Fig. 1a), node 1 computes received information from its neighbors and notice that those nodes have a high probability to become 2-hop neighbors in the near future (collision). Therefore, it sends an alert to node 3 containing information about the collision resolution process. Finally, node 3 changes its time-slot selection to avoid a merging collision (Fig. 1b).

Algorithm 1 explains in details the process of data transmission, information processing and collision avoidance.

Algorithm 1. Collision avoidance & resolution

```
Require: N: Set of time-slots
Require: Node_id: Id of the node
  Node_id.Time_Slot ← select_time_slot()        ▷ Select free time-slot (network entry)
  while True do
      receive(packet)
      process(packet)
      if packet.alert_of_collision(Node_id) then   ▷ If the the packet contains a collision alert
          Node.id ← collision_resolution()          ▷ Collision resolution
      end if
      SOT.update()
      if Current_time_slot == Node_id.Time_Slot then   ▷ Own transmission time-slot
          Packet = construct_packet()                   ▷ Packet construction
          for i in Node_id.neighbors do
              for j in Node_id.two_hop_neighbors do
                  if i.Time_Slot == j.Time_Slot then   ▷ If 3-hop neighbors share a time-slot
                      flag ← Predict_collision()        ▷ flag = 1 if predicted collision
                      if flag then
                          packet.add_alert_of_collision(i)   ▷ Add an alert to the packet
                          flag ← 0
                      end if
                  end if
              end for
          end for
          send(packet)                                  ▷ Send the packet
      end if
  end while
```

3.2 Data

To achieve collision avoidance, the nodes must share control data between each others. Control data is acquired by using packet interactions between the nodes. If one node receives a packet from another node during a time-frame, it marks the latter as neighbor in that time-frame. Control information must reach 2-hop neighbors (considered as intermediate nodes). So, every node forwards received control data to its neighbors. The packet contains the following information:

- Slot occupancy information
- Sets of 1-hop and 2-hop neighbors
- Set of interactions with 1-hop neighbors during the last f time-frames

For the initial step of control data processing, every node exploits received control data from its 2-hop neighbors to extract valuable features which will be useful to predict collisions.
The set of features is as follows:

- *was_neighbor*: if the target node was a neighbor in the past f time-frames.
- *density_difference*: the difference between neighborhood density of the intermediate and target node. The neighborhood density is a function of the number of 1-hop and 2-hop neighbors.
- *neighboring_relationship*: a function of the number of common 1-hop and 2-hop neighbors.
- *nature_of_movement*: if the node follows a group of nodes or moving alone. The nature of movement is a function of the number of interactions with neighbors in the past f time-frames.
- *staying_time*: the average time the node spent with its neighbors in the past f time-frames.
- *common_neighbor_evolution*: the evolution of number of common neighbors in the past f time-frames.

Note that it is hard to do automatic collection of data, because it is not possible to establish communication between thousands of nodes in a real life application. Therefore, we rely on a traffic simulation dataset generated with SUMO simulator [13]. Data construction is done in three stages, as shown in Fig. 2. First, we extract nodes' coordinates over time from SUMO (Fig. 2a). Then, for every time-step, we construct a neighborhood status dataset containing the list of 1-hop and 2-hop neighbors of every node in each time-step based on the radio coverage radius (Fig. 2b). Finally, for every 2-hop neighbor nodes, we extract the desired features from the neighborhood status data-set (Fig. 2c). The final dataset is not related to GPS-coordinates of the nodes, time and nodes' identity. Out of the entire data-sets, we use 70% for training and 30% for testing.

3.3 Implementation

In this subsection, we give details about the implemented algorithms as well as the performance metrics.

3.3.1 Setting of Experiment

Statistical packages "scikit-learn" and "keras" in Python were used to implement the Extreme Gradient Boosting Algorithm "XGBoost", Artificial Neural Network "ANN" and Support Vector Machines "SVM" models for classification.

	time	id	x	y
13643	293	20	4120.37	2359.47
13722	294	20	4115.52	2356.39
13801	295	20	4112.71	2352.45
13880	296	20	4111.98	2347.23
13958	297	20	4114.84	2341.33
14036	298	20	4118.18	2336.12
14115	299	20	4119.51	2329.93
14194	300	20	4117.77	2323.61
14272	301	20	4114.90	2317.82

(a) Original dataset

	time	id	list_neighbors	list_2_neighbors
77	293	20	[14, 17, 6]	[1036, 1629, 1028, 1620]
78	294	20	[14, 17, 6, 1036]	[1639, 1629, 1620, 1028]
79	295	20	[14, 17, 1028, 1036]	[1629, 1620, 6]
80	296	20	[14, 17, 1028, 1031, 1036]	[1617, 1629, 1046, 1626, 1620, 6, 1610]
81	297	20	[14, 17, 1028, 1031, 1036]	[1617, 1629, 1046, 1620, 6, 1626]
82	298	20	[14, 17, 1028, 1031, 1036]	[1639, 1629, 1620, 6]
83	299	20	[14, 1028, 1031, 1036]	[1629, 17, 1620, 6]
84	300	20	[14, 1028, 1031, 1036]	[1629, 1624, 1620, 6, 17]
85	301	20	[1028, 1031, 1036]	[17, 1620, 1629]

(b) Neighborhood status dataset

	was_neighbor	density_difference	neighboring_relationship	nature_of_movement	staying_time	common_neighbor_ev	next_state
4000	0	1.7000	0.111111	0.068966	1.0	0.0	0
4001	1	1.0625	0.111111	-0.034483	1.0	0.0	0
4002	0	1.0625	0.111111	-0.103448	0.0	-1.0	0
4003	0	1.2500	0.138889	0.137931	0.0	1.0	0
4004	0	1.0000	0.055556	0.034483	1.0	0.0	0
4005	0	2.2000	0.055556	0.206897	1.0	-1.0	0
4006	0	1.2500	0.138889	0.068966	0.0	1.0	1
4007	0	0.5625	0.055556	-0.034483	1.0	-1.0	0
4008	0	0.5625	0.055556	-0.034483	1.0	-1.0	0
4009	0	0.8750	0.055556	0.068966	1.0	0.0	0

(c) Training dataset

Fig. 2. Data-set construction process.

- XGBoost [14]: a scalable machine learning system for tree boosting that can be used for regression and classification problems. It generates a weak learner at each step and accumulates it into the total model. In this paper, we implemented the model with default values.
- ANN [15]: comprised of a node layers, containing an input layer, hidden layers, and an output layer. Each node, or artificial neuron, connects to another and has an associated weight and threshold. In this work, we created a fully-connected network structure with three layers. The first and second hidden layers have 4 nodes and uses the "*relu*" activation function. The output layer has one node and uses the "*sigmoid*" activation function. We defined the optimizer as the efficient stochastic gradient descent algorithm "adam".
- SVM [16]: also known as a discriminative classifier. It separates data points using a hyperplane with the largest amount of margin and finds an optimal hyperplane which helps in classifying new data points. We implemented the model with default values.

In the training phase, each model was evaluated by 10-fold cross-validation.

3.3.2 Evaluation of Model

The classification problem is binary (two classes); positive (collision) and negative (no collision). The prediction of the classifiers can be one of the following four states:

- *True Positive "TP"*: for correctly predicted collision
- *False Positive "FP"*: for incorrectly predicted collision
- *True Negative "TN"*: for correctly predicted no collision
- *False Negative "FN"*: for incorrectly predicted no collision

The performance of collision detection models is evaluated by four widely used measures: accuracy, precision, recall and F1-score. The definitions of these measures are based on a confusion matrix, which counts how many positive and negative in our samples that are correctly and incorrectly predicted.

- $Accuracy = \frac{TP+TN}{TP+TN+FP+FN}$: is a fraction of the correct prediction. From the two classes (positive and negative), how many of them we have predicted correctly.
- $Precision = \frac{TP}{TP+FP}$: fraction of positive predictions which are true positive. That means, from all the classes we have predicted as positive, how many are actually positive.
- $Recall = \frac{TP}{TP+FN}$: fraction of positive instances which are correctly predicted as positive. It means, from all the positive classes, how many we predicted correctly.
- $F1-score = \frac{2 \times precision \times recall}{precision+recall}$: the harmonic mean of precision and recall.

Precision can be seen as a measure of quality and recall as a measure of quantity. Higher precision means that an algorithm returns more relevant results, and high recall means that the algorithm returns most of the relevant results.

In this context, we would expect a higher recall rate in collision detection. Therefore, it is better, in many applications, to enable our classifier to return more relevant results than irrelevant ones. In other words, it will identify as many collisions (TP) as possible rather than missing some of them (FN).

We also expect good precision, which means that an algorithm returns most of the relevant results than irrelevant ones. That is to say, the algorithm will generate less unwanted alert messages and collision resolution process, which will consume bandwidth and energy.

Eventually, we will concentrate on precision and recall as the main performance metrics to evaluate our models. Unfortunately, it is not possible to maximize both these metrics at the same time, as one comes at the cost of another. Other metrics will be provided to have a full understanding of the performance of the models.

4 Evaluation Results

To evaluate the collision detection performance of the proposed models, a series of experiments were conducted on the described dataset in Subsect. 3.2. The experiments were implemented in *Jupyter notebook* with *Python* programming language. The bar charts in Fig. 3 provide a comparison between different ML models for classifications in different types of architectures.

We can observe that at least 70% accuracy was achieved by all the models in all architectures, which reflects the fairness of our models. We notice that precision is relatively low. This scenario results from the imbalance of class distribution in the dataset, as it is often the case that there are fewer collisions than the contrary.

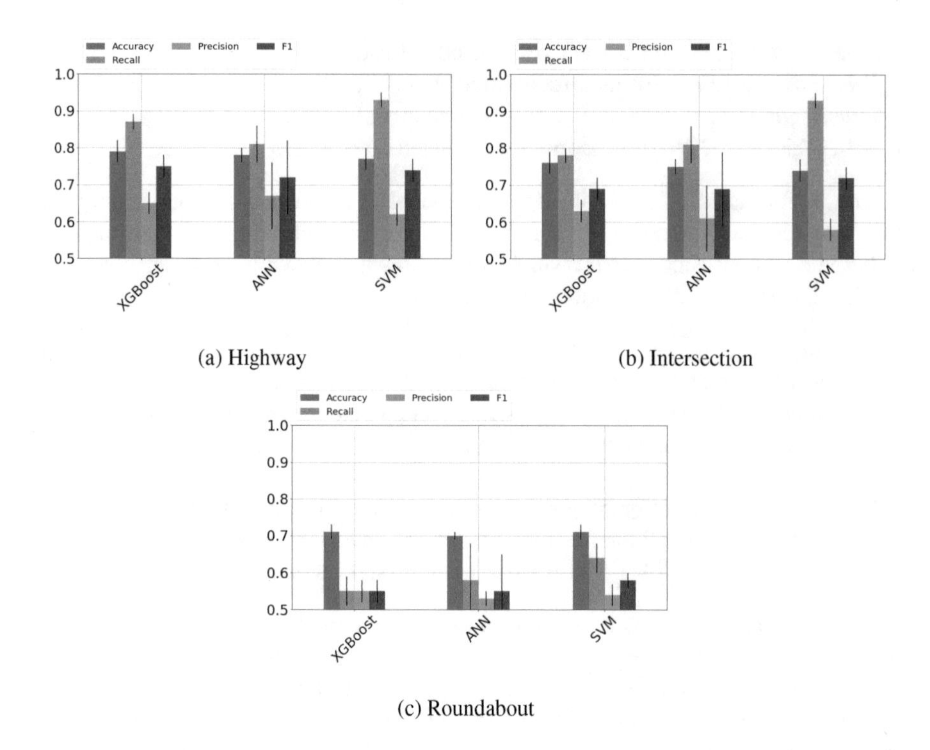

(a) Highway (b) Intersection

(c) Roundabout

Fig. 3. Performance metrics comparison between ML models for different architectures.

It can be seen that recall values of SVM are all better than the other two models for every architecture, which means it is able to detect most of the collisions in the network, hence the least energy wastage due to collision. However, precision values are better for XGBoost. That means that XGBoost causes the least false collision detection.

The choice of the best model strongly depends on the targeted application. For example, if the collision prediction model will be applied to a safety-related application, recall must be the most important performance metric. However, for applications with energy restrictions, a joint function between precision and recall (e.g. F1-score) will be the criterion to choose the best model.

More precisely, for vehicular networks, safety-related information for physical collision avoidance must be delivered in real-time manner. Each FN where a collision is not avoided will cost the network packet loss and delay. So, the better model for this application will be SVM since it gives a high ratio of TP to FN. However, for smart agriculture applications, nodes' life-time is more significant than delay and packet loss ratio. So, the model with the highest F1-score (XGBoost) will be the best for this application.

To further visualize the performance of our models, we can take a look at the confusion matrix of the different models for highway architecture. The testing data is composed of 5266 samples, of which 1896 engendered collision and 3370 did not (Fig. 4).

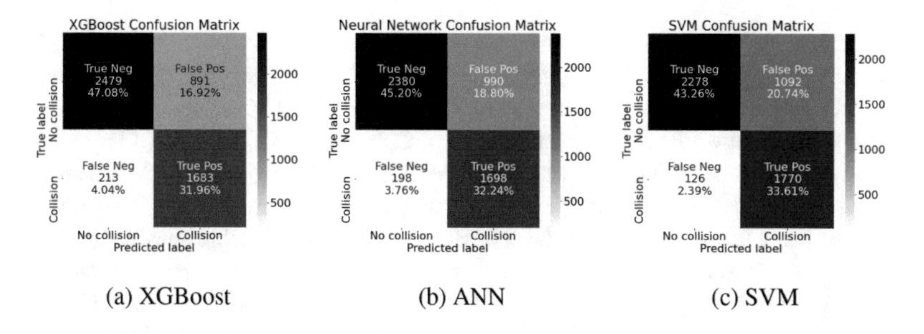

(a) XGBoost (b) ANN (c) SVM

Fig. 4. Confusion matrix of different models for highway architecture.

We can see that SVM was able to detect most of the collisions (1770) compared to XGBoost and ANN and misses only 126 ones, hence high recall. Moreover, it generates 1092 "false alarms" which is the highest compared to other models, hence its low precision.

5 Conclusion

Collision is the primary cause of delay and packet loss in wireless networks, and its prediction and resolution before it happens can improve communication quality. Therefore, the accurate prediction of collision can help to decrease PLR as well as energy consumption. In this paper we presented an initial work towards a collision prediction model designed to detect possible collision in a mobile network in some predefined architecture. The experimental results demonstrate that the collision prediction models have the characteristics of high accuracy and precision.

In the future, we will extend our research to the following two aspects: on one hand, the development of ML model capable of avoiding collisions in general architectures by searching for more comprehensive input characteristics to characterize movement direction and speed of the node. In the other hand, we could explore the impact of different speed and network density on predicting effect through the comparison of prediction performance between multi-speed and density levels.

Acknowledgements. This work was carried out and funded in the framework of the Labex MS2T. It was supported by the French Government, through the program "Investments for the future" managed by the National Agency for Research (Reference ANR-11-IDEX-0004-02).

References

1. Sinha, S.: State of iot 2021: Number of connected iot devices growing 9% to 12.3 billion globally, cellular iot now surpassing 2 billion. https://iot-analytics.com/number-connected-iot-devices/ (2021)
2. Amouris, K.: Space-time division multiple access (STDMA) and coordinated, power-aware MACA for mobile ad hoc networks. In: GLOBECOM 2001. IEEE Global Telecommunications Conference (Cat. No.01CH37270), vol. 5, pp. 2890–2895 (2001)

3. Rajandekar, A., Sikdar, B.: A survey of mac layer issues and protocols for machine-to-machine communications. IEEE Internet Things J. **2**(2), 175–186 (2015)
4. Ali, A., Huiqiang, W., Hongwu, L., Chen, X.: A survey of MAC protocols design strategies and techniques in wireless Ad Hoc networks. J. Commun. **9**(1), 30–38 (2014)
5. Huang, P., Xiao, L., Soltani, S., Mutka, M.W., Xi, N.: The evolution of MAC protocols in wireless sensor networks: a survey. IEEE Commun. Surv. Tutorials **15**(1), 101–120 (2013)
6. Doudou, M., Djenouri, D., Badache, N., Bouabdallah, A.: Synchronous contention-based MAC protocols for delay-sensitive wireless sensor networks: a review and taxonomy. J. Netw. Comput. Appl. **38**(1), 172–184 (2014). https://doi.org/10.1016/j.jnca.2013.03.012
7. Abid, K., Lakhlef, H., Bouabdallah, A.: A survey on recent contention-free mac protocols for static and mobile wireless decentralized networks in IOT. Comput. Netw. **201**, 108583 (2021). https://www.sciencedirect.com/science/article/pii/S1389128621004886
8. Xuelin, C., Zuxun, S.: An overview of slot assignment (SA) for TDMA. In: 2015 IEEE International Conference on Signal Processing, Communications and Computing (ICSPCC), pp. 1–5 (2015)
9. Yao, K., et al.: Self-organizing slot access for neighboring cooperation in UAV swarms. IEEE Trans. Wire. Commun. **19**(4), 2800–2812 (2020)
10. Sindhwal, H., Dasari, M., Vattikuti, N.: Slot conflict resolution in tdma based mobile ad hoc networks. In: Annual IEEE India Conference (INDICON), vol. 2015, pp. 1–6 (2015)
11. Jiang, X., Du, D.H.C.: PTMAC: a prediction-based TDMA mac protocol for reducing packet collisions in VANET. IEEE Trans. Veh. Technol. **65**(11), 9209–9223 (2016)
12. Bang, J.-H., Lee, J.-R.: Collision avoidance method using vector-based mobility model in TDMA-based vehicular ad hoc networks. Appl. Sci. **10**(12) (2020). https://www.mdpi.com/2076-3417/10/12/4181
13. Lopez, P.A., et al.: Microscopic traffic simulation using sumo. In: The 21st IEEE International Conference on Intelligent Transportation Systems. IEEE, pp. 2575–2582, November 2018. https://elib.dlr.de/127994/
14. Friedman, J.H.: Stochastic gradient boosting. Comput. Stat. Data Analy. **38**(4), 367–378 (2002). nonlinear Methods and Data Mining. https://www.sciencedirect.com/science/article/pii/S0167947301000652
15. Hornik, K., Stinchcombe, M., White, H.: Multilayer feedforward networks are universal approximators. Neural Netw. **2**(5), 359–366 (1989). https://www.sciencedirect.com/science/article/pii/0893608089900208
16. Cristianini, N., Shawe-Taylor, J., et al.: An introduction to support vector machines and other kernel-based learning methods. Cambridge University Press (2000)

Automatic Monitoring System for Security Using IoT Devices and Smart Contracts

Kotono Iwata[1(✉)] and Kazumasa Omote[1,2]

[1] University of Tsukuba, Tennoudai 1-1-1, Tsukuba 305-8573, Japan
s2020525@s.tsukuba.ac.jp, omote@risk.tsukuba.ac.jp

[2] National Institute of Information and Communications Technology, 4-2-1 Nukui-Kitamachi, Koganei, Tokyo 184-8795, Japan

Abstract. With the recent technological advances in the design of IoT devices and their widespread use, it has become increasingly common for alarm services that provide security for facilities to use IoT devices. However, the conventional centralized management of such IoT devices involves risks, such as data falsification that is vulnerable to cyberattacks, as well as problems pertaining to the data volume, integrity, and reliability of the IoT devices themselves. Therefore, managing the logs of a large number of IoT devices is difficult. For this reason, blockchain is often used in forensics and other applications. However, to the best of our knowledge, there has been no research to-date that can automatically determine the actual level of anomalies from the logs of IoT devices using smart contracts. In this study, we propose a method to securely record the logs of IoT devices in a blockchain and use those logs to automatically detect anomalies. We also propose a method to estimate the degree of anomalies based on the logs using smart contracts and automate the sequence of events performed by the security companies.

1 Introduction

In recent years, IoT devices have become increasingly popular and widespread. According to the International Data Corporation (IDC), by 2025 there will be 55.7 billion connected devices worldwide, 75% of which will be connected to IoT platforms [1]. In addition, the global security services market (e.g., security systems, private security guards, and security consulting services) has expanded, growing approximately 1.7 times in the 10 years from 2011 to 2020 [2]. With the proliferation of IoT devices and recent advances in IoT-relevant technologies, many security companies are now offering alarm services [3]. The alarm service is a form of security service in which sensors are placed in a facility to monitor for abnormalities, and when the log of such an abnormality is received from the sensors, security guards are dispatched to take initial action. Alarm services have expanded over the years to include a broader range of services, including temperature, fire, and medical alert monitoring, in addition to traditional burglary monitoring [3].

L. Barolli et al. (Eds.): AINA 2022, LNNS 449, pp. 205–216, 2022.
https://doi.org/10.1007/978-3-030-99584-3_18

IoT devices are often installed in large facilities, such as factories and entertainment venues. The activation of these IoT devices can provide immediate notification of abnormalities occurring in the facility, which can be effective in protecting assets and providing immediate deterrence [3]. The logs of these IoT devices also provide important evidence that can be used to identify the causes of the abnormal situation in question [4].

Typically, in an alarm service, the logs of the IoT devices are sent to the control center of the security company. Based on these logs, the security guards monitoring the alarms provide instructions and dispatch additional guards to the target facility. The logs of the IoT devices are managed by a server. Therefore, it is necessary to have a system that can manage the logs securely and is difficult to be tampered with or damaged.

Centralized management, such as a server, has several risks as it is more susceptible to cyber-attacks [5]. Therefore, the logs can become unreliable as evidence and may hinder the tracking of anomalies. Furthermore, corrupted data are difficult to recover. Therefore, it is difficult for a server to securely manage the logs of a large number of IoT devices. For this reason, blockchain technology [6] is often used for forensics and other applications which require stability and security [7–9]. The logs are stored in a blockchain to ensure integrity and traceability, and can be used as an information sharing platform. In addition, several studies have been proposed for anomaly detection using the blockchain [10–12]. However, to the best of our knowledge, there has been no research to-date that can automatically determine the actual level of anomalies from the logs of IoT devices using smart contracts.

In this study, we propose a blockchain-based system that manages the logs of IoT devices and automatically detects anomalies. Specifically, we propose storing the logs of IoT devices in the blockchain to maintain the integrity of the data and to automatically detect anomalies using the logs. In addition, the proposed method can be used to predict the possible failure of IoT devices for maintenance. Furthermore, we verify the operation and the time required for the activation of the IoT device to the start of the initial response by the security company, and demonstrate the feasibility of using blockchain. The contributions of this system are as follows:

- We propose a new method for automatically estimating the degree of anomalies based on logs using smart contracts.
- Using smart contracts, the work of judging the situation and giving instructions by guards of alarm service, which is done in the conventional method, can be done automatically, saving time and effort.

2 Predicting the Failure of IoT Devices

As the market for alarm service that uses IoT devices expands, the security companies will have to perform machine maintenance on a huge number of IoT devices. Machine maintenance can be divided into two types: corrective maintenance (CM), wherein maintenance is performed after a machine is damaged or malfunctions, and preventive maintenance (PM), where maintenance is performed in advance to prevent problems from occurring [13]. In addition, PM can be further divided into Time Based Maintenance (TBM), in which maintenance is performed periodically, and Condition Based Maintenance (CBM), in which maintenance is performed based on the condition of the machine [14]. The regular inspections that security companies generally perform are TBM and CM. However, they can only detect abnormalities and failures in the IoT devices when they are performing these activities. Therefore, in addition to the two conventional types of maintenance, it is necessary to perform predictive maintenance (PdM) that integrates the IoT technology and CBM. PdM is a method of performing maintenance by acquiring data from machines and constantly monitoring and predicting their status [15]. Therefore, based on the logs of IoT devices, we can predict the failure of IoT devices. In this study, we focus on the abnormal operation of IoT devices. We also predict failures based on the threshold values of the sensors. Typically, the sensors of IoT devices are designed to activate when they detect a sensor value that exceeds the threshold with a set limit. A temperature sensor installed in a facility is a simple and practical example. The threshold of abnormality in the facility is set to 30 °C or higher, and the threshold of failure is set to 50 °C or higher. If the temperature sensor suddenly detects a sensor value of 60 °C, it is highly probable that it may be caused due to a failure. By performing PdM, we can provide a safer alarm service and further build trust in the clients. In this study, we only focus on the use case of a method to determine the extent of abnormalities occurring in a facility.

3 Related Work

Aung et al. [16] proposed a blockchain-based system for home services that monitors and automatically controls healthcare services and home safety. When an IoT device related to health care or intrusion detection is activated, a transaction indicating an emergency situation and containing sensor information is automatically sent to the blockchain, allowing the security company to rapidly respond to the situation. A record of the process beginning at the receipt of the emergency signal from the IoT device up to the dispatch of the security team is stored in the blockchain to maintain the immutability and integrity of the record. Using smart contracts, the blockchain automatically manages the entire process from the receipt of the emergency signal to the dispatch of the security team. This reduces the amount of human labor required and therefore improves work efficiency. However, using this system, it is not possible to estimate the degree of severity of an abnormal situation based on the transactions sent from each IoT device.

Several studies have been proposed for anomaly detection using the blockchain [10,11]. Cheung et al. [12], proposed ContractGuard, the first intrusion detection system (IDS) to defend Ethereum smart contracts. It embeds a program in the blockchain to detect intrusion attacks against smart contracts. When it detects anomalous behavior, it notifies the administrator of the unauthorized action on the smart contract in the form of an alarm. However, it does not support the prediction of the culprit's intentions or the cause of the anomaly from multiple transactions.

Several blockchain-based IoT forensic frameworks [7,8] have been proposed. These frameworks can enhance the traceability, reliability, and non-repudiation of evidence provided by the blockchain. Akkaya et al. [9] proposed a forensic framework to manage the event data of IoT devices using three types of blockchains. In case of an unusual occurrence from a security point of view, investigators and police officers can access corporate databases to investigate the cause of the problem. Data from IoT devices is only stored in the blockchain as forensic data, and the degree of anomaly of the stored data cannot be ascertained.

Guizani et al. [17] proposed a blockchain-based Public Key Infrastructure (PKI) with Certification Authority (CA) functionality to manage node revocation that worked by adding the nodes to be disabled from a blockchain with CA functionality to a blacklist. This is necessary because the blockchain cannot delete node information once it is stored.

4 Proposed Method

This paper proposes a system that manages the logs of IoT devices and automatically detects abnormalities. We assume that a security company uses multiple IoT devices to guard and monitor a large facility. We also assume that the system can be operated by the security company, even without the intermediary of an alert monitoring company. The proposed method consists of five steps: (1) The IoT devices are registered in the blockchain. When an IoT device installed in the facility detects an anomaly and activates it, it sends the sensor value to the blockchain. (2) Then, the blockchain stores the data as a log and calculates the anomaly level, which estimates the degree of anomaly based on the sensor values. (3) The security company then reads the anomaly level and automatically initiates an initial response based on the anomaly level. (4) If there is any damage or loss in the facility, the log is submitted to the investigating agency as evidence. (5) If an IoT device is damaged or hijacked, the security company will revoke the IoT device. The overall diagram of the proposed method is shown in Fig. 1.

Entities

- **Security Company:** They monitor and manage the IoT devices, allowing the security companies to focus on guarding large facilities. They manage a blockchain that stores the logs of the IoT devices and automatically detects anomalies. Security companies revoke IoT devices when necessary.
- **IoT device vendor:** IoT device vendors register private keys for IoT devices and sell the IoT devices.
- **IoT device:** This is a blockchain node that sends its sensor values to the blockchain. Examples of IoT devices include temperature and infrared sensors.

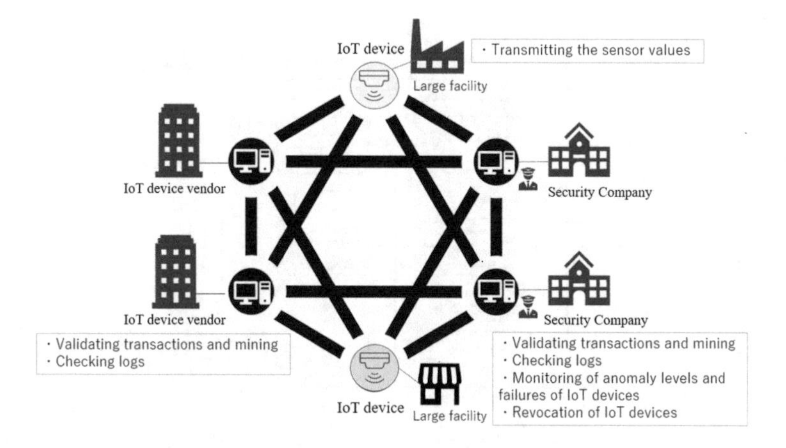

Fig. 1. Overall of the proposed method

Prerequisite

- A consortium blockchain consisting of security companies, IoT device vendors, and IoT devices. The set of nodes is closed.
- The consensus algorithm is designed to be Proof of Authority (PoA) [18], and multiple IoT device vendors and security companies validate and mine the transactions. PoA is fast consensus algorithm, wherein, it is difficult to perform fraudulent acts because the block creation process is performed only by the authorized nodes. IoT device vendors can promote the sale of IoT devices by participating in the blockchain.
- Only security companies and IoT device vendors can reference the blockchain. IoT device vendors can collect data related to the signs of failure from the logs of failed IoT devices, which can lead to improvements in IoT devices.

4.1 System Flow

The proposed system flow is illustrated in Fig. 2. First, the information of the IoT device is registered in the blockchain and the device is installed in the target facility where it is to be guarded or monitored. When an anomaly is detected, the IoT device sends its sensor values to the blockchain. The blockchain stores the logs using smart contracts, and simultaneously calculates the anomaly level using an automatic anomaly detection function (AddAlert) that estimates the degree of anomaly based on the logs activated within a certain period of time. The higher the level of the anomaly, the higher is the likelihood of damage or loss. The security company continuously monitors the blockchain, and takes an initial action based on the anomaly level detected by the system. In the event of an accident or incident, the relevant logs are submitted to the police and other investigative agencies as evidence to assist them in their investigations. If an IoT device is damaged or hijacked, the security company revokes the IoT device by adding its account address to the blockchain blacklist.

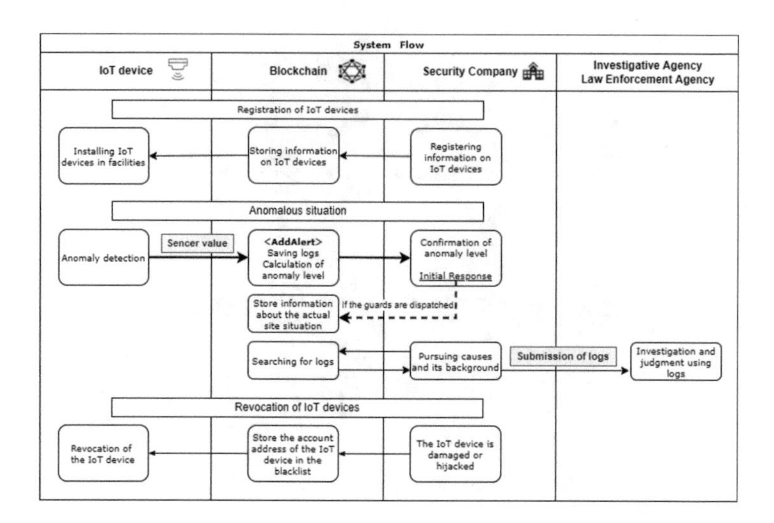

Fig. 2. System flow

Registration of IoT Devices

The IoT device vendor registers the private key to the IoT device before selling it. The purchased IoT device is then installed in the facility where the security services are provided. The security company registers the information of the IoT devices in the blockchain to use them as nodes in the blockchain. The security company stores the IoT device's identifier, account address, residence address, and detailed location within the building in the blockchain. At the same time, the account address is registered in the IoT device whitelist. Thus, only the registered IoT devices can send sensor values to the blockchain.

Saving Logs and Automatic Anomaly Detection

When the IoT device is activated, it automatically sends the sensor values to the blockchain, and the AddAlert function is executed. AddAlert stores the sensor value, anomaly level, account address, timestamp, and AlertID, which is an identifier of the log. The AddAlert function first stores the sensor value, address, timestamp, and AlertID, and then calculates and stores the anomaly level based on the sensor value. The anomaly level is calculated according to the predetermined anomaly detection rules. Specifically, the anomaly level is calculated based on which IoT devices detect sensor values that exceed the threshold value within a certain period of time. The time period, the combination of IoT devices, and the corresponding anomaly level can be determined arbitrarily according to the security purpose of the facility to be guarded.

Initial Response

The security company continuously monitors the blockchain, reading its anomaly level. The initial response is based on these readings. When a security guard is dispatched, they assess the actual site situation. Security company stores information about the AlertID, the name of the security guard in charge, and the actual situation in Blockchain. In the event of an accident or incident, the stored information can be used for investigations.

Submission of Evidence
When an anomalous situation such as an accident or incident occurs, logs that may reveal the cause and background of the situation can be searched by AlertID, timestamp, sensor value, and anomaly level. The log consists of the sensor value, anomaly level, account address, timestamp, and AlertID. The log can be submitted to the police and other investigative agencies and law enforcement agencies.

Revocation of IoT Devices
If an IoT device is damaged or hijacked, the facility manager reports it to the security company. Security company then registers the account address based on the public key corresponding to the private key of the IoT device in the blacklist, thereby revoking the IoT device.

4.2 Concrete Example

As a concrete example of the proposed system, we present a case-study involving a food factory below. We assume that three IoT devices, referred to as devices C, B, and A in order of importance, are installed in the food factory for the purpose of quality control and security of the facility, as shown in Fig. 3. We assume that IoT device A is a humidity sensor, IoT device B is a temperature sensor, and IoT device C is an infrared sensor that detects heat. The factory is maintained at a constant humidity and temperature, and if both IoT devices A and B detect sensor values that exceed T_A and T_B, respectively, the quality of the product will drop significantly.

Registration of IoT Devices
The security company registers the information of the three installed IoT devices in the blockchain.

Saving Logs and Automatic Anomaly Detection
When an IoT device is activated, it automatically sends sensor values to the blockchain. When the blockchain receives the sensor values, the AddAlert function is executed to calculate the anomaly level based on the anomaly detection rules and saves the log. An example of the rules for calculating the anomaly level is presented in Table 1. Here, we assume that the fixed time was 30 min. Therefore, the anomaly level is calculated according to which IoT device detects a sensor value that exceeds the threshold value within 30 the past minutes. For example, if only IoT Device A or only IoT Device B, but not both, detects a sensor value that exceeds the threshold value within a 30 min window, it is unlikely that someone is committing a criminal act that would deteriorate the quality of the product or an event that could cause damage, such as a fire. Therefore, anomaly level 1 is triggered. If both IoT device A and IoT device B detect sensor values that exceed the threshold values within the same 30 min window, anomaly level 2 is triggered. This is because someone may have changed the humidity and

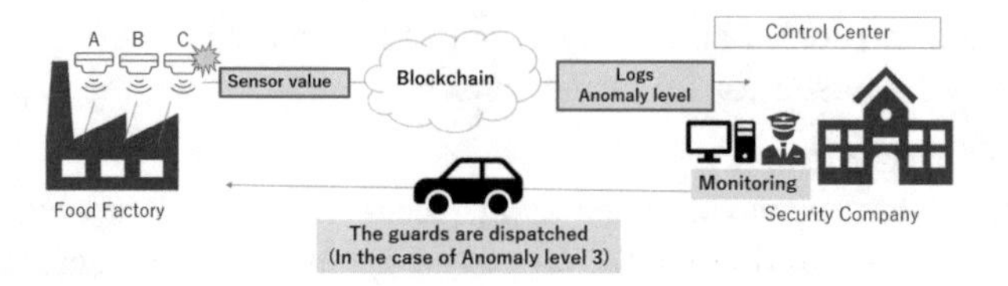

Fig. 3. IoT devices installed in food factory (In the case of Anomaly level 3)

Table 1. Example of anomaly detection rules

Anomaly level	Anomaly detection rules
Anomaly level 3	IoT device C $> T_C$
Anomaly level 2	Within 30 min IoT device A $> T_A$ **and** IoT device B $> T_B$
Anomaly level 1	Within 30 min IoT device A $> T_A$ **and** IoT device B $< T_B$ IoT device A $< T_A$ **and** IoT device B $> T_B$

temperature to lower the quality of the factory. Furthermore, if IoT device C also detects a sensor value that exceeds the threshold value, then it is extremely likely that an anomalous situation such as a fire has already occurred, and therefore anomaly level 3 is triggered immediately, regardless of the fixed time.

This AddAlert function stores the sensor values from the activated IoT devices, their account addresses, timestamps, and the AlertID (log identifier) in a blockchain, and calculates and stores the anomaly level using the sensor values. First, a hash value (Keccak256) is calculated based on the block height, timestamp, and the account address and is inserted into the AlertID. Then, the function checks whether the account address of the IoT device is registered in the whitelist and not in the blacklist. The sensor value, account address, timestamp, and AlertID are then stored in the AlertList, only if the aforementioned conditions are met. Then, the function checks whether the sensor value of the IoT device exceeds the threshold value. If it is the first time the sensor value is sent, an anomaly level 1 is triggered for IoT devices A or B, and anomaly level 3 is triggered for IoT device C. If this is not the first time that these triggers have occurred, the anomaly level is calculated based on the anomaly detection rules by searching the past logs and checking whether the timestamp of the past log was issued within the specified time before the timestamp of the most recent log. The calculated anomaly level is stored in the AlertList. Here, the threshold of the IoT device is set to T.

Initial Response

When the anomaly level is confirmed, the security company takes action based on the anomaly level. An example of the initial response of the security company is presented in Table 2. In the case of anomaly 1, the security company is alerted;

Algorithm 1. Saving logs and automatic anomaly detection: AddAlert

```
Require: SensorValue, AccountAddress, Timestamp
Ensure: AlertID, AnomalyLevel
  AlertID = Keccak256(block height, Timestamp, AccountAddress)
  if whitelist.[AcconutAddress] == 1 && blacklist[AccountAddress] == 0 then
    AlertList.push(SensorValue, AccountAddress, Timestamp, AlertID)
    if SensorValue >T then
      if Sending sensor value for the first time then
        If IoT device A or B ; AnomalyLevel=1, IoT device C: AnomalyLevel = 3
      else
        Search the past logs
        if There is an IoT device that has exceeded the threshold before a certain
        time. then
          Calculate AnomalyLevel based on the anomaly detection rules
          AlertList.push(AnomalyLevel)
        else
          Calculate AnomalyLevel based on the anomaly detection rules
          AlertList.push(AnomalyLevel)
        end if
      end if
    end if
  end if
```

Table 2. Example of initial response based on anomaly level

Anomaly level	Initial response based on anomaly level
Anomaly level 3	The guards are dispatched
Anomaly level 2	On high alert (In some cases, the guards are dispatched)
Anomaly level 1	On alert

in the case of anomaly level 2, the security company is placed on high alert; and in the case of anomaly level 3, the guards are dispatched. If the scenario involves overnight security, the guards are dispatched even in the case of anomaly level 2. When they are dispatched, the security company stores the information about the actual situation of the site.

Submission of Evidence to the Investigating Agency
The security company searches for logs related to the incident and submits them as evidence to the police or other relevant investigative agencies. It also submits the information regarding the situation at the site from the time of the anomaly that is stored in blockchain.

5 Experiment

5.1 Experimental Methods and Results

The feasibility of the proposed method was verified by conducting a series of operational experiments and verifying the time required from when the IoT device transmits the sensor values until when the security company confirms the anomaly level. We conducted a demonstrative experiment using Infura [19] and MetaMask [20] on Rinkeby [21] which uses PoA as a consensus mechanism, an Ethereum [22] test network. The smart contract was written and implemented based on the above concrete example in Solidity (Ver. 0.5.16) on the RIMIX-Ethereum IDE development platform [23]. We also implemented three programs that transmit the sensor values to mimic the IoT devices and a daemon program for security companies to confirm the anomaly level in Python on Ubuntu (18.04 LTS). In this experiment, the programs played the roles of the three virtual IoT devices and the daemon program functioned as the security company.

As a prerequisite, the IoT devices A, B, and C were registered in the blockchain. To simulate the actual operation, every time the program for sending sensor values was executed, the specific IoT device is randomly selected and the sensor value was sent to the blockchain. Because the threshold of the center value varied depending on the IoT device, the sensor value was randomly sent from 0 to 100.

The daemon program displays the anomaly level, date and time, identification number of the IoT device registered in the blockchain, residence address, and installation location. We measured the time required for the IoT activation to confirm the anomaly level. The average time for the 20 experiments we conducted was approximately 10.412 s. In the daemon program, we confirmed the update of the anomaly level and the information of the IoT device.

5.2 Discussion

Operation and Time to Confirm Anomaly Level
In a typical alarm service, a security guard monitoring the alarm at the control center reads the activation information and instructs the security guards to take the initial response. Under Japan's Security Service Law, security guards are obligated to arrive at the target facility within 25 min after the sensor is activated, and to promptly assess the situation and report the incident to the police. Therefore, immediate decision-making and a prompt initial response are required. In the proposed method, after the sensor reported its values, instructions were automatically provided, and the initial response was triggered without the intervention of the security guards who monitor the alarm. Therefore, it is possible to shorten the time required to start the initial response. In addition, control centers are typically monitored 24 h a day by multiple people, depending on the scale. In the proposed method, the number of security guards can be reduced - compared to the conventional system - because the blockchain automatically manages all the processes. Therefore, the proposed method is highly feasible for use in real-world contexts.

Reliability of Logs as Evidence
It is difficult to falsify or delete information stored in a blockchain. Furthermore, because the blockchain is decentralized, even if the information of some nodes is corrupted, it can be restored without interruption to the service. Therefore, the integrity of the logs of IoT devices stored in the blockchain is maintained. In the event of damage or loss at a guarded facility, the logs stored in the blockchain can thus provide reliable evidence to determine the background and cause of the damage.

6 Conclusion

To address the problems of managing traditional alarm services and centralized IoT devices, we proposed a blockchain-based system for managing the IoT device logs through automatic anomaly detection. By storing the logs of IoT devices in the blockchain, the integrity of the data can be maintained, and the data can be used as evidence to investigate the cause of anomalies when they occur. By implementing automatic anomaly detection, we were able to automate a series of security responses. In addition, in the Ethereum test net environment, we conducted experiments to verify the operation of the system and the time required from the time the IoT device is activated until the initiation of the initial response by the security company, to show the feasibility of the system.

In the proposed method, there is a possibility that unauthorized data will be stored in blockchain when an IoT device is hijacked. To solve this problem, we proposed to create a blacklist of logs and register the AlertID of the incorrect logs so that they cannot be used as evidence. However, this method cannot solve the problems arising from the erroneous calculation of the anomaly level due to incorrect data. We hope to investigate this topic in the future research.

Acknowledgement. This work was partly supported by the Grant-in-Aid for Scientific Research (B) (19H04107).

References

1. IDC: IoT Growth Demands Rethink of Long-Term Storage Strategies, says IDC. https://www.idc.com/getdoc.jsp?containerId=prAP46737220. Accessed 8 Dec 2021
2. Statist: Size of the security services market worldwide from 2011 to 2020, by region. https://www.statista.com/statistics/323113/distribution-of-the-security-services-market-worldwide/. Accessed 8 Dec 2021
3. Nemeth, C.P.: Private Security: An Introduction to Principles and Practice, 1st edn. CRC Press, Boca Raton (2017)
4. Mrdovic, S.: IoT forensics. In: Avoine, G., Hernandez-Castro, J. (eds.) Security of Ubiquitous Computing Systems, pp. 215–229. Springer, Cham (2021). https://doi.org/10.1007/978-3-030-10591-4_13
5. Kumar, N.M., Mallick, P.K.: Blockchain technology for security issues and challenges in IoT. Procedia Comput. Sci. **132**, 1815–1823 (2018)

6. Nakamoto, S.: Bitcoin: a peer-to-peer electronic cash system (2009). https://bitcoin.org/bitcoin.pdf
7. Le, D., Meng, H., Su, L., Thing, V., Yeo, S.L.: BIFF: a blockchain-based IoT forensics framework with identity privacy. In: 2018 IEEE TENCON, pp. 2372–2377 (2018)
8. Li, M., Qiu, M., Su, S., Sun, Y., Tian, Z.: Block-DEF: a secure digital evidence framework using blockchain. Inf. Sci. **491**, 151–165 (2019)
9. Akkaya, K., Cebe, M., Chang, M., Mercan, S., Tekiner, E., Uluagac, S.: A cost-efficient IoT forensics framework with blockchain. In: 2020 IEEE ICBC, pp. 1–5 (2020)
10. Idé, T.: Collaborative anomaly detection on blockchain from noisy sensor data. In: 2018 IEEE ICDMW, pp. 120–127 (2018)
11. Jang, J., Nang, J., Song, J.: Design of anomaly detection and visualization tool for IoT blockchain. In: 2018 CSCI, pp. 1464–1465 (2018)
12. Cheung, S., He, J., Wang, X., Xie, Z., Zhao, G.: ContractGuard: defend ethereum smart contracts with embedded intrusion detection. IEEE Trans. Serv. Comput. **13**(2), 314–328 (2020)
13. Chica, M., Tang, Q., Zhang, Z.: Maintenance costs and makespan minimization for assembly permutation flow shop scheduling by considering preventive and corrective maintenance. J. Manuf. Syst. **59**, 549–564 (2021)
14. Huang, H., Lu, Y., Wang, Y., Xu, X., Yang, L.: Digital Twin-driven online anomaly detection for an automation system based on edge intelligence. J. Manuf. Syst. **59**, 138–150 (2021)
15. Costa, C., Li, G.P., Lima, M.J., Righi, R., Trindade, E.S., Zonta, T.: Predictive maintenance in the Industry 4.0: a systematic literature review. Comput. Ind. Eng. **150**, 106889 (2020)
16. Aung, Y. N., Tantidham, T.: Emergency service for smart home system using ethereum blockchain: system and architecture. In: 2019 IEEE PerCom Workshops, pp. 888–893 (2019)
17. Guizani, N., Li, Y., Lou, C., Wang, L., Yu, Y.: Decentralized public key infrastructures atop blockchain. IEEE Network **34**(6), 133–139 (2020)
18. PoA Network: Proof of Authority: consensus model with Identity at Stake. https://medium.com/poa-network/proof-of-authority-consensus-model-with-identity-at-stake-d5bd15463256. Accessed 8 Dec 2021
19. Infura: Ethereum API | IPFS API and Gateway | ETH Nodes as a Service — Infura. https://infura.io/. Accessed 8 Dec 2021
20. MetaMask: MetaMask - A crypto wallet and gateway to blockchain apps. https://metamask.io/. Accessed 8 Dec 2021
21. Rinkeby: Rinkeby: Network Dashboard. https://www.rinkeby.io/. Accessed 8 Dec 2021
22. Ethereum: Home | ethereum.org. https://ethereum.org/en/. Accessed 8 Dec 2021
23. Ethereum: Remix IDE - Ethereum.org. https://remix.ethereum.org/. Accessed 8 Dec 2021

A Hybrid Recovery Method for Vehicular DTN Considering Dynamic Timer and Anti-packet

Minh Duc Nguyen[1], Masaya Azuma[2], Shota Uchimura[2], Makoto Ikeda[1(✉)], and Leonard Barolli[1]

[1] Department of Information and Communication Engineering, Fukuoka Institute of Technology, 3-30-1, Wajiro-higashi, Higashi-ku, Fukuoka 811-0295, Japan
s18b1016@bene.fit.ac.jp, makoto.ikd@acm.org, barolli@fit.ac.jp
[2] Graduate School of Engineering, Fukuoka Institute of Technology, 3-30-1, Wajiro-higashi, Higashi-ku, Fukuoka 811-0295, Japan
{mgm21101,mgm21102}@bene.fit.ac.jp

Abstract. In a disaster situation, the communication failures due to base station issues or increased transactions and will cause huge delay times and network outages. In such a poor environment, Delay-Tolerant Networking (DTN) with a recovery function is attracting attention. In this paper, we propose a hybrid recovery method considering dynamic timer and anti-packet for vehicular DTN. We use an urban grid road model in disaster situations. From the simulation results, we found that hybrid method can improve the storage utilization compared with conventional dynamic timer.

Keywords: DTN · Hybrid recovery · Dynamic timer · Anti-packet

1 Introduction

The concept of Delay-Tolerant Networking (DTN) [5] using automobiles and road-side units is growing rapidly as a valuable option for advertising distribution and the development of future intelligent transport systems. In DTN communication, a huge number of replicated messages has a negative effect on network resources. In [10], the authors discussed how to delete duplicated messages using timers and anti-packets.

In our previous work [2], we presented an adaptive anti-packet recovery method for shuttle buses and road-side units. However, we did not include the timer approach as a recovery method.

In [6], we proposed a recovery method with dynamic timer in Vehicular DTN. The proposed methods has timer, which can reduce the duplicated bundle messages in each vehicle. However, we did not consider anti-packet.

In this paper, we focus on the impact of hybrid recovery approaches with timer and anti-packet for replaying the replicated bundle messages. Our hybrid recovery method is based on Epidemic routing.

L. Barolli et al. (Eds.): AINA 2022, LNNS 449, pp. 217–225, 2022.
https://doi.org/10.1007/978-3-030-99584-3_19

The rest of the paper is structured as follows. In Sect. 2, we give the overview of DTN. In Sect. 3 is describe the proposed hybrid recovery method. In Sect. 4, we provide the simulation system. In Sect. 5, we provide the evaluation results. Finally, conclusions and future work are given in Sect. 6.

2 Overview of DTN

DTN is able to supply a dependable internet connection for space-related operations [4,8,12]. Space networks may have frequent link failures, disconnection and significant delays. The intermediate vehicles in Vehicular DTN store bundle messages in their storage and subsequently send them to other vehicles. The network architecture is specified in RFC 4838 [3]. The well-known DTN protocol is Epidemic Routing [7,11], which is performed by using two control messages to replicate a bundle message. Each vehicle periodically broadcasts a Summary Vector (SV) on the network. The SV contains a list of stored messages for each vehicle. When the vehicles receive the SV, they compare the received SV to their SV. The vehicle sends a request message if the received SV contains an unknown bundle message. In this method, the consumption of network resources and storage state become critical issues, because the vehicles replicate bundle messages to adjacent vehicles in their communication range. Then, the received bundle messages remain in the storage and the bundle messages are continuously replicated even if the end-point receives the messages.

As a result, recovery schemes like timer and anti-packet may delete the replicated bundle messages in the network. For the sparse network, the anti-packet deletes the replicated messages too late due to the large latency. In the case of the conventional anti-packet, the end-point broadcasts the anti-packet, which contains the list of bundle messages that are received by the end-point. Vehicles delete messages based on the anti-packet and replicate the anti-packet to other vehicles. In the case of the conventional timer, the messages have a lifetime and they are punctually deleted when the lifetime of the bundle messages has expired. The challenge is to formulate an adaptive lifetime algorithm that can be applied to different environments.

3 Hybrid Recovery Method

In this section, we explain in detail the proposed hybrid recovery method considering dynamic timer and anti-packet.

In inter-vehicle communications for disaster situation, we propose a message relaying method using a hybrid recovery approach to improve the overhead and storage utilization. In traditional protocols, the lifetime of a message is established at the moment it is generated. We consider the non-signal duration from local neighbors as part of the timer. The non-signal duration indicates that there is no vehicle in the area. A dynamic timer for messages is established for each vehicle and data is managed independently. If the lifetime has expired, the vehicle deletes the bundle message in its storage. For the hybrid recovery method, the end-point broadcasts the anti-packet, which contains the list of messages that reached the end-point. The vehicles replicate the anti-packet for other vehicles. Using the advantage of timer and anti-packet approaches, it is possible to improve storage utilization.

We present the flowchart of proposed hybrid recovery method in Fig. 1. In our method, each node periodically checks the number of received SV and measures the Non-signal Time (NT) from neighboring vehicles. If the current NT value is greater than maximum NT (NT_{max}) value, the value of NT_{max} will be updated by Hybrid Method (HM) as shown in Eq. (1):

$$HM = NT_{max} + Interval, \tag{1}$$

where *Interval* indicates the check interval, which is used for checking the number of received SVs.

Our approach resets the timer when the number of received SVs is 0. Thus, our method may sustain the bundle message and avoid a decrease in packet arrival rate even if the vehicles become disconnected from the network due to vehicle movement. Thus, by shifting the timer reset condition, our proposed approach increases network performance. In our proposed method, the reset is allowed even if the lifetime is reset more than one time. We consider the *Interval* to keep the message in the storage for checking the next SV. Even if the vehicle suddenly leaves the neighbors and there is no signal time, our method keeps the message.

4 Simulation Settings

In this section, we evaluate the proposed recovery method in urban road model considering disaster situations on the Scenargie network simulator [9].

In Fig. 2, we show an urban grid road model considering regular vehicles as message relaying terminals. In this evaluation, we deploy a maximum of 90 regular vehicles on the road and considered the following conditions.

- Regular vehicles use the proposed hybrid recovery method as a message delivery terminal.
- The message start-point and end-point are fixed.
- Some roads are destroyed in a disaster situation.
- Regular vehicles continue to move on the roads based on the map-based random way-point mobility model.

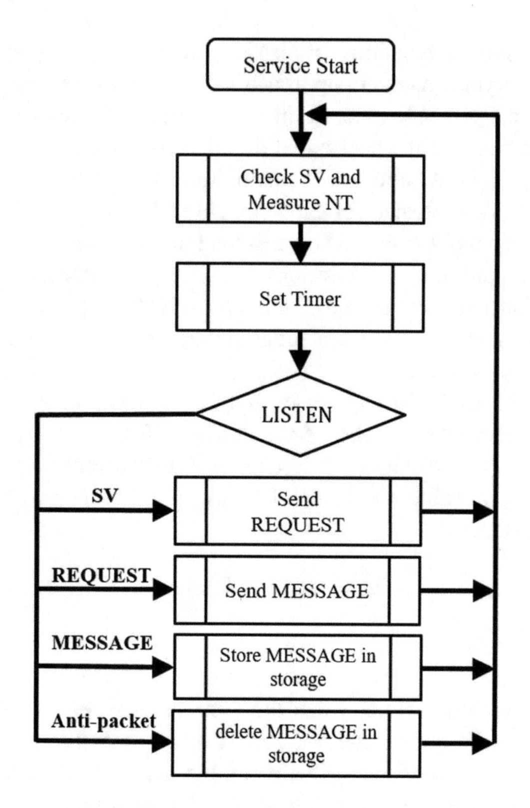

Fig. 1. Flowchart of proposed hybrid recovery method.

We consider three vehicular densities: 30, 60 and 90 vehicles. Table 1 shows the simulation parameters used for the network simulator. The start-point replicates 40 kinds of bundle messages to the relay vehicles. Then, the replicated bundle messages are delivered to the end-point. The simulation time is 600 s. The ITU-R P.1411 propagation model is used in this urban road scenario [1]. We consider the interference from buildings and other structures on the 5.9 GHz radio channel.

We evaluate the performance of storage utilization, overhead, delivery ratio and delay for different vehicles.

- The storage utilization indicates the average of the storage state of each vehicle.
- The number of replicated bundle messages in the network is indicated by the overhead.
- The delivery ratio indicates the value of the generated bundle messages divided by the delivered bundle messages to the end-point.
- The delay indicates the transmission latency of the bundle message to reach the end-point.

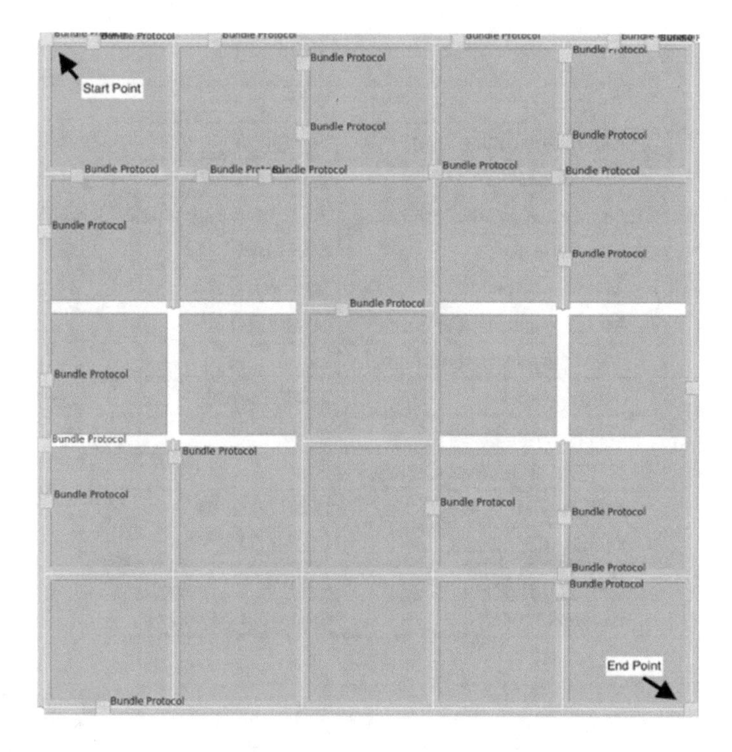

Fig. 2. Urban road model considering disaster situation.

5 Evaluation Results

We evaluate the proposed hybrid recovery method considering different vehicles compared to dynamic timer. The simulation results are shown from Figs. 3, 4, 5 and 6. In Fig. 3, we show the simulation results for storage utilization from 30 vehicles to 90 vehicles. For 30 vehicles, the storage utilization is gradually increased for both methods and is almost the same. The difference is appearing after 500 s. For 60 vehicles, we observed that the storage utilization of hybrid method is decreasing after 320 s. For 90 vehicles, the storage utilization for hybrid method is smaller.

In Fig. 4, we show the simulation results of the overhead. The overhead increases with increasing the number of vehicles. The proposed hybrid recovery method has a lower overhead than the dynamic timer regardless of the number of vehicles.

Table 1. Simulation parameters.

Parameters	Values
Simulation time ($T_{\max}$)	600 [s]
Area dimensions	500 [m] × 500 [m]
Number of regular vehicles	30, 60 and 90 [vehicles]
Minimum speed	8.333 [m/s]
Maximum speed	16.666 [m/s]
Message start and end time	10–400 [s]
Message generation interval	10 [s]
Message size	1,000 [bytes]
DT: Activated time	60 [s]
DT: Check Interval (*Interval*)	2 [s]
PHY model	IEEE 802.11p
Propagation model	ITU-R P.1411
Radio frequency	5.9 [GHz]
Antenna model	Omni-directional

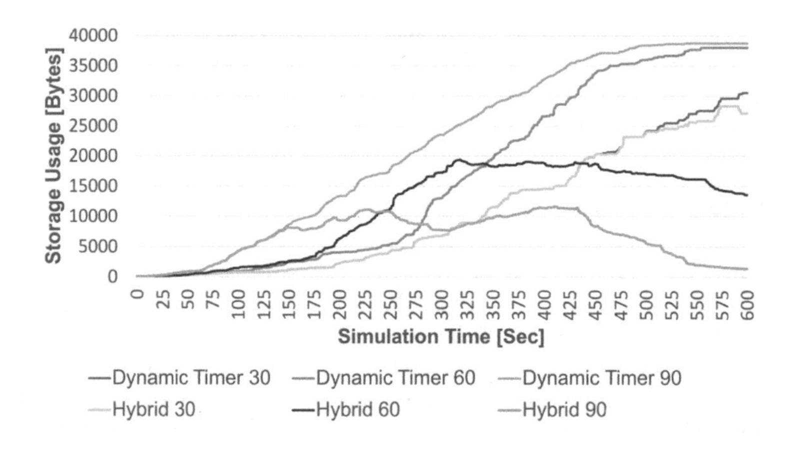

Fig. 3. Storage utilization.

In Fig. 5, we show the simulation results of the delivery ratio. For 60 vehicles, the delivery ratio for dynamic timer reached 100%. While for the hybrid method reached 100% after 90 vehicles.

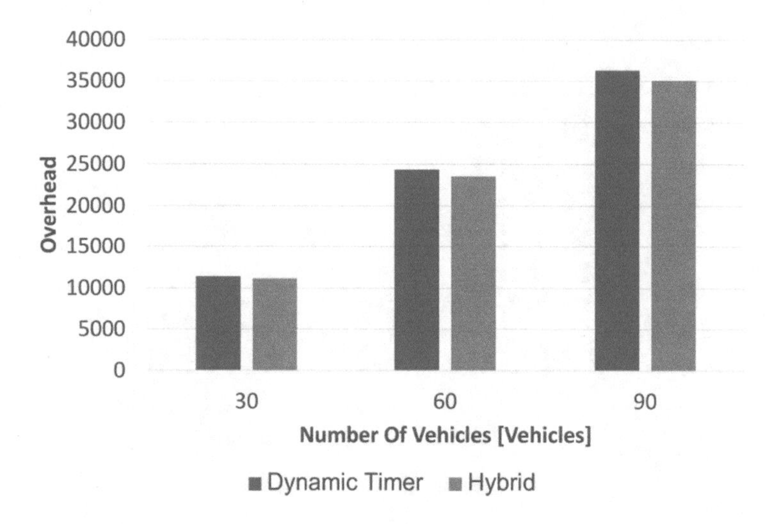

Fig. 4. Overhead.

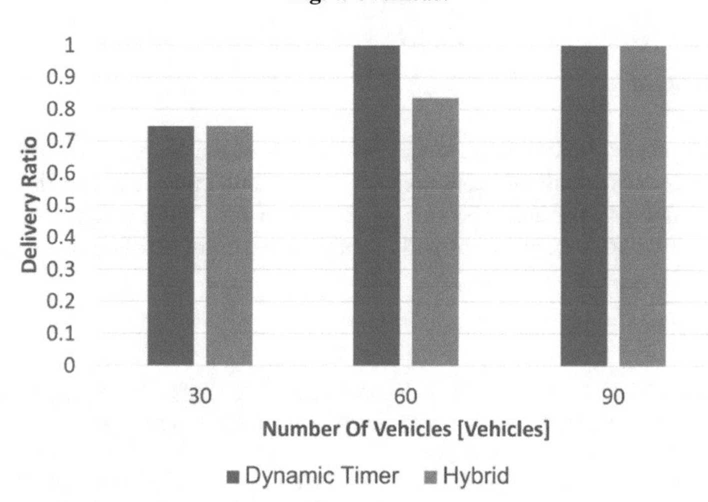

Fig. 5. Delivery ratio.

In Fig. 6, we show the simulation results of delay. The delay decreases with increasing the number of vehicles. The difference between dynamic timer and hybrid method is small for different vehicles, even the proposed method reduced the storage utilization, but did not impact the latency.

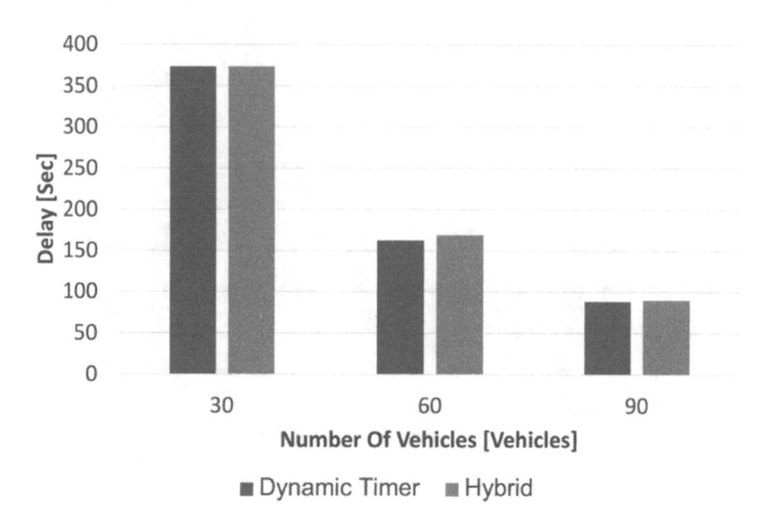

Fig. 6. Delay.

6 Conclusions

In this paper, we evaluated the network performance of the proposed hybrid recovery method in a disaster situation. From the simulation results, we found that the proposed recovery method can decrease the storage utilization for different vehicles. In future work, we would like to extend our hybrid approach and consider other parameters.

References

1. Rec. ITU-R P.1411-7: Propagation data and prediction methods for the planning of short-range outdoor radiocommunication systems and radio local area networks in the frequency range 300 MHz to 100 GHz. ITU (2013)
2. Azuma, M., Uchimura, S., Tada, Y., Ikeda, M., Barolli, L.: An adaptive anti-packet recovery method for vehicular DTN: performance evaluation considering shuttle buses and roadside units scenario. In: Barolli, L. (ed.) BWCCA 2021. LNNS, vol. 346, pp. 234–241. Springer, Cham (2022). https://doi.org/10.1007/978-3-030-90072-4_25
3. Cerf, V., et al.: Delay-tolerant networking architecture. IETF RFC 4838 (Informational), April 2007
4. Fall, K.: A delay-tolerant network architecture for challenged Internets. In: Proceedings of the International Conference on Applications, Technologies, Architectures, and Protocols for Computer Communications. pp. 27–34. SIGCOMM 2003 (2003)
5. Kawabata, N., Yamasaki, Y., Ohsaki, H.: Hybrid cellular-DTN for vehicle volume data collection in rural areas. In: Proceedings of the IEEE 43rd Annual Computer Software and Applications Conference (COMPSAC-2019), vol. 2, pp. 276–284, July 2019
6. Nakasaki, S., Ikeda, M., Barolli, L.: A message relaying method with a dynamic timer considering non-signal duration from neighboring nodes for vehicular DTN. In: Barolli, L., Nishino, H., Miwa, H. (eds.) INCoS 2019. AISC, vol. 1035, pp. 133–142. Springer, Cham (2020). https://doi.org/10.1007/978-3-030-29035-1_13

7. Ramanathan, R., Hansen, R., Basu, P., Hain, R.R., Krishnan, R.: Prioritized epidemic routing for opportunistic networks. In: Proceedings of the 1st International MobiSys Workshop on Mobile Opportunistic Networking (MobiOpp 2007), pp. 62–66 (2007)
8. Rüsch, S., Schürmann, D., Kapitza, R., Wolf, L.: Forward secure delay-tolerant networking. In: Proceedings of the 12th Workshop on Challenged Networks (CHANTS-2017), pp. 7–12, October 2017
9. Scenargie: Space-time engineering, LLC. http://www.spacetime-eng.com/
10. Solpico, D., et al.: Application of the V-HUB standard using LoRa beacons, mobile cloud, UAVs, and DTN for disaster-resilient communications. In: Proceedings of the IEEE Global Humanitarian Technology Conference (GHTC-2019), pp. 1–8, October 2019
11. Vahdat, A., Becker, D.: Epidemic routing for partially-connected ad hoc networks. Duke University, Technical Report (2000)
12. Wyatt, J., Burleigh, S., Jones, R., Torgerson, L., Wissler, S.: Disruption tolerant networking flight validation experiment on NASA's EPOXI mission. In: Proceedings of the 1st International Conference on Advances in Satellite and Space Communications (SPACOMM-2009), pp. 187–196, July 2009

Chaotic-Maps Based Access Authentication Protocol for Remote Communication Using Space Information Networks

Susmita Mandal(✉), S. S. Sravan, and Lakshmi Ramesh

IDRBT, Hyderabad, India
msusmita@idrbt.ac.in

Abstract. Space Information Networks (SIN) allow terrestrial mobile nodes to gain access to the Internet anywhere and at any time. Due to its dynamic network coverage, it has been gaining wider attention from researchers and network service providers. In comparison to terrestrial networks, the majority of data transmission in SIN is through air interface, which makes it vulnerable to adversarial attacks such as illegal data access, eavesdropping attack, etc. To prevent unauthorized user access, this research work focuses on designing a secure and lightweight access authentication mechanism using extended Chebyshev chaotic maps for resource-constrained mobile nodes as an act of the first line of defense. The proposed scheme achieves several security functionalities, including resistance to impersonation attack, session key secrecy, mutual authentication, and key leakage attack. Furthermore, the soundness of the proposed scheme is validated using the ProVerif tool. Finally, the performance analysis depicts that the proposed scheme is highly efficient compared to existing methods.

1 Introduction

The world is witnessing massive growth in the digital environment because of rapid technological advancements. The developments in digital technology bring better customer experience and make services easier for people. Even though the digital industry has advanced, there are issues pertaining to network coverage and dramatic weather conditions for less developed regions such as people residing in far ends of deserts, hilly mountains, and unconnected islands. To make use of any digitalized service, users need proper network connectivity, thus abandoning a great amount of the population from availing the privileges of modern digital facilities such as e-learning, banking at home, etc. Space Information Network (SIN) or Space-Air- Ground Integrated Network (SAGIN) can be a promising solution for last-mile connectivity. It is proposed for those regions where terrestrial network infrastructure is difficult to be established [1].

L. Barolli et al. (Eds.): AINA 2022, LNNS 449, pp. 226–237, 2022.
https://doi.org/10.1007/978-3-030-99584-3_20

Both the networks use similar architecture including Geosynchronous Earth Orbit (GEO) satellites, Medium Earth Orbit (MEO) satellites, Low Earth Orbit (LEO) Satellites, High Altitude Platform Stations (HAP), Low Altitude Platform Stations (LAP), and Terrestrial (Ground) Networks. And it has characteristics like wider coverage, dynamic network, and low transmission delay. By using this architecture, low-power remote devices can connect directly to satellites. The current LEO satellite constellation orbits 500–2000 Kms from the earth, offering better communication in terms of latency and bandwidth for a user than GEO satellites. Therefore, communication with LEO satellites can be easier. However there are still certain challenges such as (1) SIN uses open wireless channels for communication, (2) it has limited computing resources, (3) long propagation delay, and (4) lack of mutual authentication. Due to long propagation delay, the transmission of data between ground and satellite for authentication becomes at least 4 times of signaling delay. In traditional satellite-to-ground station connectivity, messages are sent in an unencrypted and insecure form where satellites are used to just forward the authentication messages. Therefore to address the openness of satellite communication it is necessary to enable mutual authentication between space and ground networks.

Existing state of art study in SIN or SAGIN environment shows that most of the applications such as [2,3,14–16] work with public-key cryptosystem. For all the remote and digital services, mutual authentication is the first and most important part of any network. After the mutual authentication phase, the entities can communicate securely with each other with the help of a secret session key. In 1996, Cruickshank [2] proposed the first authentication mechanism using SIN architecture. However, the scheme suffers from certificate management issues, replay attacks, impersonation attacks, and has high computation overhead. Later, Hwang *et al.* [3] improved the protocol and proposed a novel scheme using a symmetric key cryptosystem. His scheme cannot be applied as it fails to reduce the computational cost and suffered from impersonation attacks. With the involvement of IoT devices for direct satellite-to-ground communication, the service quality can be improved using SIN. In 2018, Meng *et al.* [14] proposed a method to authenticate the resource-constrained devices with low latency using pairing-based cryptography. However, the scheme suffers from heavy computation overhead. Recently, Xue *et al.* [15] proposed an access and handover authentication protocol for IoT in SIN by reducing the involvement of the ground station during the authentication of a user. In 2021, Guo *et al.* [16] has mentioned that Xue *et al.*'s scheme has security flaws, firstly, when the satellite is semi-trusted or has untrusted nodes. In such a case, the relay satellite functions as a node to authenticate the mobile user. Therefore, the ground station no longer validates the user's identification, allowing the satellite to interact with the ground station using a forged user identity. In addition, although Xue *et al.*'s proposed a batch handover authentication mechanism, however, the computational overhead remains high. Later, Guo *et al.* [16] presented an authentication scheme based on Elliptic Curve Cryptography (ECC), which confirms the validity of ground and space nodes to authenticate each other and designs handover schemes. However, the scheme has a high computation load.

The recent in-orbit demonstration of LacunaSat-1 nano-satellites supporting IoT devices shows that direct ground to LEO satellite communication without relying on intermediate gateways is not widely experimented with [5]. The IoT ecosystem demands more constellation of IoT-powered nano-satellites to achieve continuous coverage, which is in a nascent stage of development. Therefore, the use of Unmanned Aerial Vehicles (UAV) is a feasible way to communicate with satellites through the terrestrial network. Here the low-power IoT devices can communicate with UAVs hovering in the atmosphere. UAVs includes aircraft, drone, or balloon. Through a wireless connection, the user can connect to any nearby UAV and the UAV will transmit the data packet to a satellite. Both low earth orbit (LEO) satellites and UAVs can serve as relay nodes to provide reliable communication services for ground devices. Also, the availability of UAVs is more compared to LEO satellites, to communicate with ground nodes. Hence the wait time will be less for a ground terminal to send data packets across the SIN network. Some of the applications for SIN are military missions, disaster rescue, offline payment solutions, etc.

Several papers [6–9] proposed authentication and handover mechanisms for SIN where most of them considered the LEO satellite for only forwarding the data packets. In these schemes, the computation load is distributed between the ground station and mobile node, which increases the processing time and propagation delay. Whereas in the current SIN architecture, LEO satellites are capable of high computation. Therefore this paper proposes an efficient access authentication protocol using SIN architecture for reducing transmission delay.

The rest of the paper is organized as follows. Section 2 involves preliminaries with a mathematical background, system model, and threat model. In Sect. 3, the proposed protocol is elaborated. Section 4 discusses the formal and informal security analysis. The performance analysis is depicted in the graph in Sect. 5. Finally, the conclusion is drawn in Sect. 6.

2 Preliminaries

This section discusses the mathematical fundamentals of using Chebyshev polynomials and its computational hard assumptions. In addition, the Space Information Network model is demonstrated in Fig. 1.

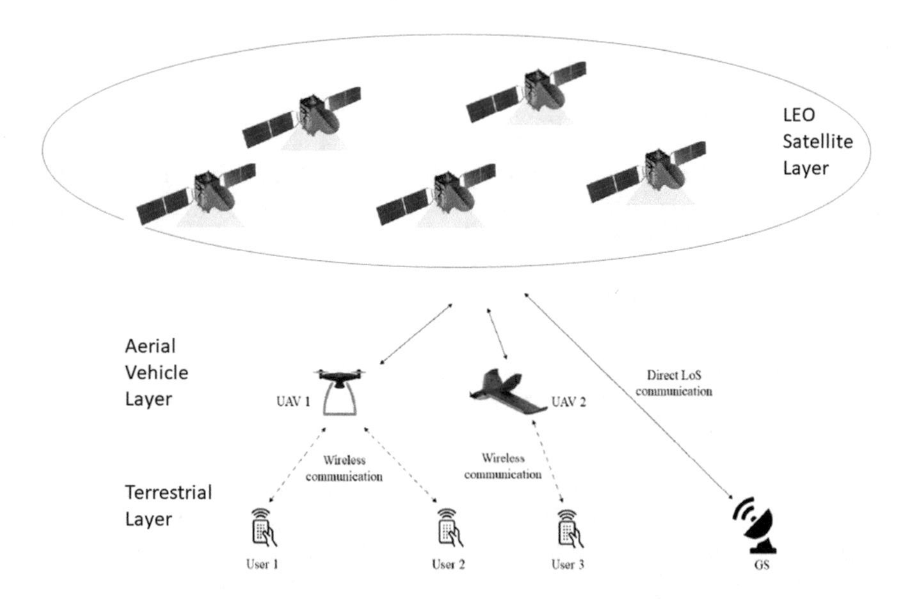

Fig. 1. System model

2.1 Chebyshev Chaotic Maps

For each natural number n, let us define a Chebyshev polynomial $Tn(x) : [-1, 1]$ $->[-1, 1]$ as,

$$T_n(x) = cos(n \times arccos(x)), \tag{1}$$

here $arccos(x))$ is regarded as a $1-1$ function from $[-1, 1]$ to $[0, \pi]$.

The recurrence of $T_n(x)$ is represented as,

$$T_n(x) = 2xT_{n-1}(x) - T_{n-2}(x), \ \ (n \geq 2), \tag{2}$$

where $T_0(x) = cos(0) = 1$, $T_1(x) = cos(arccosx) = x$.

Definition 1 (**Semigroup Property**)**:** The property states that,

$$\begin{aligned} T_a(T_b(x)) &= cos(a \times arccos(b \times arccos(x))) \\ &= cos(ab \times arccos(x)) \\ &= T_{ab}(x) \end{aligned}$$

Here a, b are positive numbers and $x \in [-1, 1]$.

Definition 2 (**Commutative Property**)**:** Chebyshev polynomials also satisfy the commutative property under composition as follows: $T_a(T_b(x)) = T_b(T_a(x))$.

In 2008, Zhang [4] proved that the semigroup property could be defined within the interval $(-\infty, +\infty)$ as,

$$T_n(x) = (2xT_{n-1}(x) - T_{n-2}(x))modp, \ \ (n \geq 2) \tag{3}$$

where p is large prime number, $x \in [0, p-1]$.

Definition 3 (**Chaotic maps discrete logarithm(CMDLP)**)**:** Given two elements $y \in [0, p-1]$ and x, a large prime p, it is hard to find an integer s, such that $T_s(x) \bmod p = y \bmod p$.

Definition 4 (**Chaotic map–based Diffie–Hellman problem(CMDHP)**)**:** Given $x \in [0, p-1]$, p is a large prime number, $T_r(x) \bmod p$, and $T_s(x) \bmod p$, it is computationally infeasible to compute $T_{rs}(x) \bmod p$.

2.2 System Model

The space information network consists of the following entities: Network authentication center (NAC), Low-earth orbit satellites (LEO), Mobile nodes (MN), and Unmanned Areal Vehicles (UAV).

- *Network Authentication Center*: It acts as a trusted third-party and is responsible for registering LEO, MN, and UAV by issuing public/private keys.
- *Low-Earth Orbit satellites*: It acts as the access point for the users to obtain the services of SIN. It is responsible for forwarding the authentication data to terrestrial controllers such as gateways stations for further processing.
- *Mobile Nodes*: Users with resource-constrained devices request access authentication through UAV through LEO satellite constellation to remote terrestrial controllers for any subscription of services.
- *Unmanned Areal Vehicles*: It is primarily responsible for forwarding the requests between terrestrial mobile nodes and LEO satellites.

2.3 Threat Model

This section outlines the widely accepted Dolev-Yao threat model [11] pursued in the paper using the following assumptions:

- An adversary is completely in control of the communication channel via the air interface.
- An adversary can eavesdrop on all the messages transmitted between mobile nodes and satellites.
- An adversary can intercept, inject, modify, and resend intercepted transmitted messages.
- All the steps of the authentication protocol are known to the adversary.

3 Proposed Protocol

In this section, the proposed chaotic map-based access authentication protocol for SIN is discussed, which involves three phases: (1) Initialization, (2) Registration, and (3) Authentication. The registration process occurs through a secure channel, with all the participating entities registering with the network authentication center.

3.1 Initialization Phase

The NAC selects a large prime number p. Then randomly selects $s_{nac} \in Z_p^*$ as its secret key and let variable $v \in (-\infty, +\infty)$ as the seed of the Chebyshev polynomial, which will be within the range of $[0, p-1]$. NAC then generates the public key $P_{nac} = T_{s_{nac}}(v) \bmod p$. Thereafter chooses one-way hash function $h_i(.) : \{0,1\}^* -> \{0,1\}^l$ where $i = \{0,1,2,3,4,5\}$. Later publicly declares the public parameters $v, T_{s_{nac}}(v), h_i(.)$.

3.2 Registration Phase

In this phase, the mobile nodes and the LEO satellites must register, providing required details to the NAC acting as a trusted authority. Before initiating the authentication process, each entity must obtain an identity-based signature as follows:

Mobile Node Registration:

1. Initially, MN chooses its identity ID_{mn} and a random number $r_m \in Z_p^*$. Then computes $W_i = h_0(ID_{mn}||r_m)$ and sends a registration request to NAC through a secure channel.
2. Upon receiving the request, NAC computes $X_i = h_1(ID_{mn}||W_i||s_{nac})$, $E_i = T_{X_i}(P_{nac}) \bmod p$ as witness, and signature as $Sig_i = h_2(E_i||ID_{mn}||ID_{nac}) \cdot X_i \cdot s_{nac}$. Later sends a message tuple as response $\langle E_i, Sig_i \rangle$ to MN.

LEO Satellite Registration:

1. Initially, the LEO satellite node chooses its identity L_{id} and generates a random number $r_i \in Z_p^*$. Then computes $W_{leo} = h_0(L_{id}||r_i)$ and sends a registration request to NAC through a secure channel.
2. Upon receiving the request, NAC computes $X_{na} = h_1(L_{id}||W_{leo}||s_{nac})$, $E_2 = T_{X_{na}}(P_{nac}) \bmod p$ as witness, and signature as $Sig_2 = h_2(E_2||L_{id}||ID_{nac}) \cdot X_{na} \cdot s_{nac}$. Later sends a message tuple as response $\langle E_2, Sig_2 \rangle$ to LEO. Finally, L_{id} securely stores received message tuple.

Similarly, UAV nodes are registered by sharing identity U_{id} to NAC. Upon receiving, NAC generates a random secret $u_r \in Z_p^*$ then computes $A_1 = h_3(ID_{nac}||U_{id}||u_r)$. Later, NAC sends $\langle ID_{nac}, A_1 \rangle$ to UAV.

3.3 Authentication Phase

Generally, the authentication phase begins when users transmit a request for access to communicate with other nodes. However, the LEO satellite nodes have a large broadcast area compared to the terrestrial network nodes. Therefore, upon completion of the registration phase, the LEO satellite nodes periodically broadcast authentication messages. The process is depicted below-

1. The LEO satellite nodes hovering over a region broadcast a tuple $\langle E_2, L_{id}, ID_{nac} \rangle$, the mobile nodes receiving the message computes a function $K_1 = (T_{Sig_i}(v) + T_{h_1(E_2||L_{id}||ID_{nac})}(E_2))$ and $F_1 = h_4(K_1||T_{st})$ then sends a message tuple including timestamp, mobile nodes identities, and witness E_i as $\langle E_i, ID_{mn}, ID_{nac}, F_1, T_{st} \rangle$ through UAV to LEO.
2. The UAV receives the message tuple and adds its registered identity then forwards it to LEO as $\langle E_i, ID_{mn}, ID_{nac}, F_1, T_{st}, A_1 \rangle$.
3. Upon receiving the message LEO node first verifies if $T' - T_{st} \leq \Delta T$ then computes $K_2 = (T_{Sig_2}(v) + T_{h_1(E_i||ID_{mn}||ID_{nac})}(E_i))$, $F_1' = h_4(K_2||T_{st})$. Using the parameters check if $F_1 \stackrel{?}{=} F_1'$. If matches, the mobile node can be proved to be a legal user.Then the LEO node computes a shared session key $SK_{ml} = h_5(F_1'||E_2||E_i||T_t)$ where T_t is the chosen new timestamp. Then sends $\langle SK_{ml}, T_t \rangle$ to MN.
4. Once the message is received at MN via UAV, then MN checks if $T' - T_t \leq \Delta T$ then computes $SK_{ml}' = h_5(F_1||E_2||E_i||T_t)$. Later checks if $SK_{ml}' \stackrel{?}{=} SK_{ml}$ matches or not else terminates the authentication (Fig. 2).

3.3.1 Proof of Correctness

Theorem 1 : The Mobile Node (MN) and LEO satellite node establishes an identical session key SK_{ml} after the mutual authentication.

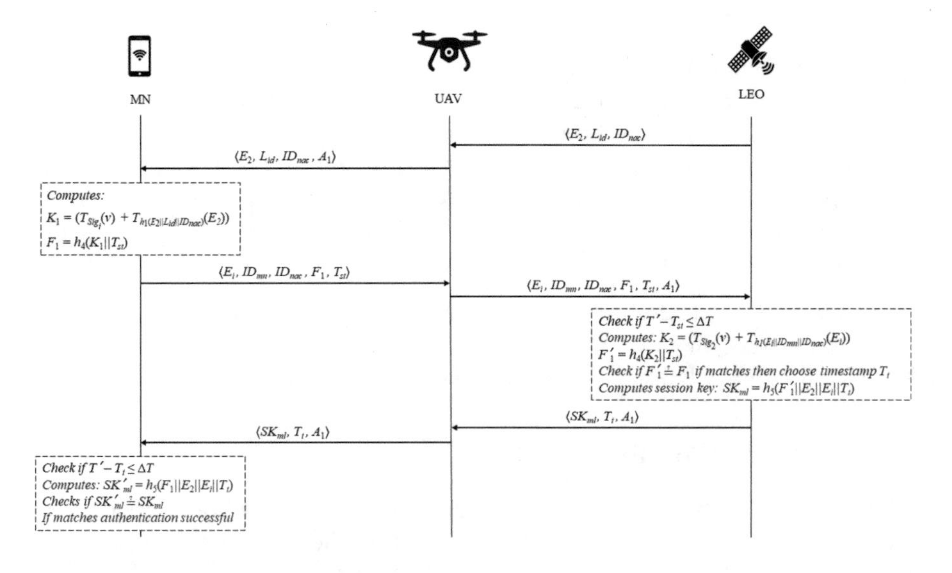

Fig. 2. Authentication phase

Proof : To prove the theorem, it is necessary to verify whether $K_1 \stackrel{?}{=} K_2$. As the session key, $SK_{ml} = h_5(E_2||E_i||T_t||h_4(K_2||T_{st})) => SK'_{ml} = h_5(E_2||E_i||T_t||h_4(K_1||T_{st}))$.

$$
\begin{aligned}
K_1 &= (T_{Sig_i}(v) + T_{h_1(E_1||L_{id}||ID_{nac})}(E_2)) \\
&= (T_{h_1(E_i||ID_{mn}||ID_{nac})\cdot X_i \cdot s_{nac}}(v) + T_{h_1(E_1||L_{id}||ID_{nac})}(T_{X_{na}}(P_{nac}))) \\
&= (T_{h_1(E_i||ID_{mn}||ID_{nac})}(T_{X_i}(T_{s_{nac}}(v))) + T_{h_1(E_1||L_{id}||ID_{nac})}(T_{X_{na}}(T_{s_{nac}}(v)))) \\
&= (T_{h_1(E_i||ID_{mn}||ID_{nac})}(T_{X_i}(P_{nac})) + T_{h_1(E_1||L_{id}||ID_{nac})\cdot X_{na} \cdot s_{nac}}(v)) \\
&= (T_{h_1(E_i||ID_{mn}||ID_{nac})}(Ei) + T_{Sig_2}(v)) = K_2
\end{aligned}
$$

4 Security Analysis

In this section, a formal analysis using the automated tool ProVerif is discussed followed by an informal (non-mathematical) security analysis of the proposed scheme.

Formal Analysis using ProVerif
In this section, the formal verification tool named ProVerif [12] is used to verify the security of the proposed protocol based on the symbolic Dolev-Yao attacker model. The tool used pi-calculus language for describing and analyzing protocols. ProVerif supports several cryptographic properties and enables session simulation to determine whether the correctness of the protocol can be proved. The security properties can be mentioned using "Query" in the code. The results can be analyzed based as follows.

- If the "RESULT [Query] is true", it depicts that an adversary fails to launch any attack.
- If the "RESULT [Query] is false", it depicts that an adversary can compromise the prototype against its claimed security property.
- If the "RESULT [Query] cannot be proved", ProVerif can not prove the state of the security prototype whether "true" or "false".

The protocol is executed between two participants the mobile node MN and the LEO node. The results for the four queries are demonstrated in Fig. 3. The events are executed in the simulation process *RESULT not attacker*(*K1*[ ]) *is true*, *RESULT not attacker*(*K2*[ ]) *is true* and, *RESULT not attacker*(*snac*[ ]) *is true*. This shows that the functions *K1* and *K2* used for compiling the session key, NAC's secret key are secured. In addition, *event*(*MNAuth*(*id*)) means that LEO satellite node authenticates MN such that *event*(*MNStart*(*id*)) must be executed before verification at LEO node end. Thus, the scheme is verified under ProVerif.

```
- Query not attacker(K1[])
Completing...
Starting query not attacker(K1[])
RESULT not attacker(K1[]) is true.
- Query not attacker(K2[])
Completing...
Starting query not attacker(K2[])
RESULT not attacker(K2[]) is true.
- Query not attacker(snac[])
Completing...
Starting query not attacker(snac[])
RESULT not attacker(snac[]) is true.
- Query event(MNAuth(id)) ==> event(MNStart(id))
Completing...
Starting query event(MNAuth(id)) ==> event(MNStart(id))
RESULT event(MNAuth(id)) ==> event(MNStart(id)) is true.
```

Fig. 3. Result of symbolic analysis using ProVerif

Informal Security Analysis

In this section, the security attributes of the proposed scheme is analyzed, considering mutual authentication, impersonation attack, session key secrecy, replay attack, and key leakage attack.

Mutual Authentication

In the proposed scheme, at the authenticated phase the LEO satellite node verifies the authenticity of Mobile Node (MN) by checking if $F_1 \stackrel{?}{=} F_1'$. Here, the function $F_1 = h_4(K_1||T_{st})$ is shared by MN which is mapped with the computed parameter $F_1' = h_4((T_{Sig_2}(v) + T_{h_1(E_i||ID_{mn}||ID_{nac})}(E_i))||T_{st})$. If it fails, then LEO will drop the communication with MN. Similarly, MN verifies the authenticity of the session key shared by the LEO node as, $SK_{ml}' \stackrel{?}{=} SK_{ml}$ where $SK_{ml}' = h_5(F_1||E_2||E_i||T_t)$. If matches, then both LEO and MN can use the session key for future communication. Hence, the proposed protocol meets the security requirement of mutual authentication.

Impersonation Attack

In the proposed scheme, it is infeasible for an adversary to impersonate the mobile using exchanged messages as he/she has to obtain E_i which is uniquely created by NAC during the registration phase of a mobile node. In addition, the identity of the mobile node, ID_{mn} is hashed with a random number r_m which is shared through a secured channel to NAC by MN. Therefore, it is difficult for an attacker to compromise the identity of any mobile node during the authentication phase.

Session key Secrecy

In the proposed scheme, the session key $SK_{ml} = h_5(F_1'||E_2||E_i||T_t)$ includes parameters E_2, E_i, and F_1'. Where E_i and E_2 are based on the CMDH problem, any adversary without the knowledge of (r_i, r_m) and s_{nac} cannot calculate the function Sig_i and Sig_2. Therefore it is infeasible for an adversary to calculate $SK_{ml}^{'}$ and $F_1^{'}$ without the prior knowledge of the above parameters.

Replay Attack
The proposed scheme is resistant against replay attacks, due to the usage of random numbers and timestamps. When MN sends the message tuple, $\langle E_i, ID_{mn}, ID_{nac}, F_1, T_{st}\rangle$, it uses the timestamp function T_{st}, random numbers r_m, and signature Sig_i. Therefore, when the LEO node receives the message tuple, it first verifies the tolerable time delay $\triangle$T then it performs the mutual authentication before accepting the authentication request from the MN node. Similarly, when the LEO sends the message tuple, $\langle SK_{ml}, T_t\rangle$ then the mobile node also performs the timestamp verification before computing the session key for future communication.

Key Leakage Attack
In 2005, Bergamo *et al.* [13] described an attack, where an adversary can obtain an integer solution b from the equation $u = T_b(v)$ if both u and v are known with $v \in [-1, +1]$. The proposed scheme resists this attack in two ways. First, $T_{X_i}(P_{nac}), T_{X_{na}}(P_{nac}), T_{Sig_i}, T_{Sig_2}, T_{h_1(E_2||L_{id}||ID_{nac})}(E_2)$, and $T_{h_1(E_i||ID_{mn}||ID_{nac})}(E_i)$ are sent securely through one way hash function. Second, the proposed scheme uses extended chaotic maps, where the random numbers are chosen in the range of [-∞, +∞]. Therefore the proposed scheme can resist Bergamo *et al.'s* attack.

5 Performance Analysis

In this section, the performance analysis of the proposed Chaotic map-based access authentication protocol is compared with existing competent schemes namely, [14,15], and [16]. The test environment to simulate the communication between MN and LEO node uses a laptop assumed to be acting like an LEO satellite running Windows 10 and 64-bit Intel(R) Core(TM) i5-8250U CPU @1.6 GHz, 8.00 GB RAM. In addition, the mobile node is assumed to be a Rasberry Pi 4 Model(B), Quad-core cortex-A72, 64-bit, CPU @1.4 GHz, 8.00 GB RAM. The cryptographic operations are compiled in both the setups for Chebyshev polynomial and elliptic safe curve x25519 for comparison with 160 bits in milliseconds (ms).

The time complexity for executing the elliptic curve scalar point multiplication in Rasberry Pi is $T_{ecc_m} = 22.56$ ms, Laptop is $T_{ecc_l} = 31.29$ ms. The time complexity for executing the Chebyshev chaotic map in Rasberry Pi is $T_{ch_m} = 4.31$ ms, Laptop is $T_{ch_l} = 11.87$ ms. The computation time for the hash is $T_h = 0.1$ ms which is negligible therefore it is omitted. The computation cost of Meng *et al.* [14] is $5T_{ecc_m} + 4T_h$ for MN whereas $3T_{ecc_l} + 2T_h$ for LEO node. Furthermore, the computation cost of Xue *et al.* [15] is $3T_{ecc_m} + 3T_h$ for MN whereas $3T_{ecc_l} + 4T_h$ for LEO node. In addition, the computation cost of Guo *et al.* [16] is $3T_{ecc_m} + 5T_h$ for MN whereas $3T_{ecc_l} + 4T_h$ for LEO node. Finally, the proposed scheme has $3T_{ch_m} + 6T_h$ for MN whereas $3T_{ch_l} + 3T_h$ for LEO node.

Therefore, it is clearly depicted that the proposed scheme has a better computation cost compared with existing ones. The comparison of computation time complexity is presented in Fig. 4 and Fig. 5 for mobile nodes and LEO nodes respectively.

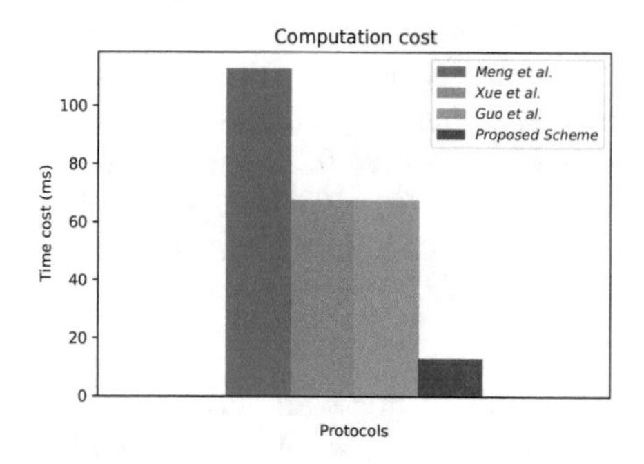

Fig. 4. Computation cost for mobile node

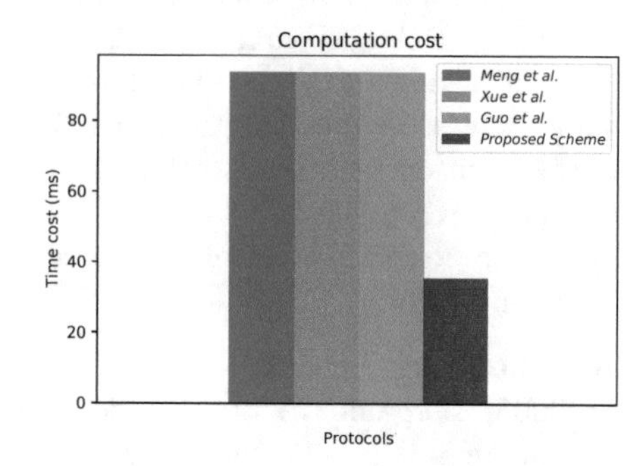

Fig. 5. Computation cost for LEO satellite node

6 Conclusion

This paper proposes a secure lightweight access authentication protocol for remote communication using SIN to prevent the challenges of illegal data access and authentication delay. The proposed scheme allows LEO satellite nodes to directly authenticate the mobile nodes without the involvement of NAC each time to access the SIN through UAV. The paper shows rigorous security analysis to prove that the scheme can withstand known attacks. The performance analysis depicts that the proposed protocol is efficient in reducing the computation overhead using Chebyshev chaotic-maps. Therefore, it is applicable for resource-constrained devices for remote connectivity.

Acknowledgement. This work is supported by Science and Engineering Research Board (SERB), a statutory body of the Department of Science and Technology, Govt. of India under Grant No. SRG/2020/002458.

References

1. Liu, J., Shi, Y., Fadlullah, Z.M., Kato, N.: Space-air-ground integrated network: a survey. IEEE Commun. Surv. Tutorials **20**(4), 2714–2741 (2018)
2. Cruickshank, H.S.: A security system for satellite networks. In: Fifth International Conference on Satellite Systems for Mobile Communications and Navigation, pp. 187–190. IET (1996)
3. Hwang, M.S., Yang, C.C., Shiu, C.Y.: An authentication scheme for mobile satellite communication systems. ACM SIGOPS Operat. Syst. Rev. **37**(4), 42–47 (2003)

4. Zhang, L.: Cryptanalysis of the public key encryption based on multiple chaotic systems. Chaos, Solitons Fractals **37**(3), 669–674 (2008)
5. Fraire, J.A., Henn, S., Dovis, F., Garello, R., Taricco, G.: Sparse satellite constellation design for LoRa-based direct-to-satellite Internet of Things. In: GLOBECOM 2020–2020 IEEE Global Communications Conference, pp. 1–6. IEEE (2020)
6. Kumar, U., Garg, M.: CCAKESC: chaotic map-based construction of a new authenticated key exchange protocol for satellite communication. Int. J. Satell. Commun. Netw. (2021). https://doi.org/10.1002/sat.1435
7. Qi, M., Chen, J., Chen, Y.: A secure authentication with key agreement scheme using ECC for satellite communication systems. Int. J. Satell. Commun. Network. **37**(3), 234–244 (2019)
8. Xu, S., Liu, X., Ma, M., Chen, J.: An improved mutual authentication protocol based on perfect forward secrecy for satellite communications. Int. J. Satell. Commun. Network. **38**(1), 62–73 (2020)
9. Altaf, I., Akram, M.A., Mahmood, K., Kumari, S., Xiong, H., Khan, M.K.: A novel authentication and key-agreement scheme for satellite communication network. Trans. Emerg. Telecommun. Technol. **32**(7), e3894 (2021)
10. Lai, C., Chen, Z.: Group-based handover authentication for space-air-ground integrated vehicular networks. In: ICC 2021-IEEE International Conference on Communications, pp. 1–6. IEEE (2021)
11. Dolev, D., Yao, A.: On the security of public key protocols. IEEE Trans. Inf. Theory **29**(2), 198–208 (1983)
12. Blanchet, B., Smyth, B., Cheval, V., Sylvestre, M.: ProVerif 2.00: automatic cryptographic protocol verifier, user manual and tutorial. Version from, pp. 05–16 (2018)
13. Bergamo, P., D'Arco, P., De Santis, A., Kocarev, L.: Security of public-key cryptosystems based on Chebyshev polynomials. IEEE Trans. Circ. Syst. I Regul. Pap. **52**(7), 1382–1393 (2005)
14. Meng, W., Xue, K., Xu, J., Hong, J., Yu, N.: Low-latency authentication against satellite compromising for space information network. In: 2018 IEEE 15th International Conference on Mobile Ad Hoc and Sensor Systems (MASS), pp. 237–244. IEEE (2018)
15. Xue, K., Meng, W., Li, S., Wei, D.S., Zhou, H., Yu, N.: A secure and efficient access and handover authentication protocol for Internet of Things in space information networks. IEEE Internet Things J. **6**(3), 5485–5499 (2019)
16. Guo, J., Du, Y., Zhang, Y., Li, M.: A provably secure ECC-based access and handover authentication protocol for space information networks. J. Netw. Comput. Appl. **193**, 103183 (2021)

Sensor Placement Strategy for SHM: Application of the Great Mosque of Sfax

Wael Doghri[1], Ahlem Saddoud[2(✉)], and Lamia Chaari Fourati[3,4]

[1] Engineering School of Sfax, University of Sfax, Sfax, Tunisia
wael.doghri@stud.enis.tn
[2] Faculty of Economics and Management, University of Sfax, Sfax, Tunisia
ahlem.saddoud@fsegs.usf.tn
[3] Higher Institute of Computer Science and Multimedia, University of Sfax, Sfax, Tunisia
lamia.chaari@isims.usf.tn
[4] Digital Research Center of Sfax (CRNS), Laboratory of Signals, systeMs, aRtificial Intelligence and neTworkS (SM@RTS), Sfax, Tunisia

Abstract. Structural Health Monitoring (SHM) is progressively developing as an essential process to insure maintenance and conservation of the structural heritage. SHM provides not only the safety of the building but also its historical and cultural value. Nowadays, Wireless Sensor Networks (WSN) are widely used for SHM presenting a prominent candidate to solve many issues such as sensors placement. Therefore, a sensor placement strategy is needed considering the fragility and the importance of the heritage buildings. In this paper, we propose sensors placement methods applied on the historical monument "The Great Mosque of Sfax". Our method is based on the Finite Element Modeling (FEM) to carry out the mesh model of the structure arches and to identify two types of the arch zones; stressed and unstressed zones. Based on FEM results, we determine the appropriate sensor positions to maximize the covered surface, given a limited number of sensor.

Keywords: Sensors placement · Structural Health Monitoring · Wireless sensors networks · Finite element modeling

1 Introduction

In civil engineering, the structure state estimation is a crucial process for existing architectural heritage. Structural Health Monitoring (SHM) is progressively developing as an essential mechanism to insure a decision-making process for maintenance and conservation of the structural heritage around the world. SHM provides not only the safety of the building but also its historical and cultural value. To ensure the safety of many existing architectural heritage, both SHM and building pathology are essential to specify the structural damage of the building. Pathology is the scientific study of diseases and anomalies according to the Cambridge dictionary. In civil engineering, Structural Pathology study (SPS) is the study of abnormal transformations or modifications of a structure or buildings in order to discover new methodologies to approach pathology

L. Barolli et al. (Eds.): AINA 2022, LNNS 449, pp. 238–248, 2022.
https://doi.org/10.1007/978-3-030-99584-3_21

problems of the structure and to make it in a safe state. Moreover, we can admit that SPS is composed of three major parts. The first part consists in identifying and determining the nature of structure defects. Then, recommending the appropriate course of action will be the second parts. The last part involves the monitoring and the supervising of the remedial work. To do these tasks, appropriate sensors are used in the purpose to extract several response parameters such as stress and displacement. Nevertheless, SHM and SPS are becoming an essential application of Wireless Sensor Networks (WSN) dealing with bridges, buildings, air-crafts and especially architectural heritage. Due WSN advantages, we believe that WSN systems are well deployed for SHM.

In [1], a combined optimal sensor placement scheme is proposed for SHM of bridge structures. This strategy named MSE-AGA is based on an hybrid method using the adaptive genetic algorithm (AGA). The objective is to determine the sensor number and its locations using AGA. The research [2] presents a sensor placement method for a multi-story structure to detect the damage locations. Using the simulated annealing algorithm, the predicted method gives estimates of the quantity and location of damage. Authors in [3] propose a method for optimal sensor placement in structural health monitoring applications based on a Bayesian experimental design approach. In order to define the optimal sensor configuration, the procedure is coupled with a Covariance Matrix Adaptation Evolution Strategy optimization scheme. Another method for detecting the appropriate positions of sensors in a building is proposed in the research [4]. The method is based on the eigenmodes of the monitored building, the mass ratio and the participation factors of each modes. Analysis results of the proposed scheme are carried based on software tools such spreadsheet and Matlab. The contribution [5] represents an application of sensor placement to the architectural heritage of Monastery of San Jerónimo de Buenavista. The proposed scheme aims to improve cost efficiency through a model identification with a limited number of sensors by applying different models dealing with the optimization sensor placement techniques. The Optimal sensor placement problem is investigated in the contribution [6]. The proposed method aims to define the optimal sensor layout ensuring the maximum dynamic information of the building with a limited number of sensors.

Our paper presents an application of the sensor placement in SHM for the heritage structure Great Mosque of Sfax. We aim to determine the suitable sensor positions, given a limited number of sensors with the maximum covered surface. The rest of this paper is organized as follows. In the next section, an overview of the Great Mosque is provided. The finite element modeling and analysis are detailed in Sect. 3. Then, sensor placement strategy is presented in Sect. 4. Finally, the conclusion and future works are drawn in Sect. 5.

2 The Great Mosque: An Overview

Structural health monitoring is the diagnosis of the state of the different parts and materiel which constitute and maintain the correct functionality of a structure [7]. The SHM main function is to remain the state of the structure as it is in designed. Thanks to the time dimension aspect of the SHM, it allowed the supervising the deformation or the states changes due to using the structure or natural causes. Like all technologies, theoretically, the main role of SHM is to save humans lives [8], that's why SHM intervene in many civil domains: towers, bridges, critical infrastructures, plane, etc. believing that saving humanity culture and popular memory as important as saving human themselves, we choose to apply our structural health monitoring system on a critical infrastructure which is an historical monument called "The Great Mosque" or "Elgamaa Elkibir". In 2012, as it is one of few medieval cities that preserve its original architecture and design and it is the only city that the wall around it is still standing according to professor MOKNI, UNESCO listed the old city of Sfax "EL Madina" in UNESCO World Heritage site.

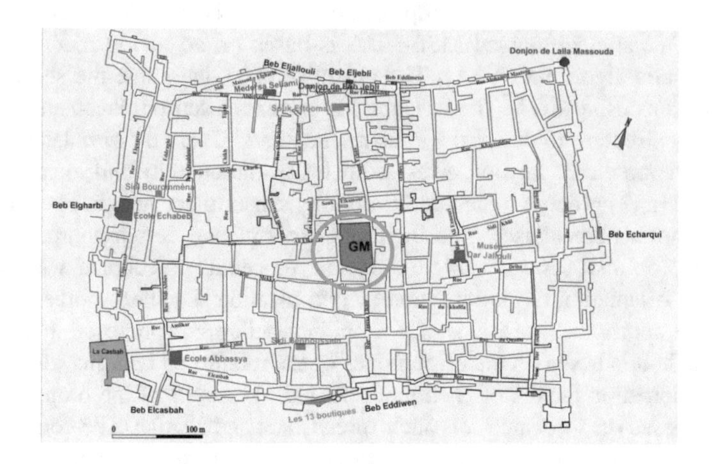

Fig. 1. The map of El Madina of Sfax

Fig. 2. A view of the Great Mosquee of Sfax and its plan

Due to its importance in the Islamic civilization and arabo-Islamic engineering, the Great Mosque (GM) mediate the old city overlooking the main road and the only exit of the walled city as shown in Fig. 2. The GM built in the ninth century with stones, clay and olive sticks based on the Islamic architecture. The GM has two domes and twelve meters minaret. As we can see in Fig. 2, The GM respects one hundred percent the Islamic architecture : a non-covered side called "esshan" and the covered side called "bayt salet" which is a combination of similar arches. Arches can be combine in a linear way or on 3d way to combine a cube. In our paper, we will deal with the elementary component and limited in studying one arch behavior when it exposed to different forces.

3 Finite Element Modeling and Analysis

Our study is based on the Finite Element Modeling (FEM) in order to get the mesh model of one structure arch and also to identify different displacement. We carry out a simulation using sketch up for designing the system and MATLAB tools to realize the FEM study.

3.1 Finite Element Modeling

The FEM is the study of system physical behavior conditions of limits points. According to Sergio R. Idelsohn [9] those points represent the result of the meshless method which is an algorithm satisfying two constraints:

- The shape functions depend only on the node positions: the whole system is related to the position of nodes
- The evaluation of nodes connectivity is bounded in time and depends only on the total number of nodes in the domain: the reason for using meshing there is no reason to use meshing without bounded evaluation of nodes.

On each finite element, we realize a linearization of the partial differential equations (PDE) the physical behavior of a system, that's means, we replace each PDE equation by a system of a linear equation, one matrix for each finite element. The system resolve simultaneously and the whole matrix is regrouped in one global matrix in the end of the calculation.

3.2 FEM Applied on Arch

Based on several studies of arches monitoring, we get different arch architectures as it is cited in reference [10]. In our study, we choose the simple round design of the structure arches.

3.2.1 Arch Meshing

To produce FEM of the specific structure, arch in our case, first of all we have to go through meshless phases. The finite element mesh used to get meshless arch is realized via MATLAB tools using *generateMesh ()* function with target maximum mesh edge length (Hmax) equal to 0.01. After meshing the arch structure, as we can see in Fig. 3, we used the triangulation as shape function to evaluate the nodal connection so we can extract the boundary node that we can use to evaluate the finite element method performances: structure displacement on different axes and stress zone.

3.2.2 FEM Results

To get the FEM results, we have to prepare our model using MATLAB tools. We decompose the arch into faces and we describe each face role using *protrusionFace ()*. In other words, we have to define the fixed faces and determine faces that will be susceptible to exerted forces using the function *applyBoundryCondition ()*. Finally, using the *solve ()* command, we produce the solution and we visualize it via *pdeplot3D ()* command and as it is shown in Fig. 4. Moreover, the fem study is based on two criteria: displacement of each dimension of the three and some stressed zone.

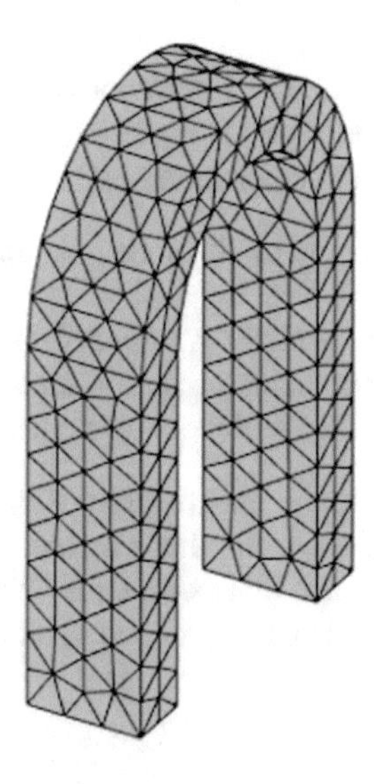

Fig. 3. Arch meshing

According to Cunlong Li et al. In [11] and based on the displacement formula $\Delta R = c\Delta\tau/2$, the SHM displacement equation could be written as follow:

$$\Delta R = \frac{c}{4\pi(f_0 + \frac{N-2B}{2N})}\Delta\phi_b \quad (1)$$

Where R is the distance from one stationary target to the sensor. τ is the round-trip time delay from transmitter to receiver and c is the propagation velocity of electromagnetic wave. f_0 is the initial frequency. $\Delta\phi_b$ is the beat signal phase b. B is the bandwidth of transmitted microwave. N is the number of recorded samples during one period T. We start with displacement [o,x) axe, as we can see the red zone which is the extremity of the arch in x direction is the most affected zone, that mean it have greater proportion to be deformed, and as we go in -x direction as the zones are less affected. However, for the prospect of displacement on y direction, we detect the far pier of the arch [12] and the intrados of the near pier [13], have more ability to be affected. However, for the probable displacement on $[o, z)$ axe is expected: the arch architecture major role is to split the forces exerted on it top (extrados) and transference those forces to the piers, that's why we find the two bases of piers are in dark red which mean the most susceptible for displacement. As for the stress probable zone, we notice that for the (o, x, y) and (o, y, z) plans the stress zone are located in the arch curve especially in the inside face (directly exposed to the exerted force). We did not present other plans result because the stress zone is evenly spread with moderate charge. Like the exerted force are parallel to z axes, it is normal that the stress zone are significantly present were we present the model projected on [o,z) axes as we can see in the Fig. 4.

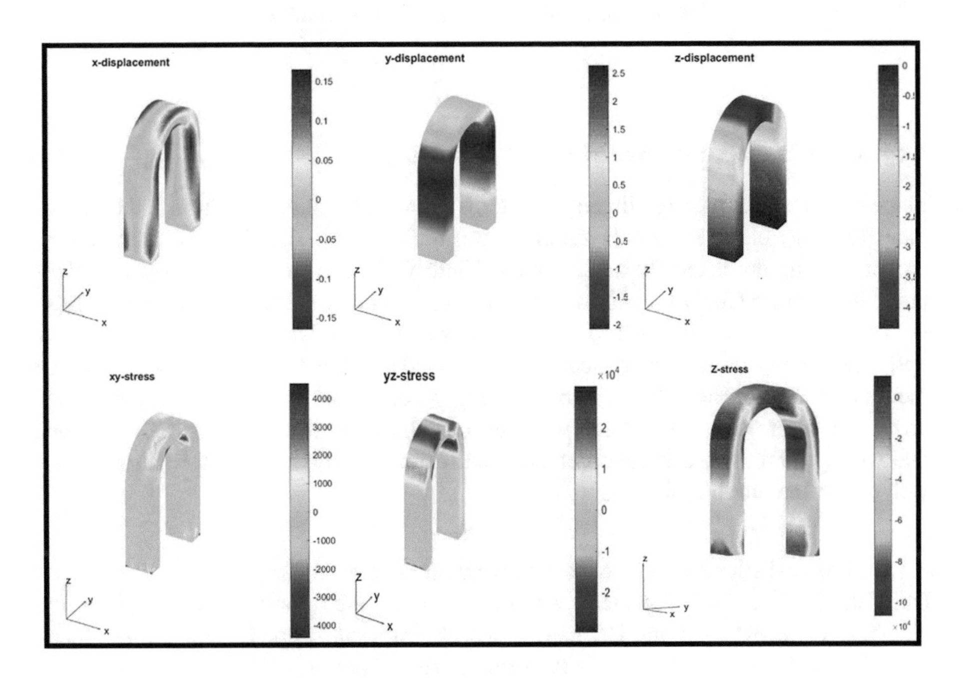

Fig. 4. FEM results

4 Sensor's Placement Strategy

We consider two zones in the arch surface named stressed and unstressed zones. In Fig. 5, the stressed zones are illustrated with red color and the rest of the arch surface is considered as unstressed zone. Two sensor's placement methods are presented in this section.

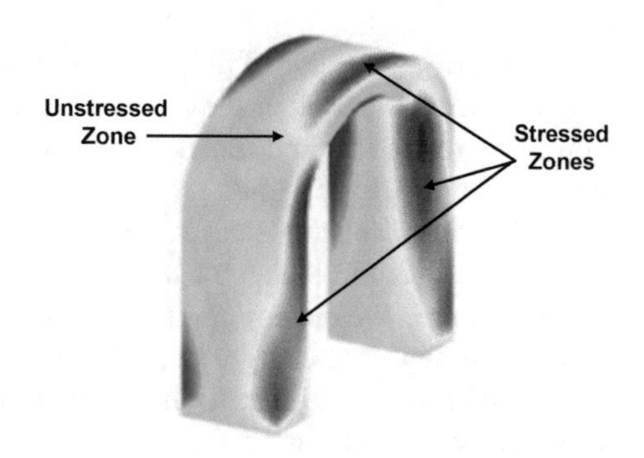

Fig. 5. Unstressed and stressed zones illustration

4.1 Sensor's Placement Method for Stressed Zone

Since the integration of intelligent monitoring in the structure monitoring, the implementation cost of the sensors is the major obstruction [14]. To reduce installation cost and preventing the detection quality and reliability, we use the optimization algorithm [15]: first step we choose an arbitrary pre-placement or a pre-placement respect a statistic low, after that we fix our objective and our constraint, based on those criteria our optimization algorithm start reduce the pre-selected placement and keep only the placement necessary for the network good working. In our paper we will admit an unorthodox method for covering the arch with sensors. Based on the FEM result, we remark that we can admit two different methods to monitor the arch shape deformation with a well determine number of sensors.

4.1.1 Parallel Elementary Sensors Placement According to the FEM output, especially the predicted displacement, we notice that the arch could be divided into five elementary pieces: right and left pier, called also vertical support, two quarter arches separated with the crown which is the fifth component and the central one. As we can see in Fig. 6, based on the FEM output (Fig. 4), we remark that there is two types of arch displacement performances: the external layer of the arch or all the deep of the arch have the same performance long of surface. That's why we modeled the arch based on 2D model and our system is limited in four sensors. In this method of placement, we used only four sensors each two of them are installed symmetrically opposed: the first pair in arch median and the second one in the left and right pier. If the two sensors lose linearity and a distance Ω appears that mean there is a change set in the arch shape. Ω could be an elementary (negligible) or functional distance: the architect and the civil engineer have to define the upper limit of the functional Ω.

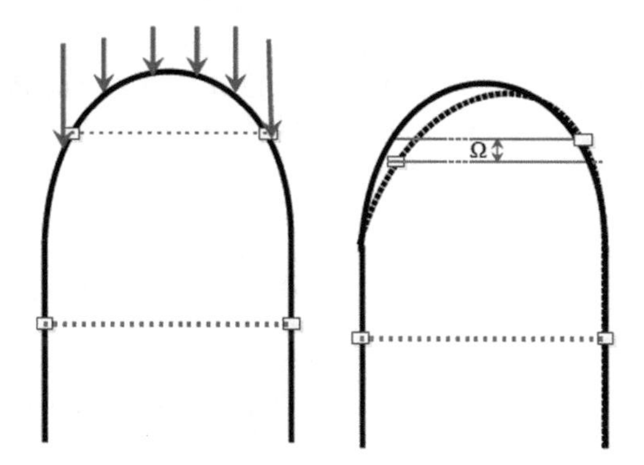

Fig. 6. Parallel elementary sensors placement

4.1.2 Angular Elementary Sensor Placement However, we detected five parts of the arch from the fem output, in this section the AESP will based on the classic geometry: according to symmetry theory our arch could be vertical divided into two equivalent parts. In the same context our sensors placement philosophy will be based on the classic geometry especially the trigonometry. In AESP our strategy is limited only on three sensor which present the vertices of the triangle. As we can see in Fig. 7, we hold the two sensors fixed in piers and the third sensor is placed in median of the crown. We suppose α the angle which is head the crown sensor and β the same angle after a shape modification on the arch, $(\beta - \alpha)$ must not surpass γ. Where γ is the tolerant value of

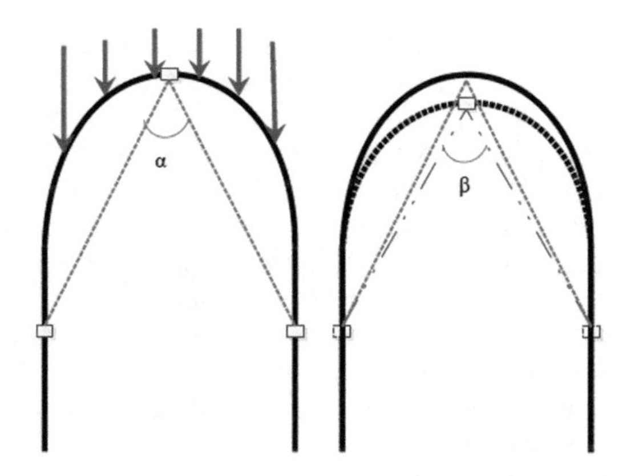

Fig. 7. Angular elementary sensors placement

the shape transforming, which is necessary for the good function of the structure, fixed by the metal and the civil engineer.

4.2 Sensor's Placement Method for Probable Unstressed Zone

The placement strategy in the probable unstressed zone is expressed by a mathematical model given by the objective function (a). Our intention is to minimize number of final sensors placement to cover the whole structure surface. Thus, the objective function and its constraints are detailed as follows:

$$\min Z = \sum_{i=0}^{k} \sum_{j=0}^{y} N_{i,j} \cdot A_{i,j} \tag{a}$$

$$s.c. \sum_{i=0}^{k} \sum_{j=0}^{y} A_{i,j} X_{i,j} \geq X_T \tag{b}$$

$$A_{i,j} \in \{0,1\}; \forall i \in \{0,k\}; \forall j \in \{0,y\} \tag{c}$$

$$\sum_{i=0}^{k} \sum_{j=0}^{y} A_{i,j} \leq 1; \forall i \in \{0,k\}; \forall j \in \{0,y\} \tag{d}$$

$$\sum_{i=0}^{k} \sum_{j=0}^{y} N_{i,j} > 0; \forall i \in \{0,k\}; \forall j \in \{0,y\} \tag{e}$$

$$\sum_{i=0}^{k} \sum_{j=0}^{y} N_{i,j} \leq \sum_{i=0}^{k} \sum_{j=0}^{y} N_{p_{i,j}}; \forall i \in \{0,k\}; \forall j \in \{0,y\} \tag{f}$$

As a beginning, we have a number of points distributed randomly over the whole structure. These points are designed by points of interest which are the potential sensors positions, each point is characterized by two coordinates (i,j). Our main objective is to determine the minimum number of final sensors placement to cover the whole unstressed zone. N_{ij} is a sensor located in the coordinates (i,j). A_{ij} is a variable that defines whether the chosen position is potential or not: A_{ij} takes 1 if the position is chosen and 0 else. In order to solve the given objective function, a set of constraints are necessary. In constraint (b), X_{ij} presents the covered area of the sensor. Having the global structure area X_T, constraint (b) ensures that the lower summation of the chosen positions must be equal or upper than the global structure area. The constraints (c) and (d) ensure that each potential position can only be affected to one sensor. Constraints (e) and (f) assure that the number of final positions cannot exceed the number of the potential sensors placement and for sure it must be upper then zero. In order to solve the sensor placement problem, the following Algorithm 1 is proposed. We assume that the number of sensor's potential positions denoted by P is equal to 50. These sensors are randomly distributed in an element space with a surface of 50×50 cm^2. The inputs $N(i)$ and $T(i)$ represent respectively x and y coordinates of 50 sensors. Furthermore, we assume that one sensor coverage is equal to 3 cm. The final sensor positions F represent the minimum number of sensors that cover the element space with coordinates $V(i)$ and $U(i)$. As a result, we obtain 22 final positions for sensors illustrated in Table 1.

Algorithm 1. SENSOR PLACEMENT ALGORITHM

```
Require: N(i), T(i) and P
Ensure: V(i), U(i) and F
  for i=2 to P do
    k ← 1
    Test ← True
    while (Test=True & K < i) do
      W ← (N(k) − N(i))^2
      Y ← (T(k) − T(i))^2
      if √(Y + W) ≥ Z then
        Test ← True
      else
        Test ← False
      end if
      k ← k + 1
    end while
    if (Test=True) then
      U(i) ← N(i)
      V(i) ← T(i)
    end if
  end for
  COM ← 0
  for (i=2 to P) do
    if V(i) ≠ 0 then
      COM ← COM + 1
    end if
  end for
  F ← COM
```

Table 1. *Sensor placement results*

Sensor	S1	S2	S3	S4	S5	S6	S7	S8	S9	S10	S11
x-coordinate	34.5	26	38	20.3	21.5	33	10.9	4.1	36.4	45	11
y-coordinate	2.1	37.9	16.2	26.1	47.6	33.4	23.7	32.5	11.5	25.1	35.5
Sensor	S12	S13	S14	S15	S16	S17	S18	S19	S20	S21	S22
x-coordinate	28	49.5	16.8	7.8	20.9	1.5	43.3	10.5	46	5.6	16.3
y-coordinate	23.6	16.2	38.2	49.8	13.3	6.7	1.8	11.5	38.8	41.6	5.6

5 Conclusion

The proposed sensor placement methods are applied on arches of an heritage structure which is the Great Mosque of Sfax. The general target of this paper is to propose a strategy to determine the suitable sensor positions, given a limited number of sensors with the maximum coverage. Based on finite element study applied on the arch of the structure, the mesh model is obtained, different displacements are produced and the stressed and untressed zones are identified. Regarding stressed zone of the arch, both angular and elementary sensor placement are proposed. Furthermore, a mathematical model is given to detail the problem of the optimization sensor placement. We propose the sensor placement algorithm as a solution process to solve the optimization sensor placement problem. Future works will be focused one models with a set of arches using sensors that gives as much information as possible.

References

1. He, C., Xing, J., Li, J., et al.: A combined optimal sensor placement strategy for the structural health monitoring of bridge structures. Int. J. Distrib. Sens. N **5–6**, 1–9 (2013)
2. Mistarihi, M., Kong, Z.: Sensor Placement Optimization and Modeling for Structural Health Monitoring **7**(6) 2349–204, ISSN(e): 2349-204X (2019)
3. Capellari, G., Chatzi, E., Mariani, S.: An optimal sensor placement method for SHM based on Bayesian experimental design and Polynomial Chaos Expansion. In: ECCOMAS Congress 2016, vol. 3, pp. 6272–6282 (2016)
4. Zhou, G.-D., Yi, T.-H., Li, H.-N.: Sensor placement optimization in structural health monitoring using cluster-in-cluster firefly algorithm. Adv. Struct. Eng. **17**, 1103–1115 (2014). https://doi.org/10.1260/1369-4332.17.8.1103
5. García, P., et al.: Evaluation of optimal sensor placement algorithms for the structural health monitoring of architectural heritage. Engineering Structures, Application to the Monastery of San Jerónimo de Buenavista (Seville, Spain) (2019). https://doi.org/10.1016/j.engstruct.2019.109843
6. Sun, H., Büyüköztürk, O.: Optimal sensor placement in structural health monitoring using discrete optimization. Smart Mater. Struct. **24**(12), 125034 (2015)
7. Balageas, D., Fritzen, C. P., Güemes, A. (eds.): Structural Health Monitoring, vol. 90. John Wiley and Sons, Hoboken (2010)
8. Doghri, W., Saddoud, A., Fourati, L.C.: Cyber-physical systems for structural health monitoring: sensing technologies and intelligent computing. J. Supercomput. 1–44 (2021). https://doi.org/10.1007/s11227-021-03875-5
9. Idelsohn, S.R., Onate, E., Calvo, N., Del Pin, F.: The meshless finite element method. Int. J. Num. Meth. Eng. **58**(6), 893–912 (2003)
10. González, G., Samper, A., Herrera, B.: Classification by type of the arches in Gaudi's Palau Güell. Nexus Netw. J. **20**(1), 173–186 (2018)
11. Li, C., Chen, W., Liu, G., Yan, R., Xu, H., Qi, Y.: A noncontact FMCW radar sensor for displacement measurement in structural health monitoring. Sensors **15**(4), 7412–7433 (2015)
12. employees.oneonta.edu/farberas/arth/ARTH-Courses.html
13. The civil engineerings.com/what-is-an-arch-components-of-arch-parts-of-arch-5/
14. Ostachowicz, W., Soman, R., Malinowski, P.: Optimization of sensor placement for structural health monitoring: a review. Struct. Health Monit. **18**(3), 963–988 (2019)
15. Tan, Y., Zhang, L.: Computational methodologies for optimal sensor placement in structural health monitoring: a review. Struct. Health Monit. **19**(4), 1287–1308 (2020)

A Comparative Analysis of Machine Learning Algorithms for Distributed Intrusion Detection in IoT Networks

Moroni N. Vieira[1], Luciana P. Oliveira[2(✉)], and Leonardo Carneiro[2]

[1] IFRN Campus Currais Novos, 773 Manoel Lopes Filho Street, Currais Novos, RN 59380-000, Brazil
moroni.neres@ifrn.edu.br

[2] IFPB Campus João Pessoa, 720 Primeiro de Maio Avenue, João Pessoa, PB 58015-435, Brazil
lpo@cin.ufpe.br, leonardo.nunes@academico.ifpb.edu.br

Abstract. The IoT devices has brought challenges in the area of information security. They have power restrictions and usually use the MQTT and CoAP protocols in plain text. This contributes to these devices being targets of malicious actions or used to attack other smart objects. Consequently, energy-efficient intrusion detection systems and procedures are essential in networks with IoT devices. An alternative for this are detection solutions based on the distribution of processing between devices in the same network domain with an artificial intelligence layer. Therefore, this article analyzed six possible algorithms (Logistic Regression, k-Nearest Neighbours, Gaussian Naive Bayes, Decision Trees, Random Forests and Linear Support Vector Machine) for the AI layer. The analysis measured the capabilities of the algorithms to identify attacks on CoAP and MQTT networks, considering the synthetic traffic of unidirectional and bidirectional flows. The metrics used were the following: energy consumption of hardware components (CPU, RAM, Package and GPU), execution time, precision, accuracy, recall and F1-Score. Finally, it was identified that the bidirectional flow is the type of traffic that was identified with greater precision and the MQTT attack was better identified by the algorithms.

1 Introduction

The three protocols most used in the application layer in security experiments in IoT networks are the CoAP (Constrained Application Protocol), the HTTP (Hypertext Transfer Protocol) and the MQTT (Message Queuing Telemetry Transport) [1]. On the other hand, comparative analysis of these protocols by [2] identified the MQTT and CoAP protocols with lower the values for power consumption, data overhead, latency and bandwidth.

Using these devices with these communication protocols requires Internet Service Providers (ISP) to support this type of traffic. For this, virtualized PoP can be an alternative, as there is a separation between control and data, allowing

L. Barolli et al. (Eds.): AINA 2022, LNNS 449, pp. 249–258, 2022.
https://doi.org/10.1007/978-3-030-99584-3_22

the scaling of services according to the needs of IoT devices. For example, the work of [3] features a virtualized PoP.

However, there are also security challenges. For example, [4] presented a study about security, MQTT, CoAP and UPnP (Universal Plug and Play) protocols in the context of IoT. The tool called honeypot was used in the study to collect attack data and it obtained numbers of attacks directed to the ports of the MQTT (1883), MQTT over SSL (8883), CoAP (5683) and UPnP (1900) protocols. The results showed that the UPnP protocol reached the first position of the analyzed attacks, the MQTT reached the second position and the CoAP reached the third position among the number of counted attacks. It is important to find ways to avoid such attacks, as they can cause interruption and unavailability of information through intrusion techniques called DoS or DDoS directed towards smart objects, these can be used to carry out attacks to other hosts on the network. Moreover, the behavior of attacks can change and it is not appropriate to use a static set of signatures to identify and block traffic.

Consequently, Machine Learning has been used to learn new occurrences of progress. These techniques allow the construction of models that facilitate the identification of classes with precision for a given dataset [5]. This has been motivating the use of ML in IDS services for virtualized PoPs [6,7].

Therefore, the general objective of this article is to compare the performance and power consumption of ML algorithms for MQTT and CoAP network problems regarding aggressive scan, UDP scan and brute force attacks on the ssh service. For this, the following research questions will be analyzed:

- Which attacks can be better detected by ML algorithms?
- Which ML algorithms are best suited for performance and power for attack detection?
- What are the recommendations for achieving a more secure IoT scenario considering MQTT and CoAP?

The organization of this paper to respond the questions is considered as follows. Section 2 presents concepts of CoAP, MQTT, attacks, AI algorithms and distributed IDS and the research methodology used to collect and analyze data. The discussion of the results is described in Sect. 3. Finally, the conclusion and future works are in Sect. 4.

2 Background

2.1 MQTT and CoAP

An element called broker is present in the MQTT network. All traffic goes through the broker, so MQTT has a centralized architecture. In the CoAP network, communication is direct between client and server, there are no messages that go through an intermediary and there is no central element. Therefore, MQTT and CoAP protocols have different architectures, but it is possible to carry out a comparative study, considering the traffic between an IoT device and the broker, which will follow the same behavior of the client/server architecture in the CoAP network.

In both networks, you can identify bidirectional and unidirectional traffic. Usually, the MQTT network uses TCP traffic which is characterized by bidirectional traffic. However, MQTT-SN is a variation of MQTT that allows unidirectional traffic with the UDP protocol. Regardless of the use of TCP or UDP, MQTT messages have a QoS field. When QoS has a value of 0, PUBLISH (publish information) and SUBCRIBE (subscribe) messages are unidirectional, as the broker does not return an acknowledgment. When QoS is 1, PUBLISH and SUBCRIBE messages are bidirectional, as the broker returns an acknowledgment. Furthermore, regardless of the QoS value, every device must first send a CONNECT message to the broker, which responds with a CONNACK message, representing bidirectional traffic.

Usually a CoAP network has UDP traffic, but there are new versions that allow the use of TCP. However, regardless of the use of TCP or UDP, CoAP messages of type CON form part of bidirectional traffic, because when a client sends a CON message, the server returns an ACK message. Unidirectional traffic consists of messages of the NON type, because when the client sends this type of message, it will not receive a reply message from the server.

In this paper, CoAP and MQTT were used in standard mode. CoAP was used with UDP in CON mode. MQTT was configured with TCP in QoS mode with zero value. However, traffic was categorized as unidirectional when a request message does not have an answer and bidirectional when the request has a reply.

2.2 Network Attacks

NMAP and Sparta tools can be used to perform attacks on devices that have an operating system, for example, Raspberry, which is widely used in IoT scenarios. Sparta allows to perform brute force attack on SSH which allows the discovery of login credentials through dictionary attacks.

NMAP has different types of parameters to attack one or more IPs. This paper will consider two types of attacks: aggressive scan and udp scan. The aggressive scan was executed by command "nmap -A target_IP" which represents the action of several existing commands in NMAP. Therefore, it send a lot packets and it is more likely to be detected, but provides a lot of valuable host information. It allows to find the operating system, version, network services, the execution of default scripts in NMAP and identification of a route to a destination. The udp scan was executed by command "nmap -sU target_IP -p port_interval". It sends UDP packets to configured ports. Depending on the port, the packet sent is a specific protocol. If it receives a response, the port is considered open. If the port is closed, Nmap will receive an "ICMP Port Unreachable" message. If Nmap does not receive any response, then the port is open and the program that is listening does not respond to Nmap's query or traffic is being filtered.

2.3 AI Algorithms

The large volume of data produced and made available by IoT scenarios are processed by artificial Intelligence (AI) algorithm [9]. Machine learning (ML) is a subset of AI that has techniques for classifying an input Xi and an unknown qualitative class Yi $\in \{1, 2, ..., L\} \mid L < \infty$s.

Therefore, an ML for classification will produce a classifier capable of generalizing the data, with the objective of classifying them later on objects with an unknown label. Next, six ML algorithms will be presented that will be analyzed in this paper.

LR (Logistic Regression) aims to find the equation that best fits the data set. This dataset can only result in two values (normal traffic or attack traffic). The result can also be used for multiclass problems, that is, it analyzes different elements or sets of variables to determine which object class is in question, through probabilistic estimates, using a logistic function.

KNN (K-Nearest Neighbors) performs the relationship between the variable K and its nearest neighbor. Thus, the value of the variable is considered as input.

GNB (Gaussian Naive Bayes) classify data based on probability, considering the frequency of variables used. In this way, it can be used to identify whether certain traffic is normal (N) or an attacker (A), for example. In this algorithm, first it calculates the occurrence probabilities in both types of traffic, then it is possible to test the received traffic, considering a conditional probability: P(N−A) = P(A−N) * P(N)/P(A).

DT (Decision Tree) builds a tree where nodes are decisions and leaves are the result of the decision, being built with a recursive divide-and-conquer approach. The purpose of the decision tree is to make several subsets of the data in such a way that the subsets get smaller (or just a subset).

RF (Random Forests) builds several decision trees and the more trees created, the better the model results will be, up to a point, where a new tree cannot lead to a significant improvement in model performance. It is important to highlight that the more trees are created, the longer the model creation time will be.

Linear SVM (Support Vector Machine) searches for a line of separation between two distinct classes by analyzing the two points, one of each group, closest to the other class. That is, the SVM chooses the line—also called the hyperplane—between two groups that are further away from each other.

2.4 Distributed IDS

IDS corresponds to a system that aims to detect a threat or intrusion on the network. IDS can be rules-based or AI-based. The first to detect known attacks. The second to recognize new attacks requiring greater energy consumption to obtain the new knowledge. For example, ML has been widely used in IDS (Intrusion Detection System) [6,7].

The IDS architecture approach can be centralized or distributed. Also, IDS in each router allows the distribution of energy consumption to analyze packets along the routing path [10]. ProNet [8] also exemplified how an intelligence layer and distributed algorithm [11] can be used to reduce energy consumption. This justifies the importance of evaluating the performance and power consumption of ML algorithms in the context of IDS and IoT performed in this article.

2.5 Dataset

CoAP and MQTT traffic were collected and generated 8 pcap files: four for each protocol. These files were read by the Tranalyzer software, which transformed the raw data from the network capture into information about the network flows between the network hosts. The next step, therefore, the Tawk and Awkf tools were used through the Linux terminal to select the data and transform them into CSV-type files with the features. That is, in each protocol group with 4 pcaps there are the following traffics:

- normal.pcap which represents the scenario with normal protocol operation
- sparta.pcap which stores the Sparta attack on the network
- scan_A.pcap which contains aggressive scan
- scan_sU. pcap which represents the UDP scan attack

When these were converted to CSV, each was subdivided into directional and bidirectional flow. The first one represents traffic from a source host to a destination host with no response in the communication. The second one represents the communication between the origin and destination with confirmation message for each request. Therefore, there are 8 CSV for each protocol and a total of 16 CVS files.

In these files there are the features that are the columns of the files. They are used by learning models in the learning phase and later to determine the behavior of network traffic. The files contains seventeen features: source IP address, destination IP, source port, destination port, transport layer protocol, number of packets, average between flow arrivals, standard deviation of time between flow arrivals, minimum time between flow arrivals stream, maximum time between stream arrivals, number of bytes, number of PSH flags, number of RST flags, number of URG flags, average packet size, standard deviation of packet size, minimum packet size, maximum packet size, identifying whether it is a packet originated by an attack and identifying whether it is a unidirectional or bidirectional flow.

Instances or occurrences are resource values found in each dataset represented by CSV files. In addition, Table 1 shows the values of these instances for MQTT and CoAP protocols. This table contains for each protocol the file size, number of unidirectional instances of normal traffic (NIUN), number of unidirectional instances of normal traffic (NIUA), number of bidirectional instances of normal traffic (NIBN) and number of bidirectional instances of attacks (NIBA).

Table 1. Table of instances in dataset

Traffic	File size MQTT	NIUN MQTT	NIUA MQTT	NIBN MQTT	NIBA MQTT	File size CoAP	NIUN CoAP	NIUA CoAP	NIBN CoAP	NIBA CoAP
Normal	192.5 MB	172313	X	167684	X	193 MB	487775	X	305522	X
Scan_A	16.2 MB	31881	20295	31553	20278	16.3 MB	73181	73006	73066	72802
Scan_U	41.3 MB	34487	22451	33539	19	41.4 MB	32135	154446	20388	48
Sparta	3.4 GB	1122002	967441	1117901	967441	3.7 GB	901355	865468	886739	864363

Thus, the lines in these files represent the amount of attacks and normal traffic. Thus, ML algorithms use this data for the learning and training process.

2.6 Repetition

Each measurement (execution) of the algorithms was composed by first executing the learning phase and then the training phase. For this, the CSV file was subdivided in a proportion of 75% for the training phase and 25% for the learning phase of each algorithm. For each execution, 250 repetitions were performed in each algorithm to perform the measurements described in the next section. The means of the values considered a confidence interval (CI) of 95%. Furthermore, severe outliers data was avoided, as some initial data were removed to avoid extreme data variations.

2.7 Metrics

The algorithms were evaluated considering five metrics: energy consumption, accuracy, precision, recall and F1-score. The energy consumption was calculated as the average energy consumption of 250 executions of the learning and testing phase using PyJoules. This is a Python language library that estimates the consumption of each of the hardware components (CPU, GPU, Package and RAM). It provides the values with the microjoule unit uJ. The accuracy, precision, recall and F1-score use the criteria of True positive, False positive, False true and False negative are in [13] and conceptualized as follows:

- True positive (PV): is the percentage of positive cases (true attacks) classified as belonging to cents to the real positive class;
- False positive (FP): is the percentage of negative cases (false attacks) classified incorrectly as belonging to the positive class;
- False true (VN): is the percentage of negative cases classified correctly as belonging to the negative class. In other words, the instance quantity that does not represent an attack;
- False negative (FN): is the percentage of positive cases classified incorrectly as belonging to the negative class. This represents normal misclassified traffic.

The accuracy is represented in the Eq. 1. It is the value obtained by the ratio between the total number of correct predictions over the total number of predictions. The precision indicates what fraction of the estimated classes were actually detected as true. It is presented in Eq. 2 and the smaller FP value, the greater the precision. The Recall is the proportion of hits when classifying elements of a certain class within the ML model. It is presented in Eq. 3 and the smaller the FN value, the higher the recall value. F1-score is a harmonic balance between the precision and recall metrics, used in order to analyze the two previous measurements in a uniform way. F1-score is presented in Eq. 4.

$$Accuracy = \frac{PV + VN}{TP + FP + VN + FN} \tag{1}$$

$$Precision = \frac{PV}{PV + FP} \tag{2}$$

$$Recall = \frac{PV}{PV + FN} \tag{3}$$

$$F1\text{-}Score = \frac{2}{\frac{1}{Precision} + \frac{1}{Recall}} \tag{4}$$

3 Results

3.1 Accuracy of ML Algorithms

The results in project site identified that most algorithms applied to MQTT traffic presented greater accuracy.

The Table 2 contains the accuracy of ML algorithms, considering MQTT and CoAP. It is possible to observe that most algorithms applied to MQTT traffic presented greater accuracy. Only LR and Linear SVM algorithms showed greater accuracy with CoAP unidirectional traffic. Analyzing only each protocol, the most accurate algorithm for MQTT was KNN and DT for CoAP.

The algorithms showed greater accuracy in the bidirectional traffic of CoAP and MQTT. A fact also observed by the work that analyzed only the MQTT in [12]

Table 2. Table of accuracy of ML models for MQTT and CoAP

Protocol	DT	KNN	GNB	RF	LR	SVM linear
Unidirectional MQTT	99.977	100.000	99.222	99.954	99.954	58.861
Bidirectional MQTT	99.854	99.951	99.848	99.951	99.946	99.928
Unidirectional CoAP	99.958	98.994	47.354	99.948	80.138	67.606
Bidirectional CoAP	99.855	99.822	98.849	99.855	99.735	99.742

3.2 Weighted Average (WA) for Accuracy, Recall and F1-Score

The Table in[1] contains full metrics information for each type of attack on MQTT and CoAP networks. These are UR (Unidirectional Recall), BR (Bidirectional Recall), UP (Unidirectional Precision), BP (Bidirectional Precision), UF (Unidirecional F1-Score) and BF (Bidirectional F1-Score). The tables also show the weighted average (WA) of attacks considering bidirectional and unidirectional traffic for MQTT and CoAP.

Analyzing WA and the MQTT network, the best algorithm was KNN, which obtained an average of 99.974 the second best was RF, which obtained an average of 99.952 and the third was DT, which obtained an average of 99.923. For a CoAP network, the best algorithm was DT, which obtained an average of 99.904 the second best was RF, which obtained an average of 99.732 and the third was KNN, which obtained an average of 99.401. Furthermore, these were the best algorithms to identify attacks on CoAP and MQTT networks when analyzing unidirectional traffic. The NB, RL and linear SVM algorithms presented a higher WA when using bidirectional traffic.

In MQTT and CoAP networks, most algorithms have identified the sparta attack. In the MQTT, only the KNN algorithm better identified the Scan_A attack. In the CoAP, the KNN and RF algorithms better identified the Scan_A attack.

3.3 Execution Time

The images in Fig. 1 present the results for the ML algorithms applied to the flow unidirectional and bidirectional. It can be observed that all algorithms had a shorter execution time to learn and detect attacks in the MQTT network regardless of whether the flow is unidirectional or bidirectional.

Analyzing separately, in unidirectional traffic, all algorithms showed similar behavior in MQTT and CoAP network. In both cases, SVM was the slowest algorithm and the fastest were decision tree and Naive Bayes. Although these two algorithms were also the fastest in bidirectional traffic over the MQTT and CoAP networks, on the other hand, the slowest in MQTT was logistic regression and in CoAP the KNN.

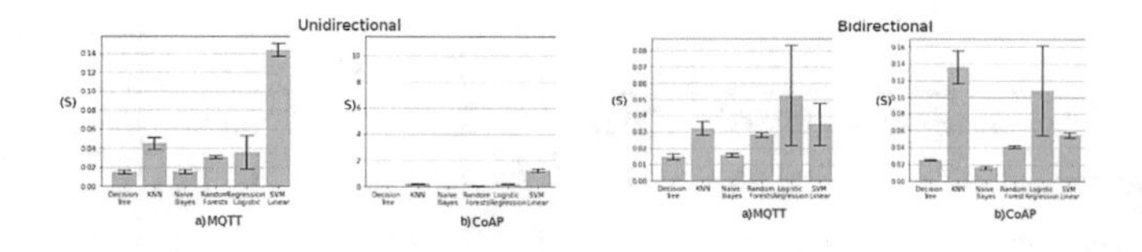

Fig. 1. Runtime average and CI of unidirectional and bidirectional flow ML algorithms

[1]https://docs.google.com/spreadsheets/d/1dvm0SD3Ym2zm7gEAc_d5L8njdZAuA8Ejc6KWEaYJUns/edit?usp=sharing.

3.4 Energy Consumption

In table[2] presents the average energy consumption of each algorithm to detect attacks on MQTT and CoAP networks, considering unidirectional and bidirectional traffic. Energy consumption is broken down into CPU, Package, RAM and GPU.

The CoAP dataset was larger than the MQTT dataset by a maximum value of 8%. However, the total power consumption of the algorithms is twice or more in the CoAP network. In both CoAP and MQTT networks, the decision tree and Gaussian Naive Bayes algorithms presented a lower energy consumption regardless of whether the traffic is unidirectional or bidirectional. Regarding the higher energy consumption, when the traffic is unidirectional, in both networks the Linear SVM algorithm presented a higher energy consumption. When traffic is bidirectional, in the MQTT network the Logistic Regression presented the highest energy consumption, whereas in the CoAP network, it was the KNN algorithm.

4 Conclusion

This work analyzed six algorithms (DT, KNN, GNB, RF, LR and SVM) of ML to identify attacks in MQTT and CoAP networks. In a drone scenario with SDN, at least one of these algorithms is extremely important so that an IDS control plane can block the attacking device and consequently avoid the interruption and unavailability of smart objects. Therefore, experiments were carried out with synthetic traffic to create two datasets (one for CoAP and the other for MQTT), considering three types of attacks: Scan aggressive, Scan UDP and Sparta.

In MQTT network, the best algorithms were DT and KNN for Scan_A type attacks and KNN and RF algorithms were better for Scan UDP attack. In the CoAP network, the DT and RF algorithms best identified the Scan_A and Scan_U attacks. In MQTT and CoAP networks, sparta was the attack with the best identification by most of the analyzed ML algorithms.

The execution of the algorithms with MQTT traffic consumed less exergy and ran faster than CoAP traffic. Highlighting the DT and GNB algorithms in the MQTT network as the best and the SVM as the slowest and most energy consuming.

Therefore, the results recommend avoiding the use of the SVM algorithm not only due to its high power consumption, but also because it was not actually satisfactory when compared to other algorithms. It is also recommended to use MQTT as the most suitable protocol to obtain a secure network, as the advertising algorithms identified it with the best values for accuracy, recall, precision and F1-Score with the benefit of shorter execution time and lower energy consumption, mainly, if the KNN and DT algorithms are used. On the other hand, if

[2]https://docs.google.com/spreadsheets/d/1CoeKSBhC0G1YCA6f3Uasm_PvkfeLOiJGOxU99RQDHMI/edit?usp=sharing.

it is necessary to build a CoAP network or a solution for both protocols (MQTT and CoAP), it is recommended to use the DT algorithm.

Acknowledgments. The authors would like to thank the Federal Institute of Paraíba(IFPB)/Campus João Pessoa for financially supporting the presentation of this research and, especially thank you, to the IFPB Interconnect Notice - No. 02/2021.

References

1. Oliveira, L.P., Vieira, M.N., Leite, G.B., de Almeida, E.L.V.: Evaluating energy efficiency and security for internet of things: a systematic review. In: Barolli, L., Amato, F., Moscato, F., Enokido, T., Takizawa, M. (eds.) AINA 2020. AISC, vol. 1151, pp. 217–228. Springer, Cham (2020). https://doi.org/10.1007/978-3-030-44041-1_20
2. Naik, N.: Choice of effective messaging protocols for IoT systems: MQTT, CoAP, AMQP and HTTP. In: ISSE (2017). https://doi.org/10.1109/SysEng.2017.8088251
3. Suarez, J., Vidal, I., Garcia-Reinoso, J., Valera, F., Azcorra, A.: Exploring the use of RPAs as 5G points of presence. In: European Conference on Networks and Communications (EuCNC), pp. 27–31 (2016). https://doi.org/10.1109/EuCNC.2016.7560998
4. Metongnon, L., Sadre, R.: Beyond telnet: prevalence of IoT protocols in telescope and honeypot measurements. In: Proceedings of the 2018 Workshop on Traffic Measurements for Cybersecurity, pp. 21–26 (2018)
5. Allahyari, M., et al.: A brief survey of text mining: classification, clustering and extraction techniques. arXiv preprint arXiv:1707.02919 (2017)
6. Aljawarneh, S., Aldwairi, M., Yassein, M.B.: Anomaly-based intrusion detection system through feature selection analysis and building hybrid efficient model. J. Computat. Sci. **25**, 152–160 (2018). Elsevier
7. Biswas, S.K., et al.: Intrusion detection using machine learning: a comparison study. Int. J. Pure Appl. Math. **118**(19), 101–114 (2018)
8. Oliveira, L.P., Sadok, D.F.H.: ProNet framework: network management using semantics and collaboration. In: INCOS (2013). https://doi.org/10.1109/INCoS.2013.78
9. Oliveira, L.P., et al.: Deep learning library performance analysis on raspberry (IoT device). In: Barolli, L., Woungang, I., Enokido, T. (eds.) AINA 2021. LNNS, vol. 225, pp. 383–392. Springer, Cham (2021). https://doi.org/10.1007/978-3-030-75100-5_33
10. Migliardi, M., Merlo, A.: Modeling the energy consumption of distributed ids: a step towards green security. In: 2011 Proceedings of the International Convention MIPRO. IEEE (2011)
11. Oliveira, L., Hadj, S.D., Gonçalves, G., Abreu, R., Kelner, J.: Collaborative algorithm with a green touch. In: Sénac, P., Ott, M., Seneviratne, A. (eds.) MobiQuitous 2010. LNICS, SITE, vol. 73, pp. 51–62. Springer, Heidelberg (2012). https://doi.org/10.1007/978-3-642-29154-8_5
12. Hindy, H., Bayne, E., Bures, M., Atkinson, R., Tachtatzis, C., Bellekens, X.: Machine learning based IoT intrusion detection system: an MQTT case study (MQTT-IoT-IDS2020 dataset). In: Ghita, B., Shiaeles, S. (eds.) INC 2020. LNNS, vol. 180, pp. 73–84. Springer, Cham (2020). https://doi.org/10.1007/978-3-030-64758-2_6
13. Batista, G.E.d.A.P., et al.: Pré-processamento de dados em aprendizado de máquina supervisionado. Ph.D. thesis - Universidade de São Paulo (2003)

Message Delivery of Nomadic Lévy Walk Based Message Ferry Routing in Delay Tolerant Networks

Koichiro Sugihara and Naohiro Hayashibara(✉)

Kyoto Sangyo University, Kyoto, Japan
{i1986089,naohaya}@cc.kyoto-su.ac.jp

Abstract. Message ferry is to provide communication service for mobile nodes in Delay-Tolerant Networks (DTNs). Various routing schemes for message ferries have been proposed so far. The efficiency of message delivery using message ferries depends on the routing scheme. Nomadic Lévy Walk, a variant of Lévy Walk, is an eligible candidate for a message ferry routing scheme. It includes homing behavior in addition to the behavior similar to Lévy walk with strategic moving the home (sink) position. This paper shows the simulation result on the message delivery probability and latency by message ferries with Nomadic Lévy Walk in DTNs. Then we compare with other variants of Lévy walk. We also discuss the influence of the sink relocation strategies of Nomadic Lévy Walk on the message delivery probability and latency.

1 Introduction

Various messaging protocols in Delay Tolerant Networks (DTNs) have been proposed so far [12]. This is because DTNs are attracted attention in various research fields such as location-awareness and emergency evacuation. In this type of network, mobile nodes which have wireless communication capability with limited communication range are located in a field, and messages sent by nodes are delivered to the destination in a Store-Carry-Forward manner because end-to-end paths do not always exist. It means that each message or data is carried to the destination by some mobile entities which repeat messages passing to each other. This type of communication is useful to send and deliver messages without a stable network infrastructure, especially in the case of disasters.

Now, we focus on the Message Ferrying based approach for data collection in DTNs. Message ferries (or MF for short) are mobile nodes that move around the field to collect messages from nodes and to deliver them to destinations properly. Whenever a node sends a message to other nodes in the network, first of all, it sends the message an MF with short-range communication capability (e.g., Bluetooth) when it gets close. The message might be reached to the destination by MFs.

Therefore, the efficiency of message delivery depends on the routing scheme of MFs. Most of the message ferry routing schemes assume a fixed route; for

L. Barolli et al. (Eds.): AINA 2022, LNNS 449, pp. 259–270, 2022.
https://doi.org/10.1007/978-3-030-99584-3_23

instance, MFs move on a circular path [11,26,27]. However, the fixed route-based routing scheme does not work if a part of the route is not available due to some event such as road maintenance and a disaster.

We focus on a random walk-based routing scheme for message ferry routing. In particular, Lévy walk has recently attracted attention due to the optimal search in animal foraging [8,25] and the statistical similarity of human mobility [13]. It is a mathematical fractal that is characterized by long segments followed by shorter hops in random directions. Lévy walk has also been used to improve the spatial coverage of mobile sensor networks [22], to analyze the properties of an evolving network topology of a mobile network [5], and to enhance grayscale images combined with other bio-inspired algorithms (i.e., bat and firefly algorithms) [7]. It is also considered to be particularly useful for message ferry routing that aims to deliver messages in DTNs [14].

Nomadic Lévy Walk (NLW) is a variant of Lévy walk, which has been proposed by Sugihara and Hayashibara [18]. An agent starts from a sink node and returns to the sink node with the given probability α. Each sink node changes its position according to the given strategy with the probability γ.

We suppose to deliver messages by using battery-driven autonomous electric vehicles as MFs. Moreover, they have an onboard camera to drive along the road and avoid obstacles. Each of them basically moves with a mobile power source vehicle. Then, it departs from the sink to visit nodes for collecting and delivering messages. It requires going back to the sink to charge its battery.

Our previous work showed the message delivery probability and latency using NLW based MFs on unit disk graphs [21]. We assumed to deliver messages among stationary nodes. This paper aims to measure the message delivery probability and latency among mobile nodes using NLW based MFs in DTNs. We conducted simulations by The ONE simulator to measure the message delivery probability and latency with several message ferry routing, including NLW. The simulation runs have been executed on the map of Helsinki. The environment of simulations is more realistic than unit disk graphs. We analyze the impact of sink relocation strategies of NLW on the message delivery probability and latency. Then, we compare NLW with other message ferry routing based on Random walk and Lévy walk. We try to confirm the practical usefulness of NLW as a message ferry routing in DTNs.

2 Related Work

We now introduce several research works related to MFs in WSNs and WSNs. Then, we explain existing movement models that can be used for message ferry routing, and existing message routing protocols.

2.1 Message Ferry in DTNs and WSNs

Tariq et al. proposed Optimized Way-points (OPWP) ferry routing method to facilitate connectivity on sparse ad-hoc networks [4]. According to the simulation result, OPWP-based MFs outperforms other ferry routing method based on Random way point (RWP).

Shin et al. apply the Lévy walk movement pattern to the routing of message ferries in DTN [14]. They demonstrated message diffusion using message ferries based on the various configuration of Lévy walk. According to the simulation result, the ballistic movement of message ferries (i.e., smaller scaling parameter of Lévy walk) is efficient regarding the message delay.

Basagni et al. proposed the notion of a mobile sink node and its heuristic routing scheme called Greedy Maximum Residual Energy (GMRE) [3]. The motivation of the work is to prolong wireless sensor nodes that are deployed in a large field and send data to the sink periodically. The mobile sink as a data collection and processing point moves around the field to save the energy consumption of sensor nodes. The simulation result showed that the mobile sink with the proposed routing scheme improved the lifetime of sensor nodes.

Alnuaimi et al. proposed a ferry-based approach to collect sensor data in WSNs [1,2]. It divides a field into virtual grids and calculates an optimal path for a mobile ferry to collect data with a minimum round trip time. It utilizes a genetic algorithm and the node ranking clustering algorithm for determining the path.

2.2 Movement Models

MFs collect data and messages and deliver them to the destination in a Store-Carry-Forward manner. Thus, the movement model of MFs is an important factor on the delivery probability and latency of message delivery in DTNs.

Birand et al. proposed the Truncated Levy Walk (TLW) model based on real human traces [5]. The model gives heavy-tailed characteristics of human movement. Authors analyzed the properties of the graph evolution under the TLW mobility model.

Valler et al. analyzed the impact of mobility models including Lévy walk on epidemic spreading in MANET [24]. They adopted the scaling parameter $\lambda = 2.0$ in the Lévy walk mobility model. From the simulation result, they found that the impact of velocity of mobile nodes does not affect the spread of virus infection.

Thejaswini et al. proposed the sampling algorithm for mobile phone sensing based on Lévy walk mobility model [22]. Authors showed that proposed algorithm gives significantly better performance compared to the existing method in terms of energy consumption and spatial coverage.

Fujihara et al. proposed a variant of Lévy flight which is called Homesick Lévy Walk (HLW) [9]. In this mobility model, agents return to the starting point with a homesick probability after arriving at the destination determined by the power-law step length. As their result, the frequency of agent encounter obeys the power-law distribution though random walks and Lévy walk do not obey it.

Most of the works related to Lévy walk assume a continuous plane and hardly any results on graphs are available. Shinki et al. defined the algorithm of Lévy walk on unit disk graphs [16]. They also found that the search capability of Lévy walk emphasizes according to increasing the distance between the target and the initial position of the searcher. It is also efficient if the average degree of a graph is small [15].

Sugihara et al. proposed the novel mobility model called Nomadic Lévy Walk (NLW) as a variant of HLW and the sink relocation strategy based on a hierarchical clustering method [18,19]. They conducted simulations to measure the cover ratio of unit disk graphs. The simulation result showed that the mobility model cover a wide area with preserving homing behavior.

2.3 Routing Protocols for DTNs

We introduce several routing protocols of opportunistic communication, which we use in the simulations.

2.3.1 Epidemic

Epidemic routing [23] is a flooding-based protocol. Every node copies and forward messages to newly encountered nodes which do not already hold a copy of the message. This protocol aims to realize a higher message delivery probability and lower latency. On the other hand, it induces a higher overhead ratio and requires a large storage space of each node as a consequence of the number of copies of messages flooded in the network. Whenever two nodes n_i and n_j exchange messages, n_i sends a Summary Vector, that contains the digest of the messages stored in n_i, to n_j and vice versa. Then, n_i and n_j send a request for messages, which are not in own storage, to each other node. Finally, they send messages according to the request. This procedure prevents to receive duplicated messages.

2.3.2 Spray and Wait

Spray and Wait protocol [17] developed by Spyropoulos et al. to optimize the resource utilization. Due to multiple copies of message, maximum resources are used in Epidemic protocol while Spray and Wait protocol limits the number of replicated messages to N. This protocol consists of two stages; the spray stage and the wait stage. When a new message arrives at a node, it generates N copies of the message and transmits them to N different nodes in the spray stage. Whenever a node receives the copy, the protocol enters in the wait stage. It holds the message copy until the destination node is come in direct communication range.

2.3.3 PRoPHET

PRoPHET [10] is a probabilistic routing protocol. First of all, it calculates delivery predictability which includes two properties; a history of encounters and transitivity. The former one indicates how many times any two nodes come across and the latter one is the probability calculated from the history in terms of message forwarding. Each node initially has an initialization constant of delivery predictability and it is updated based on encounters. The predictability is reduced if no encounters occur in a certain period of time.

3 System Model

We assume mobile nodes that move along roads in a map with a constant velocity. They could be considered pedestrians in practice. Each of them has a mobile device that has storage and a short-range communication capability such as Bluetooth, Ad-Hoc mode of IEEE 802.11, Near Field Communication (NFC), and so on.

Message ferries (MFs) are defined as special mobile nodes that help to distribute messages. They also have a short-range communication capability as the mobile nodes.

When a mobile node wants to send a message to another mobile node, it makes a copy of the message and gives it to the nearest mobile node or a MF. Messages are stored, carried, and forwarded via several mobile nodes and MFs to deliver to the destination.

This assumption is similar to the existing work [1,2,14,27]. MFs know the positions of nodes in the network. Practically speaking, an MF is a battery-driven autonomous electric vehicle which is required to charge its battery in a certain period of time. Moreover, MFs and nodes have short-range communication capability such as Bluetooth, ad hoc mode of IEEE 802.11, Near Field Communication (NFC), infrared transmission, and so on. Therefore each node is able to transmit a message by communicating with an MF when it gets close to the node. Each MF starts moving from the corresponding power source vehicle, called *sink*, and it goes back to the sink to recharge its battery.

There are two types of ferry interaction.

- *Ferry-to-Ferry interaction.* Message ferries exchange messages with each other if they are at the same node (i.e., they are in their communication range).
- *Ferry-to-Node interaction.* Each sensor node can send data piggybacked on a message to an MF when it is physically close to the node (i.e., it is at the node in the graph).

In any case, we use Epidemic protocol for message forwarding.

We assume the map (see Fig. 1) that is the Helsinki City Scenario in The ONE simulator [6]. Mobile nodes and MFs move along the roads in the map.

We also assume that each MF identifies its position (e.g., obtained by GPS), which is accessible from an MF, and it has a compass to get the direction of a walk. Each MF has a set of neighbor nodes, and such information (i.e., positions of neighbors) is also accessible. Moreover, an MF has an onboard camera to drive along the road and avoid obstacles.

The movement of mobile nodes obeys Random walk. On the other hand, MFs obeys the designated movement model such as Random walk, Lévy walk, Homesick Lévy walk and Nomadic Lévy walk.

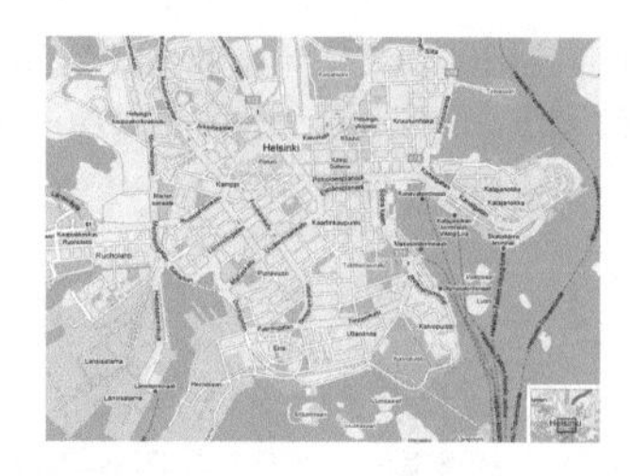

Fig. 1. Helsinki City Scenario in The ONE simulator [6].

4 Nomadic Lévy Walk

We proposed a variant of Lévy walk called Nomadic Lévy Walk in our previous work [18,19] to improve the ability of the broad area search while preserving the homing behavior. NLW is an extension of HWL [9], and holds the following properties in addition to HLW.

Nomadicity: Each sink moves its position with the given probability γ.
Sink relocation strategy: The next position of the sink is decided by a particular strategy.

The movement of each MF obeys HLW. Thus, their trajectory is radially from their sink. Moreover, they move their sink at the probability γ in Nomadic Lévy walk. In fact, the fixed sink restricts the area that each MF explores. The property of Nomadicity is expected to improve coverage by each MF.

We now explain the detail of the algorithm of Nomadic Lévy walk on unit disk graphs in Algorithm 4.

At the begging of the algorithm, each MF holds the initial position as the sink position s. In every walk, each MF determines the step length d by the power-law distribution and selects the orientation o of a walk randomly from $[0, 2\pi)$. It can obtain a set of neighbors $N(c)$ and a set of possible neighbors $PN(c) \subseteq N(c)$, to which MFs can move, from the current node c. In other words, a node $x \in PN(c)$ has a link with c that the angle θ_{ox} between o and the link is smaller than $\frac{\pi}{2}$.

In unit disk graphs, it is not always true that there exist links to the designated orientation. We introduce δ that is a permissible error against the orientation. In this paper, we set $\delta = 90$. It means that MFs can select links to move in the range ± 90 with the orientation o as a center.

In a given probability α, the MF goes back to the sink node (line 2 in Algorithm 4). In this case, it sets the orientation to the sink node as o and the distance to the sink as d.

Each MF changes its sink position with the given probability γ (line 10 in Algorithm 4) and then starts a journey from the sink.

```
 1: Initialize:
        s ← the position of the sink
        c ← the current position
        o ← 0                                  ▷ orientation for a walk.
        PN(c) ← the possible neighbors to move.
 2: if Probability: α then
 3:     d ← the distance to s
 4:     o ← the orientation of s
 5: else
 6:     d is determined by the power-law distribution
 7:     o is randomly chosen from [0, 2π)
 8: end if
 9: if Probability: γ then
10:     s ← P(c)     ▷ update the position of the sink according to the given strategy.
11: end if
12: while d > 0 do
13:     PN(c) ← {x|abs(θ_ox) < δ, x ∈ N(c)}
14:     if PN(c) ≠ ∅ then
15:         d ← d − 1
16:         move to v ∈ PN(c) where v has the minimum abs(θ_ov)
17:         c ← v
18:     else
19:         break                              ▷ no possible node to move.
20:     end if
21: end while
```

4.1 Sink Relocation Strategy

Each sink replaces its position according to the given strategy in NLW. Obviously, the sink position is one of the important factors for covering a wide area of a graph. Each sink manages the history H of the sink positions. We have proposed several sink relocation strategies for NLW [18, 19]. We now introduce those strategies as follows.

4.1.1 Lévy Walk Strategy (LWS)

The next position of the sink is changed obeying the Lévy Walk movement pattern. The orientation o of the next base is selected at random from $[0,2\pi)$ and the distance d of it from the current position is decided by the power-law distribution.

4.1.2 Reverse Prevention Strategy (RPS)

This strategy has been proposed in [18]. In this strategy, each agent assumes a set of the history H of the sink positions. It calculates the reverse orientation of each position in the set. Then, it determines the next sink so that its position is the opposite side of the past sink positions.

5 Performance Evaluation

In this section, we show the simulation result regarding the average latency of message delivery by MFs using Nomadic Lévy Walk (NLW) and Homesick Lévy Walk (HLW) as message ferry routing methods. We measure the average time between sending and delivering messages which are delivered to their destination successfully. This indicates that the efficiency of message delivery by MFs in DTNs. We conducted several simulation runs by using The ONE simulator [6] with various configurations.

5.1 Environment

In our simulation, we use The ONE simulator [6] which is a discrete event simulator on DTNs. We distribute 1,000 mobile nodes at random on the Helsinki City Scenario (See Fig. 1) in the simulator. They are assumed to be pedestrians who move with a constant velocity. Some of them send messages by using their mobile devices. These messages are stored, carried and forwarded among MFs and other mobile nodes, and are delivered to the destinations.

5.2 Configuration

We use NLW, HLW, Lévy walk, and Random walk as message ferry routing schemes. The scaling parameter λ is set 1.2 for NLW, HLW and Lévy walk. This parameter is efficient for a wide-area search in graphs [15]. The homesick probability α is common to NLW and HLW. We set $\alpha = 0.2$ for them. We also set $\gamma = 0.2$, the probability of sink relocation, for NLW.

We conducted a simulation run for 300 h. Each message has Time-to-live (TTL) for message expiration. TTL decreases by elapsing simulation steps. A message is removed from MFs that store this message when its TTL equals zero. TTL for each message is 5 h.

The number of MFs is set as $k \in [1, 50]$ and the number of mobile nodes is 50.

5.3 Simulation Result

We show the simulation results on the delivery probability and latency of message delivery by using MFs with several routing schemes.

5.3.1 Message Delivery Probability

We show the average message delivery probability result by using MFs with several routing schemes. First, we show the difference between sink relocation strategies (LWS and RPS) of NLW, which is stated in Sect. 4.1. Then, we compare other message ferry routing schemes.

NLW-based MFs

Figure 2 and 3 show the average message delivery probability of LWS strategy and RPS strategy with $\lambda = 1.2$, $\alpha = 0.2$ and $\gamma = 0.2$. RPS improves the delivery probability compared to LWS (see Fig. 2 and 3), especially with $k \geq 30$. RPS moves the sink position efficiently to cover wider area compared to LWS.

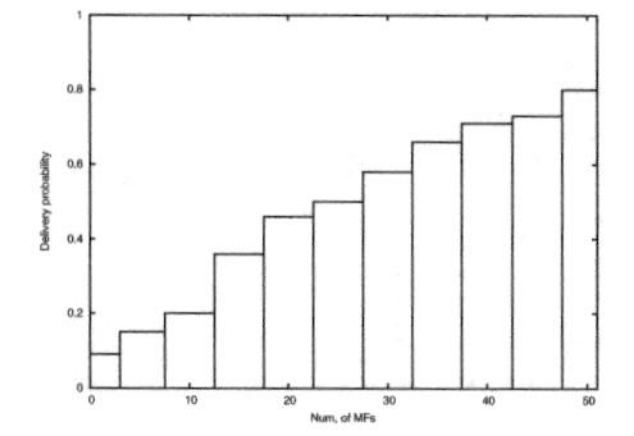

Fig. 2. Delivery probability of LWS strategy with $\gamma = 0.2$.

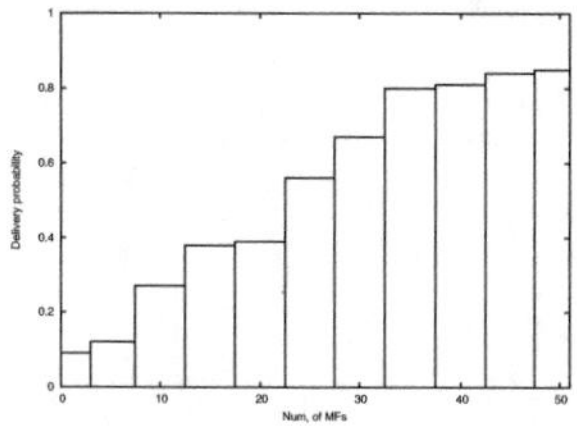

Fig. 3. Delivery probability of RPS strategy with $\gamma = 0.2$.

Comparison with other routing schemes

Figure 4, 5 and 6 show the result of HLW, Lévy walk and Random walk. According to the result, HLW-based MFs show a similar result as LWS strategy of NLW. Moreover, HLW outperforms Lévy walk and Random walk regarding the message delivery probability. On the other hand, RPS strategy of NLW is about 15% efficient compared with HLW for message delivery.

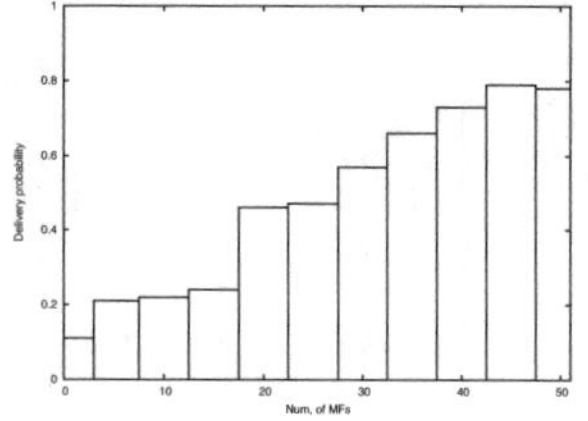

Fig. 4. Delivery probability of HLW-based MFs.

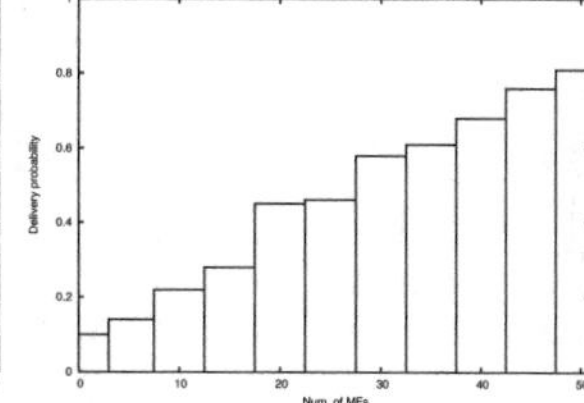

Fig. 5. Delivery probability of LW-based MFs.

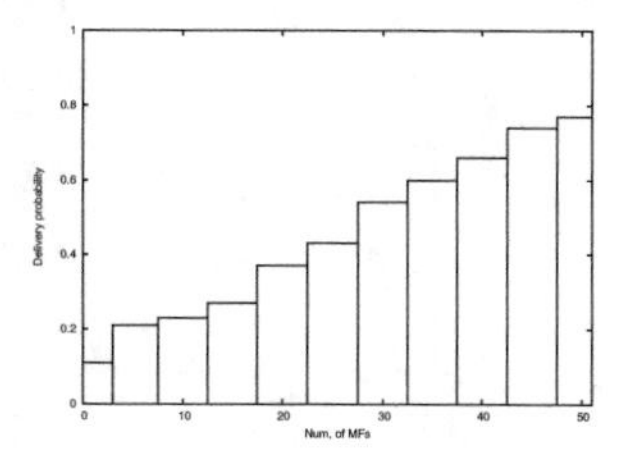

Fig. 6. Delivery probability of RW-based MFs.

5.3.2 Latency

We first show the average latency using NLW based MFs with different sink relocation strategies. Then, we compare the results with those of other message ferry routing schemes. The configuration of the number of MFs is the same as the message delivery probability. Note that each of the results is calculated based on messages that are successfully delivered to the destination nodes.

NLW-based MFs

Figure 7 and 8 show the average latency of message delivery of LWS strategy and RPS strategy with $\lambda = 1.2$, $\alpha = 0.2$ and $\gamma = 0.2$. RPS is better than LWS regarding latency on average. LWS improves latency by increasing the number of MFs. On the other hand, RPS gradually reduces latency against increasing the number of MFs.

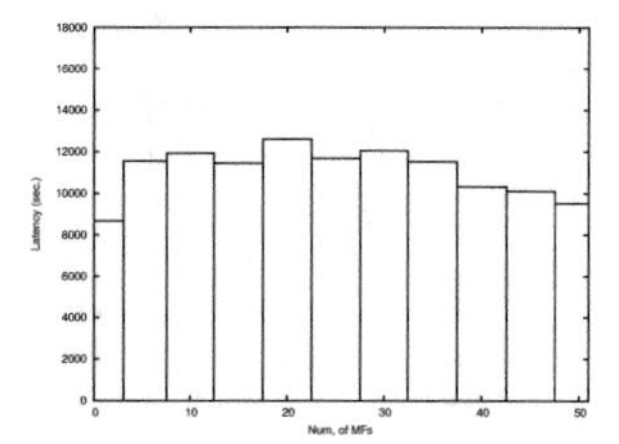

Fig. 7. Latency of LWS strategy with $\gamma = 0.2$.

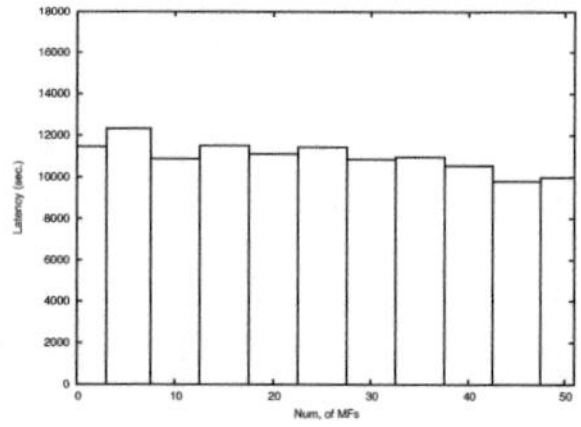

Fig. 8. Latency of RPS strategy with $\gamma = 0.2$.

Comparison with other routing schemes
NLW-RPS is 4% to 9% efficient than HLW regarding latency. HLW gradually reduces latency by increasing the number of MFs. So, both of them have a pretty similar tendency. On the other hand, Lévy walk and Random walk based MFs do not have any correlation between latency and the number of MFs.

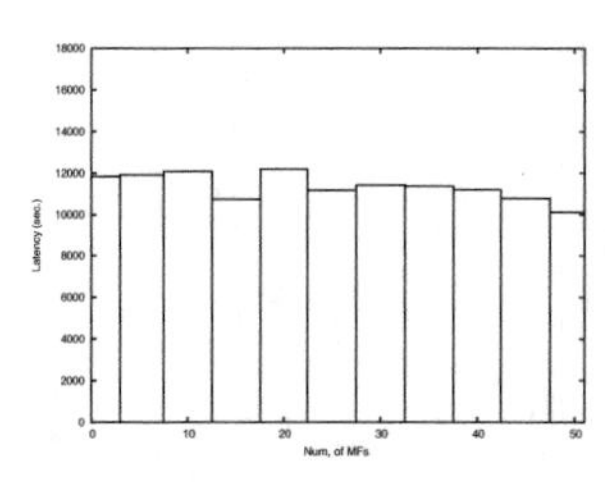

Fig. 9. Latency of HLW-based MFs.

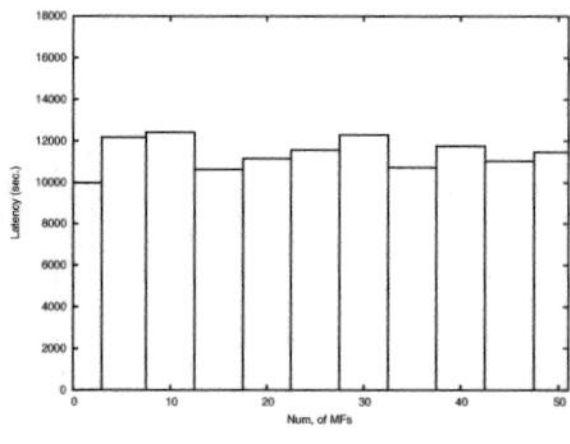

Fig. 10. Latency of LW-based MFs.

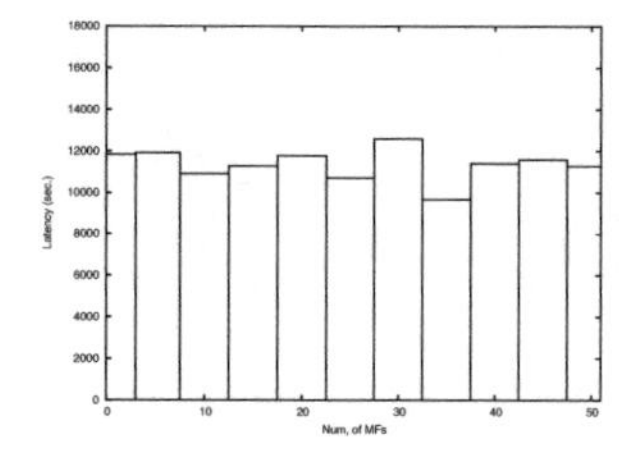

Fig. 11. Latency of RW-based MFs.

6 Conclusion

This paper measured the message delivery probability and latency on several message ferry routing schemes in delay-tolerant networks (DTNs). Then, we analyze the difference between the strategies of NLW regarding the delivery probability and latency. We also compare the results of NLW with the ones of other routing schemes.

Mobile nodes move around in the map with Random walk. On the other hand, MFs move with message ferry routing based on the designated routing schemes; NLW (RPS, LWS), HLW, Lévy walk, and Random walk. We used the Epidemic protocol as a message exchange protocol in this work.

According to the result, RPS strategy is slightly more efficient than LWS strategy in NLW. Moreover, NLW-RPS outperforms other routing schemes regarding the message delivery probability and latency.

References

1. Alnuaimi, M., Shuaib, K., Abdel-Hafez, K.A.M.: Data gathering in delay tolerant wireless sensor networks using a ferry. Sensors **15**(10), 25809–25830 (2015)
2. Alnuaimi, M., Shuaib, K., Abdel-Hafez, K.A.M.: Ferry-based data gathering in wireless sensor networks with path selection. Procedia Comput. Sci. **52**, 286–293 (2015)
3. Basagni, S., Carosi, A., Melachrinoudis, E., Petrioli, C., Wang, Z.M.: Controlled sink mobility for prolonging wireless sensor networks lifetime. Wireless Netw. **14**(6), 831–858 (2008)
4. Bin Tariq, M.M., Ammar, M., Zegura, E.: Message ferry route design for sparse ad hoc networks with mobile nodes. In: Proceedings of the 7th ACM International Symposium on Mobile Ad Hoc Networking and Computing, MobiHoc 2006, pp. 37–48. Association for Computing Machinery (2006)
5. Birand, B., Zafer, M., Zussman, G., Lee, K.W.: Dynamic graph properties of mobile networks under levy walk mobility. In: Proceedings of the 2011 IEEE Eighth International Conference on Mobile Ad-Hoc and Sensor Systems, MASS 2011, pp. 292–301. IEEE Computer Society (2011)
6. Desta, M.S., Hyytiä, E., Keränen, A., Kärkkäinen, T., Ott, J.: Evaluating (Geo) content sharing with the one simulator. In: Proceedings of the 14th ACM Symposium Modeling, Analysis and Simulation of Wireless and Mobile Systems (MSWiM) (2013)
7. Dhal, K.G., Quraishi, M.I., Das, S.: A chaotic lévy flight approach in bat and firefly algorithm for gray level image. Int. J. Image Graph. Signal Process. **7**, 69–76 (2015)
8. Edwards, A.M., et al.: Revisiting lévy flight search patterns of wandering albatrosses, bumblebees and deer. Nature **449**, 1044–1048 (2007)
9. Fujihara, A., Miwa, H.: Homesick lévy walk and optimal forwarding criterion of utility-based routing under sequential encounters. In: Proceeding of the Internet of Things and Inter-cooperative Computational Technologies for Collective Intelligence 2013, pp. 207–231 (2013)
10. Grasic, S., Davies, E., Lindgren, A., Doria, A.: The evolution of a DTN routing protocol - prophetv2. In: Proceedings of the 6th ACM Workshop on Challenged Networks, CHANTS 2011, pp. 27–30. ACM, New York, NY, USA. https://doi.org/10.1145/2030652.2030661
11. Kavitha, V., Altman, E.: Analysis and design of message ferry routes in sensor networks using polling models. In: 8th International Symposium on Modeling and Optimization in Mobile, Ad Hoc, and Wireless Networks, pp. 247–255 (2010)
12. Li, Y., Bartos, R.: A survey of protocols for intermittently connected delay-tolerant wireless sensor networks. J. Netw. Comput. Appl. **41**, 411–423 (2014)
13. Rhee, I., Shin, M., Hong, S., Lee, K., Kim, S.J., Chong, S.: On the levy-walk nature of human mobility. IEEE/ACM Trans. Netw. **19**(3), 630–643 (2011)
14. Shin, M., Hong, S., Rhee, I.: DTN routing strategies using optimal search patterns. In: Proceedings of the Third ACM Workshop on Challenged Networks, CHANTS 2008, pp. 27–32. ACM (2008)

15. Shinki, K., Hayashibara, N.: Resource exploration using lévy walk on unit disk graphs. In: The 32nd IEEE International Conference on Advanced Information Networking and Applications (AINA-2018) (2018)
16. Shinki, K., Nishida, M., Hayashibara, N.: Message dissemination using lévy flight on unit disk graphs. In: The 31st IEEE International Conference on Advanced Information Networking and Applications (AINA 2017) (2017)
17. Spyropoulos, T., Psounis, K., Raghavendra, C.S.: Efficient routing in intermittently connected mobile networks: the multiple-copy case. IEEE/ACM Trans. Netw. **14**, 77–90 (2008)
18. Sugihara, K., Hayashibara, N.: Message dissemination using nomadic Lévy walk on unit disk graphs. In: Barolli, L., Hussain, F.K., Ikeda, M. (eds.) CISIS 2019. AISC, vol. 993, pp. 136–147. Springer, Cham (2020). https://doi.org/10.1007/978-3-030-22354-0_13
19. Sugihara, K., Hayashibara, N.: Performance evaluation of nomadic lévy walk on unit disk graphs using hierarchical clustering. In: Proceedings of the 34th Int'l Conference on Advanced Information Networking and Applications (AINA-2020), pp. 512–522. Springer (2020)
20. Sugihara, K., Hayashibara, N.: Target exploration by nomadic lévy walk on unit disk graphs. Int. J. Grid Util. Comput. **11**(2), 221–229 (2020). https://doi.org/10.1504/IJGUC.2020.105536.
21. Sugihara, K., Hayashibara, N.: Message Ferry Routing Based on Nomadic Lévy Walk in Wireless Sensor Networks. In: Barolli, L., Woungang, I., Enokido, T. (eds.) AINA 2021. LNNS, vol. 225, pp. 436–447. Springer, Cham (2021). https://doi.org/10.1007/978-3-030-75100-5_38
22. Thejaswini, M., Rajalakshmi, P., Desai, U.B.: Novel sampling algorithm for human mobility-based mobile phone sensing. IEEE Internet of Things J. **2**(3), 210–220 (2015)
23. Vahdat, A., Becker, D.: Epidemic routing for partially-connected ad hoc networks. Tech. Rep. CS-2000-06, Duke University (2000)
24. Valler, N.C., Prakash, B.A., Tong, H., Faloutsos, M., Faloutsos, C.: Epidemic spread in mobile ad hoc networks: determining the tipping point. In: Proceedings of the 10th International IFIP TC 6 Conference on Networking - Volume Part I, NETWORKING 2011, pp. 266–280. Springer-Verlag (2011). https://doi.org/10.1007/978-3-642-20757-0_21
25. Viswanathan, G.M., Afanasyev, V., Buldyrev, S.V., Murphy, E.J., Prince, P.A., Stanley, H.E.: Lévy flight search patterns of wandering albatrosses. Nature **381**, 413–415 (1996)
26. Zhao, W., Ammar, M., Zegura, E.: A message ferrying approach for data delivery in sparse mobile ad hoc networks. In: Proceedings of the 5th ACM International Symposium on Mobile Ad Hoc Networking and Computing, pp. 187–198. Association for Computing Machinery (2004)
27. Zhao, W., Ammar, M., Zegura, E.: Controlling the mobility of multiple data transport ferries in a delay-tolerant network. In: IEEE INFOCOM 2005. Maiami, FL, USA (2005)

Towards Efficient Selective In-Band Network Telemetry Report Using SmartNICs

Ronaldo Canofre[1(✉)], Ariel G. Castro[1], Arthur F. Lorenzon[1], Fábio D. Rossi[2], and Marcelo C.Luizelli[1]

[1] Federal University of Pampa (UNIPAMPA), Alegrete, Brazil
{canofre,ariel.aluno,arthurlorenzon,marceloluizelli}@unipampa.edu.br
[2] Federal Institute Farroupilha (IFFAR), Alegrete, Brazil
fabio.rossi@iffarroupilha.edu.br

Abstract. In-band Network Telemetry (INT) is a promising network monitoring approach that allows broad and fine-grained network visibility. However, when a massive volume of telemetry data is reported to an INT collector, there might overload the whole network infrastructure, while still degrading the performance of packet processing at the INT sink node. As previously reported in the literature, programmable devices – in particular, SmartNICs – have strict constraints in terms of processing and memory. In this work, we propose to design and implement a lightweight Exponentially Weighted Moving Average based mechanism inside the SmartNIC data plane in order to assist the decision-making process of reporting INT data. By evaluating our solution in state-of-the-art SmartNICs, we show that our proposal can decrease the number of nonessential telemetry data sent to INT collectors by up to 16X compared to the de-facto INT approach while presenting minor overhead in terms of packet latency.

1 Introduction

In-band Network Telemetry is a promising near real-time network monitoring approach [12,17,22] that enables wide and fine-grained network visibility. In a nutshell, INT consists of instrumenting the collection of low-level network monitoring statistics directly from the data plane – allowing network operators/monitoring applications to be fed with an unprecedented level of information. Examples of such in-network statistics include data plane metadata (e.g., per-packet processing time, or queue utilization), and/or custom-made ones (e.g., network flow inter-packet gap [24]). Over the last years, INT has been successfully applied to a series of use cases [18], including the identification of short-lived network behaviors [13] and network anomalies [11].

In the classic hop-by-hop INT specification (i.e., INT-MD (eMbed Data)[1]), an INT source node embeds instructions into production network packets typi-

[1] INT specification: https://github.com/p4lang/p4-applications/blob/master/docs/INT_v2_1.pdf.

L. Barolli et al. (Eds.): AINA 2022, LNNS 449, pp. 271–284, 2022.
https://doi.org/10.1007/978-3-030-99584-3_24

cally using either unused header fields (e.g., IPv4 options) or by re-encapsulating the network traffic (e.g., using INT encapsulation). Then, INT transit nodes embed metadata to these packets according to the instructions given by the INT source. Last, an INT sink node strips the instruction out of the packet and sends the accumulated telemetry data to an INT collector. Figure 1 illustrates the whole INT procedure. In this example, a packet from network flow f_1 is used to collect INT data from forwarding devices A to F. Recently, investigations have made the first efforts to efficiently orchestrate how INT metadata are collected by network packets [2,5,11,22] in order to increase network visibility and timely detect network events. That includes, for instance, selecting the appropriate network flows/packet to collect the right network telemetry metadata in the network infrastructure. This problem has been proved to be NP-hard since packets might have different spare capacities (e.g., limited by the MTU data link) [19].

Despite these efforts, little has yet been done to efficiently and wisely report the collected telemetry data to an INT collector [23]. In the case where all the telemetry data is reported to an INT collector, there might lead to *(i) an excessive usage of network links between the INT sink and the INT collector.* For instance, if we consider a 10 Gbit/s network link sending 64-Bytes packets (i.e., 14.88 Mpps) and collecting 1 Byte per INT node transit along the way, the volume of network traffic needed to be reported per second would be 118 Mbit $*$ hops (path length). In fact, this volume of reported data can increase substantially if we assume the canonical reference architecture for programmable devices[2] where each device has at least 30 Bytes of metadata; *(ii) performance degradation on packet processing capabilities at the INT sink* due to the usage of packet cloning/recirculation primitives inside the data plane. To send the network packet to the INT collector (or part of it), programmable devices have to rely on packet recirculation/cloning primitives to duplicate the packet – which dramatically reduces the performance in terms of throughput and latency [27]; and *(iii) the overwhelm of the INT collector application* with INT data packets.

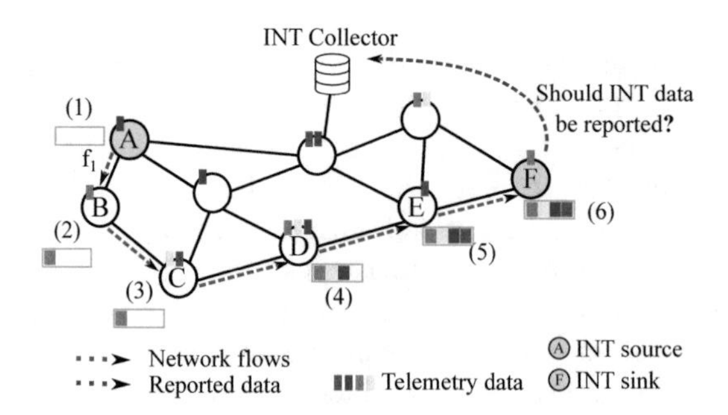

Fig. 1. Overview of the problem.

[2] https://github.com/p4lang/p4c/blob/main/p4include/v1model.p4.

To fill in this gap, in this work we propose a selective INT report mechanism entirely implemented using a SmartNIC. As previously reported by [27], programmable devices have stringent constraints in terms of processing capabilities (e.g., lack of floating-point operations) and memory usage limitations. We assume that the INT sink node is on top of a SmartNIC and, therefore, the decision whether report telemetry data or not is upon the NIC. Our proposed mechanism utilizes a lightweight Exponentially Weighted Moving Average inside the data plane. For that, we implemented it using P4 language and Micro-C routines to allow more complex operations inside the data plane. By performing an extensive performance evaluation using SmartNICs, we show that our proposed approach can reduce the amount of non-important telemetry data sent to INT collector by up to 16X (when compared to the Classical INT), while introducing negligible overhead in terms of packet latency.

The main contributions of this paper can be summarized as:

- an in-network mechanism implemented in state-of-the-art SmartNICs to wisely decide when to report INT metadata;
- a discussion of current limitations on implementing in-network computing in SmartNIC architectures; and
- an open-source code in order to foster reproducibility.

The remainder of this paper is organized as follows. In Sect. 2, we describe the SmartNIC architecture used in this work. In Sect. 3, we introduce our proposed approach. In Sect. 4, we discuss the obtained results. In Sect. 5, we overview the recent literature regarding in-band network telemetry, and. Last, in Sect. 6, we conclude this paper with final remarks.

2 Background

Cutting-edge programmable NICs (named SmartNICs) rely their architectures either on (i) multi-threaded, multi-core flow processor units or (ii) on FPGAs (Field Programmable Gate Arrays) to meet the increasing and strict demand. We concentrate our analysis on the general architectural elements of the Netronome

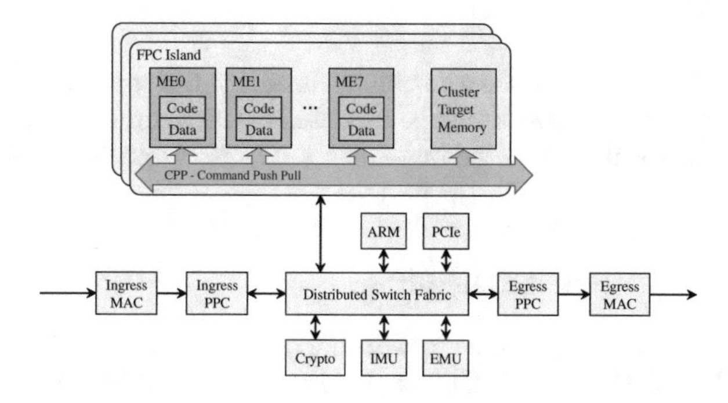

Fig. 2. An overview of the Netronome SmartNIC architecture [27].

SmartNIC architecture [21] – which is used afterward in our performance experiments – and rely on a multi-core architecture.

The SmartNIC Netronome NFP4000 architecture manages its flow processing cores (FPC) in multiple islands (Fig. 2). Each FPC includes eight Micro Engines (MEs) as a particular processor keeping its own instruction store (*code*) and local memory (*data*). Therefore, every ME in the architecture can run code with all other MEs in parallel. To support this feature, each ME holds 8 threads that can be used for cooperative multithreading in the manner that, at any given moment, at most, one thread is executing code from the same program. It means that each FPC handles at most eight parallel threads at 1.2 Ghz (one thread per ME). In each FPC, local memory comprises 32-bit registers, shared between all eight threads. Such registers are separated into: (*i*) general-purpose registers (256 32-bits registers) – used by default to store any register of up to 32-bits size; (*ii*) transfer registers (512 32-bits registers) – used for copying register over the interconnection bus (e.g., from or to other FPCs or memories); (*iii*) next-neighbor registers (128 32-bits registers) – used mostly to intercommunicate with adjacent FPCs; and (*iv*) local memory (1024 32-bit registers) – which is a little bit slower than general register. When there is a demand for more memory than available space in local FPC registers, variables are automatically and statically assigned to other in-chip memory hierarchies. Further, there are other sorts of memory available to FPCs: (*i*) Cluster Local Scratch (CLS) (20–50 cycles); (*ii*) Cluster Target Memory (CTM) (50–100 cycles); (*iii*) Internal Memory (IMEM) (120–250 cycles); and (*iv*) External Memory (EMEM) (150–590 cycles). For further details, the interested reader is referred to [21].

As packets are acquired from the network, an FPC thread picks up the packets and processes them. Extra threads are assigned to new packets as they arrive. For example, the SmartNIC NFP-4000 supports up to 60 FPCs, which enables the process of up to 480 packets simultaneously. The SmartNIC allows to program it directly using Micro-C language (i.e., a subset of C language) or using high-level domain-specific languages such as P4 [3]. The code is then compiled and statically assigned to a particular subset of FPC.

3 ETA: Early Network Telemetry Flow Analyzer Approach

In this section, we first define the model used by our proposed approach to decide whether or not network telemetry data is sent to an INT collector. Then, we describe how it is implemented in a SmartNIC and discuss existing limitations and challenges.

3.1 Model and Problem Definition

We consider that a programmable forwarding device $d \in D$ has $N \in \mathbb{N}^+$ available metadata to be collected by an INT-enabled packet p and that such packet has a limited available capacity to carry up to $M \in \mathbb{N}^+$ of such N items ($M \geq N$). For simplicity, we assume that all devices D have the same telemetry information and that an INT-enabled packet p can only collect telemetry data atomically, that is, it either collects all N metadata from $d \in D$ or none of them. Packet p can only collect once the same subset of telemetry data from the device d. We assume that the INT source instructs the packet p correctly according to a given algorithm (e.g., [19]).

Consider that a packet p has collected $M' \subseteq M$ telemetry data along its routing path – which comprises a subset of D devices. When the packet p gets to the INT sink node, the question to be answered is: *should it be sent to the INT collector or not?* To answer this question, the INT sink node computes (i) a weighted average of metadata M' collected by packet p and (ii) an exponential weighted moving average. The former tends to weigh the collected telemetry data differently according to its importance. For instance, the processing time (or the queue utilization) metadata might be more important to be considered in the decision process than the packet size. In turn, the latter tends to keep in memory the observed behavior of the latest received telemetry information over time.

Upon a received packet p in the INT sink node, it extracts the M' collected telemetry metadata and computes a weighted average per packet

$$A_p = \sum_{i=1}^{M'} w_i \cdot M_i, \tag{1}$$

where $w_i \in [0, 1]$ is the i-*th* weight given to the telemetry data. We further assume that $\sum_{i=1}^{M} w_i = 1$ and that M_i corresponds to the i-*th* collected data. Such individual averages A_p are then summing up into a accumulated weighted average metric within a given time window W (we discuss this design choice next). The time window W is defined for simplicity as a predefined number of packets. However, it can be extended to other metrics such as a time interval.

Then, the accumulated weighted average of a given window W is given by

$$A_w = \frac{\sum_{p=1}^{W} A_p}{W}. \tag{2}$$

Similarly, the exponential weighted moving average is obtained by

$$A_e = \alpha \cdot A_w + (1 - \alpha) \cdot A_e \tag{3}$$

where $\alpha \in [0, 1]$ comprises the importance given by the elements obtained in the last window W versus the historical knowledge maintained by the moving average A_e. Observe that higher values assigned to α prioritize the behavior on the latest window, while lower values prioritize the observed behavior over time. The decision-making process is made in a per-packet manner; however, the decision-making metrics (e.g., A_e) are updated in a time-window manner.

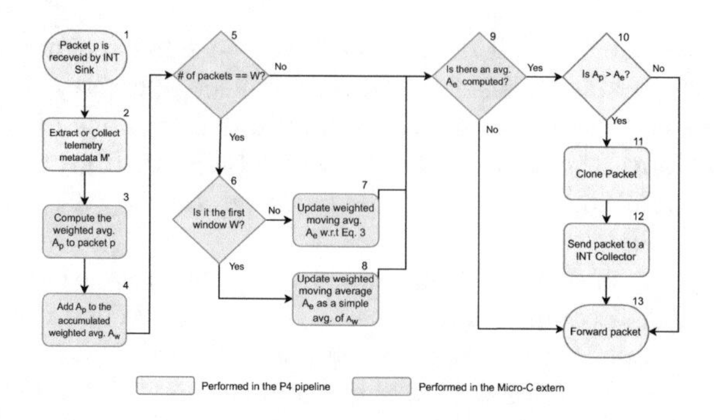

Fig. 3. Overview of the proposed P4+Micro-C pipeline approach.

3.2 Design and Implementation in a SmartNIC

Figure 3 illustrates an overview of the proposed approach pipeline implementation. Our approach utilizes the reference V1Model architecture to programmable forwarding devices as the basis to implement/add such functionalities into the Netronome SmartNIC. Also, our approach is based on P4-16 and in Micro-C languages.

Upon a packet is received by the SmartNIC, the packet is parsed accordingly. Our approach implementation resides just after the parsing step and the ingress pipeline (where the routing decision takes place), that is, in the egress pipeline. For this discussion, we assume that our forwarding device can process Ethernet frames and IP packets (or any known INT encapsulation protocol). Therefore, we omit the parsing steps since it is trivial and out of this work's scope. We then focused on the following up steps.

Our approach is implemented in the INT sink. Therefore, at this stage, all INT telemetry metadata has already been collected. The first steps of our approach consist of receiving the packet and extracting the telemetry data from it – or, in some cases, collecting them directly from the data plane (this might happen when the INT sink node acts as an INT transit node as well). This corresponds to steps 1 and 2 in Fig. 3. The extracted data is then stored in a custom-made metadata header structure – named `ETA` metadata struct. This structure comprises M 32-bit structures and a 2-bit flag used to instruct the decision-making process. For instance, if a packet needs to be sent to the INT collector, this flag is used internally. In the Netronome architecture, the existing timestamps (`ingress_timestamp` and `current_timestamp`) are 48-bit words. However, most of the applications use only the least significant bits (the last 32 bits) since they account for nanosecond differences. Further, as discussed next, the Netronome architecture also limits the results of arithmetic operations to 32 bits word (and therefore, we cannot operate on 48-bits timestamps). After extracting telemetry data into the ETA struct, our implementation calls a Micro-C extern code to handle more complex operations inside the data plane (depicted in Fig. 4). For instance, floating-point operations are not allowed in the P4-16 language reference, nor does the SmartNIC natively implement it. To allow the SmartNIC code to be partially written in P4 and Micro-C – and more importantly, to exchange data between them – we rely on ETA structs and internal P4 metadata to enable such real-time communications.

The Micro-C code starts receiving data from the P4 pipeline and locking up memory regions using a mutex (steps 1–3 in Fig. 4). Next, we calculate A_p and A_w according to Eq. 1 and Eq. 2, respectively. Figure 4 illustrates these procedures in steps 4–7. It is important to mention that neither the P4 reference architecture nor the Netronome support floating-point instructions. We implemented all these operations using fixed-point representation with 32 bits to surpass that limitation. In short, a fixed-point representation handles all real numbers as integers. The process scales up the numbers by multiplying by a constant factor C and then performs the multiplications/division as a regular integer. Finally, the scaled-up result is scaled down appropriately. In our implementation, we have used a constant C of 16 bits and performed the scale-up operation by applying a bit-shifting operation (i.e., $M_i' << 16$). After calculating A_p, we verify whether it is the case to send it out to the INT collector. We compare A_p with the observed A_e (which captures the historical behavior). In case the A_p packet value is higher than the dynamic A_e one, we mark the packet to be sent out to the collector (steps 8–12 in Fig. 4).

As previously mentioned, our approach performs per-packet decisions. However, we update the exponential weighted moving average A_e at the end of a given window W. This is done because the average A_e depends on A_w – and, the Netronome architecture does not allow arbitrary integer divisions (only by the power of 2 by performing bit-shifting – for instance $M_i' >> 16$). Our approach verifies whether it reaches or not the end of a given window W (step 14 in Fig. 4). If so, then it updates the moving average A_e (step 15–19 in Fig. 4) according to Eq. 3. Further, if it is the first time to reach the window W (i.e., $W' == W$), then A_e assumes a simple average of A_w (step 20 in Fig. 4). Then, the Micro-C sets the mutex down and allows other threads to use the blocked shared memory. Last, our Micro-C code returns the calculated values to the original P4 pipeline (step 22–23 in Fig. 4).

Back in the P4 pipeline (Fig. 3), our approach clone and recirculate the packet p. The original packet is forwarded to its destination (step 13 in Fig. 3), while the cloned one is sent to the INT collector (step 12 in Fig. 3).

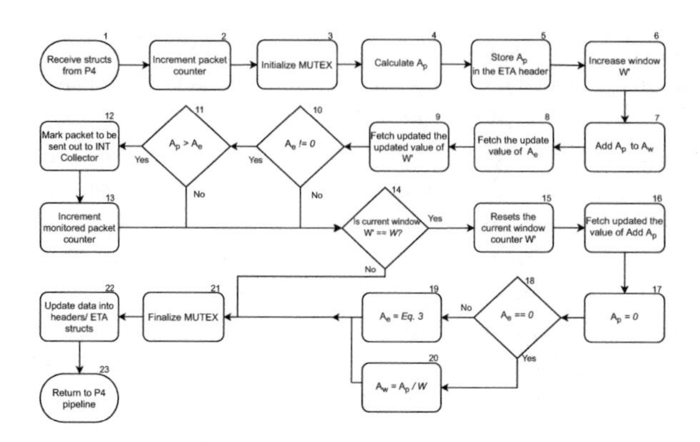

Fig. 4. Overview of the proposed Micro-C routine.

4 Performance Evaluation

In this section, we perform a performance evaluation of our proposal mechanism using a Netronome SmartNIC. We start by describing our environment setup, followed by discussing the results.

4.1 Environment Setup and Baseline Comparison

Setup. Our environment setup consists of three high-end servers. Each server has an Intel Xeon 4214R processor with 32 GB RAM. One server is our Device Under Test (DUT) – i.e., the server in which our solution (i.e., P4 program) is loaded – and the other two are used for traffic generation. All servers have a Netronome SmartNIC Agilio CX 10 Gbit/s network device with two network interfaces, which are physically connected (i.e., each traffic generator is connected to the DUT directly). We use MoonGen [10] as our DPDK[3] traffic generator. We instruct MoonGen using the Netronome Packet Generator[4]. In our experiments, we send 64B IPv4 packets at line rate (i.e., 10Gbit/s) with random source and destination prefixes. One of the traffic generator servers is used to send foreground network traffic, while the second is used to inject abrupt network traffic from time to time. This abrupt network traffic is generated with MoonGen to lead to a congested scenario. The experiment is run through 105 s, divided by time slots/epochs of 15 s each. We sent network traffic bursts in the following slots: $2nd$ (15 s–30 s), $4th$ (45 s–60 s), $6th$ (75 s–90 s). In our experiments, the SmartNIC Netronome acts simultaneously as an INT transit and an INT sink. First, it collects internal data plane metrics such as packet processing time and packet size (i.e., set M'). Then, we assume that both metrics are weighted equally (i.e., $w_1 = w_2$) for calculating A_p, A_w, and A_e. We varied the window size

[3] https://www.dpdk.org/.
[4] https://github.com/emmericp/MoonGen/tree/master/examples/netronome-packetgen.

W from 2^{20} to 2^{23} packets. Also, the parameters α in A_e varies from 0.1 to 0.9. Our solution is compiled using the Netronome's P4 compiler. The compiled code is statically assigned to a single micro engine (ME) inside the SmartNIC. This is done as there are some limitations on the Netronome's Mutex implementation when using different memories hierarchies (i.e., different from the ones inside the ME). All experiments were run at least 30 times to ensure a confidence level higher than 90%.

Baseline. We compare our ETA approach against the Full INT procedure and a fixed threshold one. In the former, all INT data are reported to the INT collector, while in the latter we use a constant value as the threshold. This constant value is obtained by calculating the average of all collected INT data in an offline manner. Our codes are publicly available[5] in order to foster reproducibility.

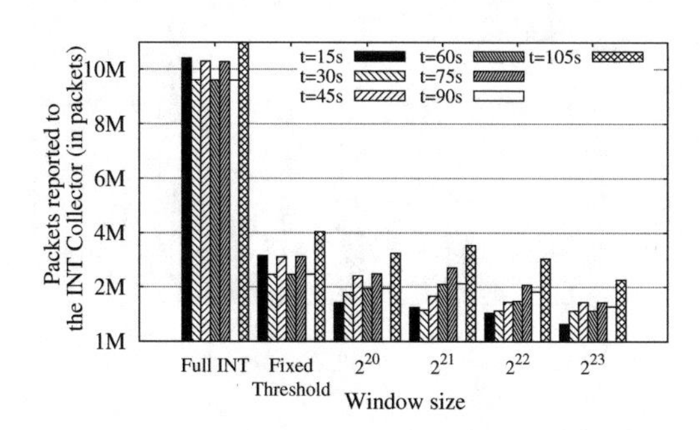

Fig. 5. Number of packets reported to an INT Collector.

4.2 Results

Number of packets sent to the INT Collector. We start by analyzing the number of packets that have been sent to an INT Collector in a given window W (Fig. 5). For this experiment, we consider that $\alpha = 0.1$ (we later analyzed its impact). We observe that our approach outperforms the Full INT and Fixed-Threshold procedures in terms of packets reported by a factor of $16X$ and $5X$, respectively. The main reason consists of the dynamic adjustment made in the threshold value over time. Further, we also observe that our approach decreases the number of reported packets as the size of the windows W increases. For instance, we note 43% less network traffic being reported using a windows $W = 2^{23}$ than $W = 2^{20}$. With larger window size, our metrics A_w tends to be smoother over time and less impacted by abrupt changes in the data plane metrics. Last, we also note a saw-tooth behavior between bars (e.g., $W = 2^{30}$), which represents the effect of sending packet bursts in a given time interval. The larger is the window size, the less is the saw-tooth behavior observed.

[5] urlhttps://github.com/canofre/mestrado/.

Impact of the window size on the computed average A_e. Next, we evaluate how the average A_e computed by the data plane evolves considering different window sizes (from 2^{20} to 2^{23}). Figure 6 illustrates the behavior when setting $\alpha = 0.2$. When the window size is small (e.g., 2^{20}), the average A_e increases over time until reaching a stable value. On the contrary, larger windows tend to decrease the average A_e value over time until reaching a similar stable value. With larger windows (i.e., 2^{22} and 2^{23}), the summation made before computing A_e within an epoch encompasses the periods where network bursts are sent. Therefore, it ends up increasing its initial value in comparison to small windows. Further, we also note that the value found in the stability tends to be more uniform in this case.

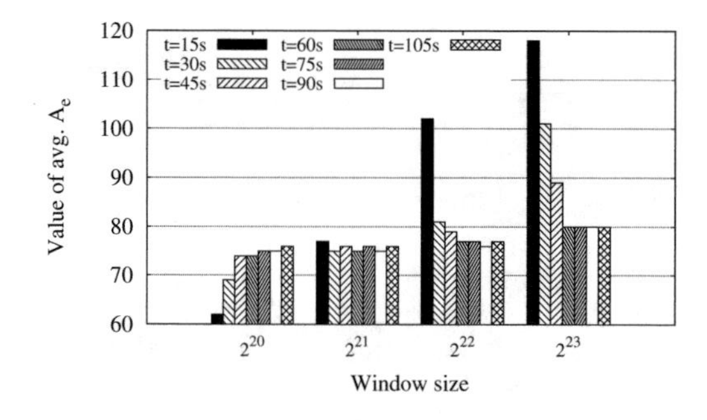

Fig. 6. Impact of window size on the average A_e with $\alpha = 0.2$

Impact of fine-tuning the parameters on the computed average A_e. We analyzed how the A_e evolves varying α from 0.1 to 0.9. For this experiment, we set the window size to $W = 2^{20}$. Note that higher values assigned to α tend to prioritize the behavior observed in the last time window (i.e., A_w), while lower values prioritize the historical behavior observed over time (i.e., A_e). As we can observe in Fig. 7, the higher the α values the higher the average A_e gets over time. In this case, network traffic bursts change data plane metrics (e.g., in-network processing time) and then propagate such values' increases with more intensity to the following-up windows. In turn, lower values of α tend to prioritize the historical behavior and, therefore, the obtained values of A_e are less susceptible to short network traffic variations.

Impact on packet processing latency and throughput. Last, we evaluate the impact of our approach in terms of packet processing and latency when processing packets in the data plane. First, Fig. 8(a) depicts the achieved throughput in packets per second. Our approach is able to outperform the Full INT strategy and the Fixed Threshold by 50% and by 10%, respectively. However, when compared to the Basic Forward – i.e., the same P4 code without any add-on – our approach is limited to run at most 60% of the maximum throughput. This occurs

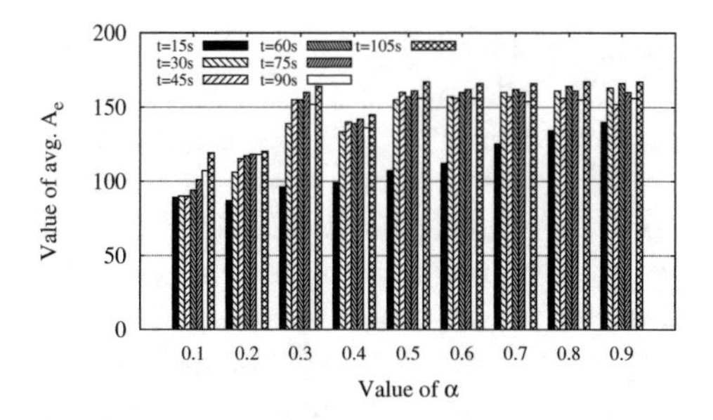

Fig. 7. Impact of the fine-tuning of α on the computed average A_e considering $W = 2^{20}$.

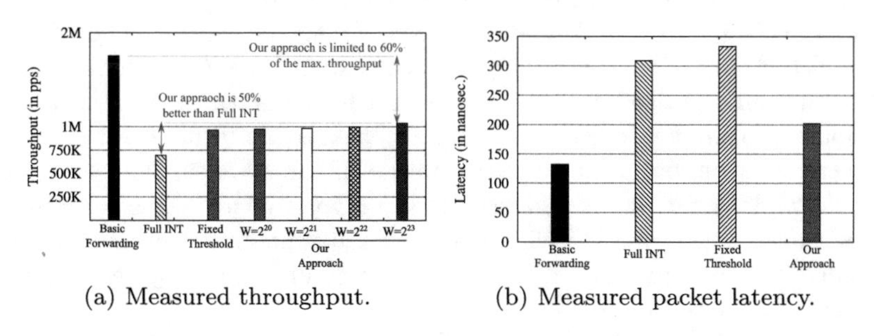

(a) Measured throughput. (b) Measured packet latency.

Fig. 8. Impact of Throughput and Latency of our proposed approach.

mostly because our approach demands more processing power and locks (due to mutex usage) and our approach is limited to running in a single ME (with 8 threads). In turn, Fig. 8(b) illustrates the incurred data plane latency. We measured the latency as the difference between the ingress and egress timestamps inside the data plane. As we observe, our proposed approach adds around 60ns – that is, 1.53X higher than the Basic Forwarding approach. In turn, the Full INT and the Fixed Threshold approaches add 2.34X and 2.52X, respectively.

5 Related Work

The data plane programmability has opened up a wide range of research opportunities to solve existing network monitoring problems such as packet priorization [7], the usage of selective telemetry to reduce the network overload [14], and the classification and analysis of network flows [4]. As previously mentioned, INT allows the network flow packets to embed network telemetry data - such as in [1,6,26]. Sel-INT [26] leverages select group tables to selectively insert INT headers into software switches based on its bucket's weight and a certain probability. Similarly, PINT [1] leverages global hash functions and randomly decides to embed INT data in a given packet, while LINT [6] implements a mechanism with adaptive telemetry accuracy, where each node in the network analyzes its impact and decides whether to send the information to the collector.

P4Entropy [9] calculates the entropy of traffic by using the information contained in the packets. The entropy results are forwarded to the controller for storage and future analysis. Lin et al. [16] perform data collection to allow the SDN controller to evaluate the behavior of network traffic to improve data routing. Likewise, [8] proposes a DDoS schema entirely in the data plane. It leverages typical DDoS metrics such as incoming flows and packet symmetry ratio and periodically triggers alarms to external controllers.

With the emergency of hardware technologies such as SmartNICs and domain-specific programming languages such as P4, more applications are being implemented directly in the data plane. The implementation of selective and dynamic monitoring with the use of additional headers [25], the classification of packets into classes with the implementation of machine learning algorithms [29] and the detection of flow events [30] are some examples of programmatic packet header manipulation. NIDS [20] consists of a machine learning-based anomaly detection algorithm for detecting anomalous packets and future error mitigation. SwitchTree [15] performs the implementation of the Random Forest algorithm in the data plane for packet analysis and decision-making while updating the rules at runtime. [28] and [15,29] present an implementation of machine learning models in the data plane using decision tree, and aim to generate network mapping for intrusion detection services.

Current research efforts in the INT domain (e.g., [1,6,26]) have mostly neglected the costs of reporting telemetry data to INT collectors. In fact, existing work have considered either a full report of information or a selective one based on static thresholds. In this work, we aim to fill this gap and offload the decision process of reporting INT data into the SmartNIC data plane in a dynamically manner. We propose a lightweight in-network mechanism based on a window-based moving average with collected INT data as input.

6 Final Remarks

In this paper, we propose a lightweight in-network mechanism to selective report in-band network telemetry to INT collectors. Our approach is based on a window-based moving average and it is tailored to SmartNIC architectures. By evaluating our approach in a state-of-the-art SmartNIC, we showed that our approach can report up to 16X less reporting statistics to INT collectors, while introducing a negligible overhead in terms of latency (1.5X higher than the baseline pipeline). As future work, we intend to explore machine learning algorithms in the data plane to decide whether to report data or not. Also, we intend to explore other offloading alternatives so that the processing workload can be split up on different SmartNICs or in eBPF/DPDK approaches.

Acknowledgements. This work was partially funded by National Council for Scientific and Technological Development (CNPq) (grant 427814/2018-9), São Paulo Research Foundation (FAPESP) (grants 2018/23092-1, 2020/05115-4, 2020/05183-0), Rio Grande do Sul Research Foundation (FAPERGS) (grants 19/2551-0001266-7, 20/2551-000483-0, 19/2551-0001224-1, 21/2551-0000688-9).

References

1. Ben Basat, R., Ramanathan, S., Li, Y., Antichi, G., Yu, M., Mitzenmacher, M.: Pint: probabilistic in-band network telemetry. In: Proceedings of ACM SIGCOMM, pp. 662–680 (2020)
2. Bhamare, D., Kassler, A., Vestin, J., Khoshkholghi, M.A., Taheri, J.: Intopt: in-band network telemetry optimization for NFV service chain monitoring. In: ICC 2019 - 2019 IEEE International Conference on Communications (ICC), pp. 1–7 (2019)
3. Bosshart, P., et al.: P4: programming protocol-independent packet processors. ACM SIGCOMM 14 **44**(3), 87–95 (2014)
4. Castanheira, L., Parizotto, R., Filho, A.E.S.: Flowstalker: comprehensive traffic flow monitoring on the data plane using P4. In: 2019 IEEE International Conference on Communications, ICC 2019, pp. 1–6. IEEE, Shanghai, China, 20–24 May 2019 (2019)
5. Castro, A.G., et al.: Near-optimal probing planning for in-band network telemetry. IEEE Commun. Lett. **25**(5), 1630–1634 (2021)
6. Chowdhury, S.R., Boutaba, R., Franċois, J.: Lint: accuracy-adaptive and lightweight in-band network telemetry. In: IFIP/IEEE International Symposium on Integrated Network Management (IM), pp. 349–357 (2021)
7. Cugini, F., Gunning, P., Paolucci, F., Castoldi, P., Lord, A.: P4 in-band telemetry (INT) for latency-aware VNF in metro networks. In: Optical Fiber Communication Conference (OFC) 2019, p. M3Z.6. Optical Society of America (2019)
8. Dimolianis, M., Pavlidis, A., Maglaris, V.: A multi-feature DDOS detection schema on p4 network hardware. In: 2020 23rd Conference on Innovation in Clouds, Internet and Networks and Workshops (ICIN), pp. 1–6 (2020)
9. Ding, D., Savi, M., Siracusa, D.: Estimating logarithmic and exponential functions to track network traffic entropy in p4. In: NOMS 2020 - 2020 IEEE/IFIP Network Operations and Management Symposium, pp. 1-9. IEEE Press (2020)
10. Emmerich, P., Gallenmüller, S., Raumer, D., Wohlfart, F., Carle, G.: Moongen: A scriptable high-speed packet generator. In: Proceedings of the ACM IMC, pp. 275-287. IMC 2015, ACM, New York, NY, USA (2015)
11. Hohemberger, R., Castro, A.G., Vogt, F.G., Mansilha, R.B., Lorenzon, A.F., Rossi, F.D., Luizelli, M.C.: Orchestrating in-band data plane telemetry with machine learning. IEEE Commun. Lett. **23**(12), 2247–2251 (2019)
12. Jeyakumar, V., Alizadeh, M., Geng, Y., Kim, C., Mazières, D.: Millions of little minions: using packets for low latency network programming and visibility. ACM SIGCOMM CCR **44**(4), 3–14 (2014)
13. Joshi, R., Qu, T., Chan, M.C., Leong, B., Loo, B.T.: Burstradar: practical real-time microburst monitoring for datacenter networks. In: Proceedings of the 9th Asia-Pacific Workshop on Systems. APSys 2018, Association for Computing Machinery, New York, NY, USA (2018)
14. Kim, Y., Suh, D., Pack, S.: Selective in-band network telemetry for overhead reduction. In: IEEE International Conference on Cloud Networking (CloudNet), pp. 1–3 (2018)
15. Lee, J.H., Singh, K.: Switchtree: in-network computing and traffic analyses with random forests. Neural Comput. Appl. (2020)
16. Lin, W.H., et al.: Network telemetry by observing and recording on programmable data plane. In: IFIP Networking Conference (IFIP Networking), pp. 1–6 (2021)

17. Liu, Z., Bi, J., Zhou, Y., Wang, Y., Lin, Y.: NetVision: towards network telemetry as a service. In: 2018 IEEE 26th International Conference on Network Protocols (ICNP), pp. 247–248, September 2018
18. Luizelli, M.C., Canofre, R., Lorenzon, A.F., Rossi, F.D., Cordeiro, W., Caicedo, O.M.: In-network neural networks: challenges and opportunities for innovation. IEEE Netw. **35**(6), 68–74 (2021). https://doi.org/10.1109/MNET.101.2100098
19. Marques, J.A., Luizelli, M.C., Tavares da Costa Filho, R.I., Gaspary, L.P.: An optimization-based approach for efficient network monitoring using in-band network telemetry. J. Internet Serv. Appl. **10**(1), 1–20 (2019). https://doi.org/10.1186/s13174-019-0112-0
20. Nam, S., Lim, J., Yoo, J.H., Hong, J.W.K.: Network anomaly detection based on in-band network telemetry with RNN. In: IEEE International Conference on Consumer Electronics - Asia (ICCE-Asia), pp. 1–4 (2020)
21. Netronome: Internet (2020). https://www.netronome.com/static/app/img/products/silicon-solutions/WP_NFP4000_TOO.pdf
22. Pan, T., et al.: Int-path: towards optimal path planning for in-band network-wide telemetry. In: IEEE INFOCOM, pp. 1–9, April 2019
23. Saquetti, M., et al.: Toward in-network intelligence: running distributed artificial neural networks in the data plane. IEEE Commun. Lett. **25**(11), 3551–3555 (2021)
24. Singh, S.K., Rothenberg, C., Luizelli, M.C., Antichi, G., Pongracz, G.: Revisiting heavy-hitters: don't count packets, compute flow inter-packet metrics in the data plane. In: ACM SIGCOMM Poster, pp. 1–4. ACM, New York, NY, USA (2020)
25. Suh, D., Jang, S., Han, S., Pack, S., Wang, X.: Flexible sampling-based in-band network telemetry in programmable data plane. ICT Express 6(1), 62–65 (2020)
26. Tang, S., Li, D., Niu, B., Peng, J., Zhu, Z.: Sel-INT: a runtime-programmable selective in-band network telemetry system. IEEE Trans. Netw. Serv. Manage. **17**(2), 708–721 (2019)
27. Viegas, P., Goes de Castro, A., Lorenzon, A.F., Rossi, F.D., Luizelli, M.C.: The actual cost of programmable smartnics: diving into the existing limits. In: Barolli, L., Amato, F., Moscato, F., Enokido, T., Takizawa, M. (eds.) Advanced Information Networking and Applications, vol. 225, pp. 381–392. Springer International Publishing (2021). https://doi.org/10.1007/978-3-030-75100-5_17
28. Xavier, B.M., Guimarães, R.S., Comarela, G., Martinello, M.: Programmable switches for in-networking classification. In: IEEE INFOCOM 2021 - IEEE Conference on Computer Communications, pp. 1–10 (2021)
29. Xiong, Z., Zilberman, N.: Do switches dream of machine learning? toward in-network classification, pp. 25-33. HotNets 2019, Association for Computing Machinery, New York, NY, USA (2019)
30. Zhou, Y., et al.: Flow event telemetry on programmable data plane. In: Proceedings of ACM SIGCOMM, pp. 76-89. SIGCOMM 2020, Association for Computing Machinery, New York, NY, USA (2020)

Energy Consumption of the Information Flow Control in the IoT: Simulation Evaluation

Shigenari Nakamura[1(✉)], Tomoya Enokido[2], and Makoto Takizawa[3]

[1] Tokyo Metropolitan Industrial Technology Research Institute, Tokyo, Japan
nakamura.shigenari@iri-tokyo.jp

[2] Rissho University, Tokyo, Japan
eno@ris.ac.jp

[3] Hosei University, Tokyo, Japan
makoto.takizawa@computer.org

Abstract. In the IoT (Internet of Things), data are exchanged among subjects and objects in devices through manipulating objects. Here, even if subjects manipulate objects according to the CBAC (Capability-Based Access Control) model, the subjects can get data which the subjects are not allowed to get, i.e. illegal information flow and late information flow occur. In order to make the IoT secure, the OI (Operation Interruption) and the TBOI (Time-Based OI) protocols are implemented to interrupt operations occurring illegal and late types of information flows. In addition, two types of capability token selection algorithms are proposed to make the protocols more useful. In our previous studies, an electric energy consumption model of a device supporting the information flow control is proposed. In this paper, electric energy consumption of each protocol is made clear based on the energy consumption model. In the simulation evaluation, it is shown that the total electric energy consumption in the OI and the TBOI protocols with capability token selection algorithms is smaller than the conventional OI and TBOI protocols.

Keywords: IoT (Internet of Things) · Device security · CBAC (Capability-Based Access Control) model · Information flow control · CoAP (Constrained Application Protocol) · Capability token selection algorithm · Electric energy consumption

1 Introduction

The CBAC (Capability-Based Access Control) model is useful to make the IoT (Internet of Things) [6,18] secure. In a CBAC model called "CapBAC model" [7], subjects are issued capability tokens which are collections of access rights. Only the authorized subjects can manipulate objects of devices only in the authorized operations. Data are exchanged among subjects and objects through manipulating objects of devices. Therefore, a subject might get data via other subjects

L. Barolli et al. (Eds.): AINA 2022, LNNS 449, pp. 285–296, 2022.
https://doi.org/10.1007/978-3-030-99584-3_25

and objects even if the subject is granted no access right to get the data, i.e. illegal information flow might occur [5,10–12,14]. In addition, a subject might get data generated out of the validity period of a capability token to get the data. Even if the time τ is not within the validity period of the capability token, the subject sb_i can get the data generated at time τ. The data are older than the subject sb_i expects to get, i.e. the data come to the subject sb_i *late* [15].

In order to solve both types of illegal and late information flow problems in the IoT, the OI (Operation Interruption) and the TBOI (Time-Based OI) protocols are implemented in a Raspberry Pi3 Model B+ [1] with Raspbian [2] which is regarded as an IoT device [14,15]. The device and a subject are a CoAP (Constrained Application Protocol) server and a CoAP client, respectively, in CoAPthon3 [20]. In the OI and the TBOI protocols, operations occurring illegal information flow and both types of illegal and late information flows are interrupted, i.e. not performed, at devices. In the evaluation, the request processing time gets longer as the number of capability tokens whose signatures are verified in devices increases in these protocols.

In order to shorten the request processing time, the MRCTSD (Minimum Required Capability Token Selection for Devices) algorithm is proposed and implemented [13]. Here, minimum required capability tokens to make authorization decisions are selected and used in devices. However, the more number of capability tokens are sent to devices, the more complex the capability token selection is. It is critical to avoid concentrating loads in devices because the devices support just low processing power and smaller size of memories. Hence, the MRCTSS (MRCTS for Subjects) algorithm to select capability tokens sent from subjects to devices is proposed and implemented [16].

In our previous studies [17], an electric energy consumption model of a device realized in a Raspberry Pi 3 Model B+ equipped with Raspbian supporting the information flow control is proposed. In the protocols, an authorization is modeled to be a computation process which uses CPU resources like scientific computation. In the paper [9], the MLPC (Multi-Level Power Consumption) and the MLC (ML Computation) models of a device to perform computation processes are proposed. Here, devices are mainly characterized in terms of the numbers of cores and threads of a CPU. This means, the power consumption of a device depends on numbers of active cores and threads. An energy consumption model of a device supporting the information flow control is proposed based on the MLPC and the MLC models.

In this paper, electric energy consumption of each protocol is made clear based on the energy consumption model. In the OI and the TBOI protocols with capability token selection algorithms, the number of capability tokens whose signatures have to be checked is reduced. Here, the execution time to authorize access requests is shortened. Hence, the electric energy consumption is smaller than the conventional OI and TBOI protocols. In the simulation evaluation with the electric energy consumption model, it is shown that the total electric energy consumption in the OI and the TBOI protocols with capability token selection algorithms is smaller than the conventional OI and TBOI protocols.

In Sect. 2, the system model and types of information flow relations are discussed. In Sect. 3, the protocols to prevent both types of information flows and the capability token selection algorithms are discussed. In Sect. 4, the electric energy consumption model of a device supporting the protocols is discussed. In Sect. 5, the protocols are evaluated in terms of the electric energy consumption.

2 System Model

In an IoT, there are the numbers dn and sbn of devices $d_1, \ldots, d_{dn}$ ($dn \geq 1$) and subjects $sb_1, \ldots, sb_{sbn}$ ($sbn \geq 1$), respectively. Each device d_k holds the number on^k of objects $o_1^k, \ldots, o_{on^k}^k$ ($on^k \geq 1$). A term, "object o_m^k" stands for a component object in the device d_k. In order to make the IoT secure, a CBAC model called "CapBAC model" [7], is discussed. Subjects manipulate data of objects in devices. Here, each subject sb_i is issued a set CAP^i which consists of the number cn^i of capability tokens $cap_1^i, \ldots, cap_{cn^i}^i$ ($cn^i \geq 1$).

A capability token cap_g^i is designed [14,15]. Let $cap_g^i.IS$ and $cap_g^i.SU$ be public keys of an issuer and a subject of the capability token cap_g^i, respectively. $cap_g^i.SG$ is a signature generated with the private key of the issuer. These signatures and keys are generated in the ECDSA (Elliptic Curve Digital Signature Algorithm) [8] and then encoded into Base64 form. The capability token cap_g^i indicates how the subject sb_i can manipulate objects in a device shown in $cap_g^i.DE$. Access rights field of the capability token cap_g^i shows object shown in $cap_g^i.OB$ can be manipulated in operation shown in $cap_g^i.OP$. The capability token cap_g^i is valid at time τ where $cap_g^i.NB < \tau < cap_g^i.NA$. Capability tokens are included in the payload field of a CoAP request.

If a subject sb_i tries to manipulate data of an object o_m^k in a device d_k in an operation op, the subject sb_i sends an access request with a capability token cap_g^i to specify the subject sb_i is allowed to manipulate the object o_m^k in the operation op to the device d_k. The access request is accepted, i.e. the operation op is performed on the object o_m^k if the device d_k confirms that the subject sb_i is allowed to manipulate the object o_m^k in the operation op. Otherwise, the access request is rejected. Since the device d_k just checks the capability token cap_g^i to authorize the subject sb_i, it is easier to adopt the CBAC model to the IoT than the ACL (Access Control List)-based models such as RBAC (Role-Based Access Control) [19] and the ABAC (Attribute-Based Access Control) [21] models.

Let a pair $\langle o, op \rangle$ be an access right. Subjects issued a capability token including an access right $\langle o, op \rangle$ is allowed to manipulate data of an object o in an operation op. A set of objects whose data a subject sb_i is allowed to get is $IN(sb_i)$ i.e. $IN(sb_i) = \{o_m^k \mid \langle o_m^k, get \rangle \in cap_g^i \wedge cap_g^i \in CAP^i\}$.

Through manipulating data of objects in devices, the data are exchanged among subjects and objects. For example, if a subject sb_i puts data got from an object o_n^l to another object o_m^k, the data of the object o_n^l flow into the subject sb_i and the object o_m^k. Objects whose data flow into entities such as subjects and objects are referred to as *source* objects for these entities. Let $sb_i.sO$ and

$o^k_m.sO$ are sets of *source* objects for a subject sb_i and an object o^k_m, respectively, which are initially ϕ. In this example, $sb_i.sO = o^k_m.sO = \{o^l_n\}$.

The validity period of each capability token cap^i_g is from the time $cap^i_g.NB$ to the time $cap^i_g.NA$. Let a pair of times $gt^i.st(o^k_m)$ and $gt^i.et(o^k_m)$ be the start and end time when a subject sb_i is allowed to get data from the object o^k_m. The time when data of an object o^k_m are generated is referred to a generation time. Let $minOT^k_m(o^l_n)$ and $minSBT^i(o^l_n)$ be the earliest generation times of data of an object o^l_n which flow to an object o^k_m and a subject sb_i, respectively.

Based on the CBAC model, we define types of information flow relations on objects and subjects as follows:

Definition 1. *An object o^k_m flows to a subject sb_i ($o^k_m \rightarrow sb_i$) iff (if and only if) $o^k_m.sO \neq \phi$ and $o^k_m \in IN(sb_i)$.*

Definition 2. *An object o^k_m legally flows to a subject sb_i ($o^k_m \Rightarrow sb_i$) iff $o^k_m \rightarrow sb_i$ and $o^k_m.sO \subseteq IN(sb_i)$.*

Definition 3. *An object o^k_m illegally flows to a subject sb_i ($o^k_m \mapsto sb_i$) iff $o^k_m \rightarrow sb_i$ and $o^k_m.sO \nsubseteq IN(sb_i)$.*

Definition 4. *An object o^k_m timely flows to a subject sb_i ($o^k_m \Rightarrow_t sb_i$) iff $o^k_m \Rightarrow sb_i$ and $\forall o^l_n \in o^k_m.sO$ $(gt^i.st(o^l_n) \leq minOT^k_m(o^l_n) \leq gt^i.et(o^l_n))$.*

Definition 5. *An object o^k_m flows late to a subject sb_i ($o^k_m \mapsto_l sb_i$) iff $o^k_m \Rightarrow sb_i$ and $\exists o^l_n \in o^k_m.sO$ $\neg(gt^i.st(o^l_n) \leq minOT^k_m(o^l_n) \leq gt^i.et(o^l_n))$.*

3 Information Flow Control

3.1 Protocols

In the CBAC model for the secure IoT, data are exchanged among subjects and objects. Here, a subject sb_i may get data of an object o^k_m flowing to another object o^l_n by accessing the object o^l_n even if the subject sb_i is not issued a capability token cap^i_g to get the data from the object o^k_m, i.e. illegal information flow from the object o^k_m to the subject sb_i occurs. In addition, although no illegal information flow occurs, a subject sb_i may get data from an object o^k_m generated out of validity period of a capability token cap^i_g to get the data. Here, the data are older than the subject sb_i expects to get, i.e. information comes to the subject sb_i *late*. In our previous studies, the OI (Operation Interruption) [14] and the TBOI (Time-Based OI) [15] protocols are implemented to prevent illegal information flow and both illegal and late types of information flows, respectively.

In order to prevent the illegal information flow, sets of *source* objects are manipulated in the OI protocol. Here, if data of an object o^k_m flow into an entity, the object o^k_m is added to a *source* object set of the entity. The sets of *source* objects are updated as follows:

1. Initially, $sb_i.sO = o^k_m.sO = \phi$ for every subject sb_i and object o^k_m;
2. If a device d_k generates data by sensing events occurring around itself and stores the data to its object o^k_m, $o^k_m.sO = o^k_m.sO \cup \{o^k_m\}$;
3. If a subject sb_i gets data from an object o^k_m, $sb_i.sO = sb_i.sO \cup o^k_m.sO$;
4. If a subject sb_i puts data to an object o^k_m, $o^k_m.sO = o^k_m.sO \cup sb_i.sO$;

On the other hand, in the TBOI protocol, the earliest generation time of data of every *source* object is also updated as follows:

1. Initially, $sb_i.sO = o^k_m.sO = \phi$ for every subject sb_i and object o^k_m;
2. If a device d_k generates data by sensing events occurring around itself and stores the data to its object o^k_m at time τ.
   a. If $minOT^k_m(o^k_m) = \text{NULL}$, $minOT^k_m(o^k_m) = \tau$;
   b. $o^k_m.sO = o^k_m.sO \cup \{o^k_m\}$;
3. If a subject sb_i gets data from an object o^k_m.
   a. For each object o^l_n such that $o^l_n \in (sb_i.sO \cap o^k_m.sO)$, $minSBT^i(o^l_n) = min(minSBT^i(o^l_n), minOT^k_m(o^l_n))$;
   b. For each object o^l_n such that $o^l_n \notin sb_i.sO$ but $o^l_n \in o^k_m.sO$, $minSBT^i(o^l_n) = minOT^k_m(o^l_n)$;
   c. $sb_i.sO = sb_i.sO \cup o^k_m.sO$;
4. If a subject sb_i puts data to an object o^k_m.
   a. For each object o^l_n such that $o^l_n \in (sb_i.sO \cap o^k_m.sO)$, $minOT^k_m(o^l_n) = min(minOT^k_m(o^l_n), minSBT^i(o^l_n))$;
   b. For each object o^l_n such that $o^l_n \notin o^k_m.sO$ but $o^l_n \in sb_i.sO$, $minOT^k_m(o^l_n) = minSBT^i(o^l_n)$;
   c. $o^k_m.sO = o^k_m.sO \cup sb_i.sO$;

Based on the sets of *source* objects and the earliest generation time of data of every object, the illegal information flow and the late information flow are detected. The OI and the TBOI protocols perform as follows to prevent illegal information flow and both illegal and late types of information flows, respectively:

[**OI protocol**] A *get* operation on an object o^k_m issued by a subject sb_i is interrupted if $o^k_m \mapsto sb_i$.
[**TBOI protocol**] A *get* operation on an object o^k_m issued by a subject sb_i is interrupted if $o^k_m \mapsto_l sb_i$.

3.2 Capability Token Selection Algorithms

In our previous studies [14,15], the OI and the TBOI protocols are implemented on a Raspberry Pi 3 Model B+ [1] whose operating system is Raspbian [2]. If a subject sb_i issues a *get* operation on an object o^k_m to a device d_k, the device d_k has to confirm the subject sb_i is issued capability tokens to get data from not only the object o^k_m but also the other objects in $o^k_m.sO$. Here, the request processing time which is the period between when a device receives a request and when the device sends the answer of the request increases as the number of capability tokens whose signatures are verified increases. The time to verify the signatures accounts for a large portion of the request processing time.

In the paper [13], the MRCTSD (Minimum Required Capability Token Selection for Devices) algorithm is proposed for the OI and the TBOI protocols to shorten the request processing time. Suppose a subject sb_i issues a *get* operation on an object o^k_m to the device d_k. Here, the device d_k selects minimum required capability tokens to confirm the subject sb_i is allowed to get data from not only the object o^k_m but also the *source* objects in $o^k_m.sO$. The OI-MRCTSD protocol and the TBOI-MRCTSD protocol are proposed and implemented. In the evaluation, it is shown that the request processing times of the OI-MRCTSD and the TBOI-MRCTSD protocols are smaller than the OI and the TBOI protocols. However, the more number of capability tokens are sent to devices, the more complex the capability token selection is in the MRCTSD algorithm. It is critical to avoid concentrating loads in devices because the devices support just low processing power and smaller size of memory.

In the paper [16], the MRCTSS (MRCTS for Subjects) algorithm to select capability tokens sent from subjects to devices is proposed. Suppose a subject sb_i tries to send a *get* operation on an object o^k_m to a device d_k. In the OI and the TBOI protocols, the subject sb_i has to indicate that the subject sb_i is allowed to get data from not only the object o^k_m but also the other *source* objects of the object o^k_m. However, when the subject sb_i starts to send a *get* operation, the subject sb_i dose not know which *source* objects are in the object o^k_m. Hence, the subject sb_i selects minimum required capability tokens to indicate the subject sb_i is allowed to get data from every object in $IN(sb_i)$ in the MRCTSS algorithm. The OI-MRCTSS protocol and the TBOI-MRCTSS protocol are proposed and implemented. In the evaluation, it is shown that the size of a UDP datagram in a *get* access request in the OI-MRCTSS and the TBOI-MRCTSS protocols are smaller than the OI, the TBOI, the OI-MRCTSD, and the TBOI-MRCTSD protocols.

4 Electric Energy Consumption Model

The IoT is more scalable than the cloud computing systems because huge number and various types of nodes are included. Hence, it is required to reduce the electric energy consumption by not only servers but also devices. In the paper [9], the MLPC (Multi-Level Power Consumption) and the MLC (ML Computation) models of a device to perform computation processes is proposed. Devices are mainly characterized in terms of the numbers of cores and threads of their CPUs. A CPU is characterized in terms of the number nc^k of cores and the number nct^k of threads on each core. Let nt^k be the total number of threads in a device d_k, i.e. $nt^k = nc^k \cdot nct^k$. A device d_k provides nt^k (≥ 1) threads th^k_0, $th^k_1, \ldots, th^k_{nt^k-1}$ on nc^k cores $c^k_0, c^k_1, \ldots, c^k_{nc^k-1}$. A thread th^k_i is *active* iff at least one process is performed on the thread th^k_i in a device d_k. A core c^k_j is *active* iff at least one thread of the core c^k_j is active. The device d_k consumes the electric power bC^k [W] if at least one core is active. The electric power consumption of a device d_k increases by values cPW^k [W] and tPW^k [W] as the numbers of active cores and active threads increase by one, respectively. Cores and threads which are not active are *idle*. A device where at least one thread is active is active. That is, at least one process is performed on an active device. A device which is not active, i.e. every thread is idle, is idle. On an idle device, no process is performed. Here, the device d_k consumes the minimum electric power $minPW^k$ [W]. The power consumption $PW^k(\tau)$ [W] of a device d_k to perform processes at time τ is given as follows:

[MLPC (Multi-Level Power Consumption) model]

$$PW^k(\tau) = minPW^k + \gamma(\tau) \cdot [bC^k + \sum_{i=0}^{nc^k-1} \alpha_i(\tau) \cdot \{(cPW^k + \beta_i(\tau) \cdot tPW^k)\}]. \quad (1)$$

If at least one core is active at time τ, $\gamma(\tau) = 1$. Otherwise, $\gamma(\tau) = 0$, i.e. an idle device d_k consumes minimum electric power $minPW^k$. If a core c_i^k is active on the device d_k at time τ, $\alpha_i(\tau) = 1$. Otherwise, $\alpha_i(\tau) = 0$. $\beta_i(\tau)$ ($\leq nct^k$) is the number of active threads on a core c_i^k at time τ. If $\alpha_i(\tau) = 1$, $1 \leq \beta_i(\tau) \leq nct^k$. '$\gamma(\tau) = 1$' means $\alpha_i(\tau) = 1$ for some core c_i^k.

Let $CP^k(\tau)$ be a set of processes which are being performed on a device d_k at time τ. In the paper [9], a process p_m^k is how much processed by a device d_k at time τ is referred to as 'computation rate $cr_m^k(\tau)$'. Let $maxC^k$ be the maximum computation rate of a core in the device d_k. The computation rate $cr_m^k(\tau)$ of a process p_m^k in a device d_k at time τ is given as follows:

[MLC (Multi-Level Computation) model]

$$cr_m^k(\tau) = \begin{cases} maxC^k & (\text{if } |CP^k(\tau)| \leq nt^k). \\ nt^k \cdot maxC^k / |CP^k(\tau)| & (\text{if } |CP^k(\tau)| > nt^k). \end{cases} \quad (2)$$

In the paper [17], an electric energy consumption model of a device d_k, Raspberry Pi 3 Model B+ [1] equipped with Raspbian [2], supporting the protocols to prevent the illegal information flow and the late information flow is proposed based on the MLPC and MLC models. The numbers nc^k, nct^k, and nt^k of a CPU in the device d_k are 4, 1, and 4, respectively. Based on measurement experiments, the parameters $minPW^k$, bC^k, and $cPW^k + tPW^k$ are decided as 2.55, 0.20, and 0.45 [W], respectively.

For $|CP^k(\tau)| \leq 4$, the computation rate $cr_m^k(\tau)$ is $maxC^k$ since only one process is performed on each thread. For $|CP^k(\tau)| > 4$, at least one process is performed on every thread. For example, if $nt^k \cdot 3$ processes are performed on the device d_k, three processes are concurrently performed on each thread. Hence, the computation rate $cr_m^k(\tau)$ of a process p_m^k is $nt^k \cdot maxC^k$ / $nt^k \cdot 3$. For $|CP^k(\tau)|$ $> nt^k$, the computation rate $cr_m^k(\tau)$ for the process p_m^k decreases as the number of processes concurrently performed on the device d_k, i.e. $|CP^k(\tau)|$, increases. Hence, for $|CP^k(\tau)| > nt^k$, the execution time of the process p_m^k increases as

$|CP^k(\tau)|$ increases. The total electric energy $EE^k(st, et)$ [J] of a device d_k from time st to time et is given as follows.

$$EE^k(st, et) = \sum_{\tau=st}^{et} PW^k(\tau). \tag{3}$$

In the papers [13–16], the protocols are implemented in the device d_k. The device d_k is manipulated by a subject sb_i which is implemented in Python. Both the device d_k and the subject sb_i are realized as a CoAP (Constrained Application Protocol) server and a CoAP client in CoAPthon3 [20], respectively. A capability token is designed as shown in the papers [14,15]. If the subject sb_i issues a *get* operation on an object o^k_m in the device d_k, the device d_k supports the authorization process composed of the following steps.

Step 1 The device d_k confirms which objects the subject sb_i is allowed to get data from.
Step 2 The device d_k verifies a signature with the public key of the subject sb_i.
Step 3 The device d_k verifies a signature of every capability token with each issuer's public key.

If the access request is accepted, the device d_k sends a CoAP response to the subject sb_i. Here, the CPU resources of the device d_k are used in all the steps. Let p^k_m be a process composed of all the steps for the object o^k_m on the device d_k. The process p^k_m is modeled to be a computation process which uses CPU resources like scientific computation.

5 Evaluation

The protocols are evaluated in terms of the electric energy consumption. In the evaluation, fifteen devices and twenty subjects are considered ($dn = 15$, $sbn = 20$). It is assumed that subjects issue only *get* and *put* operations.

Every device d_k obtains the number on^k of objects. The number on^k is randomly selected out of numbers 1, ..., 5. The type of each device d_k is randomly decided. Initially, a set of source object $o^k_m.sO$ of each object o^k_m is empty.

Every subject sb_i is issued the number cn^i of capability tokens. Every capability token cap^i_g includes the number arn^i_g of access rights. Validity period vp^i_g of the capability token cap^i_g is randomly selected out of numbers 20, ..., 50 simulation steps. The numbers cn^i and arn^i_g are randomly selected out of numbers 5, ..., 10 and 1, ..., $on^k \cdot opsn^k$, respectively. $opsn^k$ is the number of operations supported in the device d_k. Hence, if the device d_k is a sensor or actuator, $opsn^k = 1$. Otherwise, i.e. the device d_k is a hybrid device, $opsn^k = 2$. Access rights and a device d_k shown in $cap^i_g.DE$ in the capability token cap^i_g are randomly decided. The operation of an access right for an object o^k_m in a hybrid device d_k is decided to be *get* with probability 0.7. On the other hand, the operation is decided to be *put* with probability 0.3.

After the generation of devices and subjects, the following procedures are performed in every protocol:

1. Every sensor and hybrid device d_k collects data by sensing events with probability 0.5. Here, the sensing type is randomly selected so that the sensing is full one with probability 0.3. If the device d_k collects data, the data stored in an object o^k_m randomly selected.
2. For every subject sb_i, the validity period vp^i_g of every capability token cap^i_g is decremented by one. If the validity period vp^i_g gets 0, the capability token cap^i_g is revoked from the subject sb_i.
3. Every subject sb_i which has no capability token is issued capability tokens randomly generated as above.
4. Every subject sb_i issues an operation with probability 0.7. If a subject sb_i decides to issue an operation, one access right obtained by the subject sb_i is randomly selected and the subject sb_i issues an operation according to the access right. In a *get* operation, data in an object flow to the subject sb_i. On the other hand, in a *put* operation, data in the subject sb_i flow to the object. For each operation, full and partial types are randomly selected with probabilities 0.3 and 0.7, respectively.
5. The electric energy consumption of every device is calculated. After that, total electric energy consumption of all the devices is calculated.

In the simulation, we make the following assumptions:

- If $\exists o^l_n \in o^k_m.sO$ $(minOT^k_m(o^l_n) < minSBT^i(o^l_n))$, a subject sb_i does not issue a full *get* operation to a device d_k.
- If $\exists o^l_n \in sb_i..sO$ $(minSBT^i(o^l_n) < minOT^k_m(o^l_n))$, a subject sb_i does not issue a *full* put operation to a device d_k.

Let st be a simulation steps. The above procedures are assumed to be performed in one simulation step. This means, the above procedures are iterated st times. In the evaluation, one simulation step means one [s]. Here, st = 0, 600, 1200, 1800, 2400, 3000 or 3600. The sets of devices and subjects are randomly generated ten times. For a pair of given these sets, the above procedures for st are iterated five times. Finally, the average electric energy consumption of all the devices is calculated.

Figure 1 shows the differences of total electric energy consumption between the most energy-efficient protocol, i.e. TBOI-MRCTSD protocol, and the other protocols for the simulation steps st [s]. In the capability token selection algorithms, the request processing time is shorten by reducing the number of capability tokens whose signature is verified. Hence, the electric energy consumption in the OI and the TBOI protocols with capability token selection algorithms is smaller than the conventional OI and TBOI protocols.

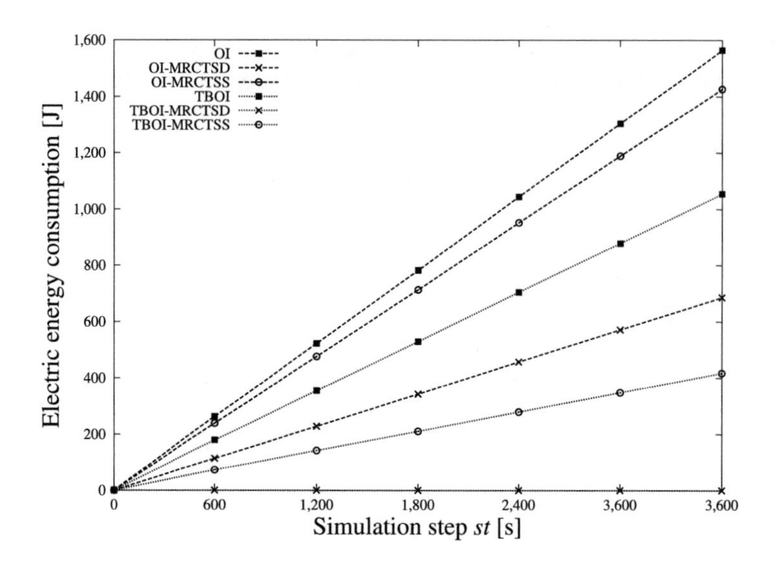

Fig. 1. Difference of total electric energy consumption.

6 Concluding Remarks

In the CBAC (Capability-Based Access Control) model proposed to make the IoT (Internet of Things) secure, capability tokens are issued to subjects as access rights. Since data are exchanged among subjects and objects through manipulating objects, two types of illegal and late information flows occur. In our previous studies, the OI (Operation Interruption) and the TBOI (Time-Based OI) protocols were proposed to prevent both types of illegal and late information flows. In addition, the MRCTSD (Minimum Required Capability Token Selection for Devices) and the MRCTSS (MRCTS for Subjects) algorithms are proposed to make the OI and the TBOI protocols more useful. In our previous studies, an electric energy consumption model of a device realized in a Raspberry Pi 3 Model B+ equipped with Raspbian supporting the information flow control is proposed based on the MLPC and the MLC models. In this paper, electric energy consumption of the protocols are made clear based on the energy consumption model. In the OI and the TBOI protocols with capability token selection algorithms, the number of capability tokens whose signatures have to be checked is reduced. Hence, since the execution time to authorize access requests is reduced, the electric energy consumption is smaller than the conventional OI and TBOI protocols.

Acknowledgement. This work was supported by Japan Society for the Promotion of Science (JSPS) KAKENHI Grant Number JP20K23336.

References

1. Raspberry pi 3 model b+. https://www.raspberrypi.org/products/raspberry-pi-3-model-b-plus/
2. Raspbian, version 10.3, 13 February 2020. https://www.raspbian.org/
3. Uwmeter (2011). http://www.metaprotocol.com/UWmeter/Features.html
4. Date, C.J.: An Introduction to Database Systems, 8th edn. Addison Wesley, Boston, MA, USA (2003)
5. Denning, D.E.R.: Cryptography and Data Security. Addison Wesley, Boston, MA, USA (1982)
6. Hanes, D., Salgueiro, G., Grossetete, P., Barton, R., Henry, J.: IoT Fundamentals: Networking Technologies, Protocols, and Use Cases for the Internet of Things. Cisco Press, Indianapolis, IN, USA (2018)
7. Hernández-Ramos, J.L., Jara, A.J., Marín, L., Skarmeta, A.F.: Distributed capability-based access control for the internet of things. J. Internet Serv. Inf. Secur. **3**(3/4), 1–16 (2013)
8. Johnson, D., Menezes, A., Vanstone, S.: The elliptic curve digital signature algorithm (ECDSA). Int. J. Inf. Secur. **1**(1), 36–63 (2001)
9. Kataoka, H., Nakamura, S., Duolikun, D., Enokido, T., Takizawa, M.: Multi-level power consumption model and energy-aware server selection algorithm. Int. J. Grid Utility Comput. **8**(3), 201–210 (2017)
10. Nakamura, S., Duolikun, D., Aikebaier, A., Enokido, T., Takizawa, M.: Read-write abortion (RWA) based synchronization protocols to prevent illegal information flow. In: Proceeding of the 17th International Conference on Network-Based Information Systems, pp. 120–127 (2014)
11. Nakamura, S., Duolikun, D., Enokido, T., Takizawa, M.: A read-write abortion protocol to prevent illegal information flow in role-based access control systems. Int. J. Space-Based Situated Comput. **6**(1), 43–53 (2016)
12. Nakamura, S., Enokido, T., Takizawa, M.: Information flow control in object-based peer-to-peer publish/subscribe systems. Concurr. Comput. Practice Exp. **32**(8), e5118 (2020)
13. Nakamura, S., Enokido, T., Takizawa, M.: A capability token selection algorithm for lightweight information flow control in the IoT. In: Proceeding of the 24th International Conference on Network-Based Information Systems, pp. 23–34 (2021)
14. Nakamura, S., Enokido, T., Takizawa, M.: Implementation and evaluation of the information flow control for the Internet of Things. Concurr. Comput. Practice Exp. **33**(19), e6311 (2021)
15. Nakamura, S., Enokido, T., Takizawa, M.: Information flow control based on capability token validity for secure IoT: implementation and evaluation. Internet of Things **15**, 100423 (2021)
16. Nakamura, S., Enokido, T., Takizawa, M.: Traffic reduction for information flow control in the IoT. In: Proceeding of the 16th International Conference on Broad-Band Wireless Computing, Communication and Applications, pp. 67–77 (2021)
17. Nakamura, S., Enokido, T., Takizawa, M.: Energy consumption model of a device supporting information flow control in the IoT. In: Proceeding of the 10th International Conference on Emerging Internet, Data, and Web Technologies, pp. 142–152 (2022)
18. Oma, R., Nakamura, S., Duolikun, D., Enokido, T., Takizawa, M.: An energy-efficient model for fog computing in the internet of things (IoT). Internet of Things **1–2**, 14–26 (2018)

19. Sandhu, R.S., Coyne, E.J., Feinstein, H.L., Youman, C.E.: Role-based access control models. IEEE Comput. **29**(2), 38–47 (1996)
20. Tanganelli, G., Vallati, C., Mingozzi, E.: Coapthon: easy development of coap-based IoT applications with python. In: IEEE 2nd World Forum on Internet of Things (WF-IoT 2015), pp. 63–68 (2015)
21. Yuan, E., Tong, J.: Attributed based access control (ABAC) for web services. In: Proceeding of the IEEE International Conference on Web Services (ICWS 2005), p. 569 (2005)

Artificial Intelligence Based Approach for Fault and Anomaly Detection Within UAVs

Fadhila Tlili[1,2,3](✉), Samiha Ayed[2], Lamia Chaari[3], and Bassem Ouni[4]

[1] National School of Electronics and Telecommunications of Sfax, Sfax, Tunisia
fadhilatlili24@gmail.com
[2] LIST3N Laboratory, University of Technology of Troyes, Troyes, France
samiha.ayed@utt.fr
[3] Digital Research Center of Sfax (CRNS), Laboratory of Signals, systeMs aRtificial Intelligence, neTworkS (SM@RTS), Sfax University, Sfax, Tunisia
lamiachaari1@gmail.com
[4] Technology Innovation Institute, Abu Dhabi, United Arab Emirates
Bassem.ouni@tii.ae

Abstract. A non-predicted defect in Unmanned Aerial Vehicles (UAVs) could occur provisionally during their deployment process. Consequently, it is critical to optimize the detection of these instances. More specifically, the deviations in normal behavior indicate the possibility of triggered attacks, failures, and flaws. Therefore, intrusion detection (ID) is mandatory for UAVs security. Meanwhile, ID performance remains an arguable problem. Most of IDSs are applied for one predefined application. There is no general model for accurately detecting both anomalies and faults. To investigate this issue, this paper presents a new dynamic approach for UAVs fault and anomaly detection to investigate this issue. To resolve the drawbacks mentioned above, we propose an attack and fault detection approach. Our method shows better performance using a large dataset, trained on only a small fraction corresponding to normal flight strategy.

1 Introduction

1.1 Comprehensive Context

Unmanned Aerial Vehicles (UAVs) [1] are designed as autonomous non-board pilots. In general, UAVs are considered as complex flying aerial vehicles that operate under a set of onboard sensors. UAVs architecture is composed of controller ground-based, remote-control operation, sensors, payloads, and a communication system. Typically, the remote operator transmits control commands through communication links to control the UAV. Internal sensors continuously monitor the behavior of the internal state. In addition, external sensors yield different data features such as temperature, pressure, position, speed, velocity, acceleration, altitude, and longitude. These features are generated during the UAV flight simulation. Generally, the flight starts by altering power engines to

L. Barolli et al. (Eds.): AINA 2022, LNNS 449, pp. 297–308, 2022.
https://doi.org/10.1007/978-3-030-99584-3_26

provide needed translational motions. Accordingly, UAVs will lift, hover, tilt left/right (*roll*), forward/backward (*pitch*) or turn on itself (*yaw*). Thus, UAV fundamental components must correlate and operate to maintain the desired state and safe flight without failures. However, the significant components of UAVs are vulnerable to failure and attacks. Based on this, it is important to counter the UAV threats efficiently. Prior works recommend intrusion detection system (IDS) [2] as an essential aspect of UAVs' security infrastructure. It (IDS) automates action in response to suspected anomalous behaviors as failures and attacks through a collection of tools, methods, and resources. Hence, IDS aims to monitor and recognize incidents in UAVs networks to detect intrusions indicators. These intrusions are defined as initiatives to violate UAVs' confidentiality, integrity, or availability properties. Given the extent of security threats, researchers adopt IDS to secure UAVs systems and networks. Hence, the challenge is to improve the detection performance of IDSs.

1.2 Motivation and Paper Organisation

Some problems within the main components could occur and block the normal UAV process. Owing to the strict control process of the management system as described in [3], failures could be hidden and amplified. Meanwhile, other problems like security issues and attacks could happen at different elements like the communication channels, the base system, and the charging system. Nevertheless, detecting UAVs anomalous behavior is time-consuming with the existing classic techniques. Therefore, to ensure a secure environment, we consider intrusion detection [3] as an essential aspect for UAVs. There are two categories of IDS: fault detection system (FDS) [4] and attacks detection system (ADS) [5]. Some actual IDSs were developed using a mixture of machine learning (ML) techniques including isolation forests [6], one-class support vector machine (SVM) [7], elliptic envelope [8], and local outlier factor [9]. Yet, ML techniques are complicated with massive and unbalanced data features. Indeed, there is a massive amount of data to process and identify during the operation. Moreover, generated false alerts and unknown attack patterns complicate the detection process. In order to solve these issues, researchers sought improved deep learning (DL) techniques with better detection performance. Considering this, building an hybrid automated system for UAVs faults and anomalies detection using Artificial Intelligence (AI) approaches could improve the detection performance.

This paper's contribution and originality consist of combining the detection of both faults and attacks. In addition, UAVs have multiple specifications in their data. Based on that, we propose a resilient approach that avoids distinctive characteristics to maintain security and safety. In this paper, we focus on:

- What makes detecting both faults and anomalies using the same method so tricky?
- How to use AI techniques to process and detect failures and attacks?

- What is the paradigm of an hybrid methodology for both faults and attacks detection?
- How to optimize features in order to enhance the detection results?

In summary, our study set a scope on UAVs faults and attacks detection using a novel approach that combines resilience and perseverance. The remainder of the paper is structured as follows: Sect. 2 provides related work to this search area. Section 3 describes our method. Meanwhile, Sect. 4 gives the details of our methodology using a deep learning approach. In order to validate our method and reach our objective, Sect. 5 describes our simulation set-up and the obtained results. Finally, Sect. 6 concludes the paper.

2 Related Work

The literature is accomplished on existing fault and attack approaches based on various UAVs aspects. Different studies were proposed and classified on faults detection approaches and attacks detection approaches. In [10], the authors proposed a model for UAVs fault detection using a stacked unsupervised learning model. It is based on a deep auto-encoder that yields low reconstruction loss. This approach was validated using the area under the curve (AUC) metric on two datasets consisting of distinct types of UAV faults and attacks. Nevertheless, this method applies the model on faults and attacks independently with several modifications. In addition, due to the high computational load, the performance is low. Authors in [11] introduced a novel approach for UAVs attacks detection that combines the Kullback-Leibler divergence (KLD) approach and artificial neural networks (ANN). By combining these two methods, the authors used the spatial and temporal correlation existing in the sensor data. The results presented an acceptable accuracy while the ANN performed better than the KLD-based approach under accuracy and f-score metrics. However, it can not present an acceptable solution to the problem. Furthermore, [12] investigated two unsupervised data mining approaches based on k-means and long short-term memory (LSTM). This approach affords suitable detection to capture time dependencies. This approach validation metric is built on the calculation of dynamic threshold distance and evaluated with high-dimensional data. However, even after using the LSTM, which enhances the detection sensitivity, results were insufficient.

Compared to the recent approaches, LSTM was the most suitable technique in attacks and fault detection. In this context, the authors in [13] proposed a method based on LSTM for UAVs attacks detection using pre-processed data. The model identifies attacks using a key log message. If the key log is not similar to the UAV safe parameter vector, the user will be alerted. The method was evaluated based on three metrics, precision, recall, and F-measure, and they reached 0.95, 0.96, and 0.96, respectively. Likely, UAVs have multiple specifications and common data overall. These data are used to train AI approaches in order to learn the safe flight state and detect potential abnormalities. The majority of studies proposed approaches that focus on one purpose, which is building a detection system for faults or attacks. Some methods aimed to detect faults and attacks but used two distinct approaches for each pattern category. FDS is remarkably similar to ADS because they both look for specific symptoms of intrusions and other security policy violations. However, failures are identified within the system. Based on that, FDS takes a static view of such signs in patterns, whereas FDS looks at the unexpected patterns dynamically. In addition, recent approaches have been developed for specific datasets or UAV types. E.g., in [12] the model received an average increase of the reconstruction error of 20% when it was tested on an unexpected attack. Table 1 provides a comparative summary of previous studies.

Table 1. Related work comparative table

Ref - year	Detection efficiency	Time efficiency	FDS	ADS	ML	DL	Resilient
[10] - 2021	Acceptable	X	✓	✓	X	✓	X
[11] - 2020	Poor	X	✓	✓	✓	✓	X
[12] - 2019	Good	✓	✓	X	X	✓	X
[13] - 2017	Acceptable	✓	X	✓	X	✓	X
Our proposal-2022	Good	✓	✓	✓	X	✓	✓

This paper seeks to overcome the limits of the previous studies. According to [12] and [13], LSTM is the most accurate approach. Generally, using time series data will affect the approach performance because more history information generates fake alerts. The LSTM controls the data in history and does not confuse the results. It can learn and detect using the knowledge and use only the recent history in a long sequence, thus ignoring the long tail. Due to gradual threshold surveillance, LSTM is potentially used in faults and anomalies detection. This fact can reduce dimensions and capture time dependencies both at once. In addition, LSTM could achieve a premature failure prediction if possible in higher confidence.

3 Proposed Scheme Overview

To better understand our work, we introduce in this section our method scheme. The design of UAVs faults and attacks detection system should be quite efficient to ensure the system's sustainability in the adversarial environment. In opposite to our novel approach, we design a resilient hybrid approach for UAVs frequent data patterns. The major supposition of our DL detection approach is that we rely on the data that exists in all UAVs categories. In our approach, we use the same DL-based approach in a novel scheme to detect both UAVs faults and attacks. Figure 1 depicts our methodology scheme. We train the network under an unsupervised learning paradigm using [safe, benign] UAVs states. We developed our detection system using the LSTM approach. There exist multiple design versions of the LSTM, we choose the most used one, namely the LSTM architecture without peephole connections [14]. This version is established to increase

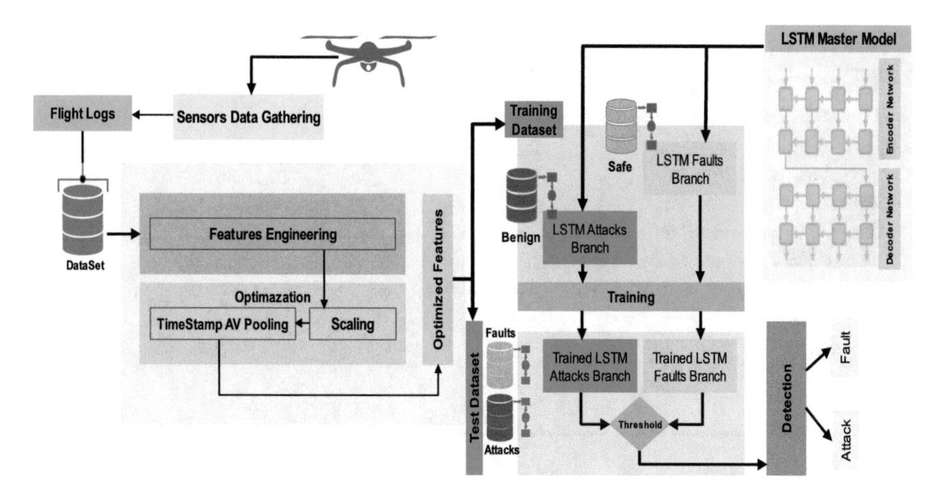

Fig. 1. Architecture of the applied methodology for faults and attacks detection

stability during the training. Our resilient method relies on having a master and two branches [Fault-Branch, Attack-Branch]. Our proposition is described as follows:

- The master is a shared-based model that yields the general LSTM encoder architecture. It is composed of an encoding and decoding network.
- The branches inherit the main structure of the master model and apply it based on the type of features which differ based on the data type.

These two branches will be trained and learned from normal [safe, benign] UAVs states to generate an hybrid global approach. These two states describe the normal flight without failures and attacks, respectively. Therefore, the trained branches will be integrated during the thresholding phase. We evaluate the network using data patterns to detect [fault, attack] states and compute the detection performance. In addition, our method includes data processing which is an essential part of generating an input feature vector to avoid the curse of dimensionality and gather key features. However, various missed or non-essential features could be incorporated into the detection process during the processing. We optimize data processing using a specific optimization algorithm to deal with this issue. For our study, the step of optimization follows Algorithm 1. This phase could be applied to different faults and attacks datasets.

Algorithm 1: Features optimization

```
Data: Flight data: CSV files
Result: Optimised Features (CSV per Cluster)
Init;
Set( Frequent Features);
Remove(NaN_values);
for CSV from faults& attacks do
    if CSV(values)≠ NULL then
        Rescaling = Min-Max_Norm(Feature(X_s c));
        Timestamp = Average(Feature_x);
    end
    ;
end
```

4 Our Hybrid Proposed Approach

This section discusses the overall proposed approach of retrieving data and generating results in detail. We go through the fundamentals of the Long Short Term Memory (LSTM) and the procedure of feature engineering and optimization. In order to optimize the selected features, a generic initial normalization method will be applied. However, we presented a new timestep average pooling method as discussed below.

4.1 Long Short Term Memory Networks Basics

In the context of time series data, we deal with a set of multiple time steps. Hence, we built a model leveraging the special architecture of Recurrent Neural Networks (RNN) named LSTM network [15]. Compared to RNN, LSTM can learn long-term relevant time-dependencies data, minimize gradient growth, and reduce vanishing problems during learning. As a result, LSTM architecture has already been successfully applied to anomaly and fault detection. The LSTM cell architecture is typically composed of three gates, The input gate i_t, The output gate o_t, and The forget gate f_t in addition to a cell, where t present the timestep. These embedded gates control the interaction with the other blocks of the network. Generally, the LSTM forget gate f_t specifies which features from the unit's state in the previous steps should be ignored and the information to be kept. Thus, LSTM receives current input X_t and previous memory cell state c_{t-1} as input using the formulas and the pipeline underneath. For each timestep t, a cell belonging to LSTM layer l and for each timestep t, the input gate is extended to incorporate multiple time sequences at a timestep t. The next step is to determine the quantity of information to learn from data. For this purpose, we calculate $\overline{c_t}$ of the new candidate values via a tanh function in order to ensure that the weights stay between −1 and 1, thus regulating the output of the neural network. Using forget and input gates, we generate new cell state data by multiplying old and added information. Therefore, the input gate

i_t regulates the stored data in the memory unit c_t and avoids unnecessary data. This process is followed by updated values decided by the sigmoid activation function used in the feed-forward network. Finally, these values will be pushed through a tanh layer and multiplied with the output gate to obtain the output of the LSTM block. As a result, it calculates the output o_t. Our LSTM aims to learn safe and benign UAV states characteristics from optimized features during the training phase and detect failure and anomalies patterns as faults and attacks during the thresholding. This version is established in order to increase stability during the training. In the context of supervised learning, the LSTM model aims to determine the mapping function accurately from the input to the output with: $outputvariables(Y) = f(X_t)$. We align the features to have the same (X) values. The objective is to predict an element of the output variables (Y) with an output sequence $Y = (Y_1, Y_2, Y_3, .., Y_i)$ from the input variables (X). The new input sequence $X = (X_1, X_2, X_3, .., X_i)$, denotes the piecewise linear representation of time series data.

4.2 Features Engineering and Optimization

In our approach, feature engineering is designed to eliminate the non-essential and unjust data patterns. To process the plotted data patterns efficiently, our model is developed using common characteristics mentioned in Subsect. 1.1. Other features have been excluded [actuators control patterns, GPS location report patterns, ...] because they are uncommon features and do not exist in all the datasets. In addition, we remove static features with indeclinable values like NULL values to clean and normalize the dataset. Furthermore, the features rescaling is fundamental to easily deploy a neural network model for faults and anomalies detection. In our pipeline, we applied a min-max normalization function for rescaling on data-frames to design an unified range of values. For the selected features via Eq. (1), this function adjusts the spectrum of features between [0, 1].

$$Feature(X_{sc}) = \frac{X - X_{sc\,Min}}{X_{sc\,Max} - X_{sc\,Min}} \tag{1}$$

Where X present the extracted features; X_{sc} refers to scaled features;

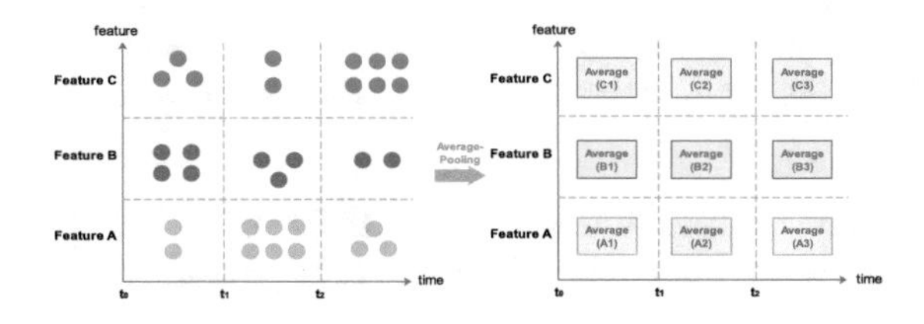

Fig. 2. Representative example of timestamp average-pooling

Following to the normalisation, we will apply timestamp average-pooling. The features are recorded in different time duration. In a particular parallel timestamp, each feature recorded a different number of values. The interval of time between $[t, t+1]$ has unequal values for different $feature(n)$. We apply a timestep pooling method to transform them into fixed sizes to be the same at a defined timestamp. In a recent work [10], the author used a timestamp pooling method to select a random single data point during a fixed time window. We assumed that the selected value could not be enough to identify anomalous states. However, we propose a timestep average-pooling method to enhance the detection efficiency and solve the unequal number of $feature(n)$ in a particular time interval $[t, t+1]$. The average of a given value in a given interval is calculated by adding the set of values and then dividing by the count of those values, as detailed in Eq. 2. Figure 2 illustrates the process of our key motivation method in an example case.

$$Average(Feature_x) = \frac{F_{x_1} + F_{x_2} + F_{x_3} + .. + F_{x_n}}{n} = \frac{1}{n}\sum_{i=1}^{n} F_{x_n} \tag{2}$$

Where Xi represents the ith value of the feature X; F(n) are the set of features from $i = 1, 2, \ldots n$

5 Performance Evaluation

5.1 Dataset

To validate our proposed model on faults and anomalies, we consider two datasets:

- **AirLab Failure and Anomaly (ALFA)** [16]: it consists of autonomous UAV flight sequences that include different control failures with recovery measures. It is available in comma-separated values (CSV) format. It contains different failures such as engine failure, aileron failure [left,right,both], rudder failure [left,right,zero], and elevator failure.
- **UAV Attack (UA)** [17]**:** it is a labeled dataset for UAV intrusion detection recorded using Quadcopter: Holybro S500 model and available in CSV format as a dataset per cluster. It consists of benign flight logs which are without malicious data. As well, it contains data related to jamming and spoofing attacks

They are open-source datasets available on IEEE DataPort [18]. In addition, both datasets are assured with their ground truth patterns.

5.2 Simulation Set-Up

Selecting the best hyper-parameter values is mandatory to design an effective DL model. We used a non-linear function (ReLU), Adam optimizer, mean-square error as the loss function. The final result consists of a dense layer with SoftMax activation. We set 800 epochs to train our LSTM with [safe, benign] states. Then it is evaluated within different [failures, attacks] states with two attacks and four types of failure.

5.3 Metrics of Evaluation

We consider accuracy the most critical performance indicator to understand faults and attacks detection performance. It represents the proportion of the total number of correct failures and attacks detection for the UAV: roll, yaw, and pitch. The accuracy is calculated using the values of the true negative (TN), true positive (TP), false negative (FN), and false-positive (FP). It is expressed as follows:

$$Accuracy = \frac{TP + TN}{TP + FP + TN + FN} \tag{3}$$

5.4 Simulation Results

We show that our proposed approach presents acceptable results through extensive simulation conducted on the two datasets. We clarify these simulation results in Tables 2 and 3 for both attacks and failure detection. Our approach gives good accuracy in the experiments for spoofing and jamming attacks. For the failures, the detection accuracy differs from one failure to another, which varies between [0.76, 0.92]. In order to evaluate our approach, we predict the roll, the pitch, and the yaw of the UAV under jamming attack, spoofing attack, and engine failure. Each input time series are re-scaled, so the values belong to the interval [0.0, 1.0]. The prediction results of the spoofing attack, jamming attack, and engine failure test mission are shown in Figs. 3, 4 and 5, respectively. The large spike towards the attack's test mission graph is flagging the flight as anomalous. To evaluate the precision of our detecting approach, we compare the prediction of the UAV flight to the ground truth of the [safe, benign] flight. These three flight major translational motions measurements show good predictive results compared to the ground truth data at the exact time. Based on this, this approach can ensure a good prediction toward both attacks and faults detection. Hence, the simulations of our resilient hybrid method present better accuracy compared to [13]. In the same variation of attacks, the KLD accuracy curve in [11] trends between [0.70, 0.95], while for the ANN is between [0.85, 0.95]. Concerning our approach results, the accuracy is similar but more effective as a resilient approach for both attacks and faults with an accuracy of 0.955 and 0.824, respectively, as shown in Table 4. In addition, Fig. 6 describes the trend of training and validation loss (test) over the number of epochs. We observe through these graphs that the loss converges after a few tens of epochs for nearly all the cases of attacks and failures. The result curves of loss emphasize the performance of our approach with a low loss.

Table 2. Representative example of timestamp average-pooling

Dataset	Attack type	Accuracy
UA	Jamming attack	0.95
	Spoofing attack	0.96

Table 3. Evaluation for faults detection results using accuracy metrics

Dataset	Failure type	Accuracy
ALFA	Engine failure	0.76
	Elevator failure	0.84
	Both_aileron failure	0.78
	Left_aileron failure	0.77
	Right_aileron failure	0.82
	Rudder_Zero failure	0.83
	Left_Rudder failure	0.87
	Right_Rudder failure	0.92

Table 4. Proposed model performance evaluation

Patterns	Accuracy
Attacks	0.955
Faults	0.824

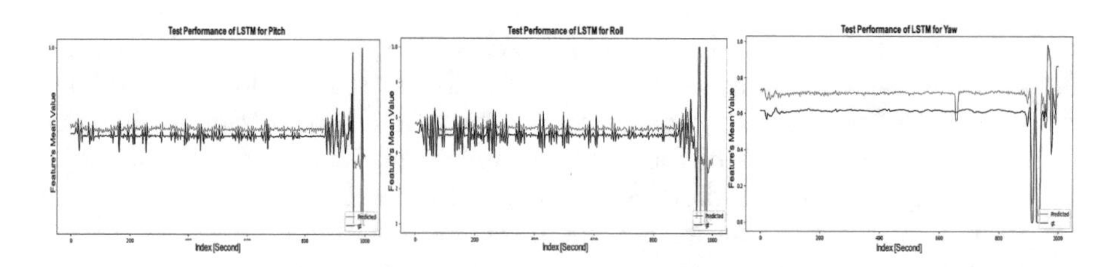

Fig. 3. Prediction detector based on pitch, roll, yaw with error spikes of spoofing attack

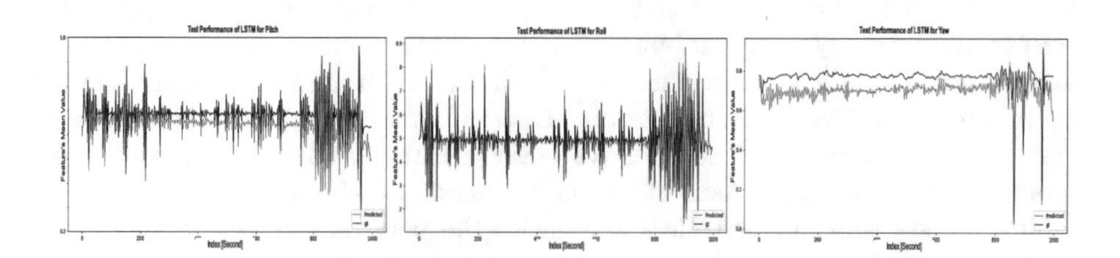

Fig. 4. Prediction detector based on pitch, roll, yaw with error spikes of jamming attack

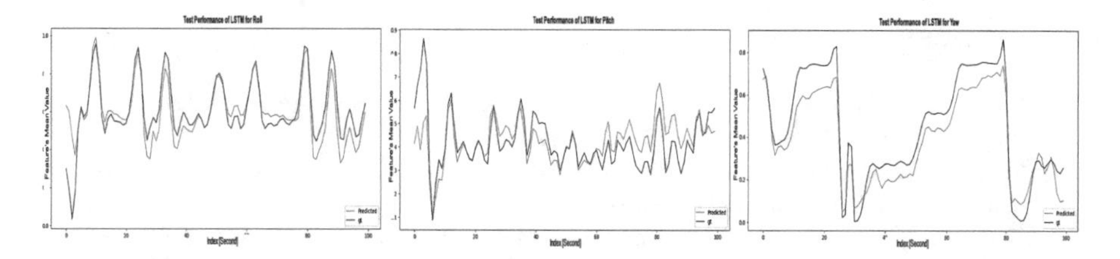

Fig. 5. Prediction detector based on pitch, roll, yaw with error spikes of engine failure

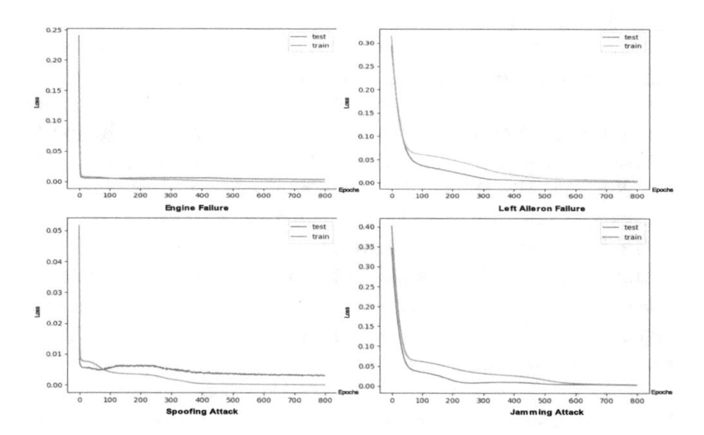

Fig. 6. Trend of training and validation loss of attacks and faults simulation over the number of epochs.

6 Conclusion and Perspectives

In this study, we proposed an hybrid resilient approach for UAVs faults and attacks detection based on AI to resolve prior approaches' limits. The contributions of our work are provided as follows. First, we proposed a series of optimizations to select the essential features. Second, we established our proposed detection approach based on the LSTM auto-encoder. The architecture consists of a master model and two branches for faults and attacks. We train the model only on the pattern of safe states. Finally, we examined our approach's promising performance using two datasets for faults and attacks. The simulation results present high accuracy along with the attacks and faults detection. Hence, future work plans to extend the architecture for multi-UAVs attacks detection. In addition, we aim to enhance the detection accuracy of some failures such as engine failure, both_aileron failure, and left_aileron failure.

References

1. Hentati, A., Fourati, L.: Comprehensive survey of UAVs communication networks. Comput. Stand. Interfaces **72**, 103451 (2020)
2. Pang, G., Shen, C., Cao, L., Hengel, A.: Deep learning for anomaly detection: a review. ACM Comput. Surv. (CSUR) **54**, 1–38 (2021)
3. Kene, S., Theng, D.: A review on intrusion detection techniques for cloud computing and security challenges. In: 2015 2nd International Conference on Electronics and Communication Systems (ICECS), pp. 227–232 (2015)
4. Baskaya, E., Bronz, M., Delahaye, D.: Fault detection diagnosis for small UAVs via machine learning. In: 2017 IEEE/AIAA 36th Digital Avionics Systems Conference (DASC), pp. 1–6 (2017)
5. Ahn, H., Choi, H., Kang, M., Moon, S.: Learning-based anomaly detection and monitoring for swarm drone flights. Appl. Sci. **9**, 5477 (2019)
6. Xu, D., Wang, Y., Meng, Y., Zhang, Z.: An improved data anomaly detection method based on isolation forest. In: 2017 10th International Symposium on Computational Intelligence and Design (ISCID), vol. 2, pp. 287–291 (2017)
7. Hoang, T., Nguyen, N., Duong, T.: Detection of eavesdropping attack in UAV-aided wireless systems: unsupervised learning with one-class SVM and k-means clustering. IEEE Wirel. Commun. Lett. **9**, 139–142 (2019)
8. Ashrafuzzaman, M., Das, S., Jillepalli, A., Chakhchoukh, Y., Sheldon, F.: Elliptic envelope based detection of stealthy false data injection attacks in smart grid control systems. In: 2020 IEEE Symposium Series on Computational Intelligence (SSCI), pp. 1131–1137 (2020)
9. Cheng, Z., Zou, C., Dong, J.: Outlier detection using isolation forest and local outlier factor. In: Proceedings of the Conference on Research in Adaptive and Convergent Systems, pp. 161–168 (2019)
10. Park, K., Park, E., Kim, H.: Unsupervised fault detection on unmanned aerial vehicles: encoding and thresholding approach. Sensors **21**, 2208 (2021)
11. Titouna, C., Na1t-Abdesselam, F., Moungla, H.: An online anomaly detection approach for unmanned aerial vehicles. In: 2020 International Wireless Communications And Mobile Computing (IWCMC), pp. 469–474 (2020)
12. Lindemann, B., Fesenmayr, F., Jazdi, N., Weyrich, M.: Anomaly detection in discrete manufacturing using self-learning approaches. Procedia CIRP **79**, 313–318 (2019)
13. Du, M., Li, F., Zheng, G., Srikumar, V.: DeepLog: anomaly detection and diagnosis from system logs through deep learning. In: Proceedings of the 2017 ACM SIGSAC Conference on Computer and Communications Security, pp. 1285–1298 (2017)
14. Ergen, T., Kozat, S.: Unsupervised anomaly detection with LSTM neural networks. IEEE Trans. Neural Netw. Learn. Syst. **31**, 3127–3141 (2020)
15. Hossain, M., Inoue, H., Ochiai, H., Fall, D., Kadobayashi, Y.: LSTM-based intrusion detection system for in-vehicle can bus communications. IEEE Access **8**, 185489–185502 (2020)
16. Whelan, J., Sangarapillai, T., Minawi, O., Almehmadi, A., El-Khatib, K.: Novelty-based intrusion detection of sensor attacks on unmanned aerial vehicles. In: Proceedings of the 16th ACM Symposium on QoS and Security for Wireless and Mobile Networks, pp. 23–28 (2020)
17. Keipour, A., Mousaei, M., Scherer, S.: ALFA: a dataset for UAV fault and anomaly detection. Int. J. Robot. Res. **40**, 515–520 (2021)
18. Whelan, J., Sangarapillai, T., Minawi, O., Almehmadi, A., El-Khatib, T.: UAV Attack Dataset (2021). https://doi.org/10.21227/00dg-0d12

Composition and Polymorphism Support in the OpenAPI Ontology

Fotios Bouraimis, Nikolaos Mainas, and Euripides G. M. Petrakis(✉)

School of Electrical and Computer Engineering, Technical University of Crete (TUC), Chania, Greece
fbouraimis@isc.tuc.gr, {nmainas,petrakis}@intelligence.tuc.gr

Abstract. An OpenAPI [5] description details the actions exposed by a REST API. A service is represented by a JSON document describing requests, responses, and security information such as authentication and authorization rules for an API action. Schema objects describe the request and response messages and can be combined to form composite or specialized objects using composed or polymorphic expressions. However, Schema properties can be vague. To clarify their meaning, Schema properties may be associated with a semantic model. Further, to resolve ambiguities in service descriptions, OpenAPI descriptions are instantiated to a reference ontology. However, the mapping of composed or polymorphed Schema objects to the ontology generated additional problems to the mapping process. The mapping of composite Schema objects to the ontology is the focus of this work.

1 Introduction

Bouraimis [1] proposed a reference ontology for OpenAPI descriptions of REST services. The ontology takes full advantage of Hydra [3] and SHACL [2]. Hydra is a promising technology towards understanding and constructing Web services that meet the HATEOAS requirement of REST architectural style. Classes and constraints on class properties are described using SHACL allowing service descriptions to be validated against the ontology.

Schema objects are mainly used to define input and output data types in request and response messages. Composition allows new Schema objects to be defined as compositions of existing ones and, Polymorphism allows Schema objects to accept more than one definition. This work (a) handles composition and polymorphism in OpenAPI Schema objects in the ontology, (b) implements the concept of inheritance on ontology entities by creating sub-class relations between the ontology classes that correspond to OpenAPI Schemas. The instantiation of all OpenAPI objects to the ontology is described in [1] and has been incorporated into a Web application[1].

Related work is presented in Sect. 2. Section 3 summarizes previous work by the authors. The instantiation of Schema objects is discussed in Sect. 4. The issues of composition and polymorphism are discussed in Sects. 5 and 6 respectively, followed by conclusions and issues for future work in Sect. 7.

[1] http://www.intelligence.tuc.gr/semantic-open-api/.

L. Barolli et al. (Eds.): AINA 2022, LNNS 449, pp. 309–320, 2022.
https://doi.org/10.1007/978-3-030-99584-3_27

2 Related Work

Syntactic description languages describe the requirements for communicating with a service and the message formats to successfully communicate with it. WSDL [10] for SOAP, WADL [9], RAML [7] for REST, are popular mainly due to their simplicity and compatibility with machine-readable formats like XML and JSON. They are not specifically designed for hypermedia-driven APIs (e.g., REST) that call for the dynamic discovery of resources at run-time (referred to as HATEOAS). Semantic approaches employ semantic models (i.e. ontologies) and are more capable of supporting automated service discovery and composition. OWL-S [6] is an upper ontology for Web services but does not support dynamic discovery of resources at runtime.

Musyaffa et al. [4] introduce annotations in Schema and Parameter objects for OpenAPI v2.0. They do not handle all causes of ambiguity nor do they handle OpenAPI v3.0 and Schema objects. The annotations appear within text properties and cannot be interpreted by a machine without pre-processing. Schwichtenberg, Gerth, and Engels [8] map OpenAPI v2.0 service descriptions to OWS-S ontology but do not deal with any causes of ambiguity in service descriptions, nor do they handle security properties. Their approach attempts to find a mapping of OAS v2.0 Schema objects to OWL-S using heuristics and name similarity matching techniques.

3 The OpenAPI Ontology

OpenAPI 3.0 comprises many parts (objects). Each object specifies a list of properties that can be objects as well. Objects and properties defined under the Components unit can be reused by other objects or they can be linked to each other (e.g. using keyword *$ref*). The Info object provides non-functional information about the service. The Servers object describes where the API servers are located. The Security object lists the security schemes of the service and, among others, it lists definitions of responses, parameters and Schemas (under the Components unit). The specification supports HTTP authentication, API keys, OAuth2 common flows or grants (i.e. ways to retrieve an access token), and OpenID Connect. The Paths object describes the operations (i.e. HTTP methods) and contains the relative paths for the service endpoints. The Request and Response objects describe the requests and responses of an operation (e.g. a HTTP status code), its message content, and HTTP headers. The Parameters object describes parameters of operations (i.e. path, query, header, and cookie parameters).

Schema objects describe the request and response messages. A Schema object can be a primitive (string, integer), an array, a model, an XML data type and may also have properties of its own. New data types are defined as a composition or a specification of existing ones using keywords *allOf*, *oneOf*, *anyOf* and *not*. The meaning of Schema properties can be ambiguous [1]. To solve the problem, OpenAPI properties are semantically annotated and associated with entities of a semantic model (e.g. an ontology) using the *x-refersTo* extension property. The *x-kindOf* extension property defines a specialization between an OpenAPI property and a semantic model (e.g. a class). The *x-mapsTo* property denotes that a Schema property is semantically equivalent with another property in the same document. Additional extension properties are defined

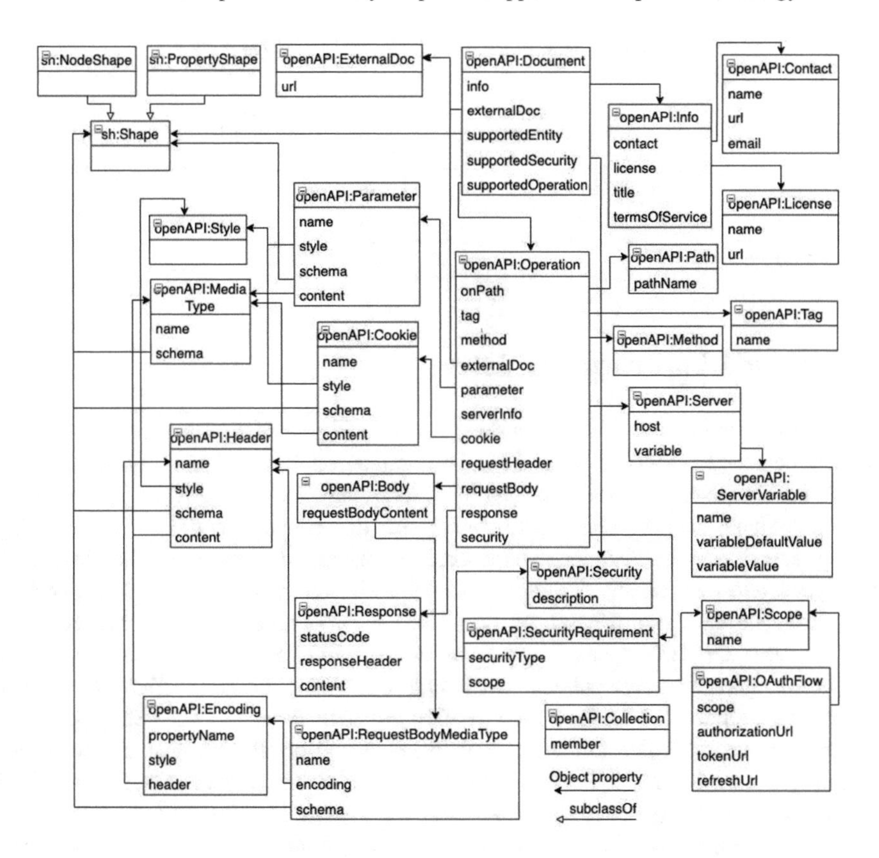

Fig. 1. OpenAPI ontology.

to clarify the meaning of the members in a collection of objects (*x-collectionOn*), for grouping Schema objects by type (*x-onResource*) and for clarifying the meaning of operations (*x-operationType*).

The semantic meaning of an OpenAPI service description is captured by the ontology of Fig. 1. This work focuses on the instantiation of Schema objects defined as composition or specification of existing ones using properties *allOf* and *oneOf*, *anyOf* and *not*). Almost every OpenAPI Object contains a Schema Object. Schema objects are mapped to classes using SHACL. SHACL can be used to describe and validate the structure of RDF data, similarly to XML-Schema or JSON Schema. It provides an RDF vocabulary of built-in types of constraints. Table 1 shows how the *allOf*, *oneOf* *anyOf* and *not* JSON Schema properties are expressed in SHACL.

Table 1. Mapping Schema properties to SHACL.

Schema object property	SHACL property
allOf	sh:and
oneOf	sh:xone
anyOf	sh:or
not	sh:not

4 Instantiating Schema Objects

Each Schema object is mapped to a Shape class using SHACL. The NodeShape class defines the properties of a class and specifies whether a class contains additional properties. The Shape class is distinguished into NodeShape class and PropertyShape class. The NodeShape class represents the classes that describe the models of an OpenAPI service description and, PropertyShape class represents the properties of a class, their data types, and restrictions. A Shape class determines how to validate a focus node (a node from the data graph) based on the values of properties and other characteristics of the focus node. Two types of Shape classes are defined in SHACL: Shapes (classes) about the focus node itself, called *node Shapes*, and Shape (classes) about the values of a particular property for the focus node, called *property Shapes*. A node Shape contains targets that specify which nodes in the data graph conform to a Shape and constraint components that determine how to validate a node. Except for *sh:targetNode*, which specifies directly the nodes to be validated, there is also a *sh:targetClass* to denote that all the nodes of a given type conform with a particular Shape. Finally, *sh:path* of a PropertyShape points at the URI of the property that is being restricted. The methodology is discussed with reference to Petstore service[2].

Listing 1. Pet object example.

```
1 Pet:
2  type: object
3  required:
4  - id
5  properties:
6   id:
7    type: integer
8    format: int64
```

Listing 1 shows the model of a Pet with property *id*. Listing 2 is the corresponding Shape class. The model becomes an instance of the NodeShape class with target class Pet. The *rdfs:label* predicate is human-readable information. The Shape node has one property (Pet_idPropertyShape in this case) of type PropertyShape of type Pet_id which is the class that is being restricted.

Listing 2. Shape and Property Shape nodes of Pet object.

```
1 <PetNodeShape> a sh:NodeShape ;
2  rdfs:label "PetNodeShape" ;
3  sh:property <Pet_idPropertyShape> ;
```

[2] https://petstore3.swagger.io.

```
4   sh:targetClass <Pet> .
5
6  <Pet> a owl:Class .
7
8  <Pet_idPropertyShape> a sh:PropertyShape ;
9   rdfs:label "Pet_idPropertyShape" ;
10  openapi:name "id" ;
11  sh:datatype xsd:long ;
12  sh:path <Pet_id> .
13
14 <Pet_id> a rdf:Property .
```

Each *sh:property* in Listing 2 becomes an instance of PropertyShape class. The values of *sh:datatype* come from the type and format of each property (i.e. *int32* has value *xsd:int* and *int64* has value *xsd:long*). The *sh:path* predicate points to the URI of the property being restricted. Each Shape property is a sub-class of *rdf:Property*.

Listing 3. Annotated Pet object.

```
1  Pet:
2   x-refersTo: https://example.com/ontology/Pet
3   type: object
4   required:
5    - id
6   properties:
7    id:
8     x-refersTo: https://example.com/ontology/Id
9     type: integer
10    format: int64
```

Listing 4. Shape models with *x-refersTo* properties.

```
1  <PetNodeShape> a sh:NodeShape ;
2   rdfs:label "PetNodeShape" ;
3   sh:property <Pet_idPropertyShape> ;
4   sh:targetClass <https://example.com/ontology/Pet>.
5
6  <Pet_idPropertyShape> a sh:PropertyShape ;
7   rdfs:label "Pet_idPropertyShape" ;
8   openapi:name "id" ;
9   sh:datatype xsd:long ;
10  sh:path <https://example.com/ontology/Id>.
```

In Listing 3 both, the Pet Schema object and its property are semantically annotated. In particular, *x-refersTo* associates the object and its property to a semantic value (class) which becomes the target and path objects of their corresponding Shapes. This is showcased in Listing 4. If the extension property is *x-kindOf* the target and path objects become sub-classes of the designated models. Listing 4 must be modified as shown in Listing 5. The *x-mapsTo* annotation property can also be used in a Schema object to designate that it points to another Schema object or property Schema with the same semantics. It is handled similar to *x-refersTo* but the target and path objects now link to the model that the *x-mapsTo* property refers to.

Listing 5. Shape models with *x-kindOf* properties.

```
1   <PetNodeShape> a sh:NodeShape ;
2   ...
3   <Pet> a owl:Class ;
4   rdfs:subClassOf <https://example.com/ontology/Pet> .
5
6   <Pet_idPropertyShape> a sh:PropertyShape;
7   ...
8   <Pet_id> a rdf:Property ;
9   rdfs:subPropertyOf <https://example.com/ontology/Id>.
```

In Listing 6, the *x-collectionOn* property is used to indicate that a Schema Object is actually a collection. Collections are represented using *openapi:Collection* class. The target class is PetCollection and will become a subclass of the general Collection class. The Shape Property class corresponds to property *pets* and declares that each member of the collection is a pet. This is showcased in Listing 7.

Listing 6. A collection of Pet objects.

```
1 PetCollection:
2  x-collectionOn: pets
3  type: object
4  properties:
5   pets:
6    type: array
7     items:
8      $ref: '#/components/schemas/Pet'
```

Listing 7. Shape models for a collection of Pet objects.

```
1  <PetCollectionNodeShape> a  sh:NodeShape ;
2   rdfs:label "PetCollectionNodeShape";
3   sh:property  <PetCollection_PropertyShape>;
4   sh:targetClass <PetCollection> .
5
6   <PetCollection> a   owl:Class ;
7    rdfs:subClassOf openapi:Collection.
8
9  <PetCollection_petsPropertyShape> a sh:PropertyShape ;
10  rdfs:label "PetCollection_petsPropertyShape";
11  openapi:name "pets" ;
12  sh:node <PetNodeShape> ;
13  sh:path openapi:member .
```

5 Composition

Schema objects can be defined using any of the keywords *allOf*, *anyOf*, *oneOf* and *not*. The *allOf* is used to define new Schemas as a composition of the Schemas referred to after the keyword. The *allOf* keyword expresses the concept of inheritance: the resulting composite schema inherits the properties of all referred Schemas. Keywords *oneOf* and *anyOf* are used to express the concept of polymorphism: they are used to define new Schemas that can take the form of one or more alternative Schemas. Finally, keyword *not* is used to restrict a Schema (i.e. it declares which type of value is not accepted as value for an object property). The Pet Schema object of Listing 8 extends the OldPet with the property *id*. The PetNodeShape is defined using SHACL predicate *sh:and* in order to express that it accepts the properties of the focus node OldPetNodeShape (i.e. the Shape of the node being extended) and the additional Shape property of a blank node. The new class must inherit the properties of the original class. The way to express this is to define PetNodeShape as a sub-class of OldPetNodeShape class.

Listing 8. Object composition with *allOf*.

```
1  Pet:
2   allOf:
3     - $ref: '#/components/schemas/OldPet'
4     - type: object
5     required:
6      - id
7     properties:
8      id:
9       type: integer
10      format: int64
11
12 OldPet:
13  type: object
14  required:
15   - name
16   - tag
17  properties:
18   name:
19    type: string
20   tag:
21    type: string
```

Listing 9. Composed Shape models.

```
1  <PetNodeShape> a  sh:NodeShape;
2   rdfs:label  "PetNodeShape" ;
3   sh:and  ( [ a sh:NodeShape;
4    sh:property  <Pet_idPropertyShape>
5    ]
6    <OldPetNodeShape>
7   );
8   sh:targetClass  <Pet>
9
10 <OldPetNodeShape> a  sh:NodeShape  ;
11  rdfs:label    "OldPetNodeShape" ;
12  sh:property  <OldPet_tagPropertyShape>, <
        OldPet_namePropertyShape> ;
13  sh:targetClass  <OldPet>
14
15  <Pet>  a   owl:Class ;
16  rdfs:subClassOf  <OldPet> .
17
18  <OldPet>  a  owl:Class .
```

The extension property *x-refersTo:* 'https://example.com/ontology/Pet' is added in Listing 8 before the *allOf* keyword. The algorithm will create a class with the name of the model referred to. This will become the target class of the new model (i.e. *sh:targetClass* <https://example.com/ontology/Pet> will replace line 8 in Listing 9). The OldPet Schema can be annotated as well by adding an extension property *x-refersTo:* https://example.com/ontology/OldPet (i.e. between lines 12 and 13 in Listing 8). The target class of the object being extended will be the class referred to in the extension property (i.e. line *sh:targetClass* <https://example.com/ontology/OldPet> will replace line 13 in Listing 9). Class *<OldPet>* in line 18 will not be created. In line 16 of Listing 10 the model https://example.com/ontology/OldPet will become superclass. Listing 10 shows the resulting composed Shape class when both models are annotated. There is also a case where either annotation property (i.e. for Pet or OldPet) has value *none*. This has the following impact to the instantiation process: (a) no target class will be created for any model annotated with *none*, (b) the Shape models for both, Pet and OldPet will be created but, (c) there will be no sub-class relationship between them.

Listing 10. Annotated composed Shape models.

```
1  <PetNodeShape>
2
3   a sh:NodeShape ;
4   rdfs:label      "PetNodeShape" ;
5   sh:and ( [ a  sh:NodeShape ;
6     sh:property  <Pet_idPropertyShape>
7    ]
8    <OldPetNodeShape>
9   ) ;
10  sh:targetClass  <https://example.com/ontology/Pet> .
11
12 <https://example.com/ontology/Pet>
13  a   owl:Class ;
14  rdfs:subClassOf  <https://example.com/ontology/OldPet> .
15
16 <OldPetNodeShape>
17  a    sh:NodeShape ;
18  rdfs:label  "OldPetNodeShape" ;
19  sh:property <OldPet_tagPropertyShape>,
20     <OldPet_namePropertyShape> ;
21 sh:targetClass  <https://example.com/ontology/OldPet> .
```

6 Polymorphism

Often, requests and responses are described by alternative Schema using keywords *oneOf* and *anyOf*. The Shape model is expressed using SHACL predicates *sh:xone* and *sh:or* respectively. Listing 11 shows an unnamed Schema for response with code "201" which can be one of "Dog" or "Cat". Because the Schema has no name, the resulting Shape node of Listing 12 is a blank Shape node (i.e. a Shape without a URI) with predicate *sh:xone* and an RDF list as object and member individuals CatNodeShape and DogNodeShape. The case with keyword *anyOf* is handled likewise; the difference is that *sh:xone* in line 2 of Listing 12 will be replaced by keyword *sh:or*.

Listing 11. Polymorphed un-named object.

```
1  responses:
2   "201":
3    description: variety
4    content: application/json:
5     schema:
6      oneOf:
7        -$ref: "#/components/schemas/Cat"
8        -$ref: "#/components/schemas/Dog"
9
10 components:
11  schemas:
12   Dog:
13    type: object
14    properties:
15     bark:
16      type: boolean
17     breed:
18      type: string
19   Cat:
20    type: object
```

```
21       properties:
22        hunts:
23         type: boolean
24        age:
25         type: integer
```

Listing 12. Polymorphed Shape object.

```
1  openapi:schema  [ a  sh:NodeShape ;
2     sh:xone   ( <DogNodeShape> <CatNodeShape> )
3    ]
4
5  <CatNodeShape>  a  sh:NodeShape ;
6   rdfs:label  "CatNodeShape" ;
7   sh:property <Cat_agePropertyShape>,<
       Cat_huntsPropertyShape>;
8   sh:targetClass  <Cat> .
9
10 <Cat>  a  owl:Class .
11
12 <DogNodeShape>  a   sh:NodeShape ;
13  rdfs:label  "DogNodeShape" ;
14  sh:property <Dog_breedPropertyShape>,  <
       Dog_barkPropertyShape>;
15  sh:targetClass  <Dog> .
16
17 <Dog>  a   owl:Class .
```

An unnamed Schema is the most common way to express a Schema Object with polymorphism. However, the response object can be named as well. This is showcased in Listing 13. This has no real impact on the instantiation algorithms: after replacing all objects referred with *$ref* the result will be the same with Listing 12.

Listing 13. Polymorphed named object.

```
1  responses:
2   "201":
3    description: variety
4    content: application/json:
5     schema:
6     oneOf:
7      - $ref: "#/components/schemas/TwoPets"
8
9  components:
10  schemas:
11   TwoPets:
12    oneof:
13     - $ref: "#/components/schemas/Cat"
14      - $ref: "#/components/schemas/Dog"
```

Polymorphism in property Schemas is supported. Listing 14 shows that property *speed* can be of any of the Schemas following keyword *anyOf* which, in turn, can be defined inline or referred to using keyword *$ref*. The resulting Shape model is shown in Listing 15 and is composed of one named Shape class (for the property referred to by *$ref* and two un-named classes for the remaining properties). Keyword *oneOf* is handled likewise by replacing predicate *sh:or* in line 4 of Listing 15 with predicate *sh:xone*.

Listing 14. Polymorphism in property Schemas.

```
1 Pet:
2  type: object
3  required:
```

```
4     - speed
5    properties:
6     speed:
7     anyOf:
8       - type: integer
9         format: int64
10      - type: string
11      - $ref: '#/components/schemas/Specified'
12
13 components:
14   schemas:
15    Specified:
16     type: integer
17    format: int32
18   ...
```

Listing 15. Polymorped property Shape model.

```
1  <Pet_speedPropertyShape>  a   sh:PropertyShape ;
2   rdfs:label    "Pet_speedPropertyShape" ;
3   openapi:name  "speed" ;
4   sh:or   ( <SpecifiedPropertyShape>
5      [ a  sh:PropertyShape ;
6       sh:datatype  xsd:string
7      ]
8      [ a sh:PropertyShape ;
9       sh:datatype  xsd:long
10     ]
11  ) ;
12  sh:path   <Pet_speed> .
13  <Pet_speed> a rdf:Property .
14
15 <SpecifiedPropertyShape>  a  sh:PropertyShape ;
16  rdfs:label  "SpecifiedPropertyShape" ;
17  sh:datatype  xsd:int ;
18  sh:path   <Specified> .
19
20 <Specified>  a  rdf:Property .
21 ...
```

Keyword *not* is used to express not acceptable values (i.e. it makes Schemas more specific). Property *byType* in Listing 16 can be anything but a string. The SHACL constraint is expressed with predicate *sh:not*. The resulting Shape model in Listing 17 denotes that the corresponding ProperyShape may not take any of the values specified in the RDF list. The restricted property cannot be annotated (e.g. by adding *x-refersTo:* https://example.com/onto\discretionary-logy/\discretionary-Specifi\discretionary-cString after line 18 in Listing 16). This would allow every other string but the one referred to by the annotation. This negates the universality of the *not* keyword and its original purpose.

Listing 16. Restricted property schema.

```
1   Pet:
2   type: object
3   required:
4     - id
5   properties:
6    id:
7     type: integer
8      format: int64
9    byType:
10    not:
11     type: string
```

Listing 17. Shape of restricted property Schema.

```
1 <Pet_byTypePropertyShape>  a  sh:PropertyShape ;
2  rdfs:label    "Pet_byTypePropertyShape" ;
3  openapi:name  "byType" ;
4  sh:not [ a  sh:PropertyShape ;
5   sh:datatype  xsd:string
6   ] ;
7  sh:path  <Pet_byType> .
8
9 <Pet_byType>  a  rdf:Property .
```

A property Schema with polymorphism accepts alternative data types. There are two cases: the first is when a semantic value is connected with different Schemas using a property; the second is when a property is connected with semantic values using property attributes (i.e. properties of properties). When a property is annotated with a semantic value, the Schema property and the semantic value are connected. If the annotation *x-refersTo:* https://example.com/ontology/Speed is inserted between lines 6 and 7 in Listing 14, the property can be of any listed after keyword *anyOf* (or after keyword *oneOf* in a similar example). The annotation connects the Schema of property *speed* with the Schemas in the RDF list in Listing 15. The second case is when the property attributes are semantically annotated instead of the property itself. If any of the properties following keyword *anyOf* (equivalently *oneOf*) is annotated, property *speed* does not acquire a semantic value but its attributes do. This is showcased in Listing 18. The Shape model of *speed* is presented along with the semantic value of each of the nodes. There are also cases that are invalid (i.e. may create inconsistencies) [1].

Listing 18. Polymorphism with annotated property Schemas.

```
1   Pet:
2   type: object
3   required:
4    - speed
5   properties:
6     speed:
7     anyOf:
8      - type: integer
9       format: int64
10      x-refersTo: https://example.com/ontology/
           Int64_property
11     - type: string
12      x-refersTo: https://example.com/ontology/
           String_property
13     - $ref: '#/components/schemas/Specified'
14
15  components:
16   schemas:
17    Specified:
18     x-refersTo: https://example.com/ontology/
         Int32_property
19     type: integer
20    format: int32
21  ...
```

Listing 19. Polymorphed Shape with annotated property Shape Schemas.

```
1   Pet_speedPropertyShape> a  sh:PropertyShape ;
2    rdfs:label    "Pet_speedPropertyShape" ;
3   openapi:name  "speed" ;
4    sh:or  ( <SpecifiedPropertyShape>
5     [ a  sh:PropertyShape ;
6      sh:datatype  xsd:string ;
7      sh:path <https://example.com/ontology/String_property>
8     ]
9    [ a  sh:PropertyShape ;
10     sh:datatype  xsd:long ;
11     sh:path <https://example.com/ontology/Int64_property>
12    ]
13   ) .
14
15 <SpecifiedPropertyShape> a sh:PropertyShape ;
16  rdfs:label   "SpecifiedPropertyShape" ;
17   sh:datatype  xsd:int ;
18  sh:path <https://example.com/ontology/Int32_property> .
19 ...
```

7 Conclusions

Instantiating composed and polymorphed Schema objects in OpenAPI to an ontology is the problem this work is dealing with. The Link and Callback Objects of the specification will be handled in a future work.

References

1. Bouraimis, F.: Instantiating OpenAPI Descriptions to the REST Services Ontology. Technical Report, Diploma Thesis, School of Electrical and Computer Engineering, Technical University of Crete (TUC), Chania, Crete (2021). https://dias.library.tuc.gr/view/88862
2. Knublauch, H., Kontokostas, D.: Shapes Constraint Language (SHACL) (2017). https://www.w3.org/TR/shacl/
3. Lanthaler, M., Gütl, C.: A Vocabulary for Hypermedia-Driven Web APIs. In: Workshop on Linked Data on the Web (LDOW 2013). Rio de Janeiro, Brazil (2013). http://www.markus-lanthaler.com/hydra/
4. Musyaffa, F.A., Halilaj, L., Siebes, R., Orlandi, F., Auer, S.: Minimally invasive semantification of light weight service descriptions. In: IEEE International Conference on Web Services (ICWS 2016), pp. 672–677. San Francisco (2016). https://doi.org/10.1109/ICWS.2016.93, https://ieeexplore.ieee.org/document/7558066
5. OpenAPI Specification v3.1.0 (2021). https://spec.openapis.org/oas/v3.1.0
6. OWL-S: Semantic Markup for Web Services (2004). https://www.w3.org/Submission/OWL-S/. W3C Member Submission
7. RAML Version 1.0: RESTful API Modeling Language (2021). https://raml.org
8. Schwichtenberg, S., Gerth, C., Engels, G.: From Open API to Semantic Specifications and Code Adapters. In: IEEE International Conference on Web Services (ICWS 2017), pp. 484–491. San Francisco (2017). https://doi.org/10.1109/ICWS.2017.56, https://ieeexplore.ieee.org/document/8029798
9. Web Application Description Language (2009). https://www.w3.org/Submission/wadl/. W3C Member Submission
10. Web Services Description Language (WSDL) 1.1 (2001). https://www.w3.org/TR/wsdl.html

Improved Road State Sensing System and Its Data Analysis for Snow Country

Yositaka Shibata[1(✉)], Akira Sakuraba[2], Yoshikazu Arai[1], and Yoshiya Saito[1]

[1] Iwate Prefectural University, Takizawa, Japan
{shibata,arai,y-saito}@iwate-pu.ac.jp
[2] Tokyo Metropolitan Industrial Technology Research Institute, Tokyo, Japan

Abstract. In this paper, the improved in-vehicle road state sensing system is introduced to identify the higher accurate road states in snow country. Various new road sensors with different sensing technologies are applied and analyzed using AI technology. The analyzed road states can be monitored for drivers to avoid the dangerous road states in advance. The system configuration and architecture of the new sensing and communication system is explained. The accuracy of decision of road state is evaluated based on the prototype of the proposed road state information platform.

1 Introduction

In order to keep safe and reliable driving, accurate monitoring and well maintenance of road states at anytime anywhere on all seasons are indespensable. There are many obstacles on the roads after serious disasters such as earthquakes, typhoons, hurricanes, heavy rains and snow. The annual road states of the snow countries are also very serious in winter. Since the roads are heavily covered with snow and the road states varies such as snowy, icy, damp, sherbet, wet and dry depending on weather conditions on both time and locations, traffic accidents are greatly increased. In fact, more than 90% of annual traffic accidents are due to slip on the snowy, sherbet and icy road states [1]. Even non-snow seasons, many traffic accidents due to fallen rocks, trees and obstacles from cargo bed from trucks occurred. Once the traffic accidents happened by those worse road states, the life of driver are fatal even to the death. In the case of autonomous driving, those slip accidents must be avoided. Those accidents can be avoided by detecting the dangerous locations and obstacles in advance and transmitting and sharing the road state information and alerting to drivers to reduce speed or stop the vehicles.

So far, we have designed and developed the road state information platform which consists of various environmental sensors and communication networks to detect the road states and transmit them and share with vehicles [2, 3]. We have implemented and evaluated our proposed the road state information platform as a social experiment [4, 5]. As results, our system could perform the accepted functions and performance as safety driving system in snow country.

In order to apply our proposed system for the autonomous driving, higher accuracy, quick and long distant communication platform is desired. For those requirements, previous version of platform should be improved. In this paper, the improved road state sensing system is proposed.

L. Barolli et al. (Eds.): AINA 2022, LNNS 449, pp. 321–329, 2022.
https://doi.org/10.1007/978-3-030-99584-3_28

In the following, the related works are introduced in Sect. 2. Next, wide area road states information platform is introduced in Sect. 3. Then, previous road state sensing system is shown in Sect. 4. After that the improved road state sensing system is precisely explained in Sect. 5. In final, conclusion and future remarks are summarized in Sect. 6.

2 Related Works

There are several related works with the road state sensing method using environmental sensors. Particularly road surface temperature is essentially important to know snowy or icy road state in winter season by correctly observing whether the road surface temperature is under minus 4 degree Celsius or over.

In the paper [6], the road surface temperature model by taking account of the effects of surrounding road environment to facilitate proper snow and ice control operations is introduced. In this research, the fixed sensor system along road is used to observe the precise temperature using the monitoring system with long-wave radiation. They build the road surface temperature model based on heat balance method.

In the paper [7–9], cost effective and simple road surface temperature prediction method in wide area while maintaining the prediction accuracy is developed. Using the relation between the air temperature and the meshed road surface temperature, statistical thermal map data are calculated to improve the accuracy of the road surface temperature model. Although the predicted accuracy is high, the difference between the ice and snow states was not clearly resolved.

In the paper [10], a road state data collection system of roughness of urban area roads is introduced. In this system, mobile profilometer using the conventional accelerometers to measure realtime roughness and road state GIS is introduced. This system provides general and wide area road state monitoring facility in urban area, but snow and icy states are note considered.

In the paper [11], a measuring method of road surface longitudinal profile using build-in accelerometer and GPS of smartphone is introduced to easily calculate road flatness and International Road Index (IRI) in offline mode. Although this method provides easy installation and quantitative calculation results of road flatness for dry or wet states, it does not consider the snow or icy road states.

In the paper [12], a statistical model for estimating road surface state based on the values related to the slide friction coefficient is introduced. Based on the estimated the slide friction coefficient calculated from vehicle motion data and meteorological data is predicted for several hours in advance. However, this system does not consider the other factors such as road surface temperature and humidity.

In the paper [13], road surface temperature forecasting model based on heat balance, so called SAFF model is introduced to forecast the surface temperature distribution on dry road. Using the SAFF model, the calculation time is very short and its accuracy is higher than the conventional forecasting method. However, the cases of snow and icy road in winter are not considered.

In the paper [14], blizzard state in winter road is detected using on-board camera and AI technology. The consecutive ten images captured by commercial based camera with 1280 × 720 are averaged by pre-filtering function and then decided whether the averaged images are blizzard or not by Convolutional Neural Network. The accuracy of precision and F-score are high because the video images of objective road are captured only in the daytime and the contrast of those images are almost stable. It is required to test this method in all the time.

In the paper [15], road surface state analysis method is introduced for automobile tire sensing by using quasi-electric field technology. Using this quasi-electric field sensor, the changes of road state are precisely observed as the change of the electrostatic voltage between the tire and earth. In this experiment, dry and wet states can be identified.

In the paper [16], a road surface state decision system is introduced based on near infrared (NIR) sensor. In this system, three different wavelengths of NIR laser sensors is used to determine the qualitative paved road states such as dry, wet, icy, and snowy states as well as the quantitative friction coefficient. Although this system can provide realtime decision capability among those road states, decision between wet and icy states sometimes makes mistakes due to only use of NIR laser wavelength.

With all of the systems mentioned above, since only single sensor is used, the number of the road states are limited and cannot be shared with other vehicles in realtime. For those reasons, construction of communication infrastructure is essential to work out at challenged network environment in at inter-mountain areas. In the followings, a new road state information platform is proposed to overcome those problems.

3 Wide Area Road States Information Platform

So far, we have designed and developed a wide area road state information platform as shown in Fig. 1. The wide area road state information platform is organized mainly by vehicles, roadside units, and cloud computing system. Vehicle includes an on-board Road Sensor Unit (RSU) and a mobile wireless node, so called Smart Mobile Box (SMB). The road side unit includes roadside wireless nodes, so called Smart Relay Shelters (SRS). The cloud computing system includes highspeed computing power and wide space of storages.

On-board RSU is consisted of various environmental sensors including including 9 axis dynamic sensor, far infrared road surface temperature sensor, air humid/temperature sensor, laser sensor, and GPS, and which are integrated to precisely and quantitatively detect the various road surface states and determine the dangerous locations on GIS map in sensor server as shown in Fig. 2. The RSU is used to detect the precise road surface states, such as dry, wet, dump, showy, frozen roads, friction rate and roughness.

The wireless communication unit in both SMB and SRS includes multiple wireless network devices with different N-wavelength (different frequency bands) wireless networks such as 0.92 GHz, 2.4 GHz, 5.6 GHz, 28 GHz, and organizes a cognitive wireless node. The SMB is used to communicate with the other SMB or SRS to exchange the sensor data and the road state information by V2R communication protocol. In our communication, the SMB can select the best link among the cognitive wireless network depending on the observed network quality, such as RSSI, bitrate, error rate by Software

Defined Network (SDN). If none of link connection is possible, those sensor data are locally and temporally stored in the database unit as internal storage until the vehicle approaches to the region where a link connection to another mobile node or roadside node is possible. When the vehicle enters the communication region, it starts to connect a link to other vehicle or roadside node and transmit sensor data by DTN Protocol. Thus, data communication can be attained even though the network infrastructure is not existed in challenged network environment such as mountain areas or just after large scale disaster areas.

On the other hand, the SRS, which is located along street, includes the same communication functions as the SMB with software and hard architecture and receives/sends the sensor data and road state information from/to the passing SMBs. In addition, the SRS also perform a Mobile Edge Computing functions (MEC) between on-board SMBs and the cloud computing system. All of the sensor data and road state information from passing vehicles on both directions are collected and sent to the cloud system. In contract, the SRS receives wider dynamic map information such as road accident, road closed, traffic jam, road regulation information in addition to wide area road state information. Then those information is sent to the passing SMBs.

Cloud computing system receives all of the sensor data and road state information from the SRSs along the various roads, processes those and organizes wide area road state map. The cloud computing system also performs road state prediction in time by collecting the statistical and predicted weather data from Meteorological Agency as bigdata and processing those and sensor data by AI technology. The predicted road state information is visualized to monitor as Web services on PCs, Tablets and Smartphones.

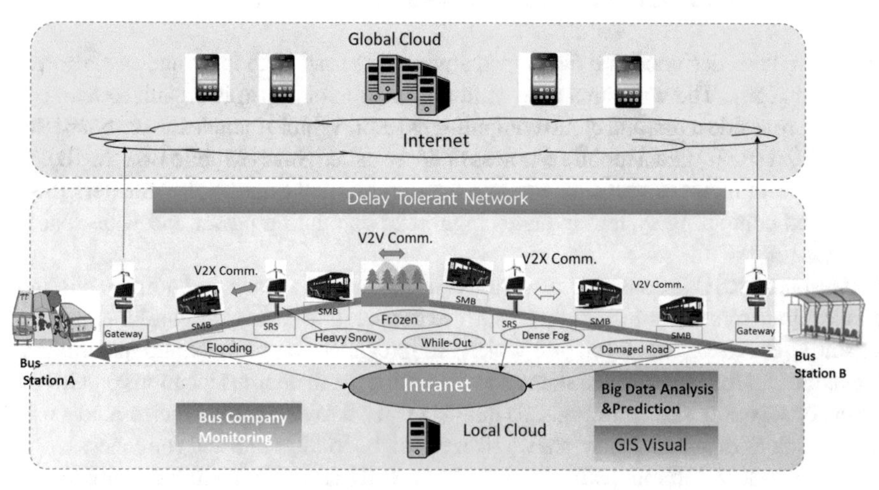

Fig. 1. Wide area road surface state information platform

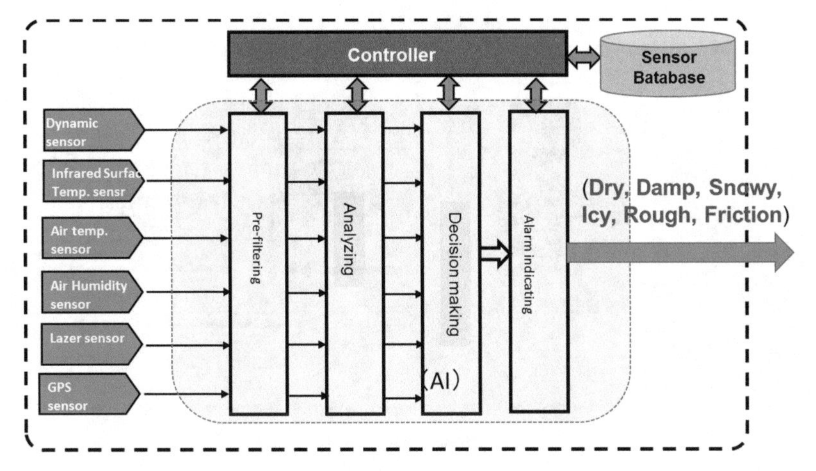

Fig. 2. On-board Road Sensor Unit (RSU)

4 Previous On-board Road Sensing System

The previous on-board road sensing system with sensor server system and communication server system of both SRS and SMB for three-wavelength communication is shown in Fig. 3 As communication components of cognitive wireless network, OiNET-923 of Oi Electric Co., Ltd. for 920 MHz band, WI-U2-300D of Buffalo Corporation for of 2.4 GHz band and T300 of Ruckus for 5.6 GHz band are used. OiNET-923 is used as control data communication link to exchange UUID, security key, password, authentication, IP address, TCP port number, socket No. On the other hand, both WI-U2-300D and T300 are used for sensor data transmission links. Raspberry Pi3 Model B+ is used for N-wavelength cognitive communication server unit to perform cognitive controller and SDN function. Intel NUC Core i7 is used for sensor database registration and data analysis by AI based road state decision.

On the other hand, in sensor server system, several sensors including BL-02 of Biglobe as 9 axis dynamic sensor and GPS, CS-TAC-40 of Optex as far-infrared temperature sensor (FIR sensor), HTY7843 of azbil as air temperature and humidity sensor and RoadEye of RIS system as near-infrared laser sensor and quasi electrical static field sensor for road surface state are used. Those sensor data are synchronously sampled every 0.1 s. and averaged every 1 s to reduce sensor noise by another Raspberry Pi3 Model B+ as sensor server. Then those data are sent to Intel NUC Core i7 which is used for sensor data storage as sensor database and data analysis by AI based road state decision. Both sensor and communication servers are connected to Ethernet switch.

From the results of performance evaluation by many social experiments [2, 3], the previous on-board road sensing system could provide good performance in the view points of sensing ability and communication facility.

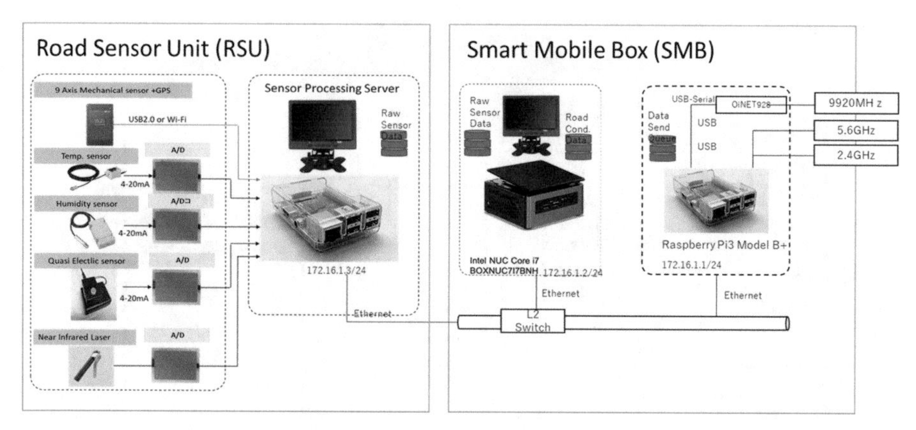

Fig. 3. Previous Road Sensor Unit (RSU-1)

5 New Road Sensing System

As next step, in order to apply our system to autonomous driving on snow road in winter, the current system is needed to be improved from the view points of the accuracy and response time as the whole system. The Fig. 3 shows an improved road sensing and communication system.

The improved system components in RSU, GPS and 9 axis dynamic sensor have been integrated into GNSS (GPS-M2) with several ten cm in distance accuracy at 1 Hz to under 5 cm at 0.1 Hz by considering multiple satellite signals including Michibiki which can provide very higher accurate positions under several cm.

As new RSU, infra-red road surface temperature sensor, air temperature and humidity sensor and near-infrared laser sensor have been integrated into one road surface sensor which can provide 4 time higher sampling rate than the previous components. Data logger is introduced to perform the faster AC/DC conversion and logging functions.

As sensor server, Raspberry Pi4 instead of previous Raspberry Pi3 model B+ is introduced to perform highspeed processing the sampled sensor data to decide the road state in realtime for higher road distance.

On the other hand, as improved system component in SMB and SRS, 5G router is newly introduced to provide higher and long distant communication between SMB and SRS and SRS and Cloud computing system. The IEEE802.11ax with 4.8 Gbps by 2.4 and 5.6 GHz are introduced instead of previous IEEE802.11ac with 867 Mbps by 5.6 GHz.

As communication server, Raspberry Pi4 instead of previous Raspberry Pi3 model B+ is also introduced to perform highspeed transmission speed and low delay.

As results, the RSU can be smaller and simpler structure and easily handled and attached to any vehicles (Fig. 4).

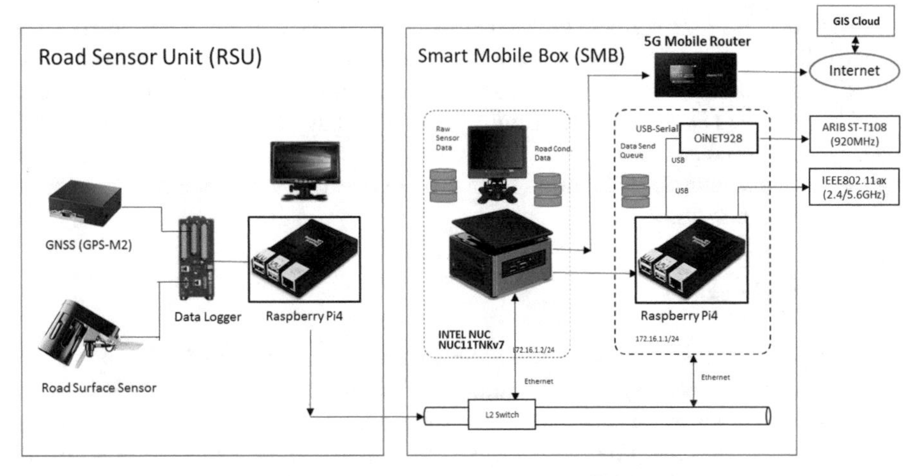

Fig. 4. New Road Sensor Unit (RSU-2)

5.1 New Data Analysis of RSU and SMB

In order to verify the performance improvement of the RSU and SMB, we organize a prototype and observe the performance of both previous and new system. The new RSU and SMB are attached to the vehicle and actually those sensors are sampled and registered in both the data logger and the sensor server while running on about 40 km/h in speed. The same trials are done by the previous RSU and SMB. The registered sensor data of both the previous and new RSUs are compared. The results for both previous and new RSU systems with various performance items are summarized in Table 1.

Table 1. Performance of RSU

	Previous RSU			
	Road Surface Temp.	Air Temp.	Air Humidity	Lazer Sensor
Sampling Rate	1 sec	1 sec	1 sec	0.1 sec
Response time	0.15sec	3min	30sec	0.1sec
Measurement Range	'-40 ~ 500℃	'-20 ~ 80℃	0 ~ 1 00%RH	0.2 ~ 0.8
Measurement Error	±3℃	±0.5℃	±5%	
Area resolution	11m	11m	11m	11m
	New RSU			
	Road Surface Temp.	Air Temp.	Air Humidity	Lazer Sensor
Sampling Rate	0.04sec	0.04sec	0.04sec	0.04sec
Response time	0.1sec	3min	30sec	0.04sec
Measurement Range	-25 ~ 40℃	-20 ~ 80℃	0 ~ 1 00%RH	0.2 ~ 0.8
Measurement Error	±0.1℃	±0.1℃	±5%	
Area resolution	3m	3m	3m	3m

From this result through the field experiment, the measurement range and error are not so different. However, since the sampling rate of the new RSU improved from 1.0 to 0.04 s, the processing time of the sensor server can be reduced, eventually area resolution could be improved about 3.7 times from 11m to 3m. Thus, the accuracy of the sensing road state increased and more precise road state decision could be attained although the registered data also increased 3 times more. The new sensor server could process those increased data in realtime.

6 Conclusions

In this paper, in order to improve the accuracy and responsibility of RSU, a new in-vehicle road state sensing system is introduced. Various new road sensors with different sensing technologies in addition to a new SMB are introduced and their performances are analyzed through the prototyped system. As results, the proposed new RSU could provide higher processing time and the accuracy of the sensing road states. This result can provide possibility of autonomous driving on snowy and icy roads in snow country. Currently, we are evaluating the functionality and performance of the new SMB which are consisted of a 5G mobile router with 4.2 GHz and 28 GHz network and several higher-speed LANs with 2.4 and 5.6 GHz.

As future work, a new wide area road state information platform which is integrated by SMBs with RSUs, SRSs as MEC and cloud computing is constructed along the public roads in a city and evaluated its performance for realization autonomous driving in snow country.

Acknowledgments. The research was supported by Japan Keiba Association Grant Numbers 2021M-198, JSPS KAKENHI Grant Numbers JP 20K11773, Strategic Information and Communications R&D Promotion Program Grant Number 181502003 by Ministry of Affairs and Communication and Communication and Strategic Research Project Grant by Iwate Prefectural University in 2021.

References

1. Police department in Hokkaido, "The Actual State of Winter Typed Traffic Accidents (Nov 2018). https://www.police.pref.hokkaido.lg.jp/info/koutuu/fuyumichi/blizzard.pdf
2. Akira, S., Yoshitaka, S., Noriki, U., Goshi, S.: Social experimental evaluation of road state information system in snow country. In: The 34-th International Conference on Advanced Information Networking and Applications (AINA-2020), Elsevier, pp. 595–604 (Apr 2020)
3. Yoshitaka, S., Akira, S., Yoshikazu, A., Yoshiya, S., Jun, H.: Road state information platform for automotive EV in snow country. In: The 34-th International Conference on Advanced Information Networking and Applications (AINA-2020), Elsevier, pp. 587–594 (Apr 2020)
4. Yoshitaka, S., Akira, S., Yoshikazu, A., Yoshiya, S., Jun, H.: Mobile sensing data analysis for Roa state decision in snow country. In: The 35-th International Conference on Advanced Information Networking and Applications (AINA-2021), Elsevier, pp. 403–413 (May 2021)
5. Akira, S., Yoshitaka, S., Noriki, U., Goshi, S.: Social experimental evaluation of road state information system in snow country. In: The 34-th International Conference on Advanced Information Networking and Applications (AINA-2020), Elsevier, pp.595–604 (Apr 2021)

6. Takahashi, N., Tokunaga, R.A., Sato, T., Ishikawa, N.: Road surface temperature model accounting for the effects of surrounding environment. J. Jpn. Soc. Snow Ice **72**(6), 377–390 (2010)
7. Akihiko, F., Tomoyuki, N., Kenji, S., Roberto, T., Naoto, T., Tateki, I.: Route-based forecasting of road surface temperature by using meshed air temperature data. In: JSSI&JSSE Joint Conference, pp. 1–34 (Sep 2018)
8. Akira, S., Kenji, S., Tomoyuki, N., Roberto, T., Sato, M.: A study of route based forecast of road surface condition by melting and feezing mass estamation method using weather mesh data. In: JSSI&JSSE Joint Conference, pp. 2–57 (Sep 2018)
9. Takumi, H., Akira, S., Tomoyuki, N., Roberto, T., Masaya, S., Kenji, S.: Basic consideration of wide-scale road surface snowy and ICY conditions using weather mesh data. In: Monthly report of Civil Engineering Research Institute for Cold Region, No. 800, pp. 28–34 (Jan 2020)
10. Shun, F., Kazuya, T., Nueraihemaitijang, A., Akira, K.: Development of a roughness data collection system for urban roads by use of a mobile profile meter and GIS. J. Jpn. Soc. Civil Eng. **69**(2), 90–97 (2013)
11. Koichi, Y.: A measuring method of road surface longitudinal profile from sprung acceleration and verification with road profiler. J. Jpn. Soc. Civil Eng. **69**(3), 1–7 (2013)
12. Hideki, M., Takashi, N., Tatsuo, S., Akira, K. A statistical model for estimating road surface conditions in winter. In: The Society of Civil Engineers, Proceedings of Infrastructure Planning (CD-ROM) (Dec 2006)
13. Saida, A., Fujimoto, A., Fukuhara, T.: Forecasting model of road surface temperature along a road network by heat balance method. J. Civil Eng. Jpn. **69**(1), 1–11 (2013)
14. Koji, O., Joji, T., Hirotaka, T., Toshimitsu, S., Tetsuya, K.: Possibility of blizzard detection by on-board camera and AI technology. Monthly report of Civil Engineering Research Institute for Cold Region, No. 798, pp. –32-37 (Nov 2019)
15. Kiyoaki, T., et al.: Trial of Quasi-electrical field technology to automobile tire sensing. In: Annual conference on Automobile Technology Association, pp. 417–540 (May 2014)
16. Casselgren, J., Rosendahl, S., Eliasson, J.: Road surface information system. In: Proceedings of the 16th SIRWEC Conference. Helsinki, 23–25 May 2012. http://sirwec.org/wp-content/uploads/Papers/2012-Helsinki/66.pdf

Multi-agent Q-learning Based Navigation in an Unknown Environment

Amar Nath[1(✉)], Rajdeep Niyogi[2], Tajinder Singh[1], and Virendra Kumar[3]

[1] Sant Longowal Institute of Engineering and Technology, Deemed-to-be-University, Longowal, India
{amarnath,tajindersingh}@sliet.ac.in
[2] Indian Institute of Technology Roorkee, Roorkee, India
rajdeep.niyogi@cs.iitr.ac.in
[3] Central University of Tamil Nadu, Thiruvarur, India
virendrakumar@cutn.ac.in

Abstract. Collaborative task execution in an unknown and dynamic environment is an important and challenging research area in autonomous robotic systems. It is essential to start the task execution in situations like search and rescue at the earliest. However, the time duration between team announcement and the arrival of team members at the location of a task delays the execution of the task. The distributed approaches for task execution assume that the path is known. However, in an environment, say, a building, the position of the doors may not be known, and some of the doors may get closed during task execution. Hence, an agent should first learn the map of the environment. The learning of the map of an unknown environment can be accelerated with multiple agents. This paper proposes a distributed multi-agent Q-learning-based approach for navigation in an unknown environment. The proposed approach is implemented using ARGoS, a realistic multi-robot simulator.

1 Introduction

Collaborative task execution in a dynamic environment, where the location and time of the task are not known in advance, is quite challenging. A mission with collaborative task execution by autonomous agents in a dynamic and unknown environment is discussed in [1–4]. The works [1–4] considered the situation where the map of the environment is known. If a search and rescue operation has to be carried out in such an environment, knowledge of the map is essential. The mission cannot be accomplished in a building if the agents do not know the rooms, the connecting doors, and the exits of the building. Some essential capabilities that allow robots to achieve a high degree of autonomy are exploration and navigation.

L. Barolli et al. (Eds.): AINA 2022, LNNS 449, pp. 330–340, 2022.
https://doi.org/10.1007/978-3-030-99584-3_29

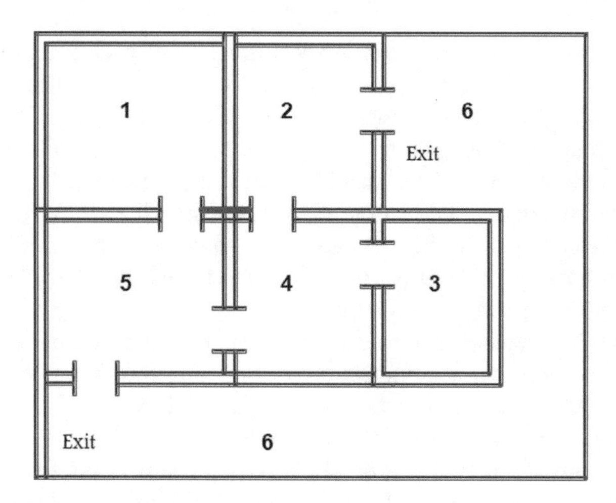

Fig. 1. Environment/Building

Consider, for example, a building (Fig. 1) that has five rooms and four connecting doors between rooms, and two exits where a search and rescue operation has to be carried out. Suppose that the positions and state of doors are not known, and some of the doors may get blocked during mission accomplishment. Hence, to carry out the search and rescue operation, it is essential to first learn the map (the locations and states of the available doors) of the building.

A multi-agent system can help in the exploration as the agents will roam around in the building to find out the doors' positions and states (closed/obstructed or open/clear). In this paper, a distributed multi-agent Q-learning is proposed for solving the aforementioned problem, where each agent periodically sends its observations to all other agents. On receipt of an observation, an agent updates its knowledge that enhances its ability to learn the map quickly. The implementation of the proposed approach is carried out using ARGoS: a modular, parallel, multi-engine simulator for multi-robot systems.

The rest of the paper is organized as of the following. Related work is given in Sect. 2. Section 3 discuss the problem formalization. The preliminary concepts and notation are given in Sect. 4. Section 5 provides the detail of the proposed approach. Implementation and results are given in Sect. 6. Conclusions are made in Sect. 7.

2 Related Work

With the advancement in robot capabilities and technologies, a multi-robot system is being used in several domains, such as search and rescue, autonomous warehouse, space, and deep-sea, where direct human intervention is impractical or impossible.

Distributed approaches for mission accomplishment with collaborative agents are suggested in [1–3,5,6]. The work [5] developed a coalition formation procedure by using a hierarchical organizational structure. In [6], a coalition formation algorithm is suggested based on the distributed computation of maximal cliques in the underlying network, where an edge connects two agents if they can communicate among themselves. The works [1–3] proposed distributed approaches for collaborative task execution in a dynamic environment. None of these works [1–3] discuss collaborative task execution in an environment where the map of the environment is not known in advance.

The work [7] discusses the static map learning techniques and sequential localization and mapping techniques to build a different kind of static map of the system like metric map and topological map. The work [7] discusses various sources of information that can help to develop the map of the environment. A technique for centralized training of Multi-Agent Deep Reinforcement Learning (MARL) using the model-free Deep Q-Network (DQN) as the baseline model and communication between agents is proposed in [8]. The work [9] presents a cooperative multi-agent exploration (CMAE): agents share a common goal while exploring the technique. They follow the standard centralized training and decentralized execution (CTDE) paradigm; at training time, the learning algorithm has access to all agents' local observations, actions, and the state. At execution time, i.e., each agent only has access to its local observation at test time.

In this paper we consider learning the navigation strategies and map of an unknown environment using multi-agent Q-learning.

3 Problem Formalization

We consider a scenario of a building (see Fig. 1) having five rooms connected with doors. The rooms are numbered $1 \ldots 5$. The outside of the building can also be thought of as one big room, numbered 6. Notice that doors 2 and 5 have exit doors outside the building. It is assumed that the map of the building is not known in advance; also, the state of the doors, i.e., open/close, is not known, and the state of the doors may change over time.

Now, suppose at some moment t, a robot $\mathcal{R}_1$ needs to search and rescue that involves the execution of a collaborative task τ in a room $\mathcal{R}_1$. The task τ requires three robots to be executed. Its means $\mathcal{R}_1$ needs to call two more robots to execute the task. Let two robots ($\mathcal{R}_2$ and $\mathcal{R}_3$) be present in rooms $\mathcal{S}_1$ and $\mathcal{S}_3$ respectively. None of the robots know the map of the building, so $\mathcal{R}_2$ may follow either

- $\mathcal{S}_2 \rightarrow \mathcal{S}_6 \rightarrow \mathcal{S}_5 \rightarrow \mathcal{S}_1$, or $\mathcal{S}_2 \rightarrow \mathcal{S}_4 \rightarrow \mathcal{S}_5 \rightarrow \mathcal{S}_1$ path.

And $\mathcal{R}_3$ may follow either.

- $\mathcal{S}_4 \rightarrow \mathcal{S}_2 \rightarrow \mathcal{S}_6 \rightarrow \mathcal{S}_5 \rightarrow \mathcal{S}_1$, or $\mathcal{S}_4 \rightarrow \mathcal{S}_5 \rightarrow \mathcal{S}_1$ path.

Now, if $\mathcal{R}_2$ and $\mathcal{R}_3$ follow the path $\mathcal{S}_2 \rightarrow \mathcal{S}_6 \rightarrow \mathcal{S}_5 \rightarrow \mathcal{S}_1$, and $\mathcal{S}_4 \rightarrow \mathcal{S}_2 \rightarrow \mathcal{S}_6 \rightarrow \mathcal{S}_5 \rightarrow \mathcal{S}_1$ respectively, it means they would take more time to reach the location of task τ. Another problem that arises here is which path to take in case some rooms get blocked during task execution.

It is essential to start the task execution in the situations mentioned above at the earliest. However, the time duration between team announcement and team members' arrival at the location of a task delays the execution of the task. The distributed approaches for task execution in dynamic environments did not study how to minimize this time duration; it is assumed that the path is known. Hence, an agent should first learn the map of the environment. The learning of an unknown environment can be accelerated with multiple agents.

The graph of the building with corresponding rewards is represented in Fig. 2. The nodes of the graph represent the rooms; node 6 represents the outside place of the building that can be reached via rooms 2 and 4. The edges represent the doors between rooms; for example, there is no door between rooms 1 and 3, 1 and 4, and 1 and 2. The rooms 2 and 5 can lead to an exit from the building. So the reward given to these edges is 20, while the other edges have 0 reward.

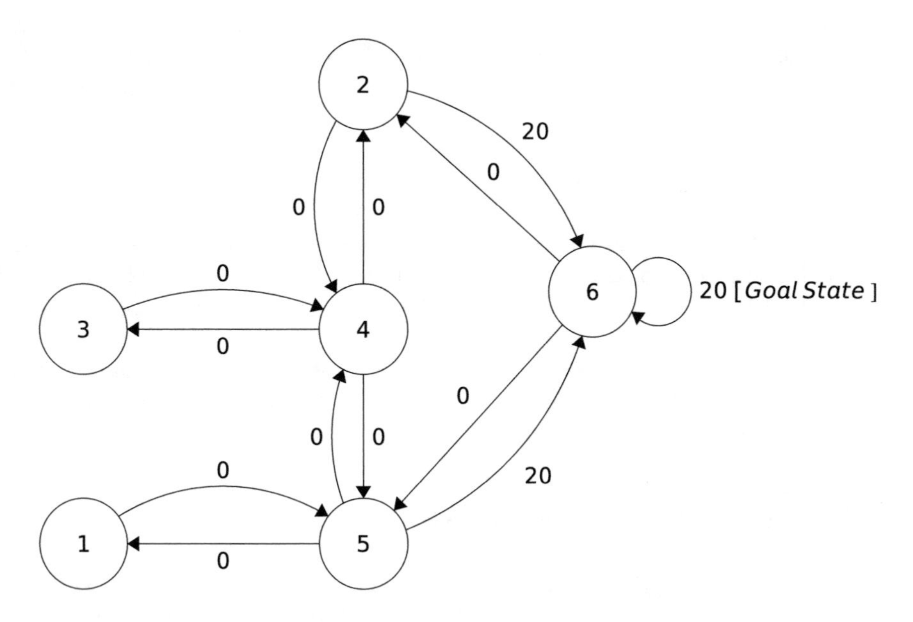

Fig. 2. Rewards for moving across the rooms of the building

The state-space contains the agents' position and the task, but the reward function is solely affected by the placement of the doors. Exploring the entire state space is far more time-consuming than locating the entrances. The rooms 1, 2, 3, 4, 5, and 6 are considered as states of the environment, while agent's movement is considered as an action. The agents initialize, populate, and modify their Q-table when they explore and re-explore the building based on the current situation. The Q-table, which maps state and action to a Q-value, is according to the structure of an environment (e.g., Fig. 1).

Let at some moment in time, a agent is in state 3 (see Fig. 2). It can go from state 3 to state 4 as state 3 is only connected to state 4. However, the agent cannot go to state 2 directly from state 3 as no door connecting room 2 and 3 (thus, no arrows). The agent has two possibilities from the state 4; it can go either to state 2 or 5 or back to 3 (look at all the arrows about state 4). There are three actions from state 5, i.e., the agent can go to state 1, 6, or 4. From state 1, it can only go back to state 5.

For this example, we will place the agents in some rooms and their goal is to exit the building, i.e., room 6. A reward of 20 is given when an agent can go directly to the goal state; otherwise reward is 0.

4 Preliminary Concepts

Q-learning is a model-free reinforcement learning for determining which action is best for a given state [10]. A cell of a Q-table gives the Q-value for a particular state-action. The initial values of the Q-table are set to zero. When an agent takes action a in state s, it receives a reward r and observes the new state s'. Then the table is updated according to Eq. (1) assuming deterministic actions and rewards [10].

$$Q(s, a) \leftarrow r + \gamma \max_{a'} Q(s', a') \tag{1}$$

γ ($0 \leq \gamma \leq 1$) is the discount factor. A high discount factor reflects the effective long-term payout, whereas a low discount factor reflects only the current benefit. The discount factor is used to strike a balance between immediate and long-term rewards.

An agent uses the Q-table as a reference to determine the optimal course of action based on the Q-values. For a state s, an agent picks the action a for which the value of $Q(s, a)$ is maximum.

5 Distributed Multi-agent Q-learning

Q-learning is concerned with how agents should take actions that maximizes the cumulative reward. The proposed Q-learning approach seeks to learn the map by exploring the environment. The advantage of the approach is that it can learn the map of an environment quickly with multiple agents. A distributed multi-agent-based Q-learning algorithm is given in the Algorithm 1.

Algorithm 1 Distributed Multi-Agent Q-learning Algorithm

```
Data: Environment
Result: Final Q-table
1  for (Agent i = 1, ..., N) do
2  |   Q_i(s,a) ← 0 ∀s,a
3  end
   /*Sender_Function_i:                                              */
4  for each episode do
5  |   set s_i randomly
   |    while (s_i ≠ Goal) do
6  |   |   if (Agent i detects a door) then
7  |   |   |   Agent i receives immediate reward r_i
   |   |   |    Agent i observes the new state s'_i
   |   |   |    s_i ← s'_i
   |   |   |    Calculate Q_i(s_i, a_i) as per Equation 1
8  |   |   end
9  |   end
10 |   Agent i broadcast its Q-table Q_i to other agents
11 end
   /*Receiver_Function_j:                                            */
12 if (Agent j receives a Q-table Q_i from Agent i) then
13 |   for (each cell c of Q_j) do
14 |   |   if (value of cell c in Q_j is less than that of Q_i) then
15 |   |   |   Update the value of cell c of Q_j with that of Q_i
16 |   |   end
17 |   end
18 end
```

5.1 Description of the Algorithm

A collaborative task has to be accomplished in an environment where agents do not know the map. So prior to task execution, the map of the environment is learned using Algorithm 1.

The agents who are present in the environment create its Q-table. The process of learning starts with initializing the Q-table with zero. The agents are placed in the building randomly in the environment (in some rooms). The agents begin exploring the environment/building, and if an agent finds a door, it crosses it and changes its state from current (s) state to next (s') state. Also, the agent receives an immediate reward, also, if the next state (s') is a goal state, then one episode of the learning process ends. Now, this agent i updates its Q-table and broadcasts its Q-table to other agents present in the environment.

Now, other agents update their Q-table on receipt of the Q-table sent by i. Suppose agent j receives the Q-table, then it updates its Q-table as follows:

- If the value of a cell c in Q_j is less than the corresponding value of received Q_i, then the value of a cell of Q_j table is updated with the value of Q_i's cell, else
- Value of the cell c in Q_j remains same.

In this way, the process of learning continues until the complete map of the environment is constructed (see Algorithm 1).

6 Implementation and Results

6.1 Experimental setup

ARGoS [12,13], a realistic multi-robot simulator, is used to simulate the proposed distributed multi-agent Q-learning algorithm. The experiments are carried out using ARGoS 3.0.0-beta49 on Intel Core i7-3600 CPU@3.40 GHz×8, 4 GB RAM and 64-bit Ubuntu operating system. We used a foot-bot, one of the ARGoS-supported robots, shown in Fig. 3. ARGoS has several sensors and actuators such as Positioning, Foot-bot gripper, and LED (summarized in Table 1).

Fig. 3. A foot-bot robot [13]

Any alternative multi-robot simulator, such as V-REP [14], ROS-Gazebo [15], can be used to implement the proposed approach. V-REP is a powerful simulation environment with a scene and model editor, a significant model library,

Table 1. Sensors and actuators used in robot

Sensors	Actuators
Colored_blob_omnidirectional_camera	Footbot_gripper
Positioning	Footbot_turret
Footbot_proximity	Differential_steering
Footbot_motor_ground	LED
Differential steering	Range_and_bearing
Range_and_bearing	

real-time mesh manipulation, and more. Complex 3D models and physics engines can be used with ROS-Gazebo. ARGoS, on the other hand, is a low-cost, open-source alternative that is well-suited to multi-robot applications. ARGoS is a multi-robot simulator with physics engine. The code that runs in ARGoS can be used on a real robot system. By adding new plug-ins, we can easily personalize ARGoS. Based on these characteristics, we chose ARGoS multi-robot simulator for implementations.

The implementation of the proposed approach necessitates addressing several difficult issues, including (i) identifying different room areas and doors, (ii) communication among robots using range-and-bearing sensors, (iii) controlling the movement of a robot when boundaries are detected using motor-ground sensors, (iv) synchronizing the robots, (v) controlling the speed and velocity of a robot, and (vi) avoiding an obstacle or an object. We were able to tweak the ARGoS features to address the difficulties indicated above.

ARGoS Setting for Algorithm's Implementation

ARGoS uses its range and bearing sensor (*RAB*) to facilitate communication. Each robot can set the sensor's bandwidth and range in the configuration file (*.argos*). In our experiments, a message format has 8 bits each for the sending address, the receiving address, and the message content (see Table 2).

Table 2. Message format

Sender address bits	Receiver address bits	Message bits
8	8	8

In a multi-robot system, collision avoidance with other robots or obstacles is a critical and challenging problem. The foot-bot's proximity sensors were used for collision avoidance. As illustrated in Fig. 4, the foot-bot contains a total of 24 proximity sensors that help in collecting data from objects that are closer to the robot.

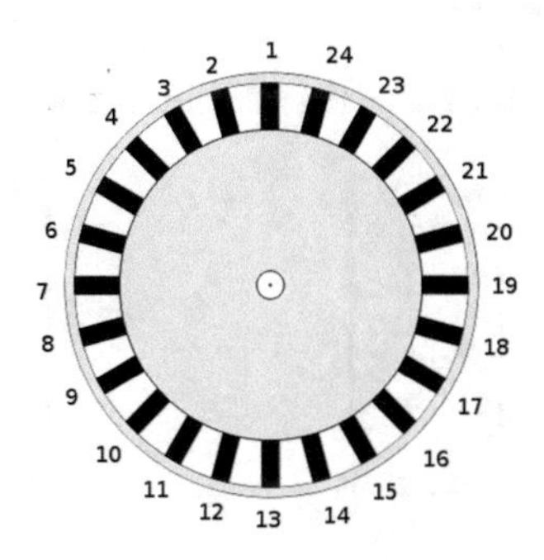

Fig. 4. Proximity sensors

6.2 Results

The initial snapshot of the simulated environment is illustrated in Fig. 5. The process of map learning starts with 2 robots, Robot1 and Robot2. The initial values of reward matrix and Q-table are given in Figs. 6a and 6b. The discount factor γ is set to 0.8. The initial states of the robots are room 1 and 4. The agents explore the building and obtain the map of the building. Throughout the exploration phase, the agent gradually gains confidence in estimating Q-values.

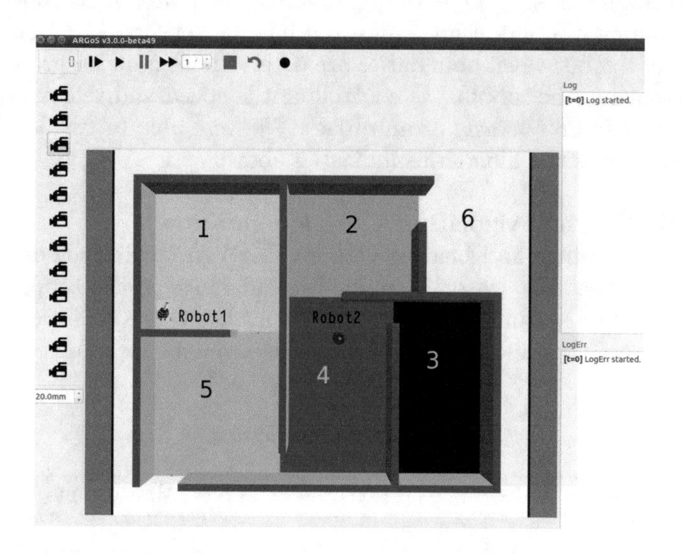

Fig. 5. Initial positions of the robots in the environment

$$\text{Q-table} = \begin{bmatrix} 0 & 0 & 0 & 0 & 0 & 0 \\ 0 & 0 & 0 & 0 & 0 & 0 \\ 0 & 0 & 0 & 0 & 0 & 0 \\ 0 & 0 & 0 & 0 & 0 & 0 \\ 0 & 0 & 0 & 0 & 0 & 0 \\ 0 & 0 & 0 & 0 & 0 & 0 \end{bmatrix}$$

(a) Initialization of Q-table

$$r = \begin{bmatrix} -1 & -1 & -1 & -1 & 0 & -1 \\ -1 & -1 & -1 & 0 & -1 & 20 \\ -1 & -1 & -1 & 0 & -1 & -1 \\ -1 & 0 & 0 & -1 & 0 & -1 \\ 0 & -1 & -1 & 0 & -1 & 20 \\ -1 & 0 & -1 & -1 & 0 & 20 \end{bmatrix}$$

(b) Initialization of r-table

Fig. 6. Initialization of Q-table and r-table

$$\text{Q-table} = \begin{bmatrix} 0 & 0 & 0 & 0 & 9.6 & 0 \\ 0 & 0 & 0 & 0 & 0 & 21.6 \\ 0 & 0 & 0 & 0 & 0 & 0 \\ 0 & 9.6 & 0 & 0 & 0 & 0 \\ 0 & 0 & 0 & 0 & 0 & 37.6 \\ 0 & 0 & 0 & 0 & 0 & 0 \end{bmatrix}$$

(a) Values of Q-table after few iterations

$$\text{Q-table} = \begin{bmatrix} 0 & 0 & 0 & 0 & 96 & 0 \\ 0 & 0 & 0 & 72 & 0 & 108 \\ 0 & 0 & 0 & 72 & 0 & 0 \\ 0 & 96 & 58 & 0 & 96 & 0 \\ 72 & 0 & 0 & 72 & 0 & 108 \\ 0 & 96 & 0 & 0 & 96 & 108 \end{bmatrix}$$

(b) Final Q-table values

Fig. 7. Intermediate and final values of Q-table

The initialization of the Q-table and reward r-table are presented in Fig. 6a and 6b respectively. The –1's in the r-table represent null values (i.e., there isn't a direct link between nodes). For example, from state 0, a robot cannot go to state 1.

The final Q-table (see Fig. 7b) is shared with all the robots present in the environment. From the final Q-table, the door's position can be identified quickly. For example, from room 3 a robot may go to either room 1 or room 4 as both have the same value 96. Now, if a robot is in a room 4, it can easily identify its next state, i.e., room 5, as it has the highest value 108. Now, suppose one exit door connecting rooms 1 and 5 gets blocked at some time. The robot will follow the gate connecting rooms 1 to 3 as the next lower Q-table value is 72. Hence in this way, the proposed approach allows for navigating an unknown and dynamic environment, as discussed in Sect.1.

7 Conclusion

This paper considers the challenging problem of collaborative task execution in an unknown environment. We developed a distributed multi-agent Q-learning-based approach for navigation of the environment. We allowed each agent to learn the map independently. However, the knowledge obtained by each agent is shared among other agents periodically by communication. This helps in avoiding an overlap of exploration, i.e., the same portion of the environment is navigated by different agents. Consequently, this results in learning the map of the environment quickly.

We developed a prototype model of the proposed algorithm in a realistic multi-robot simulation environment (ARGoS). The simulation results are quite encouraging and are as expected. As a part of our future work, we shall explore how the learning rate can be improved by also considering other complex environments.

Acknowledgements. The second author was in part supported by a research grant from Google.

References

1. Nath, A., Arun, A.R, Niyogi, R.: An approach for task execution in dynamic multi-robot environment. In: 31st Australasian Joint Conference on Artificial Intelligence (AI-2018). Wellington, New Zealand, pp. 71–76, (2018)
2. Nath, A., AR, A., Niyogi, R.: A distributed approach for autonomous cooperative transportation in a dynamic multi-robot environment. In: 35th Annual ACM Symposium on Applied Computing (SAC-2020). Brno, Czech Republic, pp. 792-799 (2020)
3. Nath, A., Arun, A.R., Niyogi, R.: DMTF: a distributed algorithm for multi-team formation. In: 12th International Conference on Agents and Artificial Intelligence (ICAART-2020). Valletta, Malta, pp. 152-160 (2020)

4. Nath, A., Niyogi, R.: Formal modeling, verification, and analysis of a distributed task execution algorithm. In: Barolli, L., Woungang, I., Enokido, T. (eds.) AINA 2021. LNNS, vol. 225, pp. 370–382. Springer, Cham (2021). https://doi.org/10.1007/978-3-030-75100-5_32
5. Abdallah, S., Lesser, V.: Organization-based cooperative coalition formation. In: ACM International Conference on Intelligent Agent Technology (IAT-2004). Beijing, China, pp. 162-168 (2004)
6. Tošić, P.T., Agha, G.A.: Maximal clique based distributed coalition formation for task allocation in large-scale multi-agent systems. In: Ishida, T., Gasser, L., Nakashima, H. (eds.) MMAS 2004. LNCS (LNAI), vol. 3446, pp. 104–120. Springer, Heidelberg (2005). https://doi.org/10.1007/11512073_8
7. Meyer, J.A., Filliat, D.: Map-based navigation in mobile robots: a review of map-learning and path-planning strategies. Cogn. Syst. Res. **4**(4), 283–317 (2003)
8. Bhalla, S., Ganapathi Subramanian, S., Crowley, M.: Deep multi agent reinforcement learning for autonomous driving. In: 33rd Canadian Conference on Artificial Intelligence (CCAI-2020). Ottawa, Ontario, pp. 67-78 (2020)
9. Liu, I.J., Jain, U., Yeh, R.A., Schwing, A.: Cooperative exploration for multi-agent deep reinforcement learning. In: 38th International Conference on Machine Learning (ICML-2021). Virtual mode, pp. 6826-6836 (2021)
10. Watkins, C.J., Dayan, P.: Q-learning. Mach. Learn. **8**(3–4), 279–292 (1992)
11. Neves, M., Vieira, M., Neto, P.: A study on a Q-Learning algorithm application to a manufacturing assembly problem. J. Manuf. Syst. **59**, 426–440 (2021)
12. ARGoS simulator www.argos-sim.info/
13. Pinciroli, C., et al.: ARGoS: a modular, parallel, multi-engine simulator for multi-robot systems. Swarm Intell. **6**(4), 271-295 (2012)
14. V-REP www.coppeliarobotics.com/
15. Gazebo //gazebosim.org/

Improving Urban Mobility with Vehicular Routing: A Parallel Approach

Fillipe Almeida Paz[1(✉)], Filipe Nascimento Almeida[1], Rubens de Souza Matos Junior[1,2], Itauan Silva Edução Ferreira[2], and Ricardo Jose Paiva de Britto Salgueiro[1]

[1] Federal University of Sergipe, São Cristóvão, SE, Brazil
{fillipe.paz,fillipe.almeida}@dcomp.ufs.com
[2] Federal Institute of Sergipe, Aracaju, SE, Brazil
{rubens.junior,itauan.ferreira}@ifs.edu.br

Abstract. The concentration of population in cities and the increase in the number of vehicles on streets pose numerous challenges to urban mobility, especially when the possibility of expanding roads is remote and social awareness is an obstacle. In this context, this work aims at analyzing the performance of routing algorithms in the periodic route planning task, in order to identify approaches that maximize quality for congestion charge, travel duration, route length, and algorithm execution time, considering smart city simulations. The results showed that routes with dynamic re-planning jointly with parallelism produced an execution time around 30000 times smaller when compared to re-planning with serial algorithms. In addition, the parallel approach managed to minimize urban mobility metrics when compared to serial algorithms results.

1 Introduction

The disorderly growth of cities has imposed many challenges on society, from aspects of sanitation to aspects of mobility. In this last example, congestion and high travel times significantly impact people's daily lives. In order to address this situation, alternatives have been sought to maximize urban mobility.

Three main challenges regarding urban mobility were identified by [10], namely: the need for standardized metrics to identify optimal routes; dynamic identification and implementation of optimal routes; in addition to simplifying the planning of investments and programming of mobility. The role of the Internet of Things in this area is also reinforced.

The use of the Internet of Things is often associated with improvements in urban mobility and Smart Cities, as seen in [13,14]. However, the concept of smart cities is quite comprehensive and assumes the use of technological infrastructure, integration of services, skills, and aspects of cities, culminating in the maximization of social well-being and economic and environmental sustainability [2].

In this context, considering the challenges associated with urban mobility and the environment provided by Smart Cities, they can support Intelligent Transport Systems, to mitigate congestion and its harmful effects. However, there is a noticeable change

L. Barolli et al. (Eds.): AINA 2022, LNNS 449, pp. 341–352, 2022.
https://doi.org/10.1007/978-3-030-99584-3_30

from the commonly used vehicular routing applications. As observed by [7], the analysis and processing infrastructure of the routes must be able to reliably capture the state of the traffic system - not just considering a limited set of possible routes - in addition to allowing observability and control by part of managers. However, to fulfill what was mentioned by [7], a dynamic re-planning of routes is necessary, considering a large space for solutions.

Potentially optimal routes will be generated from the point of view of urban mobility, but this approach might have high computational cost. In [15], a reduction of up to 45x was evidenced in the implementation of A* when compared to the serial version. However, the impacts of routing on urban traffic simulations were not surveyed, so it is not possible to measure the impacts on urban mobility metrics. Furthermore, our work proposes parallelization oriented to vehicular routing, with each thread dedicated to process the route of an active vehicle in the system, effectively exploiting GPU parallelism. In this sense, this work proposes an experimental and comparative study, based on simulation, to assess the efficiency of vehicular routes dynamic replanning associated with parallel computing, in order to reduce the processing time; in addition, the behavior of urban mobility metrics was evaluated, to verify the optimization capabilities in comparison with serial approaches.

This remaining of this work is divided into the following sections: 2) Related works, which aims at identifying papers that deal with similar issues mentioned here or that might support our approach; 3) Methodology, where the experimental process for the production of results will be addressed, as well as their analysis strategy; 4) Results, where the results will be presented and analyzed; and 5) Conclusion, which will bring considerations about the experimental evidence, as well as strengths and weaknesses associated with the treatments covered in this work.

2 Related Works

The work [8] presents a classification of different routing algorithms and the method used to improve the performance of the algorithm to be applied in a real environment. There is a suggestion of an approach to improve the performance of these algorithms through the input data; selection of metrics that optimize route quality, scalability, and computational complexity; and how the system should adapt to the dynamics of traffic data. Finally, there is an implementation of the Dijkstra algorithm to calculate and recalculate simulated vehicle routes in SUMO. In that work, there is not, in fact, a comparison between routing algorithms saying which one gives the best results, but rather a manual on how to design such algorithms.

A method is proposed in [1] to choose the best route among those available from the AODV (Adhoc On-Demand Distance Vector) algorithm. The proposed VIKOR algorithm evaluates which of the possible routes is the best one based on route length, route delay, and quality of speed signals received from vehicles. VIKOR selects the best route among all available ones based on the aforementioned parameters. Routing with AODV and VIKOR is simulated and compared, there is an advantage of the proposed method concerning the standard routing algorithm in VANET, but a low number of vehicles was used in the simulations (20 and 30), there is no information regarding the temporal and spatial complexity of the algorithms, in addition to the fact that there is no employment of statistical tests.

The work of [11] presents some algorithms used in vehicle routing and a comparison between them. Some comparisons of tests performed between the algorithms mentioned in the study are presented. The results showed that DaIBA presented better results than the others, but in some particular cases, other algorithms do better. The Holm-Bonferroni method was used to adjust the parameterization of the algorithm, however, there was no complex simulation using vehicles and a simulated real environment, nor parallelism was considered to optimize the performance of some algorithms.

3 Metodology

The objective of this work is to evaluate, the metrics travel time, travel duration, amount of congestion and execution time of the routing algorithms, the potential for maximizing urban mobility and minimizing the computational cost in the context carrying out periodic and complete re-planning of vehicles in a traffic system.

Thus, the controlled experiment will compare the aforementioned metrics for each of the treatments used, namely: Parallel A* (Fig. 1b), Serial A* (Fig. 1a), Dijkstra, and without routing. Changes in traffic load will also be considered in the form of interval between entry of vehicles into the system.

The experiment was *in virtuo* and used an HPE machine with 10 GB of RAM and Intel Xeon e3-1220 v6. The GPU used was Nvidia Super 1660 with 6 GB of RAM and 1408 CUDA Cores. In order to carry out the urban mobility simulations, SUMO (Simulation of Urban Mobility) in version 0.32.0 was used. The operating system was Ubuntu GNU/Linux 20.04. For the simulations, a map of the city of Aracaju, Brazil, with an extension of $4km^2$ was used.

The independent variables were: vehicle entry period, which defines the interval between the entry of a vehicle and its subsequent one. The assumed period values were: 0.4, 0.8, 1.2, 1.6, 2.0; state seeds to generate variability in the state of the system; hardware for carrying out the experiments; route replanning period, which was kept fixed in 240 simulation steps, with each step equaling 1 s; and, the algorithm used for the re-planning of routes.

The dependent variables were the mean values of: Duration, which reflects the average travel time of vehicles; Route Length, relative to the average length of the routes in that simulation; Jams, when cars are removed from the simulation due to traffic jams. In addition, the Execution Time samples are included in this group, which is the execution time of the route replanning, for each execution of the route replanning task.

Hypothesis Formulation

The research questions for this experiment are: do the algorithms produce the same travel times? Is the execution time of the algorithms in the route replanning task the same? Is the amount of congestion observed for each algorithm the same? Is the length of routes for each algorithm the same?

In order to answer these research questions, the experiments involved collecting metrics from the summary of simulations performed using the SUMO simulator and also collecting operating system metrics. In order to answer the first question, the Duration Route metric (λ) was used. In order to answer the second question, the execution time of the algorithm for each route replanning applied to the system was used (τ). In order to answer the third question, the Jams metric (ϕ) was used. In order to answer the fourth research question, the route length metric (δ) was used.

The experiment evaluated 4 treatments, namely: Serial A*, Parallel A*, Serial Dijkstra, No routing. This work compared them against all metrics in order to answer the research questions. In this context, let Alg be the set composed of the simulated treatments, such that Alg = {Parallel A*, Serial A*, Dijkstra, Without routing}.

Likewise, let P be the set comprising the simulated entry periods, such that P = {0.4, 0.8, 1.2, 1.6, 2.0}, with all values given in seconds.

Considering the research questions, the metrics provided by the simulations and the treatments applied, the following hypotheses can be listed: Let $(a,b) \in Alg \times Alg$ and $p \in P$. Then $\forall(a,b)$ — $a \neq b$; and $\forall p$:

Hypothesis 1: for travel time (duration)

H_0: The distributions of $\lambda_{p,a}$ and $\lambda_{p,b}$ are statistically equal ($\lambda_{p,a} = \lambda_{p,b}$) .

H_1: The distributions of $\lambda_{p,a}$ and $\lambda_{p,b}$ are statistically different ($\lambda_{p,a} \neq \lambda_{p,b}$).

Hypothesis 2: for algorithm execution time

H_0: The distributions $\tau_{p,a}$ and $\tau_{p,b}$ are statistically equal ($\tau_{p,a} = \tau_{p,b}$)

H_1: The distributions $\tau_{p,a}$ and $\tau_{p,b}$ are statistically different ($\tau_{p,a} \neq \tau_{p,b}$).

Hypothesis 3: for the number of jams

H_0: The distributions $\phi_{p,a}$ and $\phi_{p,b}$ are statistically equal. ($\phi_{p,a} = \phi_{p,b}$)

H_1: The distributions $\phi_{p,a}$ and $\phi_{p,b}$ are statistically different ($\phi_{p,a} \neq \phi_{p,b}$)

Hypothesis 4: for the length of routes

H_0: The distributions $\delta_{p,a}$ and $\delta_{p,b}$ are statistically equal. ($\delta_{p,a} = \delta_{p,b}$)

H_1: The distributions $\delta_{p,a}$ and $\delta_{p,b}$ are statistically different ($\delta_{p,a} \neq \delta_{p,b}$)

Object Selection

The experiment uses a map generated through the Open Street Map [3], from a cutout of downtown Aracaju, Brazil. 26 seeds for random number generation were also employed, concerning the positioning and origin-destination of the vehicles. In addition, 5 different vehicle entry intervals were generated in the system (0.4 s, 0.8 s, 1.2 s, 1.6 s, 2.0 s).

The serial algorithms used, A * and Dijkstra, were those already implemented in the SUMO [5] simulator. Simulations without dynamic route reprocessing were also considered, to serve as a baseline. On the other hand, the Parallel A* routing algorithm was implemented using SUMO's *traci* API methods, as well as the *pycuda* library and CUDA-C code.

The design of the experiment can be summarized in the following steps:

(1) Preparation of the experiment execution environment; (2) Map generation; (3) Generation of seeds; (4) Implementation of Parallel A* algorithm; (5) Implementation of simulation execution scripts; (6) Implementation of data collection scripts; (7) Execution of the experiments; (7a.) Execution of 26 simulation runs for each algorithm and entry period considered; (8) Collection of results; (9) Generation and analysis of charts; (10) Application of statistical tests.

Data Validation.

For validation against the hypothesis of normality, the Kolmogorov-Smirnov (KS) statistical test was employd (above 30 samples). The Shapiro-Wilks was employed for smaller amounts of samples. However, in some treatments the null hypothesis could not be accepted with 95 % confidence. Consequently, the non-parametric Kruskal-Wallis tests were adopted to identify differences or similarities between treatments. When the null hypothesis was rejected, the post-hoc Nemenyi test was employed to identify which treatments differed statistically [9].

The Wilcoxon [4] or Mann-Whitney [6] tests were used in addition to the two treatments with the lowest median for each hypothesis, in order to validate the results indicated by the Nemenyi test. In case of different results between the groups, we opted for the result of Mann-Whitney or Wilcoxon.

In order to decide on the dependence or independence of the data, a scatter plot was adopted before applying the Wilcoxon and Mann-Whitney tests. When positively or negatively correlated, Wilcoxon was applied, or Mann-Whitney otherwise.

Parallel A* Algorithm

One of the algorithms that stands out in solving routing problems is the A-star, or A*, which is usually modified to deal with specificities of the problem where it is applied. In the literature, the A* algorithm is recommended to deal with the planning of route solutions, as it is an efficient heuristic algorithm used to find a low-cost path [12]. A* makes use of an evaluation function denoted by F(n) to guide and determine the order in which the search visits the nodes of a graph representing the evaluated scenario. The function is given by: $F(n) = g(n) + h(n)$

The node visiting criterion is the smallest value of the cost F(n), a cost that is composed by the sum of a cost g(n) with a heuristic cost h(n). These costs are inherent to the problem and approach adopted, in this case the cost is related to travel time. At the end of the execution of the algorithm A*, the result must be the path that minimizes the value F(n). However, the path obtained can be suboptimal according to the quality of the heuristic function h(n) chosen for the problem. For routing problems, an admissible heuristic function is, for example, the Euclidean distance.

As a solution to the problem-focused by this work, the A* algorithm was applied to each vehicle in the traffic system of a smart city. The pseudocode of the classical A* algorithm used is shown in Fig. 1a. This application generates a high demand for computational power, due to the large number of calculations that need to be performed so that each vehicle has its planned route. To make this solution feasible, it is necessary to have a large computational infrastructure, such as clusters and supercomputers, in addition to using code parallelization techniques.

In this work, the A* algorithm was implemented with parallel programming techniques to run on GPUs (Graphical Processing Units). The algorithm was developed to be able to calculate the routes of not just one vehicle, but several vehicles simultaneously. This change was carried out from the point of view of the existence of a central entity responsible for planning the routes of all vehicles in the simulated area. Therefore, all vehicles should communicate their routes to the centralized entity. The algorithm shown in Fig. 1b has as input data: a neighborhood matrix and a matrix that stores the XY pairs. The matrix that contains the XY pairs has the same number of lines

```
Input:   vH: Matrix containing the ordered pairs XY for each point of the graph,
         the lines represent the vertices of the graph and column 1 for x and 2 for
         y.
         G: Neighborhood matrix that contains cost to leave a vertex i and reach
         a vertex j.
         Calculate: It is a function that takes the X and Y points and performs
         the distance calculation vector operation.
1:  FOR (k = 1 : N° cars)
2:    begin = BegEnd[k][0]
3:    end = BegEnd[k][1]
4:    pc[begin][k]=1
5:    point=begin
6:    WHILE (point != end)
7:      FOR (i = 1 : N° Vertex)
8:        IF ( G[point][i] != 0.0 e pc[i][k] != 1)
9:          Calculate (x , y)
10:         IF ( CG[i][k] > (G[point][i]+CG[point][k]) ou vcost[i][k]=0.0)
11:           CH[i][k] = sqrt(x*x+y*y)
12:           CG[i][k] = G[point][i]+CG[point][k]
13:           vcost[i][k]=CG[i][k] + CH[i][k]
14:           P[i][k]=point
15:         END – IF
16:       END – IF
17:     END – FOR
18:     FOR (i = 1 : N° Vertex)
19:       IF ( pc[i][k] != 1 , vcost[i][k] e vcost[i][k]!=0.0)
20:         smaller = vccost[i][k];
21:         iSmaller = i;
22:       END – IF
23:     END – FOR
24:     point = iSmaller
25:     pc[point][k]=1;
26:   END – WHILE
27: END – FOR
```

(a) Serial

```
Input:   H: Matrix containing the ordered pairs XY for each point of the graph,
         the lines represent the vertex of the graph and column 1 for x and 2 for y.
         G: Neighborhood matrix that contains cost to leave a vertex i and reach
         a vertex j.
         Calculate: It is a function that takes the X and Y points and performs
         the distance calculation vector operation.
1:  k = blockIdx.x
2:  begin = BegEnd[0+k*2]
3:  end = BegEnd[1+k*2]
4:  pc[k+begin*(N° cars)]=1
5:  point=begin
6:  WHILE (point != end)
7:    FOR (i = 1 : N° Vertex)
8:      IF ( G[i+point*(N° Vertex)] != 0.0 and pc[k+i*(N° cars)] != 1)
9:        Calculate (x , y)
10:       IF (CG[k+i*(N° Cars)] > (G[i+point*(N° Vertex)]+CG[k+point*(N° Cars)]))
11:         CH[k+i*(N° Cars)] = sqrt(x*x+y*y)
12:         CG[i+k*(N° Cars)] = G[k+i*(N° Cars)]+CG[k+point*(N° Cars)]
13:         vcost[k+i*(N° Cars)]=CG[k+i*(N° Cars)] + CH[k+i(N° Cars)]
14:         P[k+i*(N° Cars)]=point
15:       END – IF
16:     END – IF
17:   END – FOR
18: FOR (i = 1 : N° Vertex)
19:   IF ( pc[k+i*(N° Cars)] != 1 , vcost[k+i*(N° Cars)] e vcost[k+i*(N° Cars)]!=0.0)
20:     smaller = vcost[k+i*(N° Cars)];
21:     iSmaller = i;
22:   END – IF
23: END – FOR
24:   point = iSmaller
25:   pc[k+point*(N° Cars)]=1;
26: END – WHILE
```

(b) Parallel

Fig. 1. Pseudocodes A*

and vertices of the simulated area, that is, the number of intersections between paths in the simulated map. Furthermore, this matrix has two columns, with the ordered pair (x, y), for the heuristic calculation from the Euclidean distance between vertices.

Our implementation adopts a neighborhood matrix that is changed periodically, according to the state of the system. At each instant T, all vehicles can have their routes calculated taking the cost matrix as a reference. This approach enables computing routes as independent operations (each vehicle) on the same set of data (road network situation).

4 Evaluation

The results presented here come from data collected from the perspective of statistical tests. In this sense, at least one treatment for each metric did not accept the null hypothesis for the Shapiro-Wilks test, and, consequently, it was not possible to guarantee with 95% confidence that the samples followed a normal distribution. Hence, non-parametric tests were adopted.

For the execution time metric, which provided at least 416 samples, the KS test was used to assess normality, rather than the Shapiro-Wilks test. Consequently, it was not possible to guarantee with 95% confidence that the samples followed a normal distribution. Hence, non-parametric tests were adopted too.

Hypothesis 1: Average Travel Time

In evaluating this hypothesis, the parallel A* algorithm and the other treatments differed for the periods from 0.4 to 1.6. In period 0.4, the lowest median occurred in the A* Serial treatment (Fig. 2a). However, as *p-value* > 0.1 for Wilcoxon's paired test of the Serial A* and Dijkstra, the null hypothesis was not rejected and we concluded that the treatments presented similar distributions.

However, as can be seen in the graphs of Fig. 2b, 2c, 2d, the Parallel A* had a lower median compared to the other treatments, for periods 0.8, 1.2 and 1.6. The Mann-Whitiney test also confirmed the statistical difference, as *p-value* < 0.05 compared to the other treatments.

For period 2.0 (Fig. 2e) the null hypothesis was NOT rejected by neither Nemenyi nor Mann-Whitney (*p-value* > 0.2) with respect to travel time. Therefore, for the context of lower system load, the parallel and non-routing approaches obtained similar results.

It is important to note that as the system load decreased, that is, the interval between the entry of a vehicle and its successor in the system increased, the scenario without routing approached those in which periodic routing occurred. This implies that periodic routing exhibits greater effectiveness in situations of greater load.

Hypothesis 2: Algorithm Execution Time

In order to test this hypothesis, samples of the algorithms' execution times at each rerouting were analyzed. It is noteworthy that the treatment "without routing" did not have associated values for this metric since the vehicles used static routes, generated along with maps and randomness seeds.

Serial A* and Dijkstra results were in the same order of magnitude, while Parallel A* had results 10^{-5} less than the other treatments in all periods. This difference was mainly due to the massive parallelism used in the generation of vehicular routes. Furthermore, the difference between parallel A* and the other treatments was confirmed by statistical tests. Additionally, the Mann-Withiney test showed that the results of the serial A* and Dijkstra were statistically equal for periods 0.8, 1.2, and 1.6 (Fig. 3).

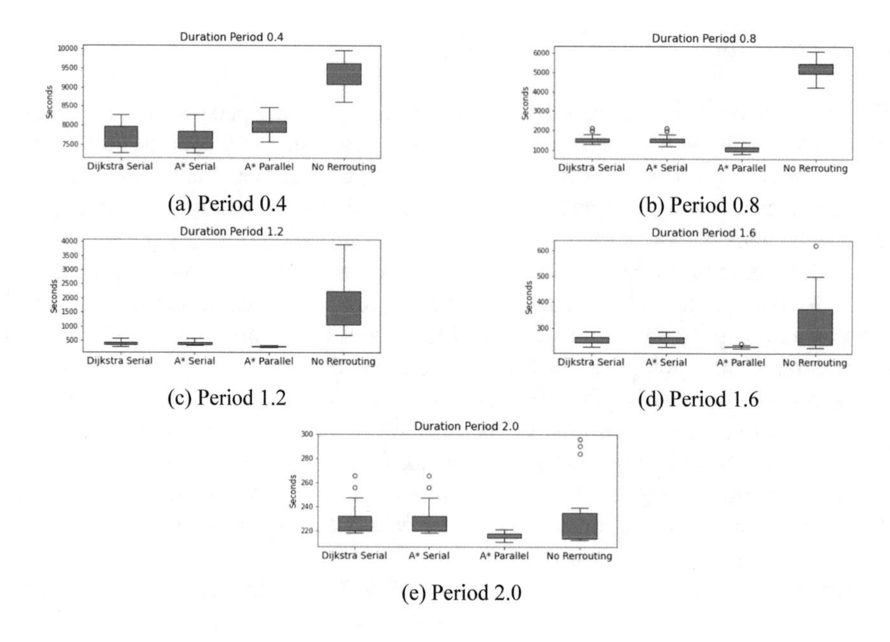

(a) Period 0.4 (b) Period 0.8 (c) Period 1.2 (d) Period 1.6 (e) Period 2.0

Fig. 2. Travel time results

Hypothesis 3: Average Amount of Congestion

It is possible to observe through Fig. 4a that the simulations submitted to the parallel A* presented a smaller amount of congestion. On the other hand, the simulations in which the rerouting was not carried out approximately double the amount observed in the simulations that used Parallel A*.

The statistical tests reaffirmed the visual findings and showed a statistically significant difference between the Parallel A* and the other treatments. It should be noted that the parallel rerouting approach showed zero jams for this metric from period 1.2 onwards, while the other treatments continued to report the occurrence of congestion (Fig. 4a, 4b, 4c, 4d and 4e).

Statistical tests showed differences in the distribution of parallel A* and other treatments for all periods (Figs. 4a, 4b, 4c, 4d and 4e), also with lower median. It was also possible to verify that the distributions with the serial A* and Dijkstra treatments behaved in a statistically similar way (*p-value* = 0.9). Meanwhile, the Mann-Withiney test applied to serial treatments compared to the non-routing approach showed that there are differences between Dijkstra and non-routing, as well as between Serial A* and non-routing for periods 0.4 to 1.6 (Fig. 4a, 4b, 4c and 4d). In the 2.0 period (Fig. 4e), as it was the one with the lowest system load, the amount of congestion was similar between these treatments.

Hypothesis 4: For the Average Length of Routes

As can be seen in the graphs of Fig. 5a, 5b and 5c, the median of route lengths in the simulations that did not perform rerouting were smaller than those which performed it. This was confirmed by statistical tests that pointed to a *p-value* < 0.05. Since the routes

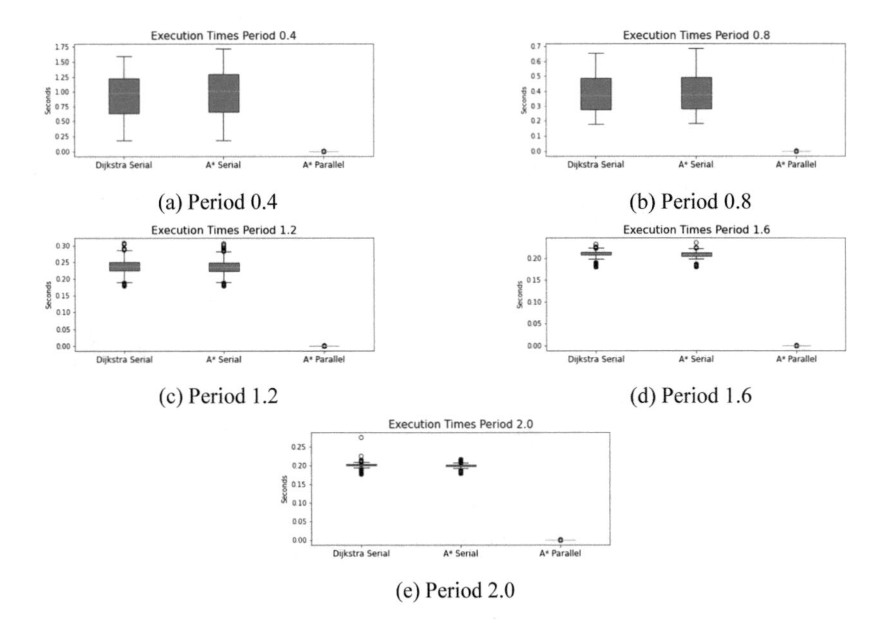

(a) Period 0.4

(b) Period 0.8

(c) Period 1.2

(d) Period 1.6

(e) Period 2.0

Fig. 3. Execution time results

generated with the simulations for the vehicles are already minimized from the point of view of route length, approximately the same average route length is observed for all simulations.

However, due to the dynamic behavior of the system, in periods 1.6 and 2.0 the best lengths of routes started to occur in simulations that performed dynamic replanning of routes using the A* Parallel. This differentiation was also supported by the Mann-Whitney paired test with parallel A* and No-Routing, as the *p-value* $< 8.10^{-3}$.

As can be seen in the graphs in Fig. 5a, 5b, 5c, 5d, 5e, the results of the serial A* and Dijkstra were very similar and correlated, which was also reflected in the statistical tests using the paired test because the *p-value* ≥ 0.1 in all periods.

Overview of the Results

In order to provide better visualization and conclusion regarding the results obtained from the experiments, the notes are presented below:

(1) The execution time of the parallel A* was in the order of 1.10^{-5} for all vehicle entry periods that were simulated, while, among the serial algorithms, the execution time is around 1.75 s; (2) The Parallel A* had the smallest medians and was statistically different concerning congestion in all simulated periods; (3) Considering the period of entry of vehicles as different load scenarios, it is possible to evidence that the parallel A* managed to minimize the urban mobility metrics in the following proportions: Jams, 5 out of 5 scenarios; Execution Time, 5 out of 5 scenarios; Route Duration, 3 out of 5 scenarios; and, Route Length, 2 out of 5 scenarios. For this comparison, the statistical difference resulting from the hypothesis tests and the median of the distributions were analyzed.

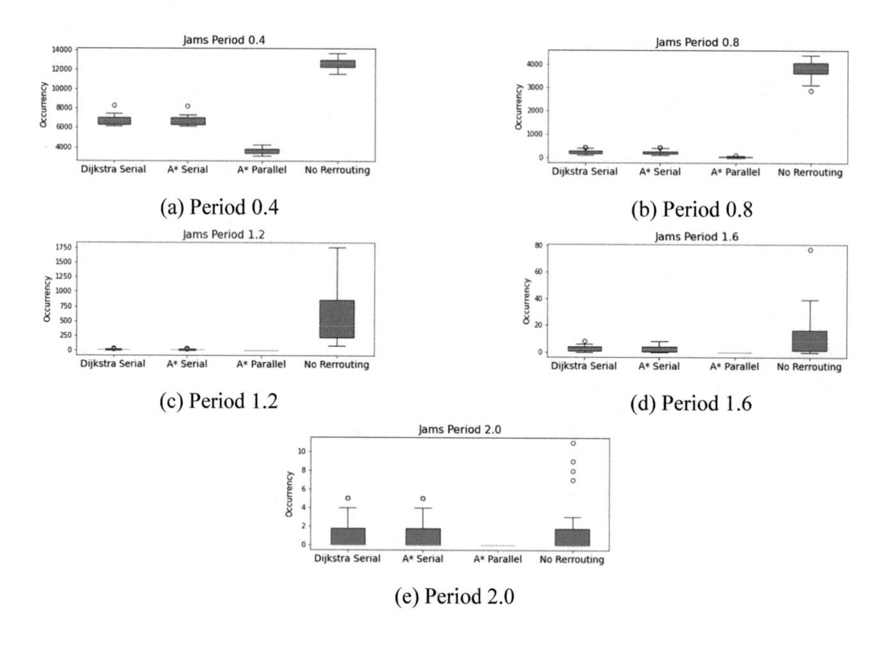

(a) Period 0.4 (b) Period 0.8 (c) Period 1.2 (d) Period 1.6 (e) Period 2.0

Fig. 4. Jams results

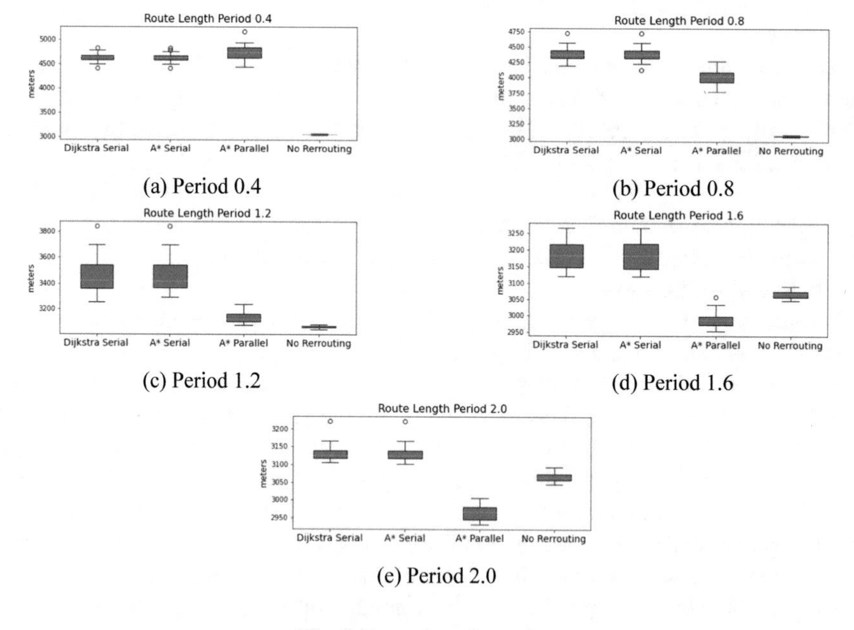

(a) Period 0.4 (b) Period 0.8 (c) Period 1.2 (d) Period 1.6 (e) Period 2.0

Fig. 5. Route length results

5 Conclusions and Future Works

Urban mobility has a significant influence on society's quality of life. In this context, the concentration of the population in urban centers and the increase in the mobility throughout the cities have intensified the occurrence of traffic jams, increasing the time lost by drivers. Consequently, research and solutions capable of proposing standardized metrics for the identification of optimal routes and the dynamic identification and implementation of optimal routes have been sought.

This work aimed to experimentally identify the behavior of different vehicle routing approaches and their optimization capabilities, under different load conditions, with respect to the execution time, route duration, route length, and jams metrics.

The experiments were carried out using the SUMO simulator, as well as scripts for collecting and processing the simulation results. Periodic re-planning of vehicular routes in the simulations were carried out to evaluate 4 treatments, namely: Parallel A*, Serial A*, Dijkstra and a scenario without route reprocessing.

Statistical tests were used to avoid bias concerning the established hypotheses – i.e., assuring conclusion validity. This statistical analysis was important to avoid selecting a treatment (algorithm) over another simply by analyzing the sample means. In order to mitigate the threat associated with evaluating the samples as dependent or independent, scatter plots were used.

Only one map and the same set of seeds were used in all experiments, which represents a low variety of scenarios and constitutes a threat to external validity. In order to mitigate this threat it is necessary to simulate with new maps and with seeds without intersection between treatments.

It was possible to evidence that the treatment that used parallel computation obtained an execution time up to 30000 times smaller than the serial algorithms. Furthermore, it was evident that the parallel A* exhibited different distributions and a lower median compared to the other treatments in the following metrics and proportions: Jams, 5 out of 5 scenarios; Route Duration, 3 out of 5 scenarios; and, Route Length, 2 of the 5 simulated scenarios.

As future work, the exploration of meta-heuristic or multiobjective optimization algorithms is intended, so they can be compared using the urban mobility and runtime metrics listed in this work.

References

1. Alizadeh, E., Lejjy, K.R., Amiri, E.: Improving routing in vehicular ad-hoc network with vikor algorithm. In: 9th International Symposium on Telecommunications (IST), pp. 337–341 (2018)
2. Dameri, R.P.: Searching for smart city definition: a comprehensive proposal. International Journal of computers & technology 11(5), 2544–2551 (2013)
3. Haklay, M., Weber, P.: Openstreetmap: user-generated street maps. IEEE Pervasive Comput. **7**(4), 12–18 (2008)
4. Hollander, M., Pledger, G., Lin, P.E.: Robustness of the wilcoxon test to a certain dependency between samples. The Annals of Statistics 2(1), 177–181 (1974)

5. Lopez, P.A., et al.: Microscopic traffic simulation using sumo. In: The 21st IEEE International Conference on Intelligent Transportation Systems. IEEE (2018)
6. MacFarland, T.W., Yates, J.M.: Mann–whitney u test. In: Introduction to Nonparametric Statistics for the Biological Sciences Using R, pp. 103–132. Springer (2016)
7. Macfarlane, J.: When apps rule the road: the proliferation of navigation apps is causing traffic chaos. it's time to restore order. IEEE Spectrum **56**(10), 22–27 (2019)
8. Nha, V.T.N., Djahel, S., Murphy, J.: A comparative study of vehicles' routing algorithms for route planning in smart cities. In: 2012 First International Workshop on Vehicular Traffic Management for Smart Cities (VTM), pp. 1–6 (2012)
9. Pohlert, T.: The pairwise multiple comparison of mean ranks package (PMCMR). R package **27**(2019), 9 (2014)
10. Porru, S., Misso, F.E., Pani, F.E., Repetto, C.: Smart mobility and public transport: opportunities and challenges in rural and urban areas. J. Traffic Transp. Eng. (English edition) **7**(1), 88–97 (2020)
11. Praveen, V., Keerthika, P., Sarankumar, A., Sivapriya, G.: A survey on various optimization algorithms to solve vehicle routing problem. In: 2019 5th International Conference on Advanced Computing Communication Systems (ICACCS), pp. 134–137 (2019)
12. Qian, J., Zhou, Z., Zhao, L., Zhang, J., Li, F.: Accelerating reconfiguration for vlsi arrays with a-star algorithm. IEEJ Transactions on Electrical and Electronic Engineering 13(10), 1511–1519 (2018)
13. Tekouabou, S.C.K., Cherif, W., Silkan, H., et al.: Improving parking availability prediction in smart cities with IoT and ensemble-based model. Journal of King Saud University-Computer and Information Sciences (2020)
14. Zafeiriou, I.: IoT and mobility in smart cities. In: 2020 3rd World Symposium on Communication Engineering (WSCE), pp. 91–95. IEEE (2020)
15. Zhou, Y., Zeng, J.: Massively parallel a* search on a GPU. In: Proceedings of the AAAI Conference on Artificial Intelligence, vol. 29 (2015)

How to Automatically Prove a Time Series Convergence to the Gumbel Distribution?

Amal Mateur(✉), Nesrine Khabou, and Ismael Bouassida Rodriguez

ReDCAD, ENIS, University of Sfax, Sfax, Tunisia
mateuramal@gmail.com, {nesrine.khabou,bouassida}@redcad.org

Abstract. As researchers are increasingly interested in natural disasters and economic problems, developing solutions to predict that is highly recommended. In this paper, we focus on the use of the Extreme Values Theory (EVT) especially the Gumbel distribution, for modelling and predicting the maximum data of extreme events in a time series. To achieve this goal, first we have to demonstrate that the time series formed by the maximum values converges to the Gumbel distribution. Then by applying several fitting methods such as the method of mayer, the method of moments and finally the method of least squares, we are able to estimate the Gumbel parameters to be used for prediction. The performance of the fitting methods was tested using specific measures of accuracy.

Keywords: Extreme value theory · Gumbel distribution · Linear curve fitting · Fitting methods · Manual fitting · Automation fitting

1 Introduction

In recent years, a great deal of research and scientific interest has been devoted to expanding the range of possibilities for predicting the strongest event in order to reduce the damage it causes. The ability to detect sudden events is frequently considered as one of the greatest scientific achievements. In that way, the Extreme Value Theory (EVT) was used to predict the probability of occurrence of rare events [1]. This approach can also be used to treat the maximum data of extreme events that are considered as peaks in non-stationary time series. As an example, in hydrology, the Extreme Value Theory can be used to predict floods of greater magnitude which has been observed in the past, in metrology to study wind speed, and finally in finance to predict crashes and extreme loss situations. Suppose $X_1, X_2, \ldots, X_n$ is a sequence of independent and identically distributed (iid) random variables with a distribution function F, the approach of extreme Value focuses on the behavior of:

$$X_i = max\{X_1, X_2, ..., X_n\} \tag{1}$$

Following the Fisher-Tippett-Gnedenko theorem, the distribution of X_i can only converge to one of the three possible distributions, namely the Gumbel distribution, the Freshet distribution, or the Weibull distribution. They can all be combined into one

L. Barolli et al. (Eds.): AINA 2022, LNNS 449, pp. 353–363, 2022.
https://doi.org/10.1007/978-3-030-99584-3_31

form: the generalized extreme value distribution (GEV) [2]. The cumulative distribution function (CDF) of the GEV distribution is:

$$F(x,\xi,x_0,S) = exp(-(1+\xi(\frac{x-x_0}{S})^{\frac{-1}{\xi}}) \tag{2}$$

The three parameters ξ, x_0 and S are the shape, the location and the scale respectively. The cumulative distribution function of the extreme values is as follows:

- When the shape parameter, $\xi = 0$ the distribution corresponds to the Gumbel (Type I):

$$F(x) = \{exp(-exp(-\frac{x-x_0}{S}))\}, x \in \mathbf{R} \tag{3}$$

  where

$$u = \frac{x-x_0}{S} \tag{4}$$

  is the reduced variable.
- When the shape parameter, $\xi > 0$ the distribution corresponds to Freshet (Type II):

$$F(x) = \begin{cases} 0, & x \leq 0; \\ exp(-\frac{S}{x})^{\xi}, & x > 0\,,\, \xi > 0; \end{cases} \tag{5}$$

- When the shape parameter, $\xi < 0$ the distribution corresponds to Weibull (Type III):

$$F(x) = \begin{cases} 1 - exp(-\frac{x}{S})^{\xi}, & x \leq 0,\, \xi > 0; \\ 1, & x > 0\,; \end{cases} \tag{6}$$

However, the shape parameter ξ of a probability distribution is often more difficult to estimate accurately than the location and the scale parameter [3].
For this reason, we propose to use the Gumbel distribution to fit the maximum values because this parameter ξ is equal to 0 unlike the Freshet and Weibull. Furthermore, the Extreme Value Theory states that the Gumbel distribution is useful for modeling extreme value of random data due to its simplicity. In hydrology, the Gumbel probability distribution is used to analyze variable such as monthly and maximum values of daily rainfall and river discharge volumes [4]. This distribution is also used to predict the strongest events such as earthquakes, floods and other natural disasters.
The mathematical formula for the Gumbel distribution is known by the equation:

$$G(x) = \frac{1}{S}e^{-u}e^{-e^{-u}} \tag{7}$$

In this case, x_0 and S are the parameters of the Gumbel distribution. x_0 denotes the location parameter and S denotes the scale one.
The Extreme Value Theory is based on the maximum block approach (MB), in which the observation is divided into periods of equal size, then limiting the attention to the maximum observation in each period. The distribution of these maximums is expected to follow an extreme value distribution [5].
To obtain the series formed by the maximum values, we should follow different steps:

1. Decomposing the original time series formed by data values (water consumption values, for example) into M blocks.
2. For each block, M, the maximum value is recorded.

$$X_i = Max\{X_1 ....X_M\} \tag{8}$$

3. Sorting the maximum values in ascending order and assigns a rank r to each value.
4. Performing the necessary calculation of the empirical frequency for each rank r. The empirical frequency is thus computed as:

$$F(x_r) = \frac{r-0.5}{n} \tag{9}$$

5. Calculating the Gumbel reduced variable U_i expressed by:

$$U_i = -ln(-ln(F(x_r))) \tag{10}$$

We finish this step by a graphical representation of pairs (U_i, X_i) of the series to be fitted. To estimate the parameters x_0 and S of the Gumbel distribution, we use a Linear Curve Fitting. In regression analysis, Linear Curve Fitting or Linear Regression, describes the process to specify the model which provides the best-fitting straight line for a data. Since the equation of a generic straight line is always given by:

$$X = US + x_0 \tag{11}$$

The question is: which S and x_0 will give us the best fitting line for the data? Previously, in order to find this straight line, a manual drawing is done by a well-trained scientist or engineer. But when we try to fit a smooth curve manually, we draw only a portion of the curve based on a relatively small number of points without taking into account the whole set of points. This problem impact the results of prediction [6]. For that reason, in this paper, we propose to automate the process of Linear Curve Fitting to the Gumbel distribution by using several mathematical methods. The rest of the paper is organized as follows: in Sect. 2, we give an overview of the different fitting methods, including the method of moments, the method of mayer and finally the method of the least squares. In Sect. 3, we detail our contribution by presenting our approach that consists on using different automated mathematical methods for finding the best-fitting curve of the time series to the Gumbel distribution. Furthermore, we present some accuracy measures to select the best fitting method and to evaluate the performance of these prediction methods. The last section provides a description of an executing scenarios of our work, followed by some results. The paper enclosed with a conclusion and gives some directions for future work.

2 Related Work: Distribution Fitting Methods Review

The process of choosing a probability model that can estimate the parameters that best fit the data is called fitting distribution. There are several statistical techniques which could be used to estimate the parameters of the Gumbel distribution, namely: (i) the method of moments, (ii) the method of maximum likelihood, (iii) the method of Regression, (iv) the method of kernel density estimation, (v) the method of the least square, (vi) The manual fitting and finally the method of mayer. For the sake of shortness, we will describe some of them.

Smail et al. [7] proposed the prediction of the Gumbel's parameters using the methods of probability weighted moments, method of moments and finally maximum likelihood method. The comparability of these methods was explored via simulations involving various sample size. The study proved that the probability weighted moments give more accurate estimates, followed by the method of maximum likelihood, but the method of moments performs less satisfactorily.

The purpose of the study performed by Hong et al. [8] is to describe the implementation of Gumbel distribution to fit annual maximum wind speed or wind velocity pressure to estimate Gumbel's parameters. Various methods were applied such as method of moments, the method of maximum likelihood (MLM), the method of L-moments and the Lieblein Blue (generalized least squares method GLSM). Results from the simulation analysis indicate that in terms of efficiency, MLM is slightly better than GLSM. In terms of Bias, GLSM and MLM are comparable, the two previous methods are better than the method of moments and the method of maximum likelihood.

The study performed by Vidal, Ignacio [9] proposed the Gumbel distribution function for modeling and predicting annual rainfall maximum intensities in different geographical zones of Chile. This article outlines the notion of return period, it gives the estimated time interval between events of a similar size or intensity. Gumbel's parameters were computed using a moment-based estimator capable of being updated automatically without the need for intervention of an expert.

To carry out the problem related to perforation and leakage accidents of oil, Xinsheng et al. used Gumbel distribution to predict the remaining life of oil and gas pipeline to ensure the safe operation of pipelines [10]. In this paper, the visual adjustment is used to estimate the parameters of Gumbel. To calculate the expected value of each point, with x_i as the abscissa and y_i as the ordinate, a manual drawn is using resultant a line will pass through the given points and will appear smooth and natural.

To conclude, the Gumbel distribution is one of the extreme value distributions used to estimate extreme values in a time series. Different types of research and studies aims to suggest several mathematical methods to improve the implementation of the Gumbel distribution through the automation in the phase of Linear Curve Fitting. Based on the previously mentioned studies, we conclude that fitting based on mathematical methods are more appropriate than the manual fitting. Ignacio Vidal has chosen to work with the method of moments, unlike the method of maximum likelihood because the moment method is fairly simple and easy to use. In our work, we choose to apply the method of moments because it is the most known method, gives a good result in some case. The method of mode gives low quality according to Varela et al. [11]. In addition to that, we decided to work with the least squares method, which is mentioned by Hong et al. [8]. But also, we have chosen the method of mayer which has not been mentioned in previous research, but has been proved that it can provide the best results in some fields.

3 An Automated Mathematical Method for Gumbel Distribution Fitting

The purpose of this paper is to perform an algorithm that studies different fitting methods to enhance the quality of prediction. Our contribution is divided into two main phases.

1. Gumbel distribution fitting methods: in this phase the main objective is to apply three mathematical methods, namely the method of moments, the method of the least squares and the method of mayer.
2. Performances of the fitting methods: Our contribution is to test the performance of different fitting methods using specific measures of accuracy.

3.1 Gumbel Distribution Fitting Method

In this section, a detailed description of the Gumbel distribution fitting method is provided.

3.1.1 Method of Moments

Many numerical methods have been derived and developed in the past to improve the quality of prediction. The method of moments was first developed by Karl Pearson in 1902; it is used to find point estimators. In statistics, the method of moments considered as a way to estimate the parameters of a frequency distribution by first computing as many moments of the distribution as there are parameters to be estimated [7]. Four moments are commonly used:

- The mean, which indicates the central tendency of a distribution.
- The standard deviation, which indicates the width or deviation.
- Skewness, which indicates any asymmetric 'leaning' to either left or right.
- Kurtosis, which indicates the degree of central 'peakedness' or, equivalently, the 'fatness' of the outer tails.

The parameters of Gumbel distribution x_0 and S can be expressed by the following formulas:

$$x_0 = \mu - S\gamma \tag{12}$$

$$S = \frac{\sqrt{6}}{\pi}\sigma \tag{13}$$

Where μ is the mean, $\gamma = 0.577215$ (the Euler-Mascheroni constant) and σ is the standard deviation. The formula to estimate the simple mean μ and standard deviation σ are as follows:

$$\mu = \sum_{i=0}^{n} \frac{x_i}{n} \tag{14}$$

$$\hat{\sigma} = \sqrt{\frac{1}{n-1} \cdot \sum_{i=1}^{n} (x - \bar{x})^2} \tag{15}$$

3.1.2 Method of the Least Squares

Like the method of moments, the method of the least squares is the most popular method used to determine the position of the trend line of a given time series [12]. In statistics, the least squares method is used to predict the true value of some quantity based on a consideration of error in an observation. In this method, the fitted value of the true (observation) variable is computed to satisfy the following condition [12]:

1. The sum of the deviations of y from their corresponding trend values is zero. i.e.,

$$\sum(y-\hat{y})=0 \tag{16}$$

2. The sum of the square of the deviations of the values of y from their corresponding trend values is the least. i.e.,

$$\sum(y-\hat{y})^2 \tag{17}$$

is least.

The equation of the fitted line can be expressed as:

$$X = SU + x_0 \tag{18}$$

To find the equation of the fitted line that best fit a set of data points (x_1, y_1), $(x_2, y_2)\ldots$, (x_n, y_n) we use the following steps:

1. Calculate the mean of the x -values and the mean of the y -values:

$$\bar{X} = \sum_{i=0}^{n} \frac{x_i}{n}; \bar{Y} = \sum_{i=0}^{n} \frac{y_i}{n} \tag{19}$$

2. Calculate the slope of the trend line:

$$S = \frac{\sum_{i=0}^{n}(x_i-\bar{X})(y_i-\bar{Y})}{\sum_{i=0}^{n}(x_i-\bar{X})^2} \tag{20}$$

3. Compute the y -intercept of the line by using the formula:

$$x_0 = \bar{Y} - S\bar{X} \tag{21}$$

4. Use the slope S and the y -intercept x_0 to form the equation of the line.

3.1.3 Method of Mayer

This fitting method is simple to be set up than the other method, it is easy to calculate and intuitive to understand. The method of mayer is sometimes called the double mean method, or the Average Point Method, consists of dividing a set of data observations into two groups of equal frequencies. The second step is to calculate the mean of the data in each group; these means will be called M1 and M2. Finally, we draw the line that joins these two points; this line passes through the center of the scatter plot.

3.2 Performances of the Fitting Methods

To verify the performance of the fitting methods, some specific measures of accuracy were used, such as the Mean Squared Error (MSE) and the Error Rate (ER).

3.2.1 Mean Squared Error (MSE)

The Mean Squared Error MSE is a measurement of how close a fitted line is to the data. In Linear Regression, when we try to find lines that best fit a set of data, we use the Mean Squared Error. However, the error in prediction is represented by the distance between the data observations and the fitted values. The Mean Squared error is calculated as the average of the squares of the errors. The line that gives minimal Mean Squared Error is considered as the best fit. The Mean Squared Error formulas is given by the equation:

$$MSE = \frac{1}{n}\sum_{i=0}^{n}(y_i - \hat{y}_i)^2 \tag{22}$$

General steps to calculate the MSE from a set of X and Y values:

1. Finding the regression line.
2. Inserting your X values into the linear regression equation to find the new values $\hat{y}$.
3. Subtracting the new $\hat{y}$ value from the original to get the error.
4. Squaring the errors.
5. Adding the errors (the $\sum$ in the formula is summation notation).
6. Finding the mean.

3.2.2 Error Rate (ER)

The Error Rate ER is a value proposed by us, which was not previously used, and serves to calculate the error margin of the fitting function. The Error Rate can be used to demonstrate if the time series formed by the maximum values converges to the Gumbel distribution. To be specific, the ER is defined as the percentage of the number of points in which the difference between the actual value and the predicted value is greater than the Percentage Error (PE) of the actual value's absolute, as indicated in the Fig. 1. The number of PE is a percentage that will be selected by the user (In our work, PE is set by an expert at 20%.). The following is the formula for the ER:

$$P = \sum_{i=1}^{n}\left[|\hat{y}_i - y_i| > |\hat{y}_i \times \frac{PE}{100}|\right] \tag{23}$$

$$ER = P \times \frac{100}{n} \tag{24}$$

Where:

- Y is the actual value
- $\hat{Y}$ is the forecast value.
- The difference between Y and $\hat{Y}$ is the prediction residue that will be compared with the 20% of the forecast value.
- P is the number of squares that have an error greater than 20% of the predicted value. The P divided by the number of fitted points n multiplying by 100% makes it an error rate.

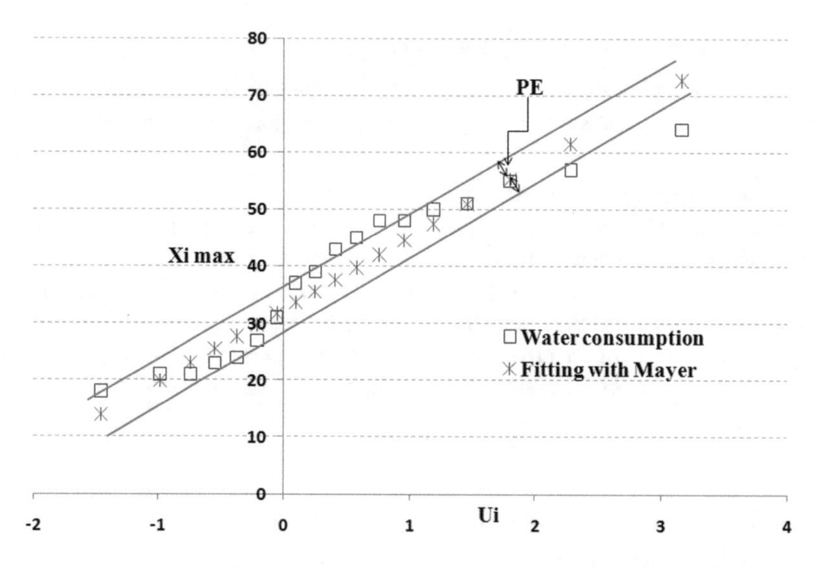

Fig. 1. Error rate illustration.

4 Illustration and Experimentation

In this section, we introduce the different steps for the Gumbel distribution fitting. We also describe our proposed work using a data collection of water consumption.

4.1 Data Collection

This study was conducted using a total of 35 years of monthly water consumption data from (1982–2017) that were obtained from National Company of exploitation and distribution of water Tunisia.

4.2 Different Steps for Gumbel Distribution Fitting

In order to illustrate our proposed work, we present the different steps to be performed in order to demonstrate that the time series formed by the maximum values converges to Gumbel distribution. Our proposed work includes the following steps:

1. Doing the necessary steps to obtain the pairs (U_i, X_i) as we explained in the introduction.
2. Performing the three fitting methods.
3. Calculating the Mean Squared Error MSE and the Error Rate ER for each fitting method.
4. Choosing the method that has the least Mean Squared Error. If the Error Rate ER for this method is less than Percentage Error PE (in our work, PE is set by an expert at 20%), so the series formed by the maximum values converges to the Gumbel distribution. Return the slope and the intercept of the line, MSE, ER, in order to use them for prediction.

4.3 Results and Discussion

In this section, we present a comparison between the results of the different fitting methods.

4.3.1 Series Fitting

When we apply different fitting methods as indicated in the Fig. 2a, Fig. 2b and Fig. 2c, we get a straight line that best fit a set of data observations. The existing three fitting methods were used to estimate the unknown parameter of the Gumbel Distribution, x_0 and S. If we compare the three results, we found that the series formed by the maximum values converges to the Gumbel distribution, as each time we obtained an error rate less than 20 % as indicated in the Table 1. However, The three results are nearly similar, so in order to select the suitable method, we choose the lowest MSE. Table 1 states that the minimum parameter of MSE is 17.42% for the least squares method. Therefore, for the prediction, we will use the following parameters:

- S=10.77
- x_0=33.53

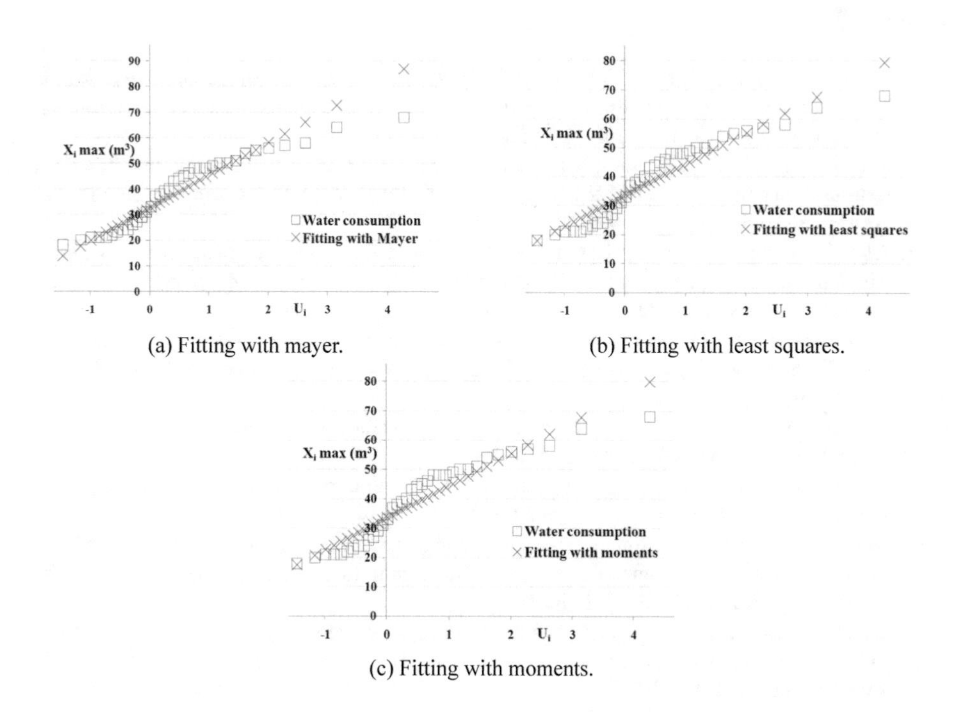

(a) Fitting with mayer.

(b) Fitting with least squares.

(c) Fitting with moments.

Fig. 2. The three fitting methods.

Table 1. Comparative table for the quality of the different results

	S	x_0	MSE (%)	ER (%)
The method of the least squares	**10.77**	**33.53**	**17.42**	**11**
The method of mayer	**12.77**	**32.39**	**23.57**	**5**
The method of moments	**10.90**	**33.36**	**17.45**	**8**

4.4 Conclusion

In this paper, we have outlined three mathematical methods, namely the method of moments, the method of the least squares and finally the method of mayer, in order to automate the process of Linear Curve Fitting for prediction purposes. To select the suitable fitting method, we have specified two accuracy measures, such as the Mean Squared error (MSE) and the error rate (ER). As future work, we are interested in studying other fitting methods, and focusing on cases in which the time series formed by the maximum values does not converge to Gumbel distribution.

References

1. Chen, S., Chai, L., Xu, K., Wei, Y., Rong, Z., Wan, W.: Estimation of the occurrence probability of extreme geomagnetic storms by applying extreme value theory to aa index. J. Geophys. Res. Space Phys. **124**(12), 9943–9952 (2019)
2. Li, M., Bi, Z., Zhou, D., Zeng, X.: Analog circuit performance bound estimation based on extreme value theory. In: 2015 IEEE 58th International Midwest Symposium on Circuits and Systems (MWSCAS), pp. 1–4 (2015)
3. Chan, J., Choy, B., Walker, S.: On the estimation of the shape parameter of a symmetric distribution. In: Stephen, On the Estimation of the Shape Parameter of a Symmetric Distribution, 1 January 2018 (2018)
4. Koutsoyiannis, D.: On the appropriateness of the gumbel distribution for modelling extreme rainfall. In: Proceedings of the ESF LESC Exploratory Workshop held at Bologna, 2003, pp. 24–25 (2003)
5. Ferreira, A., De Haan, L.: On the block maxima method in extreme value theory: Pwm estimators. Ann. Stat. **43**(1), 276–298 (2015)
6. Akima, H.: A new method of interpolation and smooth curve fitting based on local procedures. J. ACM (JACM) **17**(4), 589–602 (1970)
7. Mahdi, S., Cenac, M.: Estimating parameters of gumbel distribution using the methods of moments, probability weighted moments and maximum likelihood. Revista de Matemática: Teoría y Aplicaciones **12**(1–2), 151–156 (2005)
8. Hong, H., Li, S., Mara, T.: Performance of the generalized least-squares method for the gumbel distribution and its application to annual maximum wind speeds. J. Wind Eng. Ind. Aerodyn. **119**, 121–132 (2013)
9. Vidal, I.: A bayesian analysis of the gumbel distribution: an application to extreme rainfall data. Stoch. Env. Res. Risk Assess. **28**(3), 571–582 (2014)
10. Zhang, X.-s., Cao, N.-n., Wang, X.-w.: Residual life prediction of oil and gas pipeline based on gumbel distribution. China Safety Sci. J. (2015)

11. Gorgoso-Varela, J.J., Rojo-Alboreca, A.: Use of gumbel and weibull functions to model extreme values of diameter distributions in forest stands. Ann. For. Sci. **71**(7), 741–750 (2014)
12. Molugaram, K., Rao, G.S.: Chapter 12 - analysis of time series. In: Molugaram, K., Rao, G.S. (eds.) Statistical Techniques for Transportation Engineering, Butterworth-Heinemann, 2017, pp. 463–489 (2017). https://www.sciencedirect.com/science/article/pii/B978012811555800012X

A Machine Learning-Based Model for Predicting the Risk of Cardiovascular Disease

Chiu-Han Hsiao[1(✉)], Po-Chun Yu[2], Chia-Ying Hsieh[3], Bing-Zi Zhong[3], Yu-Ling Tsai[3], Hao-min Cheng[4], Wei-Lun Chang[4], Frank Yeong-Sung Lin[3], and Yennun Huang[1]

[1] Research Center for Information Technology Innovation, Academia Sinica, Taipei, Taiwan
chiuhanhsiao@citi.sinica.edu.tw
[2] Department of Mathematics, National Taiwan Normal University, Taipei, Taiwan
yddeuy1732@citi.sinica.edu.tw
[3] Department of Information Management, National Taiwan University, Taipei, Taiwan
{r10725059,r10725045,r10725053}@ntu.edu.tw
[4] Center for Evidence-Based Medicine, Taipei Veterans General Hospital, Taipei, Taiwan

Abstract. A growing number of medical studies have used deep learning and machine learning for the modeling and early prediction of cardiovascular disease (CVD) risk. Modern hospitals have constructed sizeable medical data sets to predict abnormal blood pressure (BP), abnormal heart vessels, and other cardiac indicators. However, hypertension has also been demonstrated to be a risk factor for cardiovascular disease and stroke. In this paper, machine learning-based and statistic-based approaches were applied to medical data to significantly identify the disease to prevent serious illness. Furthermore, lightweight BP monitoring devices that can be used at home have enabled regular BP monitoring to predict CVD risks for early treatment.

Keywords: Artificial intelligence · Machine learning · Hypertension · Cardiovascular disease

1 Introduction

Hypertension is a severe public health problem worldwide. Researchers have gained a better understanding of the pathophysiology of hypertension, and more effective treatment and prevention strategies have been implemented [1]. The Framingham Study is a prototype for the natural history of hypertension risk assessment [2]. Although major clinical guidelines advocate evidence-based treatments, these guidelines are general ones and insufficiently specific to the pathophysiology of individual patients with poorly controlled blood pressure (BP).

L. Barolli et al. (Eds.): AINA 2022, LNNS 449, pp. 364–374, 2022.
https://doi.org/10.1007/978-3-030-99584-3_32

However, novel clinical trials can improve the detection, evaluation, treatment, and prevention of hypertension and improve adherence to treatment.

1.1 Problem and Challenge

Hypertension is a common health condition that can lead to severe illnesses such as stroke, heart disease, and renal failure [3]. Risk assessment of the disease is highly complicated and must account for numerous factors and transient environmental conditions that can artificially raise BP readings [4]. Risk factors for hypertension commonly include age, gender, body mass index (BMI), obesity, stress, triglyceride levels, uric acid level, lipoprotein levels, cholesterol levels, smoking habits, and family history of the disease [5–7]. Due to the complexity of disease prediction, evaluations by clinicians may be erroneous, especially in cases with incomplete or noisy data [5]. Therefore, approaches to hypertension risk management have focused on the overall level of risk and have used conventional statistics and are based on an assessment of several risk factors instead of on identifying risk factors and observing how they interact in specific environments. Hypertension management guidelines have classified cardiovascular risk into different categories based primarily on the results of multicenter trials or meta-analyses with quality rating scales. Artificial intelligence (AI) is typically not used.

1.2 Research Motivation and Related Work

Thus, this paper proposes the use of an AI model based on the analysis of the relationships between traditional risk factors for hypertension. Hypertension is a risk factor for cardiovascular disease (CVD) [8]. Many international guidelines recommend regular BP monitoring to identify when a patient has systolic pressure (SBP) <140 mmHg and diastolic pressure (DBP) <90 mmHg [9]. Many external factors affect the accuracy of BP measurements, such as white coat syndrome. In contrast to 24-h dynamic BP monitoring, an average of six clinically collected BP readings had a low accuracy rate of 80% [10]. Moreover, the incidence of heart attack or stroke can be attributed to the onset of hypertension [11]. Therefore, over time, the average of multiple BP readings may not reflect the risk of CVD to patient with hypertension. Therefore, visit-to-visit BP variability (BPV) has also been demonstrated to be a predictor of CVD [12–15]. The standard deviation (SD) of BP readings over time is a common method of quantifying BPV. The method rests on the assumption that BP readings oscillate around a steady state; however, many studies have demonstrated that BP increases with age. To improve the accuracy of long-term BPV as a risk indicator, some studies have proposed using regression to compensate for long-term BP trends due to aging or medication. CVD includes myocardial infractions, atrial fibrillation, heart failure, and strokes. The risk factors for CVD occurrence include race, ethnicity, age, sex, weight, height, BMI, and blood test results for kidney function, liver function, and cholesterol levels [16]. Furthermore, home BP (HBP) levels were reported to be a strong predictor of future high BP and cardiovascular discomfort [17,18]. On the basis of such clear evidence, recent hypertension guidelines consistently recommend that HBP monitoring should be routinely used in clinical practice [19]. In addition to average HBP, daily HBP variability (HBPV) is increasingly recognized as an essential prognosis factor for adverse cardiovascular outcomes. Data from the J-HOP Study (Japan Morning Surge-Home Blood Pressure), which is a prospective observational study of outpatients with a prior history of or risk factors for CVD who performed HBP monitoring in the morning and evening over 14 d, was used to examine the following: the association between the maximum mean value of HBP and CVD event risks and whether the association was independent of the mean HBP or the conventional day-to-day HBPV indices [20,21]. This study hypothesized that maximum mean HBP would significantly predict future CVD events independent of the mean HBP level and conventional HBPV indices.

1.3 Research Scope

In this paper, the medical decision support systems are proposed to integrate machine learning to deliver efficient and accurate results that doctors can use to better diagnose and treat diseases. Combined with data obtained early in a patient's treatment cycle, these technologies can also enable timely interventions. It is a top priority of healthcare providers aiming to provide preventive care to these patients and to reduce overall costs. Neural networks meet stringent clinical demands and provide valuable insight. AI elucidates the factors that may affect the control of BP for prescribers and patients. One possible limitation of AI-based predictive models is that analyzing the causal relationship between risk factors and disease occurrence is challenging because AI models are usually black-box models. However, by learning complex patterns in big data, AI-based predictive models often improve the accuracy of disease predictions.

2 Methods

2.1 Study Design

This study's model is constructed by a private dataset using the Taiwan Consortium of Hypertension-associated Cardiac Disease (TCHC) data set [22], which is included by 11 medical centers in Taiwan and is a nonprofit research alliance focusing on hypertension and hypertension-related disease clinical trials and research collaboration. The dataset is based on a health study in Taiwan with nearly 2200 participants; 80% (1700 items) and 20% (500) of the data set were randomly selected and used for training and testing, respectively. Because CVD occured in only 25% of patients in the data set, 80% of the CVD-positive and 80% of the CVD-negative data sets were used for training. The remaining 20% of the items were used for testing. TCHC contained Cardiovascular disease records, blood pressure data, medication records, lifestyle interviews, basic health information, and Other disease records. It is beneficial to understand not only the impact of measurable health values on cardiovascular disease, but also whether there is an interaction between different diseases. Several approaches were used: logistic regression (LR), deep neural networks (DNNs), random forest (RF), light gradient boosting machine (LightGBM), eXtreme gradient boosting (XGboost), decision trees (DTs), k nearest neighbor (KNN), Adaboost, gradient boosting (Gboost), DT bagging (DTB), knn bagging (KNNB), and RF bagging (RFB). The characteristics of the model are presented in Table 1 for comparison [23].

Table 1. Model comparison

Model	Advantage	Limitation
Artificial neural network	• Easy to visualize • Easy to use • General use to most of the dataset	• Prone to overfitting • Can be difficult to identify causal relationship
Random forest	• Fast training and prediction speed • Scalable (i.e., very large datasets) • Good for non-linear modeling	• Prone to overfitting • Time-consuming
Decision Tree	• Fast training and prediction speed • Easy to interpret	• Prone to have poor predictive accuracy if there is an overlap problem • Prone to overfitting • Prone to ignorance of a variable if small sample size
Ensemble learning (i.e., boosting, bagging)	• Less prone to overfitting • Can be used for a combination of results from different algorithms	• Sensitive to outliers or uniform noise • Weak classifiers

2.2 Data Processing

AI is an extremely powerful tool for predictions, but its reasoning and the justification behind feature selection can be opaque. Some regression models aided the choice of good BP features that were suitable for predicting CVD. The main features are as follows.

Non-BP-related Features

- Demographics: Age, gender, weight, height, BMI, and waist circumference
- Habits: Smoking, alcohol, and exercise habits
- Disease: Hypertension (HT), diabetes mellitus (DM), hyperlipidemia (HL), panic, anger, blue, and cancer
- Treatment: (Antihypertensive drug)Anti-BP and aspirin use
- Others: GOT, GPT, glucose, cholesterol, LDL, and HDL

BP-related Features

- SBP: 7 day average, SD, coefficient of variation (CV), average real variability (ARV), and maximum and minimum values of MEave
- DBP: 7 day average, SD, CV, ARV, and maximum and minimum values of MEave
- Home central morning SBP, morning DBP, evening SBP, and evening DBP
- Office BP: Central SBP, peripheral SBP, Central DBP, and peripheral DBP

AGE, Gender, Weight, Height, BMI, Waist are easily attached and common features for medical model. The habits (Smoking, Alcohol, Exercise) are effective to ones health condition. As we mentioned, diseased may affect each other, so we included Hypertension, Diabetes mellitus, Hyperlipidemia, Cancer, Panic, Anger, Blue as features. Glutamic oxaloacetic transaminase (GOT), serum glutamic pyruvic transaminase (GPT), Glucose are risk factors to health. Low-density lipoprotein, sometimes called "bad" cholesterol (LDL) and high-density lipoprotein, or "good" cholesterol (HDL) are important parameter of Cholesterol. The goal was to predict CVD occurence by using BP data. However, BP data are complex. Therefore, the means of the SBP and DBP data are used instead. Some common CVD-related diseases were also analyzed. Age-based regression was used for missing data. Patients with and without CVD in the data set were labeled as 1 and 0, respectively. Similarly, patients with and without HT, DM, HL, and cancer were also labeled as 1 and 0, respectively. For numeric data, the original data were used. We use feature scaling method to standardize the influence of each feature. For categorical data, one-hot encoding was used to transform the features to be trainable. In addition, we use recursive feature elimination (RFE) based on logistic regression and Principle Component Analysis (PCA) to perform feature selection. The main features are listed as Table 2.

Table 2. Main features summary

#	Features	Type (Mean ± standard deviation)
1	CVD	Binary (0,1)
2	AGE	Numeric (1.27 ± 0.52)
3	Gender	Binary (0,1)
4	Weight	Numeric (70.60 ± 15.69)
5	Height	Numeric (161.50 ± 11.63)
6	Waist	Numeric (84.49 ± 11.26)
7	MEave of SBP	Numeric (129.86 ± 11.58)
8	MEave of DBP	Numeric (77.57 ± 8.45)
9	SD of the MEave of SBP	Numeric (4.98 ± 2.53)
10	SD of the MEave of DBP	Numeric (3.18 ± 1.52)
11	CV of the MEave of SBP	Numeric (0.05 ± 0.03)
12	CV of the MEave of DBP	Numeric (0.04 ± 0.02)
13	ARV of the MEave of SBP	Numeric (5.71 ± 3.11)
13	ARV of the MEave of DBP	Numeric (3.64 ± 1.92)
14	Smoking years	Categorical (1 to 12)
15	Alcohol years	Categorical (1 to 12)
16	Exercise	Categorical (1 to 5)

Some important characteristics are defined as follows: The average of morning and evening value (MEave) BP values, the SD of the MEave of home SBP and DBP, the CV of the MEave of home SBP and DBP, the ARV of the MEave of home SBP and DBP, and the variability independent of the mean (VIM) of the MEave of home SBP and DBP [20]. Despite the inclusion of BP data, features related to CVD were chosen to construct the model. Feature selection was based on past research [24]. The TCHC includes systolic and diastolic BP for each patient four times per day for seven days. Linear regression was used to reconstruct missing data [25]. The method is accurately identifies BP characteristics, variability, and trends. However, individuals may be at risk of CVD but not identified. Therefore, the BP data was used to classify CVD risk as low, medium, or high. Furthermore, silhouette coefficients, the Davies–Bouldin index, and the Calinski–Harabasz score were used to verify the accuracy of the model.

3 Experiments and Results

3.1 Model

Several models were used to predict CVD occurence. Clustering was also used to identify individual risk levels. K-means clustering was used to classify the SD, CV, and ARV of BP data; furthermore, silhouette coefficients, the Davies–Bouldin index, and the Calinski–Harabasz score were used to verify the clustering models.

Models with an accuracy of greater than 90% are listed in Table 3. A conventional LR is included for comparison.

Table 3. Model comparison with results

Model	Accuracy	Recall	Precision	F1 score	AUC
LR	0.8151 (± 0.010)	0.1218 (± 0.053)	0.6283 (± 0.118)	0.5481 (± 0.042)	0.7225 (± 0.029)
RF	0.9308 (± 0.004)	0.7470 (± 0.057)	0.8817 (± 0.042)	0.8823 (± 0.011)	0.9522 (± 0.007)
LightGBM	0.9016 (± 0.013)	0.5621 (± 0.025)	0.8933 (± 0.002)	0.8148 (± 0.030)	0.9334 (± 0.012)
XGBoost	0.9153 (± 0.010)	0.5878 (± 0.064)	0.9647 (± 0.042)	0.8389 (± 0.024)	0.9528 (± 0.011)
Decision Tree Bagging	0.91293 (± 0.004)	0.81256 (± 0.012)	0.89624 (± 0.015)	0.84546 (± 0.008)	0.92444 (± 0.008)

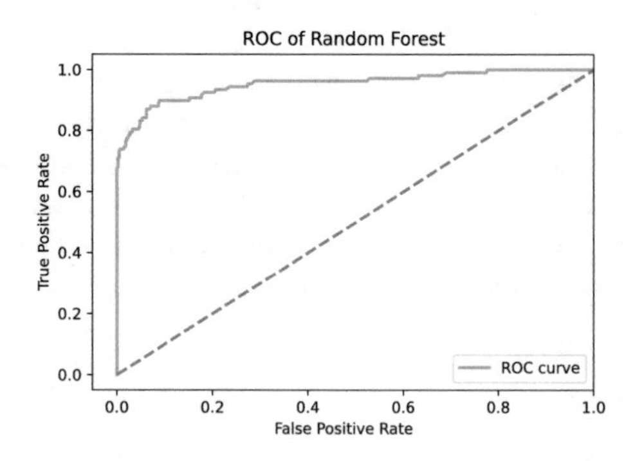

Fig. 1. RF prediction for CVD occurrence.

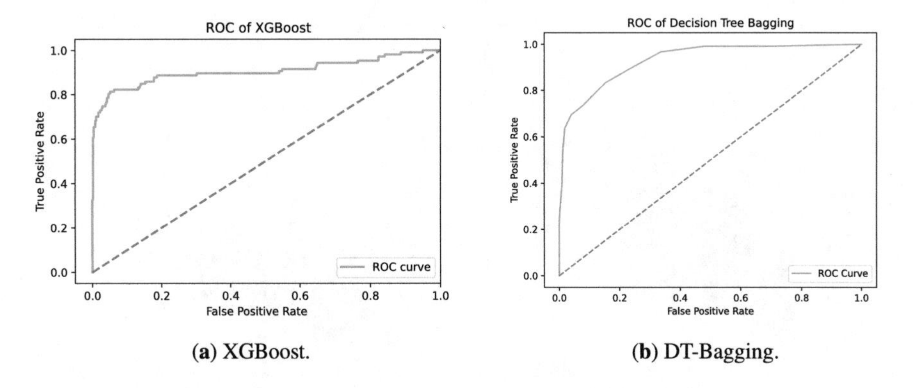

(**a**) XGBoost. (**b**) DT-Bagging.

Fig. 2. XGBoost and DT-Bagging prediction for CVD

3.2 Machine Learning-Based Approaches

The conventional logistic regression analysis had a mere 80% accuracy. The following machine learning models had a >90% accuracy: RF, LightGBM, XGBoost and DTB. RF method had the best area under the receiver curve (AUC) results, learned efficiently, and performed best overall (See Fig. 1). The results for XGBoost and DTB were also excellent shown in Fig. 2a and Fig. 2b.

3.3 Statistic-Based Approaches

Clustering methods were also used to analyze the relationship between CVD and BP data [25]. Three risk levels, low, medium, and high, were identified. The final results are presented in Table 4. Silhouette coefficient is an index to evaluate similarity within the cluster compared to other clusters, and Davies-Bouldin Index measures the data dispersion within the cluster and across the clusters. Besides, Calinski-Harabasz Index is defined as the ratio of the dispersion within-cluster and between-clusters. Therefore, low values of Davies-Bouldin Index and high values of Silhouette Index and Calinski-Harabasz index means a better clustering result to present the characteristics of the data. The results of the clustering reveal that all three clusters exhibited consistency. (See Fig. 3a, Fig. 3b and Fig. 3c). In summary, the seven-day BP ARV clustering resulted in the best BP variation.

Table 4. Model comparison on clustering models

Model	Silhouette coefficients	Davies-Bouldin	Calinski-Harabasz score
K-means (STD)	0.413978	0.833369	3046.916331
K-means (CV)	0.397395	0.8818	2515.808652
K-means (ARV)	0.513993	0.629257	4580.585313

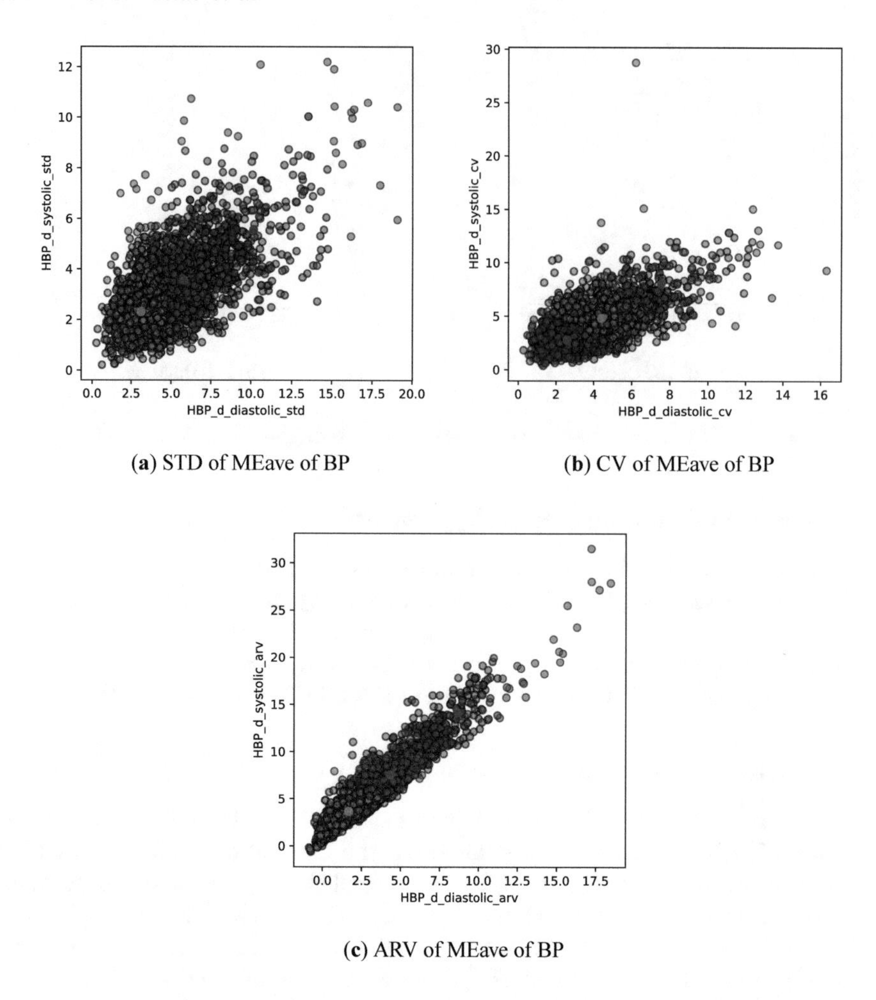

(a) STD of MEave of BP

(b) CV of MEave of BP

(c) ARV of MEave of BP

Fig. 3. K-means clustering with STD, CV and ARV.

4 Discussion

Feature selection was performed following previous studies. The results in this paper were also validated by categorical variables screening, such as the relationship between size, exercise, alcohol consumption, and medication with CVD. Moreover, it is not easy to achieve high accuracy for traditional regression models, such as LR. On the contrary, machine learning models are suitable for dealing with the biologic binary judgment model. The experiments have several suitable models for predicting CVD risks, such as RF, DTB, and xGBoost. According to the K-means result, ARV of BP is more suitable for CVD risk clustering. The goal of this paper was to identify CVD risks. However, the BP records were for only seven days based on the limited dataset size; records for an extended period of 4–6 weeks might yield more accurate results.

5 Conclusions

CVD prediction is an important goal in modern medicine. Predicting CVD risk to enable early treatment can significantly reduce the risk of serious illness. BP is a critical indicator that is measured in most health examinations. Previously, BP is only measured at hospitals or health centers; however, these measurements have become more accessible now, if lightweight BP monitoring devices that can be used at home have enabled regular BP monitoring. Based on the results shown in this paper, a relationship between CVD and BP is revealed clearly. The daily monitoring of BP, recording exercise, alcohol consumption, medication doses, and other data using wearable devices can facilitate the timely accurate treatment of CVD in the future.

Acknowledgement. This work was supported in parts by Ministry of Science and Technology (MOST), Taiwan, under Grant Number MOST 110-2222-E-001-002, 110-2221-E-002-078-MY2, and 110-2321-B-075-002.

References

1. James, P.A., et al.: 2014 evidence-based guideline for the management of high blood pressure in adults: report from the panel members appointed to the eighth joint national committee (jnc 8). JAMA **311**(5), 507–520 (2014)
2. Mancia, G., et al.: 2013 ESH/ESC guidelines for the management of arterial hypertension. Blood Press. **22**(4), 193–278 (2013)
3. Carretero, O.A., Oparil, S.: Essential hypertension. Circulation **101**(3), 329–335 (2000)
4. Pytel, K., Nawarycz, T., Ostrowska-Nawarycz, L., Drygas, W.: Anthropometric predictors and artificial neural networks in the diagnosis of hypertension. In: 2015 Federated Conference on Computer Science and Information Systems (FedCSIS), pp. 287–290 (2015)
5. Kaur, A., Bhardwaj, A.: Artificial intelligence in hypertension diagnosis: a review. Int. J. Comput. Sci. Inf. Technol. **5**(2), 2633–2635 (2014)
6. Samant, R., Rao, S.: Evaluation of artificial neural networks in prediction of essential hypertension. Int. J. Comput. Appl. **81**(12), 34–38 (2013)
7. Srivastava, P., Srivastava, A., Burande, A., Khandelwal, A.: A note on hypertension classification scheme and soft computing decision making system. Int. Scholarly Res. Notices **2013**, 1–11 (2013)
8. Dorans, K.S., Mills, K.T., Li,u Y., He, J.: Trends in prevalence and control of hypertension according to the 2017 American college of cardiology/American heart association (ACC/AHA) guideline. J. Am. Heart Assoc. **7**(11), e008,888 (2018)
9. Crim MT, et al.: National surveillance definitions for hypertension prevalence and control among adults. Circulation: Cardiovascular Quality and Outcomes **5**(3), 343–351 (2012)
10. Stergiou, G.S., Zourbaki, A.S., Skeva, I.I., Mountokalakis, T.D.: White coat effect detected using self-monitoring of blood pressure at home: comparison with ambulatory blood pressure. Am. J. Hypertens. **11**(7), 820–827 (1998)
11. Ruff, R.L., Talman, W.T., Petito, F.: Transient ischemic attacks associated with hypotension in hypertensive patients with carotid artery stenosis. Stroke **12**(3), 353–355 (1981)

12. Rothwell, P.M.: Limitations of the usual blood-pressure hypothesis and importance of variability, instability, and episodic hypertension. The Lancet **375**(9718), 938–948 (2010)
13. Rothwell, P.M., Howard, S.C., Dolan, E., O'Brien, E., Dobson, J.E., Dahlöf, B., Sever, P.S., Poulter, N.R.: Prognostic significance of visit-to-visit variability, maximum systolic blood pressure, and episodic hypertension. The Lancet **375**(9718), 895–905 (2010)
14. Muntner, P., Shimbo, D., Tonelli, M., Reynolds, K., Arnett, D.K., Oparil, S.: The relationship between visit-to-visit variability in systolic blood pressure and all-cause mortality in the general population. Hypertension **57**(2), 160–166 (2011)
15. Shimbo, D., et al.: Associations of aortic distensibility and arterial elasticity with long-term visit-to-visit blood pressure variability: the multi-ethnic study of atherosclerosis (mesa). Am. J. Hypertens. **26**(7), 896–902 (2013)
16. F. Piepoli, M.: 2016 European guidelines on cardiovascular disease prevention in clinical practice. Int. J. Behav. Med. **24**(3), 321–419 (2017)
17. Niiranen, T.J., Hänninen, M.R., Johansson, J., Reunanen, A., Jula, A.M.: Home-measured blood pressure is a stronger predictor of cardiovascular risk than office blood pressure. Hypertension **55**(6), 1346–1351 (2010)
18. Ward, A.M., Takahashi, O., Stevens, R., Heneghan, C.: Home measurement of blood pressure and cardiovascular disease: systematic review and meta-analysis of prospective studies. J. Hypertens. **30**(3), 449–456 (2012)
19. Weber, M.A., et al.: Clinical practice guidelines for the management of hypertension in the community: a statement by the American society of hypertension and the international society of hypertension. J. Hypertens. **32**(1), 3–16 (2014)
20. Fujiwara, T., Hoshide, S., Kanegae, H., Kario, K.: Clinical impact of the maximum mean value of home blood pressure on cardiovascular outcomes: A novel indicator of home blood pressure variability. Hypertension **78**(3), 840–850 (2021)
21. Umemura, S., et al.: The Japanese society of hypertension guidelines for the management of hypertension (JSH 2019). Hypertens. Res. **42**(9), 1235–1481 (2019)
22. Lin, T.T., et al.: Comparison of home and ambulatory blood pressure measurements in association with preclinical hypertensive cardiovascular damage. J. Cardiovasc. Nurs. **34**(2), 106–114 (2019)
23. Parati, G., Pellegrini, D., Torlasco, C.: How digital health can be applied for preventing and managing hypertension. Current Hypertension Rep. **21**(5), 104, 067–104,076 (2019)
24. Joo, G., Song, Y., Im, H., Park, J.: Clinical implication of machine learning in predicting the occurrence of cardiovascular disease using big data (nationwide cohort data in korea). IEEE Access **8**, 157,643–157,653 (2020)
25. Koshimizu, H., Kojima, R., Kario, K., Okuno, Y.: Prediction of blood pressure variability using deep neural networks. Int. J. Med. Inf. **136**, 104,067–104,076 (2020)

A Federated Learning-Based Precision Prediction Model for External Elastic Membrane and Lumen Boundary Segmentation in Intravascular Ultrasound Images

Chiu-Han Hsiao[1(✉)], Tsung-Yu Peng[2], Wei-Chieh Huang[3], Hsin-I Teng[4], Tse-Min Lu[3], Frank Yeong-Sung Lin[2], and Yennun Huang[1]

[1] Research Center for Information Technology Innovation, Academia Sinica, Taipei, Taiwan
chiuhanhsiao@citi.sinica.edu.tw

[2] Department of Information Management, National Taiwan University, Taipei, Taiwan
r09725024@ntu.edu.tw

[3] Division of Cardiology, Department of Internal Medicine, Taipei Veterans General Hospital and School of Medicine, National Yang Ming Chiao Tung University, Taipei, Taiwan
wchuang9@vghtpe.gov.tw

[4] Heart Center, Cheng Hsin General Hospital and School of Medicine, National Yang Ming Chiao Tung University, Taipei, Taiwan

Abstract. Deep learning technologies have been widely used in intravascular ultrasound (IVUS) image analysis. An efficient and real-time tool for IVUS image segmentation is essential during percutaneous coronary intervention (PCI). However, collecting a significant volume of data is infeasible lying to the following reasons: (1) patients' privacy concerns and (2) manually annotating data is time-consuming and laborious. Therefore, training a deep model with generalization capacity using distributed data remains a challenge. In this study, a federated learning-based deep learning model is proposed. It is a framework that enables model training on multi-source data without sharing each local dataset. A lightweight U-Net model is also adopted in the federated learning framework due to the limited volume of data. The results indicate that the lightweight model's performance can almost achieve the performance of a deeper model but significantly spends less time consumption. The proposed lightweight U-Net model parameters are less than 30 times that of the baseline model. The external elastic membrane (EEM) and the lumen boundary segmentation achieved 0.8567 and 0.8457, respectively. The efficiency of implementing such a federated learning framework on a distributed IVUS image segmentation is estimated to be increased by more than four times.

L. Barolli et al. (Eds.): AINA 2022, LNNS 449, pp. 375–386, 2022.
https://doi.org/10.1007/978-3-030-99584-3_33

Keywords: Image segmentation · Federated learning · Intravascular ultrasound

1 Introduction

Medical imaging technologies produce a vast quantity of images of various types. These images are composed in grayscale and may contain irregularly shaped lesions or other features that are difficult to detect; identifying such features manually during the diagnostic process is both challenging and time-consuming. Low-resolution images associated with some imaging technologies, such as ultrasonography, can further complicate accurate diagnoses. Therefore, an effective assistive technology that can annotate regions or biomarkers of potential interest or concern in medical images may provide more rapid and accurate diagnoses based on the data available from medical imaging studies. Convolutional neural networks are able to capture image features, and deep learning models with convolutional layers have been proposed for use in various image analysis applications [1].

1.1 IVUS for Coronary Artery Disease

This study proposes a modified U-Net model for enhanced image segmentation in intravascular ultrasound (IVUS) images. IVUS is an imaging modality used in the diagnosis of coronary artery disease (CAD) and one that can provide images containing information related to pathology in real-time [2]. Because of the low resolution of IVUS images, an automatic annotation tool to highlight arterial borders and provide clear information related to the spatial orientation of the images would result in a more effective interpretation of those images. Accordingly, adopting an area segmentation model to process these images during diagnosis would substantially increase treatment efficiency.

1.2 Related Work

To construct a segmentation model with more favorable generalization, collecting a large volume of data is required to avoid overfitting the model. However, training a model in such a supervised learning task requires paired data with raw and labeled images. Manually performing the image segmentation for such a process is time-consuming and laborious, and increasing the number of labeled images is therefore challenging. Furthermore, gathering such a large volume of data in a single location requires adequate hardware resources for storage, which is impractical. Moreover, because data in specific fields such as medicine and finance may contain private personal information, the collection of such sensitive data is necessarily highly regulated, and thereby collecting data from multiple sources or institutions is nearly impossible [3].

A novel concept called federated learning might solve the issue. Federated learning is a distributed machine learning framework that preserves data privacy during the training of machine learning models involving the use of large amounts of decentralized data. This framework enables multiple agents to be trained collaboratively through a central model without any data exchange [4]. Fundamentally, the process entails training local models on the client-side by using local data. The sets of model parameters are then delivered to the server-side and subsequently aggregated to a central model. Repeat the training and aggregation processes to optimize the central model. Because all data are stored locally, none are exposed in the cloud, and data privacy is thereby preserved.

To date, federated learning has been proposed as a promising approach for tasks with large distributed medical datasets [5]. Yan *et al.* [3] applied federated learning framework to a prostate cancer magnetic resonance image (MRI) classification task, using multi-source decentralized data. With the COVID-19 pandemic suddenly sweeping the world, Kumar *et al.* [6] train a global model using blockchain technology with federated learning to detect COVID-19 patients with computed tomography (CT) images. Zhang *et al.* [7] proposed a novel dynamic fusion-based federated learning approach for COVID-19 positive case detection. While the above studies adopted federated learning on a classification task using MRI or CT image data and presented promising results; however, IVUS image segmentation has achieved limited results. For one thing, a segmentation task provides more spatial information about lesions, which is a more advanced task than a classification task. For another, compared with CT or MRI images, IVUS has a relatively lower resolution therefore would be more challenging to be identified. Additionally, IVUS is usually used during clinical surgery, a real-time segmentation model is desired. Therefore, the method proposed in this paper adopts a federated learning-based framework with lightweight encoder-decoder architecture for real-time IVUS image segmentation.

2 Methods

2.1 Image Segmentation

2.1.1 Data Preprocessing

Every frame in IVUS imaging is a grayscale image with an intensity ranging from 0 to 255. Min-max normalization is performed to rescale the pixel value from [0, 255] to [0, 1]. In addition, data augmentation methods are applied to enhance the images. Because coronary arteries are approximately round in shape and a 2-dimensional IVUS image provides no specific directionality, flipping and rotation operations are applied to enrich the training dataset. Every image from the training dataset is randomly rotated by 0, 90, 180, or 270° and randomly flipped either vertically or horizontally. Through providing rich images from a rich training dataset by the application of such data augmentation techniques, the deep learning model is expected to learn to identify images from various angles.

2.1.2 Deep Learning Model

Because IVUS images have low resolution, marking the borders of the external elastic membrane (EEM) and the lumen is a crucial step in identifying the locations of atherosclerotic plaques. Several image segmentation techniques have been proposed in IVUS image processing. Traditional methods such as edge detection techniques have been employed to obtain borders in IVUS images through a series of image processing methods that require parameters based on statistical methods or experiences [8,9]. To date, deep learning techniques have been widely used for image analysis. U-Net, an encoder-decoder framework, is well-known for its rapid and effective development of automatic segmentation and has exhibited extremely favorable performance [1,10].

Deeper models can usually process more complex tasks. However, the multi-source data used in this study has a limited volume. To prevent the model from overfitting, a deep U-Net model would not be used, but a modified light U-Net one in this case. The modified U-Net is depicted in Fig. 1, its encoder comprised four convolution blocks, and each included a convolution layer followed by a batch normalization layer, a ReLU activation layer, and a max-pooling layer. Compared to the U-Net with a deep classifier as the encoder, the number of parameters in the modified U-Net was much smaller.

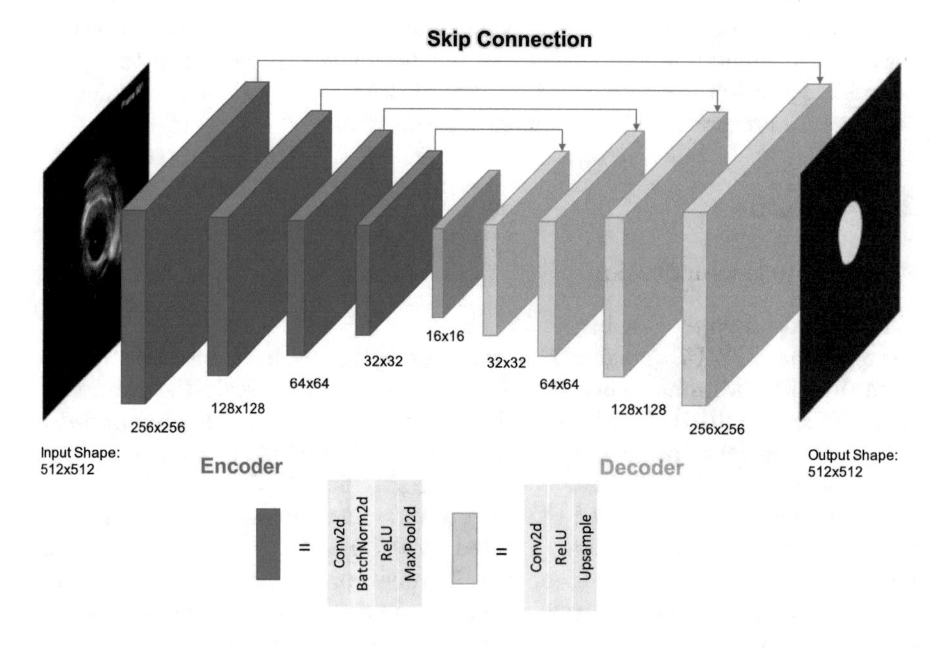

Fig. 1. The modified U-Net model architecture.

2.2 Federated Learning

2.2.1 Training Strategy

A set of experiments was performed, which involved U-Net examining our IVUS dataset. To simulate a medical environment in which a federated learning framework would be applied with N number of local clients, datasets were randomly partitioned into N sets over local clients at the patient level. For every local client in this federated learning framework, the number of patients and the U-Net architecture were identical.

The procedure of federated learning is outlined as follows. At the beginning of the federated learning process, all local clients download the model parameters from the server. Each local client then trains only with its own dataset and then uploads its model parameters to the server when local training is completed. The server aggregates all the received parameters from each local client by averaging these parameters to generate a new server model. Local clients can then download the new server model and perform the local training protocol again. Repeat this procedure until the stopping criteria are met; a global server model is eventually generated from the overall process. During the procedure, since only the local clients' model parameters, but not their data, are exchanged, ensuring that this framework satisfies data protection regulations and preserves patient privacy.

An early stopping mechanism is applied in both local and global training. The stopping criteria are met when performance no longer continues to improve. Other relative parameters settings are summarized in Sect. 3.3.

2.2.2 Differential Privacy

The basic federated learning process may be vulnerable to the risk of data leakage during communication between the server and the local clients. Differential privacy is a method that adds noise to the data transmission process to obscure data and make the process more resistant to data leakage [11]. Gaussian noise is added in the experiment when the model parameters are transmitted from the server to the clients and vice versa. Experiments on the standard deviation of Gaussian noise were conducted for sigma $= 0$, 1e$-$3, and 5e$-$3. In general, the addition of more noise to the data transmission process results in correspondingly higher levels of data protection. However, when the difference between the noise data and the original data is larger, the learning curve may tend to be more fluctuating [12].

2.2.3 Model Weights Clipping

Clipping the model weights into a given range is a common optimization technique in deep learning. The purposes of model weight clipping are to limit large gradients, provide a more stable training process, and prevent gradient explosion caused by gradient variations in each epoch. Moreover, to lower the risk of data leakage, clipping can reduce the values of those large gradients that may imply crucial features. The method of clipping the model weights may thus protect data by eliminating information gradients.

3 Experiments and Results

3.1 IVUS Datasets

The dataset was acquired from Taipei Veterans General Hospital, Taipei, Taiwan. A total of 47 patients with receiving PCI and IVUS were enrolled, and the IVUS images of each patient were in Digital Imaging and Communications in Medicine (DICOM) format; a mechanical system (OptiCross HD 60-MHz, Boston Scientific Corporation, Marlborough, Massachusetts) with a central frequency of 60-MHz was used. Images were annotated by two cardiologists who were blinded to information about each included participant.

Approximately 13,250 frames of 2-dimensional images from 47 patients were annotated at the pixel level and then classified according to two classes. Class 0 represented the image background, and class 1 represented the region of interest, which was the EEM or the lumen area. The original and annotated images were of identical dimensions, both 512 × 512 pixels. Therefore, in this study, the shape of the input and output images of the U-Net was also 512 × 512 pixels.

3.2 Evaluation Metrics

The dice score coefficient, loss value, and model training time were considered as the evaluation metrics. First, the dice score coefficient was used to measure the similarity of the two images. The definition of the dice coefficient loss is expressed in (1), where X and Y represent the ground truth image and the prediction image, respectively. This formulation means that two times the intersection of ROIs is compared to the sum of the ROIs from the two images. If the prediction perfectly matches all of the pixels in the ground truth image, the coefficient is 1, but the coefficient is 0 if the prediction matches none of the pixels in the ground truth image.

$$\text{Dice} = \frac{2 \times |X \cap Y|}{|X| + |Y|} \tag{1}$$

Furthermore, a hybrid loss function was applied to the model training. The loss function was a combination of binary cross-entropy (BCE) loss and dice score coefficient (DSC) loss, as expressed in (2). Since an image segmentation task can be considered a pixel-wise classification task, especially the binary classification task in this study, the BCE_{Loss} function was adopted. By contrast, the DSC_{Loss} is the inverse of the dice score coefficient. To align with the objective of the DSC, the DSC_{Loss} is included in the loss function. The terms α and β are the weights of the aforementioned two components and were set to 0.5 and 0.5, respectively.

$$\text{Loss} = \alpha \times BCE_{Loss} + \beta \times DCS_{Loss} \tag{2}$$

Finally, model training time is used to evaluate the efficiency of the application. This research aims to propose a real-time IVUS image segmentation. When the efficiency of a single model is improved, that of the entire federated learning framework could also be improved.

3.3 Implementation

The deep learning model was implemented using the Python programming language, version 3.9.2, with the PyTorch library, version 1.8.0. An NVIDIA GeForce RTX 2070 graphics processing unit (GPU) with 8 GB of RAM was used to speed up the processing time of the experiment. To train the adopted U-Net model, a hybrid loss function combining BCE loss and DSC loss was used. The optimizer Adam with a learning rate of 1e–5 was applied to update the model. Moreover, L2 regularization was implemented to avoid model overfitting through the addition of a penalty term to the loss function during model training. The batch size was set to 4 in every experiment due to limitations in computational power.

To evaluate the model performance in this study, 5-fold cross-validation was conducted. Each data set was randomly divided into five folds at the patient level, and the number of patients included in each fold was nearly equivalent. Furthermore, this 5-fold data set was also used to simulate a federated learning environment that assumed a case of five clients, each with its own dataset to collaboratively learn a global model.

3.4 Results

To build a real-time and accurate IVUS image segmentation model, a lightweight U-Net was adopted in this study. Table 1 presents the results of the EEM area segmentation tasks for the proposed model and a comparison of the results with those of a deeper model, namely U-Net with Resnet18 [13] as the encoder (ResUnet). Table 2 presents the same information for the lumen area segmentation task performance. Although the dice score of the proposed model was slightly lower than that of the ResUnet in both cases, the proposed model achieved a lower loss value. More crucially, the modified U-Net had only 933,281 parameters-approximately 30 times fewer than ResUnet had. The reduction of parameters led to a substantial decrease in elapsed time by more than four times in both the training and inference processes. Such a decrease is expected to improve the efficiency of clinical applications and the performance of real-time region segmentation in IVUS images. The segmentation results are shown in Fig. 2.

Table 1. Comparison of model performance (EEM area).

Model	Parameters	Loss	Dice	Elapsed time
ResUnet	28,976,321	0.1864	0.8767	22 m 42 s
Proposed	933,281	0.1553	0.8567	5 m 34 s

Table 2. Comparison of model performance (lumen area).

Model	Parameters	Loss	Dice	Elapsed time
ResUnet	28,976,321	0.1458	0.8617	22 m 38 s
Proposed	933,281	0.1335	0.8457	5 m 34 s

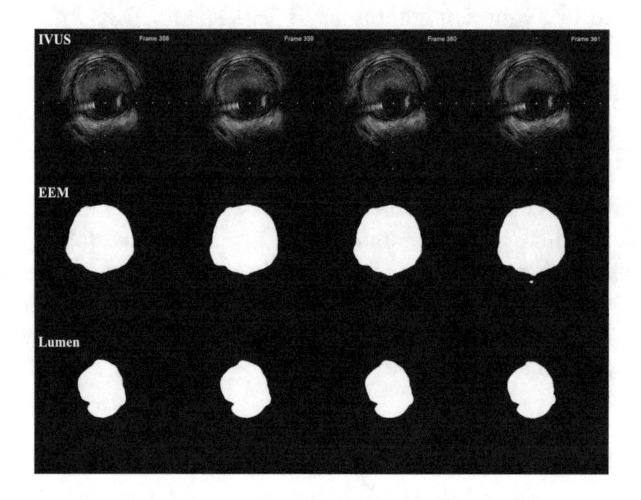

Fig. 2. IVUS segmentation results.

A federated learning framework based on the proposed model was examined. Supposing, for example, that five clients joined a federated learning system, each client and the federated server would adopt the modified U-Net model. The results of applying federated learning in lumen segmentation tasks for IVUS images are presented in Table 3. Compared with each client performing general model training independently, the federated server model collaboratively trained by all clients exhibited relatively average performance for each local dataset. The federated server model substantially improved the performance of clients on every local dataset.

Table 3. Performance of model with/without federated learning.

	Dataset1	Dataset2	Dataset3	Dataset4	Dataset5
Client1	0.8986	0.7746	0.7629	0.8278	0.7696
Client2	0.7617	0.8830	0.7929	0.8123	0.7829
Client3	0.7288	0.6836	0.8863	0.7818	0.6537
Client4	0.7741	0.7488	0.7708	0.9058	0.7717
Client5	0.7324	0.7176	0.6955	0.8003	0.8670
(FL) Server	0.8582	0.8157	0.8581	0.8608	0.8424

Table 4. Performance of model with differential privacy.

Noise Levels	Dataset1	Dataset2	Dataset3	Dataset4	Dataset5	Average
None (0)	0.9121	0.8546	0.9022	0.9007	0.8868	0.8895
Moderate (1e−3)	0.9180	0.8599	0.9023	0.9104	0.8839	0.8935
High (5e−3)	0.8747	0.8360	0.8562	0.8927	0.8514	0.8617

To improve the privacy-preserving capabilities of the federated learning frame- work, differential privacy and model weight clipping mechanisms were implemented. Gaussian noise was added to the model weights during data transmission between the server and the clients, and different standard deviations of Gaussian noise were examined through 20 rounds of federated learning. Table 4 indicates that increases in noise would affect the performance of the model. Noise with 0 standard deviations was equivalent to adding no noise, and model weights remained the same. In some cases, adding noise to model weights could be a technique to avoid overfitting, since it has a similar effect on the network as a regularizer, resulting in better generalization ability [14]. By contrast, adding noise to model weights also contributed to the interruption of model training. As indicated in Fig. 3, when moderate noise was added, federated learning performance was close to that of federated learning performance without the addition of Gaussian noise, or even better. However, greater additional noise contributed to deteriorations in the model's performance, which seemed to be associated with greater fluctuations in the learning curve.

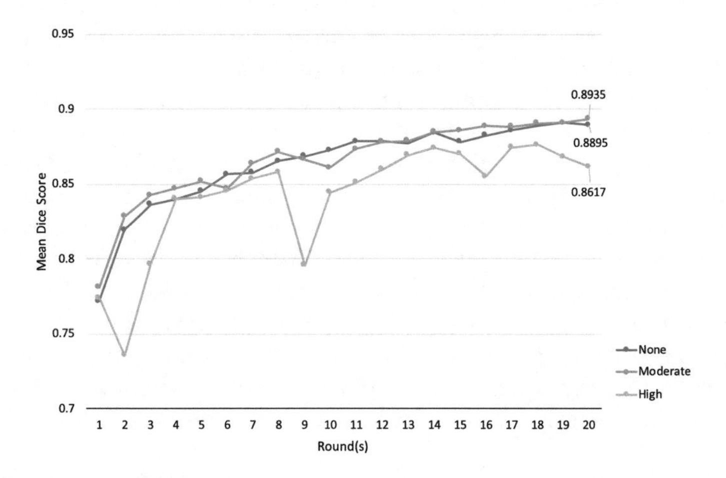

Fig. 3. Performance with respect to different degree of differential privacy.

4 Discussion

EEM and lumen border detection are essential steps in the process of concisely localizing plaques on the vessel walls, so the use of an effective real-time IVUS segmentation tool is required during surgical interventions. In this study, we proposed a deep learning model to automatically segment the EEM and the lumen areas in IVUS images. Because the collection of labeled data for model training was a substantial challenge, a federated learning framework with lightweight segmentation models was adopted. The modified U-Net model is cost-effective because the number of parameters can be reduced considerably, and this model can be executed on a single GPU with 8GB RAM. Moreover, it showed that the performance of our model could approximate that of deep models such as ResUnet (Table 2, Table 1). However, it is observed that the dice score would be varied between different fold datasets (Table 3); this implies that the discrepancy may be due to variances between patient cases, further case analysis could be conducted. In addition, each target area is segmented by a single model in this study, so the results aggregation should be performed as the post-processing, which may lead to extra process time. To put our study in practice, an end-to-end segmentation model to annotate multiple target areas would be focused on in the future.

Regarding federated learning, a federated server model was demonstrated to considerably improve the performance of each client in relation to the other local data sets, indicating that every client can benefit from the federated learning framework without sharing its own data. If concentrating on each local dataset, the server model performed worse than the local model; this might cause some clients to be reluctant to join the federated learning framework. A more comprehensive federated learning algorithm is currently understudied thus far. Given limited data availability, the modified lightweight model described herein would be suitable for federated learning to prevent all models from overfitting, and the process is also efficient. Furthermore, the capacity of federated learning to preserve privacy may be strengthened by applying differential privacy and model weight clipping. Because the model weights are obscured during communication, the risk of data leakage may be reduced, and the framework may be more resistant to malicious incursions such as those posed by model adversarial attacks [15]. On the other hand, Fig. 3 presents the tradeoff between model performance and privacy-preserving. Increasing the degree of privacy-preserving may result in a decrease in model performance. Accordingly, the optimal parameters in a federated learning algorithm could be determined according to the demands of different tasks. Alongside this, more privacy-preserving mechanisms could be examined as well.

Based on this study, a deep learning model for plaque burden segmentation is in progress. Plaque burden segmentation would be more challenging and is usually based on expertise and experience. The segmentation of the 2-dimensional plaque area can be reconstructed as a 3-dimensional volumetric measurement of plaque burden, and the distribution of calcium, lipid, and fibrotic tissue in the diseased vessel can thereby be estimated. Therefore, a federated-based precise and real-time prediction model for plaque burden analysis on IVUS images will be focused on in future work. Last but not least, the proposed approach has shown to be suitable for including but not limited to medical image segmentation tasks. It is also expected to be applied in other scenarios or data types, as long as data satisfies the characteristics of distributed or sensitive.

5 Conclusions

This study developed a federated learning-based deep learning model for IVUS image segmentation. The modified lightweight model was suitable for limited data volumes and provided real-time prediction, presenting cost efficiency compared with the U-Net with the Resnet18 encoder. The dice score can achieve 0.8567 and 0.8457 for the EEM area and the lumen area segmentation, respectively. The present study also adopted a federated learning framework to make participating clients train a more extraordinary generalizability model without sharing their local datasets. To the best of our knowledge, it is the first to employ federated learning for IVUS image segmentation. The result indicates that all clients could benefit from the federated learning framework. Using lightweight models in the proposed federated learning framework is more than four times as efficient as using deep models. Moreover, the framework is extensible. It could incorporate more privacy-preserving mechanisms to strengthen privacy protection and be applied in various scenarios with distributed and sensitive data.

Acknowledgement. This work was supported in parts by Ministry of Science and Technology (MOST), Taiwan, under Grant Number MOST 110-2222-E-001-002, 110-2221-E-002-078-MY2, and 110-2321-B-075-002.

References

1. Li, Y.C., Shen, T.Y., Chen, C.C., Chang, W.T., Lee, P.Y., Huang, C.C.J.: Automatic detection of atherosclerotic plaque and calcification from intravascular ultrasound images by using deep convolutional neural networks. IEEE Trans. Ultrason. Ferroelectr. Freq. Control **68**(5), 1762–1772 (2021)
2. Hassan, A., Dohi, T., Daida, H.: Current use of intravascular ultrasound in coronary artery disease. Clin. Med. Insights Therapeutics **8**, 45–51 (2016)
3. Yan, Z., Wicaksana, J., Wang, Z., Yang, X., Cheng, K.T.: Variation-aware federated learning with multi-source decentralized medical image data. IEEE J. Biomed. Health Inform. **25**(7), 2615–2628 (2021)
4. Hao, M., Li, H., Luo, X., Xu, G., Yang, H., Liu, S.: Efficient and privacy-enhanced federated learning for industrial artificial intelligence. IEEE Trans. Ind. Inf. **16**(10), 6532–6542 (2020)
5. Xue, Z., et al.: A resource-constrained and privacy-preserving edge-computing-enabled clinical decision system: a federated reinforcement learning approach. IEEE Internet Things J. **8**(11), 9122–9138 (2021)
6. Kumar, R., et al.: Blockchain-federated-learning and deep learning models for COVID-19 detection using CT imaging. IEEE Sens. J. **21**(14), 16,301–16,314 (2021)

7. Zhang, W., et al.: Dynamic-fusion-based federated learning for COVID-19 detection. IEEE Internet Things J. **8**(21), 15,884–15,891 (2021)
8. Zheng, M., Yubin, W., Yousheng, W., Xiaodi, S., Yali, W.: Detection of the lumen and media-adventitia borders in IVUS imaging. In: 2008 9th International Conference on Signal Processing (ICSP), pp. 1059–1062 (2008)
9. Swarnalatha, A., Manikandan, M.: Review of segmentation techniques for intravascular ultrasound (IVUS) images. In: 2017 Fourth International Conference on Signal Processing, Communication and Networking (ICSCN), pp. 1–4 (2017)
10. Ronneberger, O., Fischer, P., Brox, T.: U-Net: convolutional networks for biomedical image segmentation. In: Navab, N., Hornegger, J., Wells, W., Frangi, A. (eds.) MICCAI 2015. LNCS, vol. 9351, pp. 234–241. Springer, Cham (2015). https://doi.org/10.1007/978-3-319-24574-4_28
11. He, J., Cai, L., Guan, X.: Differential private noise adding mechanism and its application on consensus algorithm. IEEE Trans. Signal Process. **68**, 4069–4082 (2020)
12. Zhao, B., Fan, K., Yang, K., Wang, Z., Li, H., Yang, Y.: Anonymous and privacy-preserving federated learning with industrial big data. IEEE Trans. Ind. Inf. **17**(9), 6314–6323 (2021)
13. He, K., Zhang, X., Ren, S., Sun, J.: Deep residual learning for image recognition. In: 2016 IEEE Conference on Computer Vision and Pattern Recognition (CVPR), pp. 770–778 (2016)
14. Williams, T., Li, R.: Threshout regularization for deep neural networks. In: SoutheastCon 2021, pp. 01–08 (2021)
15. Kang, X., Song, B., Du, X., Guizani, M.: Adversarial attacks for image segmentation on multiple lightweight models. IEEE Access **8**, 31,359-31,370 (2020)

POTENT - Decentralized Platoon Management with Heapify for Future Vehicular Networks

Arunima Sharma(✉), Dhwani Agrawal, Nandini Roy, Sunita Bhichar, and Ramesh Babu Battula

Department of Computer Science and Engineering, Malaviya National Institute of Technology, Jaipur, India
aru.92@rediffmail.com, rbbattula.cse@mnit.ac.in

Abstract. Connected/Autonomous vehicle communication is emerging as a future commuting paradigm. Vehicle communication platoon management is vital in creating safe and sustainable mechanisms in smart cities impelling to resolve safety and security issues of private/public transportation. The existing platoon mechanisms are centralized, and most methods are designed based on predictive features, which are collected from the traveling behavior of the vehicles. Hence, the homogeneous vehicle platooning faces a significantly higher drop rate and extensive delays. Homogeneous platoons are inefficient in the smart city environment; they demand heterogeneous vehicular networks. Thus, the paper proposes a novel platoon mechanism as a decentralized approach for future vehicular networks using a heapify data structure. POTENT can successfully manage connection loss and different road conditions. In vehicular network creation, the Plexe simulator used for experimentation with C language and OMNET++. The experimental results show better performance of POTENT over the previous centralized platoon management.

1 Introduction

Vehicle platooning is a prominent mechanism to enhance the safety and improve the efficiency of connected/autonomous vehicles [5]. The vehicle platooning operations like a constant velocity keep a safe distance, collision avoidance; avoid road blockage, etc., in intelligent transportation system (ITS) that enhances the traffic behavior of commuters/society [6–8]. Moreover, platooning helps reduce fuel consumption, greenhouse emissions, etc.

Vehicle Platoon creation and management are faster and more flexible with 5G and beyond (5G/5 GB) communication [1–3]. 5 GB can provide efficient services to massive requests/responses generated by the connected/autonomous vehicles during the platoon creation/management. The platoon head is responsible for platoon movement and management. Platoon head paralleled handle requests and operations upkeep for vehicle elements (i.e., vehicle joins and left)

L. Barolli et al. (Eds.): AINA 2022, LNNS 449, pp. 387–398, 2022.
https://doi.org/10.1007/978-3-030-99584-3_34

[5–9]. The platoon head vehicle intermittently gathers every platoon member's information such as speed, area, direction, etc. The existing platoon frameworks used centralized mechanisms to perform various platoon operations [10–13].

The centralized platoon mechanisms primarily face single-point failures as a significant challenge, which may cause unwanted activities like over-speeding, accidents, heavy traffic jams, etc. [4]. In such a case, frequent connection losses cause a high packet drop rate and create heavy computational overheads in the platoon [4]. There are three major issues identified in the platoon management/creation; the first issue is that restricted platoon size causes more frequent platoon operations to provoke violations in the platoon (i.e., fewer vehicles in the platoon). These activities increase the complexity of the platoon management and make the ITS inefficient.

Second, the large size of the platoon induces higher delays in the platoon management, which creates more deplorable activities in the ITS. Finally, the third issue is proactive platoon management; in this scenario, the platoon head periodically collects its updates from the platoon members. Thus, high correspondence and calculation overheads on platoon heads consume more energy.

In this paper POTENT is proposed to resolve these issues by proposing decentralized mechanism. A dynamic platoon management and maintenance method is proposed to handle vehicle dynamics in VANET [12] (Vehicular Ad hoc Network). POTENT is a multiple lane platoon mechanism which is first in this category. The comparison of simulation results with other model shows promising solutions for real time traffic management in smart city.

This paper is divided in four sections. In Sect. 2 POTENT is explained with detailed algorithm and working procedure followed by simulation results in Sect. 3. Paper is concluded in Sect. 4 with future work and summary of POTENT.

2 Platoon Mechanism

Present mechanisms and models focus on centralized platoon safety and management [4]. They consider maintenance of vehicles in platoon as an auto process and predictive model moving in single lane [12,13]. However, these methods fail to handle multiple request together and they restrict platoon size [13]. The restriction of platoon size and centralized control of traffic system doesn't work efficiently due to several factors. Communication range limitation, connection lost, high mobility, interference, packet drop etc. are few factors affect vehicle performance in platoon.

POTENT is combination of efficient platoon operations in multiple lane system with 5G and beyond network support without limiting platoon members. The platoon in POTENT not only consist platoon head but also sub platoon heads to keep fault tolerance higher. Sub platoon heads will help to reduce load of platoon head and keep all vehicles connected in mesh network form. POTENT can work on the road segments, multiple lane, intersections, highways, urban and rural vicinity, and within dense areas. The primary parameters to create/maintain platoon in ITS under the smart city environment are used in Table 1.

Table 1. Parameters

Parameter	Specification
P	Set of Platoon
V, U	Set of all Vehicle
V_i, V_j	Vehicle
P_a, P_b	Platoon
$L_k b$	Lane k
M_P	Platoon Formation Message
pos	Position of Vehicle
pid	Platoon id
pl	Platoon Leader
β	Size of vehicles list at an instance
sz	Platoon Size
LV	List of Vehicles in a platoon
LLV	List of Leaving Vehicles
fl	Front Leader
rl	Rear Leader
ϕ	Priority of Vehicle
t	Type of Vehicle
l	Lane of Vehicle
sp	Speed of Vehicle
d	Destination
vid	Vehicle ID
sv	Splitting Vehicle

2.1 POTENT - Decentralized Platoon Management with HeaPify for FuTure vEhicular NeTworks

POTENT is divide in two phases platoon formation and platoon management. The operations are considered as group of vehicles involved or as an individual vehicle request also. These phases are explained in details as follow.

2.1.1 Platoon Formation

Platoon is created when a vehicle wants to start, find a similar vehicle, and have dense traffic conditions. Initially, as soon as the vehicle becomes active, it will start looking for other vehicles to create a platoon. In the heterogeneous scenario, the vehicle will consider different factors for selecting the platoon. If there is already an existing platoon with a similar path, the vehicle will join it else; it creates a new platoon.

The vehicle will broadcast a beacon message within a range of 1000 m (C-V2X, Cellular Vehicle-to-Everything). Those vehicles with a shared path will send the acknowledgment to the vehicle. In Algorithm 1, vehicle V_i is willing to create a platoon while traveling towards destination d. It initializes an empty platoon P_a and broadcasts a platoon formation message. The message will consist

$\{Sign_{V_i}, P_a = \{V_i\}, L_k, V_i.d, P_{ID}\}$ which are signature of V_i, Platoon P_a having V_i initially, information of lane in which V_i is present, destination of V_i and platoon ID of P_a. Vehicles who want to join the platoon will send an acknowledgment, and V_i will add them in a temporary set U. If the size of set U is ≥ 2, the platoon will be formed; otherwise, the platoon will be destroyed. After Acknowledgment, the vehicles in set U will join platoon P_a by $Join\_Maneuver(P_a, U)$.

2.1.2 Join Platoon

Vehicle v_i will initiate a platoon formation message and broadcast to the nearby vehicles available in the communication range. The vehicles going in the same direction will respond to the request message for platoon formation. There can be either an existing platoon or set of vehicles V which will answer and show interest for Platoon Formation with v_i. Based on the received response message, v_i will decide platoon formation with minimum 2 other vehicles.

$\beta$is used to calculate the size of vehicles that acknowledges and wants to be part of a platoon. When there is a change in platoon member, the priority parameter of vehicles $V_i.\phi$ get updated. The value of priority $V_i.\phi$ depends on different factors like distance to travel with the platoon, destination, power/ fuel consumption, current battery status, etc. Among these factors, distance is the most important parameter. Hence, $V_i.\phi$ is calculated as

$$V_i.\phi = distance(curr, V_i.d) + \alpha_{V_i.t} \tag{1}$$

Here $\alpha_{V_i.t}$ is used to calculate other factors that affect the priority of vehicles in the platoon. $V_i.\phi$ will help select the platoon head and find its current position in the platoon. $\alpha_{V_i.t}$ will help to reduce the number of operations required when a vehicle leaves the platoon or merge with other platoons. Function Heapify($P_a.LV$) will help to manage platoon formation structure. After forming the platoon, the leader will get updates based on the priority value (Fig. 1).

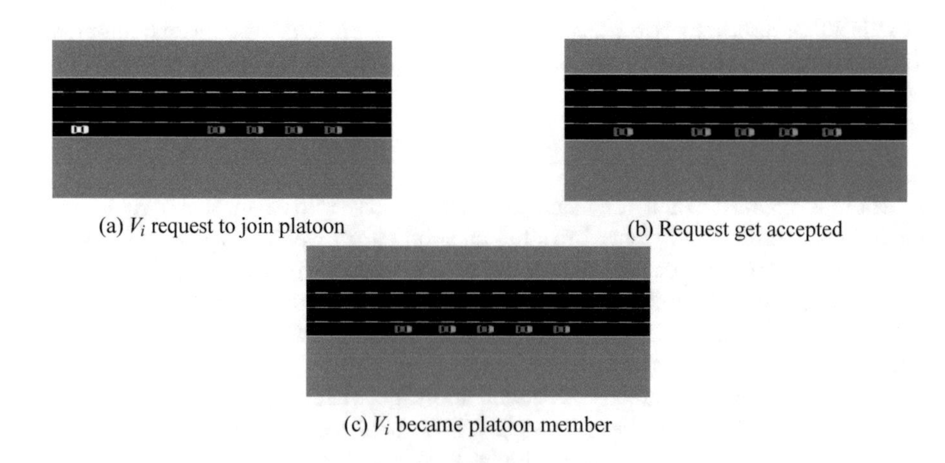

(a) V_i request to join platoon

(b) Request get accepted

(c) V_i became platoon member

Fig. 1. Join platoon

2.1.3 Platoon Leader Selection

After the platoon's formation, one vehicle needs to be the leader to take a major decision regarding the platoon. The vehicle with the highest priority ϕ will be the platoon leader *pl*. The updation of platoon leader is done whenever a platoon management operation is performed (leader leave, join platoon, merge platoon).

pl will generate the platoon ID, assign a position to fellow vehicles, distribute reward, take decisions, and control movement and speed of platoon. The vehicle will send a request to join a platoon or create a platoon is processed by *pl*.

When V_i initiates platoon formation, it will default *pl* until two other vehicles join P_a. After forming $P-a$, each vehicle's priority is calculated and updated. This updation in *pl* is broadcasted by previous *pl* as L_U leader update.

$$L\_U = H(\{(pl), C_{pl}, Sign_{pl}\}|SK_{pl})\forall P_a.LV \tag{2}$$

Here SHA 256 is used for message hashing $H(.)$ where message encrypted by a secret key SK_{pl}, signed by *pl* with public key certificate $C_{pl}, Sign_{pl}$. Platoon leader will generate a secret key for communication within platoon which is shared among only platoon members and get updated with *pl* updated. This new secret key $newSK_{pl}$ is shared with encrypted by old key in first message send by new *pl*.

$$L\_U = H(\{(pl), C_{pl}, Sign_{pl}\}, newSK_{pl}|SK_{pl})\forall P_a.LV \tag{3}$$

2.1.4 Sub-Platoon Leader Selection

The platoon leader will take all major decisions regarding the platoon, but he is unable to do so in certain conditions. Such as, when platoon members need immediate decision to be taken, platoon leader leaves without notification, loses connection with platoon leader, etc., then sub platoon head will perform the actions required.

When delay can't be considered, and platoon leader is unable to contact, then sub platoon head will manage their sub platoons. Temporary split and Un-notified actions by platoon members are initially managed by sub platoon head. And these actions are informed to the platoon head.

The sub-Platoon head, an immediate member of the platoon leader and with second-highest priority after platoon leader, will be the next leader. When an un-notified leader leaves, the sub platoon head will manage the platoon for a duration until a new platoon head gets appointed or a connection with the previous leader is reestablished. When the sub platoon head becomes a platoon leader, it gathers all information from other platoon sub-leaders and restructures the vehicular network with updated parameters.

2.2 Platoon Maintenance

The steadiness of a platoon framework is essential to ensure that every vehicle avoids accidents and can make basic moves to stay away from its impacts. In any case, the centralized platoon support technique can't meet this prerequisite. Decentralized platoon maintenance is proposed as an upkeep technique in which every vehicle communicates with its neighboring vehicles and self-decides its safe distance.

Platoon Maintenance will help keep vehicles safe by reducing the number of operations required with less impact on platoon structure. Lane change, split, merge platoons, platoon change, and leave platoon are explained in detail.

2.2.1 Lane Changing

Lane changing decision is necessary to avoid unpredictable events on the road. The lane change can be taken when a space is available to move a vehicle. This available space S_{L_k} has to be more than the length of vehicle $V_i.l$ and safe distance required among existing and entering vehicles $V_i.sd$. This safe distance can vary from vehicle to vehicle-based on length and speed. If the available space is insufficient, the followers will reduce their speed to create sufficient space.

L_k is the lane in which vehicle V_i wants to move from in current lane L. V_i, a platoon member, will notify this location modification request to its platoon leader and following vehicle members. After lane change, heapify will update platoon structure and priorities of vehicles in the platoon.

At the intersection, the platoon lane change will help avoid traffic congestion and accidents. In Lane changing at the intersection, the direction of movement will get updated, and split can occur among platoon members.

Algorithm 1: *Lane_Change*

Input: P_a, V_i, L_k
Output: P_a
1: Start
2: V_i send $Ln\_U = H(\{(V_i), C_{V_i}, Sign_{V_i}\}|SK_{V_i})$ to $P_a.pl$
3: **for** : $j = 1, j \leq |P_a.LV|, j++$ **do**
4:     **if** $S_{L_k} \geq V_i.l + V_i.sd$ && $V_j.vid == V_i.vid$ **then**
5:         $V_j.L = L_k$
6:     **end if**
7:     $V_j.\phi = distance(curr, V_j.d) + \alpha_{V_j.t}$
8: **end for**
9: $Heapify(P_a.LV)$
10: $Leader\_Selection(P_a)$
11: return (P_a)
12: End

2.2.2 Platoon Change

When the vehicle finds a better platoon during extended travel, or there will be a change in the destination, it will prefer to change any other suitable platoon. This change in the platoon will help to reduce congestion and power usage. In platoon change vehicle V_i who is a member of platoon P_a will send a request to platoon head to change and willingness to join platoon P_b. This process will get approval from both platoon heads while leaving and joining respective platoons. If V_i doesn't receive the approval from P_b it will leave P_a and travel alone till it can form a platoon.

When the change in platoon members will occur both platoon P_a and P_b will update their priorities and heapify their network. This transaction from platoon to platoon will reduce time consumption in respective of search and management (Fig. 2).

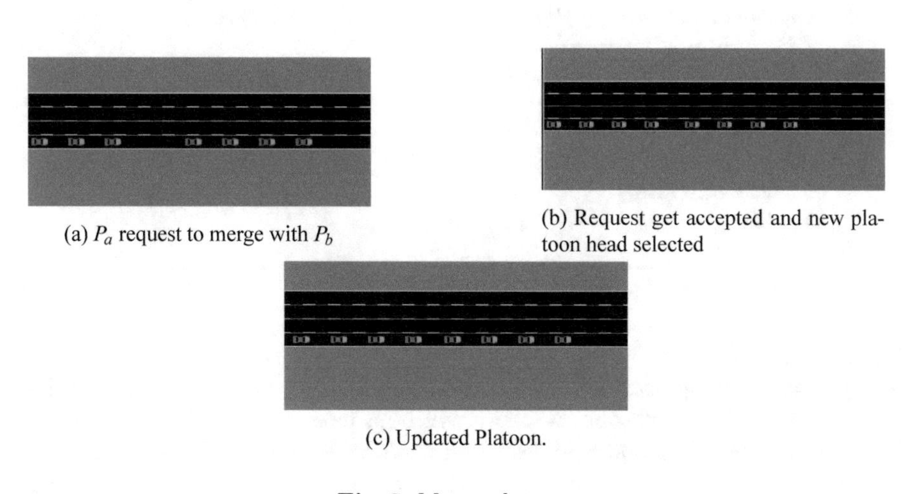

(a) P_a request to merge with P_b

(b) Request get accepted and new platoon head selected

(c) Updated Platoon.

Fig. 2. Merge platoon

2.2.3 Merge Platoons

When two platoons find some common path during traveling, they prefer to merge both platoons in one. This merging will provide more flexibility to platoon members and create flexible traveling. Any platoon present can send the merge request within communication range of each other.

In algorithm $Merge\_Maneuver$ Platoon P_b will send request to platoon P_a to merge together. If both platoons have some common destination or path during transition they will initiate merge operation. The front vehicle$fl = P_a.pl$ and rear vehicle $rl = P_b.pl$ will be the connecting point during merge. After validation and authentication both platoon get merged in P_a.

After merge is done the old platoon will get destroyed and platoon P_a will be updated and heapify.

Algorithm 2: *Merge_Maneuver*

Input: P_a, P_b
Output: P_a
1: Start
2: **if** $(d_a < d_b)$ **then**
3: return (P_a)
4: **end if**
5: $fl = P_a.pl$
6: $rl = P_b.pl$
7: rl send $Mr\_Rq = H(\{(rl), C_{rl}, Sign_{rl}\}|SK_{rl})$
8: fl decrypt Mr_Rq by using PK_{rl}
9: **if** $Sign_{rl}$ is valid **then**
10: fl send $Mr\_Ac = H(\{(fl), C_{fl}, Sign_{fl}\}|SK_{fl})$
11: $Update(rl.sp)$
12: rl broadcast $M\_U = H(\{(rl), C_{rl}, Sign_{rl}\}|SK_{rl})\ \forall\ P_b.LV$
13: $P_a = Join\_Maneuver(P_a, P_b.LV)$
14: rl send $Mr\_Dn = H(\{(rl), C_{rl}, Sign_{rl}\}|SK_{rl})$ to fl
15: $destroy(P_b)$
16: return (P_a)
17: **else**
$Discard(Mr\_Rq)$
18: **end if**
19: End

2.2.4 Split Platoon

The platoon will get split into two sections when an event occurs that can halt some vehicles in the platoon, or there will be a path change at an intersection by a few vehicles. Platoon Split will be beneficial to enhance safety features in ITS (Intelligent Transportation Services). It will help to dynamically keep a safe distance among vehicles in case of Emergency Service Braking (ESB). There are two types of split planned and sudden split.

In a platoon P_a a group of vehicles LLV will send request to split from vehicle point sv. These splitted vehicle will form another platoon P_b if they have more than 3 members in vehicle list $< V >$ of platoon.

When splitting request Sp_Rq is valid vehicles will create another platoon in different or same lane going to different destination from current location of platoon.

Algorithm 3: *Split_Maneuver*

Input: P_a, sv
Output: P_a, P_b
1: Start
2: $ld = P_a.pl$
3: Initialize List $< V > LLV$
4: **for** : $i = 1, i \leq |P_a.LV|, i++$ **do**
5:    **if** $P_a.LV_i.\phi \leq sv.\phi$ **then**
6:       $Insert(LLV, P_a.LV_i)$
7:    **end if**
8: **end for**
9: ld send $Sp\_Rq = H(\{(ld), C_{ld}, Sign_{ld}\}|SK_{ld})$
10: sv decrypt Sp_Rq by using PK_{ld}
11: **if** $Sign_{ld}$ is valid **then**
12:    **for** : $i = 1, i \leq |LLV|, i++$ **do**
13:       ld send $M\_U = H(\{(ld), C_{ld}, Sign_{ld}\}|SK_{ld})$ to LLV_i
14:       $Leave\_Maneuver(P_a, LLV_i)$
15:    **end for**
16:    ld send $Sp\_Dn = H(\{(ld), C_{ld}, Sign_{ld}\}|SK_{ld})$ to sv
17:    **if** $|P_a.LV| \leq 2$ **then**
18:       $destroy(P_a)$
19:    **end if**
20:    **if** $|LLV| \geq 3$ **then**
21:       Initialize $P_b = new_Platoon$
22:       $P_b = Join\_Maneuver(P_b, LLV)$
23:    **end if**
24:    return (P_a, P_b)
25: **else**
      $Discard(Sp\_Rq)$
26: **end if**
27: End

2.2.5 Leave Platoon

When a vehicle has to leave the platoon, if it is present at the bottom of the platoon, it doesn't cause much modification in the platoon structure, but if it is present in the middle or on top of the platoon, it becomes a complex task to manage. Heapify is used in VANET as a suitable solution. This will efficiently manage the dynamic network structure with multiple operations going on parallel.

Vehicle V_i will send request to platoon head to leave platoon P_a. If it get approved the platoon get heapify. If leaving vehicle is leader the next sub platoon head with highest priority will be the new platoon leader. When vehicle or platoon leader will leave the platoon without notifying the platoon members it is considered as connection lost situation.

3 Experimental Results

Platoon environment is created for management of multiple platoons in decentralized manner. Vehicles will communicate with each other by using Beacon Messages in C-V2X protocol. Message communication rate is 0.2 ms to 1.8 ms for one message which varies with speed and number of vehicles. In case of emergency this rate will be either 0 or 0.2 ms with nearly no delay.

Platoon formation is not limited to single lane followed by leading vehicle *pl*. One platoon will have at lease one Sub_{pl} which will be in immediate next level following *pl*. The different vehicles belong to different categories with five gears, differential ratio (>3), friction, mass, air friction coefficient, power and engine efficiency parameters.

Velocity of vehicles is between 0 to 120 km/hr while following 2 s safety distance rule. Number of platoon members are minimum three vehicles and maximum is not restricted within a range or up-to a threshold. Experimental setup is developed by using Plexe [11] simulator. Smart city environment is generated with help of Oment++ for graphical experimental setup. The road structure has multiple lanes from 1 lane upto 8 lane road. Intersections, slope, curve, one way road and known road elements are integrated together. The C language is used for management and formation of platoon.

3.1 Results

POTENT is compared with ADV [9] and CMNG [10] platoon networks. The performance is compare at run time with different situations and impact on platoon. Test bed is software based developed on AMD based Think Station P340, 64 bits windows 11 processor with Nvidia graphics.

Different operations on platoon performed simultaneously are compared on average value after multiple running of 100 rounds. Results are showing performance of POTENT, ADV [9], and CMNG [10] in graphical format. Result Fig. 3a shows comparison of average total delay and total number of vehicles in vicinity. These vehicles are part of different platoons. The average delay is counted from accepted delay to perform any operation from initiation till its completion. Figure 3b shows result between average packet drop rate at different junctions, intersections and road lanes when different number of vehicles are present in that area. Figure 3c shows how many time the platoon need to be break (violate) due to less number of vehicles, connection loss and restriction on platoon members. This result will show that in POTENT platoon violation is decreased due to multiple lane support and sub platoon heads presence. Figure 3d represents results among Average operations performed on platoon (join, merge, split, formation of platoon) vs. number of vehicles in platoons present on road.

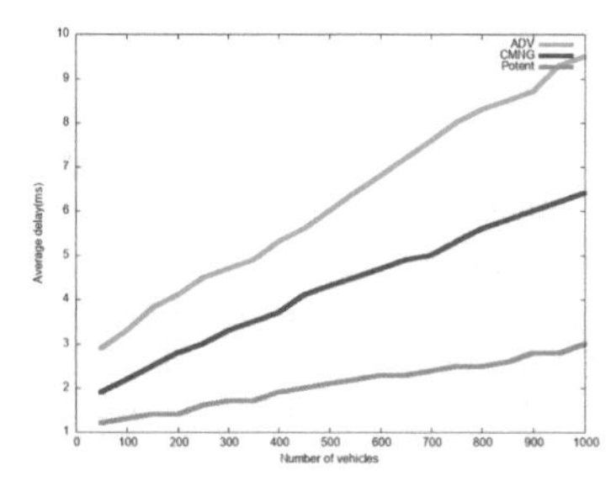

(a) Average total delay vs. number of vehicles in vicinity

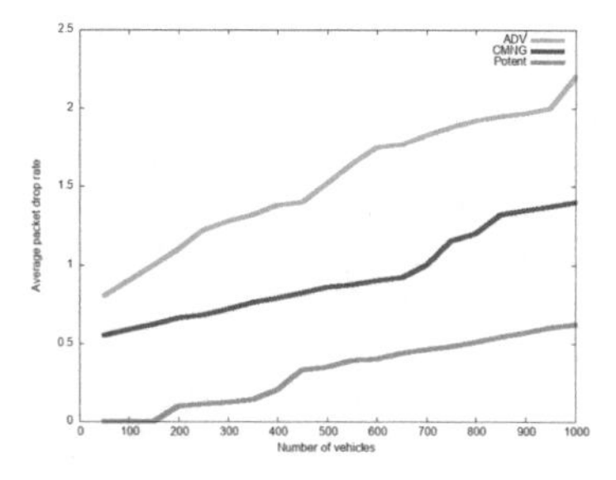

(b) Average packet drop rate vs. total number of vehicles

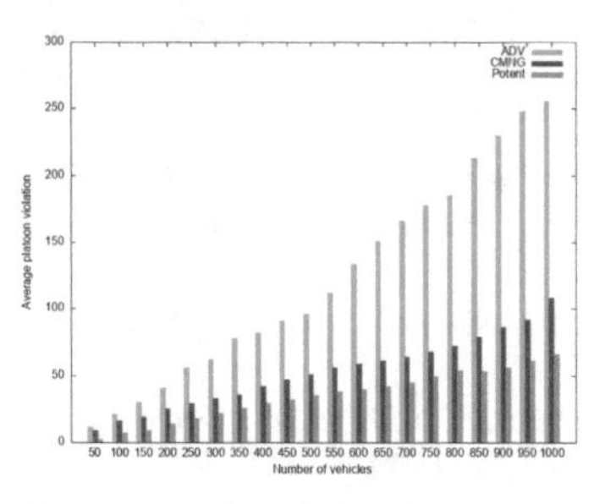

(c) Average platoon violation vs. number of vehicles on road

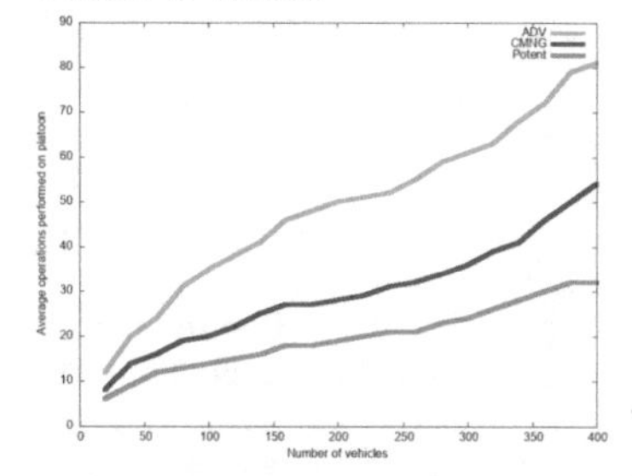

(d) Average operations performed on platoon vs. number of vehicles in platoons on road

Fig. 3. Results

4 Conclusion and Future Work

POTENT, a distributed platoon management for future vehicular networks by using heapify, is proposed in this paper. Platoon formation and maintenance algorithms are proposed in the first phase with vehicular dynamics with heapify. Connection loss and factors impacting the platoon are considered for better platoon management in smart city ITS. Restriction of platoon size is eliminated, and sub platoon heads are considered for accuracy and safety perspective. In the second phase, POTENT is compared with ADV [9] and CMNG [10]. The evaluation stated better performance of POTENT in different situations on the road. Recreation results show that without vehicle distance data, the decentralized unit support technique shows comparative execution as the unified strategy with low maintenance time and fewer platoon violations.

Future security aspects of platoon and vehicles are evaluated with different attacks possible on VANET.

Acknowledgment. This work was supported in part by SERB, the ministry of the Department of Science and Technology (DST), Government of India, Project "SWARD -Secure next-generation Wireless Access RaDio technology for smart cities in India" (grant number EEQ/2018/001482).

References

1. Ashraf, S.A., et al.: Supporting vehicle-to-everything services by 5G new radio release-16 systems. IEEE Commun. Stand. Mag. **4**(1), 26-32 (2020)
2. Sidorenko, G., et al.: Vehicle-to-vehicle communication for safe and fuel-efficient platooning. In: 2020 IEEE Intelligent Vehicles Symposium (IV). IEEE (2020)
3. Lekidis, A., Bouali, F.: C-V2X network slicing framework for 5G-enabled vehicle platooning applications. In: 2021 IEEE 93rd Vehicular Technology Conference (VTC2021-Spring). IEEE (2021)
4. Thunberg, J., et al.: Vehicle-to-vehicle communications for platooning: safety analysis. IEEE Networking Lett. **1**(4), 168–172 (2019)
5. Liu, Y., et al.: Proactive longitudinal control of connected and autonomous vehicles with lane-change assistance for human-driven vehicles. In: 2021 IEEE International Intelligent Transportation Systems Conference (ITSC). IEEE (2021)
6. Chen, Z., Park, B.B.: Preceding vehicle identification for cooperative adaptive cruise control platoon forming. IEEE Trans. Intell. Transp. Syst. **21**(1), 308–320 (2019)
7. Maiti, S., et al.: The impact of flexible platoon formation operations. IEEE Trans. Intell. Veh. **5**(2), 229–239 (2019)
8. Jin, L., et al.: Analysis and design of vehicle platooning operations on mixed-traffic highways. IEEE Trans. Autom. Control **66**, 4715–4730 (2020)
9. Hu, S., et al.: Automated discovery of denial-of-service vulnerabilities in connected vehicle protocols. In: 30th USENIX Security Symposium (USENIX Security 21) (2021)
10. Sarker, A., Qiu, C., Shen, H.: Connectivity maintenance for next-generation decentralized vehicle platoon networks. IEEE/ACM Trans. Networking **28**(4), 1449–1462 (2020)
11. Segata, M., Joerer, S., Bloessl, B., Sommer, C., Dressler, F., Lo Cigno, R.: PLEXE: a platooning extension for veins. In: Proceedings of 6th IEEE Vehicular Networking Conference (VNC 2014), Paderborn, Germany, December 2014
12. Kremer, P., et al.: State estimation for attack detection in vehicle platoon using VANET and controller model. In: 2020 IEEE 23rd International Conference on Intelligent Transportation Systems (ITSC). IEEE (2020)
13. Jiang, L., et al.: Dampen the stop-and-go traffic with connected and automated vehicles–a deep reinforcement learning approach. arXiv preprint arXiv:2005.08245 (2020)

Web Service Anti-patterns Prediction Using LSTM with Varying Embedding Sizes

Sahithi Tummalapalli(✉), Lov kumar, and Neti Lalita Bhanu Murthy

Department of Computer Science and Information Systems, Birla Institute of Science and Technology-Pilani, Hyderabad Campus, Jawahar Nagar, Hyderabad, Telangana, India
{P20170433,lovkumar,bhanu}@hyderabad.bits-pilani.ac.in

Abstract. Anti-pattern includes the concept of wrongdoing and has many options that seem to be right initially; however, it results in hassle in the long run. Studies observe that the anti-patterns in web services make them more susceptible to change proness and fault proness. Anti-patterns can easily lead to error-prone and unmaintainable solutions. This makes it essential to detect anti-patterns at the early stages of software design so that the software developers can restructure the code in the early stages itself. This will save time and effort it would require to address the issues that could stem from anti-patterns in web services. In this work, Sequence classification with LSTM is applied on the WSDL files from the repository to identify four anti-patterns by using sampling techniques. Our findings indicate that LSTM3 performs best out of three classifier models considered in our work with a mean accuracy of 91.75.

Keywords: WSDL files · Anti-patterns · Web service · Long short-term memory algorithm · LSTM · Sequence classification

1 Introduction

Service-Oriented Architecture (SOA) can facilitate companies to contour their processes to run their business more efficiently and adapt to dynamic requirements and competition, thereby realizing the concept of Software as a Service (SAAS). Web Services represents one robust approach to realizing an SOA. Though web service technology is not the sole approach to realizing an SOA, it is one that the IT industry as a whole has embraced. For web service, the industry once again faces the fundamental challenge that distributed computing has been facing: providing a uniform description, types, location, etc., and access to components or services on the network. Building a well-designed web service is crucial because it can bring cleaner services and higher-level interactions, and it can also achieve efficient software maintenance. However, Web services suffer from poor design problems called "Anti-pattern [1]". This problem frequently occurs when web services continuously evolve to fit in the ever-changing

L. Barolli et al. (Eds.): AINA 2022, LNNS 449, pp. 399–410, 2022.
https://doi.org/10.1007/978-3-030-99584-3_35

client requirements. Anti-patterns may not essentially be named as defects in the design phase, but these are very likely to develop software bugs during the maintenance phase. Studies observe that the frequency of fault-proness of the classes with anti-patterns is nine times that of the classes without anti-patterns. Likewise, the frequency of change-proness of the classes with anti-patterns is six times that of the classes without anti-patterns. This makes the detection of anti-patterns in web services a crucial step during software development. In this work, we are automating the procedure of identifying anti-patterns. We use Web Service Description Language (WSDL) files of web services, to build a set of code metrics to develop models to discover and detect anti-patterns in WSDL files. These models reduce the cost of software maintenance, helping in improving the quality of software design.

In this work, we used Long Short Term Memory Networks (LSTM) algorithm with a varying number of hidden layers and word embedding technique with different sequence padding lengths to predict the following four anti-patterns, namely, Fine-Grained Web Service (FGWS), Chatty Web Service (CWS), Ambiguous Web Service and God Object Web Service(GOWS). In this work, we have analyzed the performance of various learning algorithms in predicting anti-patterns using accuracy, the Area under the curve (AUC), and F-measure. In this paper, we have attempted to answer the following three research questions:

- **RQ1: Explore the capability of word embedding technique with varying sequence padding length to develop models for detecting anti-patterns**
  For this objective, we used embedding technique with varying sequence padding length to map each code line of the WSDL line to vector value, using which the anti-pattern detection models are developed. We compare the models performance using different feature sets (generated using various sequence padding sizes) using Area Under Curve and Statistical significance tests.
- **RQ2: Discuss the potential of Data Sampling Techniques to detect anti-patterns**
  With the original dataset, we have used two different sampling techniques to generate datasets. A total of three datasets are used for training the models in this work: 1. original dataset (ORG) 2. Synthetic Minority Over Sampling (SMOTE) 3. Borderline SMOTE (BLSMOTE). We compare the performance of the datasets mentioned above using statistical significance testing and Area under curve.
- **RQ3: Analyze the capacity of the LSTM algorithm with distinct hidden layers to predict Anti-patterns.**
  In this paper, we have used Long Short Term Memory Networks (LSTM) to detect anti-patterns using various metrics sets and datasets. We have used LSTM with varying numbers of hidden layers, i.e., LSTM with one hidden layer (LSTM1), LSTM with two hidden layers (LSTM2), and LSTM with three hidden layers (LSTM3) for detection of anti-patterns. Their perfor-

mance has been compared using Area Under the Curve analysis and statistical significance tests.

The remaining paper is organized as follows: Section 2 summarizes the related work. Section 3 gives an in-depth explanation of all the modules used in the proposed framework. Section 4 describes the proposed framework pipeline and the role of each module described in Sect. 3. Section 5 provides the experimental outcome. Section 6 shows the comparative analysis of the anti-patterns models developed using various datasets, and LSTM models. In this section, we also answer the research questions posed in Sect. 1. In Sect. 7, we conclude our research.

2 Literature Survey

Word processing context analysis and term frequency-inverse document frequency analysis (TF-IDF) were studied by Segev et al. [2]. The service repository was validated by the author for publicly accessible WSDL files and free text descriptors. The authors utilised their methodologies to demonstrate that online services may be enhanced by exploiting data accessible on the web, rather than overloading customers with service marginalization via Concept and formal explanation. Automated ORM (object-relational model) performance anti-patterns have been created by Peter Chen and his colleagues [3] to indicate performance anti-patterns in the source code. To identify web services anti-pattern, Ouni et al. [4] used genetic programming to generate detection rules using threshold values and other metrics. The five anti-patterns are tested on 310 Web services to verify the aforesaid method. Ouni et al. [5] cited the use of cooperative parallel evolutionary algorithms (P-EA) as an automated method for identifying the anti-patterns. It is their belief that the use of several detection methods in parallel optimization procedures will provide superior outcomes. Precision was measured at 0.89 when compared to random search and population-based searches.

Jaffar et al. [6] stated in his article that classes participating in anti-pattern and patterns of software designs have unvarying and modifying relationships with other classes, which may transmit difficulties to other classes. Co-change and static dependencies have been experimentally studied in this research, and their link to change proness, fault proness, and fault types has been analysed. This paper is a continuation of the work done by the authors in this area. Using static code analysis, Kumar et al. [7] offer a way to find anti-patterns automatically. It is suggested in this article that aggregated source code metrics collected at the Web service level should be used to discover anti-patterns. According to Saluja et al. [8], he and his colleagues have developed an algorithm that leverages dynamic indications in addition to static indicators for execution. The findings may be further improved using genetic algorithms. The suggested findings have a recovery rate of roughly 0.9, which is better than the current approaches. Many of the suggested methods rely on source code metrics, code analysis, or the creation of rule cards to discover anti-patterns. In order to create models for web service anti-pattern identification, we decided to apply sequence classification with LSTM on the publicly accessible web service available in the Tera-Promise repository.

3 Experimental Setup

This section describes each module used in the proposed method, i.e., Dataset, Data Sampling Techniques, Sequence Classification with LSTM.

3.1 Experimental Dataset

In this work, the database we have used comprises 226 publicly available web services that are downloaded from the Tera-Promise repository in GitHub.[1] The dataset consists of a list of web services collected from various domains such as finance, weather, education, tourism, etc., and the anti-patterns present in them. The statistics and distribution of anti-patterns by type are shown in Table 1.

Table 1. Statistics on anti-patterns distribution by type

Name of the Anti-pattern	Web services without anti-pattern	Web services with anti-pattern	% of web services without anti-pattern	% of web services with anti-pattern
Fine-Grained Web Service (FGWS)	213	13	94.25	5.75
Chatty Web Service (CWS)	205	21	90.71	9.29
God Object Web Service (GOWS)	205	21	90.71	9.29
Ambiguous Web Service (AWS)	202	24	89.38	10.62

3.2 Data Sampling Techniques

We can infer from Table 1 that there a bias between the web services which have anti-patterns and the web services that do not have anti-patterns. This is commonly referred to as the class imbalance problem. In order to mitigate the bias between the classes, we used two sampling techniques in this work to generate additional datasets. A short explanation of the sampling techniques is given below:

- **SMOTE** [9]**:** It uses the nearest neighbors of the minority class to generate the new samples artificially.
- **Borderline Smote (BLSMOTE)** [10]**:** BLSMOTE generates new examples of the minority class using nearest neighbors of the cases in the border region between classes.

[1] https://github.com/ouniali/WSantipatterns.

3.3 Sequence Classification with LSTM

Sequence classification is a predictive modeling task, for which the input is spatial or temporal sequence, and the target is to predict the sequence category. The issues here are: 1. The length of the sequence i.e. the number of words may vary. 2. The model may need to learn the long-term context or dependencies between the characters in the input sequence and, 3. The vocabulary of the input characters is very large. This sequence classification problem can be solved by using Long Short-Term Memory (LSTM) algorithm.

In this work, we used word embedding technique to encode code in WSDL file as vector values. Each word of the WSDL file is mapped to a 32 size vector. As discussed earlier, the length of the sequence may vary. Hence, we are developing models for different sequence padding sizes varying from 100 to 1000. For example, if we are developing model for padding size as 100, we will restrict each code line in WSDL file to be 100 words, diminishing lengthy code lines and padding short code lines with zero values.

Further, we develop LSTM [11] model for classifying sequence of code lines. We used Long Short-Term Memory (LSTM) classifier to predict anti-patterns. The LSTM layer consists of recursively linked blocks called memory blocks, which can be regarded as differentiable versions of memory chips in digital computers. Each memory chip contains one or more memory cells connected in different ways and three units, namely: the input, the output, and forget gateways which provide continuous write, read, and refresh operation simulations for the cells. The network can only communicate with cells through gates. The crucial adjustable hyper-parameters of LSTM are network size and learning rate. This means that the hyper-parameters can be set independently of each other. Especially the learning rate can be calibrated with a relatively small network initially, which saves a lot of experiment time. In this work, we used LSTM with varying numbers of hidden layers, i.e., LSTM with one hidden layer (LSTM1), LSTM with two hidden layers (LSTM2), and LSTM with three hidden layers (LSTM3) for training the anti-pattern prediction models. For LSTM1, the foremost layer is the Embedded layer that takes 32 size vectors (represents word) as input. Middle/ Hidden layer is the LSTM layer with memory blocks. We used a Dense output layer with single neuron and sigmoid activation function in the final layer, as ours is a classification problem with two classes: anti-pattern (1) or no anti-pattern (0). We have designed the other two techniques: LSTM2 and LSTM3 accordingly.

4 Proposed Methodology

Figure 1 shows the detailed illustration of the proposed framework. As discussed in Sect. 3.1, we conducted our experiments on 226 publicly available web services that are downloaded from GitHub. We used two data sampling techniques SMOTE, Borderline SMOTE (BLSMOTE), as stated in earlier sections, to eliminate the class imbalance problem. We used these two additional datasets and the original dataset(ORG) to train the models. Next, we applied word embedding technique with varying sequence padding sizes for feature generation. We then train the models developed using LSTM with hidden layers ranging from 1 to 3. Lastly, we applied five-fold cross-validation to repress the selection bias and overfitting problems. We then used performance metrics such as the area under the curve (AUC), F-measure, etc., and some tests like statistical significance testing to compare the performance of the various models developed.

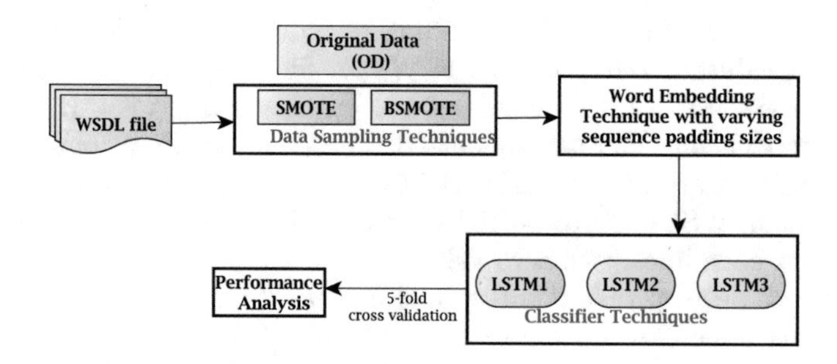

Fig. 1. Flowchart of proposed research framework

5 Experimental Outcome

Table 2 shows accuracy for all the LSTM models using feature generation techniques and data sampling techniques in the detection of GOWS and FGWS anti-patterns. The results of models for the detection of only two anti-patterns are shown in Table 2 due to space constraint. Table 3, 5 and 7 shows the descriptive statistics of different metrics used in our work. From all the above tables stated, the following observations were inferred:

- Word embedding technique with sequence padding size as 100 performs better while the embedding technique with sequence padding size as 1000 performs worst with a mean accuracy of 89.49%.
- Borderline SMOTE (BLSMOTE) performs best among the data sampling techniques with a mean accuracy of 91.75% and a median accuracy of 93.14. SMOTE has the worst performance with a mean accuracy of 87.22%.
- LSTM3 is performing better with a mean accuracy of 91.75%.

6 Comparative Analysis

RQ1: Explore the capability of word embedding technique with different sequence padding length in detecting anti-patterns
Table 3 and Fig. 2 show that the embedding technique with sequence padding size as 100 performs better than other models with varying padding size. Our

Table 2. Accuracy of all models for GOWS and FGWS anti-patterns

GOWS anti-pattern					FGWS anti-pattern				
Sampling technique	Sequence padding length	LSTM1	LSTM2	LSTM3	Sampling technique	Sequence padding length	LSTM1	LSTM2	LSTM3
OD	100	0.95	0.94	0.95	OD	100	0.96	0.96	0.92
OD	200	0.93	0.93	0.91	OD	200	0.98	0.97	0.95
OD	300	0.96	0.94	0.94	OD	300	0.98	0.98	0.96
OD	400	0.93	0.92	0.95	OD	400	0.96	0.97	0.96
OD	500	0.97	0.97	0.96	OD	500	0.96	0.96	0.96
OD	600	0.97	0.97	0.97	OD	600	0.97	0.97	0.97
OD	700	0.97	0.94	0.96	OD	700	0.95	0.95	0.94
OD	800	0.96	0.97	0.97	OD	800	0.96	0.92	0.96
OD	900	0.95	0.96	0.97	OD	900	0.96	0.96	0.96
OD	1000	0.97	0.97	0.97	OD	1000	0.96	0.95	0.96
SMOTE	100	0.91	0.97	0.91	SMOTE	100	0.98	0.95	0.95
SMOTE	200	0.86	0.92	0.90	SMOTE	200	0.85	0.87	0.86
SMOTE	300	0.75	0.85	0.87	SMOTE	300	0.85	0.88	0.95
SMOTE	400	0.79	0.89	0.90	SMOTE	400	0.94	0.85	0.88
SMOTE	500	0.88	0.86	0.87	SMOTE	500	0.87	0.91	0.89
SMOTE	600	0.89	0.92	0.91	SMOTE	600	0.89	0.89	0.84
SMOTE	700	0.87	0.86	0.83	SMOTE	700	0.76	0.80	0.86
SMOTE	800	0.89	0.88	0.91	SMOTE	800	0.87	0.85	0.80
SMOTE	900	0.90	0.84	0.87	SMOTE	900	0.76	0.78	0.80
SMOTE	1000	0.85	0.81	0.88	SMOTE	1000	0.65	0.65	0.64
BLSMOTE	100	0.94	0.94	0.92	BLSMOTE	100	0.94	0.95	0.94
BLSMOTE	200	0.85	0.88	0.86	BLSMOTE	200	0.87	0.83	0.82
BLSMOTE	300	0.84	0.87	0.88	BLSMOTE	300	0.86	0.93	0.95
BLSMOTE	400	0.84	0.90	0.84	BLSMOTE	400	0.91	0.89	0.88
BLSMOTE	500	0.97	0.93	0.94	BLSMOTE	500	0.94	0.89	0.93
BLSMOTE	600	0.97	0.96	0.97	BLSMOTE	600	0.93	0.94	0.95
BLSMOTE	700	0.97	0.96	0.95	BLSMOTE	700	0.93	0.81	0.82
BLSMOTE	800	0.97	0.97	0.98	BLSMOTE	800	0.84	0.89	0.89
BLSMOTE	900	0.95	0.96	0.95	BLSMOTE	900	0.89	0.88	0.87
BLSMOTE	1000	0.95	0.97	0.97	BLSMOTE	1000	0.72	0.89	0.86

Table 3. Statistics: sequence padding size

	Min	Max	Mean	Median	Var	Q1	Q3
100	88.61	97.61	93.50	94.03	5.98	91.12	95.45
200	81.65	97.98	90.43	90.46	22.95	86.45	94.19
300	75.14	97.98	89.56	88.51	34.24	85.66	94.53
400	79.01	97.47	90.76	91.17	15.49	88.58	93.18
500	86.46	97.47	92.07	92.87	11.56	88.83	94.95
600	83.51	97.47	93.43	94.19	14.60	90.74	96.97
700	76.06	98.48	91.62	94.19	33.15	87.46	95.71
800	80.32	97.98	92.08	94.95	27.12	87.74	96.46
900	76.06	97.47	91.73	94.19	29.18	88.90	95.45
1000	63.56	97.47	89.49	94.44	95.13	86.87	95.96

model performs better with a sequence length of 00, indicating that the size of most of the sentences given as input is around 100.

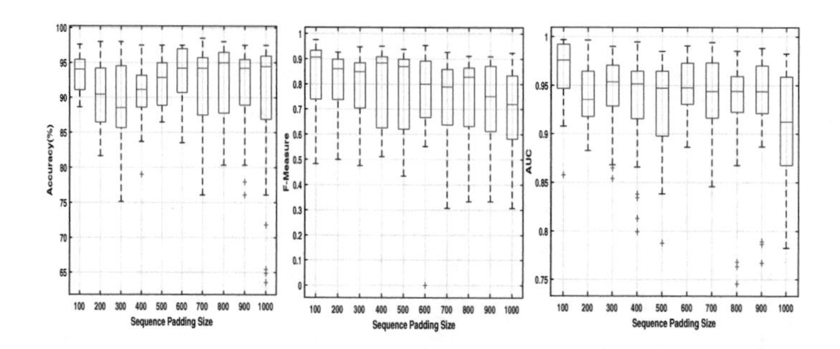

Fig. 2. Box-plot for Accuracy, F-measure and AUC: Sequence Padding Size

The Wilcoxon signed-rank test is used to compare the prediction ability of web service anti-pattern detection approaches utilising varied sequence padding sizes. If a substantial improvement is seen in the models generated using varied padding sizes, then this statistical testing will be worthwhile. In order to evaluate whether or not to accept or reject the null hypothesis, p−value is utilised in this test. "The web service anti-pattern detection models constructed using varied sequence padding sizes are substantially the same," is the investigated null hypothesis for this study. If the rank-sum test yields a p−value larger than 0.05, the null hypothesis is accepted. Wilcoxon signed-rank sum tests of vectors produced for all models with padding sizes ranging from 100 to 1000 are shown in the Table 4. Table 4 shows that most comparison points have values larger than 0.05. It follows from this that, in most circumstances, we can infer that the models built by taking into account various sequence padding lengths as input are considerably different from each other. Thus, the performance of the models produced may be determined by the duration of the padding sequences.

Table 4. Rank-sum test: Sequence Padding Size

	100	200	300	400	500	600	700	800	900	1000
100	1.00	0.00	0.00	0.00	0.00	0.01	0.01	0.00	0.00	0.00
200	0.00	1.00	0.58	0.70	0.77	0.28	0.78	0.89	0.77	0.01
300	0.00	0.58	1.00	0.79	0.28	0.55	0.91	0.41	0.72	0.01
400	0.00	0.70	0.79	1.00	0.47	0.58	0.80	0.58	0.78	0.07
500	0.00	0.77	0.28	0.47	1.00	0.15	0.37	0.88	0.55	0.15
600	0.01	0.28	0.55	0.58	0.15	1.00	0.54	0.24	0.41	0.00
700	0.01	0.78	0.91	0.80	0.37	0.54	1.00	0.64	0.86	0.02
800	0.00	0.89	0.41	0.58	0.88	0.24	0.64	1.00	0.83	0.05
900	0.00	0.77	0.72	0.78	0.55	0.41	0.86	0.83	1.00	0.04
1000	0.00	0.01	0.01	0.07	0.15	0.00	0.02	0.05	0.04	1.00

RQ2: Discuss the potential of Data Sampling Techniques to detect anti-patterns
Table 5 and Fig. 3 show that the Borderline SMOTE (BLSMOTE) is performing better across the AUC and F-measure metrics. BLSMOTE performs better than SMOTE, as SMOTE tends to ignore the instances in the outlying minority class and appear in the majority class by generating a line bridge with the majority class. However, BLSMOTE selects some random instances as border points with majority and minority instances and uses these border points to generate data synthetically. Compared to the Accuracy measure of all the models, the model developed using the original dataset (OD) shows better performance.

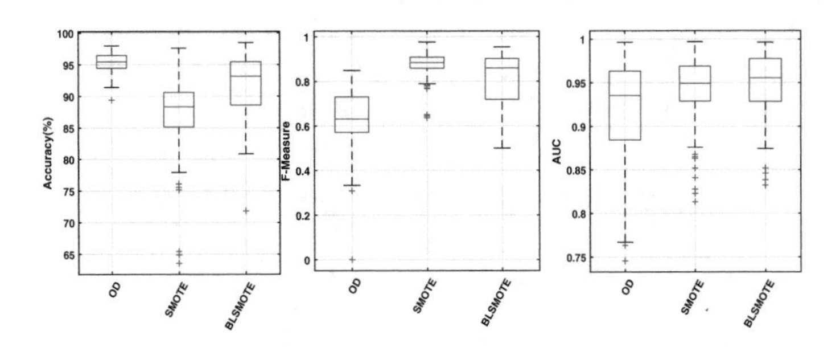

Fig. 3. Box-plot for Accuracy, F-measure and AUC: Data Sampling Techniques

Table 5. Statistics: data sampling techniques

	Min	Max	Mean	Median	Var	Q1	Q3
OD	89.39	97.98	95.43	95.45	2.58	94.44	96.46
SMOTE	63.56	97.61	87.22	88.33	32.39	85.16	90.61
BLSMOTE	71.81	98.48	91.75	93.14	21.44	88.59	95.45

Table 6 gives the results of the Rank sum test of the datasets generated using SMOTE, BLSMOTE, and OD. We inferred that the models performance trained using sampling techniques datasets varies significantly from the original dataset. We conclude that the performance values of all the datasets are highly uncorrelated. We also observed that the performance values of the models developed using SMOTE and BLSMOTE are similar.

RQ3: Analyze the capacity of LSTM algorithm with distinct hidden layers to predict Anti-patterns
Figure 4 and Table 7 show the performance of three classifier techniques developed in terms of accuracy, AUC, and F-measure. From Fig. 4 and Table 7, we infer that the LSTM3 is performing slightly better when compared to other models with a mean accuracy of 91.75. This might be because the LSTM model with

Table 6. Rank-sum test: sampling techniques

	OD	SMOTE	BLSMOTE
OD	1.000	0.008	0.000
SMOTE	0.008	1.000	0.167
BLSMOTE	0.000	0.167	1.000

three hidden layers detects more complex features while training, while LSTM1 and LSTM2 models might have failed to detect more complex features. Table 8 shows the result of the Ranksum tests on models generated using LSTM with the distinct number of hidden layers. From Table 8, we observed that the prediction models developed using various LSTM models are significantly different from each other, and the models are highly unrelated.

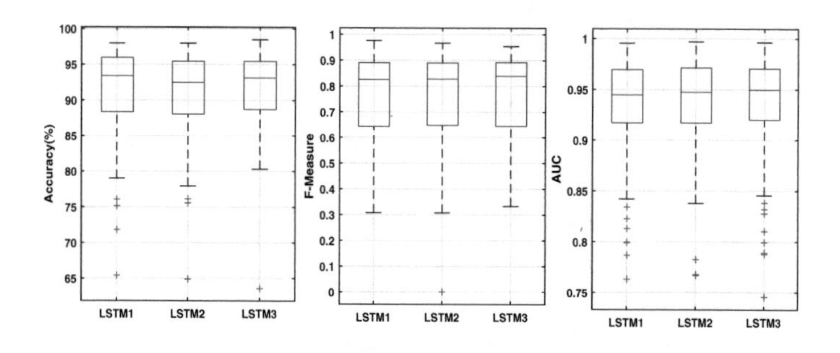

Fig. 4. Box-plot for Accuracy, F-measure and AUC: LSTM

Table 7. Statistics: LSTM

	Min	Max	Mean	Median	Var	Q1	Q3
LSTM1	65.43	97.98	91.36	93.39	34.99	88.37	95.96
LSTM2	64.89	97.98	91.29	92.49	30.25	88.06	95.45
LSTM3	63.56	98.48	91.75	93.13	25.19	88.72	95.45

Table 8. Rank-sum test: LSTM

	LSTM1	LSTM2	LSTM3
LSTM1	1.00	0.79	0.66
LSTM2	0.79	1.00	0.52
LSTM3	0.66	0.52	1.00

7 Conclusion

This paper provides the empirical evaluation of anti-pattern prediction utilizing various LSTM models, word embedding technique with distinct sequence padding sizes, and data sampling techniques. We evaluated the models using a five-fold cross-validation technique, and their prediction abilities are compared using accuracy, AUC, and F-measure. The main findings of our work are as follows:

- Word embedding technique with sequence padding size as 100 performs better when compared to other models with varying sequence padding size.
- Borderline SMOTE (BLSMOTE) performs best among the data sampling techniques.
- LSTM3, i.e., LSTM model with three hidden layers, is performing better when compared to other models.

There is much more scope for predicting the web service anti-patterns using NLP techniques. Applying different feature selection techniques and different data sampling techniques, and classifier techniques on the metric set obtained by applying various NLP techniques on WSDL files and studying the effectiveness of these models in web service anti-pattern prediction would be interesting.

References

1. Brown, W.H., Malveau, R.C., McCormick, H.W., Mowbray, T.J.: AntiPatterns: Refactoring Software, Architectures, and Projects in Crisis. Wiley, Hoboken (1998)
2. Segev, A., Toch, E.: Context-based matching and ranking of web services for composition. IEEE Trans. Serv. Comput. **2**(3), 210–222 (2009)
3. Chen, T.H., Shang, W., Jiang, Z.M., Hassan, A., Nasser, M., Flora, P.: Detecting performance anti-patterns for applications developed using object-relational mapping. In: Proceedings of the 36th International Conference on Software Engineering, pp. 1001–1012 (2014)
4. Ouni, A., Kula, R.G., Kessentini, M., Inoue, K.: Web service antipatterns detection using genetic programming. In: Proceedings of the 2015 Annual Conference on Genetic and Evolutionary Computation, pp. 1351–1358. ACM (2015)
5. Ouni, A., Kessentini, M., Inoue, K., Cinnéide, M.O.: Search-based web service antipatterns detection. IEEE Trans. Serv. Comput. **10**(4), 603–617 (2015)
6. Jaafar, F., Guéhéneuc, Y.-G., Hamel, S., Khomh, F., Zulkernine, M.: Evaluating the impact of design pattern and anti-pattern dependencies on changes and faults. Empir. Softw. Eng. **21**(3), 896–931 (2015). https://doi.org/10.1007/s10664-015-9361-0
7. Kumar, L., Sureka, A.: An empirical analysis on web service anti-pattern detection using a machine learning framework. In: 2018 IEEE 42nd Annual Computer Software and Applications Conference (COMPSAC), vol. 1, pp. 2–11. IEEE (2018)
8. Saluja, S., Batra, U.: Optimized approach for antipattern detection in service computing architecture. J. Inf. Optim. Sci. **40**(5), 1069–1080 (2019)
9. Chawla, N.V., Bowyer, K.W., Hall, L.O., Philip Kegelmeyer, W.: SMOTE: synthetic minority over-sampling technique. J. Artif. Intell. Res. **16**, 321–357 (2002)

10. Han, H., Wang, W.-Y., Mao, B.-H.: Borderline-SMOTE: a new over-sampling method in imbalanced data sets learning. In: Huang, D.-S., Zhang, X.-P., Huang, G.-B. (eds.) ICIC 2005. LNCS, vol. 3644, pp. 878–887. Springer, Heidelberg (2005). https://doi.org/10.1007/11538059_91
11. Sundermeyer, M., Schlüter, R., Ney, H.: LSTM neural networks for language modeling. In: Thirteenth Annual Conference of the International Speech Communication Association (2012)

Federated Learning with Blockchain Approach for Trust Management in IoV

Achref Haddaji[1,2,3](✉), Samiha Ayed[2], and Lamia Chaari[3]

[1] National School of Electronics and Telecommunications of Sfax, Sfax, Tunisia
haddajiachref7@gmail.com

[2] LIST3N Laboratory, University of Technology of Troyes, Troyes, France
samiha.ayed@utt.fr

[3] Digital Research Center of Sfax (CRNS), Laboratory of Signals, systeMs aRtificial Intelligence, neTworkS (SM@RTS), Sfax University, Sfax, Tunisia
lamia.chaari@isims.usf.tn

Abstract. The internet of vehicles (IoV), a variant of the traditional VANET, allows real-time data exchange between vehicles, roadside units, parking, and city infrastructure. Nevertheless, the IoV poses many security concerns due to its open nature. Traditional security solutions may not address all the IoV security risks and provide complete protection. Therefore, it is critical to establish trust and to identify dishonest nodes. As a result, trust management-based techniques are also required to improve IoV security. This paper proposes federated learning with a blockchain approach for trust management (FBTM) in IoV. Thus, a vehicular trust evaluation is designed to improve the data acquired for the federated learning model learning process. Moreover, a novel blockchain-based reputation system is developed to guarantee the storage and the share of global federated learning models. In the meanwhile, proof of reputation consensus is proposed to evaluate the roadside units operating as aggregators in the IoV network. Simulation results demonstrate that the proposed scheme is effective for IoV security.

1 Introduction

Internet of Vehicles (IoV) originated from vehicular ad hoc networks (VANET) and planned to become the internet of autonomous vehicles (IAV), is among the most active research areas, combining VANET with the Internet of Thing (IoT). It can intelligently coordinate heterogeneous vehicles, objects, humans, and environments based on the vehicle to everything communication (V2X) to build a broader network that can provide resources and services for urban areas. It also uses different types of wireless technologies (vehicular communications, mobile communication, and short-range static communication) to ensure efficient communication between vehicles and allow interaction with the infrastructure such as roadside units (RSUs) and cloud servers. However, the V2X communication

L. Barolli et al. (Eds.): AINA 2022, LNNS 449, pp. 411–423, 2022.
https://doi.org/10.1007/978-3-030-99584-3_36

makes the IoV an open-access environment that gets more exposed to attackers. In this context, numerous works have focused on security measures in the IoV. Some of these solutions are based on cryptography algorithms. However, these methods are exclusively useful against outsider attackers [1]. Recently, with the wide use of artificial intelligence (AI), more robust approaches are developed to manage IoV security and detect more complex types of threats. Security approaches based on machine learning (ML) relies on learning from data to make decisions with precision over time. Nevertheless, traditional ML necessitates centralized training, which is vulnerable to data protection and leakage issues. For data protection reasons, multi-data owners must share their training data. From this basis, federated learning (FL), a decentralized ML approach, solves the centralized training problems. All network participants may participate in the global model development without sharing data. In typical FL architecture, the centralized cloud server handles the model aggregation. Local models are collected by the server and used to achieve aggregation. RSUs act as aggregators of models. One of the most essential difficulties after the aggregation process is the secure storage and share of the model. With centralized storage, hacking and privacy leakage are prevalent problems. The most recent approaches for IoV security integrate the blockchain as a solution to tackle the aggregation threats. RSUs can work together to support blockchain to enhance the security of model storage and aggregation. The integration of FL with the blockchain promotes privacy in the IoV by enabling multiple participants to train ML models locally using their data. It also improves security by distrusting the storage of the aggregated model via the blockchain. But it is still difficult to prevent malicious nodes from participating in the FL mechanism. Furthermore, the major of existing solutions do not take into account the data gathered from untrustworthy sources and utilized for FL training, which can negatively impact the local model created and the global model aggregated. In this paper, we proposed a secure trust management solution based on the integration of FL and blockchain in IoV. Our proposed scheme consists of assigning a trust value to nodes in the network based on their interactions, to limit the effects of malicious behaviors. The addition of trust to the FL can improve the performance of the FL models. Our scheme proposes a vehicular trust evaluation, allowing only trusted vehicles to participate in the FL local training. Blockchain framework is proposed to improve the FL global model's storage and share. Using a proof of reputation (PoR) [2] consensus, only the authorized RSUs can participate in the generation of the global model. Negative reputations are assigned to RSUs with malicious behaviors in the network.

2 Related Works

Many studies have been carried out to enhance the security in IoV. Eventually, with the widespread use of ML in many areas, some researchers have already used ML algorithms on trust-security schemes in IoV. In [3], Siddiqui et al. used the ML techniques to identify malicious vehicles based on similarity, familiarity, and packet delivery ratio parameters. They used supervised learning to detect malicious vehicles and unsupervised learning to acquire an optimal trust threshold. In [4], the authors proposed a misbehavior detection system to identify and revoke internal attacks in vehicular networks. They combined ML algorithms with a reputation system to improve the proposed scheme's performance and provide high reliability of both vehicles and messages. Each message received from a trusted vehicle is evaluated before sending the report to the local authority. Then the reports are combined using the dempster-shafer theory. The reputation scores are used as belief value for the dempster-shafer theory to calculate the reputation updates. To deal with the IoV misbehavior attacks, the authors in [5] suggested a misbehavior detection system. The proposed system uses the FL algorithms to ensure the privacy of vehicles' basic safety message (BSM) dataset. The FL model can identify the misbehavior attacks using the data of each vehicle in the network. Some other works exploited the usability of the blockchain for security and trust to establish trust management in vehicular networks. The work [6] exploited the blockchain to build a distributed trust management scheme for vehicular networks. They used a Bayesian inference model to validate the messages exchanged between vehicles. Each vehicle can generate a rating for each message received from its neighbors and upload it to the RSU. Then, the RSUs compute the trust value offsets and add them to the blockchain with these ratings. Similarly, in [7], the authors proposed a decentralized trust management approach for IoV. The RSU nodes use the trust, which present the edge nodes, to maintain the update trust values of vehicles. They used the Ethereum blockchain as a platform with smart contracts to illustrate the practicality of the proposed approach. Recent work has combined blockchain and FL to develop fully distributed security systems. Chai et al. [8] integrated a hierarchical blockchain framework with a hierarchical FL in which vehicles exchange the learning knowledge in the IoV environment. The blockchain framework consists of ground chains in which vehicles gathered the data and act as FL workers, and a top chain where data is recorded data from RSUs who also act as FL workers in the top chain. We should mention that the combination of FL offers several benefits in terms of IoV security. However, training models using malicious data can significantly impact the FL model's performance. In addition, RSUs that operate as FL aggregators are also subject to several attacks. These two limitations are taken into consideration in our work. Table 1 provides a summary of the different approaches detailed above.

Table 1. Related work comparative table

Ref-year	AI techniques		Blockchain	Trust management	
	ML	FL		Vehicle	Infrastructure
[3]-2019	✓	✗	✗	✓	✗
[4]-2020	✓	✗	✗	✓	✗
[5]-2021	✗	✓	✗	✗	✗
[6]-2018	✗	✗	✓	✓	✓
[7]-2020	✗	✗	✓	✓	✓
[8]-2020	✗	✓	✓	✗	✗
FBTM-2022	✗	✓	✓	✓	✓

3 Proposed FL Based Blockchain Trust Management in IoV

3.1 Design Overview

In this section, we describe our proposed trust scheme called federated learning with blockchain for trust management (FBTM) in IoV. The general design of our proposed approach is illustrated in Fig. 1. We combine the usability of FL and blockchain with trust to develop a powerful trust management solution in IoV. Particularly, FL aims to increase privacy protection by reducing data transfer volume and speeding up the vehicle learning process. Therefore, we add the vehicular trust evaluation to provide a reliable data collection for the FL models' training. In addition, blockchain is integrated into our FBTM solution to deal with the traditional aggregation's malfunction. Our FBTM core scheme consists of provider vehicles, normal vehicles, RSU nodes, and trust authority nodes. The selection of the providers is based on the RSU logs that contains previously contacted vehicles with high trust value. In case of an empty log, we use the non-registered vehicles with reliable trust values. In our FBTM, the provider vehicles act as federated servers. Hence, to ensure a secure and trusted IoV's environment, we propose three major key contributions described as follows:

- **Vehicular trust evaluation:** Providers start with assessing the trustworthiness of normal vehicles in their transmission range. The vehicular trust evaluation consists of two types of trust: (1) Direct trust, used to compute the trust value of normal vehicles that directly communicate with the provider based on the history of interactions. (2) Recommended trust is used to calculate the trust values of vehicles that don't have direct interactions with the provider. We use the arithmetic mean algorithm (WAM) to develop trust paths.
- **Federated Learning model:** As mentioned above, providers acting as federated servers first download the global model from the RSU to train it locally. The data collected from trustworthy vehicles can improve the learning process of the local model. Then, after finalizing the train, each provider sends

the gradients of its model to RSU. Acting as an aggregator, the RSU uses the generated average technique to generate the global model and send it to the blockchain.

- **Blockchain consensus for data storage, share, and reputation estimation:** In this phase, we use a reputation-based system with the blockchain to prevent the malicious behavior of RSUs. Proof of Reputation (PoR) consensus is proposed to assign a reputation score to the RSUs participating in the global model generation. The PoR can motivate the RSUs to aggregate the FL models and generate them with high performances.

In this paper, the FBTM uses FL and blockchain technologies associated with trust management to build a decentralized and temper-proofed trust for both vehicles and infrastructure.

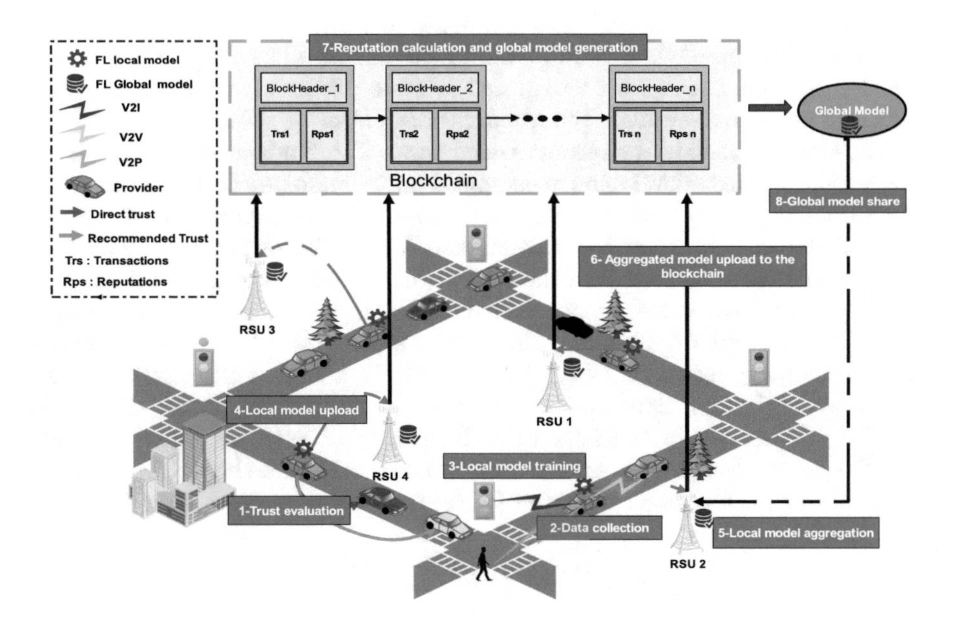

Fig. 1. Proposed network model: overall architecture.

3.2 Vehicular Trust Evaluation

The evaluation of the vehicle's trust process must not be overly complicated. The trust measurement comprises both direct trust and recommended trust. In our study, for each given provider vehicle, normal vehicles can be simply divided into nodes with direct communications with the provider and nodes without communications with the provider. The trust value of vehicles with direct communication with the provider is calculated using the direct trust evaluation method. The trust value of the rest is computed using the recommended trust.

3.2.1 Direct Trust Evaluation

The direct trust judgement must be based on the history of the interactions between vehicles. After each direct interaction, the providers will evaluate the normal vehicles. The direct trust evaluation is as follows:

$$T_{Direct}(v_p, v_n) = \frac{\sum_{i=1}^{k} t_i}{k} \tag{1}$$

Where, v_p is the provider, v_n is a normal vehicle with direct communication with v_p, t_i presents the score of the normal node evaluated by v_p after the ith transaction, and k represents the global number of direct transactions between them.

3.2.2 Recommended Trust Evaluation

The high mobility in the IoV network makes it difficult to record information, evaluate direct interactions, and maintain the history of relationships between vehicles. Recommended trust is required to solve this problem. However, many paths are possible to evaluate the recommended trust. We use the trust pathway as a metric to calculate the recommended trust. The fundamental goal of trust calculation through developing trust paths is to increase confidence in trust prediction. A recommended trust system must discover which nodes are the most trusted at delivering the recommendation of the target node. The first recommendation is derived from the vehicle directly related to the provider. The recommended trust value discovery challenge may be related to discovering a confident trust path from the provider to the target vehicle. We use the weighted arithmetic mean algorithm (WAM) to compute the recommended trust based on the average trust bath between the provider and its neighbors. We divide vehicles into the set of "vehicles with direct contact with the target vehicle" notedV_{PDT} and the set of "vehicles with direct contact with the provider" noted V_{PDP}. We define the path's weight $w_p(v_p,v_i)$ from the provider v_p to each vehicle v_i in V_{PDT}:

$$w_p(v_p, v_i) = \frac{\sum_{vj \in V_{PDP}} T(v_p, v_j) * w_p(v_j, v_i)}{\sum_{vj \in V_{PDP}} T(v_p, v_j)} \tag{2}$$

where T(v_p,v_j) presents the direct trust value of the provider v_p to the node v_j. The recommended trust value of the target node v_t is:

$$T_{rec}(v_p, v_t) = \frac{\sum_{v_i \in V_{PDT}} T(v_p, v_i) * w_p(v_i, v_t)}{\sum_{i \in V_{PDT}} T(v_p, v_i)} \tag{3}$$

After assessing all the neighbors of each provider v_p, it is now able to determine the global trust G_T in each provider's region:

$$G_T = \frac{\sum_{i}^{K} T_{Direct}(v_p, v_i)}{K} + \frac{\sum_{i}^{P} T_{Rec}(v_p, v_j)}{P} \tag{4}$$

Where K is the number of vehicles connected directly with the provider and P is the number of vehicles without communications with the provider.

3.3 Federated Learning Based Model

On the vehicle side, we use the FL to complete the computational work for the trust data request. In our proposed model, the provider nodes present the FL clients, and the RSUs take the role of the server to aggregate the local models. The providers must upload the gradients of models trained locally to the RSU, aggregating all these model parameters. The provider can gather safe data from its neighbors to construct a reliable dataset D_i used for the model's train after determining the trust degree of the provider's neighbors. The significant advantage of FL is training the model locally without sending the data to the server. For each provider, a loss function L_i is specified to measure the difference between real values and estimated values of samples in the provider's dataset D_i.

$$L_i(w) = \frac{1}{|D_i|} \sum_{j \in D_i} l_j(w, x_j, y_j) \tag{5}$$

where $l_j(w,x_j,yj)$ is the loss function on data sample (x_j, y_j) with the parameter vector w, and $|D_i|$ constitute the sample's number in a particular provider's dataset. Thus, the global loss function $L(w)$ can be defined using the formulas below.

$$L(w) = \frac{1}{|D|} \sum_{i \in PS} \sum_{j \in D_i} \beta_i . \frac{l_i(w_j, x_j, y_j)}{|D_i|} \tag{6}$$

|D| constitutes the sample's number of all provider's datasets, and β_i (where $\sum_i \beta_i = 1$) is the contribution value of provider V_i to global learning. Each provider enhances the accuracy of the global model continuously in the training process by reducing the global loss function L(w) to find the optimum global model W_T^*:

$$W_T^* = arg\ min\ L(w) \tag{7}$$

In the local training, providers must receive the global model parameters W_T from the RSU. Then, they train the model W_t^1,W_t^2,..., W_t^n with its own dataset D_i, where 1,2,...,n are the index of providers participating in the training, and t denoting the communication cycle for training. A provider v_i needs to download the update H_t^i from the RSU:

$$H_t^i = W_t^i - W_t \tag{8}$$

After the training process, the gradients of each provider's model will be sent to the RSU for aggregation. The RSU uses the federated average technique for the global model generation. Equation (9) shows the aggregation calculation:

$$W_{t+1} \leftarrow W_t + \alpha \sum_{I \in PS} \frac{|D_i|}{\sum_{k \in PS} |D_k|} H_t^i \tag{9}$$

Where PS constitutes the set of provider vehicles cooperated in training iterations t, and the learning rate is α.

3.4 Blockchain Consensus for Data Storage, Share, and Reputation Estimation

The RSUs collect all the local models uploaded by providers for the global model generation using blockchain technology. The blockchain can efficiently ensure the integrity and legitimacy of the global model. Each RSU's reputation is represented in our proposed consensus method, and it must also be saved and agreed upon by the other participants in the blockchain. To save the RSU's reputation, we modified the structure of the block. The block header contains the hash of the previous block, the timestamp, and the signature *sig* of the block creator. The transaction blocks containing the aggregated models and the reputation blocks providing the reputation scores of the RSUs participating in block validation constitutes the block body. We use the PoR consensus technique in our suggested blockchain. In the PoR protocol, the RSU with the maximum reputation value generates and publishes the block. The RSUs with higher reputations vote for the block. Then, the vote number is used to validate the block. We use rewards and reputation values to encourage RSUs to submit blocks. RSUs with high reputation values are more likely to keep the network secure. Therefore, RSU nodes must be registered using their keys saved locally to be recognized by each other. RSUs with malicious behavior are penalized, and their reputations are reduced, with the entire network keeping track of their reputation modifications.

3.4.1 Reputation-Based Consensus

The proposed Reputation procedure involves three main phases: Leader selection and block generation phase, evaluation phase, and upload phase. In the first phase, the RSU with the highest reputation value is chosen as a leader. After the global model generation, it adds the model in a block and sends it as well as its signature *sig* for the validation. RSUs with reputations more than a fixed threshold γ are selected as validators. They are in charge of confirming transactions and signatures. In the evaluation phase (described in Algorithm 1), the validators are in charge of confirming transactions and signatures. They start the verification of the block. If the transaction is illegitimate or does not come from the leader, it will be rejected. On the other hand, the validators convey a permission vote. If most validators vote in favor of a transaction, it is declared legitimate, and it will append to the blockchain. In the upload phase (described in Algorithm 2) the votes and reputation values of the validators are collected. When the overall reputation value of the validators sending the permission $R(Vot_T)$ exceeds the $\frac{2}{3}$ of the overall reputation of the validators $R(Vot_{total})$, the block is added to the blockchain. Otherwise, if the overall reputation value of the validators voting with rejection $R(Vot_F)$ is greater than the $\frac{2}{3}$ of $R(Vot_{total})$, the block will be invalid, and the value of a leader's reputation will be decreased.

Algorithm 1: Transaction Block Evaluation

```
Data: Blocks, RSU i ∈ (1..n), validators
Result: Block validation state
Block reception ;
if Block not leagal || Transaction block is not from the leader then
    if i ∈ validators then
        sendmsg(Reject, hash, sig)
    end
else
    if i ∈ validators then
        sendmsg(Accept, hash, sig)
    end
end
```

3.4.2 Reputation Estimation

The reputation value can be computed by all the nodes based on the transactions Tr, the degree of participation in the PoR consensus, the RSU's trust value and the accuracy of the aggregated model. The reputation score formula is as follows:

$$R_{RSU} = \alpha_1 R_i \ + \alpha_2 \ Rp \ + \ \alpha_3 R_{acc} \ + \ \alpha_4 R_T \tag{10}$$

Where α_1, α_2, α_3, and α_4 are the weights that vary between [0,1]. R_i is the reputation based on the RSU interaction with its neighbors. The interaction factor depends on the relationship with other RSUs, the regularity of interaction with them, and their reputations. The R_p presents the reputation based on the participation in the PoR consensus. It can be calculated as follows:

$$R_P = \sigma_1 \ C_{vot}. \sum_i^n Tr_i - \sigma_2 \ E_{vot}. \sum_i^n Tr_i \tag{11}$$

Where σ_1 and σ_2 are weight parameters that vary between [0,1], C_{vot}, and E_{vot} are respectively the number of correct voting and the number of erroneous voting of the node in recent period of time.

R_{acc} is the reputation based on the accuracy of the aggregated model. RSUs with higher model accuracy have a higher R_{acc}. R_T is the reputation based on the trust of the RSUs. The Trust of the RSU, denoted T_{RSU}, can be described as follows:

$$T_{RSU} = \sum_{i=1}^{c} G_{T_i} \tag{12}$$

Where G_{T_i}, as mentioned in Eq. (4) is the global trust assigned to each provider connected to the RSU, and c is the total number of providers connected to the RSU.

Algorithm 2: Transaction Block Upload

```
Data: Blocks, RSU i ∈ (1..n), validators
Result: Block validation state
while receive(msg) do
    if msg.type == Accept then
        sum_a = ∑ R(Vot_T)
        if sum_a ≥ 2/3 * R(Vot_total) then
            return Upload ;
        end
    else
        if msg.type == Reject then
            sum_r = ∑ R(Vot_F) ;
            if sum_a ≥ 2/3 * R(Vot_total) then
                return not Upload
            end
        end
    end
end
Result: Block validation state
```

4 Simulation Results

We use a publicly available dataset NSL-KDD [9] to perform our experimentation to evaluate our proposed approach. The NSL-KDD has already been frequently employed in IDS studies. This is owing to its enough data records and variety of attack types. It consists of 148,517 samples composed of 41 features. The 148,517 samples are divided into 125,973 samples used for the training and the rest used for the test. To address categorical variables in data preprocessing, we employ One Hot Encoding (OHE) to increase the number of features, which becomes 122. Then, we use the Attribute Ratio technique presented in [10] for features selection. The implementation of our FL model is based on PyTorch, which is an open-source ML framework enabling the creation of DL projects, and PySyft [11], which is an extension of PyTorch for DL computations that is both secure and private. Each vehicle has its own private dataset stored locally in a real scenario. However, in our simulation, The FL is released by dividing the dataset into many parts and distributing them across Vehicles, acting as virtual workers. Several virtual workers run all these simulations simultaneously in a signal machine. We use two uses cases for experimentation: the centralized learning (CL) approach and the FL approach. In the CL approach, all the set is used by a single server to train and test the model. In the FL, the data is spread across multiple clients, and the server is used to aggregate the client's models. We select respectively 2,4 and 6 clients as vehicle providers to train the FL models locally. We use the artificial neural network (ANN) model for the two approaches. The ANN consists of a 122-neuron input layer, one hidden layer of 100 neurons, and a 2-neuron output layer. It utilizes ReLU as an activation function. The comparison between the two techniques is depicted in Fig. 2(a).

The accuracy of centralized learning reaches 98.2%. Using Federated Learning, on the other hand, the highest level of accuracy is achieved when two providers train the models, up to 94,8%. We note that the FL accuracy decreases when the number of providers increases. We also indicate that the FL model's accuracy with 2 and 4 selected providers is close to the CL model's accuracy. To deal with the behavior of FL clients, Fig. 2(b) shows the performance of each provider in terms of attack detection. We have chosen 4 providers to play the role of FL clients. The first observation is that the accuracy values of the trained providers are nearly similar. The accuracy of all providers reaches high levels, exceeding 90%. In Fig. 3, we use the training time as a metric to see the difference between the CL model and the FL model. This figure shows that the FL models improved the training time. The FL model, for example, takes less than 200 s to train with 4 and 6 providers. In contrast, the CL model needs more time for the learning process, which takes more than 700 s for the training. In comparison with the related works detailed in Sect. 2, our FL model outperforms [5], which has an accuracy of 79% and [8], in which has an accuracy did not exceed 90%.

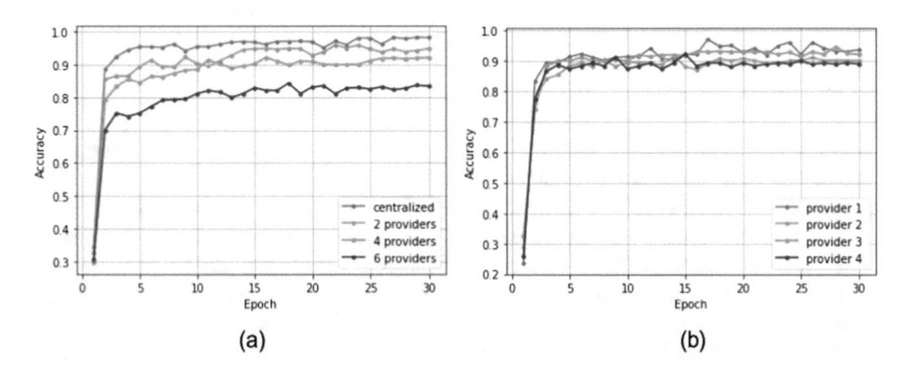

Fig. 2. Comparative results in terms of accuracy: (a) Comparative results of CL model accuracy with FL models accuracy. (b) Comparative results O FL client accuracy.

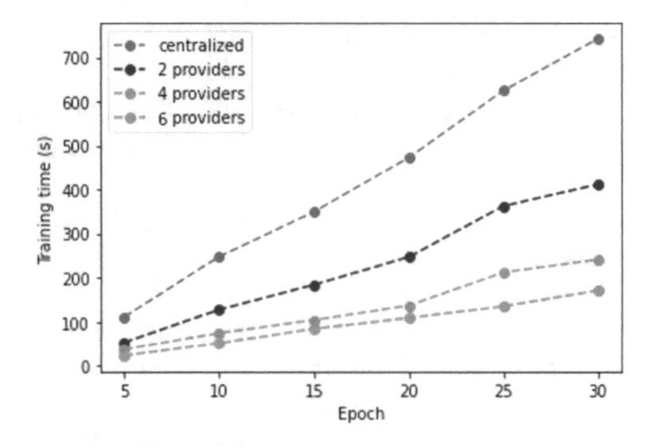

Fig. 3. Training time evaluation.

The CL models detect intrusion attacks with high accuracy. Using these models, there are privacy concerns since data must be gathered by a single node, increasing the possibility of attacks. FL models approximate the performance of the CL model in terms of attack detection with improved the real time learning.

5 Conclusion and Perspectrives

This paper proposes an approach that combines the FL and the blockchain for trust management in IoV. Using a vehicular trust evaluation, provider vehicles selected by the RSUs can estimate the trust values of their neighbors and collect accurate data. The providers also act as FL clients, and they train the FL models with their collected data. In addition, a novel blockchain is developed to maintain the secure storage and share of the FL global model. Furthermore, we suggest a proof of reputation consensus is used to evaluate the RSU acting as FL aggregators. The results demonstrate that our approach reaches an accurate performance with high accuracy values. As a future direction, we tend to add other criteria with the trust for selecting the FL participants.

References

1. Siddiqui, S., Mahmood, A., Sheng, Q., Suzuki, H., Ni, W.: A survey of trust management in the internet of vehicles. Electronics **10**, 2223 (2021)
2. Zhuang, Q., Liu, Y., Chen, L., Ai, Z.: Proof of reputation: a reputation-based consensus protocol for blockchain based systems. In: Proceedings of the 2019 International Electronics Communication Conference, pp. 131–138 (2019)
3. Siddiqui, S.A., Mahmood, A., Zhang, W.E., Sheng, Q.Z.: Machine learning based trust model for misbehaviour detection in internet-of-vehicles. In: Gedeon, T., Wong, K.W., Lee, M. (eds.) ICONIP 2019. CCIS, vol. 1142, pp. 512–520. Springer, Cham (2019). https://doi.org/10.1007/978-3-030-36808-1_56
4. Gyawali, S., Qian, Y., Hu, R.: Machine learning and reputation based misbehavior detection in vehicular communication networks. IEEE Trans. Veh. Technol. **69**, 8871–8885 (2020)
5. Uprety, A., Rawat, D., Li, J.: Privacy preserving misbehavior detection in IoV using federated machine learning. In: 2021 IEEE 18th Annual Consumer Communications Networking Conference (CCNC), pp. 1-6 (2021)
6. Yang, Z., Yang, K., Lei, L., Zheng, K., Leung, V.: Blockchain-based decentralized trust management in vehicular networks. IEEE Internet Things J. **6**, 1495–1505 (2018)
7. Singh, P., Singh, R., Nandi, S., Ghafoor, K., Rawat, D., Nandi, S.: Blockchain-based adaptive trust management in internet of vehicles using smart contract. IEEE Trans. Intell. Transp. Syst. **22**, 3616–3630 (2020)
8. Chai, H., Leng, S., Chen, Y., Zhang, K.: A hierarchical blockchain-enabled federated learning algorithm for knowledge sharing in internet of vehicles. IEEE Trans. Intell. Transp. Syst. **22**, 3975–3986 (2020)
9. Dhanabal, L., Shantharajah, S.: A study on NSL-KDD dataset for intrusion detection system based on classification algorithms. Int. J. Adv. Res. Comput. Commun. Eng. **4**, 446–452 (2015)

10. Chae, H., Jo, B., Choi, S., Park, T.: Feature selection for intrusion detection using NSL-KDD. Recent Adv. Comput. Sci. **20132**, 184–187 (2013)
11. Ziller, A., et al.: PySyft: a library for easy federated learning. Federated Learning Systems, pp. 111-139 (2021)

Detection of Distributed Denial of Service Attacks Using Entropy on Sliding Window with Dynamic Threshold

Shail Saharan(✉), Vishal Gupta, Nisarg Vora, and Mohul Maheshwari

Birla Institute of Technology and Science Pilani, Pilani Campus, Pilani, India
{p20170404,vishalgupta,f20180254,f20180229}@pilani.bits-pilani.ac.in

Abstract. The Internet has become an integral part of our day-to-day lives, from remaining connected to accessing information from any part of the world. Distributed Denial of service (DDoS) attacks disrupts the normal functioning of the Internet. Because of DDoS attacks, services over the Internet become inaccessible; regular hosts lose connectivity, etc. DDoS attacks are more dangerous because it is not always possible to differentiate whether an organization is under attack or its' just normal traffic. Therefore, an effective detection mechanism is needed that is computationally less expensive and can detect different types of attacks with good accuracy. Hence, in this paper, we propose Entropy with Dynamic Thresholds to detect DDoS attacks. A dynamic threshold helps us accurately detect an attack in different rates of traffic. To validate our approach, we have used the CICDDoS-2019 attack dataset.

1 Introduction

All the network resources have a finite amount of computational power and finite capability to handle simultaneous requests and connections. An unexpected and sudden spike in traffic can disrupt the normal functions of a network system. It can either lead to a slower service or response from the network or can even lead to complete dysfunction of the network system. Various attackers try to exploit this particular vulnerability of the network systems to disrupt the system's normal functioning. Distributed Denial of service (DDoS) attack is one such attack. DDoS attacks are malicious attempts to hijack or disturb the normal working of a networking server or infrastructure by depleting its computational and networking resources. In a DDoS attack, the attacker floods the network with a lot of unwanted traffic to deplete the server's or a host's resources, catering to this useless and artificially created traffic. The attacker does so by utilizing many compromised computer systems, Internet of Things devices (IoT), etc. The attacker can remotely control these systems. Such a network is called a botnet, and individual systems are called bots or zombies.

With the increase in the use of the Internet, IoT-based devices, cloud computing, etc., DDoS attacks are becoming a primary concern for various organizations of various scales. Furthermore, sometimes, it is also not easy to differentiate

L. Barolli et al. (Eds.): AINA 2022, LNNS 449, pp. 424–434, 2022.
https://doi.org/10.1007/978-3-030-99584-3_37

between benign traffic and attack traffic. In the first quarter of 2020, Amazon Web Services(AWS) reported one of the largest DDoS attacks at 2.3 Tbps [1]. A DDoS attack in February 2018 affected Github. The attack sent traffic at 1.3 Tbps onto the Github servers with 126.9 million packets of data each second and leading to service downtime of 20 min [2]. In another instance, Google, in 2017, reported that one of the biggest DDoS attacks on record affected the company's cloud platform. The DDoS attack lasted over a six-month campaign, peaking to 2.5Tbps in traffic [3].

From the statistics mentioned above, it can be easily said that DDoS attacks are a huge nuisance and have continuously increased over the years. Furthermore, it is not easy to always differentiate between attack traffic and benign traffic. For this, an effective technique to detect whether an attack is happening is needed so that appropriate actions can be taken afterward. One such efficient way to detect DDoS attacks is entropy. Entropy is used to measure randomness or uncertainty in a network's traffic. As soon as this randomness decreases, i.e., one type of traffic dominates the network, the entropy value decreases. This abbreviation leads to the possibility of an attack on the network. One benefit of Entropy-based algorithms is that entropy calculations require minimal computational effort. The majority of detection algorithms based on entropy use a static threshold [4,5]. However, such a static threshold would not work efficiently for DDoS attack detection and end up with many false positives. For instance, in normal traffic, the entropy values can drop due to high traffic from a host and lead to a prediction of an attack, hence false positive. To address this issue, we use a dynamic threshold approach as proposed by [6] for Software Defined Networks(SDN) to detect the attack. We have used the concept of a sliding window to get more accurate results on the dataset used. For sliding the window, our left and right pointers move by one second, to predict the attack in a particular time interval. We identify whether there is a sudden drop in the entropy value of a particular window compared to the average entropy value for some previously encountered windows. Such a drop in entropy value would increase the count of violations. To increase the effectiveness of attack detection, we look for a dynamically adjusted number of such violations in the previous fixed number of windows. It helps us to reduce the number of false positives significantly. More on this will be explained in Sect. 3. The proposed methodology is designed for real Internet scenarios, where different types of DDoS attack can take place. Hence it is tested on CICDDoS2019 attack dataset [7,15], which resembles true real-world data. We compare our results with an existing approach on adaptive threshold proposed in [8]. The main contributions are as follows-

- To detect different types of DDoS attacks and effectively differentiate between legitimate and benign traffic using entropy calculation with an adaptive or dynamic threshold.
- Sliding-Window based approach on flows for entropy calculation.
- The accuracy of the proposed approach in DDoS detection is done based on an authentic DDoS attack dataset CICDDoS-2019 [7,15].

The remaining paper is structured as follows- Sect. 2 contains related work, Sect. 3 describes the proposed methodology, Sect. 4 contains the results, and Sect. 5 concludes the paper.

2 Related Work

When DDoS attacks happen, they congest the target network's bandwidth. So monitoring traffic and analyzing the traffic is a promising way to detect these attacks. This can be done using either machine learning techniques [9,10] which classify the traffic as either attack traffic or benign traffic, or through entropy calculation [6,8,11,12].

Entropy measures the randomness in network traffic. One such type of entropy to measure uncertainty is Shannon's entropy [16]. If n is the number of packets in a window and p_i is the probability of occurrence of event x_i, Shannon's entropy H(X) is given by Eq. (1) and Eq. (2) -

$$H(X) = -\sum_{i=1}^{n} p_i log p_i \tag{1}$$

$$where, p_i = \frac{x_i}{\sum_{i=1}^{n} x_i} \tag{2}$$

Different authors have used variations of Shannon's entropy to detect DDoS attacks. These include General entropy [11], Fast entropy [8], ϕ entropy [12], etc. Sahoo et al. [11] have proposed the use of Generalized entropy (GE) to detect DDoS attacks on SDN controllers. Taking advantage of flow-based traffic in SDN controllers, GE is used to detect low rate DDoS attacks. Generalized entropy $H_\alpha(X)$ is given by [11] as shown in Eq. (3) -

$$H_\alpha(X) = \frac{1}{1-\alpha} log_2(\sum_{i=1}^{n} p_i(\alpha)) \tag{3}$$

where α is the order of general entropy, varying which different values of entropy can be calculated. When $\alpha = 1$, it becomes Shannon's entropy. Li et al. [12] have also proposed detection of DDoS attacks against SDN controllers using entropy. They have used ϕ entropy for the detection of DDoS attacks. ϕ entropy is used to widen the differences between attack traffic and normal traffic, and also adjust parameters according to network conditions. ϕ entropy, $H_\phi(X)$is given by [12] as shown in Eq. (4) -

$$H_\phi(X) = -\frac{1}{sinh(\phi)} (\sum_{i=1}^{n} p_i sinh(\phi log_2 p_i)) \tag{4}$$

The parameter ϕ is used to adjust the sensitivity of measuring the frequency of events, where $\phi > 0$. To reduce computation time in calculation of entropy, David et al. [8] proposed to use Fast entropy to detect DDoS attacks. Their approach is based upon flow-based analysis, while keeping an adaptive threshold based on traffic patterns. A flow consists of all the packets that have same Source and destination IP/Port pair for a certain amount of time. Let a random variable $x(i,t)$ represent the flow count of a particular connection i over a given time interval t. The fast entropy $H_{(i,t)}(X)$ for a particular interval and particular connection is calculated as shown in Eq. (5) and Eq. (6):

$$H_{(i,t)}(X) = -log\frac{x_{(i,t)}}{\sum_{i=1}^{n} x_{(i,t)}} + \tau_{(i,t)} \tag{5}$$

$$where, \tau_{(i,t)} = \begin{cases} \left(log\dfrac{x_{(i,t+1)}}{x_{(i,t)}}\right), if\ x_{(i,t)} \geq x_{(i,t+1)} \\ \left(log\dfrac{x_{(i,t)}}{x_{(i,t+1)}}\right), if\ x_{(i,t)} < x_{(i,t+1)} \end{cases} \tag{6}$$

To evaluate the approach proposed in this paper, the CICDDoS-2019 attack dataset has been used. Zhou et al. [13], Li et al. [14], Bulbul et al. [4], and Ali et al. [5] have also used this dataset to evaluate their approaches based on entropy. To detect anomalies caused in network traffic by botnets and DDoS attacks, Zhou et al. [13] proposed the use of Euclidean distance-based Multi-scale (EDM) fuzzy entropy algorithm. The input is taken as a time series $X = (x_1, x_2, ...x_n)$ with a time scale τ, and the output is the EDM-fuzzy entropy value. They analyzed their algorithm on the CICDDoS-2019 dataset and have shown entropy curves on different time scales. In the approach proposed by Li et al. [14], the primary focus is on volumetric DDoS attacks. For this, they proposed the use of an optimized sliding window for entropy calculation. They used Shannon's Entropy only but on a joint pair of (source IP, source port) and (destination IP, destination port), making it a joint entropy. Bulbul et al. [4] proposed DDoS detection and mitigation using Network Function Virtualization(NFV) and SDN environment. For detection of attack, they used entropy. After detecting that attack is happening, traffic is further analyzed to generate attack patterns so that attack traffic can be differentiated from legitimate traffic. Ali et al. [5] have proposed the use of entropy and Sequential Probability Ratio Test (SPRT) for DDoS detection. Flows are formed after monitoring the packets, and these flows are gathered in specific window sizes. After this entropy calculation, SPRT is used for the detection of DDoS attacks.

From above-mentioned approaches, we can say that entropy is an effective approach for DDoS attacks detection. But to detect all types of DDoS attacks, not confined to only volumetric attacks, we need to make our threshold for DDoS detection dynamic so that it can change according to attack traffic and accurately provide detection. The approach proposed in this paper uses a dynamic threshold with entropy to give better results.

3 Proposed Methodology

The entropy is used to measure the randomness in incoming packets. In the proposed approach, we have calculated entropy at fixed time intervals. We have used the concept of flows in our approach. A flow is a five-tuple entity, and a flow count is the total number of packets of a flow in a particular time interval. Flow id, source address, destination address, source port, and destination port are the five tuples of a flow. So, a flow is a uni-directional flow of packets from one source port, source IP address to another destination port, destination IP address. The entropy value would remain relatively stable without the attack, fluctuating in a specific range. During an attack, the network switch will have multiple packets with the same destination address. The Shannon's Entropy would drop significantly when a few flows would dominate others. The algorithm considers entropy for a window, and we keep on sliding this window to consider different time intervals. Table 1 describes the notations used in the proposed approach. The algorithm is explained in Algorithm 1 (EBDD). The entropy can be calculated as per Eq. (7) (8), and (9), where n is the total number of flows in the interval ΔT, left and right are pointers pointing to the leftmost and rightmost packets respectively in the current window. $Received\_Packets_i(T)$ are the packets received till time T for the ith flow.

$$X_i = Received\_Packets_i(left + \Delta T) - Received\_Packets_i(left) \tag{7}$$

$$p_i = \frac{X_i}{\sum_{i=1}^{n} X_i} \tag{8}$$

$$H+ = -p_i log p_i \tag{9}$$

We use a dynamic threshold approach as proposed by [6]. We have used the concept of a sliding window to get more accurate results on the dataset used. For sliding the window, our left and right pointers move by 1 s to predict the attack in a particular time interval. Also, the proposed methodology is designed for real Internet scenarios, where different types of DDoS attacks can take place. Hence it is tested on CICDDoS-2019 attack dataset [7], which resembles true real-world data.

3.1 Dynamic Threshold Algorithm and Attack Detection

For the current entropy calculation, if $E - H > \delta$, it would be a violation since the current entropy varies by a range of δ from the average entropy of the previous N traffic values, i.e., E. We then check if there are at least M such violations in previous S such windows, as also proposed in [6]. If that is the case, a DDoS attack is detected and reported. Checking for M such violations in S ensures that we do not report a normal increase in traffic which is not an attack. If there is no violation, we update δ and the average entropy for the previous N traffic values, i.e., E, and continue the calculations for the next time window. A constant term α_i (line 17 of EBDD) is used to give more weightage to the most recently observed entropy values as compared to the previous ones. The methodology can also be understood from the flow diagram in Fig. 1.

Table 1. Notations used in the approach

Notation	Definition
n	the total number of flows in an interval
N	For calculating average entropy, the total count of benign entropy values
D	Dictionary consisting of all the flows along with there flow packets
X_i	Flow packets corresponding to a flow D_i in a given window
H_i	Entropy of flow D_i in a given window
H	Total entropy of a window (calculated using summation of H_i)
E	Average entropy of previous N normal traffic values
M	The times violation occurs
s_len	The length of the sliding window
S	Window size to count the number of predictions in the previous history
δ	Dynamic threshold used in prediction of attack
σ	Standard deviation
λ	Threshold multiplicative factor
α	A constant value for calculation of weighted average(changes with every iteration)
$startTime$	A constant denoting starting of traffic in seconds
$endTime$	A constant denoting ending of traffic in seconds

Algorithm 1 EBDD-Entropy based DDoS detection using sliding window and flow entropy

1: $initialize: \ E, \ \delta, \ M, \ S, \ H, N, \ \lambda, \ \Delta T, \ \sigma, \ s\_len$
2: $left \leftarrow startTime$
3: $right \leftarrow startTime + \Delta T$
4: **while** $right \leq endTime$ **do**
5: $\quad H \leftarrow 0$
6: $\quad$ **for** i in all $flow$ id $between$ $left$ and $right$ $timestamp$ **do**
7: $\quad\quad X_i = Received\_Packets_i(left + \Delta T) \ - \ Received\_Packets_i(left)$
8: $\quad\quad p_i \ = \frac{X_i}{\sum_{i=1}^{n} X_i}$
9: $\quad\quad H += -p_i log p_i$
10: $\quad$ **end for**
11: $\quad H \ = \frac{H}{log\ n}$
12: $\quad$ **if** $E - H > \delta$ **then**
13: $\quad\quad$ **if** M times in S **then**
14: $\quad\quad\quad$ Report DDoS attack between window left and left + ΔT
15: $\quad\quad$ **end if**
16: $\quad$ **else**
17: $\quad\quad E = \sum_{i=1}^{N} \alpha_i \cdot H_i$
18: $\quad\quad \sigma = \sqrt{\frac{1}{N} \sum_{i=1}^{N} (E - H_i)^2}$
19: $\quad\quad \delta = \sigma\lambda$
20: $\quad$ **end if**
21: $\quad left += s\_len$
22: $\quad right += s\_len$
23: **end while**

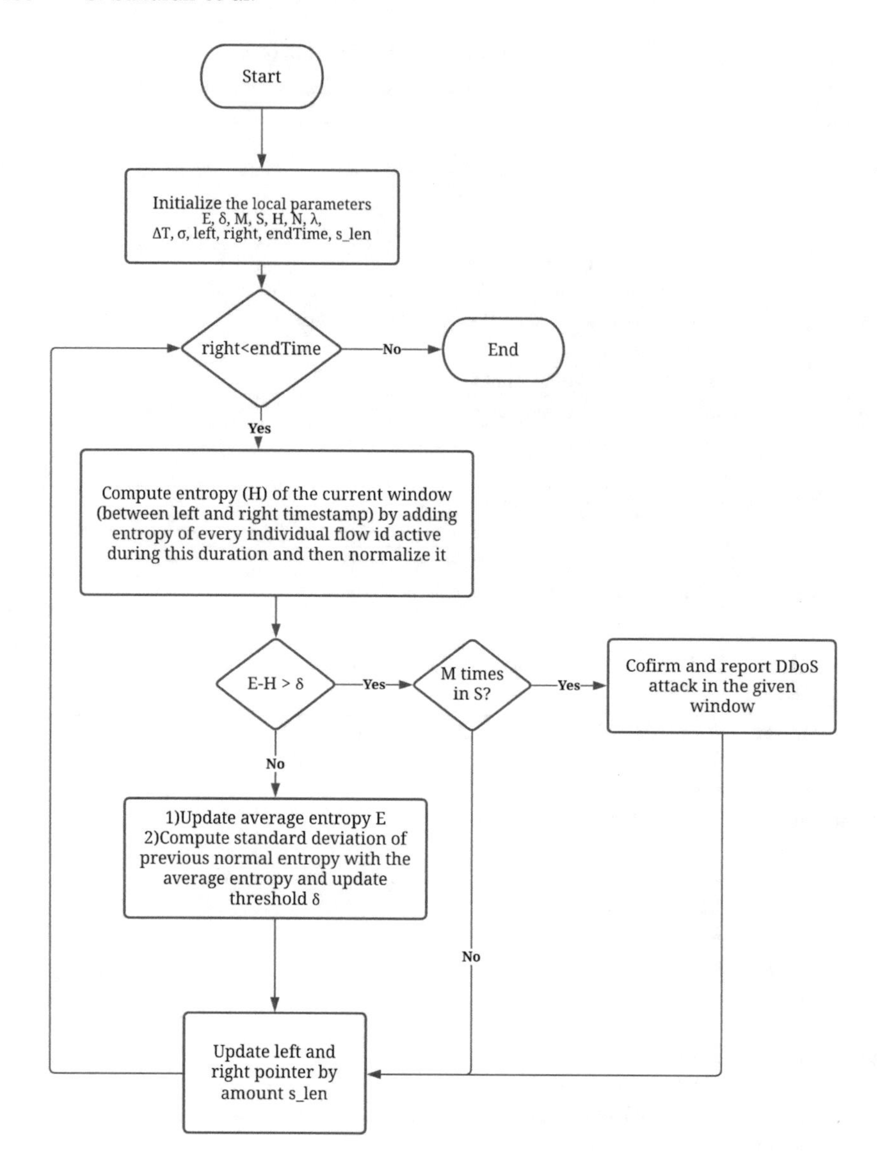

Fig. 1. Flow of the Dynamic threshold algorithm

4 Results and Discussion

4.1 Implementation Details

The CICDDoS2019 dataset contains benign and the most up-to-date common DDoS attacks, resembling actual real-world data. This authentic dataset contains both benign traffic and attack traffic. The dataset was pre-processed by removing infinity and NaN values before applying our algorithm. The features taken for entropy calculation are time-stamp, flow-ID, and flow-packets/sec. Entropy was calculated by varying ΔT and S. To test the accuracy of predicting an attack, two attack datasets of CICDDoS-2019 have been used. One is the Portmap attack dataset, which is the attack dataset of the Portmap DDoS attack, and the second is the DNS-DrDoS attack dataset which is the attack dataset of DNS amplification attacks.

4.2 Results and Analysis

The performance of our proposed algorithm is calculated using *Detection rate* and *Accuracy*. Detection rate is defined as the number of times the algorithm was able to predict an attack window during an attack. Accuracy depends upon False-Positives (FP), True-Positives (TP), False-Negatives (FN), and True-Negatives (TN), as shown in Eq. (10). FP are legitimate traffic detected as attack traffic, TN are the legitimate traffic detected as legitimate traffic, FN are attack traffic detected as legitimate traffic, and TP are attack traffic detected as attack traffic.

$$Accuracy = \frac{TP + TN}{TP + TN + FP + FN} \tag{10}$$

Figure 2 shows the comparison of accuracy with the window size S and ΔT for the Portmap attack dataset and Fig. 3 for the DNS-DrDoS attack dataset. From Fig. 2(a), we can see that for S = 3, the accuracy is maximum, and from Fig. 2(b), it is clear that accuracy starts saturating for $\Delta T \geq 10$. Similarly, in Fig. 3(a), accuracy starts increasing after S = 3, and from Fig. 3(b), it is clear that accuracy again saturates for $\Delta T \geq 10$. This observed difference in accuracy is because the attack flows are continuously present along with the benign packets in each window in the DNS-DrDoS attack dataset whereas in Portmap, the attack is there in some windows and absent in others.

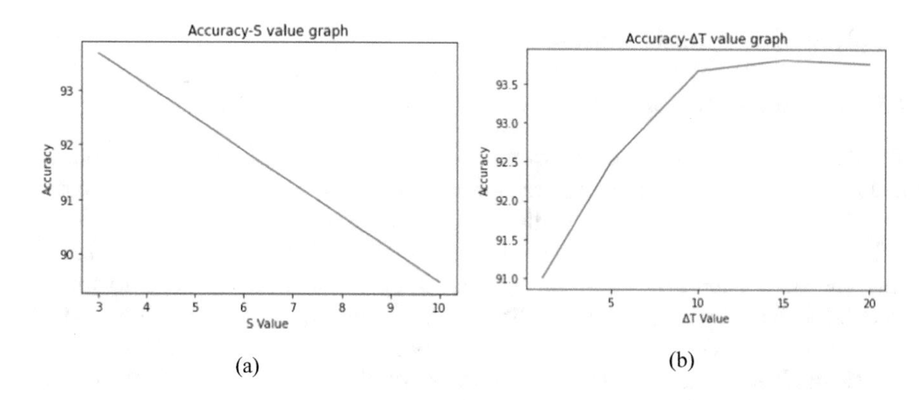

Fig. 2. Accuracy for Portmap attack dataset with varying (a) S-value, and (b) ΔT

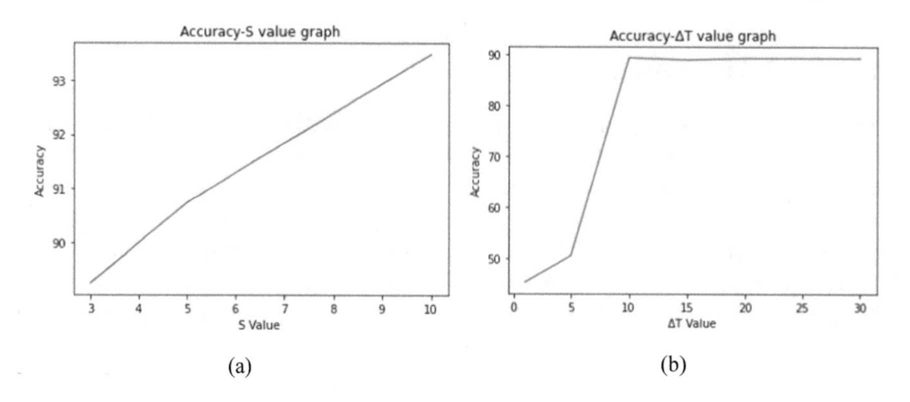

Fig. 3. Accuracy for DNS-DrDoS attack dataset with varying (a) S-value, and (b) Δ T

Based on the above-mentioned results, to predict different types of attacks with effective accuracy, we can keep S $\geq$ 3 and ΔT $\geq$ 10. Through this, for S = 3, the accuracy observed in predicting the Portmap attack was 93.67%, and that of DNS-DrDoS was 89.3%. The algorithm gave a 100% detection rate (i.e., the window in which the attack is detected) on both Portmap and DNS-DrDoS attack dataset, with only 5% false positives.

Furthermore, the dataset reports an attack during the duration of 35600 and 35670 s. As shown in Fig. 4, this attack is detected. It also aptly showcases a scenario that clearly highlights the features of our algorithm. The proposed algorithm does not consider the absolute values for predicting an attack; it considers the relative values. As shown in Fig. 4 (with ΔT = 10), since the entropy value has suddenly fallen as compared to the recent history during the window from 35657 to 35667, the algorithm detects a violation. Since there are more violations in the previous S windows and the algorithm finds M such violations (in the graph shown, there are two such violations prior to our prediction), we get a prediction for an attack.

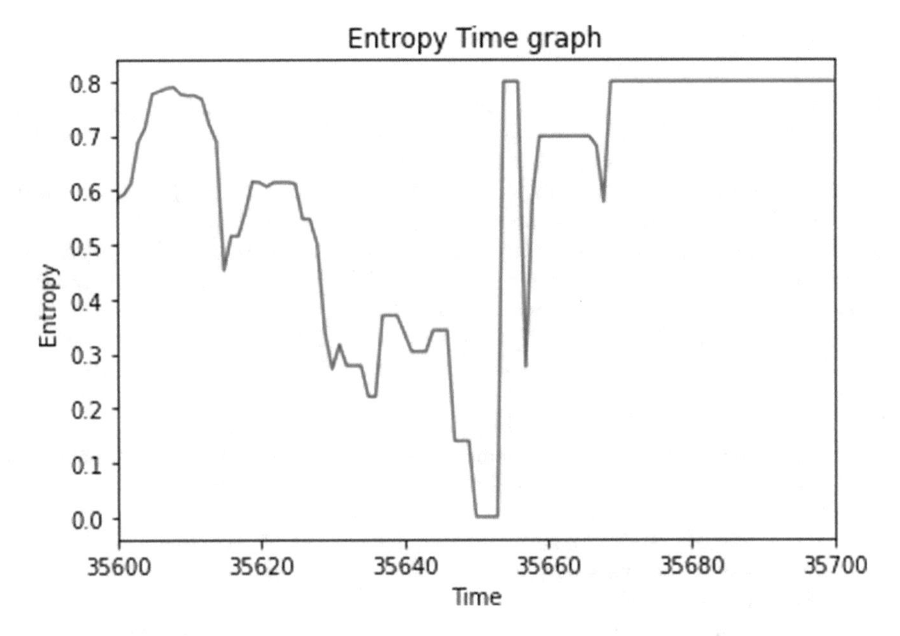

Fig. 4. Entropy as a function of time

Since we are using a dynamic threshold-based approach, we also compared our results with the ones achieved by the adaptive threshold-based fast-entropy approach in [8]. It uses the flow count of a particular flow connection between the duration t and t+1. The flow count of ith connection at duration t is denoted by x(i,t). For calculating the entropy of a flow connection, they used Eq. (5). The fast entropy approach had a detection rate of 50% and an accuracy of 78.049%. Our approach performs better due to the underlying assumption in [8] that the same connections exist over different durations, i.e., at t and t+1, which is not always the case.

5 Conclusion

According to a report by Google [3], DDoS attacks will increase more in size due to an increase in bandwidths, hence becoming more catastrophic. In this paper, an Entropy-based DDoS detection technique is proposed. It uses a dynamic threshold that helps differentiate normal traffic from attack traffic attacks. Accuracy, detection rate, and attack prediction are validated against the CICDDoS-2019 attacks dataset. The accuracy for the Portmap dataset was 93.81%, and for the DNS-DrDoS dataset was 89.3%. The attack window is predicted with a 100% detection rate, with only 5% false positives.

References

1. Cimpanu, C.: AWS said it mitigated a 2.3 Tbps DDoS attack, the largest ever (2020). https://www.zdnet.com/article/aws-said-it-mitigated-a-2-3-tbps-ddos-attack-the-largest-ever/. Accessed 11 June 2021
2. DDOS attacks and the GitHub case, Inst. Res. Internet Soc. (2018). https://irisbh.com.br/en/ddos-attacks-and-the-github-case/. Accessed 11 June 2021
3. Menscher, D.: Exponential growth in DDoS attack volumes (2020). https://cloud.google.com/blog/products/identity-security/identifying-and-protecting-against-the-largest-ddos-attacks. Accessed 11 June 2021
4. Bülbül, N.S., Fischer, M.: SDN/NFV-based DDoS mitigation via pushback. In: ICC 2020-2020 IEEE International Conference on Communications (ICC), pp. 1–6. IEEE (2020)
5. Ali, B.H., Sulaiman, N., Al-Haddad, S.A.R., Atan, R., Hassan, S.L.M., Alghrairi, M.: Identification of distributed denial of services anomalies by using combination of entropy and sequential probabilities ratio test methods. Sensors **21**(19), 6453 (2021)
6. Wang, R., Jia, Z., Ju, L.: An entropy-based distributed DDoS detection mechanism in software-defined networking. In: 2015 IEEE Trustcom/BigDataSE/ISPA, vol. 1, pp. 310–317. IEEE (2015)
7. DDoS Evaluation Dataset (CIC-DDoS2019),. Univ Brunswick n.d. https://www.unb.ca/cic/datasets/ddos-2019.html. Accessed 11 June 2021
8. David, J., Thomas, C.: DDoS attack detection using fast entropy approach on flow-based network traffic. Procedia Comput. Sci. **50**, 30–36 (2015)
9. Assis, M.V.O., Carvalho, L.F., Lloret, J., Proença Jr, M.L.: A GRU deep learning system against attacks in software defined networks. J. Netw. Comput. Appl. **177**, 102942 (2021)
10. Kshirsagar, D., Kumar, S.: A feature reduction based reflected and exploited DDoS attacks detection system. J. Ambient Intell. Humanized Comput. **13**(1), 393–405 (2021)
11. Sahoo, K.S., Puthal, D., Tiwary, M., Rodrigues, J.J., Sahoo, B., Dash, R.: An early detection of low rate DDoS attack to SDN based data center networks using information distance metrics. Future Gener. Comput. Syst. **89**, 685–697 (2018)
12. Li, R., Wu, B.: Early detection of DDoS based on φ-entropy in SDN networks. In: 2020 IEEE 4th Information Technology, Networking, Electronic and Automation Control Conference (ITNEC), vol. 1, pp. 731–735. IEEE (2020)
13. Zhou, R., Wang, X., Yang, J., Zhang, W., Zhang, S.: Characterizing network anomaly traffic with Euclidean distance-based multiscale fuzzy entropy. Secur. Commun. Netw. 2021 (2021)
14. Li, J., Liu, M., Xue, Z., Fan, X., He, X.: RTVD: a real-time volumetric detection scheme for DDoS in the internet of things. IEEE Access **8**, 36191–36201 (2020)
15. Sharafaldin, I., Lashkari, A.H., Hakak, S., Ghorbani, A.A.: Developing realistic distributed denial of service (DDoS) attack dataset and taxonomy. In: IEEE 53rd International Carnahan Conference on Security Technology, Chennai, India (2019)
16. Shannon, C.E.: Prediction and entropy of printed English. Bell Syst. Tech. J. **30**(1), 50–64 (1951)

A Detection Mechanism for Cache Pollution Attack in Named Data Network Architecture

Abdelhak Hidouri[1(✉)], Haifa Touati[1], Mohamed Hadded[2], Nasreddine Hajlaoui[1], and Paul Muhlethaler[3]

[1] Hatem Bettaher IResCoMath Lab, University of Gabes, Gabes, Tunisia
abdelhakhdr@gmail.com, haifa.touati@cristal.rnu.tn
[2] IRT SystemX, Palaiseau, France
mohamed.elhadad@irt-systemx.fr
[3] INRIA, Paris, France
paul.muhlethaler@inria.fr

Abstract. Basic Named Data Networks (NDN) security mechanisms, rely on two main key features. The first one is the caching mechanism where it manages to minimize both the bandwidth usage and the data retrieval delay all along with congestion avoidance by storing, in the intermediate routers, the contents recently demanded to quickly serve future consumers' requests. The second key feature is the NDN security which stands on its foundation by signing each Data as soon as it released by the Producer and gets verified by each requesting consumer so that it makes it resilient to most attacks that affect the integrity of such content and the privacy of its end points. However, the availability of the Data in the cache of the CS allows the malicious consumers to perform several attacks such as Cache Pollution Attack (CPA) which is easy to implement and extremely effective. As a result, it makes the data on the cache unavailable for legitimate consumers and increases its retrieval delay. In this paper, we propose a new detection mechanism of CPA called ICAN (Intrusion detection system for CPA attack in NDN architecture) based on several metrics such as Average Cache Hit Ratio, Average Interest Inter-Arrival Time, Hop Count and Prefix variation. We assess by simulation, using the NDNSim framework, the efficiency of our mechanism and the choice of the used parameters. Finally, we elaborate a qualitative comparison between our proposed solution and the state-of-the-art mechanisms.

1 Introduction

The volume of the Internet traffic is constantly increasing and shifting to content distribution. According to CISCO Annual Internet Report [1], the video content (data, streaming, etc.) that represents in 2016, 73% of the traffic consumed on the Internet, increases to 82% in 2021 and is expected to raise to 92% in 2023. However, the TCP/IP architecture is not optimal in terms of resources

L. Barolli et al. (Eds.): AINA 2022, LNNS 449, pp. 435–446, 2022.
https://doi.org/10.1007/978-3-030-99584-3_38

consumption [2] and has been always a vulnerable target for most of the security attacks, rather than the issues of not being fully capable of handling most of the new features such as massive content distribution, mobility and security. Named Data Networks (NDN) is one of the most suitable candidates for the future Internet architecture [3]. In NDN, each content is identified by a URI-like name and can be cached in intermediate NDN routers to serve subsequent requests for the same content [4].

Caching in NDN is different from the traditional web caching *"Recency-based policies"*, which is based on the idea that the content has been demanded recently in short period of time in other word *"Short time locality"* [5]. But, there is a high probability that the content will no longer be requested. NDN has different caching process, policies and metrics also NDN caching is much more optimized in which it manages to minimize both the bandwidth usage and the data retrieval delay all along with congestion avoidance.

NDN reserves three main essential components: (i) the *Pending Interest Table (PIT)* contains the interest packet entries and the corresponding list of incoming interfaces, (ii) the *Forwarding Information Base (FIB)* associates the prefix names to one or many output interfaces to which the interest packets should be forwarded and (iii) the *Content Store (CS)* holds a copy of the data packets that has been passed by the NDN router, this copy will be used to satisfy subsequent interests that request the same content [6].

In addition, NDN held a signature field inserted in each data packet where the content and the signature are binded together, to insure the verification of the data along the way of its transmission. When a data packet is received, the requesting consumer verifies the signature field [7]. Although, the verification by intervening routers is not mandatory because the verification overhead of the signature might be high, and a router needs access to multiple public key certificates to trust the public key that verifies a content signature [8].

However, since NDN is not vulnerable to a range of basic attacks that are effective on the TCP/IP model, a number of new attacks have appeared that target NDN's main components. These attacks include the CPA where the attacker tries to deplete the size of the cache presented in the CS and deny other legitimate consumers from getting the desired content from the cache.

Obviously, the CPA attack leads to a considerable deterioration in the NDN architecture's performance in terms of Cache Hit Ratio (CHR) and Average Retrieval Delay (ARD). To mitigate the effect of this attack, we propose a new CPA detection mechanism, named ICAN, that monitors the state of each NDN node using different efficient parameters, namely the Average Cache Hit Ratio (AVG-CHR), the Average Interest Inter-Arrival Time (AVG-IAT), the Hop Count and the Prefix variation. ICAN insures high accuracy of detecting CPA attack, and it outperforms the pre-existed CPA detection mechanisms that present multiple gaps such as the exhaustion of router efficiency, increasing the total bandwidth usage and the intense consumption of the NDN routers resources.

The rest of this paper is organized as follows. Section 2 overviews the security vulnerabilities in named data networking and precisely CPA. The next section details the literature review of CPA detection mechanisms. Our proposed CPA detection mechanism: ICAN is detailed in Sect. 4 followed by evaluating the efficiency of the proposed CPA detection mechanism in Sect. 5. And we end up with a conclusion and future work upon Sect. 6.

2 Cache Pollution Attack in Named Data Networks

In CPA, the attacker tries to cache unpopular content in the CS in an attempt to make the cache not available to legitimate consumers. This attack targets mostly NDN routers cache. More precisely, in CPA the attacker sends interest packets to change the priority of the content stored in the CS of nearby routers. This induces the caching of a large number of malicious packets in the CS. This behaviour changes the priority of the content and increases the popularity of these malicious contents. As a result, this attack reduces the probability of obtaining the legitimate content from the cache by the legitimate consumers. This attack confuses the router from detecting such fake content, so it rather keeps them in the router which results in *False Locality Attack (FLA)*, and that's because the attacker does not follow any specific pattern. Another type of this attack is based on multiple number of attackers with a high frequency of demanding fake content, called *Locality Disruption Attack (LDA)* [9].

In our previous work [10], we assessed via extensive simulations the extent of the damage caused by the CPA attack. Our study reveals that when the CPA is launched, the CHR is reduced by around 90%. Even with different caching strategy: LRU (Least Recently Used), LFU (Least Frequently Used), FIFO (First In First Out) and Random, the CHR still decreases which reduces the available bandwidth, overloads the network and could lead, in certain conditions, to congestion. In the same study [10], we show that the CPA attack affects also the consumers by increasing the delay needed to obtain a desired content. This behaviour results in reduced bandwidth, increased data retrieval delay and in certain cases it may result in a Time-Out.

3 Overview of Recent CPA Detection Mechanisms

Recently, several detection mechanisms have been proposed in the literature to capture the presence of a CPA attack in the network. Basically, we can classify these approaches into three main categories: (i) machine learning based approaches, (ii) statistical based solutions and (iii) probabilistic based approach. A brief literature review of these solutions is provided in the following subsections.

3.1 Machine Learning Based Approach

Adaptive Neuro-Fuzzy Inference System (ANFIS) [5] is one of the earliest solution that proposes to enhance NDN caching strategy by integrating a CPA detection mechanism based on machine learning. ANFIS works on each router independently by collecting statistics about each data packet cached in the CS and then passing this information through five layers of a fuzzy network to refine the goodness value. The goodness value is defined as the value used by the cache replacement policy for making the caching decision.

However, ANFIS detection algorithm is applied on cached contents (on the data packets present in the CS) and not on all received interests, which means that malicious interests are not detected unless the corresponding data packet is received and even more cached in the CS, i.e. after consuming network resources (bandwidth, caching resources, PIT entries, etc.). Moreover, a nearby attacker can affect the router's caching choice. Because the router has access only to local data, it may mistakenly believe the attacker's request is valid and begin caching contents requested by the attacker instead of the actual legitimate content [9].

3.2 Statistical Based Approaches

To detect the existence of CPA, CacheShield [11] monitors the received contents. When the router receives a content for the first time, it stores its placeholder in the CS. If the router receives a content whose placeholder is already stored in CS, then it computes a shield function to decide whether the content will be cached or not [9]. Like ANFIS, CacheShield is applied on cached contents and not on received interests. Moreover, this mechanism can negatively impact the caching process, mainly due to its high space usage, and its high complexity. In fact, as explained in [9], this solution induces a large overhead in terms of space to store the placeholder names and in terms of CPU to compute the shield function.

Kamimoto et al. propose in [12] a Prefix Hierarchy solution that includes a CPA detection algorithm. In the first step, this solution calculates the Weighted Request rate Variation per Prefix (WRVP) that helps to create a black-list. This black-list contains each prefix with the Request rate Variation per Prefix (RVP), then the router removes cached content from the black-list in the second step. In the final step, the router does not cache any future unpopular data, because namespace statistics are saved per namespace rather than per content. As the authors claimed, this mechanism uses less memory, but it still contains some gaps, the attacker can use less popular content of the popular prefix to perform the attack.

Guo et al. propose in [13] a path diversity based approach to detect CPA in NDN using the route of each content stored in each router. The main idea is to trace the route of each content using the information that exist in the CS, such as: the times the content object gets hit, the data structure of each content and the path of each content. Then, it calculates the Bernoulli distribution for each of the contents distinct paths, if the content has higher path value, it will be discarded from the cache otherwise it remains in the cache.

Lin Yao et al. proposed in [14] a CPA detection approach based on clustering. This solution starts by computing the number of all interests and the time interval between two consecutive interests demanding the same content in the meanwhile the NDN router computes the interest probability and its time interval then it begins the procedure of clustering by calculating the Euclidean distance between all data points and form the clusters. Based on the results of the clusters, it decides whether CPA has been launched. A CPA attack is detected if the number of requests in the non-popular cluster suddenly drops. This solution may fall in false positive and may result in wrong judgements, where some legitimate consumers can demand low popular contents.

3.3 Probabilistic Based Approach

The Randomness Checks [15] solution is based on ranking each content received by the NDN router and calculating the probability of the low rate pollution attack. When a content is received, the NDN router uses this content to construct a binary matrix. Then it analyses this matrix to rank the content. As soon as the low rate content gets captured, i.e. the matrix reaches a pre-defined threshold, the router initiates an "attack warning". Although this method is light weighted, but it is still limited to small topologies, and it can make the caching replacement policy lose its efficiency. In addition, this mechanism cannot be extended to a larger NDN topology and wider scenarios.

After studying different CPA detection approaches and identifying the limits of each one of them, we conclude that, a new efficient CPA detection mechanism should be designed to overcome these limitations. This mechanism has to match the following criteria:

- The detection mechanism needs to take into account the content popularity and should be efficient by choosing the appropriate detection parameters.
- The detection mechanism should use low computational time, less storage space usage and less time complexity.
- The detection process should not expose the identity of the consumer and the producer.
- The detection mechanism needs to be with a high accuracy and with a less detection time.
- The detection mechanism needs to be able to manage large topology and more complex environments.

4 Intrusion Detection System for Cache Pollution Attack in Ndn (ICAN)

As discussed in the previous section, despite the progress that has been made in developing a method to detect CPA in NDN, this is not sufficient because even if the CPA is detected, these solutions, in return, induces high resources' usage by detecting the CPA on the cached Data and not the interest and may don't respect the NDN philosophy by exposing the identity of the consumers. Therefore, a lightweight interest-based solution is required to detect CPA while respecting the NDN philosophy. In this section, we explain our choice of the CPA detection parameters, then we present the CPA detection process.

4.1 CPA Detection Parameters

To detect if a malicious consumer is trying to pollute the CS, we propose that each router monitors a set of parameters, that we called *CPA detection parameters*. Based on the study that we conducted in [10], to assess the impact of CPA on the NDN network, we analyse the inherent characteristics of the network when the CPA is launched, to efficiently choose which parameters could be used as CPA indicators. From this analysis, we extract four candidates CPA detection parameters: Average Cache Hit Ratio, Average Interest Inter-Arrival Time, prefix variation and Hop Count. The analysis of the variation of these parameters under CPA, will be detailed in the followings subsections.

4.1.1 Average Cache Hit Ratio (AVG-CHR)

To decide if the AVG-CHR could be chosen as a CPA detection parameter, we analyse the variation of the CHR with and without CPA in different scenarios [10]. An example of our analysis is given in Fig. 1. As illustrated in this figure, the AVG-CHR clearly decreases when the CPA is launched, and it drops to 0% in some points. The total difference between the AVG-CHR in *Normal State* and the AVG-CHR in *Attack State* reaches 40%. Therefore, we conclude that the AVG-CHR could be a good candidate to detect the appearance of a CPA attack.

4.1.2 Average Interest Inter-Arrival Time (AVG-IAT)

Inspired by previous works in the literature [16,17], that used packet inter-arrival time to detect DDoS attack in TCP/IP networks, we define a similar parameter that we called the Average Interest Inter-Arrival Time (AVG-IAT) to detect the

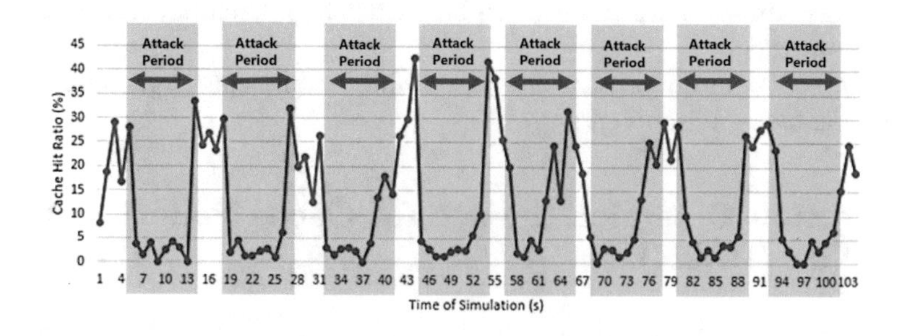

Fig. 1. Analysis of the variation of the Cache Hit Ratio under CPA.

presence of CPA in NDN networks. This parameter is defined as the average, on each time period, of the time difference between the arrival of two consecutive interests at an NDN router. Figure 2 traces an example of the variation of the AVG-IAT with and without CPA. We clearly observe that, the AVG-IAT decreases by around 40% when the attack is launched. Thus, this confirms that the Average Interest Inter-Arrival Time could be a good indicator of the presence of a CPA attack.

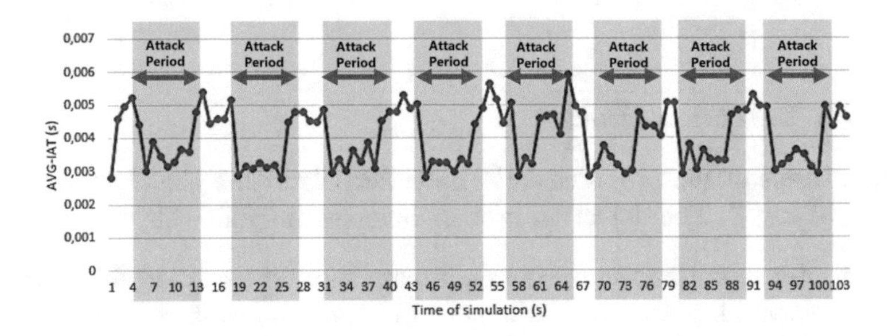

Fig. 2. Analysis of the variation of the Average Interest Inter-Arrival Time under CPA.

4.1.3 Prefix Variation

To ensure that the metrics elaborated above not falling into a false positive (i.e. considering a legitimate interest as a malicious request), we added the "Prefix Variation" parameter to ensure that such prefix is the one who is demanding such malicious content. We check the variation of the prefix in each time stamp to identify which prefix caused the decrease of the AVG-CHR and AVG-IAT and to mark it as suspected prefix.

4.1.4 Hop Count

The Hop Count parameter is defined as the number of hops between the consumer and the NDN router. As explained in Table 1, we observed that when the attack is launched, the hop count of the malicious interests stays stable. Hence, when the ratio of interests having the same Hop Count exceeds a pre-set threshold ($i = 75\%$), we suspect a CPA attack.

Table 1. Average Hop Count based Detection.

Prefix	Hop Count	State
/com/CPA.lab/xyz	7	$i = 25\%$
/com/CPA.lab/xyz	7	$i = 50\%$
/com/legitime.lab/xyz	5	$i = 50\%$
/com/CPA.lab/xyz	7	$i = 75\%$ Attack suspected
/com/CPA.lab/xyz	7	$i = 75\%$ Attack suspected

4.2 CPA Detection Process

The main steps of the proposed CPA detection algorithm are illustrated in the flowchart of Fig. 3. The detection process involves the following steps:

1. Each router monitors the upcoming interests and collects locally some statistics, like the Hop Count, the time when each interest reaches the router which will help us later to calculate the Average Interest Inter-Arrival Time, the Average of the Cache Hit Ratio and the prefix variation.
2. In each time period, the router verifies the existence of the attack by checking the state of the CPA detection parameters. If, the AVG-CHR decreases, the AVG-IAT decreases and this decrease is associated with the arrival of a new prefix (suspected prefix) and the hop count of the suspected prefix is stable, then the router concludes that it is under CPA and the algorithm marks that state as "Attack detected".

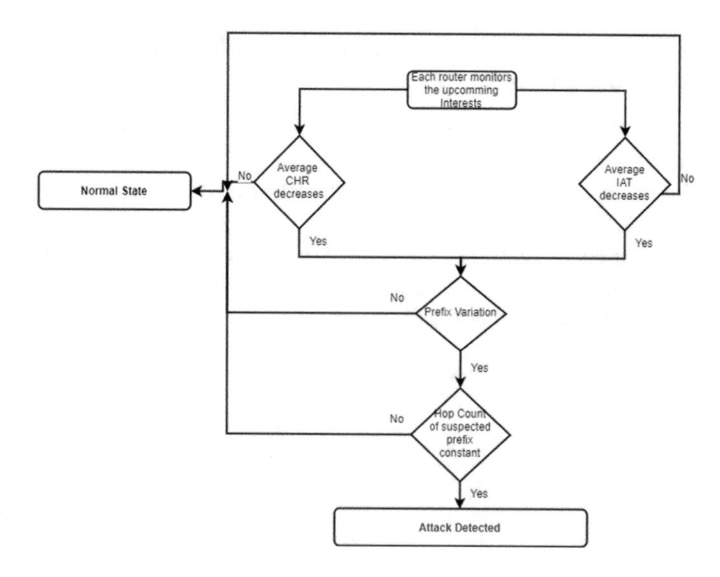

Fig. 3. Detection mechanism process.

3. At each time the detection parameters suspect an attack, the router checks the new state to a *Reference* state, where the reference point presents the optimal state of the network before the attack appears. Based on the comparison to the Reference state, the router decides to come back to the "Normal state" or to stay in the "Attack detected" state.

5 Performance Evaluation

To validate the proposed ICAN solution, we conducted a simulation study as well as a Qualitative Comparative Analysis (QCA). The details of these studies are given in the next subsections.

5.1 Simulation Based Evaluation

In this section we evaluate, through simulations, the performance of ICAN. We used the NDNSim simulator as well as the latest python version 3.10.0 to implement our solution. We used a real-world topology, namely the German Research Network (DFN) shown in Fig. 4, and we varied the CPA attack entry into 8 main ranges. Each range is composed of 7 s of attack (*Attack State*) and 6 s of non-attack (*Normal State*). Table 2 summarizes the simulation settings.

We evaluate our ICAN detection mechanism in terms of three metrics: *Accuracy, False positive ratio (FP%)* and *False negative ratio (FN%)*. Table 3 summarizes the obtained results for Router 6 and Router 10.

The obtained results, show that the accuracy of ICAN reaches 92, 307% for router R6 and 91, 346% on router R10. Moreover, the false positive ratio is limited to 4, 807% for router R6 and 6, 730% on router R10. Finally, the false negative is reduced to 2.884% in router R6 and 1.923% in router R10. These results confirm the efficiency of our CPA detection approach.

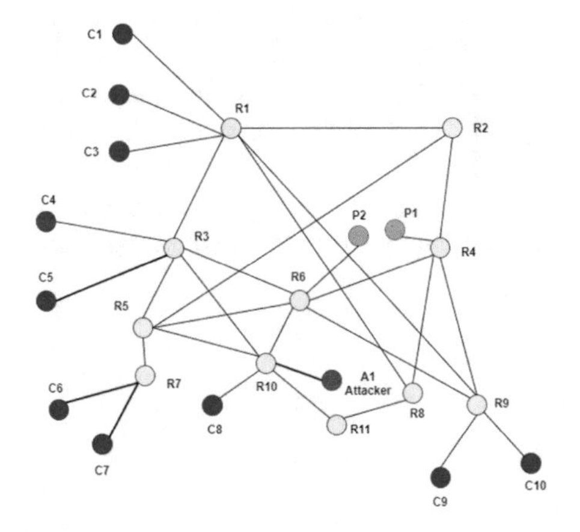

Fig. 4. The German Research Network Topology (DFN).

Table 2. Our CPA detection approach settings

Parameter	Value
Simulation time	105s
Number of legitimate consumers	6
Number of attackers	1
Consumer type	ConsumerZipfMandelbrot
Interest rate	240 Interest/s
Time of launching the attack	5–12, 18–25, 31–38, 44–51, 57–60, 67–74, 80–87, 93–100 s
Routers CS size	50
Cache policy	LRU
Topology	DFN*

* German Research Network Topology (Deutsches Forschungs Netz)

Table 3. Evaluation of the ICAN detection mechanism through simulation.

Router	Accuracy (%)	False Positive (%)	False Negative (%)
Router 6	92, 307	4, 807	2, 884
Router 10	91, 346	6, 730	1, 923

5.2 Qualitative Comparative Analysis of Our Proposed Solution

In Table 4, we elaborate a qualitative comparison based on the space usage in the router in terms of memory overhead, the complexity of the algorithm in terms of computational time and the insurance of the privacy of endpoints. Our qualitative comparison shows that ICAN requires less memory usage, conserves the identity of the consumer and insures high precision compared to state-of-the-art detection algorithms. In fact, our proposed detection algorithm ICAN is applied on interest packets, whereas most of other solutions are applied on cached data, hence our solution induces less memory usage since PIT and CS components will not be overloaded with malicious interest and cached data, respectively. Moreover, ICAN is a lightweight and simple solution, hence it induces a reduced computational time. On the other side, our proposal doesn't use the identity of the consumer or the producer to decide whether a CPA is performed or not, hence ICAN preserves a key feature of the NDN architecture, i.e. not using the identity or the addresses of the endpoints during data dissemination.

Table 4. Qualitative Comparative Analysis (QCA) of the ICAN detection mechanism.

Mechanism	Memory Overhead	Computational Time	Privacy Conservation	Accuracy	Topology
ICAN	**Low**	**Low**	**No leakage**	**High**	**DFN**
ANFIS [5]	Low	Low	No leakage	Medium	DFN
CacheShield [11]	High	High	Leak Consumers information	Medium	Self-made
Prefix Hierarchy[12]	Low	Low	No leakage	Low	DFN
Path Diversity [13]	Medium	Low	No leakage	Medium	K-ary Tree
Clustering [14]	Low	Low	No leakage	Medium	AS-1221
Randomness Checks [15]	Low	High	–	High	–

6 Conclusion and Future Works

In this paper, we begin with a literature review of CPA detection solutions in named data networks. We classified these approaches into three main categories, namely machine learning based approaches, statistical based approaches and probabilistic based approaches. We explain their detection algorithms and identify their limitations. This study reveals the need of a new robust and efficient CPA detection mechanism in NDN. Therefore, in a second step, we propose a new CPA detection mechanism called ICAN. Based on the analysis of the behaviour of the network, when the CPA is applied, we defined a set of parameters that will govern the CPA detection process. In our approach, the router continuously monitors the Average Cache Hit Ratio, the Average interest Inter Arrival Time, the Hop count and the variation of the prefix requested by the received interests to detect whether the received interests are sent by a malicious consumer or not. The proposed solution is applied on interest packets, thereby avoiding the need to load the network links as well as the PIT and the CS of intermediate routers by malicious interests. Simulation results conducted in a real-world topology illustrate the accuracy of the proposed solution.

As future work, another direction that could be explored is the design of a distributed detection mechanism where, instead of relaying on local parameters, each router communicates with other nodes to exchange related attack information.

References

1. Cisco. Cisco Annual Internet Report - Cisco Annual Internet Report (2018–2023) White Paper, 10 March 2020. https://www.cisco.com/c/en/us/solutions/collateral/executive-perspectives/annual-internet-report/white-paper-c11-741490.html
2. Yang, Z., Hua, L., Gao, N., Huo, R., Liu, J., Huang, T.: An accelerating approach for blockchain information transmission based on NDN. Future Internet **13**(2), 47 (2021). https://doi.org/10.3390/fi13020047

3. Touati, H., Aboud, A., Brahim, H.: Named Data Networking-based communication model for Internet of Things using energy aware forwarding strategy and smart sleep mode. Cluster Comput. **34**(3), e6584 (2022). https://doi.org/10.1002/cpe.6584
4. Touati, H., Mejri, S., Malouch, N., Kamoun, F.: Fair hop-by-hop interest rate control to mitigate congestion in named data networks. Cluster Comput. **24**(3), 2213–2230 (2021). https://doi.org/10.1007/s10586-021-03258-8
5. Karami, A., Guerrero-Zapata, M.: An ANFIS-based cache replacement method for mitigating cache pollution attacks in Named Data Networking. Comput. Netw. **80**, 51–65 (2015). https://doi.org/10.1016/j.comnet.2015.01.020
6. Mejri, S., Touati, H., Kamoun, F.: Hop-by-hop interest rate notification and adjustment in named data networks. In: 2018 IEEE Wireless Communications and Networking Conference (WCNC) (2018). https://doi.org/10.1109/wcnc.2018.8377374
7. Mejri, S., Touati, H., Kamoun, F.: Preventing unnecessary interests retransmission in named data networking. In: 2016 International Symposium on Networks, Computers and Communications (ISNCC) (2016). https://doi.org/10.1109/isncc.2016.7746058
8. Zhang, Z., Wong, S.Y., Shi, J., Pesavento, D., Afanasyev, A., Zhang, L.: On Certificate Management in Named Data Networking. ArXiv, abs/2009.09339 (2020)
9. Kumar, N., Singh, A.K., Aleem, A., Srivastava, S.: Security attacks in named data networking: a review and research directions. J. Comput. Sci. Technol. **34**(6), 1319–1350 (2019). https://doi.org/10.1007/s11390-019-1978-9
10. Hidouri, A., Hadded, M., Hajlaoui, N., Touati, H., Muhlethaler, P.: Cache pollution attacks in the ndn architecture: impact and analysis. In: 2021 International Conference on Software, Telecommunications and Computer Networks (SoftCOM) (2021). https://doi.org/10.23919/softcom52868.2021.9559049
11. Conti, M., Gasti, P., Teoli, M.: A lightweight mechanism for detection of cache pollution attacks in Named Data Networking. Comput. Netw. **57**(16), 3178–3191 (2013). https://doi.org/10.1016/j.comnet.2013.07.034
12. Kamimoto, T., Mori, K., Umeda, S., Ohata, Y., Shigeno, H.: Cache protection method based on prefix hierarchy for content-oriented network. In: 2016 13th IEEE Annual Consumer Communications & Networking Conference (CCNC) (2016). https://doi.org/10.1109/ccnc.2016.7444816
13. Guo, H., Wang, X., Chang, K., Tian, Y.: Exploiting path diversity for thwarting pollution attacks in named data networking. IEEE Trans. Inf. Forensics Secur. **11**(9), 2077–2090 (2016). https://doi.org/10.1109/tifs.2016.2574307
14. Yao, L., Fan, Z., Deng, J., Fan, X., Wu, G.: Detection and defense of cache pollution attacks using clustering in named data networks. IEEE Trans. Dependable Secure Comput. **17**(6), 1310–1321 (2020). https://doi.org/10.1109/tdsc.2018.2876257
15. Park, H., Widjaja, I., Lee, H.: Detection of cache pollution attacks using randomness checks. In: 2012 IEEE International Conference on Communications (ICC) (2012). https://doi.org/10.1109/icc.2012.6363885
16. Rios, V.D., Inácio, P.R., Magoni, D., Freire, M.M.: Detection of reduction-of-quality DDoS attacks using fuzzy logic and machine learning algorithms. Comput. Netw. **186**, 107792 (2021). https://doi.org/10.1016/j.comnet.2020.107792
17. Ashraf, S., Shawon, M.H., Khalid, H.M., Muyeen, S.M.: Denial-of-service attack on IEC 61850-Based substation automation system: a crucial cyber threat towards smart substation pathways. Sensors **21**(19), 6415 (2021). https://doi.org/10.3390/s21196415

Prevention of DDoS Attacks with Reliable-Dynamic Path Identifiers

Vishal Gupta, Shail Saharan(✉), and Sreetam Parida

Birla Institute of Technology and Science (BITS) Pilani, Pilani, India
{vishalgupta,p20170404}@pilani.bits-pilani.ac.in

Abstract. As per the reports of McAfee and Google, DDoS attacks are ranked the third-most dangerous network attacks, and their intensity is expected to increase in the future. To defend against these DDoS attacks, we propose a preventive approach in this paper. This approach aims to stop the DDoS attack traffic from reaching the victim's network. Our proposed technique is based upon the use of Path Identifiers (PIDs), primarily used in Information-Centric Networks, to force the forwarding of response packets on these Path IDs and not IP forwarding. As against static PIDs, we use reliable dynamic PIDs (RDPID, two for each link) to refrain the attackers from learning these PIDs and launching the attack. With the proposed RDPID technique, the PID negotiation time is reduced to 6 ms for 99% of the cases, as against 23 ms with the use of Dynamic PIDs (DPIDs) proposed in the literature. Furthermore, the attack mitigation time is reduced by (approx) 40% compared with similar techniques using DPID available in the literature.

1 Introduction

The Internet has been an incredible accomplishment since its inception. However, with the rapid increase in its users and applications, it faces many challenges, including security. It is a concern because the Internet's architecture and overall functionality were not designed with security as a focus and various kinds of attacks on the Internet still happen. One such attack category is Distributed Denial of Service (DDoS) attacks. DDoS attacks disrupt the service provider's network or user's network from providing and availing services by hogging the bandwidth and other resources. These are considered to be the third-most dangerous network attacks [1]. DDoS attacks began from a few Mbps intensity in their conception and steadily expanded to a few Gbps intensity in the immediate past [2].

Furthermore, a Google analysis predicts that they are destined to be of greater intensity in the future [3]. While unveiling the tremendous DDoS attack of 2.5 Tbps on its infrastructure, Google said the intensity would increase further in the coming years [4]. Hence, DDoS attacks are still a significant threat. The domain of activities used to tackle DDoS attacks is primarily concerned with detecting attacks and mitigating those attacks. With this, the network is susceptible to DDoS attacks and waits for them to happen. Once an attack happens, resulting in the depletion of the victim's bandwidth and other resources, attack detection and subsequent mitigation strategies follow. Instead, shifting the domain of action to attack prevention is a more promising strategy. It can

L. Barolli et al. (Eds.): AINA 2022, LNNS 449, pp. 447–458, 2022.
https://doi.org/10.1007/978-3-030-99584-3_39

happen in two ways. First, all the edge networks (also called access networks) must adhere to pre-defined policies that prevent an attacker from launching an attack. Second, make the underlying core network (i.e., network responsible for forwarding the traffic from source to destination) intelligent enough to minimize the probability of attack traffic reaching the victim. We believe it is practically impossible to stop the attacker from launching the attack since it will necessitate imposing restrictions and upgrades on each access network. However, we can prevent the attack traffic from reaching the victim's network by making the underlying core network smart enough, and this paper presents and proves the techniques to achieve it. This way, even in the presence of an attack, the victim's network is always safe.

The focus of this paper is on techniques to prevent volumetric DDoS attacks. These DDoS attacks flood the target's network with massive amounts of malicious traffic. Volumetric DDoS attacks can further be divided into the following two categories [5].

1) DrDoS (Distributed Reflection Denial of Service attacks): These attacks are made on the framework of client-server architecture. In DrDOS, the attacker constructs request packets towards a server with the source-IP address spoofed as of the victim's. As a result, the victim receives amplified response packets from the server. Domain Name Service (DNS), Light-weight directory access protocol (LDAP), and Network Time Protocol (NTP) are examples of services that attacker exploits for an attack. The amplification factor can be raised further by using bots (compromised hosts) under the attacker's control.
2) Flooding attacks- These are the ones that are generated by sending a very high number of packets targeting a victim. The attackers generate many requests and send them to victims, hogging its bandwidth.

Different authors have proposed the use of Path Identifiers (PID's) as a packet routing mechanism for Information-Centric Networks (ICNs) [6, 7]. We propose using these PIDs to prevent DDoS attacks by ensuring the response packet path is the same as that of the corresponding request packet. A novel routing mechanism in Software Defined Network (SDN) environment between all inter-domain routers is proposed to accomplish this. Different authors use PIDs to avoid DrDoS attacks [6, 7, 11]. Because of PIDs, such attacks are not possible even if the source-IP addresses are spoofed because the response packet is not forwarded using regular IP forwarding [6]. Instead, it is done based on the PIDs stored in the corresponding request packet. Luo et al. [7] propose that these PIDs should be static. Static PIDs prevent DrDoS attacks but make the network vulnerable to flooding attacks. Hence, the use of dynamic PIDs is proposed in [6]. If an attacker sniffs these PIDs (by sniffing the network packet) in legitimate packets, a DDoS attack can be launched towards the sender of these packets. Also, if there is a link break during the forwarding of the response packet, it is lost. Furthermore, the proposed approaches are prone to packet drops during PID updates.

The Reliable-Dynamic PIDs (RDPIDs) suggested in this paper address difficulties with existing technologies (as written above). RDPIDs are more secure against DDoS attacks because of the novel concept of Reserved PIDs and Open PIDs. Furthermore, the PID negotiation delays are reduced compared to [6]. The link break issue can also be resolved with the proposed technique.

Section 2 explains the existing work in the literature about the use of PIDs. The proposed methodology is defined in Sect. 3. The novelty of our approach by comparing RDPIDs and DPIDs is explained in Sect. 4. Section 5 describes simulation, results, and its analysis. Finally, Sect. 6 concludes the paper.

2 Related Work

Attackers often exploit the existing vulnerabilities of the Internet to launch attacks. One such attack is DDoS attacks. There have been approaches proposed to defend against these attacks. Once an attack has started, the primary focus is to maximize the probability of detection based on different heuristics. These heuristics vary from network traffic analysis, entropy, anomalous behavior of data traffic, etc. [8, 9]. Remedial measures can be taken once an attack is detected, like rate limiting [8], etc. Another way to defend against DDoS attacks is prevention, which is the primary focus of this paper. In this section, related work concerning DDoS prevention using PIDs is presented.

Instead of using traditional IP forwarding to transmit packets, Godfrey et al. [10] have proposed using Pathlets, which are inter-domain routing objects. The authors propose to use a standard, wireless path-vector protocol, with a pathlet declaration containing the FID (Forwarding Identifier for pathlets) and a sequence of vnode identifiers. CoLoR, an approach proposed by Luo et al. [7], says that the future Internet will continue to be organized in domains with provider/consumer/peer relationships. Every domain has a logical resource manager which maintains a registry table that stores content access information identified by SID. CoLoR also uses two local namespaces: domain-based connectors and route identifiers (PIDs). The domain in CoLoR can use different local intra-domain protocols. Inter-domain routing is based on PIDs. In CoLoR, users send GET messages to find their desired content. In the GET messages, PIDs of the inter-domain path are stored in the packet. The routing from the content provider to the content consumer is done based on these PIDs. Here, the PIDs are static, so a flooding attack is still possible. The other approach proposed by Luo et al. [6] describes how flooding attacks are still possible if the attackers can find out these inter-domain identifiers. The ways to know the PIDs are also described in detail, i.e., GET luring and Botnet cooperation. The attackers learn the PIDs between their network and the victim's network through these ways. The PIDs are static and do not change so that the attackers can launch a successful attack. That is why Luo et al. [6] proposed Dynamic Path identifiers, so even if the attackers learn the PIDs and launch the attack, the attack cannot continue as the PIDs will change after some time. We also have used Dynamic path identifiers and compared our approach with the one proposed by Luo et al. [6]. Basheer et al. [11] have also used the concept of DPID, but with Get-message logging. This approach is mainly for ICNs where users request information using Get messaging. The ICN routers log GET request messages using bloom filters. Bloom filters help in comprehensive logging, and in return, they do not even take up much space. The limitation of these techniques is that the response packet is lost if a link breaks in between the communication.

3 Methodology

The proposed technique for DDoS prevention uses PIDs described in [6] to identify the inter-domain paths. Here domain refers to an independently maintained network connected to other domains via Border Routers(BRs). A domain might be an Autonomous System (AS) or ISP's network in the Internet architecture. We assume that these domains are numbered D_1, D_2... D_n. Each domain has one or more Border Routers (BR) through which network traffic is routed to another domain. Assume that these BRs in each domain are also numbered B_1, B_2, ..., B_m. In addition, each domain has a Resource Manager (RM) whose task is to do intra-domain routing and share PIDs with the BRs. Figure 1 shows all these network elements. It shows six domains (D_1 to D_6), BRs in each domain, RMs, and hosts. Host-1 and host-2 are located in domain D_1, host-3 in D_2, and host-4 in D_6. It also shows a DNS server in D_3. For each link connecting the two BRs, a PID is generated and shared before any communication can start. These PIDs are generated and distributed by a central entity called Network Manager (NM). Although we assume NM to be a centralized entity in this paper, these can be maintained as a distributed system.

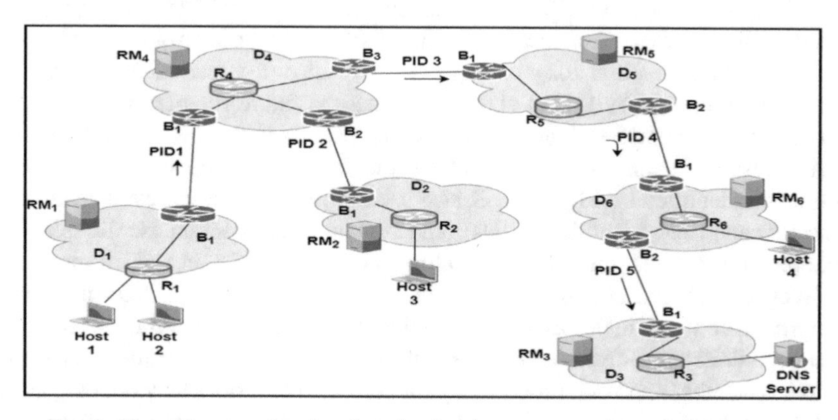

Fig. 1. Network setup showing domains, border routers, and resource managers.

For more clarity, let's take an example. Suppose, for inter-domain routing, PID1 is used between the domains D_1 and D_4, PID2 between D_2 and D_4, PID3 between D_4 and D_5, PID4 between D_5 and D_6, and PID5 between D_6 and D_3. These PIDs are negotiated and shared with the BRs before any communication can start. Also, suppose host-1 in D_1 generates a DNS request packet for the DNS server in D_3. This packet will be routed to D_3 using any inter-domain routing protocol, like that as described in [12]. The respective BRs in between will push (i.e., store) the associated PIDs in the options and padding field of the network packet's IP header. Thus, PID1, PID3, PID4, and PID5 will be stored in the DNS request packet. The corresponding response packet will follow the same path using these PIDs (rather than IP forwarding) to reach back to host-1. With this approach, even if source-IP spoofing is done by host-1 for any potential DrDoS attack, the response packet is assured to go back to the attacker itself. Thus, an attacker will attack itself.

There are two following possibilities with PID assignments.

a) As proposed in [7], PIDs can be assigned once and can be made static. This approach has a disadvantage in that an attacker can learn these PID sequences by sniffing the packets or deploying a honeypot "server" in the network. A sniffed packet (or a packet coming to the honeypot) from some host-H will have a PID sequence. The reverse of this sequence refers to the PID path back to the host-H. An attacker could use this knowledge to launch a flooding attack on H.
b) To overcome the above problem, as proposed in [6], PIDs can be made dynamic in the sense that they change after every fixed time interval T. Of course, this technique results in an increase in network traffic during PID updates. The value of T being less results in network congestion and T being more makes the source host susceptible to flooding attack, as with static PIDs.

We propose Reliable Dynamic PIDs (RDPID) to tackle the above issues. The proposed methodology is explained using three points: (a) RDPID Generation, (b) Request-Routing, and (c) Response-Routing.

3.1 Reliable-Dynamic PID (RDPID) Generation

Figure 2 shows the reserved and open PIDs (RPID and OPID) for each link interface connecting two BRs. If any two BRs (BR_1 and BR_2 as shown in Fig. 2) are connected with a link L, the RPID for an interface of BR_1 becomes OPID for an interface of BR_2 and vice versa.

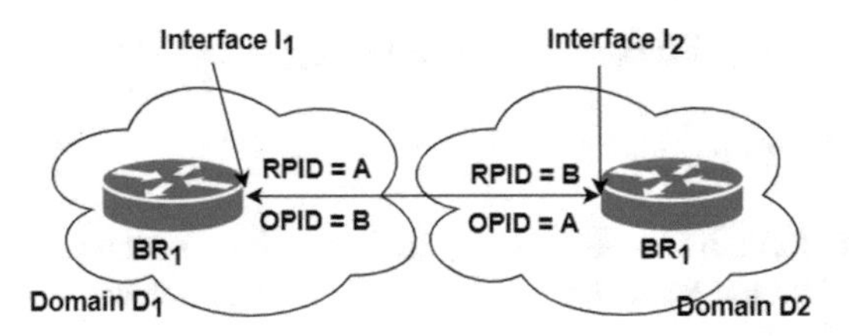

Fig. 2. Reserved and open PIDs.

The procedure PID_Generation(), shown in Fig. 3, runs periodically after every time interval T. For each domain, procedure Generate_Keys () generates reserved PIDs for each link of each of its BR. The procedure assumes a function GENERATE_UNIQUE_KEY(), which generates a unique key for a specific domain. Finally, the Network Manager communicates all the pairs for a domain with the corresponding RM, which shares the keys with its BRs. Now, these shared PIDs can be used for communication. It should be noted that while each PID is not necessary to be unique in the network, each PID must be unique inside a domain. In other words, these PIDs are used to identify neighboring devices linked to a single device; thus, they do not need to be unique across the network but only within a domain.

3.2 Request Routing

Let
D be the total number of domains, numbered 1... D
B_D: Total number of Border Routers (BR) in D^{th} domain, numbered $1 \ldots B_D$, for all D
A_{IJ}^{KM} : Link_ID between I^{th} domain's J^{th} BR and K^{th} domain's M^{th} BR ($1 \leq I \leq D$, $1 \leq K \leq D$, $1 \leq J \leq B_I$, $1 \leq M \leq B_K$)
N_{IJ}^{*} : Number of links to which J^{th} BR of Ith domain is connected to.
RPID (A_{IJ}^{KM}): Reserved PID for Link_ID A_{IJ}^{KM}
OPID (A_{IJ}^{KM}): Open PID for Link_ID A_{IJ}^{KM}
[*Note that RPID* (A_{IJ}^{KM}) = *OPID* (A_{KM}^{IJ})]

```
PID_Generation (Time interval T)
{
   Input T: time interval for PID update.
   Output: Pair [RPID (A_IJ^KM) , OPID (A_KM^IJ)] for all Link_ID's.
   Repeat after every time interval T
   {
      For d = 1 to D
           Generate_Keys (d)
   }
   Distribute all [RPID (A_IJ^KM) , OPID (A_KM^IJ)] pairs for all Links via RM's.
}
Generate_Keys (d)
{
   Input d: Domain for which key pairs are to be generated.
   Output: Unique RPID and OPID for each link interface.
   For b = 1 to B_d
   {
      For p = 1 to N*_db
           Key = GENERATE_UNIQUE_KEY();
           x = domain to which p^th link of B_d is connected to.
           y = BR to which p^th link of B_d is connected to.
           RKEY (A_db^xy) = OKEY (A_xy^db) = key
   }
}
```

Fig. 3. PID generation algorithm

While forwarding a request packet, a BR, before sending the packet to another BR, must append the open PID received from that BR within the options field of the IP header of a network packet. The options field can be considered as a stack, and the PIDs are inserted on top of this stack. The BR then consults these "stacked" PIDs to receive the response packets.

Figure 4 depicts the entire request packet routing. It shows six BRs, each assumed to be in a different domain. The link interface of each BR shows the Open PID of the next BR on the link. The request packet is to be forwarded from BR_1 to BR_6. Based on the standard IP routing, BR_1 should first send the packet to BR_2. Hence, BR_1 appends the PID value 2 within the options field of the packet and sends it. BR_2 verifies this PID for its correctness upon receiving the packet; it should be its reserved PID. If the appended PID is malicious, the packet is dropped.

Similarly, BR_2 now appends the PID value 4 to send the packet to BR_4. In this manner, the entire routing is performed. The packet contains the entire flow indicated through the PIDs at the destination. In this case, it is (2,4,5,6). This PID sequence is then used to trace the packet back to the source.

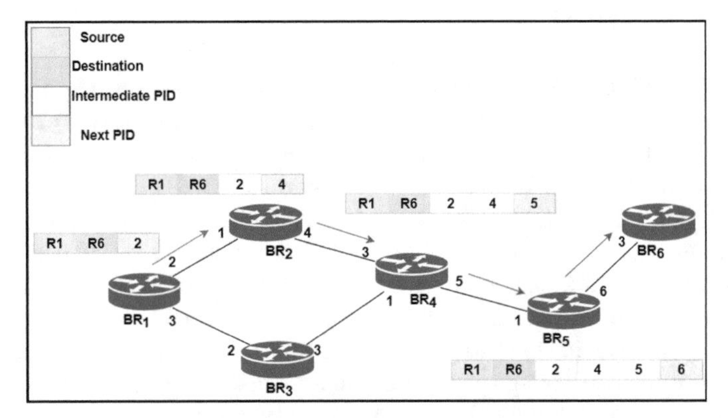

Fig. 4. Request packet routing.

3.3 Response Routing

Once the request packet has been received at the destination, the PID sequence is referred for forwarding the corresponding response packet. Each BR maintains a PID table that stores reserved and open PIDs of its interfaces. The last PID in the options field (pointed to by the top of the stack pointer) is referred to determine the subsequent BR to which the packet should be forwarded. Once determined, the PID pointed at the top of the stack is popped, and the remaining PID sequence is sent along with it.

Figure 5 portrays the flow. It shows the PID table of BR_4. When BR_4 receives the response packet with PID sequence (2, 4), with 4 being the last PID, it refers to its own PID table to find the corresponding Open PID to determine the next node on which packet would be forwarded. This last PID is then popped, and the packet with the remaining PID sequence is sent further. This exact procedure is then continued further till the packet reaches the destination.

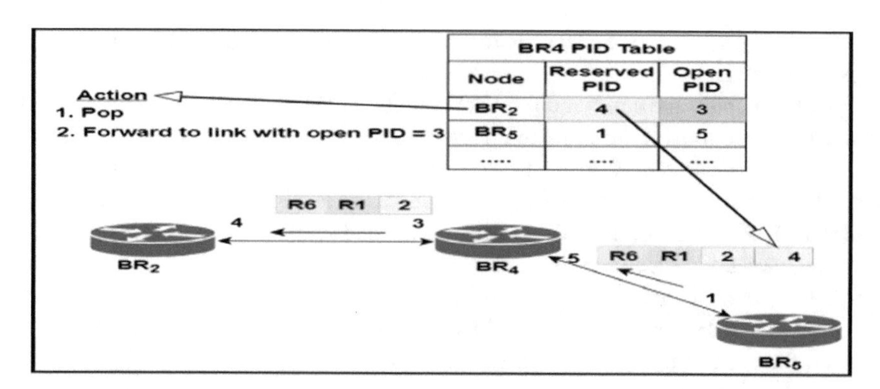

Fig. 5. Response packet routing

4 RDPID v/s DPID

In the past, single PIDs were used to combat DDoS attacks; however, they are vulnerable to other forms of attacks. Instead, using two PIDs (open and reserved) for each link interface helps secure the network further. For example, it secures the network from Syn flood type of DDoS attacks (also called as (half-open attacks)) [13]. The attacker sends a half-open connection request to the server as a client in this type. Of course, the source IP in this request packet can be spoofed. The server acknowledges the connection and waits for the final acknowledgment. Since the source-IP was spoofed, or even otherwise, this acknowledgment will not arrive; thus, server resources are wasted.

Consider the scenario depicted in Fig. 6. Assume that host h1 sends a request packet to host h2, and the underlying system uses a single PID for each link (shown within a square box on top of each link). The routers in between would append the PID's [1, 3, 5, 7] as the packet traverses through the links A, B, C and, D. Now, if the host h2 will have to send a request packet to h1, the PIDs would also be [1, 3, 5, 7], but appended in reverse order, i.e. [1, 3, 5, 7]. A sniffer at BR_5, trying to sniff incoming packets, would know this PID sequence for communication between h1 and h2. As a result, an attacker can construct a spoofed request packet somewhere between h2 and h1, thus launching a flooding DDoS attack against h1.

With two PIDs per router interface, such attacks are harder to launch. Here, the sequence remains the same for the packet from h1 to h2, i.e., [1, 3, 5, 7]. However, if h2 generates a packet for h1, the PID sequence would be [2, 4, 6, 8]. Therefore, after sniffing the request packet generated from h1 at BR_5 and getting the PID sequence [1, 3, 5, 7], when an attacker generates a spoofed request packet from h2 to h1, the packet will be discarded at BR_4; as for it to forward any request, the PID appended should be 8. But, of course, an attacker can launch a flooding attack towards host h1 using response packets. Its mitigation time depends upon the RDPID update interval. The results section shows its simulation.

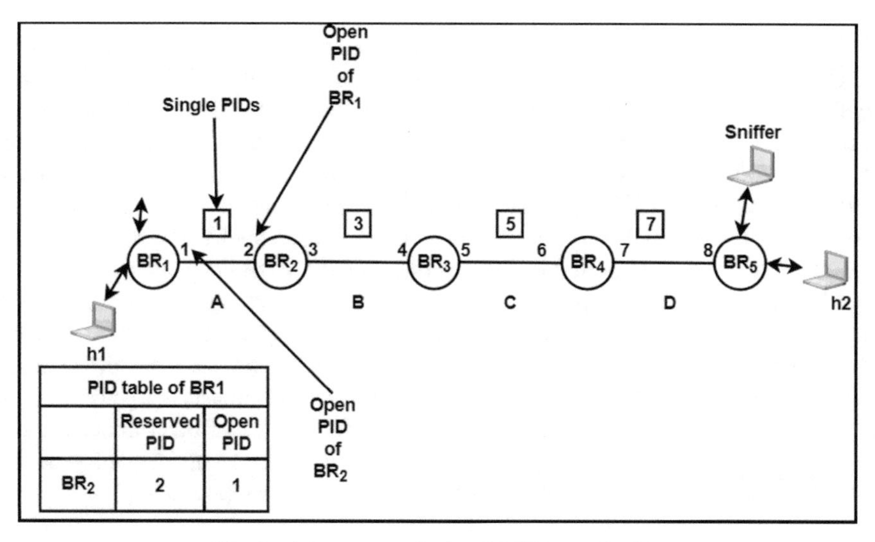

Fig. 6. Advantages of using RDPID over DPID.

5 Simulation and Results

The topology shown in Fig. 1 was emulated in mininet, using Click [14] as a router. The purpose is to demonstrate that the proposed technique can prevent DDoS attacks. Click's extensible language for deployment enables the creation of highly customizable router functionalities. This way, routers can be implemented in Linux hardware more efficiently. Also, Click achieves a very high forwarding rate per second. To do this, Click gets rid of the interrupt-driven architecture favoring polling, avoiding expensive context switches and memory accesses. Mininet is a network emulator, or more accurately, a network orchestration system. It works with a set of endpoints, switches, routers, and a single Linux kernel connection.

The proposed technique's effectiveness is shown to prevent the DrDoS attacks from reaching the victim, prevent flooding attacks, and the effect of keys (OPID and RPID) distribution on network functioning. The measured parameters are compared with the DPID technique as presented in [6].

To validate the prevention of DrDoS attacks, in the simulated network (Fig. 1), host-1 as an attacker generates DNS request packets with source IP spoofed to that of host-3. Host-1 and host-3 also generate benign traffic in the form of HTTP request packets towards the HTTP server configured in D_4. Results are measured in terms of link bandwidth v/s response delay. Without PID reverse routing in place, the DNS server sends all the amplified DNS response packets to host-3, thus choking its bandwidth and other resources. However, with RDPID routing in place, these response packets are routed towards the attacker itself, thus preventing the attack traffic from reaching the victim altogether. Results are shown in Fig. 7(a). If the HTTP response delay was greater than 100 ms (milliseconds), the packet was considered to be lost.

To test and validate the resilience of RDPID based routing against preventing flooding attacks, host-1 in Fig. 1 sends the DNS request packets to the DNS server, which is configured as a network honeypot (or otherwise) that allows an attacker to sniff the packets. Thus, an attacker in D_3 learns the PID sequence and launches a flooding attack of 20 Mbps against host-1. This attack rate was chosen to compare the effectiveness of RDPID with that of DPID, as proposed in [6]. Four different scenarios were taken. First, when PIDs are static (i.e., no RDPIDs). Second, when the RDPID routing scheme is in place and its update period is 30 s. Third, when the RDPID update period is 60 s. Fourth, reduce the RDPID update period to 10 s upon detection of an attack. Three of these four scenarios (i.e., first, second and fourth) are also considered by [6] to show the resiliency against attacks. Figure 7(b) shows the results. As expected, the victim host-1 is subjected to a full 20 Mbps attack with static PIDs. The attack gets mitigated in 33.2 s with an RDPID update period of 30 s. With an update period of 60 s, the attack gets mitigated in 70 s (as against 115 s reported in [6]). Also, with an update period of 10 s after attack detection, it got mitigated in 14 s (as against 18 s reported in [6]).

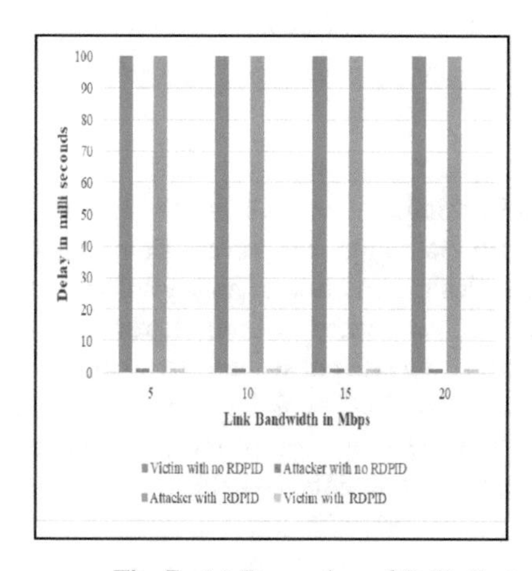

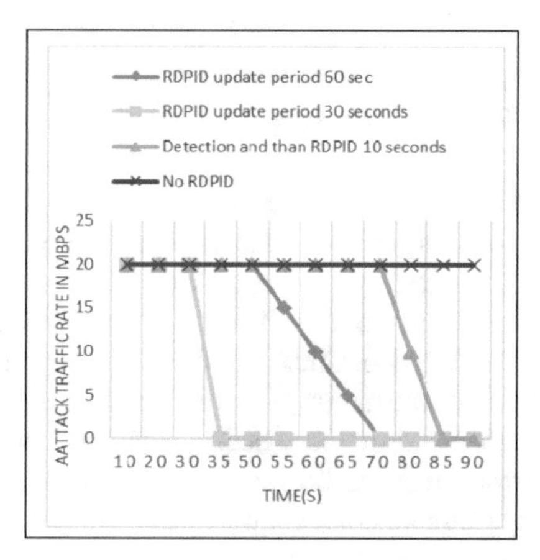

Fig. 7. (a) Prevention of DrDoS attacks (b) Prevention of flooding attacks.

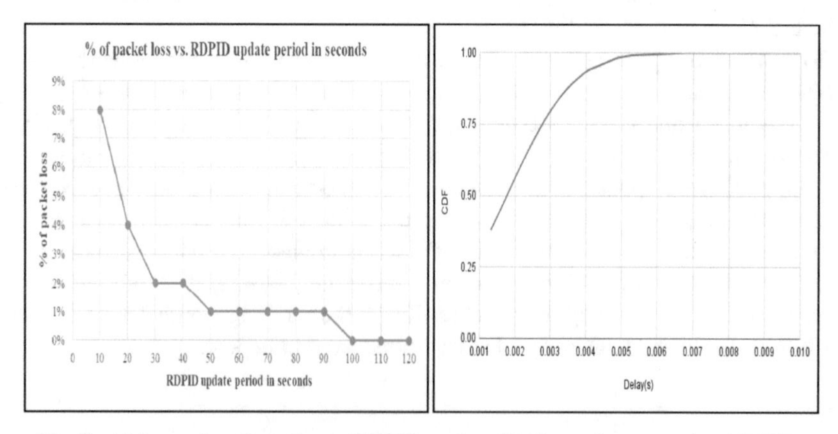

Fig. 8. (a) Loss of packets due to RDPID update (b) Time taken to update RDPIDs

The performance of RDPID was further evaluated using PID update delay. It is the sum of time taken by NM to share the RDPIDs to all RMs, and then RMs to share them with its BRs. 20000 PID update delay samples were recorded by running the proposed RDPID distribution in the topology. Figure 8(b) shows the CDF for PID update delay, which is 10 ms. In [6], the authors calculated PID negotiation delay and PID distribution delay. The former refers to the interval between when an RM sends the PID update message to its neighbor RM and receiving a corresponding acknowledgment. The latter is when an RM sends the PID distribution message to a border router and receives the corresponding acknowledgment. The sum of these two delays is the PID update delay. [6] reported an average of 20 ms for PID negotiation and 3 ms for PID distribution; a total of 23 ms. With RDPID, it was only 10 ms. In fact, as shown in Fig. 8(b), it was only 6 ms with 99% probability. This reduction can be attributed to the fact that in our approach, RMs do not negotiate RDPIDs; rather, NM's job is to update all the PIDs to RMs. The loss of packets caused due to RDPID update is shown in Fig. 8(a). As can be seen, with an RDPID update period of 50 s, the loss is 1%, and with that of 100 s, it reduces to 0%.

Further, the probability P of determining the PID depends on the number of hops (h) and the size of character space $\boldsymbol{S_{ch}}$. It is given by Eq. (1).

$$P = \frac{1}{S_{ch}^{h}} \tag{1}$$

Our experiment has a network with 4 hops for a packet to reach the DNS server and a character space of 52 characters. Even in this small network, the probability of deducing the correct PID sequence by an attacker is $\approx 10^{-7}$, as it has more than 7 million possibilities. Further, with the PID change interval of 30 s, for an attacker to sustain an attack of 1 min, the probability reduces to $\approx 10^{-14}$, having more than 50 million possibilities of different PID sequences.

6 Conclusion

This paper proposed a DDoS prevention approach using RDPIDs against DPIDs primarily used for Information-centric networks. The experiments were conducted using DNS-based request-response client-server inter-domain architecture. It is proved that with RDPID based routing mechanism in place, the victim is always prevented from DrDoS attacks. Also, a flooding attack is automatically mitigated within some constant time, depending on the PID update period. If the attacker tries to guess the used dynamic PIDs, there are approximately 50 million possible PID sequences to guess, making it computationally costly business for the attacker. Via different simulations, it is shown that with our proposed approach, the RDPID negotiation time reduces to 6 ms, as against 23 ms reported in [6]. Also, the attack gets mitigated in 70 s with a PID update period of 60 s (as against 115 s reported in [6]).

References

1. Alex, B., et al.: McAfee Labs Threats Report (2018). https://www.mcafee.com/enterprise/en-us/assets/reports/rp-quarterly-threats-mar-2018.pdf. Accessed 10 June 2021

2. Paul, N.: Five Most Famous DDoS Attacks and Then Some (2020). https://www.a10networks.com/blog/5-most-famous-ddos-attacks/. Accessed 10 June 2021
3. Menscher, D.: Exponential growth in DDoS attack volumes (2020). https://cloud.google.com/blog/products/identity-security/identifying-and-protecting-against-the-largest-ddos-attacks. Accessed 11 June 2021
4. Cimpanu, C.: Google says it mitigated a 2.54 Tbps DDoS attack in 2017, largest known to date (2020). https://www.zdnet.com/article/google-says-it-mitigated-a-2-54-tbps-ddos-attack-in-2017-largest-known-to-date/. Accessed 11 June 2021
5. Swami, R., Dave, M., Ranga, V.: Software-defined networking-based ddos defense mechanisms. ACM Comput. Surv. (CSUR) **52**(2), 1–36 (2019). https://doi.org/10.1145/3301614
6. Luo, H., Chen, Z., Li, J., Vasilakos, A.V.: Preventing distributed denial-of-service flooding attacks with dynamic path identifiers. IEEE Trans. Inf. Forensics Secur. **12**(8), 1801–1815 (2017). https://doi.org/10.1109/TIFS.2017.2688414
7. Luo, H., Chen, Z., Cui, J., Zhang, H., Zukerman, M., Qiao, C.: CoLoR: an information-centric internet architecture for innovations. IEEE Network **28**(3), 4 (2014). https://doi.org/10.1109/MNET.2014.6843226
8. Singh, J., Behal, S.: Detection and mitigation of DDoS attacks in SDN: a comprehensive review, research challenges and future directions. Comput. Sci. Rev. **37**, 100279 (2020)
9. Bhatia, S., Ahmed, I., Behal, S.: Distributed denial of service attacks and defense mechanisms: current landscape and future directions. In: Verstile Cybersecurity, pp. 55–97. Springer, Cham (2018). https://doi.org/10.1007/978-3-319-97643-3_3
10. Godfrey, P.B., Ganichev, I., Shenker, S., Stoica, I.: Pathlet routing. ACM SIGCOMM Comput. Commun. Rev. **39**(4), 111–122 (2009). https://doi.org/10.1145/1594977.1592583
11. Al-Duwairi, B., Özkasap, Ö., Uysal, A., Kocaoğullar, C., Yildirim, K.: LogDoS: a Novel logging-based DDoS prevention mechanism in path identifier-Based information centric networks. Comput. Secur. **99**, 102071 (2020).https://doi.org/10.1016/j.cose.2020.102071
12. Avramopoulos, I., Suchara, M.: Protecting DNS from routing attacks: a comparison of two alternative anycast implementations. IEEE Security and Privacy, Issue on Securing the Domain Name System (2009)
13. Imperva: TCP SYN Flood (2021). https://www.imperva.com/learn/ddos/syn-flood/. Accessed 20 Sept 2021
14. Kohler, E., Morris, R., Chen, B., Jannotti, J., Kaashoek, M.F.: The Click modular router. ACM Trans. Comput. Syst. (TOCS) **18**(3), 263–297 (2000)

Bitcoin's Blockchain Data Analytics: A Graph Theoretic Perspective

Aman Sharma, Ankit Agrawal, Ashutosh Bhatia, and Kamlesh Tiwari(✉)

Birla Institute of Technology and Science, Pilani, Rajasthan, India
{h20180137,p20190021,ashutosh.bhatia,
kamlesh.tiwari}@pilani.bits-pilani.ac.in

Abstract. Bitcoin is the first and most widely used cryptocurrency in the world. It provides a pseudonym identity to its users that is established using the user's public key, which leads to preserving the user's privacy. Each transfer of bitcoin cryptocurrency among the users makes a transaction. The pseudonym identities are considered as transaction end-points. These transactions are recorded on an immutable public ledger called Blockchain which is an append-only data structure. The popularity of Bitcoin has increased unreasonably. The general trend shows a positive response from the common masses indicating an increase in trust and privacy concerns which makes an interesting use case from the analysis point of view. Moreover, since the blockchain is publicly available and up-to-date, any analysis would provide a live insight into the usage patterns which ultimately would be useful for making a number of inferences by law-enforcement agencies, economists, tech-enthusiasts, etc. In this paper, we study various applications and techniques of performing data analytics over Bitcoin blockchain from a graph theoretic perspective. We also propose a framework for performing such data analytics and explored a couple of use cases using the proposed framework.

1 Introduction

Bitcoin is a decentralized cryptocurrency owned by pseudonyms users who make transactions that are stored on an immutable ledger called Blockchain in a distributed manner [1]. Bitcoin is also named as a peer-to-peer network having some special nodes called miners who certify and validate the transactions. Blockchain technology and bitcoin was created due to a growing distrust of existing banking systems and to provide user privacy in a digital world. Bitcoin soon became popular due to the features it provides such as security, privacy, low transaction costs, ease of access, trusted environment and least setup requirements. These features made bitcoin a popular alternative to traditional payment methods in the existing banking system. Thereby, bitcoin is accepted by several countries i.e. Germany, Japan, Mexico, and others. The government of these countries have developed regulations and allow people to use bitcoin as a legal cryptocurrency based on the developed rules.

The exchange of cryptocurrency among the users is done through bitcoin wallets that store the bitcoins and also contain some secret information i.e. private key needed to sign the transactions. According to a report given by Statista [2], the number of

L. Barolli et al. (Eds.): AINA 2022, LNNS 449, pp. 459–470, 2022.
https://doi.org/10.1007/978-3-030-99584-3_40

wallets belonging to users have been growing since the bitcoin creation in 2009 and this number has reached around 80 million wallets by the end of December 2021 that shows the popularity of cryptocurrency. In line with this rising trend, and as a result of the heavy tapping of the Internet and its services, a rise in privacy concerns has been observed among the common masses. Bitcoin is often considered a double-edged sword [3] because, on one side, it ensures the user's privacy, and on the other side, it reveals all the transactions in front of everybody. The information regarding each transaction from the time of bitcoin creation is publically available locally at every system running the bitcoin client. Such publically available data allow an individual to perform analysis on that according to the required use cases. This paper attempts to identify such use cases. The data analysis of the blockchain, particularly the bitcoin blockchain, offers to do penetration on various things such as illegal activities, economic indicators, general internet security, and others [4]. As a result, the data analysis can derive other socio-cultural trends (i.e., habits) based on transitivity or other inferencing methods. For instance, a specific type of tagged service during a period may indicate that the service has become popular, that might be indicating the changes in lifestyle [5]. An example of such services is the Illegal drug supply.

Section 2 discusses the background of cryptocurrencies. Section 3 discusses various ways of analyzing the bitcoin data. In Sect. 4, We propose a generic framework that can be followed for different analysis purposes. Then in Sect. 5, we discuss the results of certain experiments performed on the proposed framework for some considered use cases. In Sect. 6, we conclude our findings.

2 Background

Bitcoin was first introduced by Satoshi Nakamoto in the paper titled "Bitcoin: A Peer-to-Peer Electronic Cash System" [8]. The writer's identity is anonymous and has become a mystery. However, bitcoin popularity is not hidden from anyone and continuously growing. The concept of Electronic cash system is not new. DigiCash, an internet payment system founded in 1989 by David Chaum, was one of the first electronic cash services [6]. The concept of blind signature [7] is used to ensure the security and privacy of its users by restricting the third parties to access the personal information from the transactions. It also solves the double-spend issue. However, privacy is achieved as long as users don't double spend. The disadvantage of such a system is the involvement of a server handled by a centralized authority, which creates distrust among the users. Attributing value to the electronic cash was another problem. In DigiCash, one must withdraw $100 from their bank account and trade it with the issuer of ecash for $100 in order to obtain ecash worth $100. These hassles made it difficult for DigiCash to gain popularity, which led to its early demise. Other internet payment services faced the same fate at the time as well.

2.1 Bitcoin: A New Mix

Having centralized authority in the DigiCash system suffers from distrust among the users. On the other hand, Bitcoin solves this issue due to having decentralization in the blockchain. Moreover, Bitcoin provides security using the concepts of provable cryptography and offers several advantages over conventional systems such as irreversible transactions, common escrow mechanisms, continuous availability, and modest transaction fees. Irreversible transactions protect the sellers from fraud, whereas common escrow mechanisms protect the buyers. The significant components of Bitcoin are to store and validate the transactions and to protect the system from double-spending issues in a distributed manner.

Transactions
The transactions of electronic coins as a reference to Bitcoins can be considered as a chain of digital signatures. One user (owner) transfers a coin to other user (payee or receiver) by making a signature on a hash of the receiver's public key and the previous transaction and appends it to the end of the coin. In such a way, anybody can verify the chain of ownership. To prevent the system from double-spending, all transactions must be publicly available and majority of the nodes must agree and follow a single history of the received transactions order. The coin receiver should get a proof that majority of the nodes should agree on the first time arrival of each transaction. Miners group the transactions appeared at the approx same time together in a block and computes the hash of a new block from the previous hash and the current block and get published. The hash of the block is considered as a timestamp. Now anyone can check that when the hash corresponding to a block of transactions is created and published. Miners constantly do such verification that new added blocks are valid and old blocks are not tampered with. A Timestamp ensures the chronological validity of transactions to the payee much like a newspaper timestamps the events of a specific period. Once the validity is confirmed, the payee can use this transaction as a reference to spending the acquired Bitcoins.

Proof-of-Work (PoW)
Proof-of-work is a consensus algorithm used to implement a distributed timestamp in a peer-to-peer network. When users make transactions and broadcast them into the system, any other users can check the validity of those transactions. Finally, miners group the valid transactions into a block. Miners use proof-of-work to perform some computational work in order to add the blocks in the blockchain. The computational work involves solving a cryptographic puzzle, and the puzzle's solution becomes the proof for that work. The puzzle is to find a hash value starting with a few zero bits after making changes in the nonce value along with some non-changed additional fields. When the miner finds some nonce value following the restricted hash value, the miner adds that block in the blockchain. Thus, anybody in the network can verify the proof of work. Then, miners get rewarded with bitcoins as mined coins along with the transaction fee for all transactions included in the block [8].

The Process
To summarise the whole process in the Bitcoin network: A user first generates a pair of signing keys (public and private key pair) and then publishes the public key representing the user's address to send and receive the bitcoins. Individuals can create and use as many addresses as they want for transactions. As a matter of fact, the ideal number is equal to one per transaction. A user makes payment by broadcasting a transaction containing the recipient's address to its peers, each of whom forwards it to their peers.

Table 1. Block header

Field	Purpose	Size
Version	Which version of transaction data structure we're using	4 bytes
Previous Block Hash	256-bit hash of the previous block header. This is what "chains" the blocks together	32 bytes
Merkle Root	All of the transactions in this block hashed together. Basically provides a single-line summary of all the transactions in this block	32 bytes
Time	When a miner is trying to mine this block, the Unix time at which this block header is being hashed is noted within the block header itself	4 bytes
Bits	A shortened version of the Target	4 bytes
Nonce	he field that miners change in order to try and get a hash of the block header (a Block Hash) that is below the Target	4 bytes

Eventually, the broadcasted transactions reach the miners, who collect them into a block and attempt to find a hash value based on the puzzle difficulty to prove the legitimacy of the block. The miner node broadcasts the proof-of-work to all nodes when it finds it. Before accepting it into the chain, the block is checked for double-spending validity (Table 1).

3 Analysis of Bitcoin's Blockchain Data

When a node acts as a full blockchain node in the Bitcoin network, that node contains a full transaction history in its local machine. The file containing all the transaction details is encoded in hexadecimal format. Therefore, the data needs to be parsed to get ready for the analysis purposes. We used BlockSci's Parser [17] which is an open-source software platform for performing the analysis over Bitcoin blockchain data. BlockSci incorporates an in-memory blockchain database along with the analysis libraries. The BlockSci's parser performs several optimization in the form of data transformation to minimize the memory consumption so that the data can be fit in memory. The first transformation is to link the inputs of transactions with the output of the previous transactions. And the second one is to link the inputs and outputs of transactions with the addresses. The parser assign an identifier to each transaction and maps the identifier to the transaction hash. In the same way, parser assigns an identifier to the addresses and map them for deduplication (removing redundant addresses) and linking purposes. Finally, parser generates a single representation of data. Such techniques allow for efficient graph traversal.

Moreover, since the Bitcoin ecosystem is built around transactions, a graph-theoretic approach can be used for analysis. A number of different graph-centric perspectives have been proposed for Bitcoin so far [9]: 1) *Transaction Graph*: represents the flow of Bitcoins between transactions over time, each vertex is a transaction and each directed edge an output connecting two transactions with each other [10].

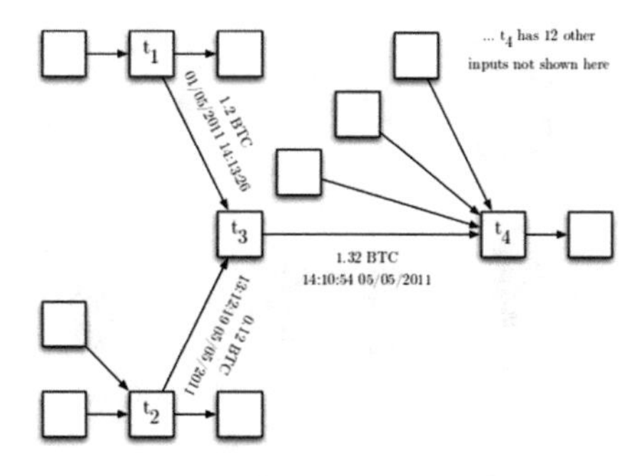

Fig. 1. An example of transaction graph

2) *Address Graph*: represents the flow of Bitcoins between addresses (public keys). The vertex represents addresses in the network and the directed edges are transaction from a source address to a destination address. 3) *Cluster Graphs*: are similar to address graphs, the only difference being vertices now represent a cluster of addresses linked by some heuristic.

Graphs have the high expressive power to model complicated structures. The above mentioned graphs are generally property- graphs where the edges are labeled and both vertices and edges can have any number of key/value properties associated with them. The graph in Fig. 1, has edges with multiple values associated with them. These properties add an extra dimension to the graph data from the analysis point of view. There are two major approaches to analyze property-graphs: 1) *Computational Graph Analytics*: involves iterating over the graph and computing properties or stats. 2) *Graph Pattern Matching*: involves querying the graph to find sub-graphs that match a given pattern.

3.1 Computational Graph Analytics

Deducing Importance of Entities

We mentioned how different entities of a Bitcoin network can be modeled as vertices in the property-graph. Deducing importance of these entities could help in understanding various dynamics of the Bitcoin's network, thus giving us an insight into much deeper structural changes. In graph theory, importance relates to the centrality of a vertex. There are various measures of centrality [11–14]: Closeness Centrality, Betweenness Centrality, PageRank, Eigenvector Centrality, and HITS (Hyperlink-Induced Topic Search). If the degree of centrality of a vertex is high, it correlates with high importance. In terms of Bitcoin, it indicates the high degree of coin flow. Such vertices generally work as service providers. For example, Satoshi Dice is a Bitcoin cash game which has high degree of centrality representing its high number of users. It has an out-degree of 9576588. However, centrality with some other heuristics is much more informative than by itself.

Traversal

Graph traversal defines a way to understand the vertex connectivity. Graph traversal can be used to achieve different objectives such as finding short distances, reachability,

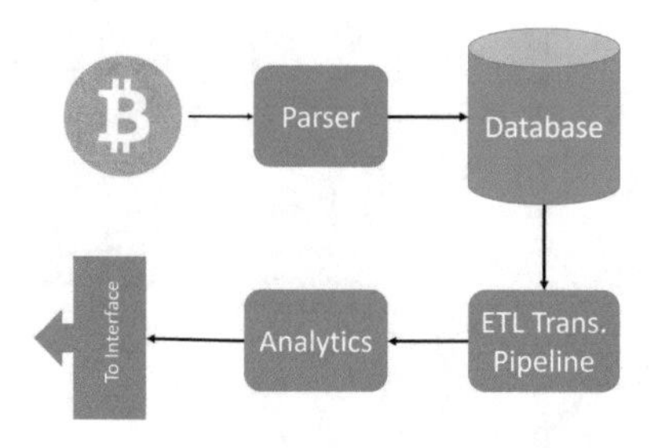

Fig. 2. A framework for bitcoin analysis

path length, etc. The problem of traversing and finding the shortest path in graph theory has been extensively studied. However, Bitcoin transactions result in large graphs with millions of nodes and billions of edges, with a highly skewed distribution of degrees. Consequently, traversal algorithms need to take into account properties such as time sequence of transactions, the cluster memberships of addresses, and the degree of node. Investigating the path from one address to another address (possibly known) is a typical use case; this can be useful for tracking the coins flow from suspicious addresses. An address must still be tagged suspicious, which is discussed later in the paper. By detecting strongly connected components of a huge dataset, such as that of Bitcoin addresses, it is possible to find addresses that are closely related to each other [15]. We can then use this information to propagate labels iteratively, thereby labeling the nodes.

3.2 Graph Pattern Matching

Analysis of big-data graphs can also be accomplished through pattern matching. Pattern matching allows analysts to query the data graph for all occurrences of a given pattern/template. Following are some possible use-cases: Fraud Detection, Anomaly Detection, and Sub-graph Extraction. For Instance, money Launderers use unregulated cryptocurrency exchange services to clean their money [16]. They accomplish this by simply trading the Bitcoin a number of times across various markets thus adding degrees of privacy similar to 'hopping' between wallet addresses. Since the number of such unregulated exchanges are few. A template matching to a launderer's trail could be used to identify all such transactions matching this template and the addresses involved could be flagged for further monitoring.

4 Analytics Architecture

In this section, we propose a generic framework, as shown in Fig. 2, to perform the analysis of Bitcoin's blockchain data with graph theoretic perspective.

- *Parser:* The parser program takes blockchain data as input and produces a sequential table of transactions As the data structure storing this representation does not include transaction hashes or addresses. A separate Indexes file is generated which maps the transaction ids to hashes.

- *Database:* The output of the parser is converted into a relational database. Even though the output of the parser can be used for the purpose of performing analysis over the parsed data, storing it in a database provides resilience as well as much flexibility in terms of a range of transformations that can be applied using various Big-data analytics tools.
- *Transformation Pipeline:* This unit is responsible for transforming data from keyspace to graphs. This stage is another reason why we had to dump our parsed data into a database. As of now, the existing parser like BlockSci [17] does not support exporting data to tools such as Apache Spark. Moreover, blocks work on a single node, unlike big-data tools that are meant for distributed computing which is required for the huge volume of Bitcoin's raw data. Therefore we used Spark to transform the parsed data into a number of keyspaces which could be queried like a graph through techniques mentioned in the previous section.

5 Experimentation

This section highlights some of the use-cases we explored and other possibilities with Bitcoin's data. At present, there are about 573,432 blocks in the Bitcoin's blockchain consisting of around 400 million transactions in total and which approximates to roughly 250 GB of data. A number which is growing almost exponentially. A regular, off the shelf machine, would be painfully slow to parse such amount of data as confirmed in our experiment. We had to restrict our analysis till 300,000 blocks which corresponded to about 35 million transactions. In this work, we used BlockSci's parser because of it's proven performance in parsing the blockchain data. We used Apache's Cassandra database to store the transaction data into a number of tables such as Complete Transactions, Block-wise Transactions, Block Statistics, etc.

5.1 Address Linking

"Bitcoin is not anonymous", is one of the points highlighted on the www.bitcoin.org website. It's true, contrary to the popular belief that Bitcoin is anonymous. However, to an extent, it's pseudo-anonymous since identities are attached to public keys which one could generate as much as they want. Nevertheless, these addresses can be linked to the user or an entity, though not completely accurately. This is achieved through various heuristics. Linking addresses into clusters/entities reduces the redundancy in the data and provides a better insight into trends over time. The two major heuristics for address linking

- **Multi-Input Transactions:** is based on the fact that people use multiple addresses for transactions, so to make a payment which is higher than the amount in any individual wallet, one can use multiple addresses that they hold to make a merged payment. This ultimately links all such addresses used as input into a single entity [21]. An example scenario is illustrated in Fig. 3. Let's assume, the owner of account C wants to make payment to the address of a service provider. But that amount exceeds the amount as UTXO (unspent transaction output) of C. However, C also

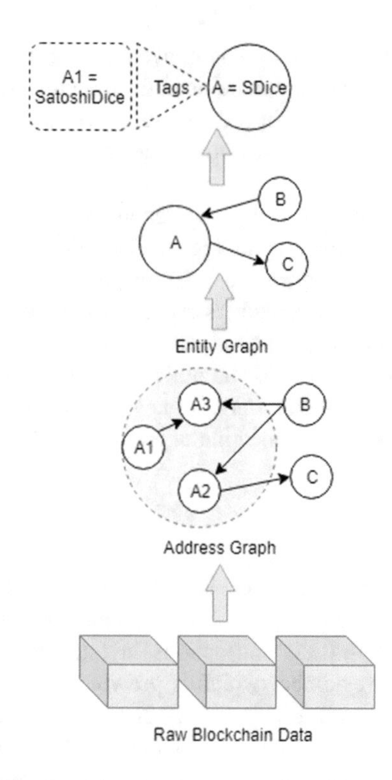

Fig. 3. Address linking as graph enrichment

owns accounts corresponding to public keys A and B. Let an outsider make a payment to C by creating a transaction that redeems to the owner of C public key only. Now the combine UTXO of C's owned accounts can compensate for the payment to the service. So, C now writes a transaction, specifying the inputs that are where the UTXOs were sent. Since C owns A and B, their outputs are redeemable at C's wallet. Hence, C is able to make the transaction to the service easily. Outsider can make out the bitcoin addresses from the transaction but can not make connect that address with a people. However, outsider can connect the addresses with a single entity by observing the input and output of multiple transactions.

- **Change Address:** Since in a transaction all of the Bitcoins of an individual are consumed and the change is returned to a new address called change address, one could link such addresses because they are hardly reused [18]. This heuristic further refines the first one for specific queries by providing a compensatory factor in certain wrong observations.

Aforementioned heuristics, when applied to the address graph, link various addresses into clusters representing a single entity. More information is attached to these clusters, by using tags attributed to one or more addresses in the entity. Once we explicitly identify some of the addresses, we could easily de-anonymize a significant amount of anonymity.

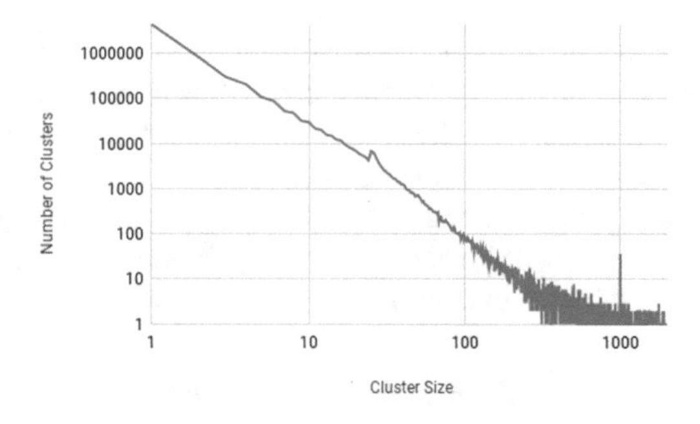

Fig. 4. Distribution of clusters with respect to sizes based on address-linking heuristic

One of the challenges is to tag addresses to real-world actors since there's no centralized infrastructure that keeps a mapping of these sorts. A fairly tested method is to scrape data off [19,20] from Bitcoin fora like BitcoinTalk.org or services like Blockchain Explorer, etc. A lot of times people reveal their addresses for business purposes or if they are looking for donations/tips, a direct relation can be obtained. Moreover, with increased internet penetration there are various passive and active attacks that can easily expose a user's identity as well. Goldfeder *et al.* [22] shows that trackers on the internet have enough information about a purchase made even through cryptocurrency, to uniquely identify the transaction on the blockchain, link it to the user's cookie, thus to the user's real identity. Furthermore, they were also able to show that if a tracker could link two transactions made by the user, it can then identify the entire cluster of addresses even if the user employs anonymity techniques like CoinJoin.

Tagging Users transacting with Known Merchants

A lot of service providers expose their wallet addresses for business purposes. So, it is not difficult to tag them. Moreover, Bitcoin is the favorite cryptocurrency of Darknet. We collected tags from various https://www.blockchain.com/btc/tagsBlockchain websites with their corresponding (seed) addresses. Then we queried the clusters for the tagged address, hence we were able to identify the cluster and tag with the help of one identity. Effectiveness of this method depends largely on the capability of the heuristic to correctly identify relations among various addresses. Considering the evolving nature of address linking techniques, and considering that different sets of heuristics may be suited to the different application, a combination of heuristics might work better.

Anlayzing Payments to Ransomware

Ransomware is a piece of malicious software that forfeits the access to a victim's data until they pay a certain sum of money "ransom" in exchange for access. In May 2017, ransomware named "WannaCry" infected about 300,000 systems worldwide. It demanded \$300–\$600 payment through bitcoin to restore access [23]. The design of ransomware requires exposing wallet addresses for collecting ransoms. These addresses can be attributed to a cluster using the multiple-input heuristic thus enabling identification of wallets of the hacker and thereby his/her activities. We carried out a similar experiment involving CryptoLocker Ransomware which was active worldwide from September 2013 to January 2014. Threat actors who demanded the ransom to be paid within 72 h through bitcoin. We used a known CryptoLocker address[1] as the seed for the clustering process. Multiple-Input heuristic was able to generate a cluster with 968 addresses, which is consistent with the results by Liao et al. [24]. Further, we analyzed that, the average in-degree (number of transaction in which said address is an output) is 1.11 and the corresponding out-degree is also approx. 1 meaning their UTXOs have been used only once. This could mean that these addresses are "Change Addresses". Figure 4 shows the distribution of cluster with respect to different sizes.

5.2 General Statistics Over the Bitcoin Transaction Graph

- **Velocity of Bitcoin Transactions:** The velocity of money is the frequency with which one unit of currency is used for purchases in a unit of time. It can provide an insight into the extent to which money is used as a medium of exchange versus a store of value.
- **Measuring Different Types of Address Use:** An insight into use of different types of transactions (see Fig. 5a): 1) *Pay to pubkey* 2) *Pay to pubkey hash* (P2PKH) 3) *Pay to script hash* (P2SH): Allows the recipient of a transaction to specify the redeem script instead of the sender. 4) *Multisignature* (multisig): refers to requiring more than one key to authorize a Bitcoin transaction. 5) *Non Standard*: All other

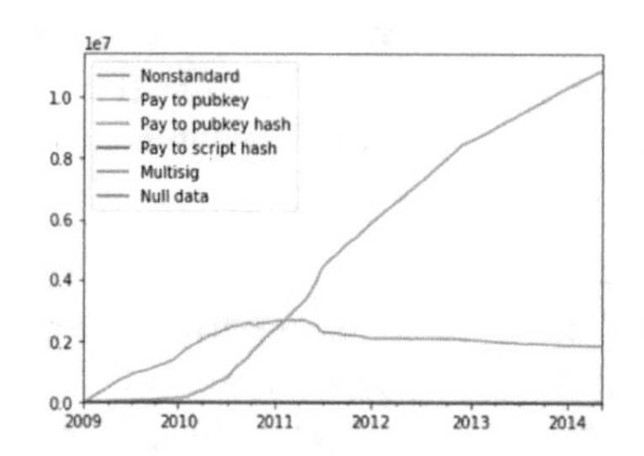

(a) Different address types for making payments

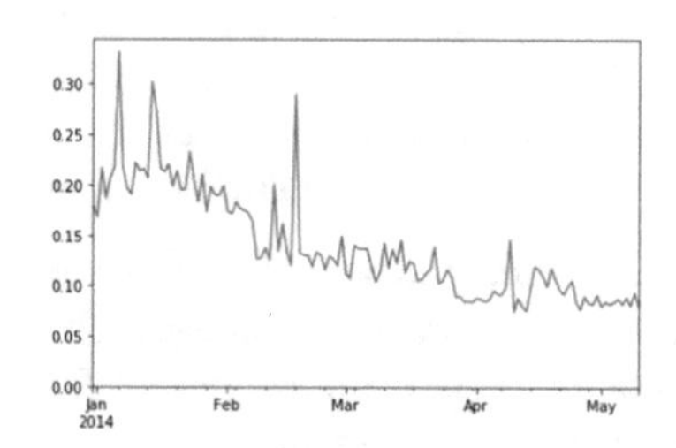

(b) Average Fee per Transaction in 2014 in USD

Fig. 5. General statistics

- **Average Fees per Transaction:** One of the advantages of Bitcoin's transaction is their meager transaction fees. The graph (see Fig. 5b) shows a positive correlation with BTC's market value at a given time. It touched its highest value in Dec 2017 at 1K Satoshis per Byte. This was the time when Bitcoin was at its highest exchange value.
- **High Value Transactions over time:** Certain transaction involving a huge transaction fees can be identified as of high value (see Fig. 6). There are over 300 high value transactions. Increase Incidentally, the highest transaction fee that has ever been paid is 291 BTC. On April 26, 2016, the creator of a transaction famously and accidentally swapped the value and the fee, losing a sum of $136,000 at the time.

Similarly, a lot of other markers can be identified that links the dynamics of Bitcoin's network with dynamics of online market and services.

[1] https://www.blockchain.com/btc/address/1KP72fBmh3XBRfuJDMn53APaqM6iMRspCh CryptoLocker Virus.

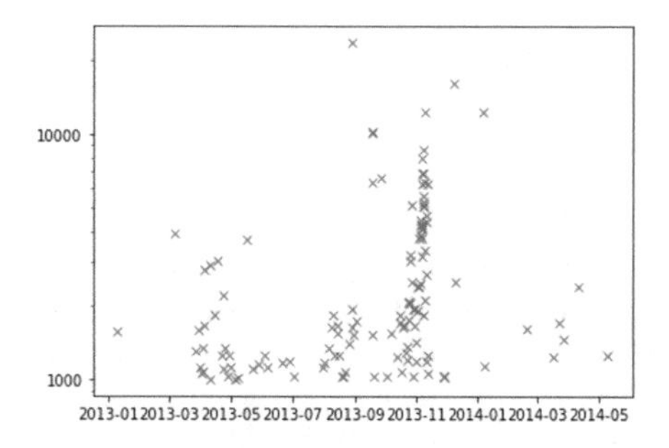

Fig. 6. Bitcoin transaction with transaction fees over $1000

6 Conclusion and Future Work

We noted various ways in which the Bitcoin's network can be analyzed. The graph-perspective of the transactions fits the purpose owing to the network/flow model of the Bitcoin's ecosystem. Also, the availability of graph-based algorithms enables proof-backed results. These algorithms can be performed with most of the Big Data analytic tools today. Such analysis is expected to get better and frequent in the future. On one hand, people would want to strengthen the system by ensuring maximum anonymity with minimal chance of linkability, for example, the coin mixing services, while on the other hand, certain actors would work towards finding out new ways to discover links between unrelated seeming transactions both for ethical or unethical purposes. However, this conflict shall reach an equilibrium just like it has for the conventional banking system. We look forward to identifying a better combination of heuristics to cluster accounts in future work. A powerful clustering mechanism that could identify and filter out false positives shall ensure better data to work on with even the same set of algorithms. The advantages of Bitcoin exceed the concerns attached with it by a large margin. Though it is required that people understand the ideal practices before transacting over Bitcoin. It is mostly through some side-channels that an attacker might gain access to one's identity/wallet. If a person is cautious enough, it is a fascinating piece of technology that has the potential to bring us closer to a universal currency someday.

References

1. Blockchains: the great chain of being sure about things. https://www.economist.com/briefing/2015/10/31/the-great-chain-of-being-sure-about-things
2. Number of Blockchain wallet users worldwide from 1st quarter 2016 to 1st quarter 2019. https://www.statista.com/statistics/647374/worldwide-blockchain-wallet-users/
3. Economics of Bitcoin: is bitcoin an alternative to fiat currencies and gold? https://nakamotoinstitute.org/research/economics-of-bitcoin/. Accessed 20 Feb 2021
4. Bouri, E., Shahzad, S.J.H., Roubaud, D., Kristoufek, L., Lucey, B.: Bitcoin, gold, and commodities as safe havens for stocks: new insight through wavelet analysis. Q. Rev. Econ. Finance **77**, 156–164 (2020)

5. Kleineberg, K.-K., Helbing, D.: A "social bitcoin" could sustain a democratic digital world. In: Dapp, M.M., Helbing, D., Klauser, S. (eds.) Finance 4.0 - Towards a Socio-Ecological Finance System. SAST, pp. 39–51. Springer, Cham (2021). https://doi.org/10.1007/978-3-030-71400-0_3
6. Prasanthi, P., Kumar, G., Kumar, S., Yalawar, M.S.: 10 Privacy and Challenges. Cyber Defense Mechanisms: Security, Privacy, and Challenges, p. 157 (2020)
7. Chaum, D.: Blind signatures for untraceable payments. In: Chaum, D., Rivest, R.L., Sherman, A.T. (eds.) Advances in Cryptology, pp. 199–203. Springer, Boston, MA (1983). https://doi.org/10.1007/978-1-4757-0602-4_18
8. Nakamoto, S.: Bitcoin: a peer-to-peer electronic cash system. www.bitcoin.org
9. (Alan) Wu, Z.: Analyzing blockchain and bitcoin transaction data as graph. https://download.oracle.com/otndocs/products/spatial/pdf/. Accessed 25 Feb 2021
10. Goldsmith, D., Grauer, K., Shmalo, Y.: Analyzing hack subnetworks in the bitcoin transaction graph. Appl. Netw. Sci. **5**(1), 1–20 (2020). https://doi.org/10.1007/s41109-020-00261-7
11. Sabidussi, G.: The centrality index of a graph. Psychometrika **31**, 581–603 (1966). https://doi.org/10.1007/BF02289527
12. Freeman, L.: A set of measures of centrality based upon betweenness. Sociometry **40**(1), 35–41 (1977). https://www.jstor.org/stable/pdf/3033543.pdf. Accessed 25 Apr 2021
13. Negre, C.F.A., et al.: Eigenvector centrality for characterization of protein allosteric pathways. Proc. Natl. Acad. Sci. **115**(52): E12201–E12208 (2018)
14. Sullivan, D.: What is Google PageRank? A Guide for Searchers & Webmasters (2007). https://searchengineland.com/what-is-google-pagerank-a-guide-for-searchers-webmasters-11068. Accessed 25 June 2021
15. Ron, D., Shamir, A.: Quantitative analysis of the full bitcoin transaction graph. In: Sadeghi, A.-R. (ed.) FC 2013. LNCS, vol. 7859, pp. 6–24. Springer, Heidelberg (2013). https://doi.org/10.1007/978-3-642-39884-1_2
16. Here's how criminals use Bitcoin to launder dirty money. https://thenextweb.com/hardfork/2018/11/26/bitcoin-money-laundering-2/. Accessed 26 June 2021
17. Kalodner, H., et al.: BlockSci: design and applications of a blockchain analysis platform. In: 29th USENIX Security Symposium (USENIX Security 20), pp. 2721–2738 (2020)
18. Meiklejohn, S., et al.: A fistful of Bitcoins: characterizing payments among men with no names. Commun. ACM **59**(4), 86–93 (2016). https://doi.org/10.1145/2896384
19. Fleder, M., Kester, M.S., Pillai, S.: Bitcoin transaction graph analysis. arXiv preprint arXiv:1502.01657 (2015)
20. Haslhofer, B., Karl, R., Filtz, E.: O Bitcoin Where Art Thou? Insight into Large-Scale Transaction Graphs. In: SEMANTiCS (Posters, Demos, SuCCESS), September 2016
21. Meiklejohn, S., et al.: A fistful of bitcoins: characterizing payments among men with no names. In: Proceedings of the 2013 Conference on Internet Measurement Conference, pp. 127–140, October 2013
22. Goldfeder, S., Kalodner, H., Reisman, D., Narayanan, A.: When the cookie meets the blockchain: privacy risks of web payments via cryptocurrencies. arXiv preprint arXiv:1708.04748 (2017)
23. Symantec Security Response "What you need to know about the WannaCry Ransomware". https://symantec-enterprise-blogs.security.com/blogs/threat-intelligence/wannacry-ransomware-attack. Accessed 5 July 2021
24. Liao, K., Zhao, Z., Doupé, A., Ahn, G.J.: Behind closed doors: measurement and analysis of CryptoLocker ransoms in Bitcoin. In: 2016 APWG Symposium on Electronic Crime Research (eCrime), pp. 1–13. IEEE, June 2016

Whole-Body Exposure to Far-Field Using Infinite Cylindrical Model for 5G FR1 Frequencies

Aymen Ben Saada[1(✉)], Sofiane Ben Mbarek[2], and Fethi Choubani[1]

[1] Innov'COM LAB, SUP'COM, University of Carthage, Tunis, Tunisia
aymen.bensaada@supcom.tn
[2] Physical Science and Engineering Division,
King Abdullah University of Science and Technology, Thuwal 23955-6900, Saudi Arabia
sofien.benmbarek@isimg.tn

Abstract. Analytic models of human body such as cylindrical and planar models are physically infinite. Hence, they can not be used to quantify whole-body SAR and Whole-body average SAR. In this paper, we proposed three averaging techniques to estimate Whole-body average SAR (WBASAR). Next, the chosen estimator was used with an infinite multilayered cylindrical (MLC) model in the first frequency range of 5G to adapt the later model to sub-6 GHz. Validating an estimator for 5G FR1 with a voxel model, has the benefit of using the estimator with 5G FR2, where voxlized model cannot be used. The adapted MLC model had matched the frequency of the first resonance.

Keywords: Body Surface Area · Specific absorption rate · SAR · Whole-body average SAR · Dosimetry

1 Introduction

In dosimetry study-field, Voxel models had become the dominant modeling tool for wave-exposure assessments. This dominance is principally related to the accuracy of voxel models in both geometry and anatomic composition of the human body. In fact, a voxel model construction is based on Magnetic resonance imaging (MRI) images. Next, these images are transformed into three dimension model. Finally, the 3D model is numerically simulated using a finite-element or FDTD computational tool to estimate the absorbed waves. The already cited workflow, has been applied in many papers to construct voxalized models [1], which have been used to assess EM fields levels up to 6-GHz [2–4]. Beyond 6 GHz, it had become technically difficult to simulate a whole-body exposure due to large computational time, and high computer-memory demand. Therefore, semi-infinite analytical models such as planar and cylindrical [5] have regained more interest for frequencies beyond 6-GHz [6].

Working with resurgent planar and cylindrical models as abstractions of the human body, will limit studies to absorbed power density (PD) and peak-SAR (pSAR) as the two intuitive metrics for absorbed energies evaluation [5,7]. PD and pSAR are both

L. Barolli et al. (Eds.): AINA 2022, LNNS 449, pp. 471–478, 2022.
https://doi.org/10.1007/978-3-030-99584-3_41

locally determined based on local E-field values, hence, they are not suitable for whole-body quantification of the absorbed energies. Expressing Whole-body SAR (WBSAR) or Whole-body-average SAR (WBASAR) for infinite-multilayered-cylindrical (MLC) and semi-infinite planar models will open the door to future comparison between analytic, semi-analytic and voxelized models for sub 6 GHz band.

In this paper, we have two objectives. The first aim is to pick the most representative metric to quantify, computationally, the whole-body absorbed energy when it becomes hard to use exact analytical solution. The second objective is to adapt an MLC model for sub- 6 GHz frequencies and try to extend the validity domain of the MLC model from above 6 GHz to much wider frequency range. In the following sections, we construct MLC model based on physical properties of human males (weight, height), then we derive the most representative WBASAR formula in order to use it as a comparative tool between voxel and MLC model below 6 GHz.

2 Model and Methods

2.1 Cylindrical Model Construction

This paper deals with two models: Voxel and MLC models. WBASAR for voxel models are extracted from Nagoaka et al. paper for 1 year, 5 years and 10 years children for an incident power density (PD) of 1 W/m^2 and an E-polarization [4]. The anatomic weights are also detailed, so each weight can be mapped into one of four tissues (Skin, Fat, Muscle and Bone). MLC model construction is based on tissues weight cited in [4].

The infinite cylindrical system formation is based on the total weight and the height of a child using body surface area formula in Eq. 1. Body surface Area (BSA) is the extern surface of the Skin. It was reported in [2], that BSA is correlated with absorption cross section in Gigahertz region. It is used in pharmacology to dose medications. Many empiric formulas have been developed to approximate BSA for various races and ages, and for both genders. In the following, we used an European male BSA formula [8].

$$BSA = 0.1225 \times Weight^{1.75} \times Height^{0.75} \tag{1}$$

Next, an outer radius is derived from BSA expression by dividing Eq. 1 by $2\,\pi \times Height$. The inner cylinders are constructed to represent one human tissue per layer. The total number of layers, in this study, is five as shown in Fig. 1. The width of the Skin layer is constant. The Fat, Bone layers thicknesses are obtained directly from [4] by dividing, consequently, the mass of the Fat and the Skeleton mass over the model weight. Muscle width is obtained by summing the mass of the heart and muscle and divide the total over the model weight. The inner layer will represent the rest of the human body tissues. MLC model thicknesses are summarized in Table 1

2.2 SAR Averaging Methods

In this section, we detail different possible techniques to estimate the WBA-SAR for an infinite cylindrical model. The first technique is a direct implementation of the average

Table 1. Thicknesses of the MLC layers for 1, 5 and 10 years old child model.

	1 year	5 years	10 years
Skin	4.2 mm	4.2 mm	4.2 mm
Fat	$0.2166 \times R_{out}$	$0.1931 \times R_{out}$	$0.1972 \times R_{out}$
Muscle	$0.2104 \times R_{out}$	$0.3205\times R_{out}$	$0.3741 \times R_{out}$
Bone	$0.1214\times R_{out}$	$0.1349\times R_{out}$	$0.14768 \times R_{out}$

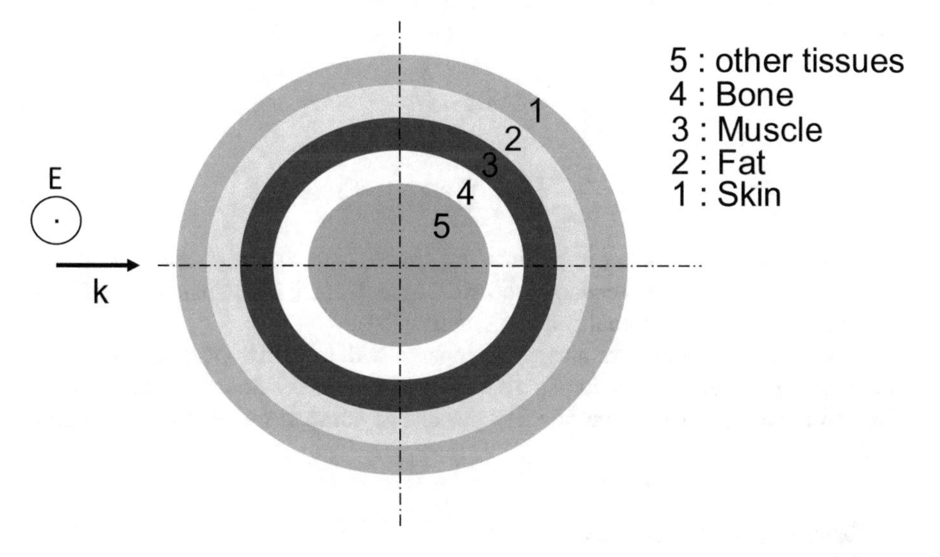

Fig. 1. 5-layers-infinite MLC model. The model is studied with an incident E-field parallel to the infinite axis of the concentric cylinders. The external radius of the MLC model is noted R_{out}.

of finite-element models. The WBASAR in this method is estimated by dividing the sum of local peak SARs (pSAR) by the total number of points, as mentioned in Eq. 2

$$WBASAR = \sum_{i=1}^{N} \frac{SAR_i}{N} \tag{2}$$

where local SAR in the ith point is calculated as follows Eq. 3

$$SAR_i = \frac{\sigma_i E_i \times E_i^*}{2\rho_i} \tag{3}$$

where σ_i is the conductivity of human tissue at the ith point in S/m. ρ_i is the mass density of human tissue in the ith point in kg.m^{-3}. E_i and E_i^* are respectively the electric field strength and its conjugate at i^{th} point.

The second technique for WBASAR prediction consists of determining the maximum electric field in each layer of MLC model. Then, the maximum local SAR (pSAR) is deduced from Eq. 3 for each layer of the model. Finally, the WBASAR is deduced by averaging the sum of pSAR for each layer over the total number of MLC layers. pSAR is deduced from Eq. 3 by replacing E_i by the maximum E-field in the i-th layers.

$$WBASAR = \sum_{l=1}^{L} \frac{pSAR_l}{L} \tag{4}$$

The third technique is derived from scattering cross-section of infinite cylinders. This technique was reported and used only by Massoudi et al. [9] to estimate average SAR. According to [9], the SAR can be estimated using the following set of formulas of Eq. 5 and Eq. 6

$$WBASAR = \frac{2\,Q_{abs}}{\pi\,R_{out}} \tag{5}$$

$$Q_{abs} = \frac{2}{k_0\,R_{out}} \left\{ Re\Big(\sum_{n=0}^{\infty} e_n D_n \Big) - \sum_{n=0}^{\infty} e_n |D_n|^2 \right\} \tag{6}$$

Where k_0 is the wave-number in vacuum. R_{out} is the outer radius of the MLC model. Q_{abs} is the absorption cross section of the MLC model, and it is computed using E-field coefficients of the total external E-field D_n and e_n [9].

These three averaging formulas, are compared with a WBASAR result from voxel model used by Nagaoka et al. in [4]. This comparison aims to figure out the most representative averaging technique in a first place. The comparison results are plotted in Fig. 2.

3 Results and Discussion

From the three averaging techniques of Fig. 2, the second estimator has the closet average to voxel WBSAR across the frequency range of this study. Averaging formulas of Eq. 2 and Eq. 4 have the same evolution, however they are shifted of 0.001 along Y axis. The stability of Eq. 2 series has been practically proved, and Eq. 2 of Fig. 2 has been traced with 839K points. Also, Fig. 2 shows small influence of the electromagnetic properties of the inner layer on WBASAR for frequencies below 200 MHz and no effect on higher frequencies despite the extreme change in permittivity and conductivity between Air and Muscle. These results will simplify the study by reducing the number of averaging techniques to one WBASAR formula of Eq. 2. It is worth noting that MLC model shows two peaks in comparison with Voxel model. It has a shifted first peak and a maximum at 6 GHz. To deeply investigate these two peaks, we add to the study another two MLC models of 1 year and 5 years and their equivalent Voxels. The results are plotted in Fig. 3.

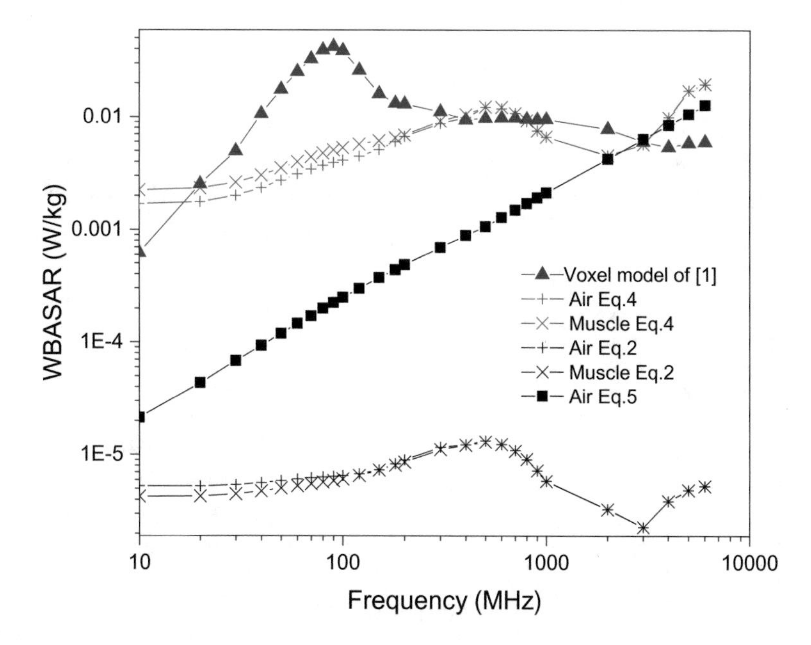

Fig. 2. WBASAR results for three methods applied on MLC model of 10 years old child and compared to its equivalent Voxel. All cylindrical models have 5 layers formed consequently from outer layer to inner layer of Skin-Fat-Muscle-Bone. The inner layer is filled with Air for the plots labeled with plus sign (+) and with Muscle for cross sign (×) labels.

The growth of the model from 1 year to 10 years is transformed into BSA increase from 0.2963 m^2 to 0.9820 m^2. This increase is traduced in shifts of the first peak toward lower frequencies. These shifts can be observed from labeled curves of Fig. 3. MLC models have conserved the order of the first peak from the highest to the smallest BSA. For E-polarization, the first peak appèars at wavelengths close to the height of the model [10]. In addition to the first peak shifts, MLC model plots have a common second peak around 6 GHz. The later peak is due to the constant thickness of the outer Skin layer of 4.2 mm.

Varying the BSA from child values (0.25 m^2 and 0.9820 m^2) to adult values (2.25 m^2 and 2.857 m^2) [2], we clearly see in Fig. 4 that the frequency of the first peak is not a linear function of BSA. Even for fictitious values of BSA (4.2 m^2), the first WBASAR peak still away from 80 MHz. Therefore, in order to attain near 80 MHz resonance with MLC model, we need to map the human height onto one of geometrical parameters of the multilayered cylindrical system (radius or diameter). Setting R_{out} equation the height of the model leads to plus-sign-label plot of Fig. 5.

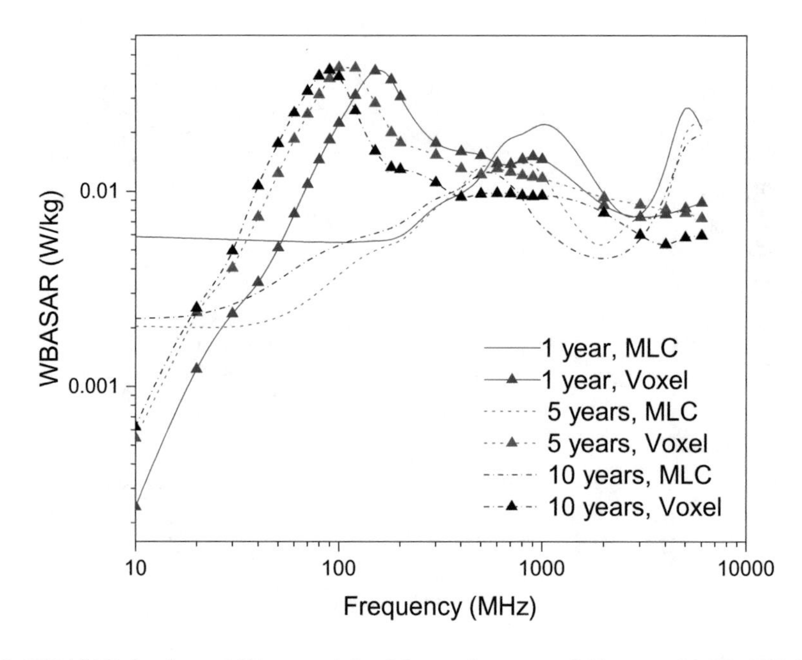

Fig. 3. WBASAR for three children models of 1 year, 5 years, and 10 years old. The WBASAR values of Voxelized models (lines with triangular labels) are extracted from [4]. WBASAR values of MLC models are obtained using Eq. 4

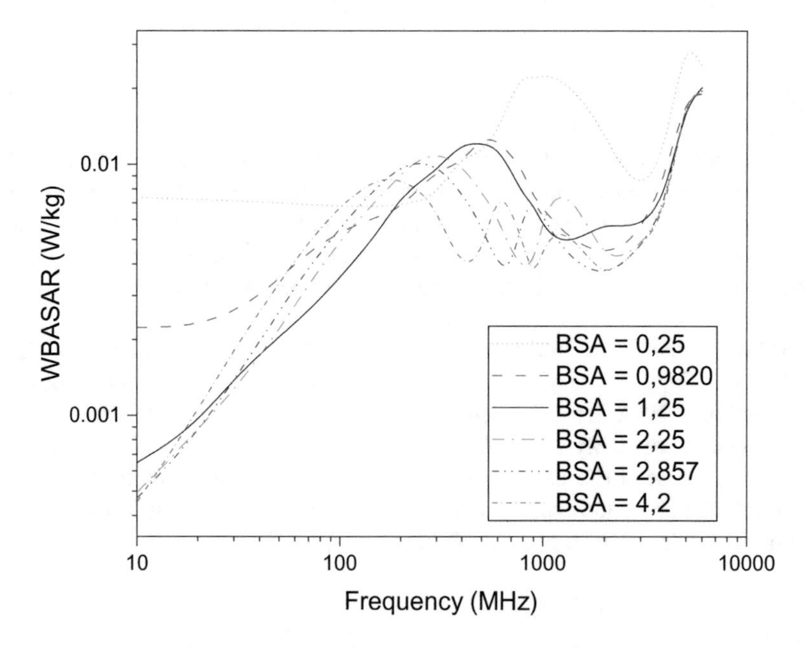

Fig. 4. Estimating WBASAR using second technique for various Body Surface Area values. BSA variations represent the Human growth from child (0.25 m^2) to adult (2.857 m^2). The 4.2 m^2 BSA value is fictitious

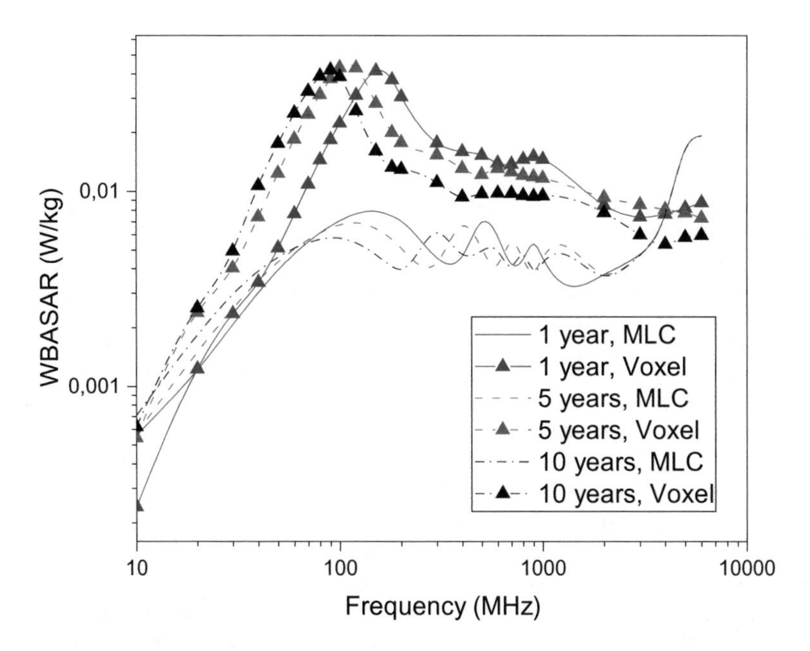

Fig. 5. WBASAR for three children models of 1 year, 5 years, and 10 years old. The WBASAR values of Voxelized models (lines with triangular labels) are extracted from [4]. WBASAR values of MLC models are obtained using Eq. 4 and an outer radius R_{out}, respectively equal to 76 cm, 109 cm and 138 cm.

4 Summary

In this study, first, we compared three WBASAR estimation techniques and we clearly visualize that the best WBASAR results for the multilayered cylindrical model can be obtained using the second averaging method of Eq. 3 and Eq. 4. Second, we vary the Body surface Area values from child level to adult level to conclude about BSA inefficiency as a cylindrical construction tool, for lower frequencies (below 6 GHz). This inefficiency is due to the non-linearity between BSA and the Height of the human model. Since WBASAR resonance occurs at wavelengths near human height, MLC's outer radius must be set equal to the human height to get the same first resonance frequency. The width of the layers still has an effect on the level of the estimated WBASAR for lower frequencies.

Despite the dominance of the voxel models for sub- 6 GHz, the semi-infinite and infinite theoretical models such as, infinite multilayered cylindrical models, are still considered effective in term of cost, computational time, and wider frequency starting from lower frequencies to near 30 GHz. In future work, we will enhance the MLC model to be applicable in the two frequency ranges of the fifth generation technology (5G).

References

1. Zhao, L., Ye, Q., Ke-Li, W., Chen, G., Wenhua, Yu.: A new high-resolution electromagnetic human head model: a useful resource for a new specific-absorption-rate assessment model. IEEE Antennas Propag. Mag. **58**(5), 32–42 (2016)
2. Hirata, A., Nagaya, Y., Osamu, F., Nagaoka, T., Watanabe, S.: Correlation between absorption cross section and body surface area of human for far-field exposure at GHz bands. In: 2007 IEEE International Symposium on Electromagnetic Compatibility, pp. 1–4 (2007)
3. Dimbylow, P., Bolch, W., Lee, C.: SAR calculations from 20 MHz to 6 GHz in the University of Florida newborn voxel phantom and their implications for dosimetry. Phys. Med. Biol. **55**(5), 1519–1530 (2010)
4. Nagaoka, T., Watanabe, S.: Development of voxel models adjusted to ICRP reference children and their whole-body averaged SARs for whole-body exposure to electromagnetic fields from 10 MHz to 6 GHz. IEEE Access **7**, 135909–135916 (2019)
5. Saada, A.B., Mbarek, S.B., Choubani, F.: Towards a new model of human tissues for 5G and beyond. In: Innovative and Intelligent Technology-Based Services for Smart Environments - Smart Sensing and Artificial Intelligence, pp. 51–55. CRC Press (2021). Num Pages: 5
6. Ziskin, M.C., Alekseev, S.I., Foster, K.R., Balzano, Q.: Tissue models for RF exposure evaluation at frequencies above 6 GHz. Bioelectromagnetics **39**(3), 173–189 (2018)
7. Saada, A.B., Mbarek, S.B., Choubani, F.: Antenna polarization impact on electromagnetic power density for an off-body to in-body communication scenario. In: 2019 15th International Wireless Communications & Mobile Computing Conference (IWCMC), pp. 1430–1433. IEEE (2019)
8. Duda, K., Majerczak, J., Nieckarz, Z., Heymsfield, S.B., Zoladz, J.A.: Chapter 1 - Human body composition and muscle mass. In: Zoladz, J.A. (ed.) Muscle and Exercise Physiology, pp. 3–26. Academic Press (2019)
9. Massoudi, H., Durney, C.H., Barber, P.W., Iskander, M.F.: Electromagnetic absorption in multilayered cylindrical models of man. IEEE Trans. Microw. Theory Tech. **27**(10), 825–830 (1979)
10. Durney, C.H.: Electromagnetic dosimetry for models of humans and animals: a review of theoretical and numerical techniques. Proc. IEEE **68**(1), 33–40 (1980)

Analysis of an Ethereum Private Blockchain Network Hosted by Virtual Machines Against Internal DoS Attacks

João H. F. Battisti[1], Guilherme P. Koslovski[1], Maurício A. Pillon[1], Charles C. Miers[1](✉), and Nelson M. Gonzalez[2]

[1] Graduate Program in Applied Computing, Santa Catarina State University, Joinville, Brazil
joao.battisti@edu.udesc.br,
{guilherme.koslovski,mauricio.pillon,charles.miers}@udesc.br
[2] IBM Watson Research Center, Yorktown Heights, USA
nelson@ibm.com

Abstract. The blockchain technology is increasingly adopted by distinct areas of knowledge and institutions filling administrative gaps related to data integrity, and audit procedures. In parallel, the adoption of resource virtualization is a reality to efficiently manage data centers. Consequently, the institutions have integrated blockchain networks into their own systems, hosting blockchain nodes atop virtual machines (VMs). In this sense, we present a resistance analysis of an Ethereum-based network hosted by VMs. The Denial of Service (DoS) attacks are used to identify the impact on the standard VM flavor. Specifically, we developed an environment for carrying out experiments using Ethereum configured with distinct consensus mechanisms: RAFT, Istanbul Byzantine Fault Tolerance (iBFT), and Proof-of-Authority (PoA). Each scenario is in-depth analyzed facing the DoS attack. Our results show that a private blockchain can suffer DoS when configured only to satisfy the default configuration defined by the application.

Keywords: Cloud computing · Blockchain · Security · DoS

1 Introduction

Blockchain consists of Peer-to-Peer (P2P) networks, cryptography, algorithms and a consensus mechanism [6]. Decentralization requires guarantees for networks to operate correctly. The consensus mechanism is a key component of a blockchain because it determines the organizations involved, how nodes interact, and their roles. Consensus mechanisms are designed for diverse and different contexts. Each application should identify the most appropriate consensus mechanism to meet its requirements while using the computing resources efficiently.

Blockchain requirements are usually identified according to the number of parties involved, types and number of transactions, how trust is managed among

L. Barolli et al. (Eds.): AINA 2022, LNNS 449, pp. 479–490, 2022.
https://doi.org/10.1007/978-3-030-99584-3_42

the parties involved, and what guarantees must be provided. Blockchains used for the general public typically have a significant number of users and transactions. As a result, they usually employ different consensus mechanisms than blockchains used internally by a company or a group of restricted parties. Furthermore, there are several ways to plan and deploy a blockchain. Our work focuses on the private model (or even consortium), in which one or more companies employ blockchain for their own use. Among the existing consensus mechanisms for this model, three stand out: RAFT, iBFT and PoA.

The use of blockchain in applications involves the configuration and deployment of nodes. These nodes allow the inclusion of new entries in the blockchain and implement the consensus process. Institutions adopting the private model create their blockchain nodes in VMs on their clouds [1,11]. The use of VMs implies defining a flavor to run the blockchain nodes. A VM for a blockchain application typically requires a modest configuration (*i.e.*, 2 vCPUS, 4 GB RAM, 20 GB storage, and 1 Gbps of networking) [4,7]. A lean flavor configuration avoids wasting resources and, according to the literature, supports most applications in terms of blockchain transactions. However, if there is an excessive number of transactions, this configuration may become a problem. Memory, processor and network resources can reach operational limits, making it difficult for the consensus engine to function.

In this context, the security of these cloud environments is a complex challenge. Users in the blockchain organization have access to the participating blockchain nodes and/or to the networks in which the VMs are hosted. In this scenario, blockchain nodes deployed in VMs are susceptible to unauthorized access, allowing a user to intentionally or accidentally initiate an attack against the institution. Regardless of motivation, this represents an attack surface for multiple security incidents. This work approaches the possibility of intentional or accidental attack by several requests which can subvert the system and generate a blockchain DoS attack. Thus, questions arise related to the environment in which the blockchain is inserted, as well as to the resistance and stability of the blockchain transactions while a DoS occurs. This work conducts a deep resistance analysis of a virtual machine based Ethereum private blockchain network against internal DoS attacks.

The work is organized as follows. Sections 2 and 3 explain the problem definition and related work. Section 4 presents our analysis proposal, as well as the experiments and scenarios. Section 5 presents the results and analysis.

2 Problem Definition

The growth of blockchain usage is remarkable [10,14], as well its applications demands and requirements [8]. Developers usually seek to improve their applications regarding efficiency and quality of the technology, but there are concerns related to the computational costs associated to performance and security. Thus, the nodes of a blockchain are deployed according to these requirements, e.g., number of transactions per minute they must process, efficiency, etc.

A blockchain consensus mechanism has a direct relationship to the amount of resources needed in the node. The choice of blockchain solution is based on the consensus mechanisms it supports. In this context, there are several consensus mechanisms (e.g., RAFT, iBFT, Practical Byzantine Fault Tolerance (pBFT), PoA, etc.) responsible for the validation processes of the blockchain network.

An important issue is the environment in which the nodes are created, highlighting computational clouds with their virtualization services. In general, the recommended flavor configurations for creating VMs, in private and consortium blockchain models, may change according to the platform/application. A platform which stands out in distributed applications and smart contracts is Ethereum. Ethereum has two types of recommendations for blockchain nodes run: *(i)* specific (4 vCPUS, 8 GB RAM and 1 Gbps network), and *(ii)* generalized (2 vCPUS, 4 GB RAM and 1 Gbps network). Each recommendation intends to meet the application needs (e.g., transactions processed per minute, latency). However, if the number of transactions submitted is above normal or if the resources are exhausted, this leads to a DoS scenario. To exemplify these situations, we define two scenarios illustrated in Figs. 1 and 2.

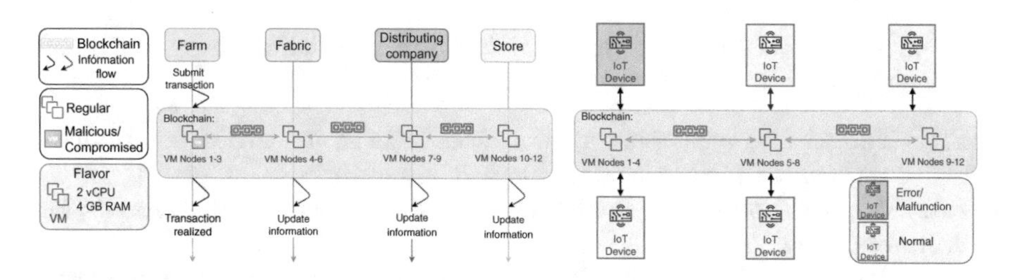

Fig. 1. Supply chain application scenario. **Fig. 2.** Internet of Things (IoT) scenario.

Initially, we present a real and simple example of a supply chain, illustrated in Fig. 1. It is possible to observe an interactive environment and message exchanges between of the VMs hosting the blockchain network. Specifically, twelve VMs maintainers of the blockchain network are observed, responsible for the transaction validation process and insertion of new blocks. In this scenario a malicious user can DoS against the VM of one or more blockchain nodes. As the flavor of these nodes is not very robust in terms of processing power, it is very likely that one or two attacking VMs may generate enough requests for some other service running on the blockchain node (e.g., SSH is normally enabled for maintenance) and exhaust its memory or compute power. This attack, while not directly against the blockchain, exhausts the VM resources needed to carry out blockchain transactions and can lead to latency or denial of legitimate requests.

Figure 2 illustrates an environment composed of twelve VMs and five IoT devices that send transactions to the blockchain network. This scenario reflects a production environment, in which at a given moment one of the IoT devices fails and sends several transactions per second to the blockchain network (e.g., instead of sending a transaction per minute, it starts sending a transaction every 2 s). The blockchain network does not have the ability to distinguish only the true transactions that are sent by the device, producing an DoS with a reduction in bandwidth. This attack occurred unintentionally due to an unexpected failure that destabilized the application and the environment. Although this scenario illustrates an unintentional DoS, this could also be done intentionally by a malicious IoT device.

The analysis of these two scenarios leads to questions related to:

(i) the proper settings of the instance flavor,
(ii) the number of transactions needed to make the technology operational; and
(iii) the metrics and criteria for detecting issues on the blockchain.

These issues are important for the development of a blockchain network, as the number of transactions required and supported is directly related to the configuration of VMs.

3 Related Work

Based on the motivation and problem definition we define three functional requirements (FR) to be addressed by this work:

- (FR1) The environment must allow transactions to be carried out through API or automated mechanisms for blockchain networks;
- (FR2) We must collect the number of transactions sent and their hashes for verification; and
- (FR3) We must collect usage metrics such as CPU, memory, networking, and transaction latency.

We used a systematic review approach to identify related work. Table 1 presents a comparison between the identified works and our requirements.

Table 1. Related work *versus* Functional requirements.

	[3]	[13]	[12]	[5]	[15]	[2]	[9]	[1]
FR1	Yes	Yes, until 5k transactions	Yes, until 10k transactions	Yes, until 10k transactions	Yes	No	Yes	Yes
FR2	No	No	No	No	No	No	No	No
FR3	Partially, related metrics: packet overloading and consumption energy	Partially, related metrics: RAM and latency	Partially, related metrics: latency	Partially, related metrics: Latency, tx rate and number of transactions	Partially, related metrics: Latency, tx rate and runtime	No	Partially, related metrics: Latency and tx rate	Partially, related metrics: latency, time and cost of transactions

Analyzing Table 1, it is perceptible that none of the identified works addresses all the FRs defined in our research. Although FR1 and FR3 are at least partially addressed in most of works, FR2 is absent. Works [1,3,5,9,12,13] assess the performance of blockchain platforms and their consensus mechanisms, carrying out on several transactions on the platforms and analyzing memory consumption, latency, transfer rate, processing time, and energy consumption. Work [2] performs a comparison between different platforms through their characteristics.

4 Proposed Approach

The examples from Sect. 2 illustrated a blockchain environment hosted by an Infrastructure-as-a-Service (IaaS) cloud. The nodes of the blockchain network are created on VMs. In a real environment in which an unintentional DoS occurs, there are two situations:

(i) The attack takes place from the accumulation of several transactions received by the platform, causing a high rate of network traffic that results in higher network latency, therefore delaying the validation and insertion of new blocks.
(ii) Similar to the previous item, but as a result of this higher latency and reduction of available network bandwidth, service unavailability and/or packets discarding may occur.

In addition, noting that in this environment the network is formed by known nodes, an intentional DoS occurs in the following situations:

(i) A malicious user belonging to the network carries out attacks against other healthy nodes to consume their resources, either to subvert the system or to reduce the number of participants in the validation process.
(ii) A malicious user sends a significant number of false/positive transactions to the network to increase traffic and latency, and consequently delays in the validation process and insertion of new blocks.

Based on these situations we propose a performance and security analysis in a private blockchain network during the execution of a DoS attack comprising the following steps:

- Analyze the number of transactions per minute that an instance can perform without service loss;
- Analyze the immutability and integrity of transactions; and
- Establish a relationship between the flavor of the instance and the intensity of DoS attacks.

Figure 3 illustrates the testbed setup. Each scenario consists of six instances running operating system (OS) GNU/Linux Ubuntu 20.04. The VMs were configured with a flavor of 2 vCPUs and 4 GB of RAM. This is the flavor recommended for generalized projects on the Ethereum platform [4]. We define three test scenarios with different consensus mechanisms: RAFT, iBFT and PoA. For each

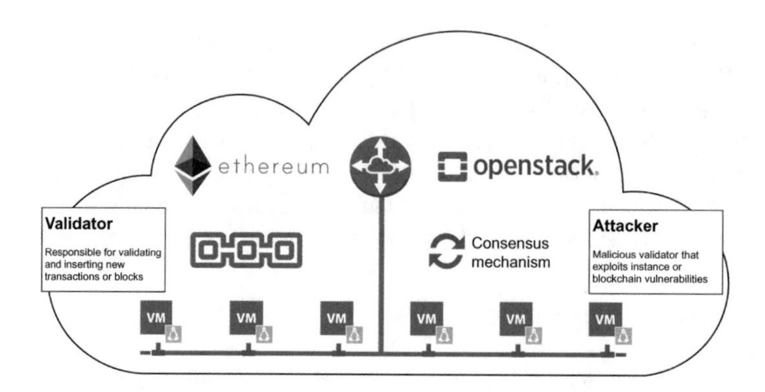

Fig. 3. Testbed used to execute Scenarios I, II, and III.

scenario we verify two situations: (1) occurrence of an unintentional DoS attack; and (2) an intentional DoS attack. In all scenarios the swap is disabled on all VMs. Only main memory is used since the objective is to exhaust the resources. In this sense, the following experiments are performed in each of the scenarios: (I) A malicious user sends transactions directly to a node of a blockchain, here the purpose is consuming the instance's resources; (II) Carrying out several transactions of a user/customer on the blockchain network, in order to reduce network traffic or unavailability of service, aiming to identify, from the monitoring of resources and transactions, the stability of the network in the event of a DoS non-malicious attack. DoS attacks selected:

- Transaction Flood: Flood attack characterized by causing a large volume of traffic on the blockchain network, so that the bandwidth becomes congested; and
- SSH Flood: A protocol exploit attack, characterized by excessive consumption of target resources, attack generates SSH packets (TCP/22) trying to establish connection to the VM.

During the attack we monitored processing and memory resources of the VMs, network traffic, number of transactions and their latency for the blockchain platform.

5 Results and Discussion

Scenario I is based on the RAFT consensus algorithm, which is an Crash Fault Tolerance (CFT) algorithm. This algorithm has a certain degree of resilience in their protocol, allowing the system to correctly reach consensus even with failures in its components. However it is not resistant to Byzantine failures. Scenarios II (iBFT) and III (PoA) have algorithms based on Byzantine Fault Tolerance (BFT), which are resistant to component failures and also to Byzantine failures.

In all scenarios, two experiments were executed. The first experiment consists of an DoS attack that seeks to exploit vulnerabilities or flood ports. The second performs a blockchain transaction flood attack. During the entire process of executing the experiments, the resources of the instances and the blockchain network were monitored (based on System Activity Report - SAR - tools) to identify possible instabilities or oscillations in the network.

For the first experiment, attacks were carried out for 240 min. For all scenarios, no changes in processes, memory or any instability that could compromise the blockchain network were detected. For the second experiment, for each of the scenarios different results were identified. Figure 4 and 5 illustrate the results of the second experiment in Scenario I, using RAFT consensus algorithm.

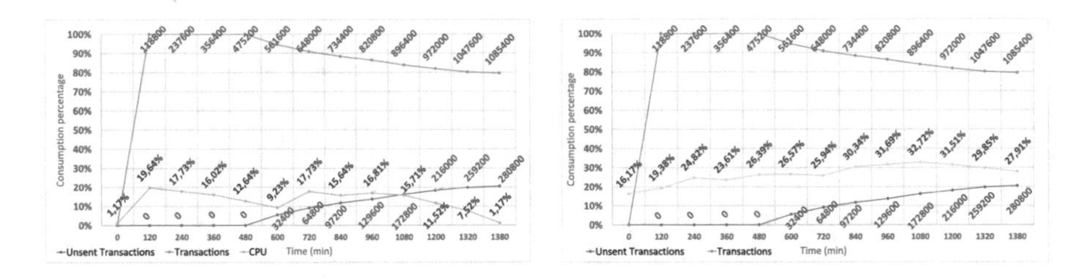

Fig. 4. Processor consumption - RAFT.

Fig. 5. Memory consumption - RAFT.

Figures 4 and 5 present three curves each. Both Figures show the number of transactions completed (and not completed) during the experiment. Figure 4 also presents the CPU consumption during the experiment. In contrast, Fig. 5 presents the memory consumption during the experiment. Analyzing CPU consumption we observe that the RAFT algorithm is relatively efficient and does not require a high computational power. During the entire execution of the attack it remained stable and with few changes in its consumption. Regarding memory consumption the main peaks of the occurrence of a reduction in the amount of transactions sent are clear, as the memory consumption in these peaks is reduced. Transactions begin to drop after 480 min of execution of the attacks.

Figures 6 and 7 illustrate the results obtained from the execution of Scenario II using the consensus algorithm iBFT.

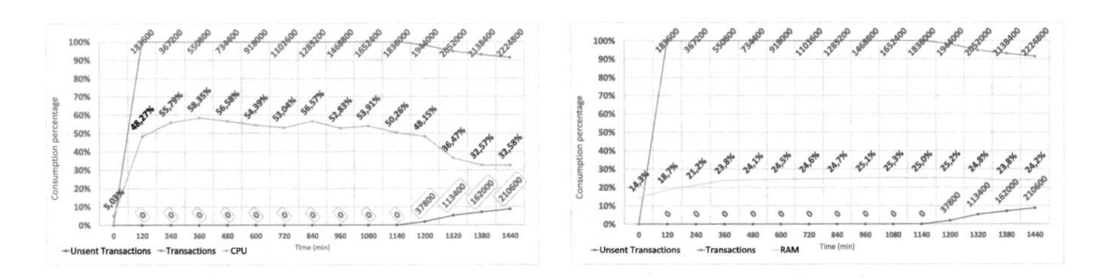

Fig. 6. Processor consumption - iBFT.

Fig. 7. Memory consumption - iBFT.

Regarding the use of processing (Figs. 6 and 7), we observe the difference between a CFT consensus algorithm compared to BFT – due to the complexity and security issues, the use of computational power is greater. However, after 1,200 min of execution of the attack the system remained stable in the use of processing, and as the reduction in processor consumption, the greater the number of transactions not performed by the platform. In a BFT consensus algorithm, the processing reduction means the transaction processing speed reduction, allowing the gradual increase of the process queue. Regarding transactions, the stability of this consensus algorithm against the Transaction Flood attack is notorious. Only after 1,200 min the transactions start to drop.

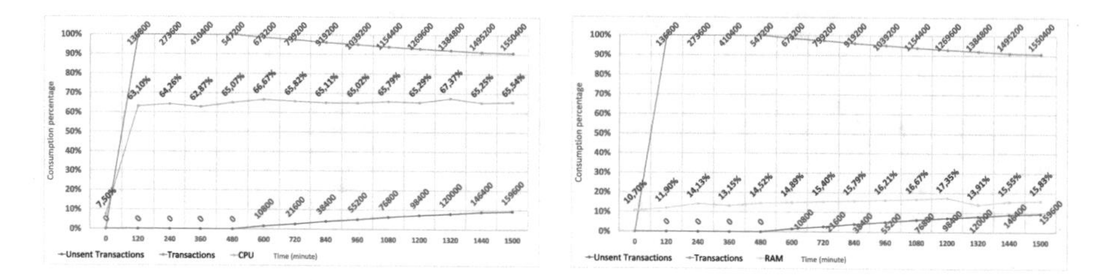

Fig. 8. Processor consumption - PoA.

Fig. 9. Memory consumption - PoA.

Figures 8 and 9 illustrate the experiments in Scenario III. Regarding processor consumption, since it is a BFT consensus algorithm we observe high processing consumption due to mathematical calculations that require greater computational power. Consumption remained stable during the experiment. Regarding transactions, we observe the stability of the consensus algorithm during the DoS attack. The first occurrence happened during the 600-min time of the attack execution but the growth of unsent transactions occurred in a contained way, compared to Scenarios I and II. In order to better understanding, we develop a compilation of these results (Figs. 10 and 11).

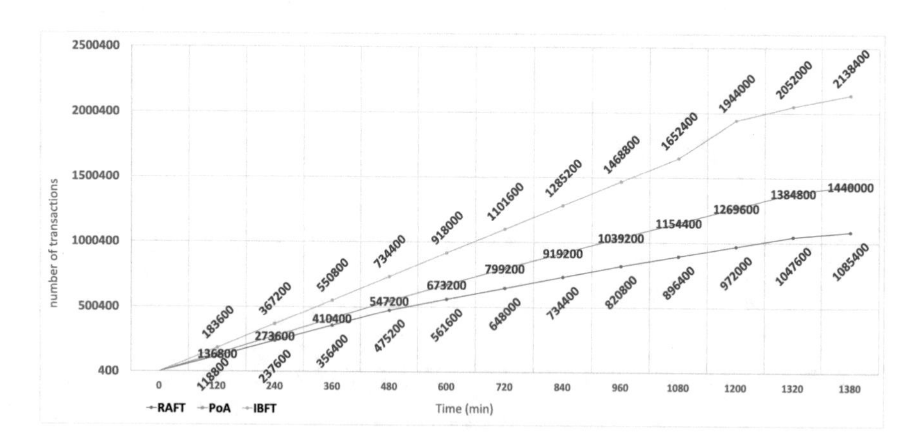

Fig. 10. Total transactions sent.

Figure 10 compares the number of transactions sent during the Transaction Flood attack for Scenario I (RAFT), Scenario II (iBFT), and Scenario III (PoA). The iBFT algorithm, which is a BFT algorithm, shows the best performance in terms of number of transactions sent. The worst case is RAFT, which is a CFT algorithm.

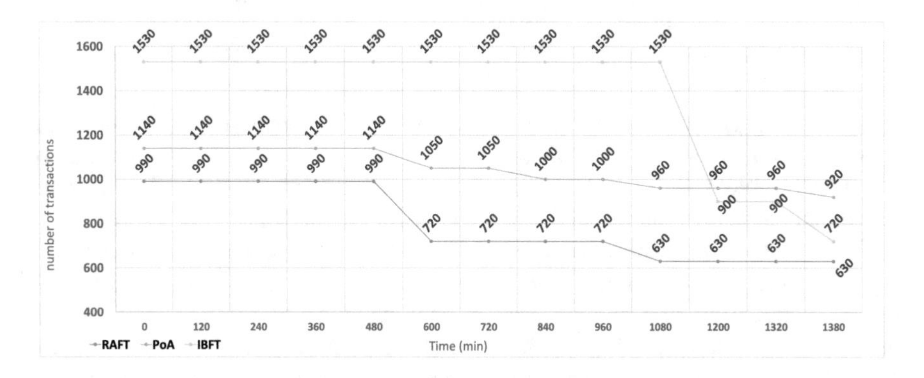

Fig. 11. Transactions per minute.

Figure 11 shows the comparison between all these scenarios in the comparison of sending transactions per minute. Regarding the RAFT algorithm scenario, since the transaction flood execution, it had the worst performance in sending transactions per minute, starting with 990 transactions per minute and ending with a sending rate of 630 transactions per minute. The iBFT consensus algorithm, from the beginning, revealed, in comparison with the others, a higher transaction sending rate, this being 1530 transactions per minute. However, with the processing loss and its high transaction buffer, these numbers started to drop, indicating a higher percentage drop in transaction per minute rates, reaching 720 transactions per minute by the end of the experiment. On the other hand, the PoA consensus algorithm showed the best behavior in relation to the amount of transactions sent per minute, which started with 1140 transactions, and since 600 min of attack execution, delays or not sending transactions started, but it kept up a good behavior, reaching the end with a rate of 920 transactions per minute, superior to RAFT and iBFT.

Table 2 presents a summary of all information collected, being separated between the consensus algorithms, the flavor of the VMs, and the number of transactions that can be supported by the instance without catastrophic impairment of the application's functioning.

Table 2. General aspects of flavor against the tested consensus algorithms.

Consensus	RAFT	iBFT	PoA
Processor	900 Transactions/min 20% of 2 vCPU	1500 Transactions/min 55% of 2 vCPU	1100 Transactions/min 67% of 2 vCPU
RAM Memory	900 Transactions/min 33% of 4 GB RAM	1500 Transactions/min 58% of 4 GB RAM	1100 Transactions/min 60% of 4 GB RAM
Transactions	900 non-continuous 700 continuous	1500 non-continuous 900 continuous	1000 continuous

Analyzing the results of Table 2, it is important to highlight some issues: (i) when BFT consensus mechanisms are applied, it is necessary to understand the needs of blockchain applications and according to the complexity and number of transactions, in parallel with the higher processing consumption, this analysis of the application becomes indicated to choose the best *flavor*; and (ii) CFT algorithms usually need more memory than BFT applications, with this it is indicated that this analysis is done so that the choice of flavor is the best possible. Furthermore, concerns about the security and proper functioning of environments and their applications have become increasingly worrying, especially with the growing number of vulnerabilities and attackers. In order to expose more comprehensively, Table 3 provides a comparison between consensus mechanisms, resources, and transactions at the time that DoS attacks started to affect blockchains and at the end of the attack.

Table 3. Flavor (2 vCPU, 4 GB RAM, 1 Gbps network): Consensus mechanisms regarding transactions/resources and DoS.

Metrics	RAFT	iBFT	PoA
DoS start time	600	1200	600
End of DoS	1380	1380	1380
Start DoS Transactions rate	720/min	900/min	1050/min
Transaction rate at the end of DoS	630/min	720/min	920/min
Percentage of processor use at the start of DoS	9,23%	48,15%	66,67%
Percentage of processor use at the end of DoS	7,52%	32,58%	65,54%
Percentage of memory usage at the start of DoS	26,57%	25,2%	14,89%
Percentage of memory usage at the end of DoS	29,86%	24,2%	15,83%

Table 3 establishes a relationship between the flavor of the instance and the intensity of the attacks DoS, in relation to the computational resources used. The results presented have a comparison between the beginning of the occurrence of DoS attacks and the final monitoring, in relation to the scenarios and their respective consensus mechanisms. Among the data, the first highlight is the RAFT algorithm which, like the PoA, had the start of DoS after 600 min of execution, revealing a greater weakness in the face of these attacks in relation to the transactions, but not about the use of your resources. The iBFT consensus algorithm showed the best performance in relation to the amount of transactions sent, however, after the floods started, there was a reduction in its processing, causing the largest differential in transaction rates. PoA showed, in general, the best performance in the ratio of the resources of its flavors vs DoS, keeping in balance, mainly in the use of its resources, with the smallest differential transaction fees. Based on these assessments, it is important to apply good security practices in a private blockchain network, monitoring network traffic, transactions and processing constantly. In addition, that all participating users and validators of the network, have the minimum confidence that is necessary for the proper functioning of the network.

6 Considerations and Future Work

The benefits of applications from a private or consortium blockchain in terms of economy, data management, and security are manifold. An important question is related to facilitating the creation of blockchain nodes in computational clouds, through virtualization. In these contexts, there is a concern with issues related to the security of the application and its development environment.

The analyzes performed by our work reinforce the concern about the need for security and continuous monitoring of a private blockchain network. Based on the results presented by our experiments, it is evident that an DoS attack, malicious or not, results in noticeable damage to the blockchain network and its users, which also results in the possibility that other vulnerabilities in blended attacks can be exploited. Regarding the flavor e.g., 2 vCPU, 4 GB RAM, and 1 Gbps network of the VMs we identified it is enough for the general needs of the blockchain platform and its application, but the choice of the consensus mechanism applied, and the number of transactions necessary to operationalize the application, are necessary for safe and optimized choices. This work allows to better understand about private/consortium blockchain applications from VMs in computational clouds, in particular those based on the Ethereum platform.

Future works includes these research to be extended in order to analyze other flavor sizes, and the use of other blockchain platforms/consensus mechanisms, applied within computational clouds.

Acknowledgements. This work was supported by Fundação de Amparo à Pesquisa e Inovação do Estado de Santa Catarina (FAPESC), Santa Catarina State University (UDESC), and Laboratory of Parallel and Distributed Processing (LabP2D).

This work was supported by Ripple's University Blockchain Research Initiative (UBRI).

This work received financial support from the Coordination for the Improvement of Higher Education Personnel - CAPES - Brazil (PROAP/AUXPE).

References

1. Abreu, A.W.d.S., Coutinho, E.F., Bezerra, C.I.M.: Performance evaluation of data transactions in blockchain. IEEE Latin America Trans. **20**(3), 409–416 (2021). https://latamt.ieeer9.org/index.php/transactions/article/view/5429. Number: 3
2. Chowdhury, M.J.M., et al.: A comparative analysis of distributed ledger technology platforms. IEEE Access **7**, 16,7930-16,7943 (2019)
3. Dorri, A., Kanhere, S., Jurdak, R., Gauravaram, P.: Blockchain for IoT security and privacy: the case study of a smart home. In: 2017 IEEE International Conference on Pervasive Computing and Communications Workshops (PerCom Workshops), pp. 618–623 (2017). https://doi.org/10.1109/PERCOMW.2017.7917634
4. Ethereum Foundation: Nodes and clients - ethereum. Technical report, Ethereum.org (2021). https://ethereum.org/en/developers/docs/nodes-and-clients/
5. Hao, Y., Li, Y., Dong, X., Fang, L., Chen, P.: Performance analysis of consensus algorithm in private blockchain. In: 2018 IEEE Intelligent Vehicles Symposium (IV), pp. 280–285 (2018)
6. Lin, I.C., Liao, T.C.: Survey of blockchain security issues and challenges. Int. J. Netw. Secur. **19**, 653–659 (2017)
7. Linux Foundation: An introduction to hyperledger (2018). https://www.hyperledger.org/wp-content/uploads/2018/07/HL_Whitepaper_IntroductiontoHyperledger.pdf
8. Loch, W.J., Koslovski, G.P., Pillon, M.A., Miers, C.C., Pasin, M.: A novel blockchain protocol for selecting microservices providers and auditing contracts. J. Syst. Softw. **180**, 111,030 (2021). https://doi.org/10.1016/j.jss.2021.111030. https://www.sciencedirect.com/science/article/pii/S0164121221001278
9. Monrat, A.A., Schelén, O., Andersson, K.: Performance evaluation of permissioned blockchain platforms. In: 2020 IEEE Asia-Pacific Conference on Computer Science and Data Engineering (CSDE), pp. 1–8 (2020). https://doi.org/10.1109/CSDE50874.2020.9411380
10. Needham, M.: Global spending on blockchain solutions forecast to be nearly $19 billion in 2024, according to new idc spending guide (2021). https://www.idc.com/getdoc.jsp?containerId=prUS47617821
11. Ogiela, M.R., Majcher, M.: Security of distributed ledger solutions based on blockchain technologies. In: 2018 IEEE 32nd International Conference on Advanced Information Networking and Applications (AINA), pp. 1089–1095 (2018). https://doi.org/10.1109/AINA.2018.00156
12. Pongnumkul, S., Siripanpornchana, C., Thajchayapong, S.: Performance analysis of private blockchain platforms in varying workloads. In: 2017 26th International Conference on Computer Communication and Networks (ICCCN), pp. 1–6 (2017). https://doi.org/10.1109/ICCCN.2017.8038517
13. Rouhani, S., Deters, R.: Performance analysis of ethereum transactions in private blockchain. In: 2017 8th IEEE International Conference on Software Engineering and Service Science (ICSESS), pp. 70–74 (2017)
14. Taylor, P.J., Dargahi, T., Dehghantanha, A., Parizi, R.M., Choo, K.K.R.: A systematic literature review of blockchain cyber security. Digital Commun. Netw. **6**(2), 147–156 (2020). https://doi.org/10.1016/j.dcan.2019.01.005. https://www.sciencedirect.com/science/article/pii/S2352864818301536
15. Vatcharatiansakul, N., Tuwanut, P.: A performance evaluation for internet of things based on blockchain technology. In: 2019 5th International Conference on Engineering, Applied Sciences and Technology (ICEAST), pp. 1–4 (2019)

A Machine Learning Approach for a Robust Irrigation Prediction via Regression and Feature Selection

Emna Ben Abdallah[1(✉)], Rima Grati[2], Malek Fredj[1], and Khouloud Boukadi[1]

[1] Miracl Laboratory, Faculty of Economics and Management of Sfax, Sfax University, Sfax, Tunisia
emnabenabdallah@ymail.com
[2] Zayed University, Abu Dhabi, UAE

Abstract. Smart irrigation has many advantages in optimizing resource usage (e.g., saving water, reducing energy consumption) and improving crop productivity. In this paper, we contribute to this field by proposing a robust and accurate machine learning-based approach that integrates feature selection techniques with several regression algorithms. To effectively determine the optimal quantity of water needed for a plant, Random Forest, Recursive Feature Elimination (RFE), and SelectKBest are used to assess the importance of the features. Different regression methods are established based on the set of effective features. The different models involved in this approach are trained and tested using a collected dataset about various crops such as tomatoes, grapes, and lemon and encompasses different features such as meteorological data, soil data, irrigation data, and crop data. The experiments demonstrated the performance of RF in analyzing the feature importance as well as the prediction of the optimal water quantity. The findings of feature selection highlight the importance level of the evapotranspiration, the depletion, and the deficit to maximize the model's accuracy.

1 Introduction

At the plot level, proper irrigation management has many benefits from several perspectives, such as saving water, generating sustainable agriculture, and decreasing environmental impacts [8]. Accordingly, improving water use is essential to ensure the sustainability of irrigated agriculture. Besides, proper irrigation that determines, precisely, and quantifies plant water needs could enhance water use efficiency, reduce energy consumption, and enhance crop productivity [1]. Therefore, many studies focused on predicting water demand at the irrigation district level [1,11,17]. However, forecasting water demand precisely is a highly complex task since it depends on many factors such as weather, soil, and water properties [8].

Nowadays, the latest sensors and weather stations allow the collection of a large amount of historical data that was unavailable in the past. According to [7],

L. Barolli et al. (Eds.): AINA 2022, LNNS 449, pp. 491–502, 2022.
https://doi.org/10.1007/978-3-030-99584-3_43

big data techniques and artificial intelligence are the best tools for handling this amount of data thanks to their capability to extract useful information and give additional value and usefulness to systems installed in the field.

Regression methods have been widely used as machine learning in agriculture fields [1,8], such as estimating the water demand [17]. Regression-based irrigation systems rely on historical data about soil, plant, and climatic sensors to predict crop water needs [1]. However, the study of [17] revealed that the regular regression models could be insufficient and do not surpass any of the machine learning models. In addition, the study of [8] indicated that an effective regression-based method depends strongly on the accuracy of measurements. The authors also revealed that the related factors should be deeply studied, and a robust model should be built to estimate the water demand accurately.

The main objective of this paper is to propose a robust and accurate approach for plant water need prediction. This approach aims to estimate daily the future water needs and maximize productivity in the irrigated areas. To fulfill this objective, we investigate the main factors influencing irrigation and evaluate their relative importance. For doing so, we rely on feature selection techniques as their capability to improve the understandability of the machine learning models and often help build models with better generalization [16]. To the best of our knowledge, this is the first attempt to explore the power of feature selection methods for feature ranking and ensemble regression methods to estimate the water need in the field of smart irrigation. Besides, we examine four ML methods to build the water need prediction model on the best effective set of features and the outperformed regression algorithm. Moreover, the proposed approach considers a crop variety such as tomatoes, grapes, and lemon for prediction water requirements for plants with a development cycle, a quantity of water to irrigate, and a different cycle duration.

The remainder of this paper is organized as follows. The related work of this paper is summarized in Sect. 2. Our proposed approach is described in detail in Sect. 3, and then experimental results with their analysis are reported in Sect. 4. Finally, the discussion and the conclusion of this study are presented in Sect. 5.

2 Related Work

Regarding the choice of machine learning solutions to optimize agricultural sustainability, we present in this section some relevant works that dealt with prediction approaches in irrigation management.

Navarro-Hellín et al. [11] proposed a machine learning program to serve as a reasoning engine to manage agricultural irrigation, thereby replacing the expert agronomist. The machine learning program employed Partial Least Squares Regression (PLSR) and Adaptive Neuro-Fuzzy Inference System (ANFIS) to calculate appropriate irrigation decisions considering the weather, crop, and soil characteristics, along with field measurements from soil sensors. Adeyemi et al. [1] presented a Dynamic Neural Network approach for modeling the temporal soil moisture fluxes. The models are trained to generate a one-day-ahead prediction of the volumetric soil moisture content based on past soil moisture, precipitation, and climatic measurements. Sayari et al. [17] also relied on artificial neural networks and other regression methods, including Multivariate Linear Regression, and Support Vector Regression (SVR), to predict the infiltrated water in the furrow irrigation system by combining them with the nature-inspired firefly algorithm. As for the input features in the different models, they used furrow length, the inflow rate, the advance time at the end of the furrow, the cross-sectional area of inflow and infiltration opportunity time. Their experiments indicated that using the combination of the SVR and FA noticeably enhances the performance of the standard SVR model.

González Perea et al. [7] propose a model for farmer's behavior and predict the irrigation events occurrence. The authors developed a hybrid heuristic method that combines Decision Trees and Genetic Algorithm to find the optimal decision tree. The proposed method was tested on a real irrigation district, and results showed that the optimal models developed have been able to predict between 68% and 100% of the positive irrigation events and between 93% and 100% of the negative irrigation events.

To conclude, we cannot deny the relevance of the machine learning-based studies, notably regression methods, in the field of irrigation management since they can run complex mathematical algorithms in real-time with low cost [8]. However, according to [8], the effectiveness of machine learning-based irrigation prediction methods depends strongly on the accuracy of measurements, reliability of the system, and the validity of the trigger threshold. Estimation errors for water need can accumulate and result in erroneous irrigation. Indeed, an accurate model prediction is highly needed in the field of irrigation. Unfortunately, most existing works ignored the significant impact of parameter selection on model accuracy and did not use reliable methods that enhance the model accuracy.

3 The Proposed Approach

The main purpose of this study is to address the accuracy problem of the water demand estimation. Thus, we propose a novel approach that ensures the accuracy of the prediction model from two perspectives. First, several types of features are studied to reveal their impact on the prediction model accuracy. Second, the best performing features are used to train several ensemble methods, such as Random Forest, CART, XGBoost and, GBR to provide accurate and robust results.

In practice, the current approach consists of three main phases: data collection and preparation, model building, and model evaluation. Figure 1 presents an overview of the proposed approach.

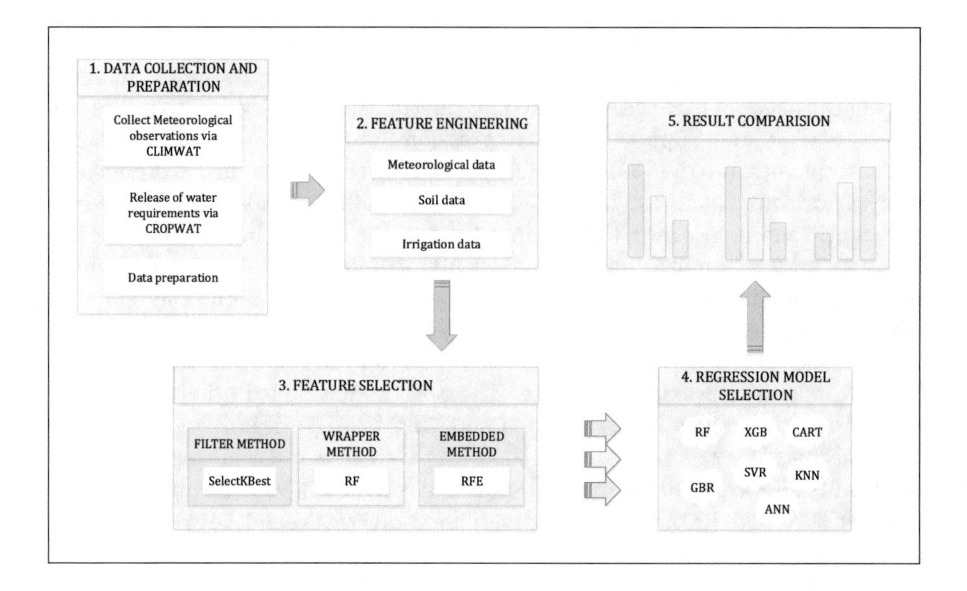

Fig. 1. A machine learning approach for water need prediction

3.1 Study Area Description

To better illustrate the proposed approach, we consider a farm located at Sfax - Tunisia. This farm cultivates peaches, nectarines, tomatoes, grapes, and lemons. As located in an arid region, well known for the lack of rain and a high variation of temperature (day/night), the farm maintains production under difficult climatic conditions characterized by hot summer (exceeding 45 °C) and a relatively cold winter reaching 1 °C. Considering these conditions (climate warning or fluctuations in temperature reaching 45 °C), there is a significant risk on the crop production/performance and the physiological behavior of the plants, which might also be damaged.

3.2 Data Collection and Preparation

Data collection is one of the main steps for implementing a smart irrigation approach. We employ CLIMWAT and CROPWAT to generate a dataset that includes plant water requirements according to several factors while considering the context of the farm. Firstly, we rely on CLIMWAT [18], a climatic database provided by the United Nations Food Organization and Agriculture (FAO), to collect climatic data related to the farm based on its location. This source provides data on weather parameters, namely temperature, humidity, precipitation, and the characteristics of the soil. Afterward, CLIMWAT can be involved with CROPWAT [4], a tool proposed by FAO to calculate a plant's water needs. Accordingly, the dataset includes data related to rainfall, temperature, humidity, sunshine, solar radiation, wind speed, and evapotranspiration. The latter is calculated using the Penman-Monteith method [2] that relies on meteorological data such as daily mean temperature, wind speed, relative humidity, and solar radiation [6]. Following data collection, we perform data preparation to employ the dataset in the later training process. Finally, the training process aims to generate a prediction model that forecasts plant irrigation requirements using a machine learning algorithm.

The data preparation aims to enhance data quality and make the dataset convenient for a training process. Indeed, many machine learning algorithms require numerical data to generate the prediction model. There, values such as string and date need to be transformed into numerical values. To do so, we employ several techniques such as transformation, cleaning, reduction, and discretization of data.

3.3 Feature Engineering

We choose the characteristics that significantly influence the use of water by crops and whose data are available throughout the growing seasons. Meteorological data, soil data, and irrigation data are used in this study to predict water demand.

- Meteorological data: This category of features is extracted from the weather dataset. We collected evapotranspiration, minimum and maximum temperature (in °C), humidity (in %), wind (in km/day), sun (in sunshine hours per day), radiation ($MJ/m^2/day$), rain (in average precipitation per month) and precipitation (in mm).
- Soil data: In this study, we consider depletion as a soil feature. The depletion presents the cumulative depth of evapotranspiration from the root zone (in mm).
- Irrigation data: This category includes the actual crop evapotranspiration (Eta), the depth of irrigation in mm (Net_Irr), the Water Stress Coefficient (KS), the deficit, the loss, and the flow. The first feature is the vapor pressure deficit which presents the difference between the saturation and the actual vapor pressure for a given period. The loss feature represents the irrigation water that is not stored in the soil, i.e., the surface runoff or the percolation (in mm). As for the flow feature, it describes the water outflow (in l/s/ha).

Besides these three categories, we introduce the date information of irrigation, including month, day, and the day of plant growth cycle [5]. We also consider the crop to irrigate, including tomato, grape and lemon, and the stage of the crop development (initial, midseason, or end season).

3.4 Feature Selection

Initially, this phase's purpose is to improve the performance of the proposed ensemble method. Feature selection consists of identifying the best subset of features that have the most critical impact on the predicted results. This subset can improve the model performance and save running time [16]. There are three categories of feature selection methods in the literature: filter, wrapper, and embedded. Feature importance is analyzed in the filter-based methods independently of the model. The wrapper-based methods examine, iteratively, the importance of features around the classification model and use the prediction accuracy to eliminate or to select a set of features. As for embedded-based methods, the search for the best subset of features is performed during the model training. This study compares the performance and robustness of three feature selection (FS) methods belonging to the above categories. The use of these methods has led to analyze the most compelling features based on the robust and the performant FS method. The details of the used FS methods are depicted as follows:

1. SelectKBest is a univariate feature selection method that belongs to the filter feature selection category. Univariate works by selecting the best features based on univariate statistical tests. It can be seen as a preprocessing step for an estimator [10].
2. Random Forest (RF) is a wrapper feature selection method [10]. The core idea of this method is to determine feature relevance by comparing the significance of the real features to that of the random probes. RF performs an implicit feature selection by creating multiple trees using regression trees, CART.
3. Recursive Feature Elimination (RFE) is an embedded feature selection method that evaluates multiple models using different procedures. These procedures try to add and/or remove predictors until finding the optimal combination that maximizes the model performance [10].

3.5 Model Development

We have used learning algorithms known by their robustness [9,15] namely Random Forest, Decision Tree, Gradient Boosting Regressor and XGBoost Regressor to constitute the model for predicting the quantity of water to irrigate. We trained the generated models by each machine learning algorithm using a set of training data made up of examples used during the learning process. In addition, we also ensured the parameters es adjustment (e.g. the weights) of a regressor using this dataset.

All observations were divided into approximately ten groups for all datasets: The different regression models were trained with the data from nine of the groups and then used to evaluate the remaining group. Therefore, around 90% of instances randomly chosen from the complete data set was used for the training process, and the remaining 10% were used as a test data set to measure the performance of all compared methods. The process of training and prediction was repeated ten times. In each of the ten rounds, the model prediction was validated against only one from the ten separate groups. Using the real values of the output estimated water, the predictive performance was measured with the regression performance metrics defined above. In this study, the 10-fold cross-validation was repeated 10 times and the overall performance measurements were computed by averaging over those 100 iterations.

Technically, all machine learning algorithms were implemented using Python 3.7.0. In particular, we used the Scikit Learn library [13] to create and adapt our models under Google Colaboratory (also known under the name Colab) [3]. Colab is a cloud service based on Jupyter notebooks which provides a fully configured environment for machine learning and free access to a robust GPU.

3.6 Model Evaluation

3.6.1 Evaluation Metrics

To evaluate the different prediction models, we rely on performance metrics. The model performance evaluation is established by plotting observed versus predicted values [14]. The metrics used for this purpose are Mean Squared Error (MSE), Mean Absolute Error (MAE), Root-Mean-Square Error (RMSE), and coefficient of determination (R^2). In this work, we use these metrics to evaluate the performance of the FS method ranking, the different base models, and the meta-model.

4 Experiments and Result Comparison

In this section, the different experiments were applied to the 216298 samples, already depicted in the previous section, to analyze and compare the performance of the models and the different features. First, series of experiments are conducted to compare the performance of the three FS methods (Subsect. 4.1), followed by the analysis of the impact of meteorological, soil, and irrigation features on the water need estimation (Subsect. 4.2). Then, the different regression models are evaluated (Subsect. 4.3).

4.1 Performance Analysis of Feature Selection

The objective of this experiment is to identify the best feature selection method that delivers the best feature subset. As illustrated in Sect. 3.4, three feature selection methods are applied to find the best one in terms of performance. These methods are incorporated in the distance-based k-Nearest Neighbors (KNN) method. KNN is a simple and intuitive one that is often considered as a powerful tool to assess the effectiveness of feature selection methods [12].

Figure 2 plots the performance of KNN in terms of MSE, MAE, and R^2 metrics averaged over 100 runs against different numbers of features considered in the water need prediction. It can be observed that RF outperforms RFE and SelectKBest by generally achieving the best values overall metrics. This characteristic suggests that RF ranks the features properly. Using the RF method, we can achieve outstanding performance (MSE = 0.1, MAE = 0.034, and R^2 = 0.99) with the top 10 features of the data set.

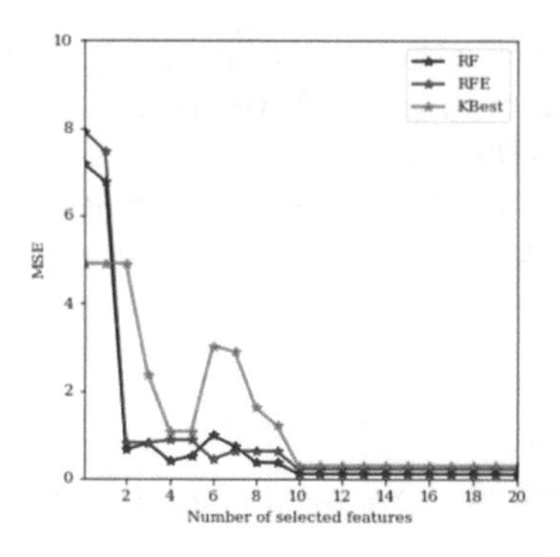

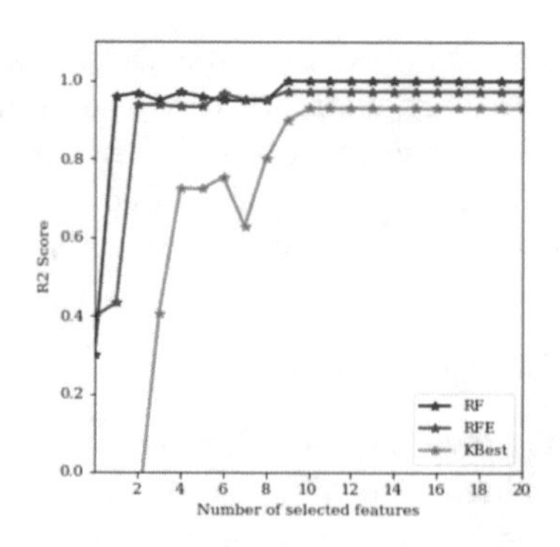

Fig. 2. Feature selection methods' performance

4.2 Feature-Wise Analysis

This section aims at investigating the importance of each feature for water need estimation. The goal of the conducted experiments is to probe the contribution of each feature to the water need. We present in Table 1 the top 10 feature ranking by the three feature selection methods. It is clear from the Table 1 that RF and RFE generate the same top 10 features. Moreover, the rankings of these two methods are more effective compared to that of SelectKBest (see Fig. 2). Alongside this, Fig. 3 illustrates the importance of each feature in the process of predicting the water need for the crop. This score is obtained using the RF model, regarding results from the training set.

From the Fig. 3 and the Table 1, it can be observed the highest importance of the feature evapotranspiration (ETo). The latter represents the amount of water transferred from the land to the atmosphere by evaporation from the soil and by transpiration from plants. A similar finding is also supported by past studies [1, 7]. According to the ranking of RF and RFE, the next most effective feature is depletion (Depl). It represents the cumulative depth of evapotranspiration from the root zone. This finding aligns with other studies such as that of Adeyemi et al. [1]. The experiments proved also the impact of the vapor pressure deficit (score≈0.12) since it is the third most important feature according to RF and RFE. In addition, the experiments illustrated the importance of the stage of the crop, notably the growth of the plant, in the estimation of water need. This explains, consequently, the significant impact of the cycle day (Day_Cyc) of the plant. From meteorological data, the minimum and maximum temperatures (Min_Temp and Max_Temp) are other important factors in the field of irrigation according to the obtained results.

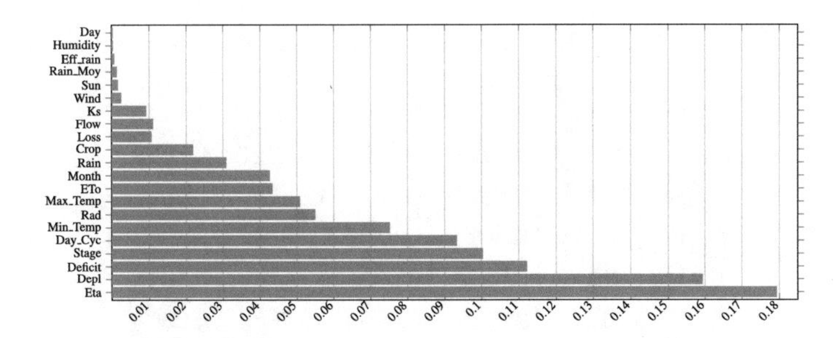

Fig. 3. RF-based ranked features with their weights

Table 1. Top 10 selected features using RF, RFE and SelectKBest.

RF	RFE	SelectKBest
Eta	Eta	Rad
Depl	Depl	Wind
Deficit	Deficit	Day_cyc
Stage	Stage	Day
Day_Cyc	Day_Cyc	Deficit
Min_Temp	Min_Temp	Depl
Rad	Rad	Ks
Max_Temp	Max_Temp	Rain
ETo	ETo	Stage
Month	Month	Eta

4.3 Results of Model Comparison

To better evaluate the obtained results, we compare in this section the different regression models while considering all features (ALL) and only the selected features (BEST). The results are given in Table 2. The results after FS showed better performance than that of before FS in all the four candidate models. Specifically, after FS, the result of R2 for GBR, CART, XGB, and RF improved by 1.27, 26.8, 3.75, 0.86, respectively, and all of them showed statistically significant differences.

Moreover, it is evident from the results that the XGB model yields the best performance compared to the other regression methods with the 10 most important features. It illustrates the evidence of the powerful predictive capacity of the XGB robust model.

Table 2. MSE, MAE, RMSE and R^2 of the different models

Model	Input	MSE	MAE	RMSE	R^2
GBR	ALL	0.0127	0.0620	0.1127	0.8894
GBR	BEST	0.0121	0.0611	0.11	0.9021
CART	ALL	0.0086	0.0399	0.0927	0.5368
CART	BEST	0.0059	0.0355	0.0768	0.8046
XGB	ALL	0.0079	0.0392	0.0888	0.9501
XGB	**BEST**	**0.0044**	**0.0354**	**0.0663**	**0.9587**
RF	ALL	0.0143	0.06238	0.1196	0.8535
RF	BEST	0.01386	0.0613	0.1177	0.8909

5 Conclusion

Improving water use is a paramount field as it ensures sustainable agriculture. Current irrigation systems demand robust and accurate predictions. Machine learning methods have been widely used, in recent years, to tackle this issue. However, the existing machine learning-based approaches still lack accuracy and robustness.

Today, feature selection and ensemble methods are often used to estimate accurate results from unseen new data. The main objective of this study was to propose a robust model to predict effectively the optimal quantity of water needed to irrigate a plant. Accordingly, this paper presents a new machine learning approach that learns from a dataset created using CLIMWAT and CROPWAT that contains meteorological data, soil data, irrigation data, and crop data. First, these data are used to select the best subset of features by relying on three FS methods: RF, RFE, and SelectKBest. Subsequently, the essential features are used to build a robust water need estimation model. Experiment results indicated that RF generates the most performed subset of features as an FS method. The experiments also showed the high impact of the evapotranspiration, the depletion, and the deficit for estimating the water need. Furthermore, other experiments are carried out to compare the different regression methods. The findings ensure that the XGB algorithm performs better than the other three model to estimate the optimal quantity of water needed to irrigate plants.

As future works, we plan to increase the data samples by considering different regions with varying characteristics (soil, weather information, etc.). Following that, we could improve the proposed approach to effectively train and estimate from diverse and massive amounts of data.

Acknowledgements. The work is carried out in the frame of the PRECIMED project that is funded under the PRIMA Programme. PRIMA is an Art.185 initiative supported and co-funded under Horizon 2020, the European Union's Programme for Research and Innovation. (project application number: 155331/I4/19.09.18).

References

1. Adeyemi, O., Grove, I., Peets, S., Domun, Y., Norton, T.: Dynamic neural network modelling of soil moisture content for predictive irrigation scheduling. Sensors **18**(10), 3408 (2018)
2. Allen, R., Pereira, L., Raes, D., Smith, M.: Fao irrigation and drainage paper no. 56. Rome: Food and Agriculture Organization of the United Nations **56**, 26–40 (1998)
3. Carneiro, T., et al.: Performance analysis of google colaboratory as a tool for accelerating deep learning applications. IEEE Access **6**, 61,677–61,685 (2018)
4. Clarke, D., Smith, M., El-Askari, K.: Cropwat for windows: user guide (2000)
5. Food, F., of the united nations, A.O.: Crop information - tomato (2020). http://www.fao.org/land-water/databases-and-software/crop-information/tomato/en/. Accessed 29 May 2021
6. Fredj, M., Grati, R., Boukadi, K.: CropWaterNeed: a machine learning approach for smart agriculture. In: ITNG 2021 18th International Conference on Information Technology-New Generations (2021)
7. González Perea, R., Camacho Poyato, E., Montesinos, P., Rodríguez Díaz, J.: Prediction of irrigation event occurrence at farm level using optimal decision trees. Comput. Electron. Agric. **157**, 173–180 (2019)
8. Gu, Z., Qi, Z., Burghate, R., Yuan, S., Jiao, X., Xu, J.: Irrigation scheduling approaches and applications: a review. J. Irrigation Drainage Eng. **146**(6), 04020,007 (2020)
9. Izzuddin, T., Johari, M., Rashid, M., Jali, M.: Smart irrigation using fuzzy logic method. ARPN J. Eng. Appl. Sci. **13**(2), 1819–6608 (2018)
10. Li, Y., Li, T., Liu, H.: Recent advances in feature selection and its applications. Knowl. Inf. Syst. **53**(3), 551–577 (2017). https://doi.org/10.1007/s10115-017-1059-8
11. Navarro-Hellín, H., del Rincon, J.M., Domingo-Miguel, R., Soto-Valles, F., Torres-Sánchez, R.: A decision support system for managing irrigation in agriculture. Comput. Electron. Agric. **124**, 121–131 (2016)
12. Neumann, J., Schnörr, C., Steidl, G.: Combined svm-based feature selection and classification. Mach. Learn. **61**(1–3), 129–150 (2005)
13. Pedregosa, F., et al.: Scikit-learn: machine learning in python. J. Mach. Learn. Res. **12**, 2825–2830 (2011)
14. Piñeiro, G., Perelman, S., Guerschman, J.P., Paruelo, J.M.: How to evaluate models: observed vs. predicted or predicted vs. observed? Ecological Modell. **216**(3), 316–322 (2008)
15. Rodriguez-Galiano, V., Sanchez-Castillo, M., Chica-Olmo, M., Chica-Rivas, M.: Machine learning predictive models for mineral prospectivity: an evaluation of neural networks, random forest, regression trees and support vector machines. Ore Geol. Rev. **71**, 804–818 (2015)
16. Saeys, Y., Abeel, T., Van de Peer, Y.: Robust feature selection using ensemble feature selection techniques. In: Daelemans, W., Goethals, B., Morik, K. (eds.) Machine Learning and Knowledge Discovery in Databases, pp. 313–325. Springer, Heidelberg (2008)

17. Sayari, S., Mahdavi-Meymand, A., Zounemat-Kermani, M.: Irrigation water infiltration modeling using machine learning. Comput. Electron. Agric. **180**, 105,921 (2021)
18. Smith, M., Nations., F., Agriculture Organization of the United Nations: CLIMWAT for CROPWAT : a climatic database for irrigation planning and management (1993). http://books.google.com/books?id=SCEtAQAAMAAJ

An Energy Efficient Scheme Using Heuristic Algorithms for 5G H-CRAN

Hasna Fourati[1,2,3(✉)], Rihab Maaloul[1,2], Lamia Chaari[1,2], and Mohamed Jmaiel[1,4]

[1] Digital Research Center of Sfax, Sfax, Tunisia
[2] Laboratory of Signals, systeMs, aRtificial Intelligence and neTworkS, Sfax, Tunisia
hasna.fourati21@gmail.com, rihab.maaloul.abid@gmail.com, lamiachaari1@gmail.com
[3] Faculty of Economics and Management of Sfax (FSEGS), University of Sfax, Sfax, Tunisie
[4] Research Laboratory on Development and Control of Distributed Applications, Sfax, Tunisia
mohamed.jmaiel@redcad.org

Abstract. The explosion of traffic demand combined with increasingly complex heterogeneous network (HetNet) infrastructures, has presented ever more challenges for mobile network operators such as QoS, autonomous network management and energy efficiency (EE). In this paper, we present an energy-efficient solution considering the Self-Organizing Network (SON) paradigm in 5G. Turning off small base stations (SBS) while maintaining a satisfactory level of QoS is a promising solution to enhance EE. Our solution is based on two heuristic algorithms: Particle Swarm Optimization (PSO) and Genetic Algorithm (GA) to empower the Energy Saving SON function in order to increase EE. We model the problem as linear program that aims to find the optimal set of SBS can be turn off. Numerical results show the efficiency the heuristic algorithms, which effectively serve users's traffic requirement while reducing energy consumption.

1 Introduction

The last decade has experienced a never-ending growth in global mobile data traffic. The International Telecommunication Union (ITU) predicted that the overall mobile data traffic will reach 5 zettabytes (ZB) per month [1]. Thus, this trend is pushing network operators to consider more rigorous elements in the QoS. The key technologies to achieve 5G goals were network densification, millimetre wave, and massive MIMO architecture. The 3rd Generation Partnership Project (3GPP) proposes multitier heterogeneous networks (HetNet), in which small base stations (SBSs) are organized under macro base stations (MBSs) to fulfill QoS and traffic volume requirements. That said, the number of small cell deployments in 5G has been densely rising in radio access networks (RANs) to improve the network coverage and capacity. The HetNet concept overlays low-power and low-cost base stations with conventional macro cellular networks. SON is a key enabler in 5G, which allows to autonomously configure, optimize and heal the RAN performance. SON promises operators to enhance the QoS as well as to reduce CAPEX and OPEX costs in an autonomous way. Thus, SON mechanisms deliver the easy management of network operations, resources and optimization [3]. The

L. Barolli et al. (Eds.): AINA 2022, LNNS 449, pp. 503–515, 2022.
https://doi.org/10.1007/978-3-030-99584-3_44

minimization of energy consumption is considered as a primary problem in the resource optimization. The base station consumes a large part of the energy generating a large amount of electricity bill. Thus, not only the operator's but also the consumer's concern, obtaining EE has significant economic benefits [4]. The Energy Saving mechanism is implemented to deliver SON function that aims to reduce consumption and enhance EE by allowing cells to go into Sleep mode, resulting in low energy consumption. To minimize the energy consumption and satisfy user's requests, a typical used solution is to turn into saving mode the underutilized BSs. We consider a centralized SON architecture in which the Energy Saving algorithm is mainly run in central network management system, it offsets the SDN controller functions presented in wired topology [5]. Artificial Intelligence (AI) techniques such as Machine Learning (ML), bio-inspired algorithms and fuzzy neural networks could be applied to design and optimize SON operations. We consider in this work to use bio-inspired algorithms, in particular Particle Swarm Optimization (PSO) and Genetic Algorithm (GA) for managing the Energy Saving SON function. These algorithms are suitable to manage complex system as HetNet since they have relatively low complexity, enabled by recursive feedback-based learning and local interactions [2]. The remainder of this paper is organized as follows. Related works are discussed in Sect. 2. The system model is presented to describe the scenario implementation, as well as a Mixed-Integer Linear Program (MILP) formulation of the problem in Sect. 3. Section 4 presents the obtained results by PSO and GA compared to the CPLEX. Finally, Sect. 5 concludes the paper and gives future directions.

2 Related Work

This section provides an overview of the work related to Energy Saving in 5G cellular network. Three techniques should be considered to achieve high EE in 5G, i.e., the design of the network architecture, the resource management and the power states management. Energy can be saved by switching some BS components into sleep mode during low-traffic periods. Typically, a BS has three power states (e.g., OFF, ON, Sleep) [6,7,12,13]. Considering only two ON/OFF states [11,14] can degrade the EE in the system network. Different 5G network architectures in the existing literature are proposed through which BS can switch to Sleep state and achieve the best Energy Saving: HetNet [7,10–14]. The most of these studies did not consider the implementation of SON. H-CRAN takes the advantages of two approaches HetNet and C-RAN to provide an efficient allocation of energy resources. The HetNet off-loads the charge of traffic demands from MBS to SBSs in order to minimize the MBS energy consumption. In C-RAN, the cloud assumes the responsibility of taking decisions and execution of algorithms. This responsibility off-loads the energy of calculation from BSs to cloud servers. Shuvo et al. [10] formulate the total power consumption as the sum of power consumption in macro and small cells. In addition to [10], Kang et al. [7] add the power consumption due to switching between different states to the sum of the power consumption by an MBS and the power consumption by SBSs under the coverage of an MBS. To optimally solve Energy Saving problem, several AI-based (e.g., ML, PSO, GA) solutions are selected to intelligently perform the switching between power states. Kang et al. used PSO algorithm [7] to provide the optimal user-BS association according to the best BS mode. Salem et al. [8] proposed the ML Markov Decision Process to model the system problem and used the Value iteration algorithm to decide the optimal Sleep policy. Shuvo et al. [10] used Meta-heuristic GA solution to optimize the number of activated small cells in a 5G and satisfy the users minimum required services. Son et al. [11] proposed an ML Long short-term memory (LSTM) solution to decide the best mode of BSs in ultra-dense networks. The studies [8, 11] applied their solutions in dynamic environment whose the conditions remain variant during the time. In addition, their dynamics programming methods can not be used in static environment. Unlike ML solutions proposed in [8, 11], the studies [7, 10, 12–14] have proven the ability of the heuristic PSO and GA algorithms to optimize the Energy Saving in static and dynamic environments. In addition, PSO and GA are easy in implementation and they present minimal computational cost compared to other ML and other AI solutions. Compared to all above studies, we choose an innovated 5G H-CRAN architecture concatenating C-RAN with slight modification of the HetNet architecture proposed in [7]. We propose a centralized implementation of the heuristic algorithms into an SON controller to operate its Energy Saving function. In addition, we have chosen to minimize the sum of three powers consumption. The minimization of power consumption due to switching between different states is done by switching between ON, OFF and Sleep modes. Unlike [7, 10, 12–14], we have realized a comparison between the AI solutions PSO and GA which have proven their efficiency to achieve a high Energy Saving. Moreover, we have compared the performance of heuristic algorithms with MILP model. Both GA and PSO provide feasible solutions within reasonable computation times. The results highlight the efficiency of GA solution which is closer to MILP than PSO in terms of Energy Saving and delay aggregation.

To the best of our knowledge, there is no paper that considers all these points (i) three power states (ii)H-CRAN architecture (iii) 5G in Static environment (iv) Comparison between two meta heuristic solutions PSO and GA (iiv) Comparison between PSO-GA and MILP results.

3 System Model and Problem Formulation

We consider a system model similar to that presented in [12]. We consider H-CRAN architecture [15], as depicted in Fig. 1, (i.e., combines HetNet and C-RAN). H-CRAN comprises one macro cell that includes several small cells. Each cell uses its Base Station (BS) to serve the demands of their users (e.g., Small Base Station (SBS) for small cell and Macro Base Station (MBS) for macro cell). Each MBS and SBS is composed of BaseBand Unit (BBU) and less intelligent Remote Radio Head (RRH). BBU is placed in the cloud and RRH remains in the BS. The cloud has a fully centralized intelligence system which is composed of two parts. First, a centralized SON server which controls the different BS states will decide the best power state. Second, BBU pool which virtually manages the users' data. Thus, our architecture consists of three principles separated planes (control, data and user planes). Centralized SON is an extension of SDN's control/data plane architecture in the cellular radio access network (RAN). The data are sent from the user plane to the cloud by a Fronthaul that implement logical one-to-one and one-to-many mapping between BBUs and BS infrastructure [5]. The decision of BS state depends of the traffic load. We contemplate one MBS and m SBSs with n users who are randomly distributed in small cells with $m<n$. We consider that all users are associated to the MBS. The user can request two types of data traffic d: high and low data rate traffic. In addition, *involved SBSs* is the number of SBS that serve data traffic demands from users belonging under its small cell coverage. We modify the architecture of [7] by adding high and low transmitted traffic conditions. During low traffic period, if *involved SBSs* is higher than the half of m, the users will directly be associated to MBS which will always be in ON state. Moreover, all SBSs will be switched to OFF state and MBS will manage both control and data traffic. Otherwise, MBS will always be in Sleep state. Each user will be connected to the best SBS maintaining best SBSs states and the minimum total energy consumption in the system. During high traffic period, the MBS will always be in ON state to manage the control signals and serve the low data rate traffic. In fact, the centralized SON orders the MBS to manage the control and low data rate signals while the SBSs serve high data rate traffic. We consider a_{ji} as the connection between the User Equipment (UE) UE_i and BS_j, w_j is the state ON of BS_j, e_j is the state Sleep of BS_j, f_j is the state switching between ON and OFF of BS_j, s_j is the state switching between ON and Sleep of BS_j, d_i is the data traffic demand of a UE_i, r_i is the type of required data traffic (high or low rate data traffic). The problem can be formulated as mixed integer linear program:

$$min\ \ EC_{Total} = P_T \times T_{agg} \tag{1}$$

This objective function EC_{Total} minimizes the total energy consumption of our system, P_T is the total power consumption and T_{agg} is aggregate delay or the time needed by a UE_i to get service from a BS_j. $P_{switching}$ is the power consumption due to switching between different states. $T_{switching}$ is the average time of switching between BS_j's states and $T_{connect}$ is the average connection initiation time between a UE_i and BS_j.

$$P_T = P_{macro} + \sum_{j=0}^{m} P(j)_{small} + P_{switching} \tag{2}$$

$$T_{agg} = T_{switching} + T_{connect} \tag{3}$$

$$T_{switching} = \sum_{j=0}^{m+1} (1-w_j) \times ((1-e_j) \times f_j \times T_{ON/OFF} + e_j \times s_j \times T_{ON/Sleep})$$

$$T_{connect} = \sum_{i=0}^{n-1} (d_i \times T_c)$$

Subject to

$$P(j)_{small} \leq P_S^{max}; (0 \leq j \leq m) \tag{4}$$

$$P_{macro} \leq P_M^{max} \tag{5}$$

$$C_{macro} \leq C_M^{max} \tag{6}$$

$$C(j)_{small} \leq C_S^{max}; (0 \leq j \leq m) \tag{7}$$

$$\sum_{j=0}^{m+1} a_{ji} \leq 1; (0 \leq i \leq n) \tag{8}$$

$$a_{ji} \leq w_j; (0 \leq i \leq n), (0 \leq j \leq m+1) \tag{9}$$

$$r_i \leq d_i; (0 \leq i \leq n) \tag{10}$$

$$w_j + e_j \leq 1; (0 \leq j \leq m+1) \tag{11}$$

$$f_j + s_j \leq 1; (0 \leq j \leq m+1) \tag{12}$$

The first and second constraints (4–5) ensure that the power consumption of one SBS $P(j)_{small}$ must be lower than maximum power consumption of SBS P_S^{max} and power consumption of MBS P_{macro} must be lower than the maximum power consumption of MBS P_M^{max}. If $P(j)_{small} > P_S^{max}$, the UE_i will be associated to MBS. We adopt the same idea regarding the capacities (C_{macro}, $C(j)_{small}$) and maximum capacities of MBS C_M^{max} and SBS_j C_S^{max} in (6–7). The constraint (8) says that UE_i can have one or no association with either MBS or any SBSs. The constraint (9) says that the UE_i can be associated to BS_j only when the BS_j is active. The constraint (10) determines the type of data rate traffic (high or low) only if the user requires a data traffic. The constraints (11–12) consider that the activation of BS_j must be only either ON or Sleep and the switch between state must be either ON-OFF or ON-Sleep. We compare the efficiency of two heuristics PSO and GA results with CPLEX solver results. Note that CPLEX is a solver that uses exact methods of resolution to solve integer, mixed integer and quadratic programs [16]. The algorithms 1 and 2 describe the different steps of PSO and GA. In PSO, we assume that the particle $p[p\_index]$ is the a_{ji} connection *new_connection* in each swarm *p_index*, *z* is swarm size, *v* is the current velocity of particle at swarm in iteration, which is included between v_{min} and v_{max}, *Pb* is the local optimum and the *Pbest* is the value of the objective function applying *Pb*. *wg* is the inertia weight, *c*1 is the cognitive coefficient, *c*2 is the social coefficient, *r*1 and *r*2 are the magnitudes of the cognitive and social vectors. To obtain the next *Pbest*, the next velocity v' must be calculated, *PSO_iteration* is the number of iteration, *Gb* is global optimum, *Gbest* is the best value of *Pbest* which is stored in *Gb*. In GA, we assume that the population *new_population* is a set of chromosomes. Each chromosome is composed of set of genes. The number of genes is *num_weights*. The *new_population* is solution matrix with *pop_size* size that composed of *num_weights* × *sol_per_pop*. The *num_parents_mating* is the number of parents and the *num_generation* is the number of generation of *new_population*. We consider that the *sol_per_pop* is number of chromosomes. In our architecture, *new_population* is a set of new connections between UEs and BSs and each chromosome in population presents the possible associations a_{ji}. The number of UE in each small cell is *num_weights*. The GA objective is to minimize the fitness function by finding the best a_{ji}.

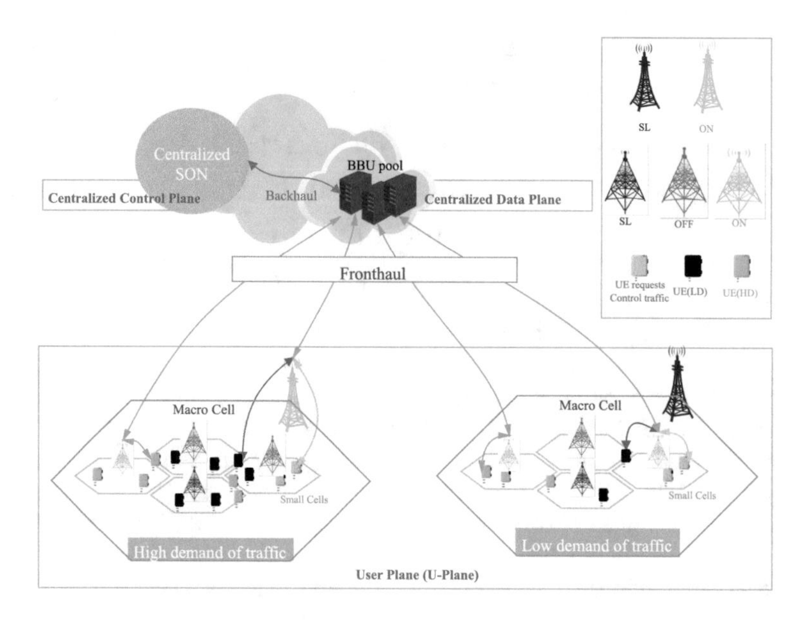

Fig. 1. System architecture

4 Evaluation Performance

In this section, we compare the PSO and GA solutions with MILP model. We can state that the GA provides Energy Saving values better than to those obtained by PSO. For high traffic, Figs. 2 and 3 report the obtained results by GA and PSO in terms of number of iterations. Figures 4 and 5 report the obtained results by GA and PSO in terms of number of iterations for low traffic. We consider $n = 500$ users and $m = 50$ SBSs, $involvedSBSs = 25$, 5 swarms for PSO and 5 sol_per_pop for GA. We consider that the users can request one of the data traffic types: low data rate (Medical Application (10 Kbps) or Image Browsing (128 Kbps)) and high data rate (Video data (1024 Kbps)). We assume an uniform distribution of the traffic time service. The low data rate traffic can be served with 0.01 Mbps and high data rate traffic can be served with 1 Mbps. The maximum power consumption of an SBS P_S^{max} is 27 W, the maximum power consumption of an MBS P_M^{max} is 1350 W. The maximum capacity of SBS is 25 Mbps and the maximum capacity of MBS is 500 Mbps. For low traffic, we assume that the number of users who exchange traffic with the system does not exceed 150 users scattered in small cells, thus, the percentage of traffic transmitted in the system is low than 30%. For high traffic, the number of users who exchange traffic is between 300 and 500 users, thus, the percentage of traffic transmitted in the system is high than 60%. The time service of traffic is U[0,2.5] min.

Algorithm 1: PSO for Energy Saving

```
   Input: a, d, r, w, e, f, s, n(u_{r,j}), n(u_{I,j}), PSO_iteration, wg, c1, c2, r1, r2, Pb, v, Gb, z, p,
          Pbest, Gbest
   Output: EC_Total
1  p=a
2  for iteration in range (0, PSO_iteration) do
3      for p_index in range (0,z) do
4          /* Calculate the velocity */
5          v' =
            1 ÷ (1 + exp(wg × v[p_index] + c1 × r1 × (Pb[p_index] − a) + c2 × r2 × (Gb − a)))
6          /* Algorithm3: Adjust the new connection and state w of BS. */
7          [p[p_index], w'] = Adjust_the_new_connection_and_state_of_BS(v', d, r).
8          /* Algorithm4: Find State Information of BS.*/
9          [e', f, s] = Find_State_Information_of_BS(w', e, w).
10         /* Calculate Energy consumption */
11         EC_Total = P_T × T_agg
12         /* Update v, Pbest and Pb, if EC_Total is lower than Pbest */
13         if EC_Total < Pbest then
14             Pb[p_index] = p[p_index];
15             Pbest[p_index] = EC_Total;
16             v[p_index] = [v'];
17         /* Update w', e',Gbest and Gb, if the updated Pbest is lower than the best
            value of Gbest */
18         if Pbest < Gbest then
19             Gb = Pb[p_index];
20             Gbest = Pbest[p_index];
21             BSstate = [w', e'];
```

Our proposed model solution can satisfy users by providing SBSs and MBS capacities more than needed to treat all traffic transmitted between all UEs and BSs. For low traffic, we evaluate two cases: first, when the $involvedSBSs < 25$ (e.g., $involvedSBSs = 14$), the capacity of one SBS can treat a maximum of 24 users (e.g., 24.576 Mbps < 25 Mbps) who are all required video data rate as well 14 SBSs (e.g., 350 Mbps) can treat more than 150 users (e.g., 153 Mbps). Second, when $involvedSBSs > 25$, MBS stays always in ON state to serve users, all SBSs are in OFF state. The maximum capacity of MBS (e.g., 500 Mbps) can treat more than 150 users who require video data rate. For high traffic, the number of UEs who are scattered in the entire network is high. SBSs serve only high data rate (video data). If all users require only video data, all SBSs in the system can serve more than 500 users (e.g., 512 Mbps < 1250 Mbps). If all users require only Image Browsing data (low data rate), MBS can serve more than 500 users (e.g., 64 Mbps < 500 Mbps).

Algorithm 2: GA for Energy Saving

```
Input: a, d, r, w, e, f, s, n(u_{r,j}), n(u_{I,j}), num_weights, sol_per_pop, num_parents_mating,
       num_of_generation, pop_size
Output: EC_Total
1  new_population = a
2  for generation in range (0,num_generations) do
3      /* Selecting the best parents in the population mating */
4      parents = ga.select_mating_pool(new_population, fitness, num_parents_mating) ;
5      /* Generating next generation using crossover.*/
6      offspring_crossover = ga.crossover(parents, offspring_size =
         (pop_size[0] − parents. shape[0], I, num_weights));
7      /* Adding some variations to the offspring using mutation. */
8      offspring_mutation = ga.mutation(offspring _crossover, num_mutations = 2);
9      /* Creating the new connection based on the parents and offspring. */
10     new_population[0 : parents.shape[0], :] = parents;
11     new_population[parents.shape[0] :, :] = offspring_mutation;
12     for sol in range (0,sol_per_pop) do
13         /* Algorithm3: Adjust the new connection and the state of BS. */
14         [new_population, w'] = adjust_the_new _connection_and_the_state_of_BS(d, r);
15         /* Algorithm4: Find State Information of BS. */
16         [f, s, e'] = Find_State_Information_of_BS(e, n(u_{r,j}), n(u_{I,j}), P_thr, I_thr, w, w');
17         EC_Total = P_T × T_agg;
18         fitness[sol] = EC_Total;
19     Best result = min(fitness);
```

For energy consumption, we evaluate GA and PSO results compared to MILP model results. We observe that GA always achieves Energy Saving better than PSO for low and high traffic scenarios. For low traffic, when $involvedSBSs < 25$, the minimum energy consumption for PSO is 20.350 KJ while for GA is 19.745 KJ. GA is closer to MILP results (e.g., 10.782 KJ). When $involvedSBSs > 25$, only MBS is active in the system. Thus, the energy consumption will not be changed. For high traffic, the energy consumption for PSO is minimised from 43.912 KJ to 38.727 KJ which is higher than GA energy consumption (e.g., 37.638 KJ). We observe that the energy consumption for low traffic is much less than high traffic, which highlights the efficiency of our modification in the system model compared to [7]. In addition, the difference of energy consumption minimisation for low traffic is less than the difference of energy consumption minimisation for high traffic, it is because the number of the *involvedSBSs* which is less than *involvedSBSs* in high traffic.

Algorithm 3: Adjust the new connection and the state of BS

```
   Input: d,r
   Output: new_connection, w'
 1 for n' in range (0,n) do
 2     if Percentage of traffic in system < 30% then
 3         if involvedSBSs > 25 then
 4             w'[m+1] = 1 ;
 5             new_connection[m+1][n'] = 1;
 6         else
 7             /* Power of small cell condition*/
 8             if P(j)_small > P_S^max then
 9                 new_connection[m+1][n'] = 1;
10                 w'[m+1] = 1;
11             else
12                 for j in range(0,m) do
13                     /* If the user exists in the overlapping areas commonly covered
                         by the BS and the neighbor BSs, select among neighbouring
                         cells condition*/
14                     new_connection[j][n'] = 1;
15                     w'[j] = 1;
16     else if Percentage of traffic in system > 60% then
17         /*Set association if a UE demands low rate data service.*/
18         if (d[n'] == 1 and r[n'] == 0) or d[n'] == 0 then
19             new_connection[m+1][n'] = 1;
20             w'[m+1] = 1;
21         else if d[n'] == 1 and r[n'] == 1 then
22             /* Power of small cell condition*/
23             if P(j)_small > P_S^max then
24                 new_connection[m+1][n'] = 1;
25                 w'[m+1] = 1;
26             else
27                 for j in range(0,m) do
28                     /* If the user exists in the overlapping areas commonly covered
                         by the BS and the neighbor BSs, select among neighbouring
                         cells condition*/
29                     new_connection[j][n'] = 1;
30                     w'[j] = 1;
```

As well energy consumption and system capacity, GA outperforms PSO in terms of aggregate delay. For low traffic, GA presents $T_{agg} = 9.057$ s compared to PSO which offers 10.13 s. For high traffic, the T_{agg} in GA is 21.397 s while the PSO offers 22.256 s. The efficiency of the separation between low and high traffic is emphasized in aggregate delay. In fact, with low traffic, the delay represents the half of high traffic aggregate delay. According to objective function (1), we aim to minimize the energy consumption which depends on T_{agg} and P_T. We mention that T_{agg} depends on $T_{switching}$. When the BS_j

Algorithm 4: Find State Information of BS

```
Input: e, n(u_{r,j}), n(u_{I,j}), P_thr, I_thr, w, w'
Output: f, s, e'
1 for j in range (0,m+1) do
2     if n(u_{r,j})[j] > P_thr and n(u_{I,j})[j] > I_thr and w'[j] =!1 then
3         e'[j] = 1;
4     /*Find state switching of BS.*/
5     if w[j] != w'[j] and e[j] == e'[j] then
6         f[j] = 1;
7     if w[j] == w'[j] and e[j] != e'[j] then
8         s[j] = 1;
```

is in ON state, $T_{switching}$=0. Otherwise, $T_{switching} > 0$. Thus, when the power consumption in the system is high, the delay aggregate in the system is low. We take high traffic scenario, P_T of PSO is minimized from 1764 W to 1740 W while the minimum P_T of GA is 1759W. T_{agg} of GA is 21.397 s while T_{agg} of PSO is 22.256 s. GA provides Energy Saving values better than PSO due to its low aggregate delay despite the slight height P_T compared to PSO.

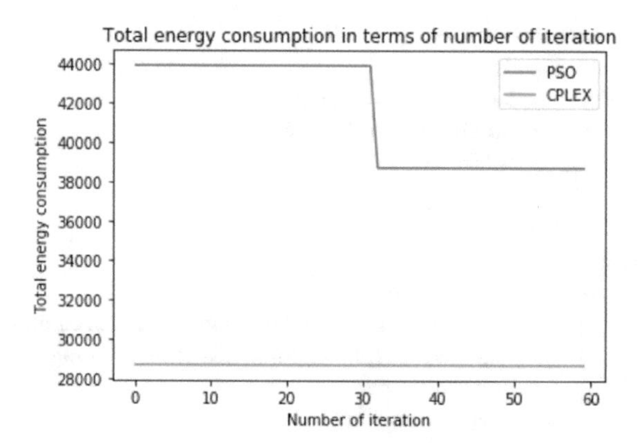

Fig. 2. High traffic comparison between PSO and CPLEX results.

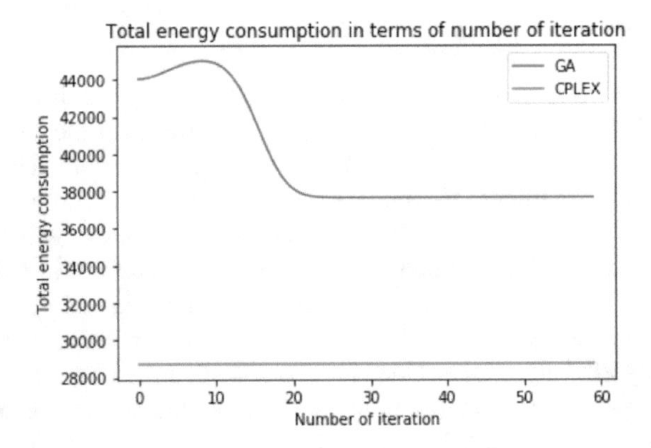

Fig. 3. High traffic comparison between GA and CPLEX results.

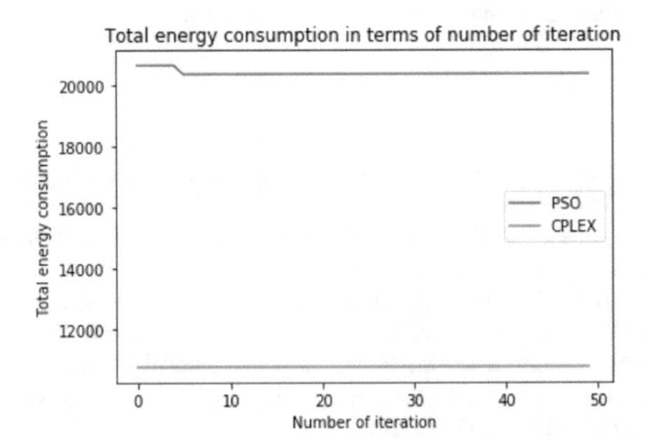

Fig. 4. Low traffic comparison between PSO and CPLEX results with s = 14.

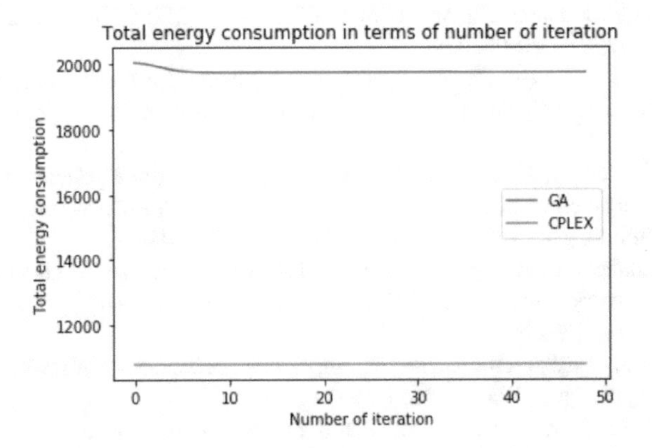

Fig. 5. Low traffic comparison between GA and CPLEX results with s = 14.

5 Conclusion

In this paper, we present an energy efficient solution to deliver an Energy Saving SON function 5G H-CRAN. We choose to save energy by turning off the components of the BS reactively to traffic demand. The proposed solution is implemented in static environment with constant conditions. We model the problem as MILP formulation. Then, we propose to solve the problem using GA and PSO heuristic. These algorithms don't need for historical results and they are able to achieve optimal solutions in polynomial time. We evaluate the efficiency of these heuristics with results obtained by CPLEX solving the MILP. Experiment results show that GA provided Energy Saving values better than those obtained by PSO algorithms. Compared to objective values generated by CPLEX, GA provided near-optimal solutions in terms of Energy Saving. The aggregate delay to serve the user with GA is lower than the aggregate delay with PSO. As a future work, it would be interesting to expand the number of MBS and SBS and shift to dynamic environment using Markov decision Process and Reinforcement Learning. Moreover, we can use more than three states to save energy in BSs.

References

1. Tariq, F., Khandaker, M., Wong, K., Imran, M., Bennis, M., Debbah, M.: A speculative study on 6G. IEEE Wirel. Commun. **27**, 118–125 (2020)
2. Kuribayashi, H., et al.: Particle swarm-based cell range expansion for heterogeneous mobile networks. IEEE Access **8**, 37021–37034 (2020)
3. Fourati, H., Maaloul, R., Chaari, L., Jmaiel, M.: Comprehensive survey on self-organizing cellular network approaches applied to 5G networks. Comput. Netw. 108435 (2021). https://www.sciencedirect.com/science/article/abs/pii/S1389128621003960
4. Abrol, A., Jha, R.: Power optimization in 5G networks: a step towards GrEEn communication. IEEE Access **4**, 1355–1374 (2016)
5. Arslan, M., Sundaresan, K., Rangarajan, S.: Software-defined networking in cellular radio access networks: potential and challenges. IEEE Commun. Mag. **53**, 150–156 (2015)
6. Post, B., Borst, S., van den Berg, H.: A self-organizing base station sleeping and user association strategy for dense cellular networks. Wirel. Netw. **27**(1), 307–322 (2020). https://doi.org/10.1007/s11276-020-02383-3
7. Kang, M., Chung, Y.: An efficient energy saving scheme for base stations in 5G networks with separated data and control planes using particle swarm optimization. Energies **10**, 1417 (2017)
8. Salem, F., Chahed, T., Altman, E., Gati, A., Altman, Z.: Optimal policies of advanced sleep modes for energy-efficient 5G networks. In: 2019 IEEE 18th International Symposium On Network Computing And Applications (NCA), pp. 1–7 (2019)
9. Salem, F., Chahed, T., Altman, Z., Gati, A.: Traffic-aware advanced sleep modes management in 5G networks. In: 2019 IEEE Wireless Communications And Networking Conference (WCNC), pp. 1–6 (2019)
10. Shuvo, M., et al.: Energy-efficient scheduling of small cells in 5G: a meta-heuristic approach. J. Netw. Comput. Appl. **178**, 102986 (2021)
11. Son, J., Kim, S., Shim, B.: Energy efficient ultra-dense network using long short-term memory. In: 2020 IEEE Wireless Communications And Networking Conference (WCNC), pp. 1–6 (2020)

12. Sigwele, T., Hu, Y., Susanto, M.: Energy-efficient 5G cloud RAN with virtual BBU server consolidation and base station sleeping. Comput. Netw. **177**, 107302 (2020)
13. Farooq, H., Asghar, A., Imran, A.: Mobility prediction-based autonomous proactive energy saving (AURORA) framework for emerging ultra-dense networks. IEEE Trans. Green Commun. Network. **2**, 958–971 (2018)
14. Chang, K., Chu, K., Wang, H., Lin, Y., Pan, J.: Energy saving technology of 5G base station based on internet of things collaborative control. IEEE Access **8**, 32935–32946 (2020)
15. Gonçalves, G., et al.: Flying to the clouds: the evolution of the 5G radio access networks. The Cloud-to-Thing Continuum, pp. 41–60 (2020)
16. ILOG, I. cplex-optimizer. (https://www.ibm.com/analytics/cplex-optimizer)

A Multi-agent Based Framework for RDF Stream Processing

Wafaa Mebrek[1,2] and Amel Bouzeghoub[1](✉)

[1] Samovar, Télécom SudParis, Institut Polytechnique de Paris, 19 place Marguerite Perey, 91120 Palaiseau, France
{Wafaa.Mebrek,Amel.Bouzeghoub}@telecom-sudparis.eu
[2] EEDIS, Djillali Liabes University of Sidi Bel Abbes, Sidi Bel Abbes, Algeria
w.mebrek@esi-sba.dz

Abstract. When a large amount of data is generated from multiple, heterogeneous and continuous data streams, the need for continuous processing and on-the-fly consumption of the overwhelming flow of data is crucial. In this context, the W3C RDF Stream Processing (RSP) Community Group has defined a common model for continuous querying RDF Streams, giving rise to a plethora of RSP engines. However, their main limitation is that, depending on the application queries, one RSP engine may be more appropriate than another, or multiple engines are required to address complex queries. In this paper, we propose a multi-agent based framework for distributed continuous processing that gives the opportunity to use several RSP engines in the same framework in order to benefit from their advantages and to offer the possibility to use them at the same time or in a sequence to answer complex queries. A preliminary experimental evaluation with a real-world benchmark shows promising results when compared to an existing work.

1 Introduction

Recent years have seen an increasing interest in data stream analysis systems that correlate data generated by applications, sensors, and smart devices enriched with background knowledge for processing heterogeneous data streams. Therefore, stream processing has become an important paradigm for processing data at high speed and large scale.

As a result, the Resource Description Framework (RDF) was recognized as the most suitable model to represent these data streams, thus enabling the integration of diverse data sources across heterogeneous domains. A large number of RDF Stream Processing (RSP) systems have been proposed and implemented over the last decade. They differ with respect to syntax, expressiveness, scalability, reasoning support, underlying semantics, and the supported query languages. In practice, querying data streams requires diverse query operators, such as selection, aggregation, filtering, in addition to explicit temporal features. In particular, we can point out the following main limitations: (1) Window management operators or temporal operators are only carried by few systems; (2)

L. Barolli et al. (Eds.): AINA 2022, LNNS 449, pp. 516–528, 2022.
https://doi.org/10.1007/978-3-030-99584-3_45

Aggregates functions are not always implemented; (3) Their output streams are not well designed to be an input for another RSP engine. If no R2S (relation-to-stream) operator is implemented, the output is a relation or a stream. In this case, three operators are considered Rstream (streams out all data in the last step); Istream (streams out what is new); Dstream (streams out what is old). Depending on the desired type of output, some engines are more appropriate than others. C-SPARQL[1] [12] implements Rstream while CQELS[2] [17] implicitly supports IStream); (4) They are centralized solutions that cannot execute several queries in parallel and it is not always possible to process multiple data streams in different time windows. Thus, given that (i) no standard language exists to address these shortcomings, we believe we are making a step forward on this issue by providing a framework that allows RSP engines to coexist. Hence, the main contributions of this paper are as follows: (i) A new extensible and distributed framework for data streams processing that integrate multiple RSP engines enabling the expression of more complex queries by taking advantage of their benefits and capabilities; (ii) Definition of query patterns for automatic selection of the appropriate engine. The rest of the paper is organized as follows: Sect. 2 presents a running example to identify the requirements for our proposed framework. Section 3 details the state of the art on RSP and distributed approaches. Our proposed system architecture is described in Sect. 4 and evaluated in Sect. 5. Section 6 concludes and gives some perspectives.

2 Motivating Example

Citybench [6] is a benchmark that assesses two RSP engines (C-SPARQL and CQELS) in smart city applications using a set of data streams collected from the city of Aarhus (Denmark). This benchmark provides a set of simple continuous queries that can express aggregation over RDF streams. However, neither temporal ordering over triple streams nor temporal patterns expression over RDF graphs are possible. Besides, complex queries that require involving several engines in parallel or sequence cannot be expressed. Let's consider the following example: "*Bob is driving and wants to know which cultural events are happening around his current location and for a particular event, to be notified if traffic conditions are increasing in order to switch his trajectory or find the nearest parking lots*". This query poses a threefold issue: First, it is necessary to compare the level of congestion of two consecutive road segments. However, both of the aforementioned RSP engines lack suitable expressive temporal operator to find out the sequence of events that determines the congestion level of the two road sections near Bob's location. Even though C-SPARQL supports the timestamp function to partially express the events sequence, the query construction is not intuitive and the validity of the results cannot be determined due to the missing of semantics that show the usefulness of a more appropriate engine.

[1] C-SPARQL: A Continuous Query Language For RDF Data Streams.
[2] CQELS: Continuous Query Evaluation over Linked Streams.

Second, multiple streams must be aggregated (Traffic, Parking, and User Location). Third, static knowledge for cultural events is needed. All these features cannot be expressed with C-SPARQL and CQELS. Based on the aforementioned observations, we highlight the following requirements:

R1 - Need for processing multiple streams: As showed in the previous query, only events belonging to the same geographical scope are of interest. Streams for Traffic, Parking and location must be aggregated. R2 - Need for cascade queries: Given the complex nature of the query, it is necessary to split it into subqueries, each dedicated to one engine. Thereby, the result of a query (or subquery) is an input stream for another RSP engine. R3 - Need for temporal operators: The aforementioned query requires comparing a sequence of congestion level events. R4 - Need for n-window queries: some queries may need comparing events in windows of different sizes. R5 - Need to separate aggregation functions from temporal operators: Using a query language only for aggregation functions and another for temporal operators would guarantee the query results and help for the cohabitation of RSP engines. R6 - Need for additional knowledge: Enriching stream processing with background knowledge (e.g. point of interest or cultural events) may improve the result efficiency. R7 - Need for extensible architecture: An RSP framework should enable to easily add or delete data sources or RSP engines and facilitate settings and extensions. In addition, several RSP engines need to collaborate.

3 Related Work

Several solutions were proposed to process RDF streams each with its own properties in terms of supported query language, expressiveness, reasoning abilities or scalability.

C-SPARQL represents the well-known extended SPARQL language. It follows two main steps to guarantee the processing task: the execution of continuous queries over RDF streams with Esper to produce a sequence of RDF graphs; then the execution of a SPARQL query against each RDF graph in the sequence. Each C-SPARQL query can therefore be divided into a static part dealing with knowledge (the semantic part) and a dynamic part dealing with data (the stream part) and can combine triples from multiple RDF streams. C-SPARQL fulfills R1 and R6 requirements. On the contrary, CQELS [17] natively uses some heuristics during the query execution to reorder operators, implements the query evaluation, and recompiles the query plan to reduce the processing overhead and delay. It supports queries with multi-stream, allows the definition of two windows in the same stream (while other languages do not) and can query static knowledge. CQELS satisfies R1, R4, and R6. SPARQLstream [7] follows the same path and adopts OBDA-like (Ontology-Based Data Access) approach to rewrite SPARQLstream queries in relational algebra expressions, optimizes and converts them in the language of a target RSP. SPARQLstream satisfies only R1. In the same path, INSTANS [10] has been shown to support expressive CEP, and is the only system that allows data to be persisted to support stateful operations. However, it does not support windows over streams natively, and removing outdated events to reduce the volume of data to be processed is not a trivial task. With regards to our requirements, INSTANS satisfies R1.

Aggregation functions are supported by all these languages but most of them suffer from a strong limitation as they do not support temporal operators except EP-SPARQL [2] which allow temporal ordering over triple streams. It represents the only system with reasoning capabilities to evaluate SPARQL queries.

However, it cannot deal with data streams from multiple sources and also the time window mechanism is not well defined. EP-SPARQL satisfies R3.

SPAseq [8] too enables temporal patterns expression. Even this language supports the RDF graph-based events and multiple RDF graph streams, it is not possible to express different time windows on each stream. SPAseq satisfies R3.

To the best of our knowledge, a recent version of CQELS proposed the BEFORE operator but it is still not supported in the released version and still lack of support for the optional clause and the union pattern in addition to arithmetic operations that prevent CQELS to calculate the spatial distance.

On the other side, several RSP benchmarks have been proposed: SRBench [9], LSBench [14], CSRBench [3], CityBench [6], etc. Despite the existence of all these benchmarks, comparing their performances remains out of sight due to both syntactic and semantic differences. For example CQELS, EP-SPARQL, and INSTANS use a reactive execution strategy, where queries are matched as soon as new data becomes available. This property reduces the latency in real-time applications and minimizes unnecessary query execution. On the contrary, SPARQLstream, and C-SPARQL execute queries periodically. As a consequence, execution rates are adjusted to optimize the performances. This is useful when the reactive evaluation strategy may not have time to return all results before the next execution.

All in all, R1, R3, R4 and R6 are fulfilled by one or more RSP engines while R2, R5 and R7 are not satisfied.

Most of these solutions are black-boxes. They are centralized and do not support a high processing capacity. In the literature, some distributed solutions exist.

[1] proposes WeSP. However, this infrastructure lack flexibility and it is not easy to add or remove an RSP engine or to set different behavior. Calbimonte in [11], offers the first distributed infrastructure for RSP that focuses on connecting different engines in digital rehabilitation scenarios. However, in this work, they do not take into consideration the expressiveness level of the query that defines the behavior of the RSP agents. Besides, agents communication is ensured through a centralized RSP agent that contains a shared mailbox. Thus, having this intermediate layer may impact system latency and performance when agents are duplicated and a high number of messages is exchanged.

In another hand, some infrastructure-oriented distributed RSP systems were developed: CQELS Cloud and C-SPARQL on S4 [13], Strider [16] and DSECP [4]. They provide, in real-time, a processing capacity of large volume of RDF data in parallel and distributed mode. These distributed RSP systems represent

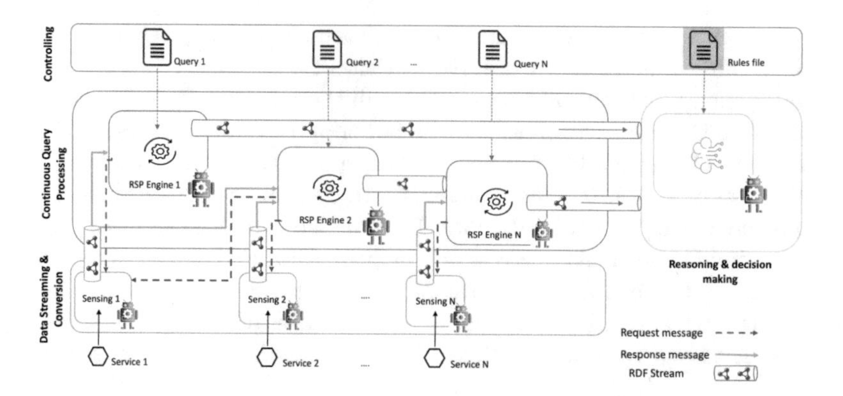

Fig. 1. MAS-based RDF stream processing architecture overview

an great limitation in windowing operators associated with the query language and propose a re-evaluation of queries by micro-batch (Strider). In addition, the distributed join between static and dynamic data as well as the intermediate results generated during the join operations (CQELS Cloud and C-SPARQL on S4) considerably degrade the performance. Although these solutions offer a distributed architecture, they are not extensible. R7 is not satisfied.

In this paper, we propose a framework whose main objective is to satisfy the seven aforementioned requirements.

4 General Architecture

In summary, the architecture sketched in Fig. 1 is composed of a set of modules having different roles: controlling, processing, sensing, and reasoning. The reasoning task is out of the scope of this paper. It is worth mentioning that any component can be duplicated, removed, or added easily. Moreover, a component can either be executed independently or in collaboration with others.

Controlling: It represents the core component of the model and is in charge of managing all the agents that communicate and exchange RDF streams. More precisely, we proposed query patterns that must be checked by the controller before deciding which agent to launch for the stream processing task. These query patterns are engine-specific, support the syntax and data models of each particular engine. The continuous - user - query is given as an argument to the controlling module and if this latter matches a query pattern, a new processing agent that embeds the corresponding engine is automatically created. The query of the motivating example is split in two sub-queries Query 1 and Query 2, each dedicated to a specific engine.

By executing these sub-queries in cascade, we fulfill the R2 requirement. We present an example of a SPAseq query (Listing 1) expressing the Query 1 that obtains the congestion level in two consecutive roads from the resulted stream of

Query 2 (Listing 2) and compare their values to determine which road is more congested.

Listing 1 : SPAseq continuous query over Traffic stream observations

```
PREFIX ssn: <http://purl.oclc.org/NET/ssnx/ssn#>
PREFIX sao: <http://purl.oclc.org/NET/sao/ssn#>

SELECT ?congestion1 ?congestion2 WITHIN 10 seconds
FROM STREAM S1<http://www.insight-centre.org/dataset/SampleEventService#Query2>
WHERE SEQ(A; B){
 DEFINE GPM A ON S1 {
        ?obId1 ssn:observedProperty ?p1;
               sao:hasValue ?congestion1;
               ssn:observedBy <http://www.insight-centre.org/dataset/SampleEventService#
                      AarhusTrafficData>.
 }
 DEFINE GPM B ON S1 {
        ?obId2 ssn:observedProperty ?p2;
               sao:hasValue ?congestion2;
               ssn:observedBy <http://www.insight-centre.org/dataset/SampleEventService#
                      AarhusTrafficData>.
         filter (?congestion1 > ?congestion2)
 }}
```

In the same way, (Listing 2) illustrates a typical C-SPARQL query for the application of Query 2. We aggregate Traffic, Parking, and user location streams using (AVG) aggregation function with C-SPARQL engine in 10 s time window. We add a static knowledge for cultural events information and register the constructed graph as Query2, then the resulted stream within a IRI[3]. This latter is processed by the SPAseq engine (Listing 1) that uses "followed-by (;)" temporal operator to compare two congested roads segments needed for Query 2[4].

Listing 2 : C-SPARQL continuous query over Traffic, Parking, and User location streams.

```
PREFIX ssn: <http://purl.oclc.org/NET/ssnx/ssn#>
PREFIX sao: <http://purl.oclc.org/NET/sao/ssn#>
PREFIX ct: <http://www.insight-centre.org/citytraffic#>
PREFIX rdfs: <http://www.w3.org/2000/01/rdf-schema#>
PREFIX rdft: <http://www.w3.org/1999/02/22-rdf-syntax-ns#type>

REGISTER STREAM Query2 AS
CONSTRUCT {
        _:c0 sao:hasValue ?congestion .
        _:c0 ssn:observedProperty _:c1 .
        _:c0 rdft: <http://www.insight-centre.org/citytraffic#congestionLevel> .
        _:c1 ct:hasStartLatitude ?lat1 .
        _:c1 ct:hasStartLongitude ?lng1 .
        _:c2 rdft: <http://www.insight-centre.org/citytraffic#userPosition> .
        _:c2 ct:hasLatitude ?avglat .
        _:c2 ct:hasLongitude ?avglng .
        _:c3 sao:hasValue ?parkingCapacity .
        _:c3 ssn:observedProperty _:c4 .
        _:c3 rdft <http://www.insight-centre.org/citytraffic#ParkingVacancy> .
        _:c3 ct:hasStartLatitude ?lat4 .
        _:c3 ct:hasStartLongitude ?lng4 .
        _:c3 rdfs:label ?parkingId.
}
 FROM STREAM <http://www.insight-centre.org/dataset/SampleEventService#AarhusTrafficData> [
        range 10s step 10s]
 FROM STREAM <http://www.insight-centre.org/dataset/SampleEventService#UserLocationService> [
        range 10s step 10s]
 FROM STREAM <http://www.insight-centre.org/dataset/SampleEventService#AarhusParkingData> [
        range 10s step 10s]
WHERE {
?p1   a <http://www.insight-centre.org/citytraffic#CongestionLevel>.
?p1 ssn:isPropertyOf ?v2.
?v2 ct:hasStartLatitude ?lat1. ?v2 ct:hasStartLongitude ?lng1.
?v2 ct:hasEndLatitude ?lat2. ?v2 ct:hasEndLongitude ?lng2.

?p4   a <http://www.insight-centre.org/citytraffic#ParkingVacancy>.
?p4 ssn:isPropertyOf ?foi4.
?foi4 ct:hasStartLatitude ?lat4. ?foi4 ct:hasStartLongitude ?lng4.
```

[3] http://www.insight-centre.org/dataset/SampleEventService#Query2.

[4] Due to space constraint, the full query is available on Github link.

```
?foi4 rdfs:label ?parkingId.
{       ?obId1 ssn:observedProperty ?p1.
        ?obId1 sao:hasValue ?congestion.
        ?obId1 ssn:observedBy <http://www.insight-centre.org/dataset/SampleEventService#
              AarhusTrafficData>.
}
{ SELECT ( AVG (?lat3 ) AS ?avglat ) ( AVG (?lng3 ) AS ?avglng )
        WHERE
{       ?obId3 a ?ob.
        ?obId3 sao:hasValue ?v3.
        ?v3 ct:hasLatitude ?lat3. ?v3 ct:hasLongitude ?lng3.
        ?obId3 ssn:observedBy <http://www.insight-centre.org/dataset/SampleEventService#
              UserLocationService>.
        }
}
{       ?obId4 ssn:observedProperty ?p4.
        ?obId4 sao:hasValue ?parkingCapacity.
        ?obId4 ssn:observedBy <http://www.insight-centre.org/dataset/SampleEventService#
              AarhusParkingData>.
}
Filter (((?avglat-?lat1)*(?avglat-?lat1)+(?avglng-?lng1)*(?avglng-?lng1)) < 0.1 && ((?avglat-?lat4)
      *(?avglat-?lat4)+(?avglng-?lng4)*(?avglng-?lng4)) < 0.1)
}
```

Continuous Query Processing: the agents ingest incoming RDF data streams, continuously process them and generate new RDF data streams as output. We propose several agent models and with the help of their autonomy and flexibility, we can implement and deploy different mechanisms responsible of several variety of functionalities (e.g. adapting input streams to the RSP engines syntax constraints or adapting outputs so that the resulting stream could be processed by other engines). Each of these agents embeds one RSP engine instance, it takes as input a continuous query sent by the controlling module. This latter contains the identifier of the source (IRI), indispensable for data streaming and processing. Each agent in the continuous query processing module may act as a sender or receiver and interacts with other agents in two different ways: (1) With sensing agents that control the data sources. RSP agents ask for a data stream; (2) With other RSP agents in order to reply collectively to a query. By design, RSP agents produce continuous results in RDF so that they can be retrievable/reused by other RSP agents after processing Therefore, from both possible interactions, the RDF streams output can reach their final destination and could be sent to the Reasoning & Decision making module. This latter uploads its rule file from the controlling module and proceeds for reasoning. Hence, different agents model and multiple instances of RSP engines can easily be set up. Consequently, our architecture is able to handle more queries in a time window and set parallel or sequence processing. Furthermore, we accord importance to asynchronous messages that we use for exchange, share information, and interactions between agents. This is defined on existing standards, such as FIPA protocol [15] for mutli-agent cooperation and delegation. These exchanged messages support and use RDF as data model. Consequently, all agents in Fig. 1 exchange RDF stream elements that are accessible through time windows, but after switching from one window to another, all the old messages are expired.

Data Streaming and Conversion: Sensing agents in data streaming and conversion module aim to manage data flows and convert data for integration purpose. They control the sensors that monitor the environment, generate and share a large amount of data that is used for processing. The smart environment is composed of multiple smart areas that deploy heterogeneous sensors (e.g. Traffic, Parking, Weather, etc.). Each sensor produces continuously raw data of one specific source that are collected later by the sensing agents that convert the data on-the-fly to RDF streams. When the sensing agent receives these raw data, it parses them and transforms their format to RDF triples. At this stage, different ontologies are used for data annotation and semantic enrichment such as SAO (Stream Annotation Ontology)(sao: StreamData to annotate sensory observations or sao: Segment to describe temporal features in Listing 1), and CT (City Traffic ontology) (ct:hasStartLongitude/ct:hasEndLongitude to annotate the beginning and the end of Longitude for the city roads in Listing 2). The conversion to RDF streams as a unified data model allows first to semantically enrich the raw data coming from sensors. Second, it helps to add knowledge using vocabularies and ontologies. Third, it provides interoperability in heterogeneous environments. Consequently, the outcome of the conversion is an RDF graph stream. Finally, the sensing agent takes the output of the conversion, declares an RDF stream with an IRI in the service directory where all source identifiers are stored, then, pushes these RDF streams to the RSP agent consumer that asks for it through a message.

To conclude, the proposed architecture can handle several queries in parallel. For each query, sensing agents are launched on-the-fly and with query patterns, the corresponding RDF engine is automatically activated. In addition, it easily integrates multiple engines and through agent cooperation, they can work together to achieve the same goal. The easy addition and removal of RSP engines and data sources ensure that our solution meets the R7 requirement. Complex queries can be split into subqueries and processed in cascade and if necessary, temporal constraints and aggregation functions can be treated with different RSP engines. Hence, R2 and R5 are also met. As a result, our framework meets all the requirements R1–R7.

5 Experimental Results

In this section, we present in detail our experimental evaluation that illustrates (i) the analysis and comparison with the CityBench benchmark that evaluate RSP engines within smart city applications; (ii) the effect of data streams rates on query execution time and system latency, (iii) the memory consumption of RSP engines during the execution of the same queries. We first describe our experimental setup, and then we analyze both experiments and their results.

5.1 Experimental Setup

Real-world datasets already proposed in City-pulse project scenarios [5] and their associated queries in CityBench [6] were used. The evaluation started with three variations of query Q10 that generates an immediate notification about polluted areas in the city depending on three configurations for a number of input data streams (2,5 and 8). We run these queries on a standard PC with a 2.6 GHz core processor and 16 Gb RAM. We conduct extensive experiments to verify the performances of a centralized against a distributed solution. To this end, we compared our framework with Citybench firstly with a single RSP agent (C-SPARQL or CQELS) and secondly with the full distribution of several instances of RSP engines. As a proof of concept we have implemented our model with Jade (version 4.5). The code is open-source[5].

5.2 Results and Analysis

Our first assumption was that a distributed architecture based on the multi-agent paradigm positively impacts the speed up of RSP and proves the enhancement of both C-SPARQL (version 0.9) and CQELS (version 0.9) engines. To validate it, we followed the same series of evaluations as in CityBench and we measured the three following metrics in 15 min time execution: First, we executed $Q10_2$, $Q10_5$ and $Q10_8$ with C-SPARQL embedded in one agent and compared the evaluation results (cf. Fig. 2) to the C-SPARQL. We also evaluated the same queries with CQELS in order to perform the scalability test and compare the results with those obtained with CityBench (cf. Fig. 4). Second, we monitored and observed the use of system memory during the execution of these queries in both C-SPARQL (cf. Fig. 3) and CQELS (cf. Fig. 5) with different variations in order to evaluate their memory consumption and compare the results with CityBench.

In Fig. 2, using a single C-SPARQL encapsulated in one agent with 2, 5 input data streams do not affect the system latency comparing to the Citybench one. On the contrary, most of the time, there is a considerable improvement in the throughput efficiency and system latency with one C-SPARQL agent of our framework. However, when increasing the stream variation to 8 in Citybench, the C-SPARQL engine has an abnormal state and suffers in the first minute execution compared to our single C-SPARQL agent. Therefore, we took advantage of our architecture and split the query into two subqueries ($Q10_8a$ and $Q10_8b$) and executed each with a C-SPARQL agent in the same time window. We observe that the latency of the subqueries significantly decreases (about 50% less than one single query's latency). This is due to our extensible framework (R7).

The next experiment in Fig. 4 is set in exactly the same conditions, except that it uses CQELS. From the execution time of $Q10_2$ and $Q10_5$, one single CQELS agent results are still very close to the CityBench one. However, splitting the query provides higher throughput efficiency. This experiment shows how a set of CQELS can handle multiple queries significantly faster than a single engine. Moreover, CQELS in our architecture is able to process 8 streams within a single query. Besides, it is quite easy to set up two agents with CQELS instances, each of them processing 4 variation streams for $Q10_8a$ and $Q10_8b$. In Fig. 4, the increase in performance is impressive and shows the efficiency in terms of scalability and latency (R7).

[5] https://github.com/anonymSma/sma_processing.

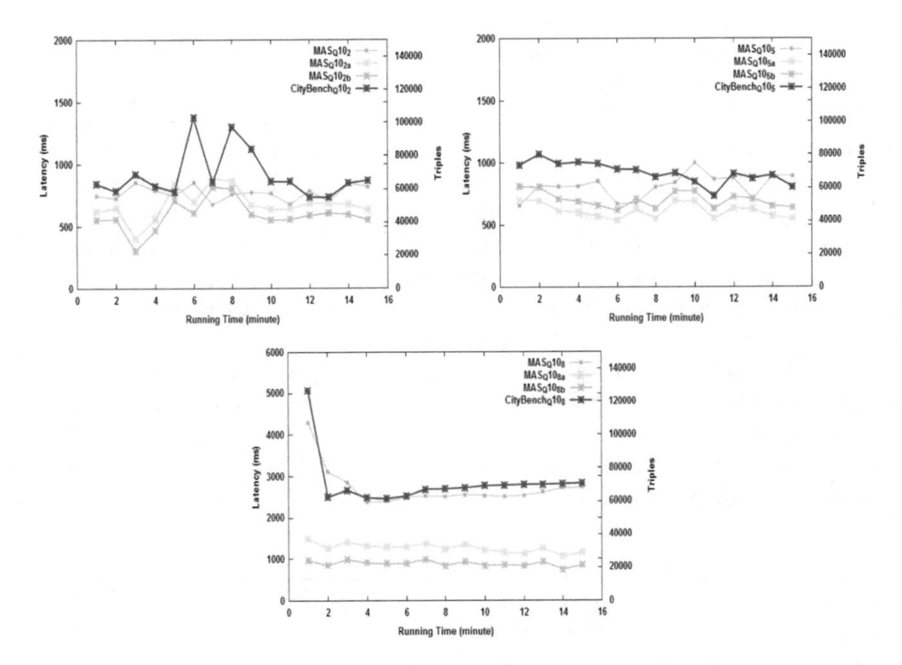

Fig. 2. Latency for centralized vs distributed C-SPARQL ($Q10_2$, $Q10_5$ and $Q10_8$)

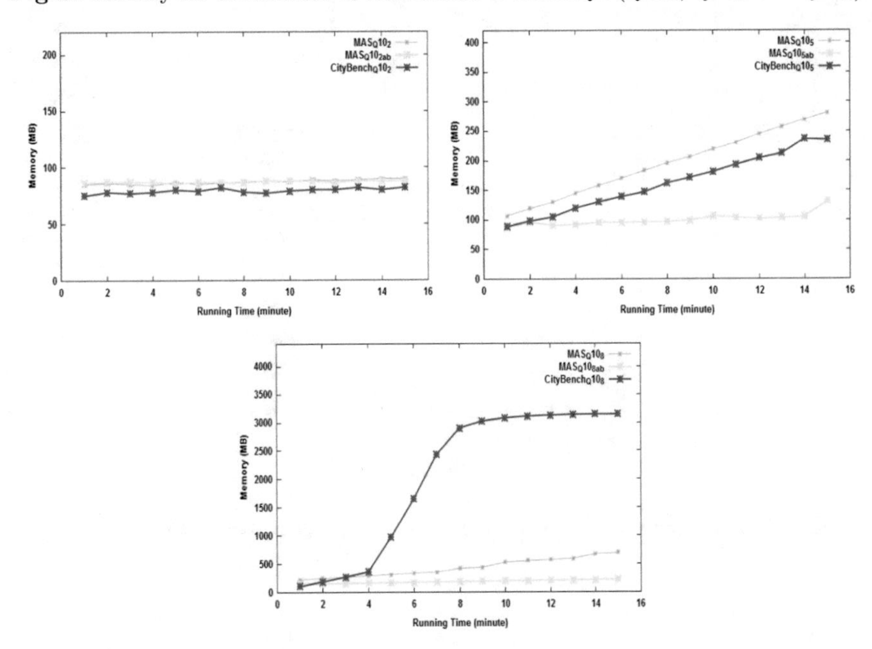

Fig. 3. Memory consumption for centralized vs distributed C-SPARQL ($Q10_2$, $Q10_5$ and $Q10_8$)

As for the memory consumption, Fig. 3 shows that it is roughly equal with $Q10_2$ for all the implementations. Later, it starts to increase with $Q10_5$ when the number of C-SPARQL agents increases, as a side effect, this explains that agents are sometimes memory costs. However, from this experiment, we notice that when we duplicate C-SPARQL agents for distribution, this operation has a minimal impact on memory consumption comparing to CityBench where we observe that the memory consumption of $Q10$ with 5 streams is decreasing. In parallel, in case of $Q10_8$ there is a slight margin between the memory consumption of multiple agents compared to one single agent. This proves that agents deployment may give rise to the same use of resource and does not have a real effect on the memory consumption. In the next experiment depicted in Fig. 5, we evaluate the memory consumption of CQELS with similar query variations. As we can see, the memory consumption of CityBench with $Q10_2$ and $Q10_5$ is greater than with MAS. However, CityBench was unable to process a query with 8 stream variations with CQELS engine.

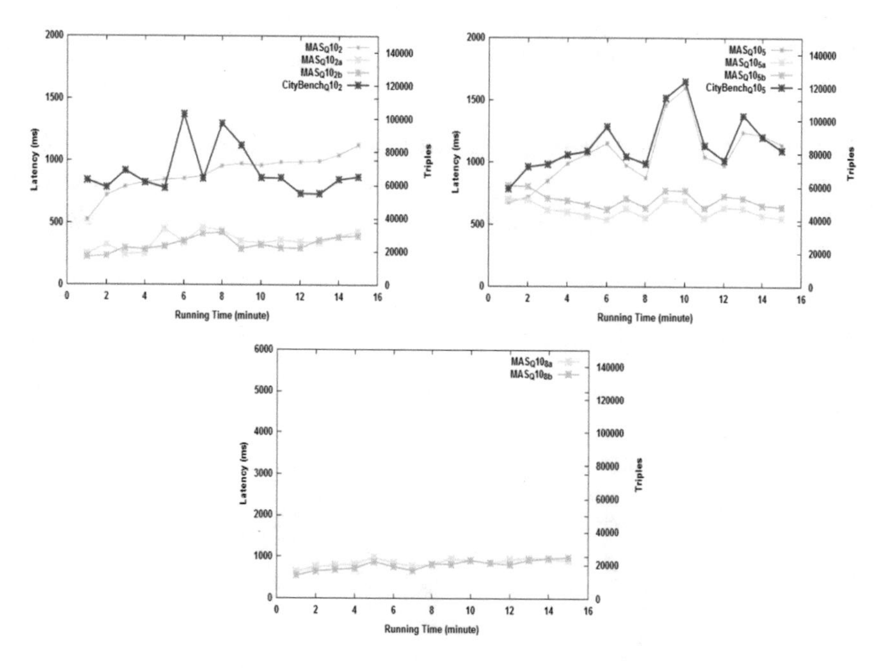

Fig. 4. Latency for centralized vs distributed CQELS ($Q10_2$, $Q10_5$ and $Q10_8$)

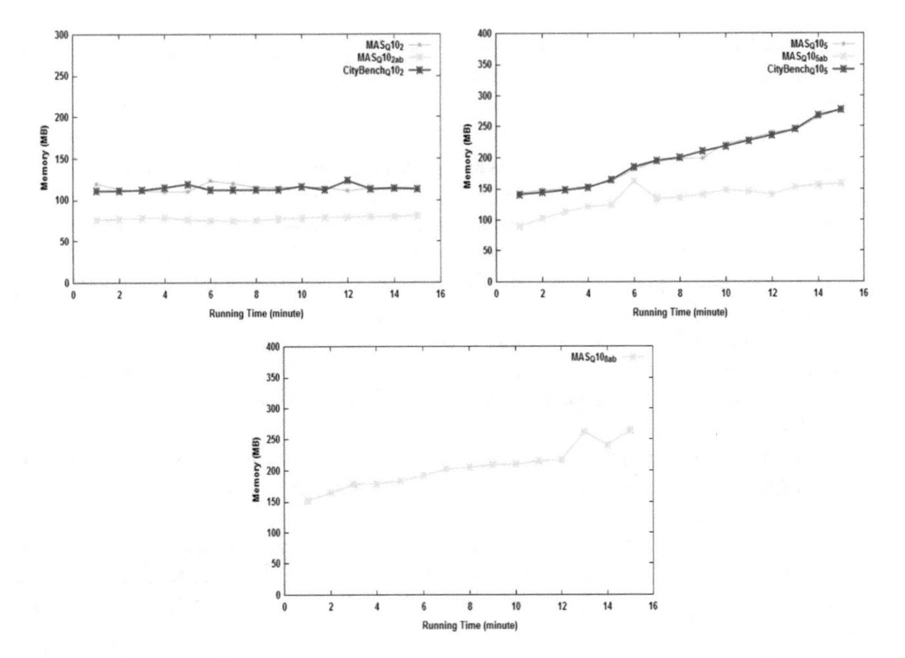

Fig. 5. Memory consumption for centralized vs distributed CQELS ($Q10_2$, $Q10_5$ and $Q10_8$)

6 Conclusion and Future Work

In this paper, we proposed a framework for RDF stream processing based on multi-agent architecture. The main benefit of this approach that it can easily work with multiples data sources and RSP engines. It simplifies the execution of complex queries involving temporal operators, aggregation functions, or background knowledge. Comparing to CityBench queries, the results show a considerable improvement of the throughput efficiency. We are working on the automatic split of complex queries Having demonstrated the validity of our framework, we are working on the reasoning and decision making module.

References

1. Dell'Aglio, D., Le Phuoc, D., Le-Tuan, A., Intizar Ali, M., Calbimonte, J.-P.: On a web of data streams (2017)
2. Anicic, D., Fodor, P., Rudolph, S., Stojanovic, N.: EP-SPARQL: a unified language for event processing and stream reasoning. In: Proceedings of the 20th International Conference on World Wide Web, WWW 2011, Hyderabad, India, 28 March–1 April 2011, pp. 635–6441 (2011)
3. Dell'Aglio, D., Calbimonte, J.-P., Balduini, M., Corcho, O., Della Valle, E.: On correctness in RDF stream processor benchmarking. In: Alani, H., et al. (eds.) ISWC 2013. LNCS, vol. 8219, pp. 326–342. Springer, Heidelberg (2013). https://doi.org/10.1007/978-3-642-41338-4_21

4. de Almeida, V.P., Bhowmik, S., Lima, G.F., Endler, M., Rothermel, K.: DSCEP: an infrastructure for decentralized semantic complex event processing. In: IEEE International Conference on Big Data, Big Data 2020, Atlanta, GA, USA, 10–13 December 2020, pp. 391–398 (2020)
5. Puiu, D., et al.: CityPulse: large scale data analytics framework for smart cities. IEEE Access **4**, 1086–1108 (2016)
6. Ali, M.I., Gao, F., Mileo, A.: CityBench: a configurable benchmark to evaluate RSP engines using smart city datasets. In: Arenas, M., et al. (eds.) ISWC 2015. LNCS, vol. 9367, pp. 374–389. Springer, Cham (2015). https://doi.org/10.1007/978-3-319-25010-6_25
7. Calbimonte, J.-P., Corcho, O., Gray, A.J.G.: Enabling ontology-based access to streaming data sources. In: Patel-Schneider, P.F., et al. (eds.) ISWC 2010. LNCS, vol. 6496, pp. 96–111. Springer, Heidelberg (2010). https://doi.org/10.1007/978-3-642-17746-0_7
8. Gillani, S., Zimmermann, A., Picard, G., Laforest, F.: A query language for semantic complex event processing: syntax, semantics and implementation. Semant. Web **10**(1), 53–93 (2019)
9. Zhang, Y., Duc, P.M., Corcho, O., Calbimonte, J.-P.: SRBench: a streaming RDF/SPARQL benchmark. In: Cudré-Mauroux, P., et al. (eds.) ISWC 2012. LNCS, vol. 7649, pp. 641–657. Springer, Heidelberg (2012). https://doi.org/10.1007/978-3-642-35176-1_40
10. Rinne, M., Nuutila, E., Torma, S., Glimm, B., Huynh, D.: INSTANS: high-performance event processing with standard RDF and SPARQL. In: Proceedings of the ISWC 2012 Posters and Demonstrations Track, Boston, USA, 11–15 November 2012, vol. 914 (2012)
11. Calvaresi, D., Calbimonte, J.-P.: Real-time compliant stream processing agents for physical rehabilitation. Sensors **20**(3), 746 (2020)
12. Barbieri, D.F., Braga, D., Ceri, S., Valle, E.D., Grossniklaus, M.: C-SPARQL: a continuous query language for RDF data streams. Int. J. Semant. Comput. **4**(1), 3–25 (2010)
13. Le-Phuoc, D., Nguyen Mau Quoc, H., Le Van, C., Hauswirth, M.: Elastic and Scalable Processing of Linked Stream Data in the Cloud. In: Alani, H., Kagal, L., Fokoue, A., Groth, P., Biemann, C., Parreira, J.X., Aroyo, L., Noy, N., Welty, C., Janowicz, K. (eds.) ISWC 2013. LNCS, vol. 8218, pp. 280–297. Springer, Heidelberg (2013). https://doi.org/10.1007/978-3-642-41335-3_18
14. Le-Phuoc, D., Dao-Tran, M., Pham, M.-D., Boncz, P., Eiter, T., Fink, M.: Linked stream data processing engines: facts and figures. In: Cudré-Mauroux, P., et al. (eds.) ISWC 2012. LNCS, vol. 7650, pp. 300–312. Springer, Heidelberg (2012). https://doi.org/10.1007/978-3-642-35173-0_20
15. Greenwood, D.A.P., Lyell, M., Mallya, A.U., Suguri, H.: The IEEE FIPA approach to integrating software agents and web services. In: 6th International Joint Conference on Autonomous Agents and Multiagent Systems (AAMAS 2007), Honolulu, Hawaii, USA, 14–18 May 2007, p. 276 (2007)
16. Ren, X., et al.: Strider: an adaptive, inference-enabled distributed RDF stream processing engine. Proc. VLDB Endow. **10**(12), 1905–1908 (2017)
17. Le-Phuoc, D., Dao-Tran, M., Xavier Parreira, J., Hauswirth, M.: A native and adaptive approach for unified processing of linked streams and linked data. In: Aroyo, L., et al. (eds.) ISWC 2011. LNCS, vol. 7031, pp. 370–388. Springer, Heidelberg (2011). https://doi.org/10.1007/978-3-642-25073-6_24

VINEVI: A Virtualized Network Vision Architecture for Smart Monitoring of Heterogeneous Applications and Infrastructures

Rodrigo Moreira[1(✉)], Hugo G. V. O. da Cunha[2], Larissa F. Rodrigues Moreira[2], and Flávio de Oliveira Silva[2]

[1] Institute of Exact and Technological Sciences (IEP), Federal University of Viçosa, Rio Paranaíba 38810-000, Brazil
rodrigo@ufv.br

[2] Faculty of Computing (FACOM), Federal University of Uberlândia, Uberlândia 38400-902, Brazil
{hugo.cunha,larissarodrigues,flavio}@ufu.br

Abstract. Monitoring heterogeneous infrastructures and applications is essential to cope with user requirements properly, but it still lacks enhancements. The well-known state-of-the-art methods and tools do not support seamless monitoring of bare-metal, low-cost infrastructures, neither hosted nor virtualized services with fine-grained details. This work proposes VIrtualized NEtwork VIsion architecture (VINEVI), an intelligent method for seamless monitoring heterogeneous infrastructures and applications. The VINEVI architecture advances state of the art with a node-embedded traffic classification agent placing physical and virtualized infrastructures enabling real-time traffic classification. VINEVI combines this real-time traffic classification with well-known tools such as Prometheus and Victoria Metrics to monitor the entire stack from the hardware to the virtualized applications. Experimental results showcased that VINEVI architecture allowed seamless heterogeneous infrastructure monitoring with a higher level of detail beyond literature. Also, our node-embedded real-time Internet traffic classifier evolved with flexibility the methods with monitoring heterogeneous infrastructures seamlessly.

1 Introduction

Understanding how Internet services and resources are used is essential to support the user experience within Service-Level Agreement (SLA) [1]. Among the Internet resources, the cloud computing infrastructures, which use virtualization and handle vast amounts of data generated by users, require the entire stack monitoring from the hardware to the virtualized applications. In 2014 [2], the cloud computing paradigm evolved to Multi-Acces Edge Computing (MEC) to address the challenges of having centralized computing capabilities geographically distant from users [3].

L. Barolli et al. (Eds.): AINA 2022, LNNS 449, pp. 529–541, 2022.
https://doi.org/10.1007/978-3-030-99584-3_46

Monitoring these infrastructures, especially cloud computing, has become essential for maintaining the service's operation, yielding users' service level agreements [4]. Also, monitoring is critical to support visibility regarding their resource consumption behavior, enabling the prediction of outages, perform performance diagnosis, and Service Level Agreement (SLA) violation [5–7]. Furthermore, due to the large amount of data that monitoring tools generate, it becomes challenging to find methods, frameworks, or tools that detail the status of infrastructure entities, especially bottlenecks, without causing a significant overhead on the system [8].

Besides, monitoring heterogeneous infrastructures Methods, technologies, and monitoring strategies found in the literature are predominantly integrated with cloud provider tools or tied to a specific SLA [9–11] and do not take into account container-native monitoring [12].

Hence, this work proposes the **VI**rtualized **NE**twork **VI**sion architecture (VINEVI) framework for seamless monitoring of heterogeneous infrastructures and services. VINEVI provides a set of entities and technological enablers that allow monitoring cloud infrastructures such as heterogeneous bare-metal (x64) and low-cost (AArc64) [13] architectures. VINEVI advances state of the art with a monitoring method based on Artificial Intelligence (AI) that monitors the infrastructures, considering their resources and services. VINEVI enables the monitoring of network traffic volume by application class for each monitored entity and service. Additionally, this paper innovates with a framework that monitors hosted or virtualized services.

Among the contributions of this work, the following stand out:

- A seamless monitoring framework for network entities, hybrid infrastructures, and hosted and virtualized services;
- A traffic volume counter customizable by application class for hybrid architectures;
- An assessment of the performance of CNNs as enabling technologies for real-time sampling network traffic classification.

This work is organized as follows. Section 2 presents the current state of the art on network monitoring. Section 3 presents the VINEVI architecture proposed in this paper, whereas the experimental setup as described in detail in Sect. 4. Section 5 reports results and analysis of the experimental evaluation. Section 6 concludes the paper.

2 Related Work

Currently, the literature presents different monitoring solutions for virtualized networks, and infrastructures [14–16]. Some approaches focus on networks, others on infrastructures, but none rely on seamless monitoring of low-cost and high-performance infrastructures. Also, the well-known monitoring solutions lack monitoring both running on top of virtualized infrastructures or non-virtualized.

This section describes some related work considering the seamless infrastructures and network monitoring capabilities.

Borylo et al. [17] proposed and evaluated a portable monitoring module that combines monitoring capabilities to Network Function Virtualization (NFV), Software-defined Networking (SDN), and Cloud. Their solution architecture receives monitoring statistics through well-known and universal interfaces with SDN controllers, Virtualized Infrastructure Manager (VIM), host, and tenants. Unlike our work, we proposed a seamless infrastructure monitoring that considers a virtualized and non-virtualized infrastructure and its services. Although, we provide fine-grained statistical monitoring of the entire network and infrastructure.

In [18] contains the description of a seamless platform based on Prometheus and Grafana for the deployment and monitoring of containers over infrastructures. Unlike our proposal, the authors focused on monitoring containers to achieve availability and OPEX reduction. Although our monitoring proposal can monitor container-based services, we go further by intelligently monitoring network traffic. VINEVI monitors bare-metal, low-cost infrastructures, network entities, and services transparently, unlike previous approaches. Similarly, we find other monitoring architectures in [18] which take into account logs [19] while [20] takes into account flow rules.

Won and Kim [21] believe that open source tools like Prometheus and Zabbix require configuration and rely on empirical knowledge about failures, requiring administrators to set accurate thresholds for each situation. They proposed an intelligent Multi-Layer monitoring architecture based on Prometheus and machine learning to address these challenges. The authors evaluated the efficiency of some machine learning models such as Random Forest (RF), Support Vector Machine (SVM), and Deep Neural Network (DNN) for monitoring tasks. The results suggest that DNN proved to be a promising technology with reasonable accuracy for predicting CPU, RAM, and network failures. Our paper also deals with statistics gathering and proactive submission to related endpoints. However, the VINEVI framework thoroughly monitors the infrastructure, which can be bare-metal, low-cost, network entities, virtualized, or hosted services on top of those infrastructures.

In [22] we find DynAMo, an alternative monitoring approach to Prometheus that causes low computational overhead on monitored entities and services. The use case used in the evaluation refers to the communication service integrated into a railway company simulated in a software environment. Among the results reported, the low memory consumption compared to monitoring based on Prometheus stands out. Unlike the authors, VINEVI, in addition to monitoring containers in production environments, monitors the consumption of network resources of containers.

3 VINEVI Architecture

This work proposes a method for seamlessly monitoring bare-metal, low-cost infrastructures and network elements. In addition to dealing with the infrastructure, our solution can deal with the services that run on these infrastructures, directly monitoring hosted, virtualized, and nested-virtualized services.

VINEVI monitors the network elements and the traffic volume detailed by application class, services, and computing infrastructure. We present the conceptual diagram of the VINEVI framework in Fig. 1 and read it from left to right, and we see network resources, data center infrastructure, and low-cost infrastructures.

In the upper flow, the blue arrow refers to the collection of information about network traffic of resources and services running on the infrastructures. In Fig. 1 the *s0/3* interface denotes the monitoring of the traffic volume of the router network interface, which can occur live or by samples. These samples feed the previously built [23] Packet Vision component for traffic classification based on *Convolution Neural Network (CNN)*.

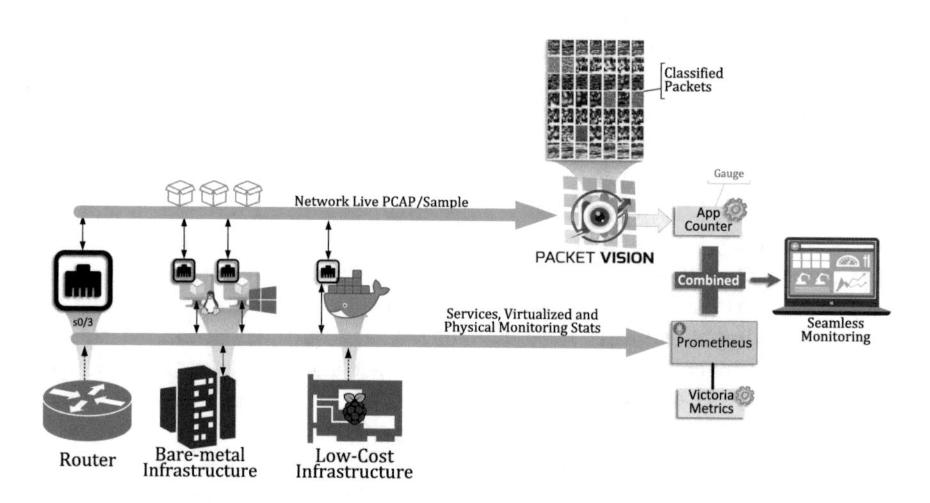

Fig. 1. VINEVI monitoring schema.

In the lower flow, information regarding the physical resources of the infrastructure feeds the event-driven statistics collector based on Prometheus [24]. In addition to physical information, the metrics collector receives information related to the service, depending on the service operator specifying what will be monitored from the application and pointing to the corresponding end-point. In addition to monitoring the infrastructure, VINEVI can monitor the service orchestrators of these infrastructures and the Virtualized Infrastructure Managers (VIMs).

The VINEVI framework combines service and resource monitoring metrics with specified traffic volume by application class. When combined, we bring novelty with a seamless monitoring method that allows administrators and the infrastructure operations team to monitor their resources adaptively and granularly.

A critical component of the VINEVI framework is PacketVision [23], which receives sampled or live stream network packets and classifies them according to their application class. In VINEVI, we build a gauge-type counter for each predominant Internet traffic class. This counter decreases or increases over time depending on the current state of the network. The traffic classes for our VINEVI Proof-of-Concept (POC) are detailed in Sect. 4.2. Another fundamental component of VINEVI is Prometheus or Victoria Metrics which receive metrics by end-points and store them for temporal analysis.

For the VINEVI framework to consolidate the various monitoring metrics, each monitored entity must have a statistics publisher that runs as a daemon. This monitoring entity sends the metrics to the corresponding endpoints in Prometheus.

4 Experimental Setup

To functionally validate VINEVI, we deploy it in an experimental testbed. This testbed consisted of four (4) distinct hosts: the Monitor Server, Experimental Server, Orchestrator Server, and AI Server as in Fig. 2. Among these hosts represent the infrastructure in the VINEVI framework, the Experimental Server, a bare-metal with four vCPUs and 8 GB RAM with Ubuntu 18.04 LTS. Another type of Experimental Server admitted by VINEVI is the low-cost one. A Raspberry Pi4 is hosting virtual machines and is managed by Orchestrator Service.

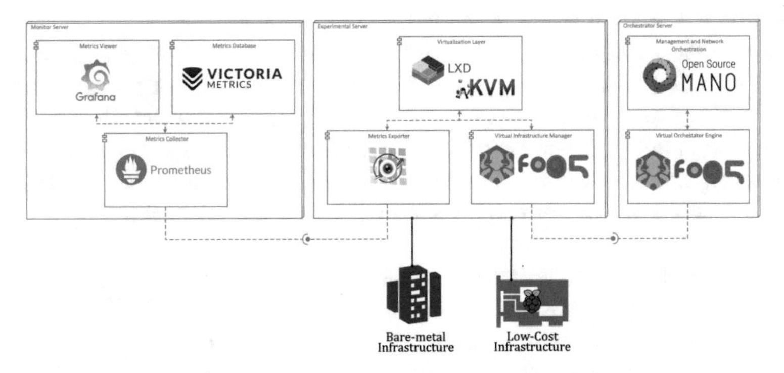

Fig. 2. Intelligent monitoring testbed overview.

The Monitor Server hosts the platform for viewing and supporting the monitoring data. We configure Grafana, Prometheus, and Victoria Metrics services to handle monitored entities. Besides, we configure a Virtual Private Network (VPN) service so that metric collectors publish metrics to endpoints correctly, even when they are outside the network domain. Part of our VINEVI experimental setup was deployed on the Microsoft Azure cloud computing platform with Ubuntu 18.04 LTS with a flavor 2vCPU and 8 GB RAM.

On the other hand, Orchestrator Server runs on a bare-metal server. Among the roles of this server, the implementation of virtualized services on bare-metal and low-cost infrastructure stands out. To seamlessly deploy virtual machines on bare-metal (x86_x64) and low-cost (AArc64) infrastructures, we configured the open-source Eclipse fog05 virtual Orchestrator. This Orchestrator deals directly with the Fog5 agent running on top of Experimental Servers enabling them to launch virtual machines or containers.

4.1 Smart Traffic Monitoring

For VINEVI to monitor the volume of network traffic by application class, we incorporated the capabilities of CNNs into the framework through Packet Vision. Thus, VINEVI has an intelligent node agent for monitoring network traffic. We train CNNs to classify network traffic into seven (7) typical Internet classes: Bittorrent, Browsing, DNS, IoT, RDP, SSH, and VoIP. In this experimental setup we considered three (3) CNNs to classify traffic on the VINEVI framework: SqueezeNet [25], ResNet-18 [26] and MobileNetV2 [27].

Among these three CNNs, the SqueezeNet and MobileNetV2 architectures were explicitly designed for use in mobile and edge devices, so we hypothesized that they are good candidates to compose the VINEVI monitoring agent. SqueezeNet aimed to reduce the number of parameters through fire modules, which use the strategy of compression and expansion of activation maps in the convolution layers [25]. MobileNetV2 [27] uses separable depth convolutions, which consists of factoring the standard convolution into a depth convolution followed by a 1×1 convolution, called a point convolution.

The literature claims that it is not promising to embed large models such as AlexNet or VGG-16 in small devices because they demand high computational load [28]. However, we decided to investigate how the ResNet-18 architecture, which is the smallest network in the ResNet family (composed of $\approx$11 million parameters), behaves when embedded. ResNet is composed of residual blocks that allow accelerating convergence and better deal with the [26] overfitting problem.

A premise of the VINEVI framework is a previous CNN training to enable traffic prediction by application class for the monitoring agent. Once the model has been trained, the CNN has uploaded to the VINEVI intelligent traffic monitoring agent for future on-the-fly predictions on the infrastructures. The Torch framework allows loading previously trained models, avoiding training bottlenecks. Furthermore, due to the hardware restrictions of low-cost infrastructures, embedding the previous trained CNN model proves to be advantageous and functionally correct.

We adopted this strategy to enable the smart traffic monitoring agent on RPi4 to enrich infrastructure monitoring with application class details on traffic volume. We configured PyTorch 1.10.0 to run on AArc64 and x64 hardware, thus enabling the loading of an already trained CNN and for the monitoring agent to predict the sample packages' application class regardless of the architecture or underlying architecture infrastructure.

4.2 Dataset Description

The dataset images evaluated in this paper were built from the Packet Vision [23] component, which considered the raw information carried in the packet, including header and payload. About 9645 images were obtained from Wireshark traces (*pcap*) from different sources. These images are categorized into seven (7) classes, as summarized in Table 1. Figure 3 shows some images from the dataset for each class. All images are in PNG format with 224×224 pixels size.

Table 1. Descriptive summary of the images.

Class	Samples	Source
Bit torrent	1217	UPC data [29]
DNS	1412	
Browsing	1225	ISCXVPN2016 [30]
RDP	1271	
SSH	1352	
IoT	1848	IoT Sentinel [31]
VoIP	1320	NASOR [32]
Total	**9645**	

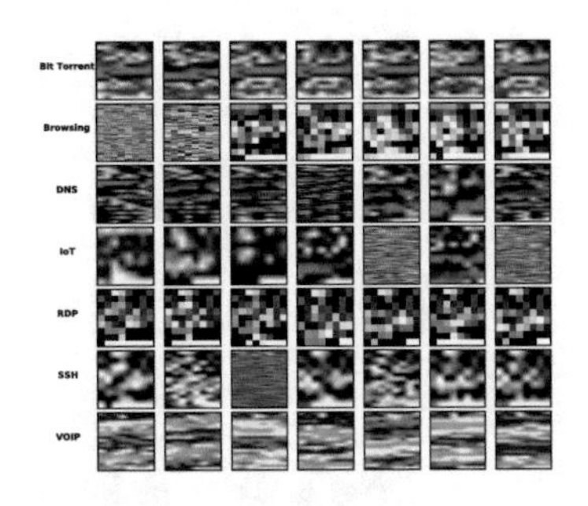

Fig. 3. Examples of images for each class.

5 Results and Discussion

Considering the testbed presented above, we carried experiments to validate the VINEVI framework functionally. The objectives of the experiments were to answer the following questions:

1. For the VINEVI testbed, which CNN outperforms regarding accuracy for real-time network traffic prediction by application class?
2. For a real-time traffic prediction environment, where the prediction time of packets is essential, which CNN is suitable to compose the intelligent monitoring agent in AArc64 architectures?
3. When does VINEVI intelligent monitoring agent run on x64 architectures, which CNN is best suited for traffic prediction by application classes?
4. Is there differentiation in CPU consumption in the traffic class prediction process depending on CNN type and traffic class?

For VINEVI's experimental setup, and considering the dataset described above, the CNN that performed best regarding accuracy was MobileNet. The numerical results of the learning and testing process denoted that the accuracy of this CNN was 99.90%. The learning and prediction behavior, according to Fig. 4, implies the generalization and aptitude of the model to compose the VINEVI smart traffic monitoring agent.

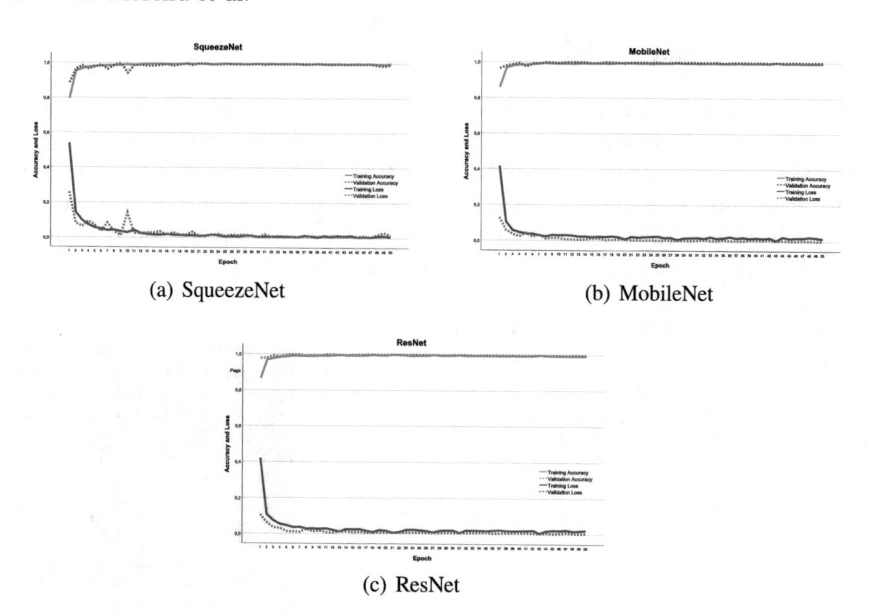

(a) SqueezeNet

(b) MobileNet

(c) ResNet

Fig. 4. Training and loss values evolution for each CNN model.

When we implemented the intelligent traffic classification agent running on low-cost and bare-metal infrastructures, we noticed a different behavior for the same CNN. Intelligently monitoring the traffic class of applications on low-cost infrastructures (AArc64) can be done by pooling sampling, not in real-time because of the large amount of data in a production network. To predict a single packet regarding its application class on AArc64 architectures required ≈811 ms using ResNet. Thus, ResNet was the CNN that required the shortest time to predict the application class of a given package in AArc64.

When the intelligent traffic prediction agent ran on bare-metal (x64) infrastructures, ResNet demanded the longest prediction time ≈77 ms, behaving differently than AArc64 architecture. In the x64 architecture, the CNN that consumed the least amount of time to predict the application class of a given package was SqueezeNet, demanding around ≈62 ms, according to Fig. 5. In Fig. 6 it can be seen that CNN ResNet took about 9.5× less time than SqueezeNet to predict the traffic class of the same sampled network packet.

Accordingly, for VINEVI's testbed, the CNN outperforms prediction time depending on the architecture where the intelligent network traffic monitoring agent stands out. If deployed over low-cost infrastructures, the best CNN to be used will be ResNet; if deployed on bare-metal, the most recommended is SqueezeNet.

We investigate how the placement of intelligent traffic monitoring agents over low-cost infrastructure impacts the consumption of computational resources. In this investigation, we consider how and to what extent the prediction of different classes of applications by the three (3) CNNs considered in the VINEVI testbed

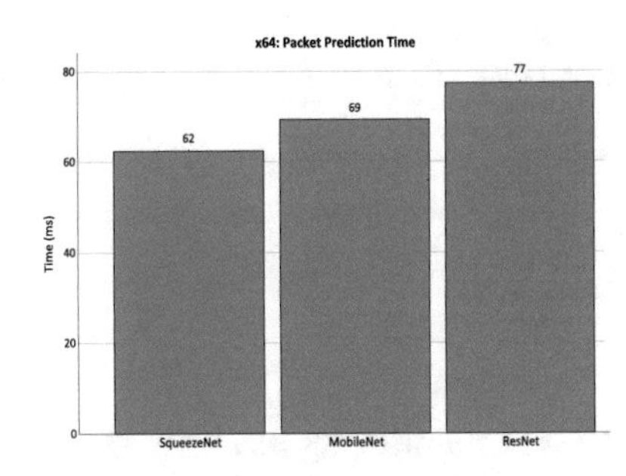

Fig. 5. x64: prediction time.

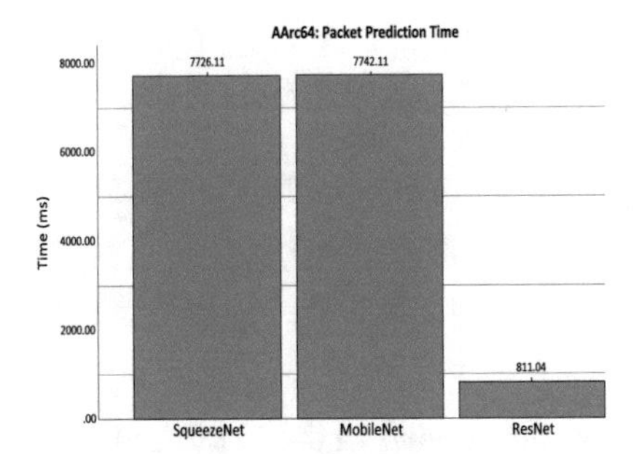

Fig. 6. AArch64: prediction time.

affects the CPU consumption of low-cost infrastructures. According to Fig. 7 it is possible to see that MobileNet and SqueezeNet consumed on average and within the confidence interval the same amount of CPU for all application classes.

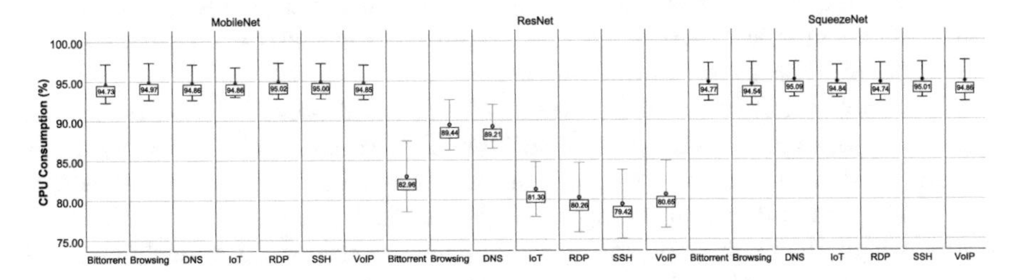

Fig. 7. Comparison of CPU consumption by classification agents.

However, with ResNet, it was observed that the average CPU consumption was ≈82.49%, being ≈13% lower compared to other CNNs. We believe that this happened because in the process and prediction, having the CNNs already been trained, the last SoftMax layer is responsible for the forecast. Since this layer is less complex, it implies lower CPU consumption. Therefore, incorporating this CNN in the intelligent traffic monitoring agent to monitor low-cost infrastructures is proved to be more suitable.

We also assess the complexity of each CNN, and according to [33] the complexity of a CNN is measured by the amount of multiplication and addition operations that are required for the underlying architecture to perform a computation [34]. According to Table 2, the last layer of ResNet is less complex in terms of FLOPS compared with others in the same experiment. Due to this lesser complexity, we argue that the ResNet consumed less CPU than others. Also, this result suggests that embedded systems can provide better performance during feedforward prediction [35].

The VINEVI framework showed to be adaptable to monitor network traffic and heterogeneous infrastructure resources seamlessly, not impeding various infrastructures that exist on the Internet. Thus, when we combine the traffic

Table 2. Complexity of the last layer of CNNs.

CNN	No. of parameters	Complexity (%)
SqueezeNet	0.004 M	0.082
MobileNet	0.009 M	0.003
ResNet	0.004 M	0.000

class predictor with well-established state-of-the-art monitoring mechanisms, we arrive at an AI-based solution that seamlessly monitors heterogeneous infrastructure and services.

6 Conclusion

This work introduced the VINEVI framework for seamless monitoring of network traffic, hybrid infrastructures, and hosted or virtualized services. By combining the monitoring capabilities of well-established state-of-the-art tools [13] with artificial intelligence technologies, we enrich state-of-the-art with detailed monitoring of hybrid entities and services.

We proposed and functionally evaluated a counter gauge for monitoring network traffic volume by class of applications. This counter relies on CNN-based network traffic classification. In addition, we assessed the placement of the network traffic prediction agent on top of possible VINEVI monitoring architectures. We found that the traffic monitoring module must consider the underlying architecture type to load the CNN model, which takes the least time and consumes the least CPU to predict the application class of a given packet. Furthermore, seamless monitoring of infrastructures or services requires flexible solutions that are adaptable to the environment regardless of vendor, hardware, or software.

For future work, urges to try and validate other AI techniques for traffic prediction. In addition, we consider it essential to create standardized interfaces for infrastructure operation and automation solutions to use metrics monitored by VINEVI to positively impact metrics such as Mean Time to Failure (MTTF) or Mean Time to Recovery (MTR). Furthermore, we consider it necessary to study intelligent pooling and network sampling mechanisms for estimating traffic volume, class of applications, and others.

Acknowledgements. This study was financed in part by the Coordenação de Aperfeiçoamento de Pessoal de Nível Superior - Brasil (CAPES) - Finance Code 001. And we would like to thank National Education and Research Network (RNP) for financial support under the CT-Mon call.

References

1. Singh, M., Baranwal, G.: Quality of service (QoS) in internet of things. In: 2018 3rd International Conference On Internet of Things: Smart Innovation and Usages (IoT-SIU), pp. 1–6 (2018)

2. ETSI Mobile-edge Computing (MEC) Industry Initiative: Mobile Edge Computing-introductory Technical White paper. etsi2014mobile (2014)
3. Porambage, P., Okwuibe, J., Liyanage, M., Ylianttila, M., Taleb, T.: Survey on multi-access edge computing for internet of things realization. IEEE Commun. Surv. Tutor. **20**(4), 2961–2991 (2018)
4. Birje, M.N., Bulla, C.: Commercial and open source cloud monitoring tools: a review. In: Satapathy, S.C., Raju, K.S., Shyamala, K., Krishna, D.R., Favorskaya, M.N. (eds.) Advances in Decision Sciences, Image Processing, Security and Computer Vision. LAIS, vol. 3, pp. 480–490. Springer, Cham (2020). https://doi.org/10.1007/978-3-030-24322-7_59
5. D'Alconzo, A., Drago, I., Morichetta, A., Mellia, M., Casas, P.: A survey on big data for network traffic monitoring and analysis. IEEE Trans. Netw. Serv. Manag. **16**(3), 800–813 (2019)
6. Kontsek, M., Moravcik, M., Jurc, J., Sterbak, M.: Survey of the monitoring tools suitable for CC environment. In: 2020 18th International Conference on Emerging eLearning Technologies and Applications (ICETA), pp. 330–335 (2020)
7. Calzarossa, M.C., Massari, L., Tessera, D.: Performance Monitoring Guidelines, pp. 109–114. Association for Computing Machinery, New York (2021)
8. Popiolek, P.F., dos Santos Machado, K., Mendizabal, O.M.: Low overhead performance monitoring for shared infrastructures. Expert Syst. Appl. **171**, 114558 (2021)
9. Dawson, D., Desmarais, R., Kienle, H.M., Müller, H.A.: Monitoring in adaptive systems using reflection. In: Proceedings of the 2008 International Workshop on Software Engineering for Adaptive and Self-Managing Systems, SEAMS 2008, pp. 81–88. Association for Computing Machinery, New York (2008)
10. Müller, C., et al.: Comprehensive explanation of SLA violations at runtime. IEEE Trans. Serv. Comput. **7**(2), 168–183 (2014)
11. Cedillo, P., Insfran, E., Abrahão, S., Vanderdonckt, J.: Empirical evaluation of a method for monitoring cloud services based on models at runtime. IEEE Access **9**, 55898–55919 (2021)
12. Taherizadeh, S., Jones, A.C., Taylor, I., Zhao, Z., Stankovski, V.: Monitoring self-adaptive applications within edge computing frameworks: a state-of-the-art review. J. Syst. Softw. **136**, 19–38 (2018)
13. da Cunha, H.G.V.O., Moreira, R., de Oliveira Silva, F.: A comparative study between containerization and full-virtualization of virtualized everything functions in edge computing. In: Barolli, L., Woungang, I., Enokido, T. (eds.) AINA 2021. LNNS, vol. 226, pp. 771–782. Springer, Cham (2021). https://doi.org/10.1007/978-3-030-75075-6_63
14. Kreutz, D., Ramos, F.M.V., Veríssimo, P.E., Rothenberg, C.E., Azodolmolky, S., Uhlig, S.: Software-defined networking: a comprehensive survey. Proc. IEEE **103**(1), 14–76 (2015)
15. Tsai, P.-W., Tsai, C.-W., Hsu, C.-W., Yang, C.-S.: Network monitoring in software-defined networking: a review. IEEE Syst. J. **12**(4), 3958–3969 (2018)
16. Zhou, D., Yan, Z., Fu, Y., Yao, Z.: A survey on network data collection. J. Netw. Comput. Appl. **116**, 9–23 (2018)
17. Borylo, P., Davoli, G., Rzepka, M., Lason, A., Cerroni, W.: Unified and standalone monitoring module for NFV/SDN infrastructures. J. Netw. Comput. Appl. **175**, 102934 (2021)
18. Mfula, H., Ylä-Jääski, A., Nurminen, J.K.: Seamless kubernetes cluster management in multi-cloud and edge 5G applications. In: International Conference on High Performance Computing and Simulation (HPCS 2020) (2021)

19. Sukhija, N., et al.: Event management and monitoring framework for HPC environments using ServiceNow and Prometheus. In: Proceedings of the 12th International Conference on Management of Digital EcoSystems, MEDES 2020, pp. 149–156. Association for Computing Machinery, New York (2020)
20. Bali, A., Gherbi, A.: Rule based lightweight approach for resources monitoring on IoT edge devices. In: Proceedings of the 5th International Workshop on Container Technologies and Container Clouds, WOC 2019, pp. 43–48. Association for Computing Machinery, New York (2019)
21. Won, H., Kim, Y.: Performance analysis of machine learning based fault detection for cloud infrastructure. In: 2021 International Conference on Information Networking (ICOIN), pp. 877–880 (2021)
22. Moeyersons, J., Kerkhove, S., Wauters, T., De Turck, F., Volckaert, B.: Towards cloud-based unobtrusive monitoring in remote multi-vendor environments. Softw. Practice Exp. (2021)
23. Moreira, R., Rodrigues, L.F., Rosa, P.F., Aguiar, R.L., d. Oliveira Silva, F.: Packet vision: a convolutional neural network approach for network traffic classification. In: 2020 33rd SIBGRAPI Conference on Graphics, Patterns and Images (SIBGRAPI), pp. 256–263 (2020)
24. Sukhija, N., Bautista, E.: Towards a framework for monitoring and analyzing high performance computing environments using kubernetes and prometheus. In: 2019 IEEE SmartWorld, Ubiquitous Intelligence Computing, Advanced Trusted Computing, Scalable Computing Communications, Cloud Big Data Computing, Internet of People and Smart City Innovation (SmartWorld/SCALCOM/UIC/ATC/CBDCom/IOP/SCI), pp. 257–262 (2019)
25. Iandola, F.N., Moskewicz, M.W., Ashraf, K., Han, S., Dally, W.J., Keutzer, K.: SqueezeNet: alexNet-level accuracy with 50x fewer parameters and <1 mb model size. CoRR, abs/1602.07360 (2016)
26. He, K., Zhang, X., Ren, S., Sun, J.: Deep residual learning for image recognition. CoRR, abs/1512.03385 (2015)
27. Sandler, M., Howard, A., Zhu, M., Zhmoginov, A., Chen, L.-C.: MobileNetv2: inverted residuals and linear bottlenecks. In:Proceedings of the IEEE Conference on Computer Vision and Pattern Recognition (CVPR), June 2018
28. Alippi, C., Disabato, S., Roveri, M.: Moving convolutional neural networks to embedded systems: the alexnet and VGG-16 case. In: 2018 17th ACM/IEEE International Conference on Information Processing in Sensor Networks (IPSN), pp. 212–223 (2018)
29. Carela-Español, V., Bujlow, T., Barlet-Ros, P.: Is our ground-truth for traffic classification reliable? In: Faloutsos, M., Kuzmanovic, A. (eds.) PAM 2014. LNCS, vol. 8362, pp. 98–108. Springer, Cham (2014). https://doi.org/10.1007/978-3-319-04918-2_10
30. Draper-Gil, G., Lashkari, A.H., Mamun, M.S.I., Ghorbani, A.A.: Characterization of encrypted and VPN traffic using time-related features. In: Proceedings of the 2nd International Conference on Information Systems Security and Privacy - Volume 1: ICISSP, pp. 407–414. INSTICC, SciTePress (2016)
31. Miettinen, M., Marchal, S., Hafeez, I., Asokan, N., Sadeghi, A.-R., Tarkoma, S.: IoT sentinel: automated device-type identification for security enforcement in IoT. In: 2017 IEEE 37th International Conference on Distributed Computing Systems (ICDCS), pp. 2177–2184 (2017)
32. Moreira, R., Rosa, P.F., Aguiar, R.L.A., de Oliveira Silva, F., Nasor: a network slicing approach for multiple autonomous systems. Comput. Commun. **179**, 131–144 (2021)

33. Zhang, X., Zhou, X., Lin, M., Sun, J.: ShuffleNet: an extremely efficient convolutional neural network for mobile devices. In: Proceedings of the IEEE Conference on Computer Vision and Pattern Recognition (CVPR), June 2018
34. Ma, N., Zhang, X., Zheng, H.T., Sun, J.: ShuffleNet V2: practical guidelines for efficient CNN architecture design. In: Proceedings of the European Conference on Computer Vision (ECCV), September 2018
35. Dundar, A., Jin, J., Martini, B., Culurciello, E.: Embedded streaming deep neural networks accelerator with applications. IEEE Trans. Neural Netw. Learn. Syst. **28**(7), 1572–1583 (2017)

Parallel IFFT/FFT for MIMO-OFDM LTE on NoC-Based FPGA

Kais Jallouli[1(✉)], Azer Hasnaoui[1], Jean-Philippe Diguet[2], Alireza Monemi[3], and Salem Hasnaoui[1]

[1] University of Tunis El Manar, Tunis, Tunisia
kais.jallouli@enit.utm.tn, salem.hasnaoui@enit.rnu.tn
[2] CNRS, IRL CROSSING, Adelaide, Australia
jean-philippe.diguet@cnrs.fr
[3] Barcelona Supercomputing Center, Barcelona, Spain
alireza.monemi@bsc.es

Abstract. The evaluation of wireless communication systems over the last decades has led to a growing demand for more advanced high-speed communication systems. In this paper, we propose a hardware workflow developed for implementing the Long Term Evolution (LTE) communication system. This work studies the Multiple-input, multiple-output orthogonal frequency-division multiplexing (MIMO-OFDM) LTE system. The main focus of this work is on implementation of OFDM modulation/demodulation functions as they are the main contributors to the processing time and latency in high speed communication systems. To achieve this goal, a multicore low latency OFDM LTE system is proposed. The multicore RTL code is generated using the ProNoC tool. The main contribution of this system the archived speed up in OFDM LTE computation using parallel processing techniques on an NoC based multicore system. The speed-up comparison for systems having different numbers of cores computing the IFFT task are reported in this paper. The proposed multicore system is also compared with a single-core system as a reference design. Systems having different LTE OFDM configurations are synthesized, implemented and verified using Altera Stratix V GX FPGA. The application execution time and FPGA resource utilization are used as compassion metrics. The proposed multicore LTE OFDM systems having 2 and 16 processing tiles computing IFFT tasks on different LTE channel bandwidths, the execution time is reduced by 24% and 76%, respectively compared to a conventional LTE OFDM system that is running on a single-core system.

1 Introduction

Digital communications are based on various important pillars including big data, mobility, Internet of Things (IoT). With the advent of 5G, there has been a growing interest in the study of connected objects [1]. IoT hinges on the current and evolving interoperable information and communication technologies. Its success is due to the evolution of hardware devices and communication technologies, most notably wireless. In this case, developing a communication system for these objects has become

L. Barolli et al. (Eds.): AINA 2022, LNNS 449, pp. 542–553, 2022.
https://doi.org/10.1007/978-3-030-99584-3_47

paramount [2]. Thus, In IoT, we need wireless communication systems providing very low latency, high bandwidth, and low energy consumption as well as the establishment of high-performance processing units offering both low computing power and short development time. Orthogonal Frequency Division Multiplexing (OFDM) is a multicarrier modulation method used especially for high data rate mobile transmission systems such as LTE and obviously 5G [3]. OFDM modulation/demodulation functions are based on Inverse Fast Fourier Transform (IFFT) and Fast Fourier Transform (FFT) [4], respectively. In this work, we use the Long Term Evolution (LTE) which is one of the wireless communication systems offering high-speed data transmission based on OFDM technology [5]. LTE communication aims to respond adequately to growing demands for video conferencing, video streaming, gaming applications, and fast transfer in mobile applications. However, its complex protocol imposes strict constraints on the lower communications layers where the implementation of operations with high demand in computation is a challenge that applicable for both IFFT and FFT. These two building blocks are the most computationally intensive modules compared to other modules in LTE system [6]. These two crucial modules are not only computationally complex, but also the major latency bottleneck. The IFFT and FFT operations must be repeatedly performed in very precise calculation time to respect the LTE standard [7]. The need for real-time execution and the hardware complexity continues to increase with the evolution of wireless communication systems. Therefore, designing communication systems that meet the requirements of speed, latency, power consumption and that are less complex in terms of hardware implementation has become a necessity for IoT technology. Therefore, the computation time of IFFT algorithm should be reduced [8]. By relying on a distributed programming model, multicore architectures are obviously more effective than a single core architecture. Subsequently, we can exploit that IFFT is a parallelizable algorithm. The contribution of this paper is the development of a multicore low latency OFDM LTE system that parallelizes the IFFT task on a NoC-based multicore platform that is an advantage over the conventional OFDM system.

2 Related Works

In digital signal processing, Discrete Fourier Transform (DFT) has always played a significant role in several technological fields such as wireless communication networks based on OFDM system [9]. The DFT of a discrete signal $x(n)$ can be directly calculated by the equation

$$X(K) = \sum_{n=0}^{N-1} x(n) W_N^{Kn} \tag{1}$$

with $x(n)$ and $X(k)$, respectively, input sequence and output sequence. The two sequences $x(n)$ and $X(k)$ are sequences of complex numbers. N is the length of the DFT. $W_N^{Kn} = e^{-j2\pi kn}$ are the Twiddle Factor.

The calculation of DFT needs an important number of complex operations. The time complexity of DFT is $O(N^2)$. It becomes expensive for large N. A faster algorithm for the calculation of the DFT is the Fast Fourier Transform. It demands $N \log_2 N$ operations instead of N^2 for direct calculation of the DFT where N is the length of IFFT. An enormous and serious problem for the real-time application MIMO OFDM LTE is still posed due the very high time complexity of IFFT. It is crucial to efficiently reduce the time complexity of IFFT. IFFT is a parallelizable algorithm but its parallelization needs communications. An efficient implementation of IFFT/FFT modules is required. A solution consists in the parallel execution of IFFT/FFT modules on p processors. Besides, NoC is the most adequate solution given the interconnection requirements of advanced MPSoC [10, 11]. However, NoCs pose numerous challenges related to their architectures. Communications between processing tiles must be performed in a reliable and transparent manner with minimal latency, high data throughput, minimal silicon surface area and low power consumption. The aim of this paper is to overcome these challenges in order to design and validate a NoC architecture for the MIMO OFDM LTE type radio telecommunications standard. A NoC FPGA prototype platform is necessary to evaluate the performance of the target application MIMO-OFDM LTE system. FPGA devices are used for emulation and testing architectures in several applications requiring rapid prototyping of complex digital circuit such as digital communications. The development and validation of a SoC is made easier thanks to these devices. Nowadays, the implementation of a SoC with a limited number of IPs and the interconnection between them via a NoC is within reach. Our main contribution in this paper consists in the implementation of OFDM LTE system where the parallel execution of IFFT is our main challenge. Execution time is our major concern in our implementation choice of OFDM system. The rest of the paper is structured as follows: Sect. 3 depicts our hardware workflow based on a heterogeneous NoC prototype platform. Section 4 presents our proposed parallelization approach for the OFDM LTE system. Section 5 presents our implementation results. Finally, Sect. 6 gives a conclusion of our paper.

3 Hardware Workflow

Designing our own NoC for the target MIMO-OFDM LTE system seems unnecessary as NoCs are already available in synthesizable hardware description language (VHDL, Verilog). Consequently, the reuse of a dedicated NoC and the use of a platform for the design and generation of a NoC will be integrated as a hardware workflow given that NoCs are flexible, generic and scalable communication architectures. In this paper, we have chosen the ProNoC prototyping tool as a hardware workflow. This tool presents the desired characteristics of future NoCs and is available in Verilog synthesizable and can be easily parameterized. It has enabled us to automate tedious tasks and thus increase productivity in terms of NoC design.

The hardware workflow of ProNoC [12] platform as illustrated in Fig. 1 consists of 4 steps: Interface generator, IP generator, PT generator, and NoC-based MPSoC Generator. Each step corresponds to every layer in the NoC design architecture. An Interface Generator consists of a combination of several ports providing specific functionality. An IP Generator consists of making a library for every intellectual property (IP) such as processor, bus, timer, memory. A processing tile consists of a combination of different IPs that can be connected together. This tool generates the complete RTL code of the processing tile which can be used subsequently in the final step. Finally, the NoC-based MPSoC generates a complete RTL code (Verilog) of a NoC (routers, network interface, Links) and Processing Tiles.

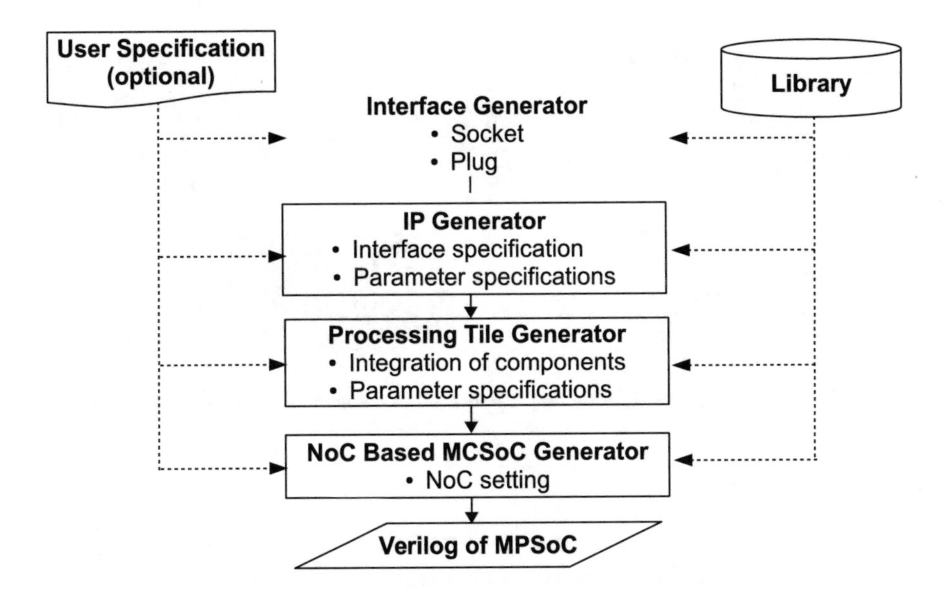

Fig. 1. ProNoC hardware design flow

3.1 SoC Generation

We have generated a SoC architecture implemented in FPGA device for the conventional OFDM LTE system in a single core using ProNoC tool as illustrated in Fig. 2. It contains a mor1kx processor [13], a memory, a wishbone bus, a timer and an UART. It is used to compare the execution time of conventional OFDM with multicore OFDM LTE systems having a different number of processing tiles that compute IFFT task applied to NoC. Then, we write the C code of the conventional LTE OFDM system, the mor1kx GNU tool chain is used to compile it. After that, we simulate the SoC architecture system using verilator when it runs the conventional LTE OFDM system. Using software tool Quartus Prime 18.1, the SoC architecture is synthesized so as to generate the SOF programming file and emulation with Altera Stratix V GX FPGA. After validation of the conventional OFDM LTE system, we move to the design methodology to generate a NoC architecture for a multicore OFDM LTE system.

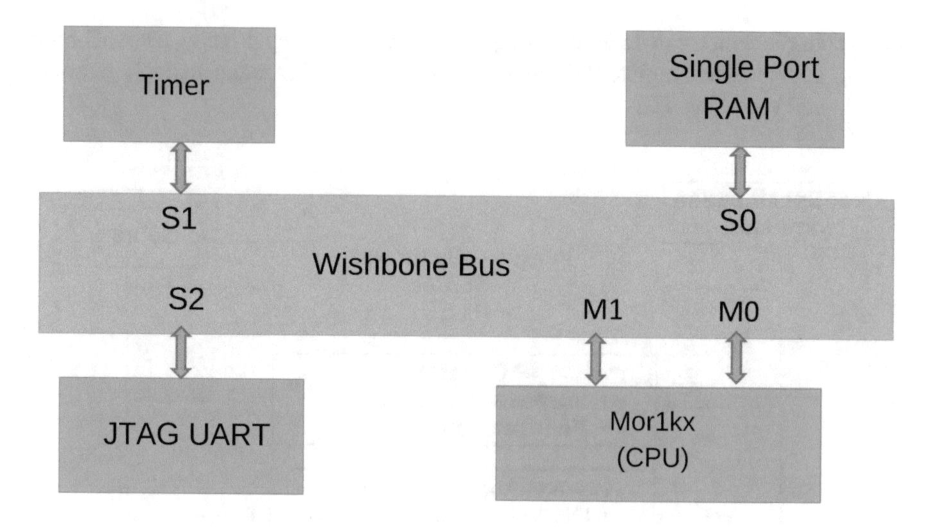

Fig. 2. SoC architecture used for the implementation of a conventional OFDM LTE system.

3.2 NoC Generation

There are 8 steps in the design methodology for the generation of a NoC design architecture implemented in FPGA device for our multicore LTE OFDM system which are (1) the tile generation process, (2) network-on-chip configurations (3), network-on-chip generation, (4) parallelization of the LTE OFDM system, (5) compilation of parallel LTE OFDM system, (6) NoC design verification (7), NoC design synthesis and (8) NoC implementation. Firstly, we have generated the processing tile that can be connected to our NoC architecture as illustrated in Fig. 3. Then, we configure our NoC architecture such as topology, number of tiles, number of virtual channel per port, routing algorithm. In the next step, we generate the complete RTL code of the NoC architecture. Then, we parallelize the OFDM LTE system by considering our application as a mixture of serial and parallel modules in which IFFT module is running in p processing tiles based on our parallel IFFT algorithm detailed in Sect. 5. Then, we compile the C code of each processing tiles using the processor mor1kx GNU toolchain. Then, we simulate the whole NoC using Verilator simulator [14]. In the next step, the NoC design architecture is synthesized using the Intel Quartus prime 18.1. Finally, the complete NoC is implemented to Altera Stratix V GX FPGA using the USB blaster II. According to performance evaluation, in which we mainly focus on execution time and resource occupation in FPGA, the number of processing tiles computing IFFT task can be varied.

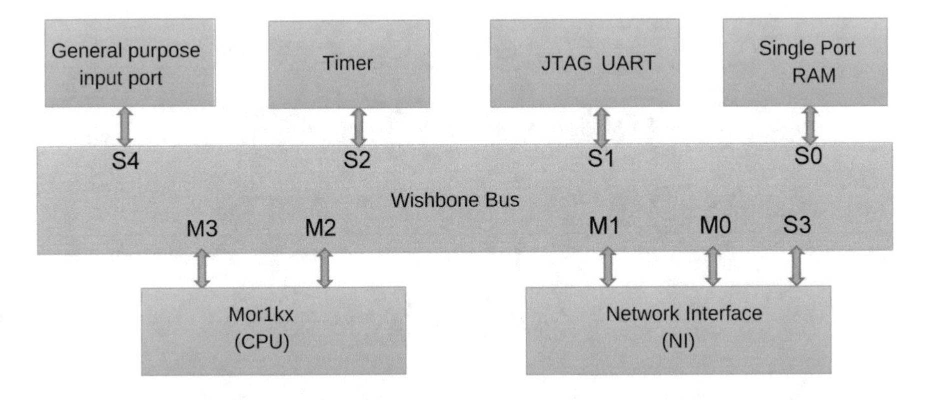

Fig. 3. :Processing tile architecture used in the NoC design.

4 Proposed Parallelization Approach

Figure 4 illustrates an overview of our proposed parallelization approach for a multicore OFDM LTE system. Our application can be considered as a mixture of serial and parallel blocks. Based on a NoC architecture, the multicore OFDM system is decomposed into two sequential blocks and one parallel block. Each of two sequential blocks should be running in one processing tile such as packing input data, inserting DC component, adding cyclic prefix. A parallel block needs to be performed using p processing tiles to compute the IFFT algorithm. We can improve the overall system performance of our OFDM by increasing the number of processing tiles computing the IFFT algorithm to reach the high computational requirement of real-time application MIMO-OFDM. To achieve this, we propose a parallel IFFT/FFT algorithm used for the implementation of OFDM system.

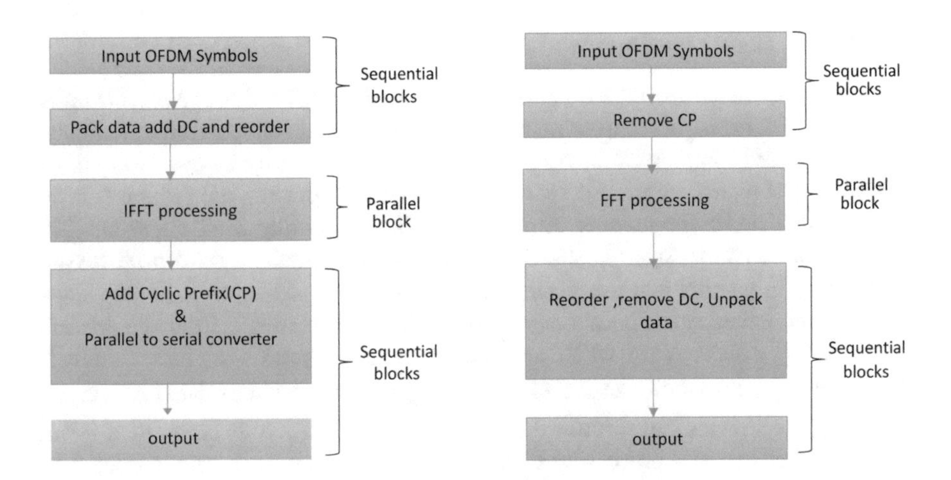

Fig. 4. OFDM LTE system.

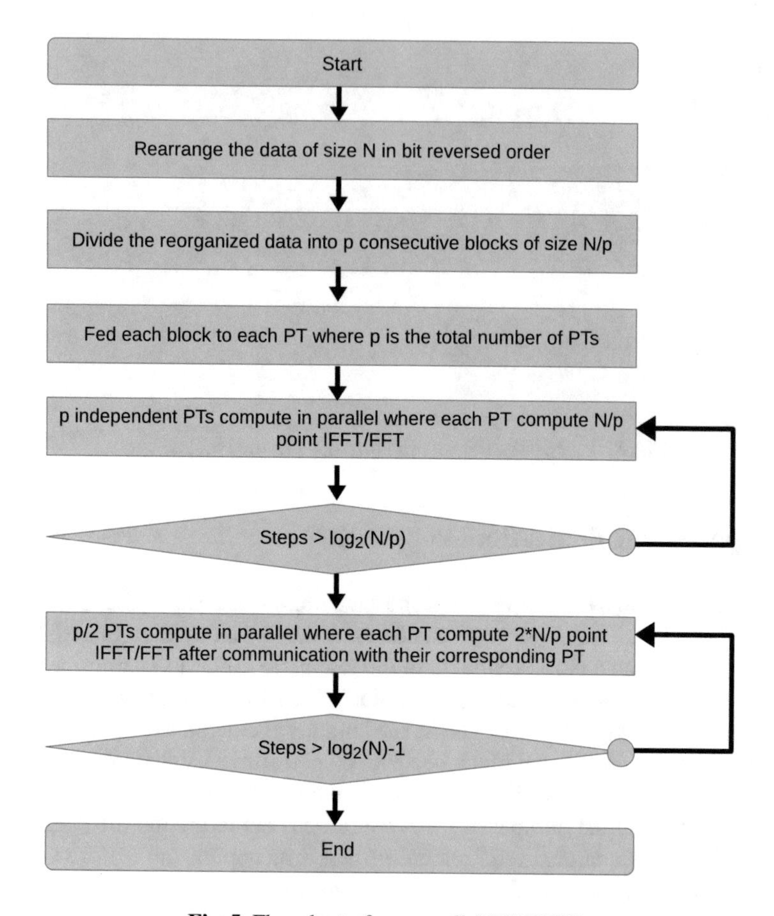

Fig. 5. Flowchart of our parallel FFT/IFFT

4.1 Parallel IFFT Algorithm

We can decompose our IFFT/FFT algorithm [15] into three different stages as illustrated in Fig. 5. Firstly, the input complex numbers are reorganized in bit reversed order. Then, the N reorganized complex numbers are divided into p consecutive blocks. Every block containing $\frac{N}{p}$ complex numbers is sent to its respective processing tiles where N is the length of IFFT and p the number of processing tiles computing IFFT algorithm. The next stage presents the $\log_2(\frac{N}{P})$ steps of IFFT/FTT algorithm, we have p processing tiles running in parallel and every tile computes an IFFT of length $\frac{N}{P}$. In this stage, there is no communication between processing tiles. Finally, the third stage presents the rest of $\log_2 p$ steps of IFFT/FFT algorithm. We have $\frac{p}{2}$ processing tiles running in parallel. Every tile computes an IFFT of length $2\frac{N}{p}$ after receiving its data from its corresponding tile. At this stage, index distance between a butterfly is larger than $\frac{N}{p}$ and communication between processing tiles is required, as a result. Both the operations count and the process time are reduced by our parallel IFFT/FFT. Consequently, compared to the time complexity of serial FFT/IFFT $O(N\log_2 N))$, the time complexity of our parallel algorithm is reduced to $O(N\log_2 \frac{N}{p}))$ which results in latency decrease of our OFDM system.

5 Implementation Results

In this paper, we explore a variety of architectural designs of OFDM LTE implemented in FPGA device using the hardware workflow ProNoC. This is meant to analyze the impact of increasing the number of processors calculating the IFFT. We report a comparison of performance evaluation of the OFDM LTE system having a different number of processing tiles including a single processing tile result as reference. Table 1 enumerates the specific parameters of the network design for the implementation of our multicore LTE OFDM system. For the different architectures designs, we adopt mesh topology with deterministic routing algorithm, wormhole packet switching technique. In this work, the execution time and resource FPGA utilization are analyzed for the different OFDM LTE system where 1, 2, 4, 8 and 16 are the number of processing tile computing IFFT task.

Table 1. NoC parameters for the implementation of OFDM system

Parameters	Value
Number of tiles computing IFFT task	2, 4, 8, 16
Topology	Mesh
Payload width	32
Number of virtual channel per port	4
Buffer flits per VC	4
Packet switching technique	Wormhole
Routing algorithm	XY

5.1 Execution Time Versus Number of Processing Tiles

Applying the design methodology adopted by ProNoC tool enables us to evaluate the execution time performance of our implemented multicore LTE OFDM system. The different channels bandwidth are 1.4 MHz, 3 MHz, 5 MHz, 10 MHz, and 20 MHz give us the execution time results. The number of processing tiles computing the IFFT block of LTE OFDM system is represented by the horizontal axis whereas the execution time is depicted by the left vertical axis. The speed-up metric is represented by the right vertical axis. Compared to the conventional OFDM system, the execution time of our proposed multicore OFDM system at all examined channel bandwidth decreases importantly with the increase in the number of processing tiles as shown in Fig. 6. The multicore LTE OFDM system is more performant than conventional OFDM system which is ranging from 24% for 2 processing tiles computing IFFT task to 76% for 16 processing tiles computing IFFT task. For a larger number of processing tiles, the OFDM

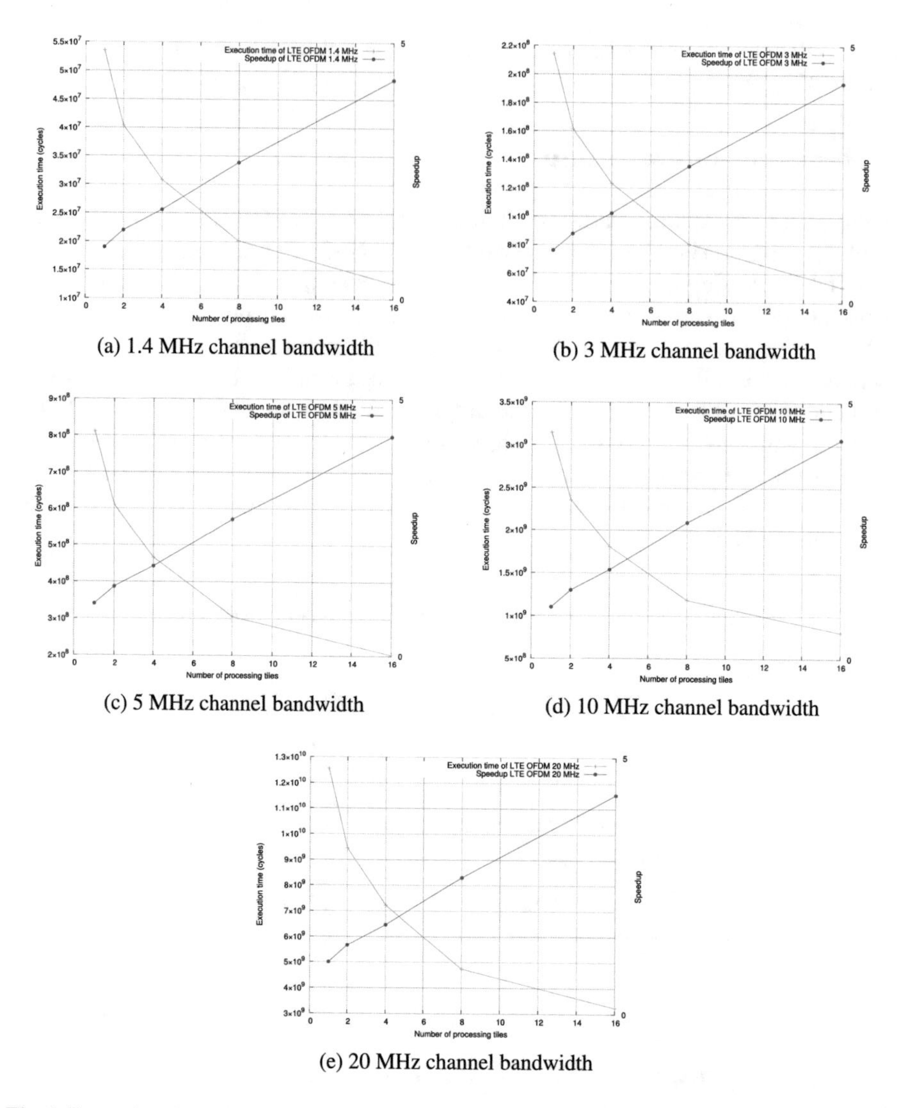

(a) 1.4 MHz channel bandwidth

(b) 3 MHz channel bandwidth

(c) 5 MHz channel bandwidth

(d) 10 MHz channel bandwidth

(e) 20 MHz channel bandwidth

Fig. 6. Execution time of our multicore OFDM system in NoC design architecture using the fixed parameters in Table 1 versus number of processing tiles computing IFFT at different channel bandwidths.

system requires a more limited execution time. Movers, the speed-up metrics of our proposed OFDM system varies from 1.33 times faster for 2 processing tiles computing IFFT task to 4.27 times faster for 16 processing tiles computing IFFT task contrary to a conventional OFDM.

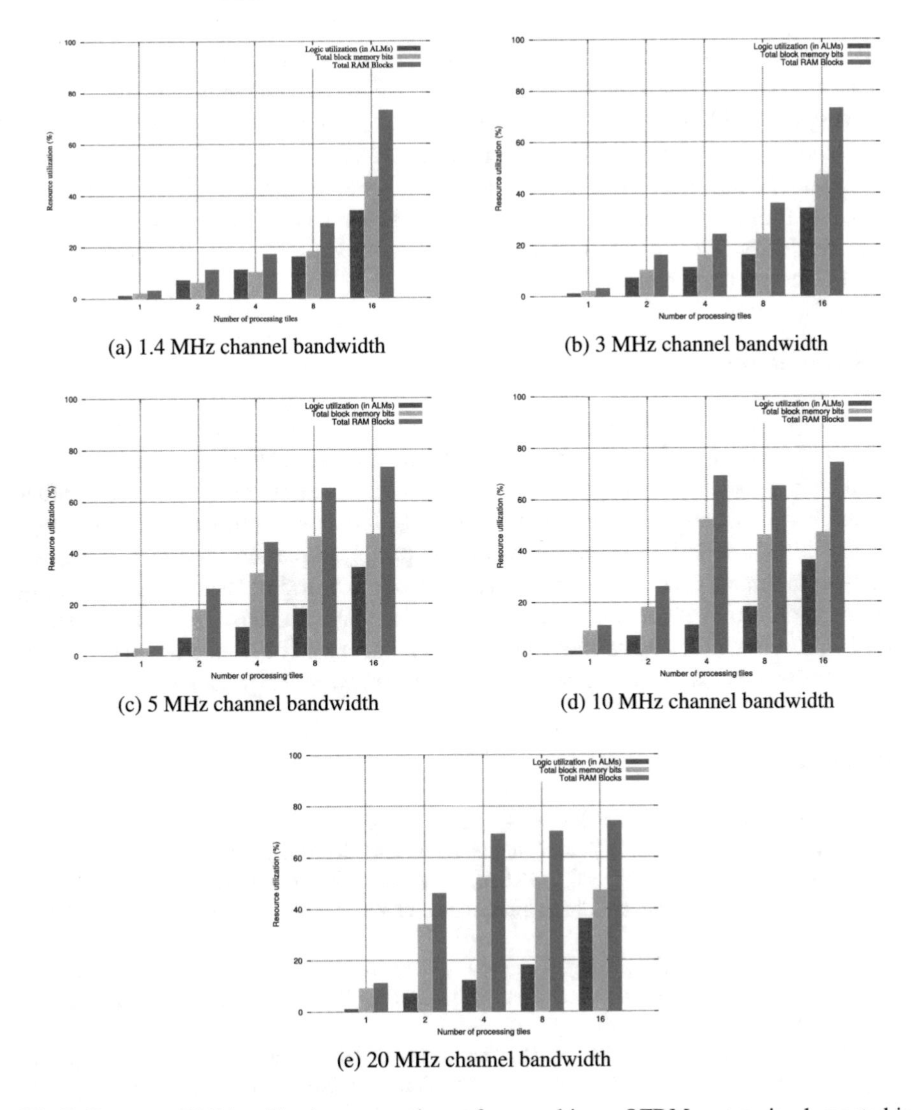

(a) 1.4 MHz channel bandwidth

(b) 3 MHz channel bandwidth

(c) 5 MHz channel bandwidth

(d) 10 MHz channel bandwidth

(e) 20 MHz channel bandwidth

Fig. 7. Resource FPGA utilization comparison of our multicore OFDM system implemented in NoCs using the fixed parameters in Table 1 versus number of processing tiles computing IFFT at different channel bandwidths.

5.2 FPGA Resource Utilization Versus Number of Processing Tiles

The second performance evaluation that we conducted is the Resource Utilization of the NoC. Figure 7 presents the FPGA resource utilization comparison required to implement the different configurations of our multicore LTE OFDM system in NoC design with a different number of processing tiles computing IFFT task including 1 single-core result at different channel bandwidth 1.4 MHz, 3 MHz, 5 MHz, 10 MHz, and 20 MHz. The different hardware architectures (Verilog code) are synthesized on Altera Stratix

V GX FPGA. As expected, the increase of the number of tiles leads to faster acceleration and a negative impact on the occupied area. Although, generating a multicore LTE OFDM system with more than 16 processing tiles computing IFFT is not possible due to the memory constraints and to the fact that the FPGA Altera Stratix V GX does not contain enough resources to implement larger NoC. For an LTE OFDM system with 16 processing tiles computing IFFT tasks in different channels bandwidth leads to an area occupation of 34% for logic utilization in ALMs, 47% for total blocks memory bits, and 74% for total ram blocks.

6 Conclusion

In this work, we presented a hardware design flow based on a prototype platform to design and evaluate different designs architectures for our multicore OFDM LTE system. It chiefly focuses on the parallelization of IFFT which can be generalized to other wireless communication systems based on OFDM like 5G. Given that our application is a combination of serial and parallel tasks, this prototype platform is used to explore a variety of multicore architectures by varying the number of processing tiles computing IFFT task. Our implementation results demonstrate that a 20 MHz OFDM LTE system with 16 processing tiles computing IFFT task conduct to a 76% reduction in execution time against a conventional OFDM system implemented in one processor.

References

1. Hsieh, H.C., Chen, J.L., Benslimane, A.: 5G virtualized multi-access edge computing platform for IoT applications. J. Netw. Comput. Appl. **115**, 94–102 (2018)
2. Younus, M.U., ul Islam, S., Ali, I., Khan, S., Khan, M.K.: A survey on software defined networking enabled smart buildings: architecture, challenges and use cases. J. Netw. Comput. Appl. **137**, 62–77 (2019)
3. Liu, S., Huang, S., Li, M., Dong, X., Qiu, D., Yang, R.: Novel low-complexity low-latency orthogonal frequency division multiplexing transmitter. Wuhan Univ. J. Nat. Sci. **24**(1), 71–78 (2019)
4. LaSorte, N., Barnes, W.J., Refai, H.H.: The history of orthogonal frequency division multiplexing. In: 2008 IEEE Global Telecommunications Conference, IEEE GLOBECOM 2008, pp. 1–5. IEEE (2008)
5. Ferreira, M.L., Barahimi, A., Ferreira, J.C.: Dynamically reconfigurable LTE-compliant OFDM modulator for down-link transmission. In: 2016 Conference on Design of Circuits and Integrated Systems (DCIS), pp. 1–6. IEEE (2016)
6. Nouri, S., Hussain, W., Nurmi, J.: Evaluation of a heterogeneous multicore architecture by design and test of an OFDM receiver. IEEE Trans. Parallel Distrib. Syst. **28**(11), 3171–3187 (2017)
7. Huang, J., Ruan, F., Su, M., Yang, X., Yao, S., Zhang, J.: Analysis of orthogonal frequency division multiplexing (OFDM) technology in wireless communication process. In: 2016 10th IEEE International Conference on Anti-counterfeiting, Security, and Identification (ASID), pp 122–125. IEEE (2016)
8. Dali, M., Guessoum, A., Gibson, R.M., Amira, A., Ramzan, N.: Efficient FPGA implementation of high-throughput mixed radix multipath delay commutator FFT processor for MIMO-OFDM. Adv. Electr. Comput. Eng. **17**(1), 27–38 (2017)

9. Kumar, G.G., Sahoo, S.K., Meher, P.K.: 50 years of FFT algorithms and applications. Circ. Syst. Sig. Process **38**(12), 5665–5698 (2019). https://doi.org/10.1007/s00034-019-01136-8
10. Gaur, M.S., Laxmi, V., Zwolinski, M., Kumar, M., Gupta, N., et al.: Network-on-chip: current issues and challenges. In: 2015 19th International Symposium on VLSI Design and Test, pp. 1–3. IEEE (2015)
11. Bhaskar, A.V., Venkatesh, T.: Performance analysis of network-on-chip in many-core processors. J. Parallel Distrib. Comput. **147**, 196–208 (2021)
12. Monemi, A., Tang, J.W., Palesi, M., Marsono, M.N.: ProNoC: a low latency network-on-chip based many-core system-on-chip prototyping platform. Microprocess. Microsyst. **54**, 60–74 (2017)
13. mor1Kx: an OpenRISC processor IP core. http://github.com/openrisc/mor1kx. Accessed Dec 2021
14. Snyder, W., Wasson, P., Galbi, D.: Verilator - Convert Verilog code to C++/SystemC. http://www.veripool.org/wiki/verilator. Accessed Dec 2021
15. Jallouli, K., Mazouzi, M., Ahmed, A.B., Monemi, A., Hasnaoui, S.: Multicore MIMO-OFDM LTE optimizing. In: 2018 International Conference on Internet of Things, Embedded Systems and Communications (IINTEC), pp. 166–170. IEEE (2018)

Software-Defined Overlay Network Implementation and Its Use for Interoperable Mission Network in Military Communications

Shuraia Khan(✉) and Farookh Khadeer Hussain

School of Computer Science, University of Technology Sydney, 15 Broadway, Ultimo, NSW 2007, Australia
Shuraia.khan@student.uts.edu.au, Farookh.hussain@uts.edu.au

Abstract. The extensive joint force modernization program for defense requires integrating various modern network systems that operate globally connected networked systems. This advancement of the battlefield is developing very complex and striving to adapt with ever-increasing capacity challenges at the tactical edge. Military network in revolutionary Software Defined Network (SDN) architecture with Service Oriented Architecture (SOA) principle offers an intelligent centralized network with advanced management features. However, interoperability between heterogeneous communication equipment of a battlefield network remains an ongoing challenge. This paper argues that the overlay feature in an SDN network can resolve the interoperability issues referenced from Federation Mission Network (FMN). We have used two simple scenarios to realize the simplicity of configuring the overlay network over SDN and discuss the usability of the overlay approach in the military network in terms of interoperability.

1 Introduction

Mr. Michael Street from NATO Communications and Information (NCI) Agency handed the second keynote where he illustrated that "*Has the NATO community determines their position in terms of operations, standards development and interoperability, and how missions and their demands on the networks have progressed*" [1]. These keynotes left the researcher with various focus points that need to address. One of the critical points is interoperability. Besides, a Tactical network is progressing rapidly to deliver efficient operational practices and situational awareness for data exchange and information sharing among the military force. The current progression demands the capability to work together for joint operations effectively. Moreover, It requires facilitating network-centric warfare's command and control (C2) capabilities [2]. From here, the term "Interoperability" emerges.

Tactical communication network capabilities and their variations depend on several factors such as; landscape or ground status, operational mobility, availability of the resources of media communication, and characteristics of the warfare network [3]. Suitable application and network configurations must consider facilitating interaction and

L. Barolli et al. (Eds.): AINA 2022, LNNS 449, pp. 554–565, 2022.
https://doi.org/10.1007/978-3-030-99584-3_48

interoperability at different tactical data networking levels. NATO has defined interoperability as conducting joint operations of different military organizations [1]. Interoperability requires the equipment to share common facilities and communicate with other types of equipment from a different network.

Over the last decade, service-oriented architecture (SOA) has been found as a critical enabler within NATO and NATO nations for military communication and information infrastructures. However, the tactical networks' military battlefields face new challenges that classical SOA systems and technologies initially developed for well-provisioned enterprise environments cannot cope well. Tactical networks are mobile, low-bandwidth, and prone to lengthy delays and frequent interruptions. To address these underlying challenges in the Tactical domain, NATO defines a Software Defined Network (SDN)-based Tactical network with service-oriented principles to facilitate interoperability within the network [4]. In software-based SDN, Network intelligence is (logically) centralized, and the controller provisions the network managers quickly configure, manage, secure, and optimize network resources via dynamic, automated SDN programs.

The motivation of this paper is to find insight into the current military network challenges they are facing. Additionally, we envisioned a future heterogeneous tactical network on an SOA-based SDN architecture that can provide a protected, integrated, and supportable sovereign software-defined battlefield network.

This research shows a Mininet based implementation established on two scenarios: a) OpenFlow protocol forward packets and update the flow entry in an OpenFlow switch. b) VXLAN tunnel between two virtual machines using Open Flow switch. In this paper, we have reviewed a list of existing literature to identify research in terms of interoperability of the tactical domain discussed in Sect. 2. An overview of the Overlay network in SDN in Sect. 3, Two scenario-based implementations and lessons learned from the findings discussed in Sect. 4. Discussions of the above scenarios in Sect. 5 and a list of benefits of implementing an overlay network in the SDN-based mission network in Sect. 6. We have concluded the paper with future research directions in Sect. 7, followed by an implementation appendix.

2 Literature Review

In the literature review, we have broadly looked at an interoperable mission-critical network in the tactical domain. We also have looked at service-oriented architecture (SOA)-based SDN networks that focus on interoperability in the military network. High mobility fighting vehicles with advanced communication characteristics introduce to enhance combat capabilities. However, these advancements require interoperability between heterogeneous tactical network equipment [5]. A 3-tiered SDN architecture-based tactical network is proposed to abstract the physical layer, control layer, and management layer as Tactical land Network (LTN), battlefield Tactical Network (BTN), and Joint Task Force Headquarters (JTFH) consecutively. This research guarantees the interoperability between various communication equipment across multiple platforms. However, minimizing delay and congestion control of the network is still needed to experiment.

NATO introduces a Web-services notification approach in military networks with an SOA baseline that defense researchers are currently experimenting with to achieve full interoperability in the military network [6]. Another research proposes a numerical simulation-based approach to achieve efficiency in the Tactical ad hoc network. The method evaluates mobile ad hoc networking interoperability and cooperation (MANIAC) datasets [7]. Another research proposes a proof of concept-based framework for integrating various network technologies to eliminate interoperability issues for emergency operations in the Military [8]. The researchers developed a prototype that includes a programmable radio integration unit using a backhaul device that can integrate with multiple radios and switches on the ground, aligning with the SDN concept. The researchers picturing the benefits by including Software Defined Radio (SDR) with the proposed prototype; however, a standard approach is required to integrate all legacy devices, not only SDR or modern radio devices.

[9] This paper examined the application of SDN across military coalition operations, proposed a method for facilitating dynamic Communities of Interest (CoI) within these environments, and evaluated such interoperability architectures following essential performance metrics. The authors consider the Software-Defined Coalitions (SDC) as the mechanism capable of facilitating the operation of dynamic tactical CoIs. Consequently, they relate three interoperability levels: a) network, b) network and storage, and c) network, storage, and compute. They simplify the process by merging the SDN and Software-Defined Storage processes while constructing upon SDN Environments' concept. The presentation and discussion of the proposed mechanism are comprehensive, even though additional scenario-based validation/verification can be desirable.

Overlay network usage in the tactical domain concept is primarily theoretical, except Federation Mission Network evolved by NATO. Therefore, we have not included literature in this paper. The literature above demonstrated that the research to achieve an interoperable military network is immature and remains an issue. Interoperable intelligent and centralized SDN-based tactical network incorporated with the service-oriented principle for Command and Control (C2) warfare is an overarching demand for the military network.

3 Overlay Network in Software-defined Network (SDN)

SDN introduces a sovereign model that can intelligently support various applications with dynamic demands while lessening operating costs by simplifying hardware and software resources [10]. Researchers have explored various possible features to make the best use of SDN for disadvantaged networks, such as the tactical domain for military communications. Thus, service-oriented architecture (SOA) based SDN, an evolutionary approach, is introduced to receive the best outcome.

3.1 SDN Architecture

Software-Defined Network (SDN) model is an advancement of the traditional network and a remarkable game-changer. The static architecture with the decentralized nature of the conventional network causes numerous constraints to dynamically adapt the various application needs. SDN has introduced itself as an evolutional of the network technology that can address the limitations of the traditional network. SDN decouples the data and control planes and centralizes the whole network's controlling process [10]. SDN performs the operations and manages the network in a centralized or distributed manner using the following key elements. 'Forwarding device' is the hardware or software-based data plane device that obtain instructions or well-specified rules set to take the appropriate actions when a packet appears. In contrast, the 'data plane' comprises interconnecting forwarding devices.

Similarly, the 'control plane' refers to the network brain, the control logic to the forwarding device through the Southbound interface (SI). At the same time, the Northbound interface (NI) offers an API to develop the application and abstract the low-level instructions. On the other hand, the Management plane consists of applications that leverage the functions provided by NI to implement network control and operation logic.

3.2 OpenFlow Protocol

OpenFlow protocol plays a vital role in the possibility of SDN OpenFlow is a groundbreaking communication protocol that allows deploying routing and switching and can implement enormous functionalities in various application features such as virtual machine mobility, high-security networks, and next-generation IP-based mobile networks.

Traditional network, packet forwarding (data path), and routing decision (Control path) ensure by network devices (Router, switch) themselves. However, the OpenFlow switch brings the concept of separating these two functions of packet forwarding (data path) with a device and high-level routing decisions in the control component. We will look at the detailed functionality of this protocol in Sect. 4.

3.3 VXLAN Overlay Network

An overlay network is an overarching layer between two networks and increases the mobility of virtual machines. Virtual eXtensible LAN (VXLAN) is one of Layer 2 overlay schemes over a layer-3 network proposed by the Internet Engineering Task Force (IETF). This approach creates multiple separate, dedicated, discrete virtualized layers of abstraction within the network for deploying specific applications or security reasons. Overlay networking may combine Virtual Local Area Networks (VLANs), Peer-to-Peer networks, and IP networks using various networking protocols and standards. The most widely held protocols are Virtual Extensible Lan (VXLAN, Virtual Private Networks (VPNs), and IP multicast. This encapsulation mechanism supports up to 16 million virtual overlay tunnels over a physical later 2/3 underlay network to provide multitenancy within a network [2].

NATO S&T research task group IST-124 has performed a comprehensive experiment to achieve interoperable and efficient routing in a tactical network [3] and configured explicit multicast-unicast groups for military communication. The researchers mentioned that 16M overlay networks or segments could resolve the 4K limitation of VLANs [11].

4 Scenario-Based Implementation Using Mininet

In this section, we develop two scenario-based implementations of SDN to realize how this groundbreaking technology leverage various innovative features for future military network. We simulate the scenarios using the Mininet simulation tool to realize the functionality of the OpenFlow protocol and discuss the potential serviceability of implementing this concept in military communication networks.

4.1 Scenario 1: How SDN Forwards the Packets Using OpenFlow

We experiment the scenario to realize the packet forwarding process applying the OpenFlow protocol. Figure 1 demonstrates a simple SDN network simulation to accomplish the following insights; a) How packets move through an OpenFlow enabled network. b) Packet walkthrough of an example HTTP request and reply in an OpenFlow enabled network.

We use Mininet 2.2.1 on Ubuntu 14.04 – 32 bits in Windows and configure the above scenario with four hosts related to Open vSwitch S1 and S1 related to reference controller C0. Host 4 is configured as a simple Web Server. Configuration details are available in the appendix. Receiving successful pings reply from h4 using port number 80 shows the webserver is successfully connected and running. A bunch of OpenFlow packets was captured to analyze the traffic flow.

5 Findings and Lessons Learned:

When host h1 makes HTTP: web request to web server h4, the following process happens in the OpenFlow enabled network.

1. Transport Control Protocol (TCP) starts sending a message from h1 to switch s1 using port 80. In a traditional network, S1 forwards a packet to the web server by knowing the MAC address of h4. However, the OpenFlow switch sends packets following the flow table.
2. If there is no matching flow in the flow table, the default process sends the packet to the controller. The packet is called the 'Packet-IN' message. The Packet-IN message encapsulates the TCP message inside its referencing buffer-ID. Switch buffer the entire packet with reference (buffer-ID) and send it to the controller for instructions.
3. The controller sends back a 'Packet-OUT' message, sending a flow modification message to the switch. The Packet-OUT message contains the whole message or the message's header with instructions about what to do with this packet referencing the Buffer-ID.
4. In this Scenario, The Packet-OUT message contains instructions for the switch as "forward to port 4". Alternatively, the controller sent a flow modification message for adding a new flow entry in the flow table. Flow modification message says, any TCP packet source an IP in MAC of h1 destination an IP in MAC of h4, send all of those to port 4. The exact process continues for an acknowledgment message from h4 to h1

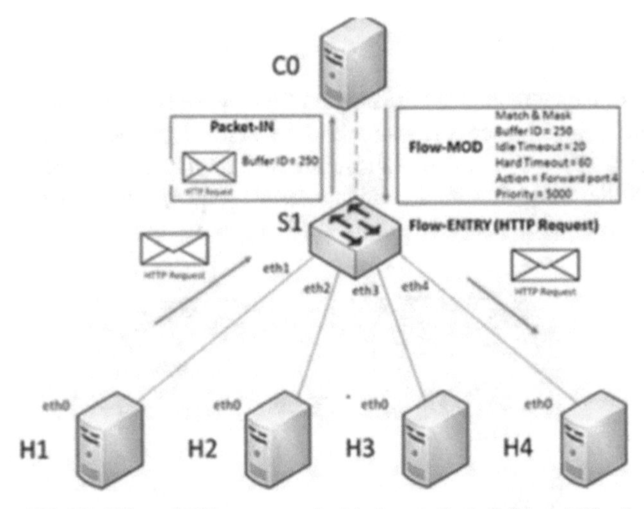

Fig. 1. 4 hosts (H1, H2, H3, and H4) connected with Open vSwitch S1 and S1 relate to reference controller C0

The controller can instruct various actions through flow modification messages. The actions may include: a) flooding out all ports, b) dropping the packets, c) matching packets to the controller. In addition, two types of time-out are present in the inflow modification messages to tell how long to cache Flow to flow entry. a) 'Idle time out' = 30 means removing this flow entry if there is no match of an HTTP request for 30 s. b) 'Hard time-out' = 60, which means after 60 s, no matter it is a living packet or not, remove the flow entry. Moreover, priority is another entity of the flow modification message. If two flow entry matches in the flow table, that refers to higher priority; therefore, the controller ignores the lower priority. Based on the above procedure, we analyze the output of this scenario in the result section.

5.1 Scenario 2: VXLAN Overlay Network Using Open vSwitch

In this scenario, we configure an overlay network (discussed in Sect. 3.4) in SDN using Open vSwitch. We create two networks in two virtual machines using Mininet and then create a VXLAN tunnel between these two networks (Fig. 2). We also ensure that two VMs are successfully communicating using reference controller instruction (Fig. 3).

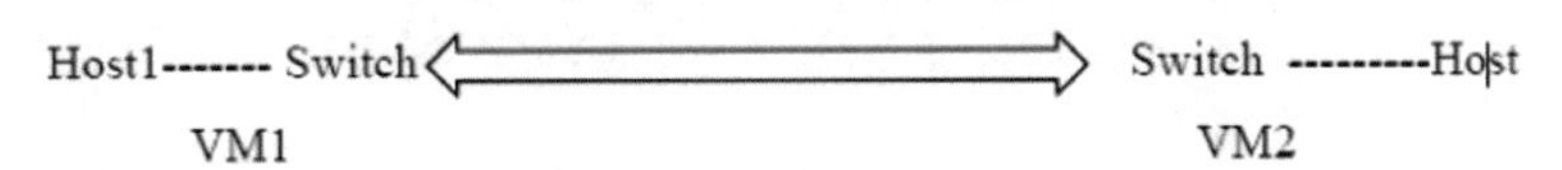

Fig. 2. VXLAN Overlay Network topology design.

After that, we manually terminated the reference controller and ensured the two VMs were not communicating anymore. Employing the learning from scenario 1, we Add the flow entry to the Open vSwitch to create tunnels between two VMs. We test the overlay network by receiving successful ping commends.

6 Discussion

We capture traffic data to analyze the output of the above two scenarios. A 'packet-IN' message shows (Fig. 3) where the source and the destination address are corresponding host1 and host4 address recorded with Buffer-ID 285. The switch receives instruction from the controller using flow modification message (Fig. 4) that any packet from Source host1 and the destination host4 should forward to port 80, referencing 285.

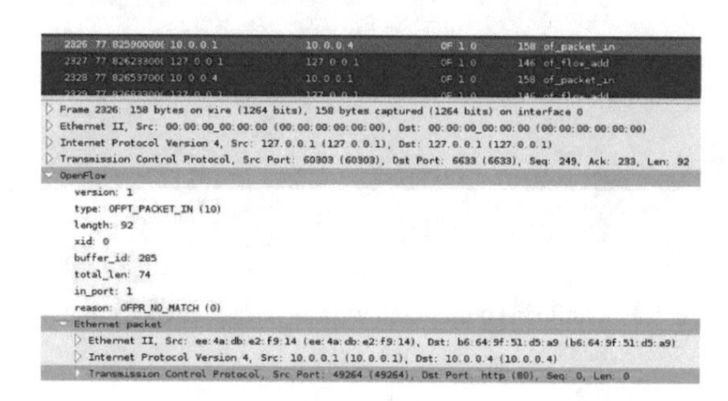

Fig. 3. Packet-IN message capture

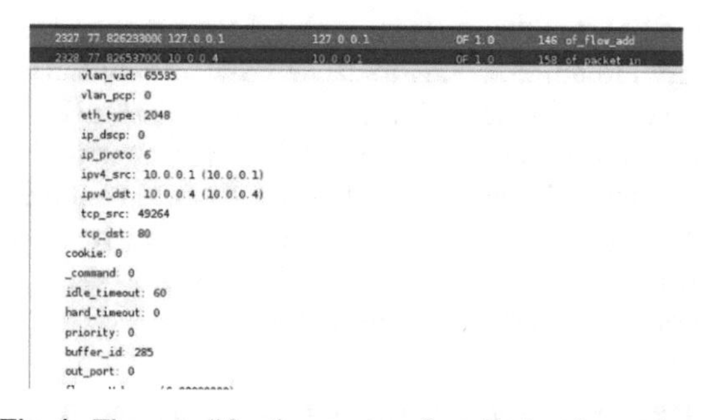

Fig. 4. Flow modification capture from Packet-In message.

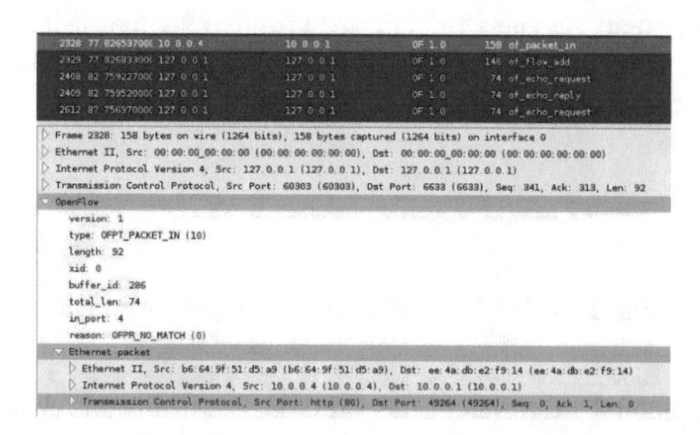

Fig. 5. Packet in a reply message

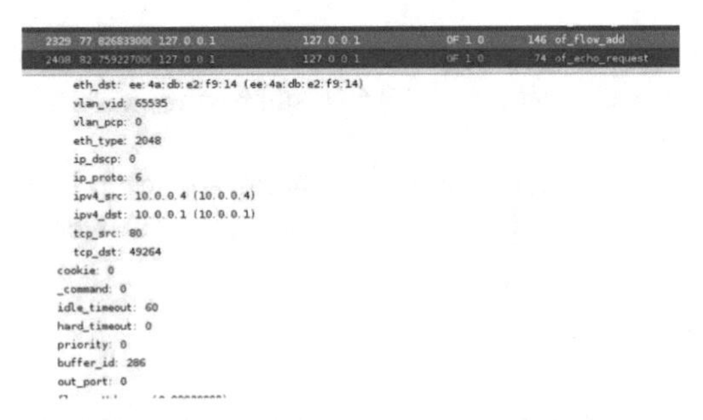

Fig. 6. Flow Modification to the flow table.

An acknowledgment message (Fig. 5) shows (ack) packet from h4 forward to h1 with buffer id 286, and we add this instruction to the flow table (Fig. 6) with buffer id 256.

7 Benefits of Overlay Network in SDN Based Military Network with SOA Principal in Tactical Domain

NATO introduces the Federated Mission Networking (FMN) concept, the foundation for instituting future mission networks. The FMN concept describes and realized in the NATO FMN Implementation Plan (NFIP) [12]. NFIP builds on the concept of service-oriented architecture (SOA) on the information infrastructure to achieve interoperability in the mission network [12] and standardize through NATO Network Enabled Capabilities (NNEC). The NNEC recognizes several Core Enterprise Services (CES) on SOA Baseline, representing the standard functionality demand for building an interoperable service-oriented infrastructure in a federation network. The questions raised of "*How to enhance additional network features(robustness, efficiency, security, and QoS especially for disadvantaged networks) needed for military applications*"? Based on the above question, we have discussed a simulation-based implementation of an overlay network in an SDN architecture and overseen the overlay network as an efficient technique to facilitate advanced features of the future SOA-based mission networks. The possibilities of these advancements are following:

- Advancing from the Overlay concept, NATO has introduced an Interconnect-overlay architecture that bears a resemblance to an interconnect-flat architecture to achieve the goal of interconnecting many network segments that are running separate routing protocol domains and span the whole heterogeneous network [3].
- Improve the flexibility by allowing network managers to move around network endpoints using software management.
- Capable of achieving a high level of robustness by enhancing routing [13]. Advanced routing features can overcome high and variable wireless bit error rates by sending messages over multiple disjoint paths without altering the underlying IP routing tables.
- Enable QoS routing according to delay requirements, bandwidth requirements, or both.

- Enhances network processing capability and assists in traffic aggregation and redundant Forward Error Correction (FEC) [9], improving network resource utilization efficiency and reliability.
- Enable to deploy very critical and protected special applications into a secured channel and improve the security and robustness of the military network.

8 Conclusion

Software-Defined Network (SDN) has emerged as a revolutionary technology capable of eliminating the limitations of current network infrastructure by separating the control plane from the data forwarding plane. SDN expansions receive more and more acceptance, the controller of SDN can be able to fulfill the expectation and can be able to close the gaps that remain to develop an intelligent tactical network. In addition, SDN with SOA principle brings a new groundbreaking concept of service-oriented network. Tactical network researchers investigate the best utilization of this new concept of the SOA principle and identify its benefits. An interoperable network is a crucial enabler for developing an intelligent and dynamic military network. Overlay network can bring a bridge in this research to establish an interoperable network for future The **C4 (Command Control Communications and Computers) Edge (Evolutionary Digital Ground Environment)** network that will provide a Protected, Integrated, and Supportable Sovereign system. An interoperable SDN-based tactical network with SOA principles is the future direction of this research, and currently, we are working to develop a testbed implementation of the interoperable heterogeneous radio network in SDN architecture.

Acknowledgments. I am grateful to Dr. Farookh Khadeer Hussain and Adam Wilson for helping and supervising me for this research and providing continuous guidance.

Appendix:

The commands to create a simple SDN network.

```
$ sudo mn -topo=single,4
```

The “dump” command shows the available nodes, IP address, OVS-Switch, usable ports, and the reference controller. Open flow reference distribution includes Wireshark dissector, which analyzes the OpenFlow messages and sends to OpenFlow default port (6633) conveniently readable. To capture the OpenFlow packets, we set up capture interface loopback 0 as Mininet uses loopback to communicate with the host. Filter ‘of’ is set up to grab the OpenFlow packet. We configure host4 as a simple Web and ping from h1 to h4.

```
H4 python -m SimpleHTTPServer 80 &

H1 wget 10.0.0.4
```

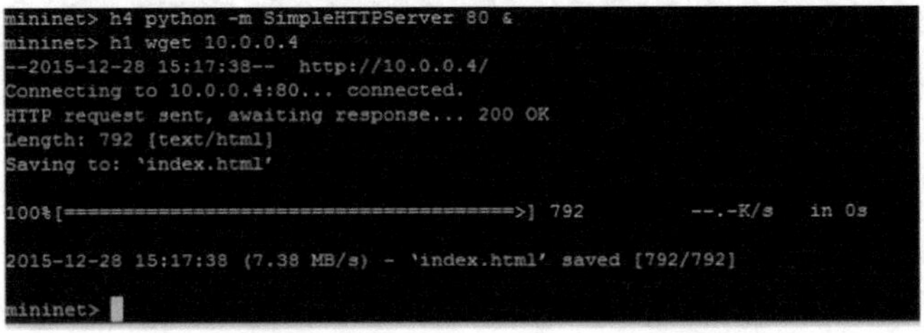

```
mininet> h4 python -m SimpleHTTPServer 80 &
mininet> h1 wget 10.0.0.4
--2015-12-28 15:17:38--  http://10.0.0.4/
Connecting to 10.0.0.4:80... connected.
HTTP request sent, awaiting response... 200 OK
Length: 792 [text/html]
Saving to: 'index.html'

100%[======================================>] 792         --.-K/s   in 0s

2015-12-28 15:17:38 (7.38 MB/s) - 'index.html' saved [792/792]

mininet>
```

Fig. 7. Ping reply from the webserver.

The above screenshot (Fig. 7) shows the successful ping reply from h4 using port number 80. While running this process, Wireshark captures a bunch of OpenFlow packets.

VXLAN Overlay Network Using Open vSwitch

In this scenario, we implement a VXLAN overlay network using Open vSwitch. We create two networks in two virtual machines using Mininet and then create a VXLAN tunnel between these two networks. Firstly, we collect the eth0 interface address of each VM. The addresses are respectively mn1: 192.168.159.133 and mn2: 192.168.159.132. The first line command creates a simple topology in each VM. In the next phase, we create a VXLAN port in each switch and direct it with remote IP. The program codes are following:

```
$ sudo mn --topo single,1

Mininet 1> sh ovs-vsctl add-port s1 vxlan
Mininet 1> sh ovs-vsctl set interface vxlan type=vxlan
option:remote_ip=192.168.159.132option:key=5566
ofport_request=9

Mininet 2> sh ovs-vsctl add-port s1 vxlan

Mininet 2> sh ovs-vsctl set interface vxlan type=vxlan
option:remote_ip=192.168.159.133option:key=5566
ofport_request=9
```

Using the ping command, we ensure that hosts of two VMs are successfully communicating using reference controller instruction. In the next stage, we manually stopped the reference controller of both VMs shown in Fig. 8 (mn1).

```
login as: mininet
mininet@192.168.159.133's password:
Welcome to Ubuntu 14.04 LTS (GNU/Linux 3.13.0-24-generic i686)

 * Documentation:  https://help.ubuntu.com/
Last login: Sun Jan  3 10:07:14 2016 from 192.168.159.1
mininet@mininet-vm:~$ ps -aux | grep controller
root      1596  0.0  0.0   2320   708 pts/1    S+   10:34   0:00 controller -v p
tcp:6633
mininet   1711  0.0  0.0   4676   828 pts/5    S+   11:07   0:00 grep --color=au
to controller
mininet@mininet-vm:~$ sudo kill 1596
mininet@mininet-vm:~$ 
```

Fig. 8. Stop the reference controller manually from mn1

Ping commands ensuring that they are communicating anymore. Then we manually add the flow entry of the switches to create tunnel between two VMs. The commands are following:

```
Mininet 1> sh ovs-ofctl add-flow s1
'in_port=1,actions=set_field:5566->tun_id,output:9'

Mininet 1> sh ovs-ofctl add-flow s1
in_port=9,tun_id=5566,actions=output:1

Mininet 2> sh ovs-ofctl add-flow s1
'in_port=1,actions=set_field:5566->tun_id,output:9'

Mininet 2> sh ovs-ofctl add-flow s1
in_port=9,tun_id=5566,actions=output:1
```

We ping the VMs from each end and receive a successful ping reply.

References

1. Dahel, S.: Technical Evaluation Report. Real Time Intrusion Detection (2003)
2. Cheung, L.Y., Yin, C.W.: Designing tactical networks-perspectives from a practitioner. dSTA Horizons Tech. Rep. 122–133 (2013)
3. Treaty, N.A.: Heterogeneous Tactical Networks–Improving Connectivity and Network Efficiency
4. Division, NPD, Interoperability for joint operations, NPD. Division, Editor. 2006, BGR2-INTEROP-ENG-0706 © NATO 2006: Belgium
5. Elgendi, I., Munasinghe, K.S., Mcgrath, B.: A heterogeneous software defined networking architecture for the tactical edge. In: 2016 Military Communications and Information Systems Conference (MilCIS). IEEE (2016)
6. Dehghan, M., et al. Optimal caching and routing in hybrid networks. In: 2014 IEEE Military Communications Conference. IEEE (2014)
7. Yan, X., Hu, X., Liu, W.: SDN controller deployment for QoS guarantees in tactical Ad Hoc networks. Wireless Commun. Mobile Comput. **2021**, 1–8 (2021)
8. Mihailescu, M., Nguyen, H., Webb, M.R.: Enhancing wireless communications with software defined networking. In: 2015 Military Communications and Information Systems Conference (MilCIS). IEEE (2015)

9. Mishra, V., et al.: Comparing software defined architectures for coalition operations. In: 2017 International Conference on Military Communications and Information Systems (ICMCIS). IEEE (2017)
10. Khan, S., Hussain, F.K., Hussain, O.K.: Guaranteeing end-to-end QoS provisioning in SOA based SDN architecture: a survey and open issues. Futur. Gener. Comput. Syst. **119**, 176–187 (2021)
11. Nakagawa, Y., Hyoudou, K., Shimizu, T.: A management method of IP multicast in overlay networks using openflow. In: Proceedings of the First Workshop on Hot Topics in Software Defined Networks (2012)
12. Brannsten, M.R., et al.: Toward federated mission networking in the tactical domain. IEEE Commun. Mag. **53**(10), 52–58 (2015)
13. Elaoud, M., et al.: Self-initiated and self-maintained overlay networks (SIMONS) for enhancing military network capabilities. In: MILCOM 2005–2005 IEEE Military Communications Conference. IEEE (2005)

Fault Tolerant Multiple Dominating Set Constructions for Wireless Ad-hoc Networks

Khaleda Akther Papry(✉) and Ashikur Rahman

Bangladesh University of Engineering and Technology, Dhaka, Bangladesh
papry.05084@gmail.com, ashikur@buet.ac.bd

Abstract. In wireless ad-hoc networks, broadcasting is the most common communication method. To reduce redundancy, traffic and collision induced by broadcasting, different virtual backbones are used on top of the physical topology and Connected Dominating Set (CDS) is one of those. However, constructing minimum connected dominating set (MCDS) containing minimum number of nodes is an NP-complete problem. Although some approximation algorithms are available, the CDS or its approximation has poor fault tolerance. In this work, we present two heuristics, one centralized and the other distributed, for constructing multiple connected dominating sets providing enhanced fault tolerance of the network. Both algorithms are intended to maximize network lifetime involving minimal nodes. Moreover, both the algorithms also ensure load balancing over the network. Finally, we simulate our heuristics to show the improvement of network lifetime and system fault tolerance.

1 Introduction

Wireless ad-hoc networks consist of some wireless nodes that communicate over the networks without the existence of any fixed infrastructure. Broadcasting is the most common communication method in such networks where each node over the network receives the message from a source node. Among different approaches, uncontrolled flooding is the easiest approach for broadcasting where each node unconditionally distributes its incoming packets to each of its neighbors. Therefore, it causes too much traffic, contention and collision resulting in broadcast storm problem [1]. This problem can be minimized by creating a virtual backbone and Connected Dominating Set (CDS) is one of them [2].

A CDS is the subset of a graph where all nodes within the set are connected and the other nodes are 1-hop neighbour of at least one of the members of CDS. A CDS in an ad-hoc network can serve as a backbone for packet routing over the network. Figure 1 shows a network with seven nodes where connectivity means their transmission range. Here, some possible CDSs are: {A, C}, {A, B, F}, {A, B, C}, {A, C, E}, {A, B, C, D}, and so on. However, a small size CDS is desirable in many applications. The less nodes in a CDS, the more efficient a network is as along with routing redundancy, the number of forwarding packets are also decreased. For example, {A, C} is the most desirable CDS for the graph.

L. Barolli et al. (Eds.): AINA 2022, LNNS 449, pp. 566–578, 2022.
https://doi.org/10.1007/978-3-030-99584-3_49

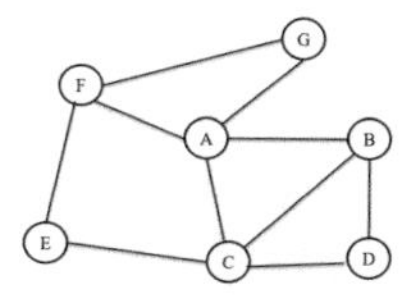

Fig. 1. A random ad-hoc network topology with 7 nodes

A CDS with minimum number of nodes is called minimum connected dominating set (MCDS). However, finding an MCDS is an NP-complete problem [3]. Different heuristics have already been proposed to this end. Moreover, nodes are usually battery operated. Therefore, load balancing among the nodes ensure proper utilisation of energy over the network and increases network lifetime as well. Another serious issue is fault tolerance ability of a network. There may contain some nodes in a network which may fail to forward packets or communicate due to power failures or other errors. If there is a backup CDS for routing, the system might continue to work properly.

In this paper, we aim at finding multiple minimum connected dominating sets (MMCDS) using minimal nodes to make system more fault tolerant and to increase the network longevity. We devise both centralized and distributed algorithms where we allow some minimum overlaps when fully disjoint CDSs are not possible to generate. Moreover, we can apply our algorithm in both static and dynamic scenario with some trivial changes.

The main contributions of this work are enumerated below:

(a) We develop a centralized and a distributed approach for MMCDS construction using overlapping boundary K.
(b) Finally, we validate the algorithms using simulation results, and compare the performance of the two algorithms with existing state-of-the-art algorithms.

2 Related Work

Many broadcasting algorithms have been proposed over the decades to overcome the broadcast storm problem. Lim and Kim provide a new flooding methods [4] for efficient broadcasting named dominant pruning. In dominant pruning algorithm, each node uses its 2-hop neighbor information to reduce redundant transmissions. To reduce broadcast redundancy in ad-hoc wireless networks, Lou and Wu [5] proposed two improved algorithms based on dominant pruning. Rahman et al. proposed enhanced dominant pruning based [6] and partial dominant pruning based [7] broadcasting in untrusted ad-hoc wireless networks.

Ephremides et al. [8] first proposed the idea of using a CDS as virtual backbone for broadcasting. Butenko et al. [2,9] proposed both distributed and centralized greedy algorithms for constructing minimum CDS (MCDS). In paper [10], the authors propose a greedy algorithm for MCDS in unit-disk graphs based on Maximal Independent Set (MIS) which has been used in 3D heterogeneous network [11] later.

Fault tolerance is one of the major issues in broadcasting algorithm. A lot of researches have already been performed to address fault tolerant for broadcasting in ad-hoc networks. Papers [12,13] provide fault tolerant CDS models as k-connected m- dominating set where the CDS is k-connected, and each node not in CDS is dominated (adjacent) by at least m nodes in CDS. However, to achieve this mechanism, each node must be connected with at least m nodes, where m $\geq$ k. The idea of connected minimum secure dominating sets has been first proposed by Barnett et al. in [14]. However, this works in cylindrical or toroidal grid networks only. In [15,16], the authors derive multiple set covers for directional sensor networks which is a special case of wireless ad-hoc networks.

None of the above algorithms have considered fault tolerance and network longevity jointly for wireless ad-hoc networks which is our main focus in this paper.

3 Preliminaries

Here, we discuss network model and some definitions used throughout the paper.

3.1 Network Model

A wireless ad-hoc network can be represented as a graph like a wired network. We represent the connectivity between two nodes of a network if they are within transmission range. We represent the ad-hoc network with a graph G(V, E) using the idea of unit disk graphs. Figure 1 represents a wireless ad-hoc network with seven nodes with similar transmission range.

3.2 Important Definitions

- **Multiple MCDS (MMCDS)**: MMCDS refers to generating different possible MCDSs from a network. As fully disjoint sets might not be achieved always, here we consider overlapping up to a certain amount denoted as K which refers to the boundary value of a node's presence in multiple sets.
- **Overlapping Boundary (K)**: For creating multiple MCDSs (MMCDSs), we consider overlapping among sets bounded by K. The parameter denotes that no node can participate in MCDSs more than this given upper bound value. For example, if K is 2, then a node can be present up to two MCDSs among MMCDSs.
- **Cardinality (C)**: It is the measure of participation in different CDS of node i. For each node the value is bounded by K, that means $C_i \leq K$.
- **Dominant Pruning**: For distributed system, we use the procedure of creating the forwarding list followed by dominant pruning [4]. Suppose, u sends a message to v. Now, v will create its forwarding list from it's one-hop neighbor B_v that are not in $N(u)$. For constructing the new forwarding list, node v needs all uncovered two-hop neighbours U_v. Node v selects nodes from B_v to cover all the nodes in U_v and added to forward list F_v. To create (F_v), we calculate U_v and B_v from the following formulas [4]:

$$U_v = N(N(v)) - N(v) - N(u) \text{ and } B_v = N(v) - N(u)$$

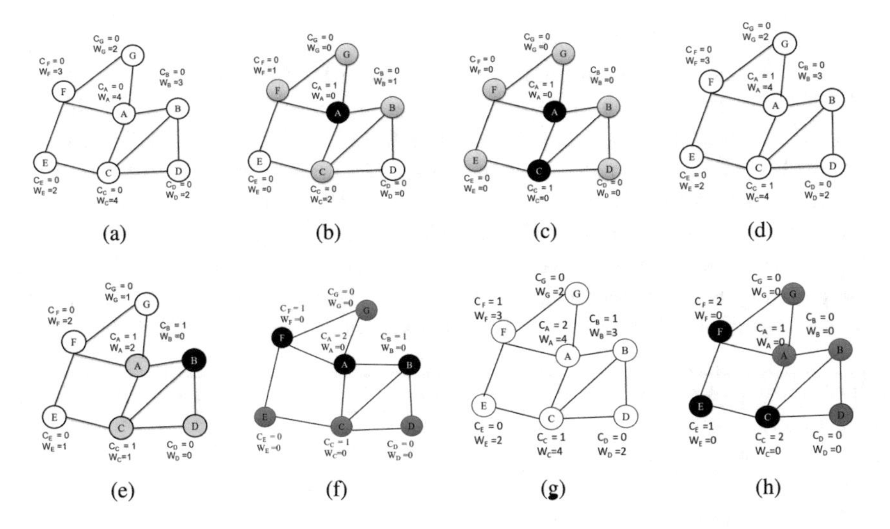

Fig. 2. Basic Greedy Centralized Construction of MCDS (a)–(c) and multiple MCDS construction (d)–(h)

4 Multiple Minimum Connected Dominating Sets (MMCDSs)

In this section, we represent our both centralized and distributed multiple connected dominating sets (MMCDSs) constructions respectively.

4.1 Centralized Algorithm for MMCDSs

In this approach, we run a centralized greedy algorithm to generate multiple CDSs with minimum nodes and minimum overlaps (up to K). It follows the basic MCDS construction [2] at each iteration. After a complete iteration a CDS is constructed with minimum nodes. In the next generation of MCDS, the algorithm tries to select nodes from minimum cardinality (unused nodes of the previous iterations). If there is no new node, it can select a previously used node up to K times which is its maximum cardinality. The process continues until no new sets can be generated. After the generation of each CDS set, another optimization algorithm is run to remove redundant nodes from the set. Figure 2 represents how our algorithm works. We keep the overlapping boundary value K = 2. We can see that 2(a)–(c) constructs the first MCDS. Then all the nodes are initialized again except their cardinality values (2 (d)). Another new set is generated by node F, A and B (2(e)–(f)). At third iteration (2(g)–(h)), C,E, F constructs a new set although node D was selected first. D is discarded using optimization function as redundant node.

The centralized algorithm of MMCDSs is shown in Algorithm 1. Initially, we provide a graph representation of the network $G(V, E)$ and overlap limit K. The outer loop (line 3–23) generates the desired MMCDSs and runs until no new set is found. The inner loop (lines 5–17) creates a single MCDS after a complete iteration. In the inner while loop, a node s from V (first iteration) or $GraySet$ (other iteration) with minimum cardinality and maximum white neighbors is selected (line 6–10). The selected node is added to CDS set and this procedure runs until $WhiteSet$ is empty or no new node can be selected with cardinality less than K. After a complete iteration, the set is optimized by removing redundant nodes using optimization algorithm in [16], if there is any and added it to CDS_n set (line 20). If there is any redundant nodes in CDS that is not in CDS_n, then its cardinality is decremented by 1, as the node is not really used in the set (line 21). The value of n is incremented after a successful creation of CDS.

Algorithm 1. Centralized Multiple Dominating Sets construction Algorithm

Require: $G(V, E), K$
Ensure: A Set of MCDSs CDS
1: $CDS = \emptyset$, $n \leftarrow 0$
2: Set $C_i \leftarrow 0$, for all i $\in V$
3: **while** a new set can be generated **do**
4: $CDS = \emptyset$, $WhiteSet = V$, $GraySet = \emptyset$
5: **while** $WhiteSet \neq Null$ **do**
6: **if** $CDS == \emptyset$ **then**
7: Find the node $s \in V$ with maximum W_s and minimum C_s
8: **else**
9: Find the node $s \in GraySet$ with maximum W_s and minimum C_s
10: **end if**
11: **if** $C_s < K$ and $W_s > 0$ **then**
12: $CDS = CDS \cup \{s\}$, $C_s = C_s + 1$
13: $GraySet = GraySet \cup N(s) - \{s\}$, $WhiteSet = WhiteSet - N(s) - \{s\}$
14: **end if**
15: **end while**
16: **if** $CDS \neq Null$ **then**
17: $n = n + 1$
18: Optimize CDS by removing redundant nodes and add to CDS_n
19: Update C_k by $C_k - 1$ for all $k \in (CDS - CDS_n)$
20: **end if**
21: **end while**

In this algorithm we have two while loops. The outer while loop generates multiple sets which is $O(V)$ times in the worst case. The inner while loop generates a single CDS in $O(V^2)$. There is an optimization function after the inner while loop and within the outer while loop which also costs $O(V^2)$. Therefore, the total time complexity of our Centralized MMCDS is $O(V^3)$ times.

4.2 Distributed Algorithm for MMCDSs

In distributed algorithm, for creating multiple MCDSs, we use the dominant pruning based MCDS construction multiple times. Given 2-hop neighbour information, each node tries to contribute to multiple CDS with minimum overlapping over sets.

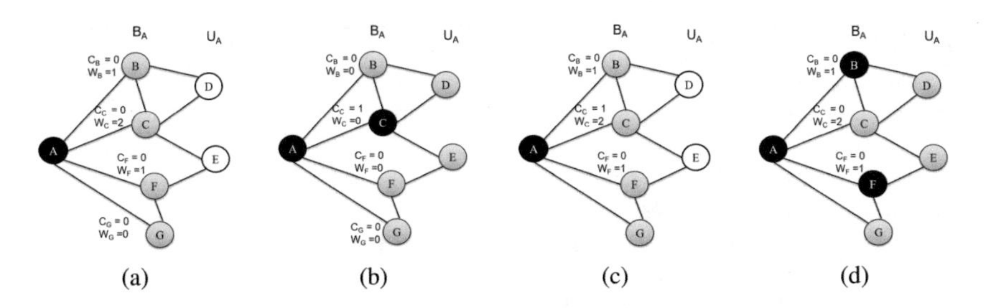

Fig. 3. Basic Greedy Distributed Forwarding List Construction of MCDS (a)–(b) and multiple MCDS (c)–(d) constructions for static scenario

Initially, we consider the algorithm for static ad-hoc network and generate multiple forwarding lists for a node receiving packets. Here, a node can be maximally used up to K forwarding lists as like the centralized one. For this algorithm, a node from its one-hop neighbor list is selected for forwarding that has minimum cardinality and maximum number of node coverage of its uncovered two-hop neighbors. The process is repeated until all uncovered two-hop neighbors are covered. To generate multiple sets, the above process is repeated until no new set is generated. Figure 3 represents how node A creates its forwarding lists for static system. Figure (3(a)) illustrates one-hop (B_A) neighbors and two-hop uncovered neighbors (U_A) of node A. From B_A, node C is selected first with maximum white neighbors from U_A. Here, only C constructs the first forwarding list as it covers all nodes from U_A (3(b)). As, we aim to create multiple forwarding lists, after first iteration, configuration are initialized as initial values except the cardinality values of C (3(c)). Again, another new forwarding list (B and F) is created using the same procedure (3(d)–(e)).

For dynamic scenario, as the environment changes, we generate only a set for each time and store last $K - 1$ used sets. For the next time, it creates its new forwarding list with minimal overlap with last $K - 1$ sets.

Algorithm 2 represents a set of n forward lists (F_{vn}) creation algorithm for static scenario of a node v. The outer while loop generates multiple forwarding lists (line 3–14). In each iteration it selects node s from B_v with minimum cardinality and maximum uncovered 2-hop neighbours (line 4). All neighbors of s are discarded from U, the cardinality of s is incremented by 1, and s is added to F_{vn} (line 6). If it's cardinality reaches to K then it is discarded from B_v (line 7–8). This procedure runs until there is no node remaining in U. Whenever, U becomes null, it is again initialized to U_v to create next forwarding list along with necessary changes.

Suppose, Δ is the maximum degree of the graph. Hence, the size of B_v and U_v can be at most Δ and Δ^2 respectively. Therefore, the run time complexity of creating a forwarding list is Δ^3. For static network, multiple forwarding lists could be at most Δ. Hence, total time complexity is Δ^4. On the other hand, for dynamic scenario, a single set is generated comparing with previous $K - 1$ lists with $K\Delta^3$ complexity .

Algorithm 2. Forward Lists Creation of a node v for Distributed Algorithm

Require: B_v, U_v
Ensure: A Set of forwarding lists
1: $F_{v1} = \emptyset$, $U = U_v$, $n = 1$
2: Set $C_i \leftarrow 0$ for all node $i \in B_v$
3: **while** a new set can be generated **do**
4: Select a node $s \in B_v$ with minimum C_s and maximum $N(s) \cap U$
5: **if** $(N(s) \cap U) > 0$ **then**
6: $F_{vn} = F_{vn} \cup s$, $U = U - N(s)$, $C_s = C_s + 1$
7: **if** $C_s == K$ **then**
8: $B_v = B_v - s$
9: **end if**
10: **end if**
11: **if** U is $NULL$ **then**
12: $U = U_v$, $n = n + 1$, $F_{vn} = \emptyset$
13: **end if**
14: **end while**

5 Experiments

Finally, in this section, simulation results of the algorithms and the improvement of system fault tolerance and network life time will be shown in performance metrics.

5.1 Experimental Setup

We simulate our fault tolerant MMCDS algorithms along with basic MCDS and other algorithms. Here, we consider a network over an area of 100 × 100 units followed by N wireless nodes. Initial battery power is kept fixed at 100 unit and transmission range is kept fixed at 25 units. Overlapping boundary K is set 1 to 5 for different simulations. Additionally, to observe the impact of K on different performance we vary K from 1 to 20. In our simulations, we vary N from 20 to 200 with an increment of 20 to simulate the network performance. We simulate all the algorithms in Java programming language in Netbeans IDE.

5.2 Performance Measurements

In this section, we present the performance of our algorithms considering number of MMCDSs evaluation, size of MMCDSs with and without optimization function, average packet passing, network life time and fault tolerance of the network.

5.2.1 Multiple MCDSs Evaluation

An individual CDS is capable of communicating whole network or keep the network active. However, generating multiple MCDS indicates that the network gets more options to choose for communication. However, generating multiple disjoint CDSs always might not be feasible. Therefore, we allow some overlapping among the CDSs bounded by K. If $K = 1$, only disjoint CDSs are generated. When $K = 2$, we allow any node to be present at maximum two CDSs. Therefore, with the increase of K, the number of MCDSs increases as well. However, it reaches to a saturated value after a certain value of K as no new sets can be generated. For simulation, we use $K = 1$ to 20 with an increment of 1 and find out a suitable value of K for upper boundary of overlapping among generated MCDSs.

5.2.2 Network Lifetime

The network lifetime means how long the network remains active. Generating multiple CDSs can increase network longevity by using all the sets in round robin scheduling. For calculating network life time, we assume that all the nodes have similar battery power with T time unit. If only one CDS is generated and all the nodes are activated for the entire time to communicate over the network, then the lifetime of the network becomes $NL = T$. On the other hand, with the increase of the number of CDSs, network lifetime also increases as each CDS is scheduled at different time periods. If there are n disjoint CDSs, then the network lifetime becomes, $NL = nT$ [16]. However, if there exists at most K boundary overlaps of anode among all CDSs, then the network lifetime becomes as follows [16]:

$$NL = \sum_{i=1}^{n} t_i \qquad where, \qquad t_i = \frac{T}{\max(C_{ij} : j = 1, 2, ..., \|CDS_i\|)} \tag{1}$$

Here, C_{ij} is the cardinality of node j in the i-th CDS.

5.2.3 Network Fault Tolerance

Fault tolerance of a network can be calculated as up to how many node failures it can tolerate or handle. For example, if any node of the system fails but the network still remains operative, then it has fault tolerance of 1. Moreover, if any two nodes of the system fails and the system still works, it has fault tolerance value of 2. For our system, fault tolerance is very high compared to a single MCDS network. If our system generates n disjoint sets, then it can tolerate up to n-1 node failures. However, as we consider some overlapping up to K, the fault tolerance value decreases and it depends on total number of sets creation and the overlapping boundary K. We calculate fault tolerance F for the system which satisfies the following equation:

$$n - \sum_{i=1}^{F} \max(C_i) = 0 \tag{2}$$

where, C_i is cardinality of i-th node failure.

5.2.4 Average Forwarding Nodes

The number of forwarding nodes can be defined as the total number of nodes (forward nodes) who forward or rebroadcast the broadcast packet plus node (for the source node). The equation of number of nodes forwarding can be defined as [17]:

$$\text{NFN} = \text{Number of nodes forwarding} + 1(\text{source node}).$$

As, our algorithm runs multiple CDSs, we consider average forwarding nodes (AFN) where it is the average of total forwarding nodes (TFN). Suppose, our algorithm runs n CDSs in round robin fashion. Hence, The equation can be defined as:

$$AFN = \frac{TFN}{n}; \qquad where, \qquad TFN = \sum_{i=1}^{n} NFN_i \tag{3}$$

5.3 Experiment Results

In this subsection, we present the results of our algorithms based on the performance measurement parameters along with other algorithms.

5.3.1 Number of MMCDSs

We evaluate total number of CDSs varying network size with $N = 20$ to 200 with an increment of 20. Figure 4(a) illustrates the results of CDSs construction applying optimization step. The result shows that with the increase of density of nodes the total number of MCDSs increases. Moreover, with the increment of overlapping boundary it also increases, as each node contributes to more new sets. For our MMCDSs construction, we use a step to minimize redundant nodes, hence the size of the CDSs also decreases. Figure 4(b) shows how total number of MCDSs changes with the increase of overlapping boundary K. It is clear that with the increase of overlapping boundary number of MMCDSs increases. However, it moves to a saturation point when there is no new node to generate a new set. For example, when we consider network size N = 20, for $k = 4$ the set construction is almost in saturation, whereas, for N = 30, the saturation point moves to $K = 18$.

Finally, Fig. 5 illustrates how the sizes of MCDSs construction decreases for different network size with optimization step. This happens because when we apply optimization, unnecessary nodes are removed from a set which reduces the sizes of MCDSs. Therefore, the total number of MCDSs also might increase as removed nodes might be used for further set construction.

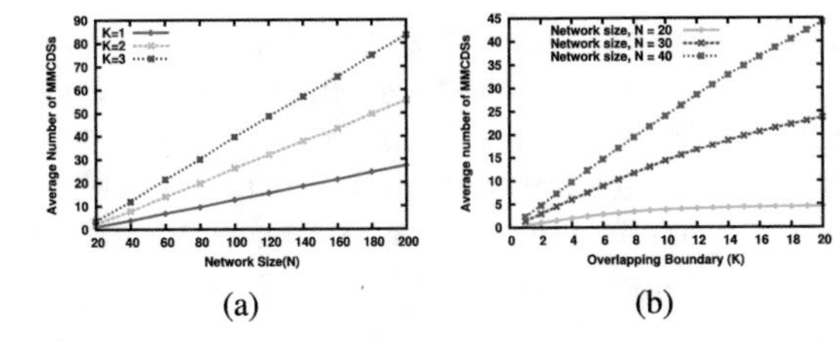

Fig. 4. Number of MMCDSs construction (a) for different overlapping boundary varying network size, (b) for different network size varying overlapping boundary

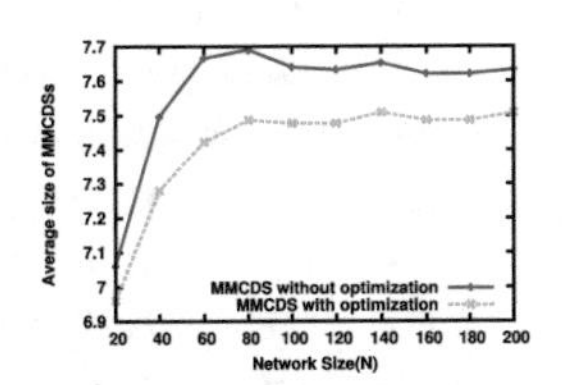

Fig. 5. Size of MMCDS construction for K = 2 with optimization and without optimization

5.4 Network Lifetime

Here, we present how our algorithms increase network lifetime in an ad-hoc network. Figure 6 illustrates the network lifetime by varying network size 20 to 200 with an increment of 5 for overlapping boundary $K = 1$ to 5 with an increment of 1 and keeping other parameters fixed. From figure, we can see that network lifetime increases for K $\geq$ 2 than $K = 1$. However, the lifetime remains almost same for higher values of K. Therefore, if we use overlapping boundary $K = 2$ or close to 2, then we can achieve maximum network life. Additionally, we represent here in Fig. 7, the relation of network lifetime with K more elaborately for two types of graph: sparse (number of nodes are minimum) and dense graph (number of nodes are maximum). Here, we represent the values of average network lifetime for overlapping boundary $K =$ 1 to 20. From figure, we can observe that for both types of graph give higher values for $K = 2$ or near values of 2.

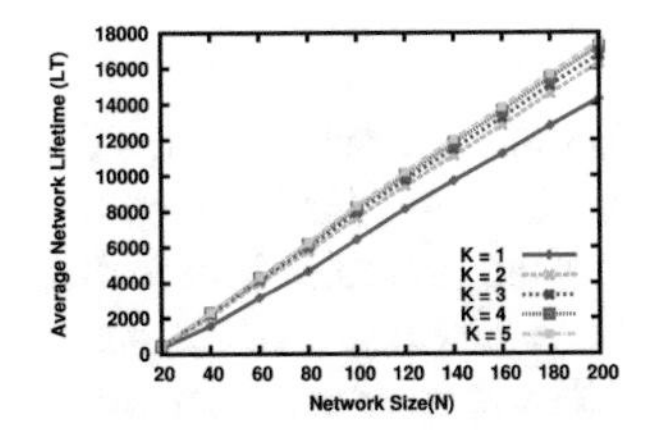

Fig. 6. Average network lifetime for different overlapping boundary varying network size

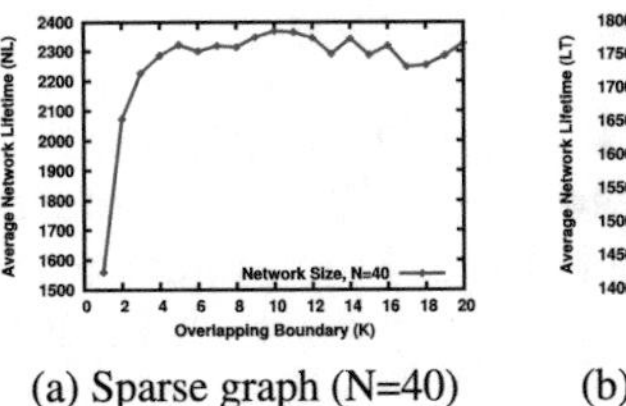

(a) Sparse graph (N=40)

(b) Dense graph (N=200)

Fig. 7. Results of average network lifetime for different overlapping boundary values on (a) Sparse graph and (b) Dense graph

5.5 Network Fault Tolerance

Here, we present how network fault tolerance changes if we apply our algorithms in an ad-hoc network varying the overlapping boundary K. Figure 8 illustrates the network fault tolerance for overlapping boundary $K = 1$ to 5 with an increment of 1. Other parameters also remain fixed here. From figure, we can see that network fault tolerance for $K = 1$ increases with the increase of network size. For, network size of 200, it can tolerate approximately 20 faults on the average. Although the number of MMCDSs increases gradually for other values of $K \geq 1$ (Fig. 4), the fault tolerance values don't increase with that similar proportion. For example, for network size 200, when $K = 2$, the network can handle almost 24 node failures. However, for same network size, the network can handle nearly 27 node failures for $K = 5$. Therefore, we can say that more overlapping boundaries although generate more MCDSs, the fault tolerance doesn't improve that much. Figure 9 shows how K changes the fault tolerance for different network. It can be observed that, K has similar effects for all types of network.

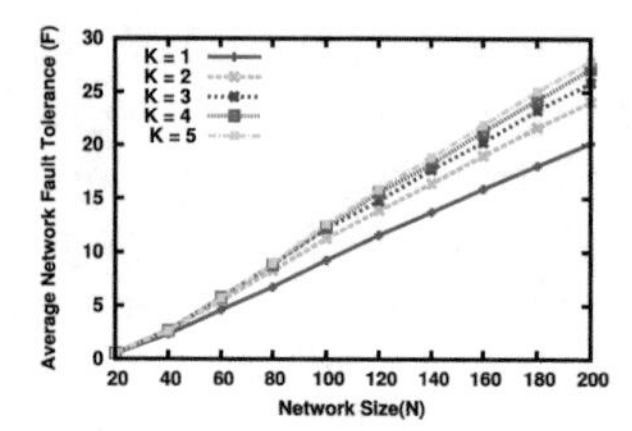

Fig. 8. Average Network Fault Tolerance for different overlapping boundary varying network size

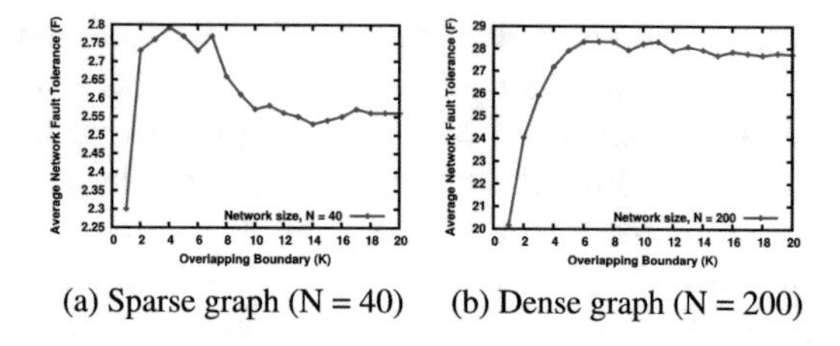

Fig. 9. Results of average Fault Tolerance for different overlapping boundary values on (a) Sparse graph and (b) Dense graph

5.6 Average Forwarding Nodes

Figure 10 represents average forwarding nodes for different network sizes for different algorithms. Here, in Fig. 10(a) we compare our centralized MMCDSs algorithm with $K = 1$ and $K = 2$. From, figure we can see that, average forwarding nodes increase almost linearly for each algorithm. It is obvious that a single MCDS has less nodes forwarding than any other algorithms. However, our algorithm has better result than 1–2 CDS which is a general case of k-m CDS. Additionally, Fig. 10(b) shows the average node forwarding values of our algorithm along with the basic dominant pruning algorithm for packet forwarding. Although we consider here multiple sets construction, our algorithm performs nearly the basic one.

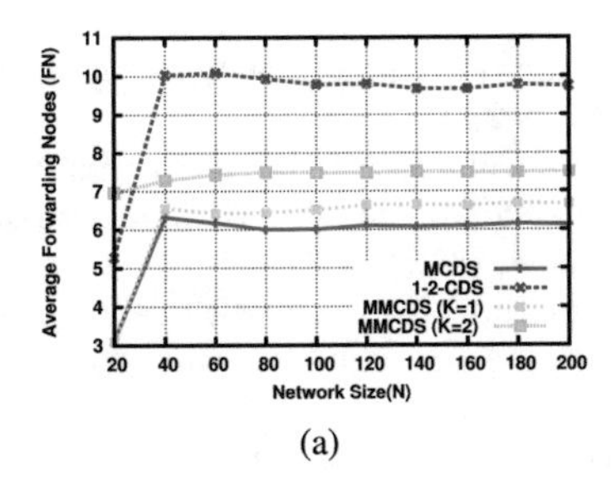

(a)

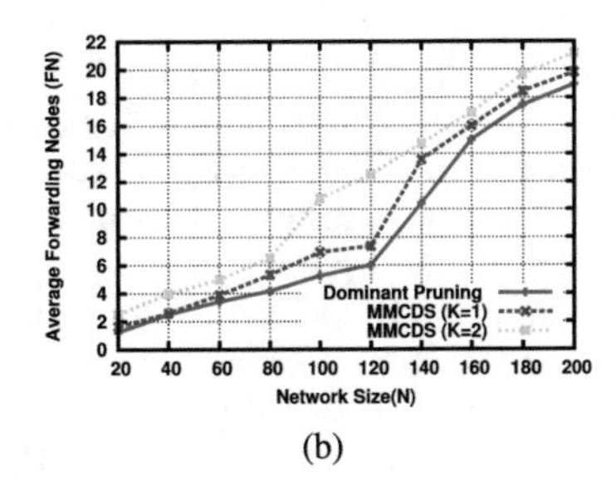

(b)

Fig. 10. Average Forwarding Nodes of (a) centralized and (b) distributed algorithms along with our MMCDSs with $K = 1$ and $K = 2$

6 Conclusion and Future Work

In this paper, we concern about network lifetime and fault tolerance of wireless ad-hoc networks. Therefore, for efficient communication among nodes over the network, we construct multiple connected dominating sets using possible minimum nodes. We can use those sets in round robin fashion to enhance network lifetime or keep as back up of active sets to handle system fault tolerance. However, always disjoint sets constructions might not be possible. Therefore, we introduce a user defined overlapping boundary which indicates in how much sets a node can be present. We apply the strategy both in centralized and distributed version of our algorithm. A comprehensive simulation result is presented to analyse the behaviour of the developed algorithms. However, when we consider overlapping boundary $K \geq 2$, we only consider the worst case for calculating average fault tolerance. If we could consider all possible cases of node failures, the fault tolerance would improve more than our calculated values. Therefore, our future challenge is to provide a mathematical probabilistic model for analyzing system fault tolerance for all possible node failures. In future, we plan to develop analytical model(s) for finding out overlapping boundary based on network pattern.

References

1. Tseng, Y.-C., Ni, S.-Y., Chen, Y.-S., Sheu, J.-P.: The broadcast storm problem in a mobile ad hoc network. Wirel. Netw. **8**(2), 153–167 (2002)
2. Butenko, S., Cheng, X., Du, D.-Z., Pardalos, P.M.: On the construction of virtual backbone for ad hoc wireless network. In: Butenko, S., Murphey, R., Pardalos, P.M. (eds.) Cooperative Control: Models, Applications and Algorithms, pp. 43–54. Springer, Boston (2003). https://doi.org/10.1007/978-1-4757-3758-5_3
3. Blum, J., Ding, M., Thaeler, A., Cheng, X.: Connected dominating set in sensor networks and MANETs. In: Du, D.Z., Pardalos, P.M. (eds.) Handbook of Combinatorial Optimization. Springer, Boston (2004). https://doi.org/10.1007/0-387-23830-1_8
4. Lim, H., Kim, C.: Flooding in wireless ad hoc networks. Comput. Commun. **24**(3), 353–363 (2001)

5. Lou, W., Wu, J.: On reducing broadcast redundancy in ad hoc wireless networks. IEEE Trans. Mob. Comput. **1**(2), 111–122 (2002)
6. Rahman, A., Gburzynski, P., Kaminska, B.: Enhanced dominant pruning-based broadcasting in untrusted ad-hoc wireless networks. In: 2007 IEEE International Conference on Communications, pp. 3389–3394. IEEE (2007)
7. Rahman, A., Hoque, M.E., Rahman, F., Kundu, S.K., Gburzynski, P.: Enhanced partial dominant pruning (EPDP) based broadcasting in ad hoc wireless networks. J. Netw. **4**(9), 895–904 (2009)
8. Ephremides, A., Wieselthier, J.E., Baker, D.J.: A design concept for reliable mobile radio networks with frequency hopping signaling. Proc. IEEE **75**(1), 56–73 (1987)
9. Butenko, S., Cheng, X., Oliveira, C.A., Pardalos, P.M.: A new heuristic for the minimum connected dominating set problem on ad hoc wireless networks. In: Butenko, S., Murphey, R., Pardalos, P.M. (eds.) Recent Developments in Cooperative Control and Optimization, pp. 61–73. Springer, Boston (2004). https://doi.org/10.1007/978-1-4613-0219-3_4
10. Cheng, X., Ding, M., Chen, D.: An approximation algorithm for connected dominating set in ad hoc networks. In: Proceedings of the International Workshop on Theoretical Aspects of Wireless Ad Hoc, Sensor, and Peer-to-Peer Networks (TAWN), vol. 2 (2004)
11. Bai, X., Zhao, D., Bai, S., Wang, Q., Li, W., Mu, D.: Minimum connected dominating sets in heterogeneous 3D wireless ad hoc networks. Ad Hoc Netw. **97**, 102023 (2020)
12. Shi, Y., Zhang, Y., Zhang, Z., Wu, W.: A greedy algorithm for the minimum k-connected m-fold dominating set problem. J. Comb. Optim. **31**(1), 136–151 (2016)
13. Zhou, J., Zhang, Z., Tang, S., Huang, X., Mo, Y., Du, D.-Z.: Fault-tolerant virtual backbone in heterogeneous wireless sensor network. IEEE/ACM Trans. Netw. **25**(6), 3487–3499 (2017)
14. Barnett, J., Blumenthal, A., Johnson, P., Jones, C., Matzke, R., Mujuni, E.: Connected minimum secure-dominating sets in grids. AKCE Int. J. Graphs Comb. **14**(3), 216–223 (2017)
15. Farzana, S., Papry, K.A., Rahman, A., Rab, R.: Maximally pair-wise disjoint set covers for directional sensors in visual sensor networks. In: Wireless Days (WD), pp. 1–7. IEEE (2016)
16. Saha, S., Zishan, A.A., Rahman, A.: On target monitoring in directional sensor networks by jointly considering network lifetime and fault tolerance. In: Proceedings of the 6th International Conference on Networking, Systems and Security, pp. 68–76 (2019)
17. Akter, M., Islam, A., Rahman, A.: Fault tolerant optimized broadcast for wireless ad-hoc networks. In: 2016 International Conference on Networking Systems and Security (NSysS), pp. 1–9. IEEE (2016)

Forecasting the Number of Firemen Interventions Using Exponential Smoothing Methods: A Case Study

Roxane Elias Mallouhy[1(✉)], Christophe Guyeux[2], Chady Abou Jaoude[3], and Abdallah Makhoul[2]

[1] Prince Mohammad Bin Fahd University, Khobar, Kingdom of Saudi Arabia
reliasmallouhy@pmu.edu.sa
[2] Univ. Bourgogne Franche-Comté, Belfort, France
{christophe.guyeux,abdallah.makhoul}@univ-fcomte.fr
[3] Antonine University, Baabda, Lebanon
chady.aboujaoude@ua.edu.lb

Abstract. Predicting the number of firemen interventions to size the appropriate workload of firefighters to the appropriate need is vital for reducing material and human resources. Therefore, it will have a great impact on reducing the financial crisis resulting from global warming and population growth. The database in this research includes interventions recorded hourly from "1 January, 2015 00:00:00" to "31 December, 2019 23:00:00" in Doubs, France. The data were processed, decomposed, outliers were detected and replaced. Thenceforth, optimal smoothing values were selected and then three different models of Exponential Smoothing were deployed. Experiments have shown that Holt-Winters' method has the best accuracy comparing to the baseline and other Exponential Smoothing techniques. The results are promising and would optimize the number of firefighters' resources.

1 Introduction

Time series forecasting is a data science analysis used in a variety of fields. It has a substantial impact on future decisions with a generous range of complexity and applications. In agriculture, predicting the weather of the coming days helps to plan sowing or harvesting. In retail, forecasting daily sales gives guidance for inventory decisions. In finance, more successful investments are possible after predicting stock prices. Also, forecasting the number of firemen brigades would help optimize human and material resources.

Over time, the number of firefighters' interventions in France has steadily increased, reaching more than 4.65 million operations in 2017 [1]. Predicting the future number of firefighting missions would help the emergency response system to estimate the possible flow of events in the coming hours and days, and then be able to optimize human and material resources. This would have a remarkable impact on protecting people, the environment, and property from

L. Barolli et al. (Eds.): AINA 2022, LNNS 449, pp. 579–589, 2022.
https://doi.org/10.1007/978-3-030-99584-3_50

damage, critical incidents, and disasters. Consequently, this would increase the efficiency of emergency operations while reducing operational costs.

On top of that, the use of machine learning to predict the number of firefighters' operations seems to be efficient as it can be assumed that the number of operations is affected by climate, time, and other events such as New Year's Eve, holidays. Following this principle, it will be possible to analyze past observations using historical values, and associated patterns to predict future deployments. These properties are well represented in time series forecasting approaches that consider the trend, level, and seasonality of a time-ordered series. They analyze and forecast data observations and the result can lead to better decisions. To date, several models of time series forecasting models have been created, one of which is Exponential Smoothing, which gives more weight to recent values but gives less exponential importance to older observations. This is particularly the case with the firefighter dataset, as it is reasonable that the number of deployments in the upcoming hour tends to be influenced by the number of deployments in the previous hour.

The objective of this study is to predict the number of firemen brigades by applying the techniques of forecasting Simple Exponential Smoothing, Holt and Holt-Winters to a concrete dataset over the period 2015–2019. In this paper, we will show how we prepare the data and how we build each model to realistically predict the future. This paper is organized as follows: Sect. 2 presents an overview of Exponential Smoothing techniques and related work to our research; Sect. 3 demonstrates how the data was prepared; Sect. 4 provides the optimal choice of values of each model and experiments. Results are interpreted in Sect. 5 and last section provides conclusions and future work.

2 State of the Art

This section reviews various literature studies based on Exponential Smoothing methods in many fields. A large number of researches have been done on the powerful ES methods for time series prediction in many fields such as tourism, weather, COVID-19 pandemic, finance, economics, and medicine.

Singh et al. [2] implemented in his research Exponential Smoothing method to predict the number of tourists for 2018 of an Indian state using Java programming language for the years 2008 to 2017. He applied different values of smoothing constant to find the best accuracy of the model. Zafar et al. [3] analyzed and studied temperature data and variability of two major regions using the Simple Smoothing Technique and concluded that SES gives the best predictive values compared to other models.

Moreover, Argawu [4] predicted the number of COVID -19 new cases in the 10 most infected African countries by applying regression, ARIMA, and Exponential Smoothing Models. Yasar and Kilimci [5] emphasized how to mix Time Series Forecasting methods with Financial Sentiment Analysis data collected from Twitter, Instagram, and Facebook. In their case study, they employed ARIMA, Holt, and Holt-Winters to provide a more consistent exchange rate prediction to any user wishing to exchange Turkish Lira /US dollars, and ended their study with the best-observed performance belonging to the Holt-Winters' method.

Anggrainingsih et al. [6] analyzed time series data of website visitors using the Triple Exponential Smoothing method. Their results showed the optimal alpha, beta, and gamma for the best prediction accuracy. Lai et al. [7] proposed a hybrid methodology by integrating Neural Network with Exponential Smoothing for financial time series prediction. Their experimental results considered the accuracy and directional predictions and showed that the hybrid-integrated method performs better than the two benchmark models. Jones et al. [8] examined and evaluated the use of SARIMA, Exponential Smoothing, time series regression, and ANN to predict daily patient volume in the emergency department, compared the results with the multiple linear regression model previously performed and concluded that the regression-based model provided the most consistent accuracy.

Incidentally, few studies on time series forecasting using skills in Artificial Intelligence have considered the topic of our research about fire department operations. Nahuis et al. used Long Short-Term Memory and demonstrated the possibility to build a neural network from scratch and the ability to predict the number of deployments in 2017 from those from 2012 to 2016 [9]. Couchot et al. [10] worked on a learning process based on real but anonymous data and conducted the study using the Extreme Gradient Boosting technique. In [11], Guyeux et al. applied an ad-hoc Multi-Layer Perceptron in which hyper-parameters were selected using a supercomputer, and their work gave a reasonable prediction of firefighters' operations. Furthermore and Cerna et al. [12] compared between XG-Boost, Gradient Boosting, and AdaBoost, as these techniques are considered very effective in modeling nonlinear systems. Arcolezi et al. [13] focused on local-differential privacy-based data in their study. Their approach was to predict the number of firefighters in specific locations by applying differential privacy along with XG-Boost techniques. None of these methods are classified as exponential smoothing techniques, which is the added value of this research compared to the aforementioned previous studies.

3 Data Preparation

3.1 Data Acquisition

The data considered in this study contains the number of firemen interventions collected from "January 1, 2015, 00:00:00" to "December 31, 2019, 23:00:00" by the Fire and Rescue Department SDIS25, in Doubs, France. In this paper, two datasets are analyzed: the first one (hourly-dataset) contains the number of interventions per hour, while the second one (daily-dataset) contains the average of interventions per day. Therefore, the hourly dataset consists of 43824 interventions, while the daily dataset carries 1826 interventions.

3.2 Outliers Detection

As can be seen in Fig. 1, there are black dots outside the blue box: these are the outliers.

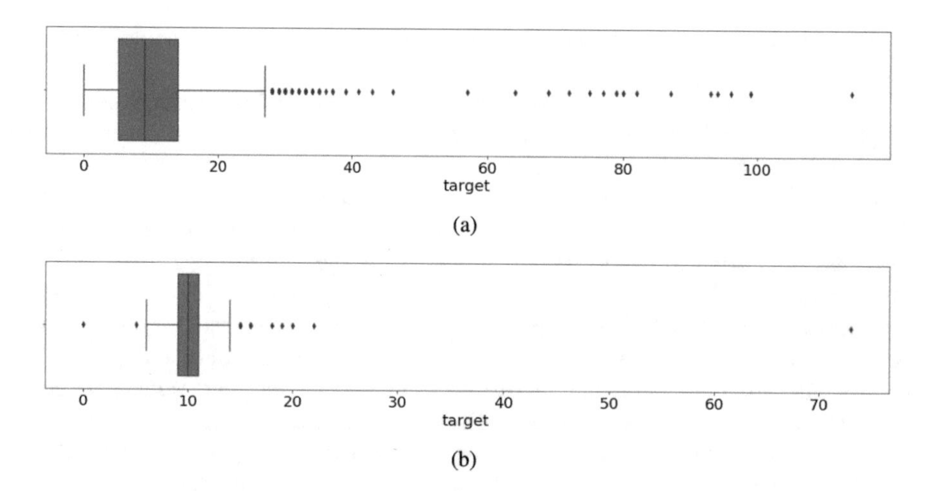

Fig. 1. Graphical visualization for outliers detection for (a) hourly-dataset and (b) daily-dataset

We consider everything above 25 for the hourly-dataset and everything below 6 and above 14 for the daily-dataset as anomalies. The boxplot detects 327 outliers for the hourly-dataset and 39 outliers for the daily-dataset.

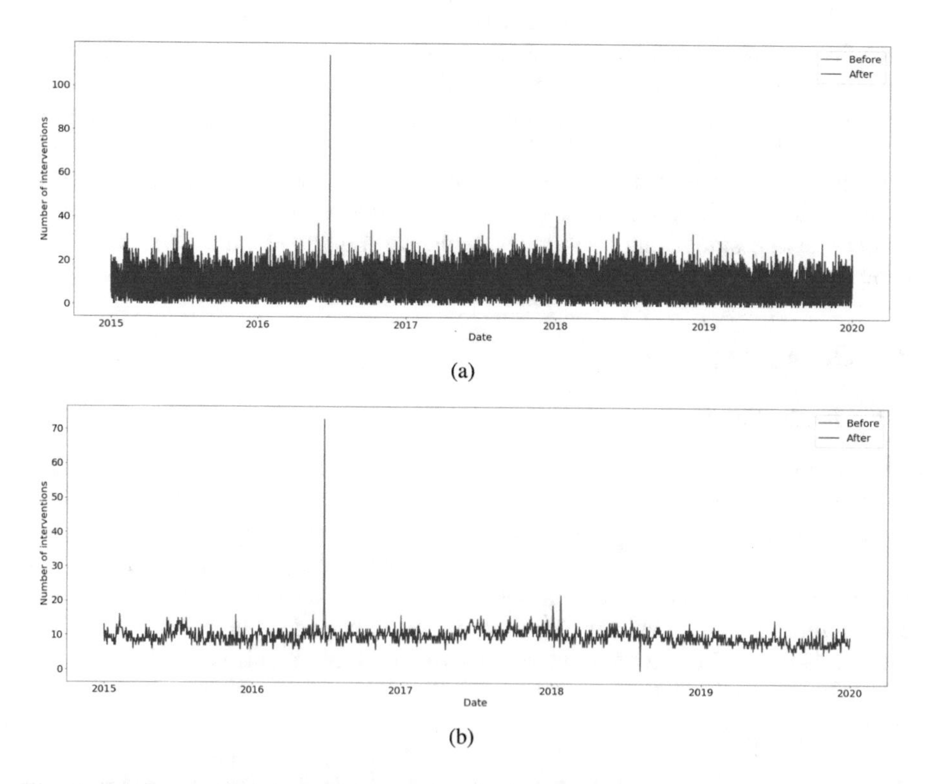

Fig. 2. Number of firefighters' interventions before and after replacing outliers for (a) hourly-dataset and (b) daily-dataset

To remove what can be considered as anomalies, we replaced the outliers with the lower or upper whisker, and consequently, the datasets have been updated, as can be viewed in Fig. 2.

3.3 Datasets Decomposition

It is important to perform the decomposition of the datasets to get a structured view about the components used in the Exponential Smoothing methods, such as:

- trend: increasing/decreasing tendencies of firefighters' interventions;
- seasonality: repeating cycle;
- residual: random variation of the dataset.

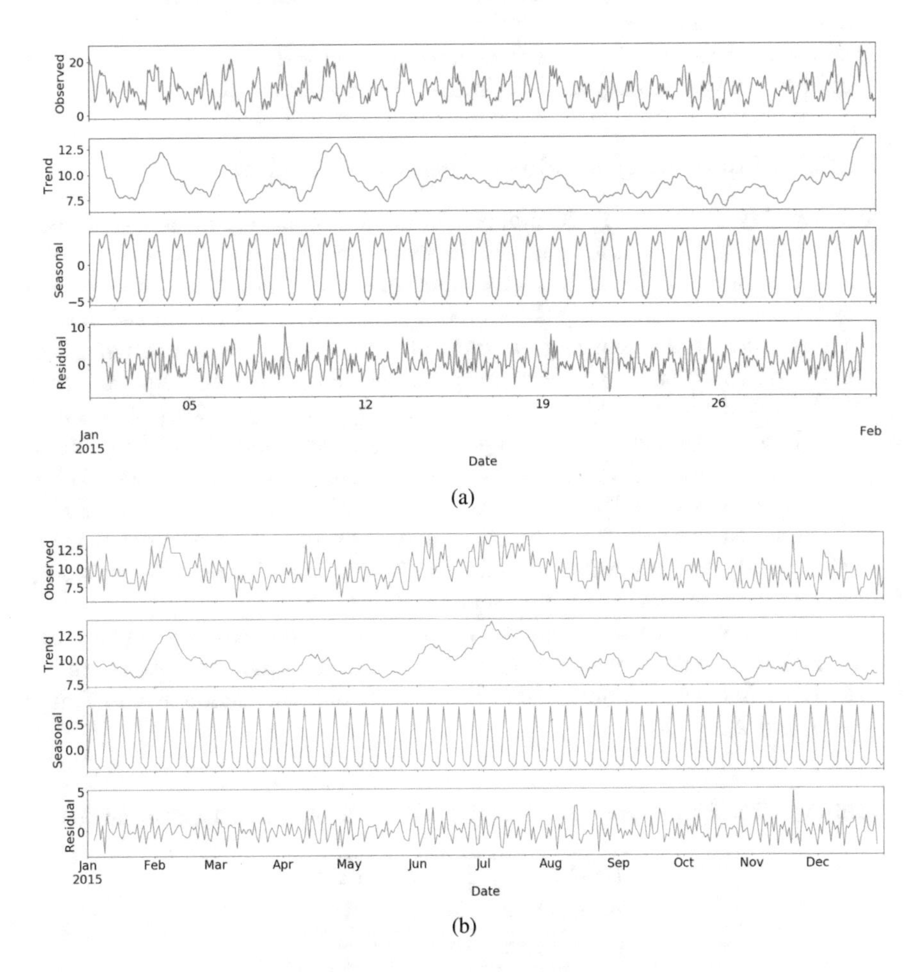

Fig. 3. Decomposition charts for (a) hourly-dataset and (b) daily-dataset

In this study, both the daily and hourly datasets show an interesting seasonality and trend: the cycle repeats every day/24 h for the hourly-dataset and every week/7 days for the daily-dataset, as shown in Fig. 3. Moreover, the residuals are also reasonable and show different variability over time.

4 Experimental Methods and Results

4.1 Datasets

Each dataset is divided into training and testing, and to back test the three Exponential Smoothing methods, this study uses the walk-forward validation technique by repeating the following steps:

1. training the model using the minimum number of samples in the window;
2. prediction for the unique next step;
3. evaluation of the predicted value against the real value;
4. expansion of the window to include the known value.

4.2 Selection of Appropriate Smoothing Constants

The sensitivity of the predictions depends on the smoothing constants. Larger values of alpha (α) form a forecast that is more sensitive to recent observations, while in contrast, smaller values give a dampening result. The same concept applies to beta (β), which emphasizes recent trends over older observed values.

In this study, the smoothing parameters were selected as a function of the minimum values of mean absolute error (MAE) and root mean squared error (RMSE) so that the forecasts are more accurate. Different values of α and β were tried on the datasets. The concept is to repeat the loop 99 times for simple Exponential Smoothing, whereas for Holt's method the loop is repeated 99 * 99 times because it has two smoothing parameters. α and β are ranged from 0 to 1 and the loops increase the value of α and β by 0.01 at each iteration. On the other hand, the seasonality used for the Holt-Winters' method is picked up depending on the seasonal curve presented in Sect. 3.3.

The obtained optimal constants for the hourly-dataset are alpha = 0.9, beta = 0.05 and the seasonality is 24 h. On the other side, the optimal values for the daily-dataset are alpha = 0.1, beta = 0.05 and the seasonality is 7 days (Tables 1 and 2).

Table 1. The RMSE measures using different value of smoothing constant (α)

Daily-dataset		Hourly-dataset	
Alpha	RMSE	Alpha	RMSE
0.1	1.409	0.1	4.775
0.5	1.464	0.5	3.476
0.9	1.624	0.9	3.172

Table 2. The RMSE measures using different value of smoothing constants (α) and (β)

Daily-dataset			Hourly-dataset		
Alpha	Beta	RMSE	alpha	beta	RMSE
0.1	0.05	1.408	0.9	0.05	3.175
0.5	0.05	1.451	0.9	0.5	3.246
0.1	0.5	1.656	0.4	0.5	3.961

4.3 Forecasting Results

To check the efficiency of the models used, forecasting was made for different time intervals in the future, and to compare the obtained results with a reference, the persistence model is considered as the baseline and MAE and RMSE are calculated. Moreover, the prediction of the number of firemen for 300 h in 2019 for the hourly-dataset is shown in Fig. 4 and the result of prediction for the daily-dataset in the whole year 2019 is displayed in Fig. 5.

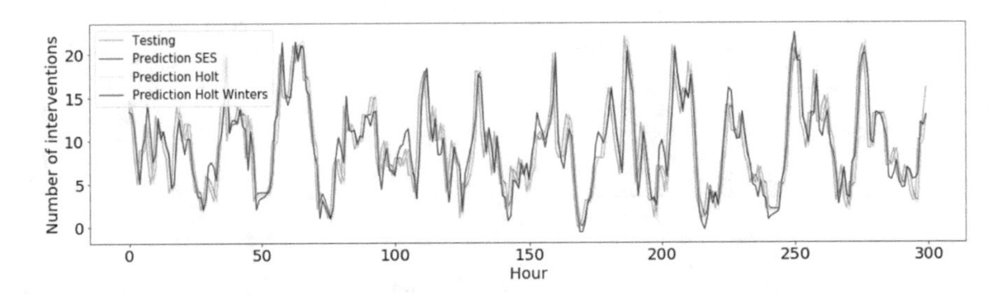

Fig. 4. Various models to predict the number of firefighters' interventions during 300 h in January 2019 for hourly-dataset

To show the results for multiple periods and since seasonality is evident in this study for both the daily and hourly datasets, we calculated the average number of firefighters' interventions on an hourly basis from "00:00:00" to "23:00:00" for the hourly-dataset, and on weekly basis from Monday to Sunday for the daily-dataset. Subsequently, the same concept is processed by applying Simple Exponential Smoothing, Holt's, Holt-Winters', and persistence models. The results for the different methods used are illustrated in Figs. 6 and 7.

5 Discussion

This research aimed to develop three time forecasting methods using hourly and daily data from 2015 to 2019, split them into training/testing, implement walk-forward validation to back test datasets, and select optimal values of smoothing parameters and seasonality.

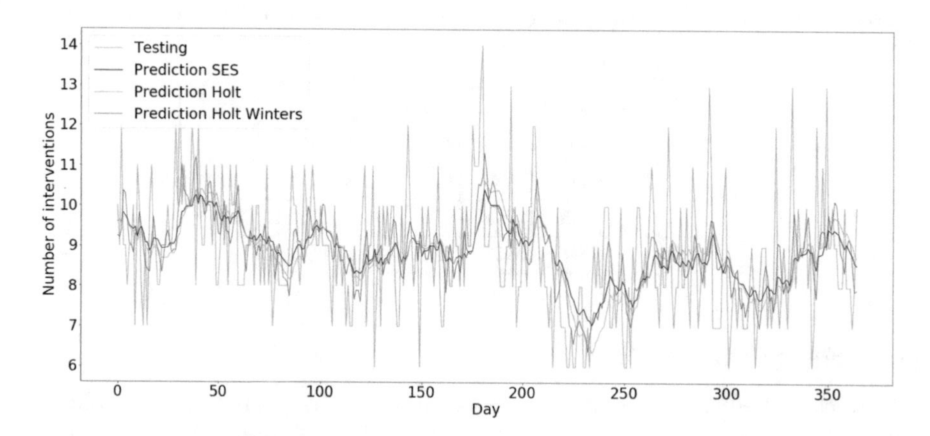

Fig. 5. Various models to predict the number of firefighters' interventions during the whole year 2019 for daily-dataset

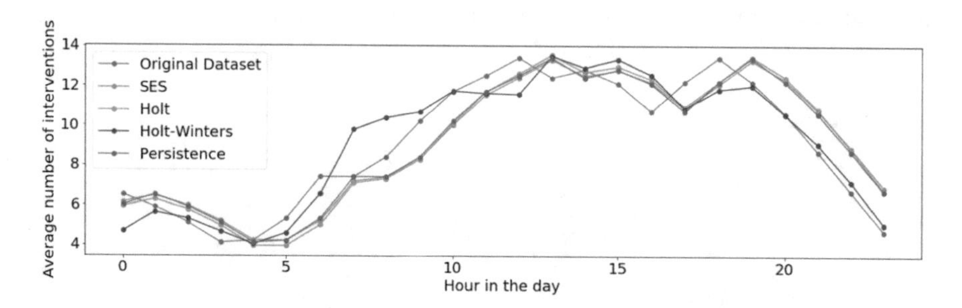

Fig. 6. The average number of firefighters' interventions over the hours of a day for hourly-dataset

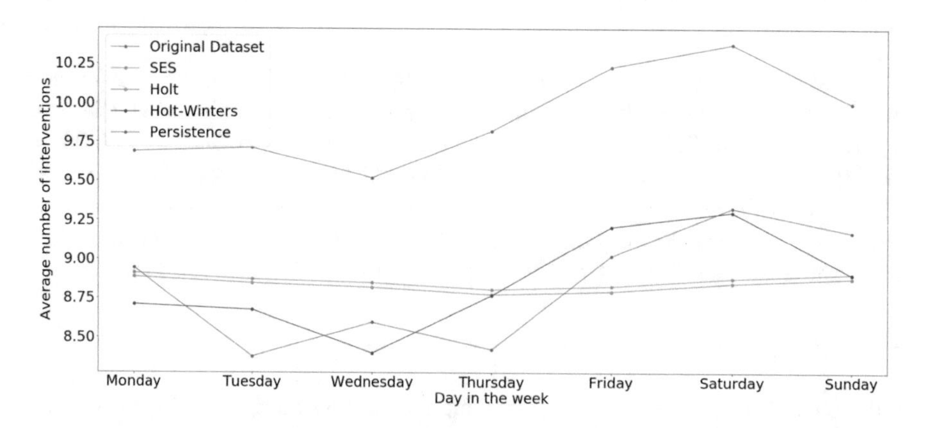

Fig. 7. The average number of firefighters' interventions over the days of the week for daily-dataset

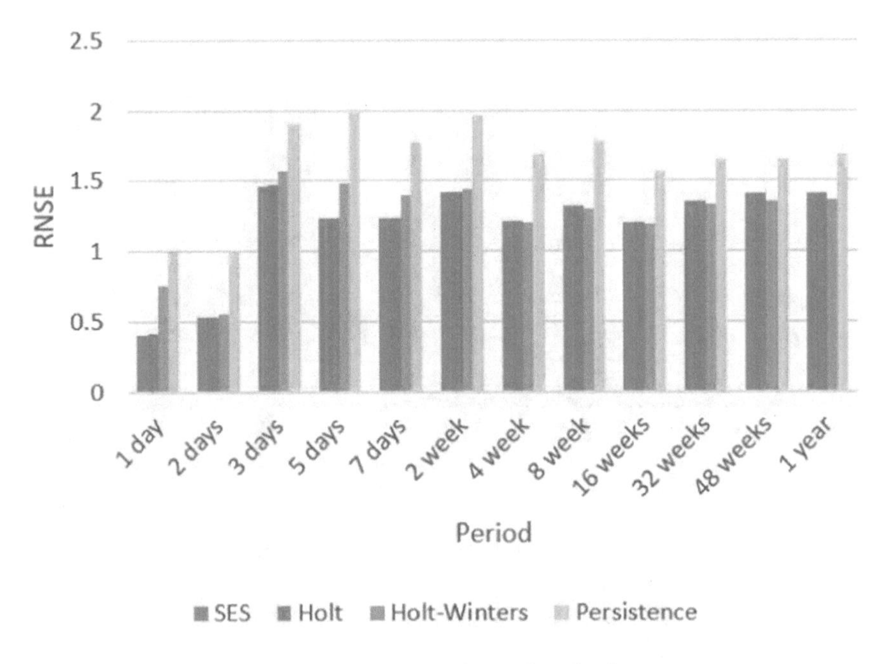

Fig. 8. Prediction results on hourly-dataset

From Figs. 8 and 9, it can be seen that the value of RMSE for hourly-dataset is increased with the increase of alpha. However, the opposite is observed for the daily-dataset as the RMSE decreases with the increase of alpha. The selected optimal values resulted from the minimal RMSE. These results reflect that Exponential Smoothing assigns larger weights to the recent observations in the hourly-dataset and fewer weights in the daily-dataset. This means that α has a smaller effect and gives more importance to the recent observations in hourly-dataset, while the data in the daily-dataset is less sensitive to the recent changes. Furthermore, the optimal value of beta is 0.05, which is very close to zero. This means that more weight has been given to the past trends in the estimation of current trends.

After selecting the smoothing constants that produce less error, MAE and RMSE are calculated for each prediction period for the hourly and daily datasets to measure the analysis performed during the forecasting process. The use of single and double Exponential Smoothing in this work is not effective compared to the Holt-Winters' method, which gives the lowest prediction error over time. In other words, as can be seen in Figs. 8 and 9, when Holt-Winters' method is used, the RMSE decreases as the prediction period increases. This result reveals that Triple Exponential Smoothing is a feasible technique used in this research because both the daily and hourly datasets are seasonal.

Finally, the compilation of the average number of firefighters' operations during the days of the week for the hourly-dataset and during each hour of the day for the daily-dataset expose many relevant facts. Figures 6 and 7 show that

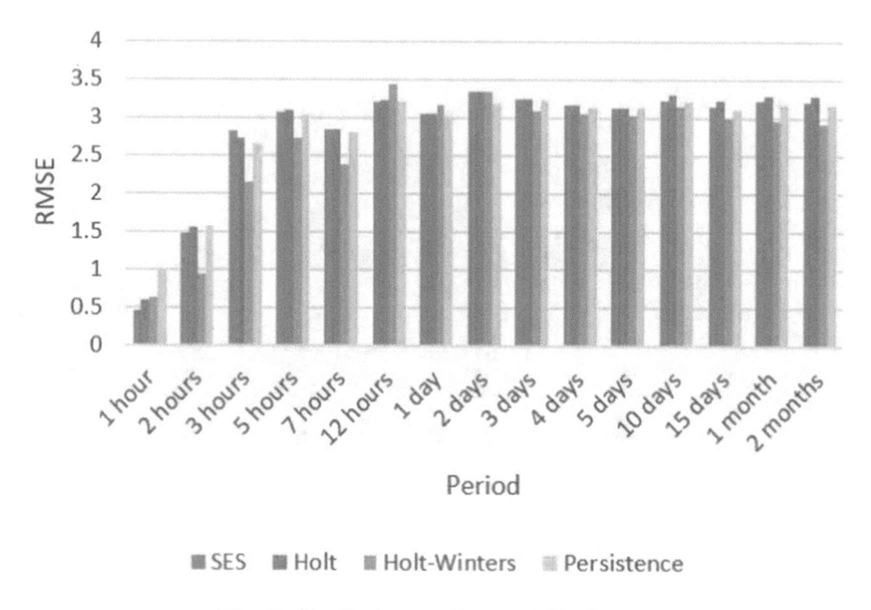

Fig. 9. Prediction results on daily-dataset

Holt-Winters' method has the most accurate values of prediction comparing to the original values of interventions. Additionally, it is observed in Fig. 6 that the firemen services increase from 5:00 am, tended to peak, and remain broadly stable throughout the day before gradually decreasing at 7:00 pm. This is very reasonable because the number of vehicles and flow of the people out of their homes is higher during the day causing rush hours and thus more risk of damage and incidents. On the other hand, it can be seen in Fig. 7 that the firefighters have the highest number of interventions from Friday to Sunday with less fluctuations on weekdays. Unsurprisingly, weekends are the riskiest days for fatal crashes.

6 Conclusion

It is very discernible that the prediction of the number of firefighters' interventions is not a simple random process, but is influenced by hourly and weekly changes related to human activities. Exponential Smoothing is a prominent tool to use in such a study that provides reliable forecasting. Statistics characteristics and graphical exploration of data have been presented to find the best Exponential Smoothing technique and three models were developed and then compared to each other and the baseline. Based on the prediction error, the less variation measures were found for the method Holt-Winters' since it takes into account seasonality, which is the main component of hourly and daily datasets.

For future work, new machine learning techniques will be explored on this dataset. Testing new and larger time steps is planned, such as ambulatory transportation. Finally, adding integrated explanatory inventories to the dataset is a possible idea to predict not only the number of interventions but also the type of interventions.

Acknowledgement. This work has been supported by the EIPHI Graduate School (contract ANR-17-EURE-0002) and is partially funded with support from the Hubert Curien CEDRE programme n 46543ZD.

References

1. Statista Research Department: Number of firefighting operations in France between 2007 and 2017
2. Singh, K., et al.: Implementation of exponential smoothing for forecasting time series data. Int. J. Sci. Res. Comput. Sci. Appl. Manag. Stud, **8** (2019)
3. Zafar, M.K.S., Khan, M.I., Nida, H.: Application of simple exponential smoothing method for temperature forecasting in two major cities of the Punjab, Pakistan
4. Argawu, A.: Modeling and forecasting of Covid-19 new cases in the top 10 infected African countries using regression and time series models. medRxiv (2020)
5. Harun Yasar and Zeynep Hilal Kilimci: US dollar/Turkish lira exchange rate forecasting model based on deep learning methodologies and time series analysis. Symmetry **12**(9), 1553 (2020)
6. Anggrainingsih, R., Aprianto, G.R., Sihwi, S.W.: Time series forecasting using exponential smoothing to predict the number of website visitor of Sebelas Maret university. In: 2015 2nd International Conference on Information Technology, Computer, and Electrical Engineering (ICITACEE), pp. 14–19. IEEE (2015)
7. Lai, K.K., Yu, L., Wang, S., Huang, W.: Hybridizing exponential smoothing and neural network for financial time series predication. In: Alexandrov, V.N., van Albada, G.D., Sloot, P.M.A., Dongarra, J. (eds.) ICCS 2006. LNCS, vol. 3994, pp. 493–500. Springer, Heidelberg (2006). https://doi.org/10.1007/11758549_69
8. Jones, S.S., Thomas, A., Evans, R.S., Welch, S.J., Haug, P.J., Snow, G.L.: Forecasting daily patient volumes in the emergency department. Acad. Emergency Med. **15**(2), 159–170 (2008)
9. Nahuis, S.L.C., Guyeux, C., Arcolezi, H.H., Couturier, R., Royer, G., Lotufo, A.D.P.: Long short-term memory for predicting firemen interventions. In: 2019 6th International Conference on Control, Decision and Information Technologies (CoDIT), pp. 1132–1137. IEEE (2019)
10. Couchot, J.-F., Guyeux, C., Royer, G.: Anonymously forecasting the number and nature of firefighting operations. In: Proceedings of the 23rd International Database Applications and Engineering Symposium, pp. 1–8 (2019)
11. Guyeux, C., et al.: Firemen prediction by using neural networks: a real case study. In: Bi, Y., Bhatia, R., Kapoor, S. (eds.) IntelliSys 2019. AISC, vol. 1037, pp. 541–552. Springer, Cham (2019). https://doi.org/10.1007/978-3-030-29516-5_42
12. Cerna, S., Guyeux, C., Arcolezi, H.H., Couturier, R., Royer, G.: A comparison of LSTM and XGBoost for predicting firemen interventions. In: Rocha, Á., Adeli, H., Reis, L., Costanzo, S., Orovic, I., Moreira, F. (eds.) WorldCIST 2020. AISC, vol. 1160, pp. 424–434. Springer, Cham (2020). https://doi.org/10.1007/978-3-030-45691-7_39
13. Arcolezi, H.H., Couchot, J.-F., Cerna, S., Guyeux, C., Royer, G., Al Bouna, B., Xiao, X.: Forecasting the number of firefighter interventions per region with local-differential-privacy-based data. Comput. Secur. **96**, 101888 (2020)

Mechanisms to Avoid the Unavailability of Points of Presence: A Systematic Review

Maria Camila Lijó and Luciana Pereira Oliveira(✉)

IFPB Campus João Pessoa, Av. Primeiro de Maio, 720 - Jaguaribe, João Pessoa, PB, Brazil
maria.lijo@academico.ifpb.edu.br, luciana.oliveira@ifpb.edu.br

Abstract. Typically, there are service providers teams that need to travel to resolve equipment failures present in one or more PoPs (Points of Presence). The biggest problem is when there is a shift to corrections due to service unavailability. It is believed that this can be avoided when monitoring the following variables: energy, internal and external PoP temperature, battery bank charge level and climatic changes (rain, fog and strong winds). Therefore, this work performs a systematic review with a start tool to find evidence of how these variables are worked on in the articles to ensure a better operation of the PoP without the need for face-to-face operators.

1 Introduction

The Internet is indispensable for obtaining information that can be found anywhere in the world. ISPs (Internet Service Providers) carry out preventive actions and maintenance to ensure the availability and scalability of the Internet service. The intention is for the end user to have access to the service in a transparent way to any failure that may occur in the ISP's network [1].

The PoP (Point of Presence) is a place where the ISP has devices that make connections with customers or with other PoPs [2]. Each ISP has a set of PoP's distributed in commercially strategic locations in order to serve customers [1,3].

In this ISP scenario, the SLA (Service Level Agreement) is associated with the control of the quality of service between the ISP and the customer. The SLA contains the specification of variables, ranges of acceptable variable value and penalties when the values obtained are not within what was agreed. So it is difficult to define the unique SLA for any location. For example, when the PoPs are far from the teams, there is an additional cost to perform device maintenance [2]. In addition, failures in PoP may require the presence of a technician and result in long downtime, increasing the possibility of penalties according to the SLA. In order to avoid non-compliance with the SLA, the ISP must look for:

- Variables and solutions to prevent failures
- Strategies to extend the useful life of network assets in PoP
- The remote activities in network infrastructure that contains a great diversity of assets in the network

L. Barolli et al. (Eds.): AINA 2022, LNNS 449, pp. 590–601, 2022.
https://doi.org/10.1007/978-3-030-99584-3_51

For example, temperature and energy of each device can be variables. They must be within predetermined ranges by the manufacturer. The operation of PoP device at temperatures and energy outside the predetermined ranges can result in irreversible damage to the asset. So, in addition to monitoring, action must be taken so that these variables are in accordance with what was determined by the manufacturer. Furthermore, the temperature produced by the assets during operation must be removed at a rate that ensures the proper functioning of the device in question [4,5].

In this way, the SLA and prevention are keys to guarantee the quality of the ISP, as it is directly linked to the availability of the customer's service. Failure prevention and contingency operations shall be planned and managed in such a way that service management is in compliance with the results of the service delivered.

Therefore, the purpose of this research based on Systematic Literature Review (SLR) method is to comprehend the diversity of PoP following two lines of study: in the first, the survey of works that addressed the context of PoPs to understand the structure, evolution and problems; in the second, the identification of solutions to avoid unavailability of services in PoPs. Such solutions were analyzed in the context of monitoring, automation and IoT (Internet of Things) that can be applied to PoPs' network assets. Information was extracted from a total of 162 papers, 89 on PoPs and 73 solutions that can be applied to PoPs.

The organization of this research based on systematic review is considered as follows. Section 2 presents the related works. The Sect. 3 describes the protocol used in the research will be discussed. The Sect. 4 presents the results of the research. Finally, the conclusion and future works are in Sect. 5.

2 Related Works

Demands regarding the availability and quality of PoP are increasing, due to the importance of the Internet. For this, the ISPs has some activities to keep the PoPs: design, planning, implementation, operation and support. However, the support team performs the activities that must deal with the challenge of several problems to avoid processing overload, inefficient management and failures.

However, this SLR did not find any secondary work in terms of PoP. On the other hands, reviews, survey and systematic review in context of automation were found. They can help the PoP support activities.

In terms of processing overhead, [6] presents a systematic review with IoT device. They can be used to improve efficiency and address the challenges of storing and processing a large volume of data. The paper [6] used data mining that is a method to explore data and look for consistent patterns, relationships between variables and detection of new data. The work [7] also addresses the context of IoT, complementing [6] with mechanisms to collect information from various sensors and cameras in order to control autonomous devices. These papers can be used to remotely automate and control PoPs.

Table 1. Secondary works in the context of PoP and support activities

Ref	Review P. type	Year	PoP	Main topic
[6]	S.Review	2018	No	Over process
[7]	Survey	2015	No	Over process and Management
[8]	S.Review	2018	No	Management
[9]	S.Review	2016	No	Faults
[10]	S.Review	2020	No	Faults
[12]	Survey	2018	No	Faults

In terms of inefficient management, the work [8] is a systematic review that presents Industry 4.0 trends with a variety of technologies to enable value chain development. It proposes an alternative to reduce production times and better product quality and organizational performance. Although they do not address the concept of PoP, these technologies can also be used to provide efficient management of PoPs. In this context, the survey [7] contributes to the study of ways to carry out activities without human intervention and, thus, saving people's routine connected to the collection and processing of information.

Failures and the displacement of technicians must be avoided. When some fault occur, the support team must be resolved quickly and looking for mechanism to solve remotely. In this context, [9] presents a systematic review that deals with an in-depth study of how to avoid DDoS attacks through IP traceback that can be solved remotely. The work [10] is also a systematic review to describe how SDN can be used to avoid or resolve failure remotely. The paper [12] is a survey that explores several issues related to failures in the area of networks, including issues related to earthquakes, terrorist attacks, DDoS attacks, cable disruptions and others that will require the displacement of technicians.

Therefore, this paper is the first SLR study in terms of PoP. The Table 1 presents the resume of the secondary papers and the existing deficiencies propose that is important to do a comprehensive literature review to address these weaknesses as follows:

- The present studies do not provide an in-depth study of PoP in terms of evolution and new services.
- Some papers presents review about faults, but they did not identify the problems and faults in PoP context.
- The works did not identify solutions (algorithms and prototype) to PoP problems.

3 Methodology

The SLR is a study based on the method of analyzing published works, and analyzing the most relevant works [13]. This method can be used by manual process or by tools. For this work, the start tool [11] was used in two steps.

In the first stage of this research, the keyword point-of-presence was applied to IEEE source. After the exclusion and inclusion criteria were applied to answer a set of questions related to the context of PoP, with 89 articles selected. These papers were used to categorize and extract the characteristics and problems related to PoP, as well as to select set of keywords related to problems with failures that require the presence of a technician in the PoP.

In the second step, the keywords (energy and temperature extracted from the first step) and considering monitoring, automation and IoT were applied in IEEE, ACM and Science Direct. After, the inclusion and exclusion criteria were applied to answer a set of questions to preventively solve problems. The objective is to find solutions to avoid the displacement of a technician and, thus, reduce the unavailability of services in the PoP.

Therefore, as shown in Fig. 1, first an SLR was performed to generate a set of keywords were defined to seek solutions to the problems. The final graphics about characteristics and problems associated with PoP. Finally, the results contain graphics in the context of PoPs. Moreover, at each phase, the review was carried out with three steps: Data planning; Execution; and Analysis.

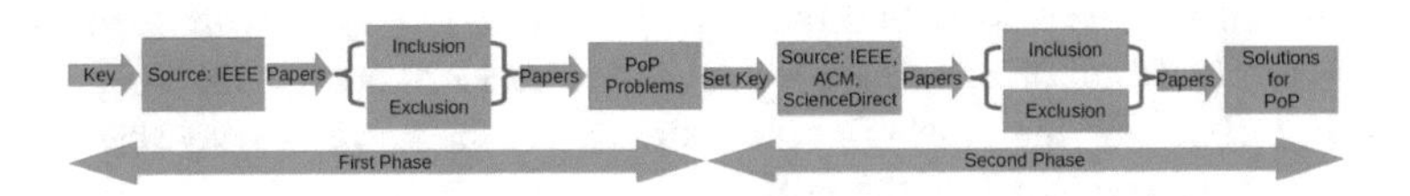

Fig. 1. Methodology

3.1 Research Questions (RQs)

In order to understand the evaluations of studies about PoP, this SLR paper formulated two set of the questions that will answer in the last phase of the research.

The first set of RQs referring to the understanding of the characteristics and problems related to the PoP:

- Do PoPs only offer connectivity service?
- What is the importance of the temperature variable for PoPs?
- How important is the energy variable for PoPs?
- What variables should be monitored to avoid having to travel a technician to troubleshoot PoPs?

The second group of questions related to solutions to avoid the displacement of technicians to the PoP site:

- What sensors and actuators are found as solutions to avoid downtime in PoPs?
- What algorithms were presented to avoid unavailability of device in the PoP?
- What protocols (at the application and transmission level) were presented in the solutions that may be applicable to PoPs?
- What programming languages and equipment were used in the prototypes that might be applicable to troubleshoot PoPs?

4 Search Results

This section presents and discusses the answers to eight RQs in Sect. 3.1.

4.1 RQ1: Do PoPs Only Offer Connectivity Service?

Initially, PoPs only offered connectivity services (data transmission), as described in [14] that was the first paper about PoP in 1989. This article also demonstrates the concern with the evolution and growth of the infrastructure of PoPs. At that time, there was no network integration with voice services that were provided by another infrastructure (telephone network). In 1998 and 1999 the integration of the data and voice network appears in the works [15] and [16]. They consider the voice and content service located within the PoP. On the other hand, when evaluating the interval between 1998 and 2021, shown in Fig. 2, there was a tendency for services to be in customer networks between 1998 and 2011. In 2011, the first article with the concept of virtualization considered the virtual routing as new service. Later, there was a tendency to have services (CDN, security and others services) within the PoP. A total of 42 papers with services inside of the PoP, while a total of 21 articles are related to services outside the PoPs.

So, currently, PoPs don't just offer the connectivity service. PoPs have been offering new services managed inside of the PoP.

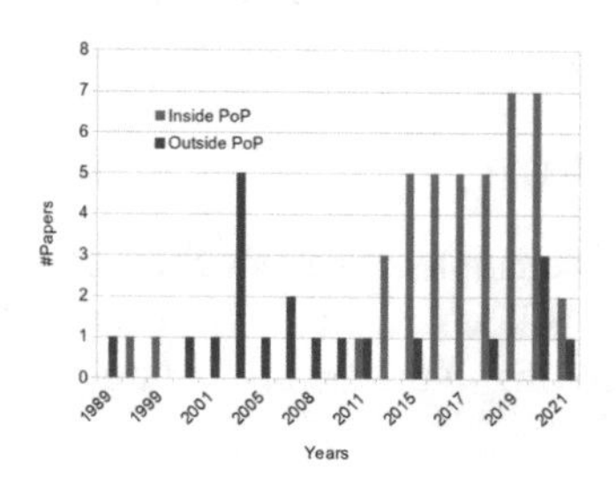

Fig. 2. Services and PoP

4.2 RQ2: What Is the Importance of the Temperature Variable for PoPs?

The emergence of the concept of virtualization brought flexibility to the growth of PoPs and also the responsibility of an infrastructure that offers different services. Although the first reference to the use of virtualization in PoP occurred in 2011, only in 2015 the first paper that addresses NFV (Network function virtualization) is identified. It decouples the network functions from the underlying hardware. This decouples the physical location of the virtualized function resources from the physical location of the equipment where the function is hosted. Essentially the function can be instantiated anywhere in the available infrastructure (i.e. PoPs) while it maintains the same logical location in the service chain. The functions are the virtualization of services that can be CDN, security, firewall, proxy and others.

A total of 24 works that referenced NFV. In 2017, the name Network Function Virtualization-Infrastructure Point of Presence (NFVI-PoP) for PoP infrastructure appears, which uses the technique of virtualization. A total of eight papers [17–21] used such a name that commonly calls the use of virtualization in PoP. In 2019, [18] considers NFV to run on Virtual Machines (VM) located NFVI-PoP. However, some articles consider NFV running in virtual machine or container or at the controller level on SDN networks. This is a consequence of choosing how to implement virtualization in PoPs, with SDN being the most referenced as shown in Fig. 3(a).

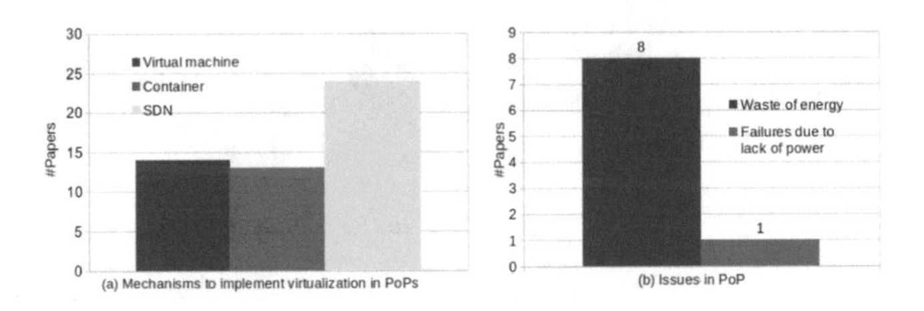

Fig. 3. Temperature variable X Energy variable

So, due to the tendency for there to be virtualization in PoPs, the greater the number of virtual machines and containers to be processed in a PoP. There is also a need for more devices, as well as a rise in temperature to carry out the processing of services. Consequently, the variable temperature is extremely important. It not only of the device, but also the temperature of the room where it is located, considering cities with temperatures above 20°. For example, the paper [1] presents the importance of remotely monitoring and controlling the room temperature of a PoP.

4.3 RQ3: How Important Is the Energy Variable for PoPs?

The articles about energy as important variable also referred battery, electricity and electricity. It was possible to find a total of 9 paper relating this variable and PoP.

Furthermore, the studies can be grouped into one of the following issues: failures due to lack of energy and cost or waste of energy. As shown in Fig. 3(b), it can be seen that most papers in the articles address issues of how to reduce the energy consumption in PoP with 8 article. Furthermore, virtualization can contribute to solving this problem, because it allows dynamically to activate and deactivate programs and devices and only two related articles related energy and NFVI-PoP.

Therefore, the variable energy and its synonyms found that they present the importance of further studies addressing the issue of PoP failure due to lack of energy, as only one article presented such a problem. In addition, this issue may require the presence of a technician if it matches the need for battery replacement; device failures (start and final time return) and other facts related to power failure.

4.4 RQ4: What Variables Should Be Monitored to Avoid Having to Travel a Technician to Troubleshoot PoPs?

The analysis of the papers allowed the identification of six types of PoP, as shown in the Fig. 4(a): Legacy PoP, PoP Tower, PoP CDN, PoP Datacenter, PoP Drone and PoP NFVI.

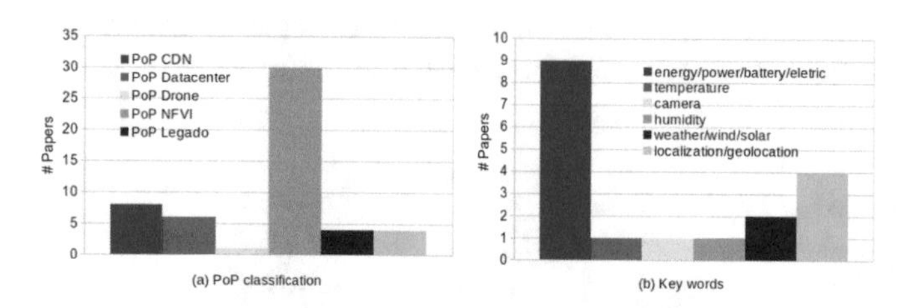

Fig. 4. Variables extracted from PoP Classification

Legacy PoP corresponds to an infrastructure that only provides the internet connectivity service. The PoP Tower is somewhat similar to the legacy PoP in that it also has the main objective of providing the connectivity service only. The difference in PoP Tower is the possibility of failures caused by environmental factors (rain when the transmission medium is by radio, lightning strikes and others).

As described above, it is possible to observe a growth in PoP studies that contain other types of services within the PoP, being possible to highlight two main ones: PoP CDN and PoP Datacenter. These PoPs have a greater concern with energy consumption, outages caused by power outages and the need for temperature control due to high data processing.

The PoP Drone is a more recent type with only one work found in the context of disasters, despite being a type already announced by facebook [23].

The NFVI PoP has been emerging since 2011 and is applicable to all other identified types of PoP.

These types of PoP reinforce the importance of the variables temperature, energy and their synonyms (power, electric and battery). In addition, PoP Tower also demonstrates the importance of the weather/wind/solar variable, as winds, rains and other phenomena can explicitly influence availability.

Furthermore, even if there is only one article on PoP Drone, this type will probably depend on these variables presented and other variables such as camera, location and humidity (since it may be more exposed to rain).

Therefore, the Fig. 4(b) summarizes the variables that must be monitored to avoid outages in PoPs. These variables will be used as keywords along with monitoring and automation to look for solutions to be applied to PoPs.

4.5 RQ5: What Sensors and Actuators Are Found as Solutions to Avoid Downtime in PoPs?

The total of 106 papers were found with proposals for solutions to monitor variables and automations that can be applied in PoP to avoid downtime. In the Fig. 5(a), the articles are distributed 93 works that addressed six types of sensors. It is important to note that only one paper has explicitly solution for PoP.

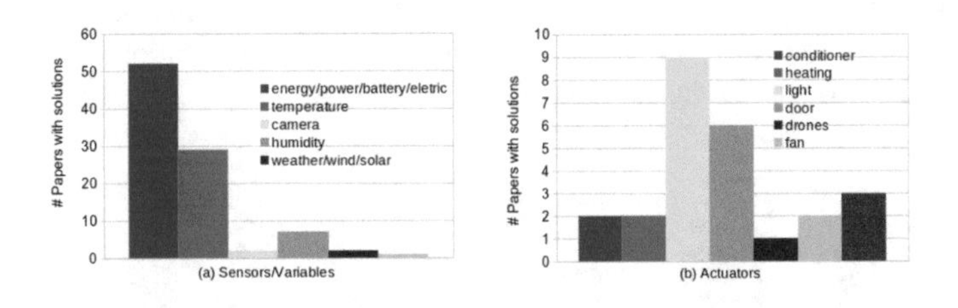

Fig. 5. Sensors and actuators in papers with solutions

Regarding actuators, 7 types were found, shown in the Fig. 5(b). For example, papers were identified that used lamps and doors as actuators that are activated according to the analysis of camera images. These works can offer a solution for PoP security in order to avoid unavailability of services due to equipment theft.

Another example of an actuator is the fan. For example, the paper [22] presents an artificial intelligence algorithm in order to control and reduce the temperature and humidity, being possible to be applicable without PoP rooms.

Air conditioning and heating are also important actuators to control the temperature of PoP rooms, despite contributing to increased energy consumption.

Robots and specifically drones were found as actuators that can even replace certain technical functions. For example, [23] developed a robot to run fiber optic wires on poles, resulting in a lower cost as it does not require human labor.

4.6 RQ6: What Algorithms Were Presented to Avoid Unavailability of Device in the PoP?

Several artificial intelligence algorithms were found to monitor the electrical network, temperature and other variables. These algorithms found can be grouped into one of the following classes: supervised (regression or classification or neural networks), semi-supervised and reinforcement (optimization or genetic algorithm or ant), being possible to visualize the quantity in the Fig. 6.

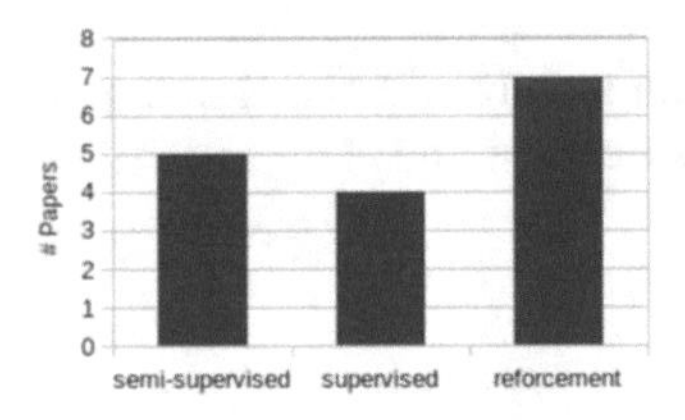

Fig. 6. Sensors and actuators in papers with solutions

4.7 RQ7: What Protocols (at the Application and Transmission Level) Were Presented in the Solutions that May Be Applicable to PoPs?

The papers [24] and [25] presented proposals for a new protocol related to a solution applicable to PoP to avoid downtime. At the application level, solutions were found that used MQTT, CoAP or HTTP, as shown in the Fig. 8(a).

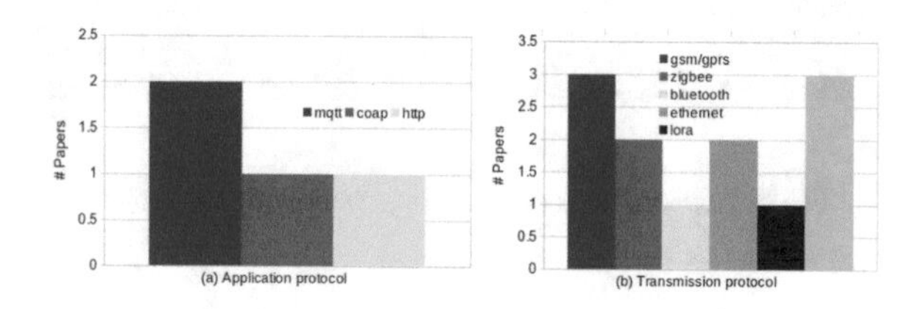

Fig. 7. Protocols

Regarding data transmission technology, six types were found, presented in the Fig. 8(b). Some articles used more than one transmission technology option.

4.8 RQ8: What Programming Languages and Equipment Were Used in the Prototypes that Might Be Applicable to Troubleshoot PoPs?

Prototypes developed with one of the following programming languages presented in the Fig. 8(a). The C++ being the most used language.

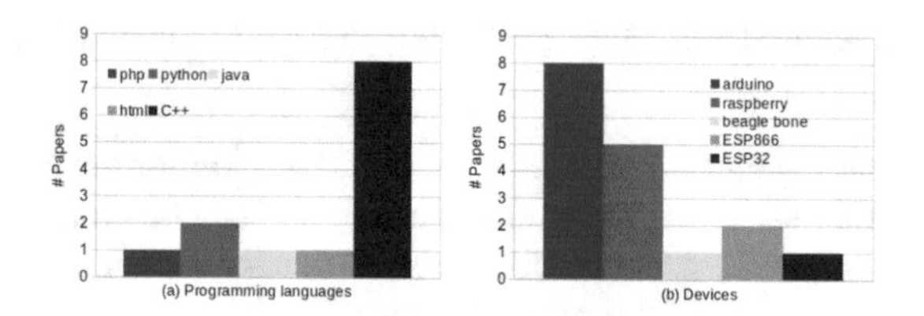

Fig. 8. Tools to build a prototype

The prototypes were built mainly with the Arduino microcontroller. The following Fig. 8(b) shows other hardware identified in the articles.

5 Conclusion

The presented results of this SLR aimed at investigating the characteristics of PoP problems and solutions to avoid failures that require the displacement of technicians by two phases.

The first phase helped to understand the current state-of-the-art, the evolution and feature of PoP. This phase identified a strong trend for PoP to contain no only internet service and to use virtualization in order to provide robustness, reduce energy consumption, minimize costs and more. The selected papers were analized and extracted six classifications: PoP CND, PoP Datacenter, PoP Drone, PoP NFVI, PoP legacy and PoP Tower. Moreover, these class guided the identification of some variable that must be monitored like as temperature; energy; its sinonimous (power, electric e battery), camera, localization and humility.

The second phase analized solutions for PoP extracted by words identifies in first step. The results suggest seven actuator to be used in solution to avoid failures in PoP: conditioner, heating, light, door, drones, fan and robot. Moreover, the selected paper contains prototypes and algorithms to handle with identified variables. The algorithms are based supervised, semi-supervised and reinforcement to monitor and automate the PoP. The major number of prototype used MQTT as application protocol; RFID and GSM/GPRS as transmission protocol; C++ as program language and arduino as microcontroler.

This SLR found only one prototype as solution to avoid failures in PoP and to recover the system remotely. Therefore, this paper encourages researches to continually propose new algorithms and prototype to avoid failures in PoP. Moreover, it was possible to find gaps in PoP drone and robots to provide new services.

Acknowledgments. The authors would like to thank the Federal Institute of Paraíba(IFPB)/ Campus João Pessoa for financially supporting the presentation of this research and, especially thank you, to the IFPB Interconnect Notice.

References

1. Anwar, H., Santoso, H., Khameswara, T.D., Priantoro, A.U.: Monitor-PoP-ISP's PoP room temperature and humidity web based monitoring using microcontroller. In: IEEE 8th Control and System Graduate Research Colloquium (ICSGRC), pp. 212–216. IEEE (2017)
2. Portela, F.A.: Posicionamento de servidores com minimização delatência em redes de operadoras de telecomunicações (2017)
3. Al Rosas, J.C.D., Constante, J.H., Oliveira, M.C.: Cálculo de radioenlace terrestre. In: SOMI XXXI Congreso de Instrumentación (2016)
4. Chooruang, K., Meekul, K.: Design of an IoT energy monitoring system. In: 2018 16th International Conference on ICT and Knowledge Engineering (ICT KE), pp. 1–4, November 2018
5. Case, J.D., Fedor, M., Schoffstall, M.L., Davin, J.: Simple network management protocol (SNMP). Request for Comments, vol. 1098 (1990). Shadroo, S., Rahmani, A.M.: Systematic survey of big data and data
6. Shadroo, S., Rahmani, A.M.: Systematic survey of big data and datamining in Internet of Things. Comput. Netw. **139**, 19-47 (2018)
7. Khorov, E., Lyakhov, A., Krotov, A., Guschin, A.: A survey onieee 802.11 ah: an enabling networking technology for smart cities. Comput. commun. **58**, 53-69 (2015)
8. Kamble, S.S., Gunasekaran, A., Gawankar, S.A.: Sustainable industry 4.0 framework: a systematic literature review identifying the current trends and future perspectives. Process Saf. Environ. Protect. **117**, 408–425 (2018)

9. Singh, K., Singh, P., Kumar, K.: A systematic review of IP traceback schemes for denial of service attacks. Comput. Secur. **56**, 111–139 (2016). https://www.sciencedirect.com/science/article/pii/S0167404815000930
10. ur Rasool, R., Wang, H., Ashraf, U., Ahmed, K., Anwar, Z., Rafique, W.: A survey of link flooding attacks in software defined network ecosystems. J. Netw. Comput. Appl. **172**, 102803 (2020). https://www.sciencedirect.com/science/article/pii/S1084804520302757
11. Fabbri, S., Silva, C., Hernandes, E., Octaviano, F., Di Thommazo, A., Belgamo, A.: Improvements in the StArt tool to better support the systematic review process. In: Proceedings of the 20th International Conference on Evaluation and Assessment in Software Engineering (EASE 2016). Association for Computing Machinery, New York, NY, USA, pp. 1–5 (2016). Article 21. https://doi.org/10.1145/2915970.2916013
12. Aceto, G., Botta, A., Marchetta, P., Persico, V., Pescapé, A.: A comprehensive survey on internet outages. J. Netw. Comput. Appl. **113**, 36–63 (2018). https://www.sciencedirect.com/science/article/pii/S1084804518301139
13. Ravindran, V., Shankar, S.: Systematic reviews and meta-analysis demystified. Ind. J. Rheumatol. **10**(2), 89–94 (2015)
14. Hearst, S.: Intelligent multiplexers with digital cross-connect capability-a requirement in network evolution. In: 1989 IEEE Global Telecommunications Conference and Exhibition Communications Technology for the 1990s and Beyond, vol. 1, pp. 527–532 (1989). https://doi.org/10.1109/GLOCOM.1989.64027.
15. Schoen, U., Hamann, J., Ugel, A., Kurzawa, H., Schmidt, C.: Convergence between public switching and the Internet. IEEE Commun. Mag. **36**(1), 50–65 (1998). https://doi.org/10.1109/35.649328
16. Skelly, P., Li, M.: EIPMon: an enhanced IP network monitoring tool. In: Proceedings 1999 IEEE Workshop on Internet Applications (Cat. No.PR00197), pp. 20–27 (1999). https://doi.org/10.1109/WIAPP.1999.788013.
17. Carapinha, J., et al.: Deployment of virtual network functions over multiple WAN interconnected PoPs. In: 2017 IEEE Conference on Network Function Virtualization and Software Defined Networks (NFV-SDN), pp. 252–257 (2017). https://doi.org/10.1109/NFV-SDN.2017.8169883.
18. Eramo, V., Lavacca, F.G.: Proposal and investigation of a reconfiguration cost aware policy for resource allocation in multi-provider NFV infrastructures interconnected by elastic optical networks. J. Lightwave Technol. **37**(16), 4098–4114 (2019). https://doi.org/10.1109/JLT.2019.2921428
19. Rashid, A.H., Shah Muhammad, S.: Traffic intensity based efficient packet schedualing. In: 2019 International Conference on Communication Technologies (ComTech), pp. 88–101 (2019). https://doi.org/10.1109/COMTECH.2019.8737826.
20. Vilalta, R., et al.: End-to-end network service deployment over multiple vims using a disaggregated transport optical network. In: 2019 21st International Conference on Transparent Optical Networks (ICTON), pp. 1-4 (2019). https://doi.org/10.1109/ICTON.2019.8840222.
21. Eramo, V., Lavacca, F.G., Catena, T., Giorgio, F.D.: Reconfiguration of optical-NFV network architectures based on cloud resource allocation and QoS degradation cost-aware prediction techniques. IEEE Access **8**, 200834–200850 (2020). https://doi.org/10.1109/ACCESS.2020.3035749
22. Bushnag, A.: Air quality and climate control Arduino monitoring system using fuzzy logic for indoor environments. In: 2020 International Conference on Control, Automation and Diagnosis (ICCAD), pp. 1-6 (2020). https://doi.org/10.1109/ICCAD49821.2020.9260514
23. Facebook built a new fiber-spinning robot to make internet service cheaper (2020). Cnet. short- url.at/jlIK7

24. Ellinidou, S., Kontogiannis, S., Kokkonis, G.: RF low energy monitoring protocol and system architecture for location based sensor measurements. In: Proceedings of the SEEDA-CECNSM 2016. Association for Computing Machinery, USA, pp. 60-65 (2016). https://doi.org/10.1145/2984393.2984397
25. Díaz-Reséndiz, L., Guerrero-Sánchez, J., Toledano-Ayala, A.E., Rivas-Araiza, E.A.: IoT based ambient monitoring system for intelligent buildings. In: 2018 IEEE International Conference on Automation/XXIII Congress of the Chilean Association of Automatic Control (ICA-ACCA), pp. 1–6 (2018). https://doi.org/10.1109/ICA-ACCA.2018.8609862.

A Provably Secure User Authentication Scheme Over Unreliable Networks

Toan-Thinh Truong[1(✉)], Minh-Triet Tran[1], Anh-Duc Duong[2], and Anh-Duy Tran[1]

[1] Vietnam National University Ho Chi Minh City, University of Science, Ho Chi Minh City, Vietnam
{ttthinh,tmtriet,taduy}@fit.hcmus.edu.vn

[2] Ho Chi Minh City Government, Ho Chi Minh City People's Committee, Ho Chi Minh City 700000, Vietnam
ducda@uit.edu.vn

Abstract. Wearable devices not only are the advancement in reduction of computation-systems, but also convenient for indirect interactions. Financial or medical data are exchanged between these devices through unreliable networks must be kept secret. Hence, lightweight and secure user authentication schemes always receive special attentions of many works in information security. Recently, Saleem et al. and Kandar et al. proposed a provable biometrics-based scheme providing the users with guarantee of secure authentication. Although elliptic curve cryptosystem-based combined with cryptographic-hash function, their schemes fail to satisfy session-key perfect forward secrecy and not applicable in practice. In this paper, we provide an improved version of their scheme to overcome these limitations.

1 Introduction

Authentication scheme plays an important role because it is the first step of initiation of transactions exchanged between mobile devices and service providers.

Lamport [1] utilized the user-password and repeatedly applied the cryptographic hash to conceal the true password. However, his approach's drawbacks are that the authentication depends on the requirement of user-verification-table stored at server-side, and offline-password guessing attack. Shamir [2] used the identity combined with server's master-key to generate the long-term keys. This idea helps to easily verify log-in-message because this mechanism depends on the identity's legitimate. Besides, it also helps reduce the computational time in comparison with public certificates. All [3–6] applied Shamir's idea in their schemes to eliminate the password-table, but they used plain-identity. Therefore, anyone can know who are using which services, and this limits the practical usage. Yoon [7] proposed the usage of smart-card combined with time-stamp in their scheme. This is a convenient method, but they fail to impersonation attack due to stolen smart-card. Sood [8] proposed a similar scheme, but they used the random values to create a different long-term keys at different time. This idea helps to

L. Barolli et al. (Eds.): AINA 2022, LNNS 449, pp. 602–613, 2022.
https://doi.org/10.1007/978-3-030-99584-3_52

reduce all the kinds of attack related to these keys. In summary, the works with the approach of hash function have some common points: distribute the common long-term key to all users, and lack of session-key agreement phase. Nevertheless, hash function is time-efficient and plays an inevitable role in user authentication scheme. The approach of hard-problems is a reasonable method to provide the security foundation for the schemes. Rivest [9] and Elgamal [10] proposed Public-key cryptography-PKC. These systems need Public-key infrastructure-PKI, so when the number of users increase, the storage and computation-time will be complex. Notably, elliptic curve cryptography-ECC proposed by Miller [11] and Koblitz [12] may be an reasonable alternative for PKC because it achieve the same security-level with smaller size of keys, for example 160-bit ECC is the same as 1024-bit RSA [13]. Yang [14] used elliptic curve discrete logarithm problem-ECDLP and related problems to proposed the lightweight authentication scheme. However, Yoon [15] proposed the improved version because they found that Yang's scheme failed to provide session-key perfect-forward secrecy. Islam [16] found that Yoon's scheme is vulnerable to similar attacks as Yang's scheme. Also, they proposed the scheme using random values combined with three-way handshake technique. However, their scheme is vulnerable to known session-specific temporary information attack [17]. In 2020, Wu [18] proposed a scheme in multi-server environment. This scheme needs three parties when another user authenticates with the server, and we find that it will increase the time of message-exchange. Especially Saleem [19] proposed a provable biometric-based scheme for secure communication over unreliable networks. This is really a lightweight scheme with ECC and suitable for quickly authentication. Also, Kandar [20] proposed a similar biometric-based one in multi-server architecture. This is efficient in time and reasonable cost. However, we find these are vulnerable to three-factor attack and cannot provide session-key perfect-forward secrecy, an important requirement. In this work, we will clarify these disadvantages and provide a modification to overcome these limitations.

The rest of our paper is organized as follows. Section 2 analyzes the Saleem et al.'s scheme and Kandar et al.'s scheme. In Sect. 3, we present the improved scheme of previous works. Section 4 provide an security and efficiency analysis. Finally, our conclusion is presented in Sect. 5.

2 Related Works

This section analyzes the Saleem et al.'s and Kandar et al.'s schemes. Although designed in different architectures, they are used to protect the secure communication over an unreliable networks.

2.1 Cryptanalysis of Saleem et al.'s Scheme

In this section, we only present our cryptanalysis on Saleem et al.'s scheme (refer [19] for more details). We find that Saleem et al.'s scheme is vulnerable to three-factor attack and cannot provide session-key perfect forward secrecy.

2.1.1 Three-Factor Attack

At first, we assume that client's identity ID_c is not hard to know. Because if the adversary is one of the legitimate users in this service. Furthermore, this kind of attack needs two of three factor leaked to the adversary, and we choose a biometric B_c and SC_c. Now, the adversary performs as follows:

- Create the expression $E_c = h(G_c \parallel ID_c)$, where ID_c and E_c are known.
- Analyze $G_c = H_c \oplus \bar{PW}_c$, where H_c is known.
- Analyze $\bar{PW}_c = h(ID_c \parallel PW_c \parallel a \parallel B_c)$, where ID_c, B_c are known.
- Analyze $a = Z_c \oplus h(ID_c \parallel PW_c \oplus B_c)$, where Z_c, ID_c, and B_c are known.
- Finally, the adversary achieve a final expression $E_c = h((H_c \oplus h(ID_c \parallel PW_c \parallel <Z_c \oplus h(ID_c \parallel PW_c \oplus B_c)> \parallel B_c)) \parallel ID_c)$.

Of course, it is not easy to know the client's SC_c and B_c. However, this requirement is an essential condition to evaluate the biometric-based scheme. So, this scheme cannot resist the offline-password-guessing attack.

2.1.2 Session-Key Perfect-Forward Secrecy

This kind of attack assumes the long-term keys of both server and user are leaked. So, it is important to protect the previous sessions. Now, assume the adversary caught another previous messages: $\{M_c, PID_c, Auth_c, N_c, t_s\}$ and $\{O_s, Auth_s, t_s\}$.

- Recover $ID_c = PID_c \oplus s^{-1} \times M_c$ and re-computes $G_c = h(ID_c \parallel s)$, where s is a leaked key of S_s.
- Then, the adversary computes $r_c = N_c \oplus h(h(G_c \parallel ID_c) \parallel t_c)$. With r_c, the adversary computes $r_s = r_c \oplus O_s$.
- With r_c, r_s, t_c, t_s and G_c, the adversary recovers $SK = h(G_c \parallel r_c \parallel r_s \parallel t_c \parallel t_s)$

Clearly, their scheme fails to provide the session-key perfect forward secrecy.

2.2 Cryptanalysis of Kandar et al.'s Scheme

Similar to previous section, we only present our analysis on Kandar et al.'s schemes (refer [20] for more details). We see their scheme is vulnerable to server-impersonation attack and cannot provide session-key perfect forward secrecy.

2.2.1 Server Impersonation Attack

We see any valid user can obtain all the $Skey_j$ of all service-providers in session-key agreement phase. Also, SID_j is also easily known because it is not a secret information. Now, we assume in another session, RC sends $\{M_7, M_8\}$ to a service-provider and it is caught by another adversary. He/she will do as follows to cheat the U_i and RC:

- Recover $N_3 = M_7 \oplus h(SID_j \parallel SKey_j)$.
- Choose a nonce N_4 and computes $M_9 = h(Skey_j \parallel N_3) \oplus N_4$ and $M_{10} = h(Skey_j \parallel N_3 \parallel N_4)$. Then, sending $\{M_9, M_{10}\}$ to RC and waiting.
- When the adversary receives $\{M_{14}, M_{15}\}$ from RC and $\{M_{16}, M_{17}\}$ from U_i, he/she recovers $h(N_2 \parallel N_3) = M_{14} \oplus h(Skey_j \parallel SID_j \parallel N_4)$, $V_{ij} = h(SKey_j \parallel h(N_2 \parallel N_3))$, $B_i = M_{16} \oplus h(Skey_j \parallel V_{ij})$.
- Finally, the adversary computes $SK = h(V_{ij} \parallel Skey_j \parallel h(N_2 \parallel N_3) \parallel B_i)$.

Clearly, the malicious user can collect all the service-provider by requesting services to RC. Especially, the design of this scheme reveals all the information of user's smart-card when authentication phase happens.

2.2.2 Session-Key Perfect-Forward Secrecy

If the RC's X_S is leak, the adversary can compute all the past session-keys and recover all the messages encrypted by them. Assuming that another adversary has all the necessary message-packages of another session, she/he can do as follows to compute session-key of this session.

- Computes $N_1 = h(D_i \oplus h(C_i \parallel X_S))$, $N_2 = M_3 \oplus h(h(C_i \parallel X_S) \parallel N_1)$, and $N_3 = h(D_i \oplus h(C_i \parallel X_S)) \oplus M_{11}$.
- Next, he/she computes $Skey_j = M_{12} \oplus h((D_i \oplus h(C_i \parallel X_S)) \parallel N_1 \parallel N_3)$, $V_{ij} = h(Skey_j \parallel h(N_2 \parallel N_3))$ and $B_i = M_{16} \oplus h(Skey_j \parallel V_{ij})$.
- Finally, she/he computes session-key $SK = h(V_{ij} \parallel Skey_j \parallel h(N_2 \parallel N_3) \parallel B_i)$.

3 Proposed Scheme

Our scheme inherits the advantages of [19,20], and overcomes their drawbacks. Although designed with client-server architecture, proposed scheme may be easily extended in multi-server environment. Main contributions are as follows:

- The design is based on mathematical models, for example BAN-logic and random-oracle. Therefore, it can be proven formally.
- Our scheme can be extended in multi-server environment by using cryptographic hash-function combined with ECC approach.
- The design can resist the popular kinds of attacks at authentication stage.
- Our scheme integrates the user-biometrics to enhance the security.

3.1 Registration Phase

User C_c selects identity ID_c, a random value $a \in Z_p^*$ and imprints biometric B_c.

- Then, C_c computes $<R_c, P_c> = Gen(B_c)$ and $P\bar{W}_c$, when Gen is an algorithm for biometric information. Next, C_c sends $\{ID_c, P\bar{W}_c\}$ to server S_s.
- Once receiving registration-message of C_c, S_s checks ID_c and stores $<h(ID_c), ID_c>$ into its database for further reference.

- Then, S_s computes $G_c = h(ID_c \parallel s)$, $H_c = G_c \oplus P\bar{W}_c$ and $E_c = h(G_c \parallel ID_c)$, where s is a secret master-key of S_s.
- Finally, S_s sends smart-card $SC_c = \{H_c, E_c, Pub\}$ to C_c. Once receiving SC_c of S_s, C_c computes $Z_c = a \oplus h(ID_c \parallel R_c)$ and inserts $\{P_c, Z_c\}$ into SC_c.

We use an algorithm to generate a template P_c for later reproduction to solve the problem of inconsistency of biometric. Furthermore, we eliminate the password so can resist all password-related attacks and reduce the computational cost.

3.2 Authentication Phase

C_c inserts its SC_c, provides ID_c and imprints $B_c^{'}$ into the terminal. Then:

- SC_c computes $R_c = Rep(P_c, B_c^{'})$, $a = Z_c \oplus h(ID_c \parallel R_c)$, $G_c^{'} = H_c \oplus h(ID_c \parallel R_c \parallel a)$ and $E_c^{'} = h(G_c^{'} \parallel ID_c)$, where Rep is a reproduction algorithm of R_c from $B_c^{'}$.
- Then, SC_c checks if $E_c^{'} ?= E_c$. If this does not hold, it terminates the session. Otherwise it selects $r_c \in Z_p^*$ and computes $Ran_c = r_c \times P$, $M_c = r_c \times Pub$, $PID_c = h(ID_c) \oplus h(Ran_c)$ and $Auth_c = h(ID_c \parallel G_c^{'} \parallel Ran_c \parallel t_c)$, where t_c is the current time-stamp. Next, SC_c sends $\{M_c, PID_c, Auth_c, t_c\}$ to S_s via a common channel.
- Once receiving login-message of C_c, S_s checks if $t_s - t_c ?\leq \Delta t$, where t_s is S_s's current time-stamp. If this does not hold, it terminates the session. Otherwise it continues. Then, S_s computes $Ran_c^{'} = s^{-1} \times M_c$, and $h(ID_c) = PID_c \oplus h(Ran_c^{'})$.
- So, S_s extracts ID_c according to $h(ID_c)$ in its database and checks ID_c. Next, S_s computes $G_c = h(ID_c \parallel s)$ and $E_c = h(G_c \parallel ID_c)$. Then S_s checks if $Auth_c ?= h(ID_c \parallel G_c \parallel Ran_c^{'} \parallel t_c)$. If this does not hold, it terminates the session. Otherwise it selects $r_s \in Z_p^*$ and computes $O_s = Ran_c^{'} + r_s \times P$, session-key $SK = h(G_c \parallel r_s \times Ran_c^{'} \parallel t_c \parallel t_s)$, $Auth_s = h(SK \parallel E_c \parallel ID_c)$.
- Finally, S_s sends reply-message $\{O_s, Auth_s, t_s\}$ to C_c through a common channel.
- Once receiving reply-message of S_s, C_c checks if t_k-t_s $?\leq \Delta t$. If this does not hold, it terminates the session. Otherwise, it re-computes $Ran_s^{'} = O_s - Ran_c$ and $SK = h(G_c^{'} \parallel r_c \times Ran_s \parallel t_c \parallel t_s)$. Then C_c checks if $Auth_s ?= h(SK \parallel E_c \parallel ID_c)$. If this does not hold, it terminates the session. Otherwise, C_c and S_s successfully authenticate with each other.

Also, we apply DDHP in message O_s and CDHP in message SK. With these improvements, our scheme can provide session-key perfect forward secrecy.

3.3 Biometric-Update Phase

For simple, we ignore the 1st step of this phase because it is the same as log-in step of authentication phase.

- C_c inserts SC_c, provides ID_c and imprints $B_c^{'}$ into another terminal.
- SC_c computes $R_c = Rep(P_c, B_c^{'})$, $a = Z_c \oplus h(ID_c \parallel R_c)$, $\bar{PW}_c = h(ID_c \parallel R_c \parallel a)$, $G_c^{'} = H_c \oplus \bar{PW}_c$, $E_c^{'} = h(G_c^{'} \parallel ID_c)$. Then, SC_c checks if $E_c^{'} ?= E_c$. If this does not hold, it terminates the session. Otherwise it selects $r_c \in Z_p^*$ and requests C_c to provide B_c^{new}.
- Once receiving B_c^{new} of C_c, SC_c computes $<R_c^{new}, P_c^{new}>$, $\bar{PW}_{new} = h(ID_c \parallel R_c^{new} \parallel a)$, $H_{new} = H_c \oplus \bar{PW}_c \oplus \bar{PW}_{new}$, $Z_{new} = Z_c \oplus h(ID_c \parallel R_c) \oplus h(ID_c \parallel R_c^{new})$.
- Finally, SC_c updates $Z_c \longleftarrow Z_{new}$, $H_c \longleftarrow H_{new}$ and $P_c \longleftarrow P_{new}$.

3.4 Experimental Extended-Registration Phases

We present the ability to extend the registration phase into two ones: a service-provider side and a client-side. Of course, we need a registration center *RC*. Also, we do not have to modify the authentication and biometrics-update phases, and this design helps to reduce the number of transmitting message-package as in [20].

3.4.1 Server Registration Phase

Another server S_j provides its identity ID_S to RC through a secure channel. When receiving this message, RC checks ID_S and chooses a random value r_j. RC stores $\{h(ID_S), ID_S, r_j\}$ into its database. Then, it computes $ASID_j = h(ID_S \parallel k)$, where k is a secret master-key of *RC*. Finally, it returns $\{ASID_j, r_j\}$ to S_j through a secure channel. Of course, if there are many S_j registering to *RC*, there will be many values of $\{ASID_j, r_j\}$, where $j \in \{1, ..., n\}$.

3.4.2 Client Registration Phase

C_C provides $\{ID_C, \bar{PW}_C\}$ to *RC*. When receiving these messages, RC checks ID_C and stores $\{h(ID_C), ID_C\}$ into its database. Then RC computes $G_C = h(ID_C \parallel ASID_j)$, where $ASID_j$ is a S_j's authentication key which C_C wants to register, $H_C = G_C \oplus \bar{PW}_C$, $E_C = h(G_C \parallel ID_C)$. Finally, RC sends back $SC_C = \{H_C, E_C, r_j \times P\}$ to C_C. When receiving SC_C, C_C does the same thing as in previous registration phase. Of course, if C_C registers many service-providers, the SC_C will contain many values of $r_j \times P$, where $j \in \{1, ..., n\}$.

4 Proposed Scheme's Analysis

As for formal proof and security model in [21,22], we only summarize some main points. Also, we re-use their execution time of cryptographic operation to provide a comparison of computational cost.

4.1 Security Analysis

Firstly, we present an informal security analysis on common kinds of attacks and Table 1 summarizes all the comparison. Then we review a formal model of semantic security on proposed scheme.

4.1.1 Informal Analysis

This section presents only some typical common kinds of attack our scheme resists:

- User-anonymity: In this scheme, PID_c contains $h(ID_c)$. So if Ran_c is leaked, attacker cannot know the true ID_c. Furthermore, with various random value r_c, our scheme can provide strong-anonymity.
- Impersonation attack: as for user-impersonation, attacker must know the G_c to generate a valid $Auth_c = h(ID_c \parallel G_c \parallel Ran_c \parallel t_c)$. Clearly, this is impossible. As for server-impersonation, attacker must know the Ran_c and G_c of another victim. Also, this is impossible because it needs master-key s to know user's Ran_c. In short, our scheme can resist this kind of attack.
- Perfect forward secrecy: in this kind of attack, if the master-key s and all long-term keys G_c are leaked, attacker can derive Ran_c or Ran_s of any session. However, it cannot derive the $r_s\times Ran_c$ or $r_c\times Ran_s$ due to ECDDHP.
- Three-factor attack: our scheme uses biometric information, so we assume one of two factors is leaked. Firstly, R_c is leaked, attacker cannot use it to impersonate or compute session-key. Secondly, $SC_c = \{H_c, E_c, P_c, Z_c, Pub\}$ is leaked, attacker cannot extract long-term key G_c due to lacking of random value a and R_c.

Table 1. The comparison in resisting the various common attacks

	Saleem's	Kandar's	Ours
Mutual-authentication	✓	✓	✓
User-anonymity	✓	✓	✓
Impersonation	✓	✗	✓
Replay	✓	✓	✓
Off-line password guessing	✓	✓	✓
Insider	✓	✓	✓
Perfect forward secrecy	✗	✗	✓
Three-factor	✗	†	✓

† Untouched

4.1.2 Formal Analysis

This section re-uses all Saleem et al.'s the security model including: a threat model and seven standards. However, we have some modification. Firstly in $Corrupt(\Pi^i_{C_c}, v)$, there is no $v = 2$ in this query because our scheme eliminate the user's password. Furthermore, PW_c is replaced with B_c in position of long-lived key of each client instance C_c. As for **Game**$_z$ $(0 \leq z \leq 6)$, we re-analyze the **Game**$_5$ and **Game**$_6$.

- **Game**$_5$: in this game, there is only $Corrupt(\Pi^i_{C_c}, 3)$ to guess B_c because there is no PW_c. So this only results in $\mathbf{Max}\{q_s \times (\frac{1}{2^{l_b}}, \epsilon_b)\}$, where q_s is the number of *Send* queries, l_b is the bit-length of biometrics, and ϵ_b is the probability of false. Finally, we got $|\mathbf{Pr}[Suc_5] - \mathbf{Pr}[Suc_4]| \leq \mathbf{Max}\{q_s \times (\frac{1}{2^{l_b}}, \epsilon_b)\}$.
- **Game**$_6$: in this game with the second case, we let *Pollard*'s *Rho* be the algorithm of adversary, and guessing r_c or r_s takes $O(\sqrt{2^l})$. Finally, we got $|\mathbf{Pr}[Suc_6] - \mathbf{Pr}[Suc_5]| \leq \frac{q_h}{2^l} + \frac{1}{O(\sqrt{2^l})}$, where q_h is the number of *Hash* queries and l is the prime number p's bit-length.

Eventually, we conclude that $Adv^{\Pi}_{A} \leq \frac{q_h^2+(q_s+q_e)^2}{2^l} + \mathbf{Max}\{q_s \times (\frac{1}{2^{l_b}}, \epsilon_b)\} + \frac{q_h}{2^l} + \frac{1}{O(\sqrt{2^l})} + \frac{2 \times q_s}{2^l} + \frac{q_h}{2^l}$, where q_e is the number of *Execute* queries.

4.2 Correctness Analysis

This subsection uses BAN-logic model [23] with objectives in [24] to prove the correctness of our scheme. Next are some assumptions being initial beliefs of S_S and C_C. For instance, A_1 means that another C_C shares its identity with S_S at registration phase, or A_2 means that another C_C and S_S successfully agree authentication-key.

- A_1: $C_C \mid\equiv (C_C \stackrel{ID_C}{\leftrightarrow} S_S)$. C_C believes C_C shares ID_C with S_S
- A_2: $C_C \mid\equiv (C_C \stackrel{G_C}{\leftrightarrow} S_S)$. C_C believes C_C shares G_C with S_S
- A_3: $C_C \mid\equiv (S_S \Rightarrow (C_C \stackrel{SK}{\leftrightarrow} S_S))$. C_C believes S_S controls the share of SK between them
- A_4: $S_S \mid\equiv (C_C \Rightarrow (C_C \stackrel{ID_C}{\leftrightarrow} S_S))$. S_S believes C_C controls the share of ID_C between them
- A_5: $S_S \mid\equiv (C_C \Rightarrow (C_C \stackrel{SK}{\leftrightarrow} S_S))$. S_S believes C_C controls the share of SK between them
- A_6: $S_S \mid\equiv (S_S \stackrel{G_C}{\leftrightarrow} C_C)$. S_S believes S_S shares G_C with C_C
- A_7: $C_C \mid\equiv \#(r_S \| G_C)$. C_C believes the random messages of S_S is fresh, where $\|$ is only a symbol standing for relationship of random value and authentication-key
- A_8: $S_S \mid\equiv \#(r_C \| G_C)$. S_S believes the random messages of C_C is fresh

All messages transmitted in our scheme are normalized. We let G_C denote long-term key $h(ID_C \parallel s)$, C_C/S_S denotes client/server, ID_C denotes client's identity and SK denotes session-key. This step explicitly shows all messages transmitted between S_S and C_C. For instance, $Auth_C$ contains random challenge-message $r_U \| G_C$ and G_C, or $Auth_S$ contains challenge-message $r_S \| G_C$ and G_C. Also, we let $< C_C \overset{ID_C}{\leftrightarrow} S_S,\ C_C \overset{G_C}{\leftrightarrow} S_S,\ r_C \| G_C >$ denote $\{PID\}$, $< r_C \| G_C,\ C_C \overset{G_C}{\leftrightarrow} S_S >$ denote $\{Auth_C\}$, $< r_S \| G_C,\ C_C \overset{G_C}{\leftrightarrow} S_S >$ denote $\{O_S\}$ and $< C_C \overset{G_C}{\leftrightarrow} S_S,\ C_C \overset{SK}{\leftrightarrow} S_S >$ denote $\{Auth_S\}$

Next are seven lemmas re-organized from [24] and our proofs.

Lemma 1: If S_S believes G_C is shared with C_C and its messages encrypted with G_C are fresh, S_S believes C_C believes its ID_C is successfully shared with S_S

Proof: Using message-meaning rule with PID and A_6, we have $\frac{S_S|\equiv(S_S \overset{G_C}{\leftrightarrow} C_C), S_S \triangleleft PID}{S_S|\equiv(C_C|\sim PID)}$. Using freshness rule with A_8 to have $\frac{S_S|\equiv\#(r_C\|G_C)}{S_S|\equiv\#PID}$. Then, using nonce-verification rule to have $\frac{S_S|\equiv C_C|\sim PID, S_S|\equiv\#PID}{S_S|\equiv C_C|\equiv C_C \overset{ID_C}{\leftrightarrow} S_S}$. Finally, using believe-rule to have $\frac{S_S|\equiv C_C|\equiv PID}{S_S|\equiv C_C|\equiv C_C \overset{ID_C}{\leftrightarrow} S_S}$. So, with A_6 and A_8 we complete Lemma 1.

Lemma 2: If S_S believes C_C believes its ID_C is shared successfully and C_C controls this ID_C's sharing, S_S believes C_C's ID_C is shared successfully.

Proof: Using jurisdiction-rule with Lemma 1 and A_4 to have $\frac{S_S|\equiv C_C|\equiv C_C \overset{ID_C}{\leftrightarrow} S_S, S_S|\equiv C_C \Rightarrow C_C \overset{ID_C}{\leftrightarrow} S_S}{S_S|\equiv C_C \overset{ID_C}{\leftrightarrow} S_S}$. So, we complete Lemma 2.

Lemma 3: If C_C believes G_C is shared with S_S and its messages encrypted with G_C are fresh, C_C believes S_S believes its ID_C is shared successfully.

Proof: Using jurisdiction-rule with A_2 and O_S to have $\frac{C_C|\equiv C_C \overset{G_C}{\leftrightarrow} S_S, C_C \triangleleft O_S}{C_C|\equiv S_S|\sim O_S}$. Then, using freshness-rule with A_7 to have $\frac{C_C|\equiv\#(r_S\|G_C), C_C|\equiv S_S|\sim O_S}{C_C|\equiv\#O_S}$. So, combining two results with nonce-verification-rule to have $\frac{C_C|\equiv S_S|\sim O_S, C_C|\equiv\#O_S}{C_C|\equiv S_S|\equiv O_S}$. Finally, applying believe-rule to have $\frac{C_C|\equiv S_S|\equiv O_S}{C_C|\equiv S_S|\equiv C_C \overset{ID_C}{\leftrightarrow} S_S}$. Therefore, we prove how proposed scheme satisfies Lemma 3, and claim both S_S and C_C successfully share their identities. Next are the similar proofs for session-key, so we ignore these for compactness.

4.3 Efficient Analysis

This sub-section uses the same desktop-system [19] for convenient comparison (Fig. 1):

- Communication cost: The total communication cost of Saleem et al.'s scheme including $\{M_c, PID_c, Auth_c, N_c, t_c\}$ and $\{O_s, Auth_s, t_s\}$ is $160 \times 5 + 256 \times 3 = 1568$ bits. The total communication cost of Kandar et al.'s scheme including $\{M_1, M_2, C_i, D_i\}$, $\{M_5, M_6\}$, $\{M_7, M_8\}$, $\{M_9, M_{10}\}$, $\{M_{11}, M_{12}, M_{13}\}$, $\{M_{14}, M_{15}\}$ and $\{M_{16}, M_{17}\}$ is $256 \times 19 = 4864$ bits. As for proposed scheme, its cost of $\{M_c, PID_c, Auth_c, t_c\}$ and $\{O_s, Auth_s, t_s\}$ is $160 \times 4 + 256 \times 3 = 1408$ bits.

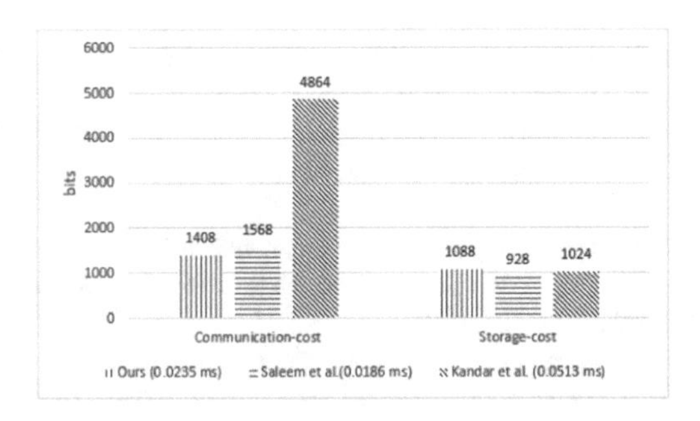

Fig. 1. Efficiency comparison of proposed scheme and Saleem et al.'s

- Computation cost: The total computation cost of Saleem et al.'s scheme is: registration phase costs $4t_{owh}$ ($\approx$0.0036 ms) and authentication phase costs $13t_{owh} + 3t_{mu}$ ($\approx$0.015 ms), where t_{owh} and t_{mu} are the execution time of hash and point multiplication. The total computation cost of Kandar et al.'s scheme is: all registration phases costs $11t_{owh}$ ($\approx$0.0099 ms) and all authentication phases costs $46t_{owh}$ ($\approx$0.0414 ms). As for proposed scheme, its cost is: registration phase costs $5t_{owh}$ ($\approx$0.0045 ms) and authentication phase costs $14t_{owh} + 2t_{ad} + 4t_{mu}$($\approx$0.019 ms), where t_{ad} is the execution time of point addition. Our cost isn't much higher than previous scheme, and this is an essential thing for a stronger scheme.
- Smart-card storage cost: The storage-cost of Kandar et al.'s scheme is 1024 bits. As for our scheme, it only adds a 160-bit P_c. Therefore, our cost is almost the same as Saleem et al.'s and Kandar et al.'s scheme.

5 Conclusions

In this paper, we survey the current works in this area, and then analyze two recent typical-works, Saleem et al.'s and Kandar et al.'s schemes. These scheme are time-efficient and has a good provable security foundation. However, we discover some limitations and proposed a modified version. Our scheme overcomes the problems and inherits all the advantages of previous ones. In the near future, we propose the stronger version in multi-server environment.

Acknowledgements. This study was funded by Vietnam National University, Ho Chi Minh City (VNU-HCM) under grant number C2021-18-21

References

1. Lamport, L.: Password authentication with insecure communication. Commun. ACM **3468**, 770–772 (1981)

2. Shamir, A.: Identity-based cryptosystems and signature schemes. In: Blakley, G.R., Chaum, D. (eds.) Advances in Cryptology, CRYPTO 1984. LNCS, vol. 196, pp. 47–53. Springer, Heidelberg (1984). https://doi.org/10.1007/3-540-39568-7_5
3. Tsaur, W.J.: A flexible user authentication scheme for multi-server internet services. In: Lorenz, P. (ed.) ICN 2001. LNCS, vol. 2093, pp. 174–183. Springer, Heidelberg (2001). https://doi.org/10.1007/3-540-47728-4_18
4. Hwang, M., Lee, C., Tang, Y.: A simple remote user authentication scheme. Math. Comput. Model. **36**, 103–107 (2002)
5. Shen, J., Lin, C., Hwang, M.: A modified remote user authentication scheme using smart cards. IEEE Trans. Consum. Electron. **49**(2), 414–416 (2003)
6. Lee, C., Hwang, M., Yang, W.: Flexible remote user authentication scheme using smart cards. IEEE Trans. Neural Netw. **36**(3), 46–52 (2002)
7. Yoon, E., Yoo, K.: A flexible user authentication for multi-server internet services. In: Debruyne, C., et al. (eds.) OTM Workshops. LNCS, vol. 4277, pp. 499–507. Springer, Cham (2006)
8. Sood, S., Sarje, A., Singh, K.: A secure dynamic identity based authentication protocol for multi-server architecture. J. Netw. Comput. Appl. **34**(2), 609–618 (2011)
9. Rivest, R., Shamir, A., Adleman, L.: A method for obtaining digital signatures and public-key crypto-systems. Commun. ACM **21**(2), 120–126 (1978)
10. Elgamal, T.: A public key crypto-system and a signature scheme based on discrete logarithms. IEEE Trans. Inf. Theory **31**(4), 469–472 (1985)
11. Miller, V.S.: Use of elliptic curves in cryptography. In: Williams, H.C. (ed.) CRYPTO 1985. LNCS, vol. 218, pp. 417–426. Springer, Heidelberg (1986). https://doi.org/10.1007/3-540-39799-X_31
12. Koblitz, N.: Elliptic curve crypto-systems. Math. Comput. **48**(177), 203–209 (1987)
13. Hankerson, D., Menezes, A., Vanstone, S.: Guide to Elliptic Curve Cryptography, 1st edn. Springer, New York (2004). https://doi.org/10.1007/b97644
14. Yang, J., Chang, C.: An ID-based remote mutual authentication with key agreement scheme for mobile devices on elliptic curve cryptosystem. Comput. Secur. **28**(3–4), 138–143 (2009)
15. Yoon, E., Yoo, K.: Robust ID-based remote mutual authentication with key agreement scheme for mobile devices on ECC. In: International Conference on Computational Science and Engineering, pp. 633–640 (2009)
16. Islam, S.H., Biswas, G.: A more efficient and secure id-based remote mutual authentication with key agreement scheme for mobile devices on elliptic curve cryptosystem. J. Syst. Softw. **84**(11), 1892–1898 (2011)
17. Canetti, R., Krawczyk, H.: Analysis of key-exchange protocols and their use for building secure channels. In: Pfitzmann, B. (ed.) EUROCRYPT 2001. LNCS, vol. 2045, pp. 451–472. Springer, Heidelberg (2001). https://doi.org/10.1007/3-540-44987-6_28
18. Wu, H., Chang, C., Zheng, Y., Chen, L., Chen, C.: A secure IoT-based authentication system in cloud computing environment. Sensors **20**(19), 5604 (2020)
19. Saleem, M., Islam, S., Ahmed, S., Mahmood, K., Hussain, M.: Provably secure biometric-based client-server secure communication over unreliable networks. J. Inf. Secur. Appl. **58**, 102769 (2021)
20. Kandar, S., Pal, S., Dhara, B.C.: A biometric based remote user authentication technique using smart card in multi-server environment. Wireless Pers. Commun. **2021**(120), 1003–1026 (2021)

21. Bellare, M., Rogaway, P.: The security of triple encryption and a framework for code-based game-playing proofs. In: Vaudenay, S. (ed.) EUROCRYPT 2006. LNCS, vol. 4004, pp. 409–426. Springer, Heidelberg (2006). https://doi.org/10.1007/11761679_25
22. Blake-Wilson, S., Johnson, D., Menezes, A.: Key agreement protocols and their security analysis. In: Darnell, M. (ed.) Cryptography and Coding 1997. LNCS, vol. 1355, pp. 30–45. Springer, Heidelberg (1997). https://doi.org/10.1007/BFb0024447
23. Burrows, M., Abadi, M., Needham, R.: A logic of authentication. ACM Trans. Comput. Syst. **8**, 18–36 (1990)
24. Tsai, J., Wu, T., Tsai, K.: New dynamic id authentication scheme using smart cards. Int. J. Commun. Syst. **23**(12), 1449–1462 (2010)

Event-Triggered Based Distributed Agreement Algorithm to Ensure the Cohort Stability

Imen Zidi[1(✉)], Abir Ben Ali[1], and Farouk Kamoun[2]

[1] CRISTAL Laborarory, ENSI, Manouba, Tunisia
zidi.imene@gmail.com, abir.benali@isetr.rnu.tn
[2] SESAME University, Tunis, Tunisia
farouk.kamoun@sesame.com.tn

Abstract. Intelligent Vehicular Networks (IVN) are considered a promising solution to mitigate the road traffic's problems and improve its safety and efficiency. Due to the initial unbounded-size mesh topology, IVN has suffered from the channel congestion and unreliable message dissemination issues. So to deal with these issues we proposed, in this work, to break down the IVN into a fully distributed bounded-size cyber-physical construct called cohort, to guarantee the timeliness and the validity of the Safety-critical (SC) data. In addition, a distributed event-triggered agreement algorithm based on the Neighbor to Neighbor (N2N) communication is proposed, here, to help the cohort, making collaborative decisions to ensure its safety and stability in the face to external perturbation. Finally, we studied the variation of the cyber space delay in purpose to define its upper bound, depending on several factors.

1 Introduction

The serious situation of the road traffic, caused by the increasing number of vehicles and their related social and environmental issues, has motivated the industrial and academic community, around the world, to push up toward the next generation of the transportation system [3]. The next generation of the transportation system exploits the proliferation of the embedded communication, networking and sensing technologies to build the automation of the driving features and set up the vehicles' connectivity to form the IVN [1]. Further, the connected automated vehicles are considered a promising solution to mitigate the road traffic's problems [14] and are expected to achieve important results in improving its safety and efficiency [13].

In the early stages, IVN was characterised by an unbounded-size mesh topology. This structure leads to a crucial problem of channel congestion that causes additional latency and reliability's loss. Such a limitations could not be tolerated in a real-time critical environment. Several studies aimed to overcome this situation. Breaking down the network into strings of vehicles, is a contemplated

L. Barolli et al. (Eds.): AINA 2022, LNNS 449, pp. 614–626, 2022.
https://doi.org/10.1007/978-3-030-99584-3_53

technique since SC data are only useful in the vicinity of the hazard, and its dissemination far away is wasting of the network limited resources.

Platooning approaches [2], aiming to structure the traveling vehicles into linear strings depending on their direction and lane of motion, have attracted a lot of consideration since their appearance in 1978 [21]. Maintaining the platoon's stability is an active field of research. Firstly, sensing-based strategy has been proposed to control the speed and the gap between two consecutive vehicles. Since sensors are failure prone, so, alternative solutions base-on Inter-Vehicular (IV) communication are required to ensure direct information exchange within the platoon.

Based on the standard IEEE 802.11p, normalized in 2011 [5,18,20] to enable the so called Vehicle to everything (V2X) communication, a new platoon's stability control mechanism appealed Cooperative Adaptive Cruise Control (CACC) has been proposed [6]. Despite the several attempts [2,4,6] to improve its behavior, the strategy still suffers from fundamental limitations, reviewed on a non realistic assumptions about the fixed communication pattern and constant delay. These assumptions could not fit the real world where the communication topology is variable [14] and the time delay variation is considered the source of platoon instability [19,25]. In addition, the study [22] affirmed that, CACC schemes lose their performance once the communication pattern changes. Then, more realistic solutions are required to deal with this situation.

In recent years, consensus alternatives are considered as a promising solutions to ensure the string's safety and stability and a variety of consensus algorithms have been reported. Most of them focused on solving the limitations of CACC algorithms. So, considering the issues related to the time-varying communication delays, consensus control algorithms have been introduced in [16,17,19]. In addition, the authors in [14] proposed a consensus control algorithm, counting the time-varying communication and the input saturation issues. Taking into consideration the platoon's problems related to the network imperfections and external disturbances, Elahi et al. [3] have introduced a consensus control algorithm for vehicles platoon to mitigate the serious effects of the communication topology uncertainty and the time-varying delay.

However, the consensus schemes cited ahead are interpreted difficult to implement in realistic scenario [15] and may waste the network's resources [25] due to their uninterrupted input and output. To overcome these limitations, researchers turn to event-triggered consensus solutions due to its resource-saving features. Several event-triggered leader-follower consensus control algorithms have been proposed to address the problem of the network saturation [15,24]. Due to the lack of space, we could not detail these approaches in the present work, so for more details, interested readers may refer to [21,23,24,26,27].

It is clear that platooning approaches help reduce the network load. But the fact of being an unbounded-size centralized construct, where all the members' actions depend on the platoon's leader information and instruction, is considered in some cases a crucial limitation. Depending on this consideration and the limitations of the communication based on 802.11p, a new vision is proposed by

G. Le Lann [7–12]. The cornerstone of this approach is to divide the IVN into a cyber-physical construct called *cohort*. Briefly, a cohort is a fully distributed bounded-size, n, vehicular string circulating in the same lane within a similar velocity ν_0 and keeping a same IV spacing s. Every cohort's member is characterized by its rank r and noted M_r with $r \in [1-n]$. Intra-cohort communication is based on the N2N communication paradigm. More details about the cohort concept could be find in [7,9]. In addition, Le Lann has invoked the important role that could play the consensus algorithms [11] to improve the road traffic safety and also to ensure the cohort stability. In accordance with this vision, the present work proposes a distributed event-triggered agreement algorithm to control the stability of the cohort in the face to a SC maneuver, and in this context we select the use case of deceleration maneuver.

The remained of this paper is organized as follow: in Sect. 2 we introduce our proposition, experimentation is presented in Sect. 3, and finally, we end with a conclusion in Sect. 4.

2 Agreement Scheme for Safety Critical Longitudinal Maneuver

Every cyber-physical maneuver, distinguished by a unique identifier, is a combination of a cyber part and a physical part. In our study, we are only concerned with the cyber space, a set of c automated phases carried out to ensure the safe execution of the physical space. Indeed, the time required to finish the cyber space might have an important effect on the execution of the entire maneuver. In this context, the author in [12] indicated that the major part of the SC maneuver delay must be affected to the physical space. Then it is mandatory to reduce as much as possible the cyber space delay. In this section, we ensure that the adopted use case is executed on the highway and all the members M_r in the cohort Γ are supposed to be highly automated and have similar size, l_r.

2.1 Assumptions

1st *Assumption:* Directional short range N2N communication is used, as a major key to: (i) ensure the cohort's members' cooperation within SC event appearance, and, (ii) exchange cohort control messages, principally, adapted to maintain neighboring awareness.

2nd *Assumption:* Time division-based deterministic MAC layer algorithms are crucial to schedule the channel access in between the cohort members. In this study, we adapt the MAC layer scheme invented by G. Le Lann in [10]. According to this algorithm, time is divided into several frames, named δ, and every cohort member M_r acquires two slots per frame: one slot is used to communicate with its successor M_{r+1}, called *downStream slot*, and the second one is used to cooperate with its predecessor M_{r-1}, called *upStream slot.* To mitigate the slots' allocation problem the algorithm proceeds as next: (i) the frame is divided fairly into two equal lots of h slots, one reserved for the *downStream slots*, and the other for *upStream slots*, (*ii*) every node M_r calculates the index of its will-be-used slots based on its position r in its cohort. Here we offered a brief description of this algorithm, more details about its main concept are presented in [10], although, its implementation, and evaluation are the subject of an ongoing publication.

3rd *Assumption:* Agreement algorithms are necessary for the cohort members coordination in hazardous situation. Therefore, we extend our MAC layer with an agreement module. Such a module makes sure that all the cohort's members, involved in a SC maneuver, participate in the decision making. The trend towards collective decision allows to avoid the catastrophic consequences of individual decision made in the absence of consensus algorithms. Details about our proposed agreement scheme, are given in the next subsection.

4th *Assumption:* The agreement delay is established from the MAC layer delay. Here we are going to explain, briefly, how the dissemination delay is determined according to the adopted MAC algorithm. So, every SC message is able to run through h hops per frame. Consequently, to travel the entire cohort, from the first member to the last one, in ideal condition, our message needs $\lceil \frac{n-1}{h} \rceil$ frames. In addition, we consider that the cohort is a fault-tolerant system, that tolerates $\kappa \leq \kappa^{\bullet}$ omissions per N2N link l. Actually, every omission gives rise to an additional delay equal to δ. Thus, to travel the entire cohort, the message exploits $\left(\sum_{i=1}^{n-1} \kappa_{l_i} + \lceil \frac{n-1}{h} \rceil\right)$ frame.

Finally, we assume that the arrival rate of the SC messages is equal to 1 message per second. Then, every SC message is going to wait up to δ to get access to the channel. Consequently, the aggregate delay, needed to disseminate a SC message over an entire cohort, is equal to: $\delta.\left(1 + \sum_{i=1}^{n-1} \kappa_{l_i} + \lceil \frac{n-1}{h} \rceil\right)$.

2.2 Agreement Scheme

In this subsection we describe our distributed event-triggered agreement scheme, presented on Algorithm 1. Within our proposed scheme, and as depicted on Fig. 1 and Fig. 2, every member M_r in Γ might get 3 major different states as following; the *initiator state*, the *collector state* and the *decider state*. Hereafter, we are going to detail all of these states.

1. ***Initiator Node***: is the cohort member, called M_i and located at x_i, who has detected a hazard O in x_O, at a time t_0. Then, it generates the message m_{SC} to inform the rest of Γ about the emergent situation. Depending on Algorithm 1, M_i should go through the following instructions:
   - Measure the distance, d_i, to O, using the method $distanceToHazard(x_o, x_i)$,
   - Calculate its rate of deceleration, γ_{SC_i}, using the method $calcul\gamma_{SC}()$,
   - Generate m_{SC}, which includes: $[x_i, \gamma_{SC_i}, d_i, x_o]$,
   - Send m_{SC} to its successor at its *downStream slot*,
   - Switch to the *waiting* state.

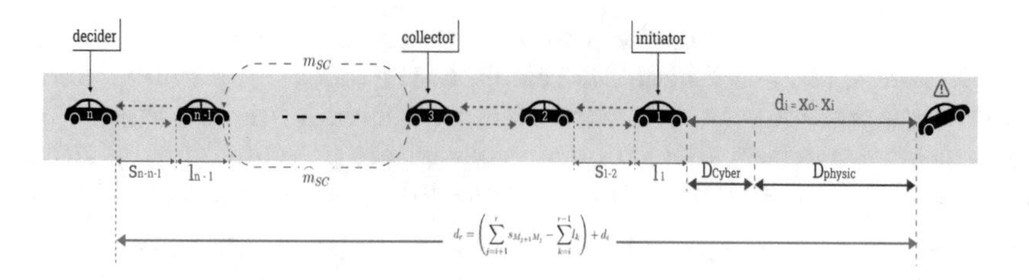

Fig. 1. Global operation of the distributed event-triggered agreement scheme

The *waiting* state refers to the state in which the node is waiting for the agreement decision.

2. ***Collector Node***: is a cohort member who has received m_{SC}, from its predecessor, treat and forward it to its successor, after adding, its proposed γ_{SC_r}. At the reception of m_{SC}, the collector acts as following:
   - Check the maneuver code,
   - Check its eligibility of making decision, by verifying if $r == \widehat{r}$ is true,
   - Measure the distance d_r, to the hazard O,
   - Calculate its deceleration rate, γ_{SC_r},
   - Add $[d_r, \gamma_{SC_r}]$ to m_{SC}, by the method *upDateSCMsg*(),
   - Send m_{SC} to the next neighbor at its *downStream slot*,
   - Switch to *waiting* state.

   Remember that m_{SC} is the message created by M_i and updated, by every M_r to include the necessary data helping the *decider* node to make decision. Otherwise, m_{SC} is containing all γ_{SC_r}, proposed from the initiator to the decider, coupled with all the other necessary information.
3. ***Decider Node***: is the cohort member who is elected to make decision. First of all, when m_{SC} arrives, this node behaves like the *collector* and goes through all the instructions indicated above until the determining γ_{SC_r}, then it behaves as following:
   - Calculate $\widehat{\gamma}_{SC}$, the most suitable deceleration rate to the current situation,
   - Generate $\widehat{m}_{SC}$, that will contain $\widehat{\gamma}_{SC}$
   - Send $\widehat{m}_{SC}$ to its predecessor at its *upStream slot*,
   - Switch to *executing* state.

The *executing* state is used to mark up the end of the cyber space ($i == c$) and hand over the control to the physical space.

Hereafter, we are going to explain the operation of the method *calcul*γ_{SC}(), that allows the nodes to calculate their deceleration rate γ_{SC}, taking into account its current ν_0, as follows:

(i) Firstly, we estimate the upper bound of the time needed to receive the decision, according to the 4th *assumption.* This latency matches the cyber space delay, we called ρ^r, which depends on the size of the hazard zone, established on the number of vehicles invoked on the maneuver, called ξ. Within our use case,

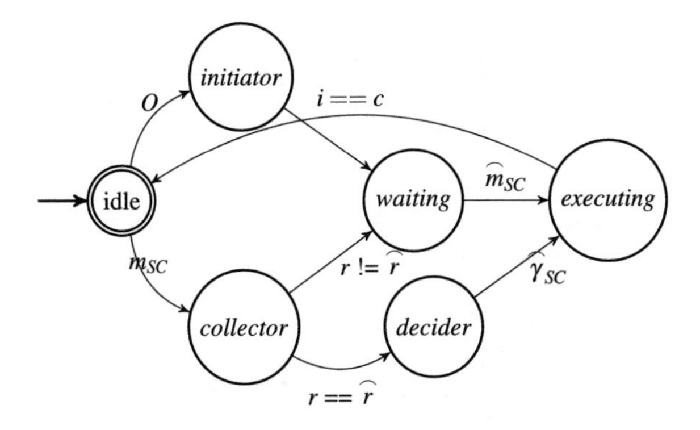

Fig. 2. Cohort's member sate variation according to the role played on the maneuver

when it determines $\widehat{\gamma}_{SC}$, the *decider* switches to the *executing* state and sends $\widehat{m}_{SC}$ to $M_{\widehat{r}-1}$ immediately. Then, $\widehat{m}_{SC}$ is keeping been disseminated toward the *initiator*. Then, the delay required by any member M_r to start braking is expressed as next:

$$
\begin{aligned}
\rho^{\widehat{r}} &= t_{m_{SC}}^{i\rightarrow\widehat{r}} + t_{\widehat{\gamma}_{SC}}^{\widehat{r}} + 0 \\
\rho^{\widehat{r}-1} &= t_{m_{SC}}^{i\rightarrow\widehat{r}} + t_{\widehat{\gamma}_{SC}}^{\widehat{r}} + t_{\widehat{m}_{SC}}^{\widehat{r}\rightarrow\widehat{r}-1} \\
&\dots \\
\rho^{i} &= t_{m_{SC}}^{i\rightarrow\widehat{r}} + t_{\widehat{\gamma}_{SC}}^{\widehat{r}} + \Delta_{\widehat{m}_{SC}}^{\widehat{r}\rightarrow i}
\end{aligned}
$$

So generally, the cyber delay is expressed according to this formula:

$$\rho^{r} = t_{m_{SC}}^{i\rightarrow\widehat{r}} + t_{\widehat{\gamma}_{SC}}^{\widehat{r}} + t_{\widehat{m}_{SC}}^{\widehat{r}\rightarrow r} \tag{1}$$

where $t_{m_{SC}}^{i\rightarrow\widehat{r}}$ is the time needed to transmit m_{SC} to $\widehat{r}$, $t_{\widehat{m}_{SC}}^{\widehat{r}\rightarrow r}$ the dissemination delay of $\widehat{m}_{SC}$ and $t_{\widehat{\gamma}_{SC}}^{\widehat{r}}$ is the time needed to calculate $\widehat{\gamma}_{SC}$. In accordance with the 4th *assumption* such a formula is translated to the following equation:

$$\rho^{r} = \delta.\left(1 + \sum_{i=1}^{n-1}\kappa_{l_i} + \left\lceil\frac{n-1}{h_m}\right\rceil + \left\lceil\frac{n-r}{h_m}\right\rceil + \sum_{j=1}^{n-r}\kappa_{l_j}\right) \tag{2}$$

(ii) Afterwards, we calculate the distance travelled by M_r during ρ^r, and we call it $\psi^r = \rho^r.\nu_0$. Then we can deduct the distance left to the hazard, we call $\varphi^r = d_r - \psi^r$. It is clear that φ^r depends on d_r. Depending on kinematic rules d_r is the total stopping distance available for M_r, measured from the moment M_r notices the hazard till the moment it completely stops. Here we explain how to express this distance. Let us start with the distance separating the *initiator*,

Algorithm 1: Longitudinal Agreement Scheme

Data:

x_r, l_r, s, ν_0 // current member situational information

begin

- **switch** *event* **do**
  - **case** *event*$_1$ **do**
    - //hazard detection
    - switchto(*initiator*)
    - d_i = distanceToHazard(x_o, x_i)
    - γ_{SC_i} = calculγ(d_i, ν_i)
    - m_{SC} = generate(γ_{SC_i}, d_i, (x_o) //create a new m_{SC}
    - send(m_{SC}) //at its downSlot
    - switchTo(*waiting*) //switch to waiting state
  - **case** *event*$_2$ **do**
    - //m_{SC} arrival
    - **if** (r != $\widehat{r}$) **then**
      - switchTo(*collector*)
      - checkManeuverCode(C)
      - d_r = distanceToHazard(x_o, x_r)
      - γ_{SC_r} = calculγ_{SC}(d_r, ν_r)
      - m_{SC} = upDateSCMsg(m_{SC}, γ_{SC_r}, d_r) //update initial m_{SC}
      - send(m_{SC})
      - switchTo(*waiting*)
    - **else if** (r ==$\widehat{r}$) **then**
      - switchTo(*decider*)
      - checkManeuverCode(C)
      - $d_{\widehat{r}}$ = distanceToHazard(x_o, ($x_{\widehat{r}}$)
      - $\gamma_{\widehat{r}}$ = calculγ_{SC}($d_{\widehat{r}}$, $\nu_{\widehat{r}}$)
      - $\widehat{\gamma}_{SC}$ = decide$\widehat{\gamma}_{SC}$($\boldsymbol{\gamma_r}$, $\boldsymbol{d_r}$)
      - $\widehat{m}_{SC}$ = generate($\widehat{\gamma}_{SC}$)
      - send($\widehat{m}_{SC}$) //at its upSlot
      - switchTo(*executing*)
  - **case** *event*$_3$ **do**
    - switchTo(*executing*) // After $\widehat{m}_{SC}$ arrival

located at x_i, and O located at x_o, already called d_i and given by the next equation:

$$d_i = x_o - x_i(t_0) \tag{3}$$

Accordingly, the distances separating the member behind M_i and O is established, in terms of d_i, as following:

$$d_{i+1} = s_{M_{i+1}M_i} - l_i + d_i$$
$$d_{i+2} = s_{M_{i+1}M_i} + s_{M_{i+2}M_{i+2}} - (l_i + l_{i+1}) + d_i$$
$$\ldots$$

Generally, the distance from any member M_r to O is given by the next equation:

$$d_r = \left(\sum_{j=i+1}^{r} s_{M_{j+1}M_j} - \sum_{k=i}^{r-1} l_k \right) + d_i \quad (4)$$

(iii) Finally, we select to right deceleration rate, γ_{SC}, that could avoid the collision with the O. According to the physic rules γ_{SC} is calculated as follows: $\gamma_{SC} = -\frac{\nu_0^2}{2.\varphi}$.

3 Safety Critical Deceleration Maneuver

Consider the following scenario, where a cohort Γ of n highly automated vehicles, travelling on a crowded highway, at a velocity ν_0 and maintaining a similar IV spacing s_{xy}. Assume also that all of Γ's members have a similar size l_r. For every M_r we consider the next situational information: the location x_r, the velocity ν_r, the size l_r and the deceleration rate γ_{SC_r}. Working with highly automated vehicles means that the human drivers are completely out of the deceleration task. Then, the time required by the human driver to recognize the danger is replaced by the m_{SC}'s dissemination delay, and the time needed to react is replaced by the time needed to receive the agreement decision message $\widehat{\gamma}_{SC}$. The sum of the m_{SC} and $\widehat{m}_{SC}$'s dissemination delays matches the cyber delay we are looking up to study.

In purpose to establish the upper bound of the cyber delay, we called $\rho^{\bullet}$, we suppose that the hazard is located ahead of the entire cohort. Then at t_0, M_1 realizes that, they are approaching a hazard zone O, located at x_o. Consequently, it is crucial to inform the rest of Γ about this danger by triggering a collaborative cyber-physical maneuver of deceleration. So, M_1 states itself the *initiator* M_i and proceeds as pointed out by Algorithm 1. Since the entire cohort is invoked in the maneuver, so M_n (the Γ's tail) will be the *decider*. Unlike the scenario without agreement, within our solution no member from the concerned ones has the ability to start braking before the agreement procedure gets ended. The thing is, M_r could start braking only after receiving the agreement decision from its neighbor M_{r-1}.

In accordance with the assumptions in Sect. 2.1, M_r are configured to use N2N communication, based on a deterministic MAC layer extended by an agreement module, to share information and collaborate to overcome critical situation. Under serious condition losing these messages or even their delayed reception might lead to harmful consequences, like rear-end collision within our use case.

So, depending on 4th *Assumption* in Sect. 2.1, every SC message loses its validity up to $\rho^{\bullet}$.

Consequently, to ensure a reliable and delay-free dissemination of SC messages, we study in this section the impact of the cyber delay variation on the execution of the entire SC deceleration maneuver and afterward figure out $\rho^{\bullet}$. Previously in Sect. 2.1, we indicated that ρ^{r} increases when r decreases. Accordingly, ρ^{i} matches $\rho^{\bullet}$ and ρ^{n} represents the lowest value, we called ρ°, so $\rho^{\circ} < \rho^{r} < \rho^{\bullet}$.

Then, studying the variation of upper bound of ρ^{r} means studying the variation of ρ^{i}. Actually, during $\rho^{\bullet}$ Γ loses a distance, $\psi^{\bullet}$, towards O, as indicated in Sect. 2.2, and the condition is getting more critical, particularly for the members located at the head of Γ. Wherefore, we indicate that the braking must start from the Γ's tail, to reduce as much as possible the number of vehicles implied in the rear-end collision, if any.

Within our configuration the cyber delay is influenced by the following parameters: (i) the hazard zone size ξ, (ii) the frame duration δ and (iii) the messages omission's number κ that our system could tolerate. All the parameters used in this study are reviewed in the following table.

Table 1. Configuration parameters

Parameter	Description	Value
ν_0	Initial velocity	$[9, \text{m.s}^{-1}$–$37\,\text{m.s}^{-1}]$
n	Cohort size established on vehicle number	[10–50]
ξ	Hazard zone size established on vehicle number	n
h	Interference range established on vehicle number	[3–5]
θ	Time slot duration	1 ms
κ	Number tolerated omission per N2N link	[1–3]
s_{xy}	Inter-vehicular gap	$\nu_0.2\,\text{s}$
δ	Frame duration (ms)	$2.h.\theta$

We divided our study into three tests depending on the value of h as depicted on Fig. 3. As mentioned in Table 1, we keep varying $n \in [10\text{–}50]$ and $\kappa \in [1\text{–}3]$. On one hand, the different configurations of h have indicated that the $\rho^{\bullet}$ increases significantly when n and/or κ increase. Let us consider the case of $h = 3$, illustrated by Fig. 3a, the results show that for $n = 50$ and $\kappa = 1$, $\rho^{\bullet} \approx 500\,\text{ms}$, while $\rho^{\bullet} > 1000\,\text{ms}$ when $n = 50$ and $\kappa = 3$. Similarly when $h = 4$, see Fig. 3b, we obtained $\rho^{\bullet} > 360\,\text{ms}$ when $n = 30$ and $\kappa = 1$, $\rho \approx 600\,\text{ms}$ for $n = 30$ and $\kappa = 2$ and $\rho > 800\,\text{ms}$ for $n = 30$ and $\kappa = 3$ and so on. In addition, the results highlight that the values of $\rho^{\bullet}$ obtained for $\kappa = 1$ are smaller than the others regardless of n. On the other hand, by comparing the results issued from the different h we find that these obtained for $h = 3$ are smaller then those obtained for $h = 4$ or $h = 5$.

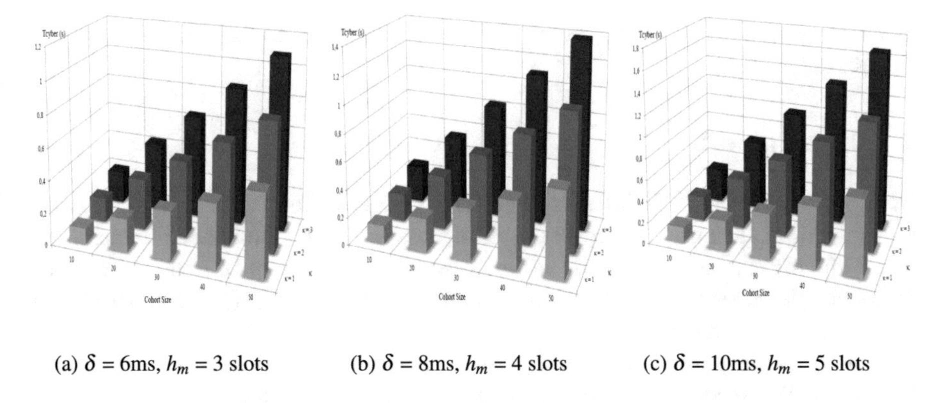

(a) δ = 6ms, h_m = 3 slots (b) δ = 8ms, h_m = 4 slots (c) δ = 10ms, h_m = 5 slots

Fig. 3. Cyber delay upper bound variation according to δ and κ

Author in [12] has indicated that the cyber delay should not exceed 500 ms. Based on this assumption and considering $\kappa = 1$ and $h = 3$ it is remarkable that the best values of $\rho^\bullet$ are given when $n < 50$. But if we consider the IEEE 802.11p assumption that the SC dissemination delay should not exceed 100 ms then the best values of $\rho^\bullet$ are given when $n < 20$.

Hereafter, we pursue with the following set up $h = 3$, $\kappa = 1$, $n \in [10\text{–}27]$ and $\nu_0 \in [9\,\text{m.s}^{-1}\text{–}37\,\text{m.s}^{-1}]$ to determine the distance $\psi^r = \nu_0.\rho^r$ travelled during ρ^r. The results are presented on Fig. 4.

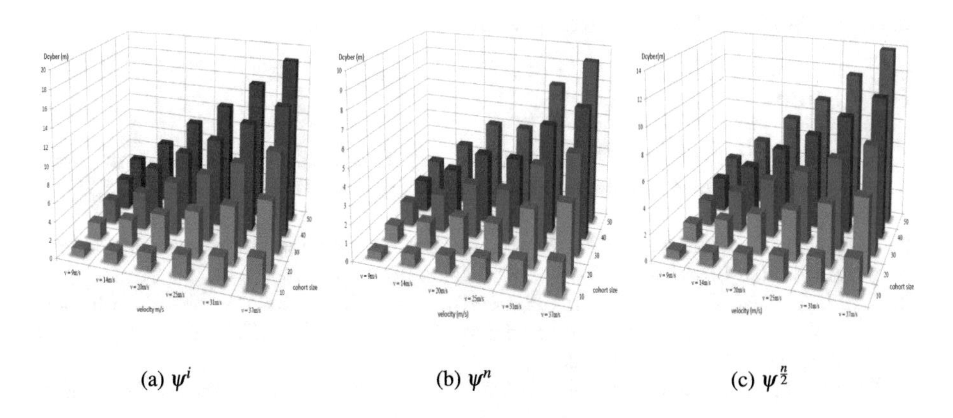

(a) ψ^i (b) ψ^n (c) $\psi^{\frac{n}{2}}$

Fig. 4. Cyber distance ψ^r variation according to n and ν_0

Three cases of ψ^r are highlighted here; (i) ψ^i travelled by the *initiator*, ψ^n travelled by the *decider* and $\psi^{\frac{n}{2}}$ representing the distance travelled by the node situated in the middle of Γ. The results show that this distance increases when ν_0 and n increase. In accordance with the previous results, the largest distance is ψ^i and the narrowest distance is ψ^n and that remains to the fact that this distance is tightly related to the cyber delay value. $\psi^{\frac{n}{2}}$ represents the average of this distance.

4 Conclusion

Breaking down the IVN into size-bounded stings, called cohorts, is considered a promising solution to improve the timeliness and the validity of SC messages, and reducing the network congestion. Like platoons, maintain the cohort stability face to external perturbation, like a brusque cyber-physical maneuver of deceleration, is a necessity.

Then, we proposed a distributed event-triggered agreement algorithm that, (i) allows all the concerned vehicles to participate on the decision making, to (i) ensure the choice of the most suitable deceleration rate. Afterwards, we studied the variation of the variation of the cyber delay according to different configurations of the parameters, in purpose to determine the upper bound of this delay and the right configuration. The results indicate that the upper bound depends, generally, on the size of risk zone, in our use case setted up at n. According to previous study [12], that indicated that more the cyber delay is small more the execution of the entire SC maneuver might be safe. Then we recommend to choose small cohort size, less then 20 vehicles, to satisfy this hypothesis.

In our future work we aim to focus on the multi-lane maneuvers, also know under the name of lateral maneuvers, where the execution of the maneuver is done over several lanes in the road, which is more complex than these taking place on one lane like our use case in this paper. So, we are going to propose agreement algorithms that cover this kind of situation.

References

1. Kajackas, A.A., Žuraulis, V., Sokolovskij, E.: Influence of VANET system on movement of traffic flows in emergency situations. Promet Traffic Transp. **27**, 237–246 (2015)
2. Peters, A.A., Middleton, R.H., Mason, O.: Leader tracking in homogeneous vehicle platoons with broadcast delays. Automatica **50**(1), 64–74 (2014)
3. Elahi, A., Alfi, A., Modares, H.: H_∞ consensus of homogeneous vehicular platooning systems with packet dropout and communication delay. IEEE Trans. Syst. Man Cybern. Syst. **51**(10), 1–12 (2021)
4. Flores, C., Milanés, V.: Fractional-order-based ACC/CACC algorithm for improving string stability. Transp. Res. C Emerg. Technol. **95**, 381–393 (2018)
5. Raviglione, F., Malinverno, M., Casetti, C.: Characterization and performance evaluation of IEEE 802.11p NICs. In: ACM Workshop on Technologies, mOdels, and Protocols for Cooperative Connected Cars (TOP-Cars), Catania-Italy, pp. 13–18, July 2019
6. Naus, G.J.L., Vugts, R.P.A., Ploeg, J., van de Molengraft, M.J.G., Steinbuch, M.: String-stable CACC design and experimental validation: a frequency-domain approach. IEEE Trans. Veh. Technol. **59**(9), 4268–4279 (2010)
7. Le Lann, G.: Cohorts and groups for safe and efficient autonomous driving on highways. In: IEEE Vehicular Networking Conference (VNC) (2011)

8. Le Lann, G.: On safety in ad hoc networks of autonomous and communicating vehicles: A rationale for time-bounded deterministic solutions. In: CoRes 2017–2ème Rencontres Francophones sur la Conception de Protocoles, l'Évaluation de Performance et l'Expérimentation des Réseaux de Communication, Quiberon, France, May 2017 (2017)
9. Le Lann, G.: On the power of cohorts - multipoint protocols for fast and reliable safety-critical communications in intelligent vehicular networks. In: International Conference on Connected Vehicles and Expo (ICCVE), pp. 35–42 (2012)
10. Le Lann, G.: A collision-free MAC protocol for fast message dissemination in vehicular strings. In: IEEE Conference on Standards for Communications and Networking (CSCN), Berlin-Germany (2016)
11. Le Lann, G.: Fast distributed agreements and safety-critical scenarios in VANETs. In: International Conference on Computing, Networking and Communications, Santa Clara, CA, USA, Jan 2017, p. 7 (2017)
12. Le Lann, G.: On the power of cohorts-multipoint protocols for fast and reliable safety-critical communications in intelligent vehicular networks. In: International Conference on Connected Vehicles and Expo (ICCVE), Beijing, China, December 2012, pp. 35–42 (2012)
13. Bae, I., Moon, J., Seo, J.: Toward a comfortable driving experience for a self-driving shuttle bus. Electronics **8**, 943 (2019). http://www.mdpi.com/journal/electronics
14. Chen, J., Liang, H., Li, J., Lv, Z.: Connected automated vehicle platoon control with input saturation and variable time headway strategy. IEEE Trans. Intell. Transp. Syst. **22**(8), 4929–4940 (2021)
15. Zhang, L., Sun, J., Yang, Q.: Distributed model-based event-triggered leader-follower consensus control for linear continuous-time multiagent systems. IEEE Trans. Syst. Man Cybern. Syst. **51**(10), 6457–6465 (2020)
16. di Bernardo, M., Salvi, A., Santini, S.: Distributed consensus strategy for platooning of vehicles in the presence of time-varying heterogeneous communication delays. IEEE Trans. Intell. Transp. Syst. **16**(1), 102–112 (2015)
17. di Bernardo, M., Falcone, P., Salvi, A., Santini, S.: Design, analysis, and experimental validation of a distributed protocol for platooning in the presence of time-varying heterogeneous delays. IEEE Trans. Control Syst. Technol. **24**(2), 413–427 (2016)
18. Muhammad, M., Kearney, P., Aneiba, A., Kunz, A.: Analysis of security overhead in broadcast V2V communications. In: International Conference on Computer Safety, Reliability, and Security, Turku, Finland, 10–13 September 2019, pp. 251–263 (2019)
19. Yang, P., Tang, Y., Yan, M., Zhu, X.: Consensus based control algorithm for nonlinear vehicle platoons in the presence of time delay. Int. J. Control Autom. Syst. **17**(X), 1–13 (2019)
20. Baek, S., Lee, I., Song, C.: A new data pilot-aided channel estimation scheme for fast time-varying channels in IEEE 802.11p systems. IEEE Trans. Veh. Technol. **68**(5), 5169–5172 (2019)
21. Shladover, S.E.: Longitudinal control of automated guideway transit vehicles within platoons. J. Dyn. Syst. Meas. Control **100**(4), 302–310 (1978)
22. Santini, S., Salvi, A., Valente, A.S., Pescapè, A., Segata, M., Lo Cigno, R.: A consensus-based approach for platooning with inter-vehicular communications. In: IEEE Conference on Computer Communications (INFOCOM), Hong Kong, China, 26 April–1 May 2015, pp. 1158–1166 (2015)
23. Wen, S., Guo, G., Chen, B., Gao, X.: Event-triggered cooperative control of vehicle platoons in vehicular ad hoc networks. Inf. Sci. **459**, 341–353 (2018)

24. Dolk, V.S., Abdelrahim, M., Heemels, W.P.M.H.: Event-triggered consensus seeking under non-uniform time-varying delays. IFAC-PapersOnLine **50**(1), 10096–10101 (2017)
25. Wang, W., et al.: Nonlinear consensus-based autonomous vehicle platoon control under event-triggered strategy in the presence of time delays. Appl. Math. Comput. **404**, 1 (2021)
26. Zhu, W., Jiang, Z.-P.: Event-based leader-following consensus of multi-agent systems with input time delay. IEEE Trans. Autom. Control **60**(5), 1362–1367 (2015)
27. Yin, X., Yue, D., Hu, S.: Adaptive periodic event-triggered consensus for multi-agent systems subject to input saturation. Int. J. Control **89**(4), 653–667 (2016)

Game Theory-Based Energy Efficient Routing in Opportunistic Networks

Jagdeep Singh[1(✉)], Sanjay Kumar Dhurandher[2], and Isaac Woungang[3]

[1] Department of Computer Science and Engineering, Sant Longowal Institute of Engineering and Technology, Longowal, Punjab, India
jagdeep@sliet.ac.in, jagdeepknit@gmail.com

[2] Department of Information Technology, Netaji Subhas University of Technology, New Delhi, India

[3] Department of Computer Science, Ryerson University, Toronto, ON, Canada
iwoungan@ryerson.ca

Abstract. This paper proposes a novel optimized routing protocol called Game Theory-based Energy Efficient Routing (*GTEER*) in Opportunistic Networks that uses game theory for selecting the best possible next hop to forward data packets efficiently. In this protocol, the best strategy for the selection of the energy efficient next hop is dependent upon a non-zero sum cooperative game of two players considering the context information, energy, and successful deliveries of the corresponding node from the destination as vital attributes in framing the game. Simulation results are provided, showing that the proposed protocol *GTEER* performed better than the benchmark protocols *E-EDR, E-ATDTN,* and *GT-ACR* in terms of delivery ratio, average latency and average residual energy by using the *haggle-infocomm-2006 real data trace.*

1 Introduction

Delay Tolerant Networks (*DTNs*), also known as Opportunistic Networks [1], are those in which the sender and target nodes do not have an end-to-end connection. DTN is an example of a network that tries to provide best-effort and low-cost access to locations such as underwater networks, space communication networks, satellite communication networks, war zones, or sparsely populated areas. In DTN sometimes the nodes act as selfish. First, nodes may not store or copy the data, or it will drop the packet. Secondly, even if the nodes agree to store data, they may refuse to participate in relaying data to various other nodes. Both cases raise concerns for energy consumption, bandwidth consumption and storage space occupation. Efficient usage of energy, simply called energy efficiency, is the intent to minimize the amount of energy necessary to impart service in the network. Seeing that it is conventionally inconvenient as well as expensive to substitute faulty nodes once they are established, mitigating the consumption of energy is of utmost significance in order to augment the system

L. Barolli et al. (Eds.): AINA 2022, LNNS 449, pp. 627–639, 2022.
https://doi.org/10.1007/978-3-030-99584-3_54

lifespan. The lifespan of a network is elucidated as the stretch until sensors in the network die.

Game theory [2] is a handy tool in examining present day wireless systems, as it yields methodical tools to represent interactions amongst units with clashing concerns that contend for the finite system resources. Just like nodes that are resource-restricted, that may opt not to relay packets so that they can conserve their own battery. In some instances, nodes may attempt to hone the overall system performance. There exists nodes in the network who are not willing to cooperate and drop the packets before they reache the destination resulting in loss in message delivery ratio, wastage of energy and increase in overhead [3,4]. One of the major issues in DTN is the enforcement of cooperation. An efficient lightweight game theory is proposed that is based rewarding the node that make positive move in the direction of message delivery and enforces the node to participate the network for network gains. Through game theory [5,6], it can be made sure that the nodes not willing to contribute to message forwarding are discarded or excluded from the network thereby leading to enhancement of the overall network performance.

In this proposed game theory, game is played between the initiator node and the intermediate nodes. The destination node is not a part of this game. So, the game starts when the source node relays the packet to one of the intermediate nodes. Now, the decision is in the hands of that intermediate node whether to further relay or drop the packet. If it drops, then the game is ended but if it decides to relay then the decision lies in the hands of the next intermediate node. If the packet reaches successfully to the destination, then the communication is considered as a success else, it is considered as a failure. This critical decision whether to relay or drop the packet is taken by the nodes is based on their past behavior and the reward that's evaluated. As the messages are not passed on to the less trusted nodes which might drop the packet, so it is saving the energy of the network. This proposed game theory has led to the conservation of energy of the network.

In this work, we follow the routing procedure trend by using game theory to study the intelligent interactions and behaviors of nodes since the nodes that wish to share the common summary vectores may have no incentive to cooperate with each other and may prefer behaving selfishly.

1.1 Paper Contributions

The main contribution of the paper is given below:

1. Energy efficient routing in opportunistic networks using game theory is proposed.
2. The proposed protocol *GTEER* ensures the energy efficient nature of routing by running the simulations using the haggle INFOCOM 2006 real data traces.

The rest of the paper is organized as follows. In Sect. 2, the game theory based routing protocols for OppNets are discussed. In Sect. 3, the proposed

GTEER scheme is described. Section 4 provides the simulation results and analysis. Section 5 concludes the paper.

2 Related Work

Various routing protocols for OppNets have been proposed in the literature. Representative ones include: Prophet [7], PBMT [8], FQLRP [9], GT-ACR [10], CGOPP [11], CTIDM [12], and E^2MOOR [13]. In addition to these benchmark protocols, some research work have used game theory as a strategic tool for designing routing protocols for OppNets. The main objective of these protocols has been to increase the utilization of the network resources, maximize the delivery probability, or minimize the average delay.

In Guo et al. [14], if one node requires to be the intermediary node then it shall primarily dispatch its reference value to its circumambient proximate nodes which is used to note or mark that a particular node has taken part in the routing. This technique is referred to as a Misbehavior Detection System. It is used to safeguard the safety of the hybrid network.

Lin et al. [15] propose a protocol of game theory based on real time fault-tolerance with emphasis on regulating selfish behavior of nodes, considered a major cause of loss of packets during data transmission or network traffic.

Jiang et al. [16] talks about DTNs which can be thought of as a better solution to help users connect with each other with the help of various hops. Unlike a traditional scenario, real time DTNs make use of battery operated nodes which may or may not be functional at all times giving rise to the selfish issue of these networks. Present along are comparisons of various methods incentive mechanisms which attempt to analyze the problem and stimulate the network nodes.

Most of the above discussed dissemination-based routing techniques suffer from high network congestion, hence, require more network resources. In addition, in terms of next hop selection, extra time is involved in calculating the context information. Due to this, the discussed context-based routing protocols do not perform well in terms of delivery probability. Furthermore, the above-mentioned techniques did not consider any other parameter apart from security, when formulating the game theory based solution for dissemination and context based scenarios. The work proposed in this paper considers the context information of nodes to formulate a strategic energy efficient optimized game in such a way that a message can be routed towards its destination with minimum latency, and maximum possible message delivery probability.

3 Game Theory-Based Energy Efficient Routing in Opportunistic Networks

Game theory has been used for modelling the consequences of the mutual and possibly conflicting actions of the decision makers [5], and for modeling the complex interactions between nodes in networks [6]. In this work, we follow the same trend by using game theory to study the intelligent interactions and behaviors of nodes since the nodes that wish to share the common summary vectores may have no incentive to cooperate with each other and may prefer behaving selfishly.

The proposed *GTEER* protocol assumes an opportunistic network scenario composed of N mobile nodes. A sense of cooperation exists among all the unbiased mobile nodes in the given OppNet environment.

In addition, it is also assumed that these nodes have sufficient buffer size to store their respective context information. Furthermore, once the source or intermediate node has generated and broadcast the HELLO packets, every node dynamically calculates it's transmit energy and the successful deliveries with respect to the corresponding destination.

These values are assumed to be stored in the node's buffer. This node specific context information is further used in formulating a strategic game in order to find the best strategy from the available ones, for next hop selection purposes.

Here, the strategic game for two players is assumed to be a finite, cooperative and non-zero sum game. Finally the proposed work also assumed that each node participating in the message forwarding has enough energy level and no node behaves maliciously.

Successful deliveries means the number of successful message transfers from the start of simulation to current time, between the two given nodes. Also, successful deliveries values of these N nodes with respect to the destination is calculated dynamically, also is represented as SD.

Transmit energy is the energy which are required to transfer a message from one node to another node. This energy utilization is called transmit energy. In opportunistic environment, the transmit energy is calculated dynamically and is represented as TE. For one player, the distance for different messages may be different. Messages may have different order as distance travelled by nodes. For a node, smaller distance the message has, low transmission energy is required in transmission. In general, if distance of path is more, than transmission energy required by the node after successful transmission of message will be high.

3.1 System Model

The *GTEER* protocol tends to define the game as a Tuple $G = (N, S, U)$, where N is the set of mobile nodes/players, $S = S_i$ is the set of available strategies, and $U = U_i$ is the set of utility functions for the same game. Let TE be the transmit energy of node required for message transmission to neighboring nodes.

In the formulated strategic game, the game is being played between the source or intermediate node s and the neighbouring nodes i, $i = (1, 2, 3, \ldots, m)$.

Here, the neighbouring nodes are the ones that lie in the communication range of the source or intermediate node s. The outcome of the strategic game gives the best strategy from the already available set of strategies S. Based on the findings of the formulated strategic game, a source or intermediate node will select the next hop from the neighbouring nodes i to forward the message towards its destination.

In this section, we construct two player game for message transmission among nodes in opportunistic networks and construct the energy efficient utility function by using transmit energy and successful deliveries.

Suppose for instance, a source wants to send the message to destination. The source node S creates a message M and wants to send this message to destination D. Initially, S checks that, the particular destination node is available in transmission range. If D is available in transmission range, then S directly transfer the message to D. If D is not available in transmission range, then S tries to find suitable node with the help of utility function among neighboring nodes. The utility function is calculated by using the calculation of successful deliveries done by node before simulation starts and calculation of transmit energy required by node. The nodes may be available in transmission range but at different spatial distance. Then Sender/Intermediate node computes their utility function and sorts the available messages in ascending order according to spatial distance. Then each node can get the message priority. The spatial distance can only determine the priorities of message for single node i. In fact, there are many neighboring nodes around node i, each of them has a spatial distance for a certain message of node i. Hence, the priority of message for each player will be determined by utility function. λ is utility function, which play key role in message forwarding. Mathematically, the utility function is given below:

$$\lambda = TE * SD \tag{1}$$

The utility function is calculated iteratively according to the message priority. When the sum of utility functions on sender node equals to that of utility functions on intermediate node. Then, these nodes will reach equilibrium. And the pivot point is also determined. After determining the pivot point, it is natural that the solution is to assign the messages to corresponding nodes. For summary vector exchange, this pivot point is used. The messages which are left to pivot point, all those message are assigned to sender node and rest message are assigned to intermediate node.

From the Fig. 1, we can see that S transmit the message m by using shortest path. If S transmit the message to node 2, then because of larger distance between nodes, the transmit energy required high. In the current scenario, the energy exhausted by node is less due to lesser distance between source to destination by taking path.

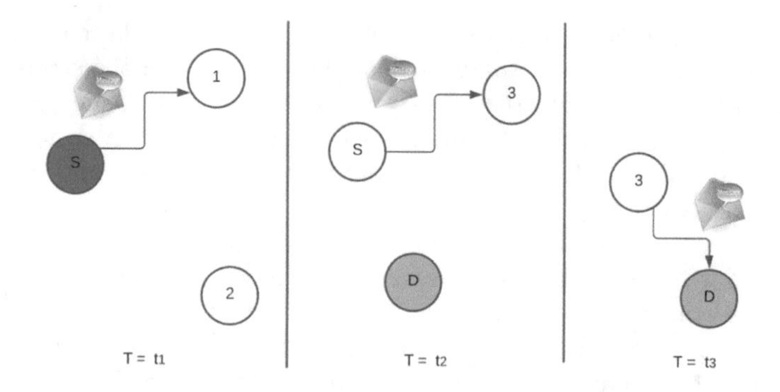

Fig. 1. Message transmitted on the basis for two player game at different time stamps

Algorithm 1. Game Theory-based Energy Efficient Routing

Input: Source node S, Destination node D, Successful Deliveries, Transmit Energy, Distance and Number of Nodes
Output: List of best relay nodes P
1: **Begin** N nodes (S_0, S_1, S_2, S_3,......S_n) in Opportunistic network.
2: Source node S generated a message m for sending to destination D.
3: Repeat steps 4 to 19 while message do not reach destination.
4: **for** each encounter between source/intermediate node pair **do**
5: Case 1:
6: **if** node D is avilable in the transmission range of node S **then**
7: transfer the message m directly to node D.
8: **end if**
9: Case 2:
10: **if** node D is not in the transmission range of node S **then**
11: Search for neighboring nodes, which are avilable in the transmission range of node S.
12: follow the two player game method of all the neighboring nodes.
13: for each neighbor node N, calcualate SD and TE
14: Calculate utility function λ

$$\lambda = TE * SD$$

15: Calculate pivot point for determination about message transmission.
16: Exchange the messages among nodes as per the pivot point.
17: Forward the message m to destination D.
18: **end if**
19: **end for**
20: Message delivered successfully to destination node D.

3.2 Illustrative Example

Suppose in an opportunsitic network environment, There are 2 players and they are having 7 messages, where node n_1 has the message 1, 2, 3 and 4. Node n_2 has the message 5, 6, and 7. The total messages denoted by P = {1, 2.......7}. When two nodes come in to the contact or chance to meet, they exchange their message list, and will know the total message. Then two nodes compute their utility function by using the parameters of successful deliveries and residual energy; and sort these 7 messages in ascending order according to spatial distance. Then each node can get the message priority. We cumulate utility function for each message iteratively according to the message priority. When the sum of utility functions on node n_1 equals to that of utility functions on node n_2. These two nodes will reach an equalibrium. And the pivot point is also determined. After determining the pivot point, it natural that the solution is to assign the messages to corresponding nodes. Messages which are left to pivot element, all those message are assigned to node n_1 and rest message are assigned to node n_2. Hence, node n_1 will forward their messages 2 and 3 to node n_2. And still keeps messages 1 and 4. The node n_2 forwards the messages 5 and 6 to node n_1, keeps the message 7.

4 Evaluation

In this section, the performance of the proposed *GTEER* protocol is evaluated using the Opportunistic Network Environment (ONE) [19] simulator. The area of the simulation is in a rectangle form in which the position of a node is represented by Cartesian x and y coordinate system. The values of various parameters considered in this simulation are shown in Table 1. The performance of the proposed *GTEER* scheme is compared against that of the E-EDR [17], E-ATDTN [18] and GT-ACR routing protocols, chosen as benchmark schemes, under varying TTL values from 100 min to 300 min. The performance metrics used for comparing the protocols are delivery probability, average latency and average residual energy by varying TTL, and buffer size. The simulations are done on real mobility data *haggle-one-infocom 2006* [20].

Figure 2 shows the relationship between delivery probability and TTL (Time To Live), showing that as the TTL increases, the delivery probability of *GTEER*, *E-ATDTN*, *GT-ACR* and *E-EDR* decreases. As a result, as the TTL increases, the time allocated to each message increases, and as more messages are kept in the node's buffer, the message delivery probability drops. The delivery probability of a message reaching it's destination increases with the application of game theory in the network. This happens so because with the application of the game, messages are forwarded only to nodes with a high coopertiveness and not to any node without checking it's corresponding context parameters. *GTEER* is 18.25% times better than *E-ATDTN*, 14.7% times better than *GT-ACR*, and 24.7% times better than *E-EDR* in terms of delivery probability.

Table 1. Simulation parameters

Parameter	Value
Real mobility data trace	Cambridge-haggle-one-infocom 2006
Trace format	Standard events reader
Trace fields in each line	5
Communication interface	Bluetooth
Transmission range	10 m
Number of nodes	98
Number of contacts	170601
Simulation time	337418 s
Transmission speed	250 Kbps
Message size	500 Kb up to 1 Mb
Buffer capacity	5 Mb
Message Time-to-live (TTL)	100–300 min

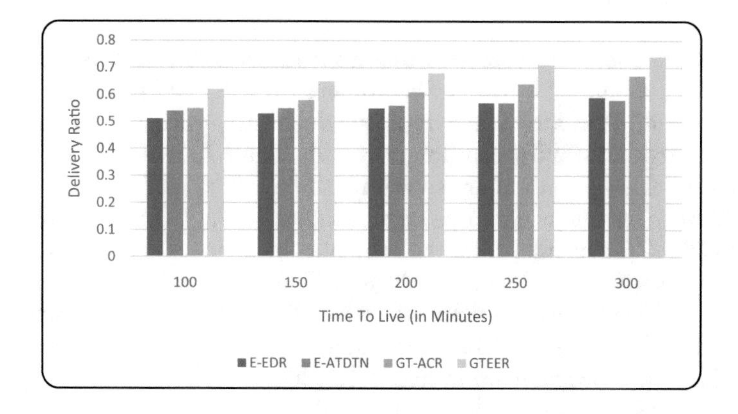

Fig. 2. Delivery ratio vs. time to live

Next, the TTL is altered, and the effect on latency is measured. Figure 3 depicts the results. It has been found that as the TTL is increased, the average latency increases as well. This is due to the fact that a large TTL value extends the time that the message stays in the node's buffer. In fact, *GTEER* performs 5.07% better than *E-ATDTN*, 4.41% better than *GT-ACR* and 6.65% better than *E-EDR* in terms of latency.

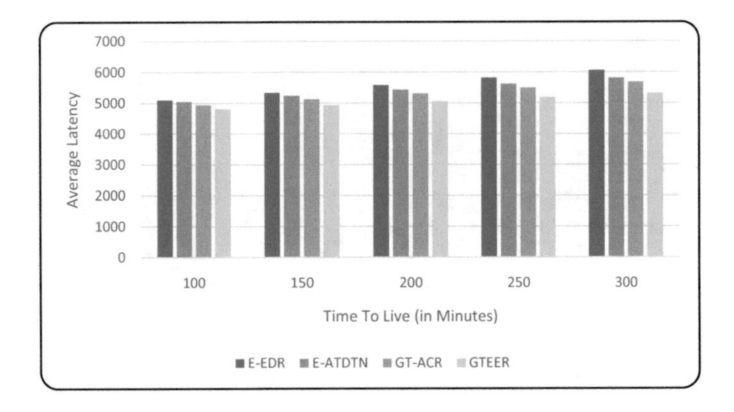

Fig. 3. Average latency vs. time to live

Thereafter, the buffer capacity is changed, and the effect of this modification on the probability of delivery is examined. Figure 4 depicts the performance of the routing strategies. It has been found that as buffer capacity grows, so does the probability of delivery. This is because when a node's buffer capacity grows, the number of messages stored in that buffer grows, resulting in more messages being transmitted to the receiver node. *GTEER* outperforms *E-ATDTN*, *GT-ACR* and *E-EDR* in terms of delivery probability by 11.72%, 8.05%, and 9.42%, respectively.

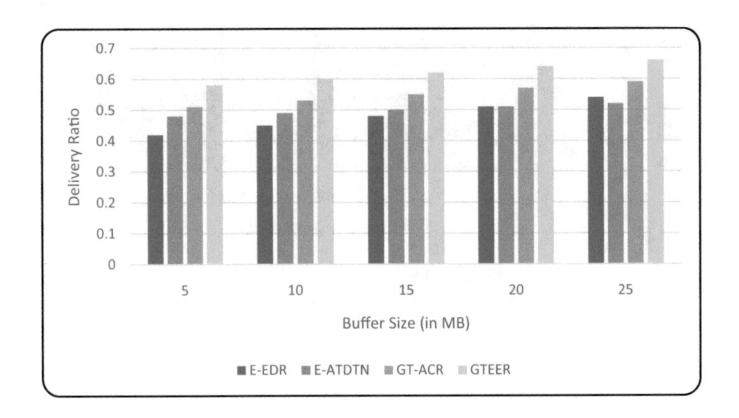

Fig. 4. Delivery ratio vs. buffer size

Next, the buffer capacity is changed, and the impact on the average delay is evaluated. Figure 5 depicts the performance of the buffer policies. *GTEER*, as expected, produces the lowest latency when compared to *E-ATDTN*, *GT-ACR*, and *E-EDR*.

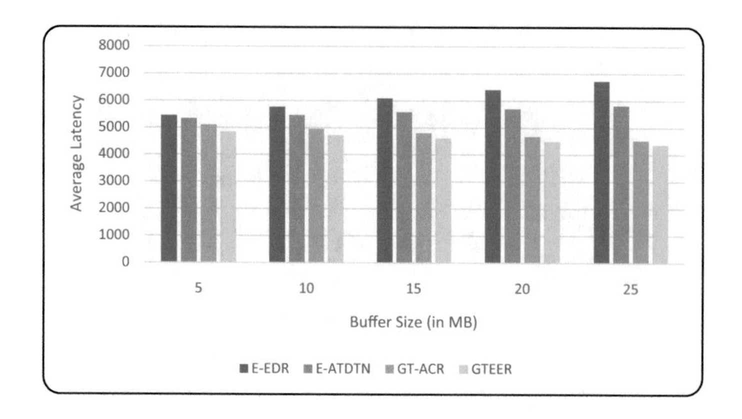

Fig. 5. Average latency vs. buffer size

Next, the main focus of implementing game theory into the network was so that it can be energy efficient. So, the message life time is changed, and the impact on the average residual energy is evaluated. Figure 6 depicts the performance of the remaining energy available after simulation. *GTEER*, outperforms to *E-ATDTN*, *GT-ACR*, and *E-EDR* routing protocol.

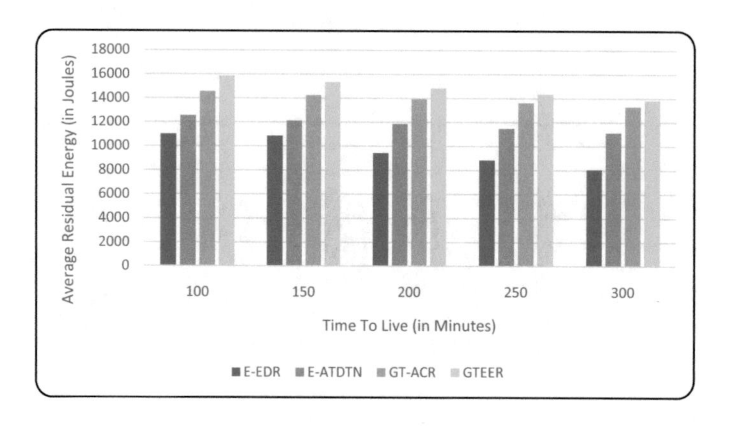

Fig. 6. Average residual energy vs. time to live

Figure 7 depicts the performance of the average residual energy available after simulation when varied buffer size. *GTEER*, outperforms to *E-ATDTN*, *GT-ACR*, and *E-EDR* routing protocol.

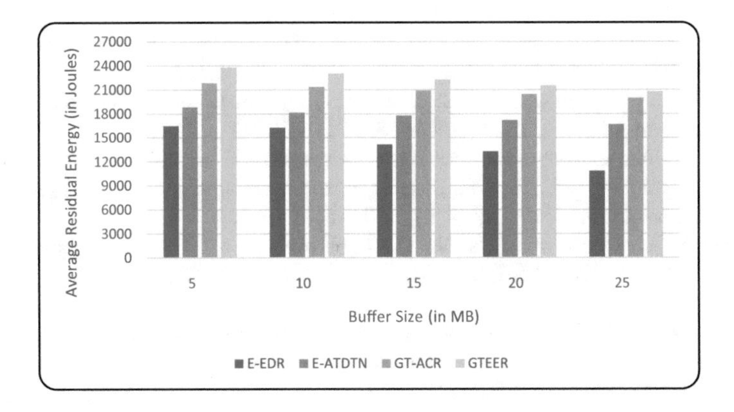

Fig. 7. Average residual energy vs. buffer size

5 Conclusion

We have addressed the issue of making the system energy efficient and to do the same, have implemented game theory in the system which enforces the nodes in the system to cooperate with each other. This approach has been designed and implemented in the *GTEER* protocol. In this paper, a novel routing protocol for OppNets *GTEER* has been proposed, which relies on the identification of the most significant attributes of a node. In this protocol, the selection of the best message forwarder for a given node is determined by means of a non-zero-sum cooperative game technique. Using the best response analysis approach, it has been shown that the proposed *GTEER* scheme can efficiently identify and select the next energy efficient best message forwarder for a given node from the set of its available neighbors. By using real data trace, simulation results have shown that *GTEER* yields a superior performance than *GT-ACR*, *E-EDR* and *E-ATDTN* routing protocols, in terms of average latency, and delivery probability, under varying TTL, and buffer size. As future work, we plan to investigate the security aspects of the proposed *GTEER* scheme with real traces mobility models.

References

1. Dede, J.: Simulating opportunistic networks: survey and future directions. IEEE Commun. Surv. Tutor. **20**(2), 1547–1573 (2019)
2. Qin, X., Wang, X., Wang, L., Lin, Y., Wang, X.: An efficient probabilistic routing scheme based on game theory in opportunistic networks. Comput. Netw. **149**, 144–153 (2019)
3. Wu, F., Chen, T., Zhong, S., Qiao, C., Chen, G.: A bargaining-based approach for incentive-compatible message forwarding in opportunistic networks. In: 2012 IEEE International Conference on Communications (ICC), Otawa, Canada, pp. 789–793. IEEE (2012)

4. Zhang, C., Zhu, Q., Chen, Z.: Game-based data-forward decision mechanism for opportunistic networks. J. Comput. **5**(2), 298–305 (2010)
5. Li, L., Wang, H., Liu, Z., Ye, H.: GIR: an opportunistic network routing algorithm based on game theory. IEEE Access **8**, 201158–201172 (2020)
6. Vazintari, A., Cottis, P.G.: Mobility management in energy constrained self-organizing delay tolerant networks: an autonomic scheme based on game theory. IEEE Trans. Mob. Comput. **15**(6), 1401–1411 (2015)
7. Lindgren, A., Doria, A., Schelén, O.: Probabilistic routing in intermittently connected networks. ACM SIGMOBILE Mobile Comput. Commun. Rev. **7**(3), 19–20 (2003)
8. Dhurandher, S.K., Singh, J., Woungang, I., Rodrigues, J.J.: Priority based buffer management technique for opportunistic networks. In: 2019 IEEE Global Communications Conference (GLOBECOM), Hawaii, USA, pp. 1–6. IEEE (2019)
9. Dhurandher, S.K., Singh, J., Obaidat, M.S., Woungang, I., Srivastava, S., Rodrigues, J.J.: Reinforcement learning-based routing protocol for opportunistic networks. In: ICC 2020-2020 IEEE International Conference on Communications (ICC), Dublin, Ireland, pp. 1–6. IEEE, June 2020
10. Borah, S.J., Dhurandher, S.K., Woungang, I., Kumar, V.: A game theoretic context-based routing protocol for opportunistic networks in an IoT scenario. Comput. Netw. **129**, 572–584 (2017)
11. Singh, J., Dhurandher, S.K., Woungang, I., Takizawa, M.: Centrality based geocasting for opportunistic networks. In: Barolli, L., Takizawa, M., Xhafa, F., Enokido, T. (eds.) Advanced Information Networking and Applications, AINA 2019. AISC, vol. 926, pp. 702–712. Springer, Cham (2020). https://doi.org/10.1007/978-3-030-15032-7_59
12. Gupta, N., Singh, J., Dhurandher, S.K., Han, Z.: Contract theory based incentive design mechanism for opportunistic IoT networks. IEEE Internet Things J., 1–11 (2021)
13. Singh, J., Dhurandher, S.K., Woungang, I., Diwakar, S., Chatzimisios, P.: Energy efficient multi-objectives optimized routing for opportunistic networks. In: ICC 2021-IEEE International Conference on Communications, Montreal, Canada, pp. 1–6. IEEE (2021)
14. Guo, H., Wang, X., Cheng, H., Huang, M.: A routing defense mechanism using evolutionary game theory for Delay Tolerant Networks. Appl. Soft Comput. **38**, 469–476 (2016)
15. Lin, C., Guowei, W., Pirozmand, P.: GTRF: a game theory approach for regulating node behavior in real-time wireless sensor networks. Sensors **15**(6), 12932–12958 (2015)
16. Khalid, W., et al.: A taxonomy on misbehaving nodes in delay tolerant networks. Comput. Secur. **77**, 442–471 (2018)
17. Borah, S.J., Dhurandher, S.K., Tibarewala, S., Woungang, I., Obaidat, M.S.: Energy-efficient prophet-PRoWait-EDR protocols for opportunistic networks. In: GLOBECOM 2017-2017 IEEE Global Communications Conference, Singapore, pp. 1–6. IEEE (2017)
18. Dhurandher, S.K., Woungang, I., Singh, J., Borah, S.J.: Energy aware routing for efficient green communication in opportunistic networks. IET Netw. **8**(4), 272–279 (2019)

19. Keranen, A., Ott, J., Karkkainen, T.: The ONE simulator for DTN protocol evaluation. In: Proceedings of SIMUTools, Rome, Italy, 2–6 March 2009, pp. 1–9 (2009)
20. Scott, J., Gass, R., Crowcroft, J., Hui, P., Diot, C., Chaintreau, A.: CRAWDAD dataset cambridge/haggle (v. 2009-05-29), CRAWDAD wireless network data archive (2009)

Accurate Modelling of A-MPDU Aggregation Technique with Markovian Techniques and M/M/1/k Queues

Kaouther Mansour[1(✉)] and Issam Jabri[2]

[1] IResCoMath Laboratory, ENIG, University of Gabès, Gabès, Tunisia
mansour.kaouther@enit.utm.tn
[2] CoEA, Yamamah University, Riyadh, Kingdom of Saudi Arabia
issam.jabri@enig.rnu.tn

Abstract. The 802.11ac amendment specifies that all Physical Protocol Data Units (PPDUs) are transmitted in aggregated MAC Protocol Data Units (A-MPDU) frames. Yet, when the medium is free all the eligible packets in the transmission queue are aggregated and transmitted. Hence, the size of A-MPDU aggregated frames does not depend only on the Block Acknowledgement window size but also on the number of eligible PDUs available in the transmission buffer. Even though that this specific may have a potential impact on the A-MPDU aggregation behavior, the existing analytical modelling approaches do not consider it. This may incurs inaccurate estimations of network performance. In this paper, we use markovian techniques and queuing theory to derive a mathematical model that provides accurate estimation of the A-MPDU aggregation size. An M/M/1/K queue is also considered, to reflect the impact of the network load on A-MPDU aggregation behavior. The accuracy of our model is well benchmarked by ns-3 simulator. Besides, compared to existing models, our proposal presents more precise estimation of the A-MPDU aggregation size especially in low network loads.

1 Introduction

Frame aggregation technique was defined for the first time by the 802.11n amendment. The key idea is to transmit multiple frames at a single access to the channel at the aim of optimizing channel utilization efficiency [14]. Two forms of aggregation are defined: A-MPDU aggregation and A-MSDU aggregation. The former consists in accumulating and encapsulating multiple MSDUs of the same access category addressed to the same destination. An MSDU is composed of a Link Layer Control (LLC) header, an IP header and the data payload. The latter offers the possibility to deliver up to N mpdus to the same destination using a unique PHY header. N presents the size of the Block Ack window and corresponds to 64 and 128 with 802.11n and 802.11ac amendments, respectively. A-MPDU aggregation is accomplished at the MAC layer. Firstly, MAC Packets Data Units are logically aggregated. Next, padding bits are added such that the length of each subframe is a multiple of four bytes. An MPDU delimiter is inserted at the end of each subframe. The latter is necessary for aggregated subframes depiction

L. Barolli et al. (Eds.): AINA 2022, LNNS 449, pp. 640–650, 2022.
https://doi.org/10.1007/978-3-030-99584-3_55

by the receiving node. The Block Akcnowledgment technique is used jointly with A-MPDU aggregation technique, to ensure individual error control over the aggregated subframes [14]. Indeed, the sequence number of the oldest mpdu in the aggregated frame is recorded in the Starting Sequence Number subfield. Then each bit in the bitmap subfield ensures the acknowledgement of a specific mpdu in the aggregated frame such that the bit in position k is assigned to the subframe of sequence number $(SSN + k - 1)$. New research studies have revealed that this acknowledgement policy causes A-MPDU aggregation size fluctuation in lossy environments, since the aggregation size depends on the position of the first lost subframe [5].

Besides, the 802.11ac amendment specifies that all Physical Protocol Data Units (PPDUs) are transmitted in MAC Protocol Data Units (A-MPDU) frames [13]. Yet, when the medium is free all the eligible packets in the transmission queue are aggregated and transmitted. Hence, the size of A-MPDU aggregated frames does not depend only on the Block Acknowledgement window size but also on the number of eligible PPDUs available in the transmission buffer.

In this paper, we propose an analytic model for 802.11ac A-MPDU aggregation scheme. Our model provides accurate estimation of A-MPDU aggregated frames size. An M/M/1/k queue model is considered to outline the fact that A-MPDU aggregated frames are not forcedly composed of the maximum supported number of subframes. In other words, once a node gains access to the medium, the eligible mpdus in the transmission queue are aggregated and transmitted, there's no gathering time. Using queueing theory, we investigate the behavior of A-MPDU aggregation technique face to network load variation. Markovian techniques are further used to reflect the impact of the legacy acknowledgement policy on the aggregated frame size. The effectiveness of our model is validated using ns-3 simulations [4]. Besides, compared to the analytic model proposed in [11], our proposal presents more precise estimations of the average A-MPDU aggregation size especially in low network loads.

The paper is structured as follows. The most relevant related works to the area of our concern are briefly reviewed in Sect. 2. Our analytic model is presented in Sect. 3. Section 4 is dedicated to analysis and discussions. And finally the paper is concluded in Sect. 5.

2 Related Works

In the research literature several mathematical model have been developed for 802.11 enabled-aggregation MAC layers. In [1] and [8], authors have designed mathematical models to estimate the achieved throughput and the end-to-end delay in unsaturated networks conditions. These two models are limited to error-free channel conditions. More recently, authors in [7] have proposed an analytic model for A-MPDU aggregation-enabled MAC layers in lossy environments. In [9], authors have designed a novel aggregation scheme that estimates the optimal aggregation size in presence of channel errors. They have, further, elaborated an analytic model to evaluate the performance of their proposal. In [3], authors have designed an analytic model for A-MPDU aggregation-enabled WLANs. Numeric estimations of networks key performance indicators (KPI) are used to permit dynamic adjustment of the aggregation size face to channel conditions variation. In [2], Hajlaoui et al. have used markovian techniques to derive an accurate model for the conservative A-MPDU aggregation scheme that takes into consideration the anomalous time slots. In [10], Cong et al. have elaborated a precize end-to-end delay model for unsaturated networks. They have, further, conducted extensive analysis on delay distribution to investigate the impact of the aggregation level on QoS performance for real time multimedia applications.

Recently, new analytic models have focused on the issue of the Block Ack window limit for the greedy A-MPDU aggregation scheme [6,11,12]. Hence, based on markovian techniques, authors in [12] have proposed an enhanced analytical model of aggregation-enabled WLANs with compressed Block Ack. Extensive analysis have been conducted to explain the impact of the compressed Block Ack operation on the maximum supported aggregation size. As far as, a three-dimensional DTMC model has been elaborated in [6] to provide accurate estimation of the aggregation level for the conventional A-MPDU aggregation scheme for saturated networks. Assessments revealed that the Block Ack window limit has a significant effect on the aggregated frames size, besides ommiting this specific induces estimation errors that raise up to 55% in error-prone networks. A more sophisticated version of this analytic model was proposed in [11].

3 Analytic Model

3.1 Model Assumptions

For our analysis we consider a 802.11-ac station transmitting saturated uplink traffic. We suppose that all mpdus are of the same length. Further, we assume that preambles and control frames are transmitted without errors, since they are transmitted at the lowest basic rate, furthermore they are very short. Finally, the binary symmetric channel model with independently distributed bit errors is considered. This channel model has been widely used in the literature [12].

3.2 DTMC Model Description

We consider a two dimensional DTMC model; $T_{u,v}$, $(0 \leq u \leq N$ and $0 \leq v \leq N)$. Being u refers to the number of subframes aggregated for the first time and v presents the number of retransmitted ones. To elaborate our model, 3 categories of states are considered according to the form of the aggregated frame. The first category includes the states $T_{u,0}$ with $(1 \leq u \leq N)$; an aggregated frame composed of u new subframes. The second category includes all the states of aggregated frames containing only retransmitted mpdus. In other words, the states $T_{0,v}$ with $1 \leq v \leq N$. And finally, the third category includes the states $T_{u,v}$; $1 \leq u \leq N-1$ and $1 \leq v \leq N-u$, a frame composed of new and retransmitted mpdus.

In a next step, the possible one step transitions between the different states are determined and the expressions of their corresponding probabilities are computed. For each category, 3 types of transitions are defined, according to the category to which belongs the next state. The markov chain is then designed. Finally, the asymptotic distribution at the steady state is determined. The probability generating function (PGF) of the aggregation size random variable (A) is further computed and the corresponding moment of first order is calculated.

3.2.1 One-Step Non Null Transition Probabilities

As aforementioned, the transition probabilities are computed by category. Assume $p_e = PER$ and $p_s = 1 - PER$ the probabilities that an mpdu is lost due to channel errors and delivered with success respectively.

For the first category, a successful transmission of the u new mpdus ensures remaining at the same category with a frame composed of u' new subframes with the following probability:

$$P[T_{(u',0)}/T_{(u,0)}] = p_s^u \, p^{u'} \quad 1 \le u \le N,\ 1 \le u' \le N. \tag{1}$$

Here $p^{u'}$ gives the probability that the waiting queue contains u' mpdus. This probability is calculated using the predefined equations from the queuing theory. However the corruption of the first aggregated subframe induces a transition to one of the states of the second category, according to the total number of corrupted subframes.

$$P[T_{(0,v')}/T_{(u,0)}] = p_e \, C_{u-1}^{v'-1} \, p_e^{(v'-1)} \, p_s^{(u-v')} \quad 1 \le u \le N,\ 1 \le v' \le u. \tag{2}$$

The first term p_e presents the probability of corruption of the first aggregated mpdu. The second term gives the probability that $(v'-1)$ mpdus among the remaining $(u-1)$ are corrupted while $(u-v')$ mpdus are transmitted with success.

And finally, the loss of one or more subframes different from the first aggregated subframe provoques a transition to a state of the 3rd category.

$$P[T_{(u',v')}/T_{(u,0)}] = p^{u'} \, p_s^{u''} \, p_e \, C_{u-(u''+1)}^{v'-1} \, p_e^{(v'-1)} \, p_s^{(u-(u''+v'))}$$
$$1 \le u'' \le u-1,\ 1 \le u' \le u'',\ 1 \le v' \le u-u''. \tag{3}$$

The term $p_s^{u''}$ refers to the probability of the successful transmission of the first u'' subframes. p_e is the probability of corruption of the subframe of sequence number $(SSN + u'')$. Similarly to the precedent equation, the 3rd term gives the probability that $(v'-1)$ mpdus among the remaining $u-(u''+1)$ are corrupted while $(u-(u''+v'))$ mpdus are transmitted with success. Finally, $p^{u'}$ corresponds to the probability that u' mpdus are present in the waiting queue.

We proceed similarly to compute the non-null one-step transition probabilities for the states of the second category.

We note that, for states of the second category, not only a successful transmission of the v retransmitted mpdus provoques a transition to the first category. But also, the dropping of subframes that have reached the maximum retry limit.

$$P[T_{(u',0)}/T_{(0,v)}] = p^{u'} \sum_{i=0}^{v} p_{RL}^{i} \, (p_{NRL} \, p_s^{(v-i)}) \quad 1 \le v \le N,\ 1 \le u' \le N. \tag{4}$$

$p^{u'}$ refers to the probability that there exists u' mpdus in the waiting queue. The second term gives the probability that i subframes have reached the maximum retry limit. Then if they are not transmitted with success they will be rejected. The last term corresponds to the probability that the remaining $(v-i)$ mpdus have not yet reached the maximum retry limit and are transmitted with success.

The corruption of the oldest subframe within the aggregated frame provoques a transition to another state of the second category according to the total number of corrupted mpdus.

$$P[T_{(0,v')}/T_{(0,v)}] = p_{NRL}^{v} p_e C_{v-1}^{v'-1} p_e^{(v'-1)} p_s^{(v-v')} \quad 1 \leq v \leq N,\ 1 \leq v' \leq v. \tag{5}$$

The term p_{NRL}^{v} refers to the probability that none of the v retransmitted subframes has reached the maximum retry limit. Thus this transition can not occur in the other case. p_e presents the probability of corruption of the oldest subframe in the aggregated frame. The second term gives the probability that $(v-v')$ mpdus among the $(v-1)$ remaining ones are transmitted successfully while $(v'-1)$ are lost.

If the oldest subframe is transmitted with success or rejected while one or more other subframes among the remaining ones are corrupted then a transition to a state $T_{u',v'}$ of the third category occurs.

$$\begin{aligned} P[T_{(u',v')}/T_{(0,v)}] =& p^{u'} \frac{1}{N-(v-1)} \left[\sum_{i=i_{min}}^{i_{max}} \sum_{j=0}^{i} p_{RL}^{j} \left(p_{NRL}\, p_s\right)^{(i-j)} \right] \\ & \left[p_{NRL}^{(v-i)}\ p_e \right] \left[C_{v-(i+1)}^{v'-1}\ p_e^{(v'-1)} p_s^{(v-(v'+i))} \right] \\ & 2 \leq v \leq N,\ 1 \leq u'' \leq N-1, \\ & 1 \leq u' \leq u'',\ 1 \leq v' \leq min\,(v-1, N-u''). \end{aligned} \tag{6}$$

The term $\frac{1}{N-(v-1)}$ accounts for the fact that the $(v-1)$ retransmitted mpdus are dispersed over the $N-1$ available positions in the aggregated frame. The term $(\sum_{i=i_{min}}^{i_{max}} \sum_{j=0}^{i} p_{RL}^{j} \left(p_{NRL}\, p_s\right)^{(i-j)})$ defines the probability that the i aggregated subframes older than the first corrupted subframe are either dropped or transmitted with success. The argument i depends on the number of available positions before and after the first corrupted subframe. A detailed explication of how to compute the value of the argument i will be given lateral. The third term $(p_{NRL}^{(v-i)}\ p_e\ )$ presents the probability that the first corrupted subframe and the retransmitted subframes with sequence number greater than the sequence number of the first corrupted subframe have not yet reached the retry limit. p_e is the probability of corruption of the first lost subframe. The last expression gives the probability that $(v-(v'+i))$ mpdus among the remaining $(v-(i+1))$ mpdus are transmitted without errors, whereas $(v'-1)$ mpdus are corrupted.

Computation of the argument i

Assume s the number of subframes transmitted with success; then $s = v - v'$.

- $if\ \ u'' = 1$

$$i_{min} = i_{max} = 1$$

- $if \;\; N-u'' \geq v-1$

$$i_{min} = 1 \text{ and}$$
$$i_{max} = min(s,u'')$$

- $if \;\; N-u'' < v-1$

$$i_{min} = s-(N-u''-v') \text{ and } i_{max} = min(s,u').$$

Likewise, three types of transitions are defined for the states of the third category. Firstly, a successful transmission of the u new subframes and the retransmitted subframes that have not yet reached the retry limit permits to go to a state $T_{u,0}$.

$$P[T_{(u',0)}/T_{(u,v)}] = p^{u'} \left[\sum_{i=0}^{v} p_{RL}^{i} \, (p_{NRL}\, p_s)^{(v-i)}\right] \, p_s^u$$
$$1 \leq u \leq N-1, \; 1 \leq v \leq N-u, \; 1 \leq u' \leq N. \tag{7}$$

$p^{u'}$ defines the probability that the transmission queue contains u' mpdus. p_s^u gives the probability that the u new mpdus should be transmitted with success.The term $(\sum_{i=0}^{v} p_{RL}^{i} \, (p_{NRL}\, p_s)^{(v-i)})$ presents the probability that the v retransmitted mpdus have either reached the maximum retry limit or retransmitted with success.

If it has not yet reached the maximum retry limit, the corruption of the first aggregated subframe prevents the insertion of new subframes in the next transmitted frame. Thus a transition to a state of the second category occurs.

$$P[T_{(0,v')}/T_{(u,v)}] = p_{NRL}^{v} p_e \, C_{(v+u)-1}^{v'-1} \, p_e^{(v'-1)} \, p_s^{((v+u)-v')}$$
$$1 \leq u \leq N-1, \; 1 \leq v \leq N-u, \; 1 \leq v' \leq u+v. \tag{8}$$

Similar to Eq. 5, except that in the second term we should take into consideration the u new subframes in addition to the v retransmitted ones.

Finally, a transition to another state of the third category happens in case of loss of at least one subframe different from the oldest one in the aggregated frame. For simplicity reasons, we differentiate two cases according to the nature of the first lost subframe: If it is among the u subframes transmitted for the first time the transition probability is calculated as follows.

$$P[T_{(u',v')}/T_{(u,v)}] = p^{u'} \left[\sum_{i=0}^{v} p_{RL}^{i} \, (p_{NRL}\, p_s)^{(v-i)}\right] \left[p_s^{(u-(N-u''))} \; p_e\right]$$
$$\left[C_{v'-1}^{(N-u''-1)} \; p_e^{(v'-1)} \; p_s^{(N-u''-v')}\right]$$
$$1 \leq u \leq N-1, \; 1 \leq v \leq N-u,$$
$$N-u \leq u'' \leq N-1, \; 1 \leq u' \leq u'', \; 1 \leq v' \leq N-u''. \tag{9}$$

This equation is composed of two main parts. The first part serves to compute the probability that the v retransmitted subframes are either delivered successfully or dropped while the second part concerns the u new subframes. The latter is divided into three terms. The first term gives the probability that the first $(u-(N-u'))$ new subframes are transmitted with success. We note that the number of these subframes is determined based on the structure of the aggregated mpdu. The second term gives the probability that the following subframe is lost. Finally, the third term presents the probability that $(v'-1)$ among the $(N-u'-1)$ remaining ones are lost while $(N-u'-v')$ are transmitted successfully.

In the case that the first lost subframe is a retransmitted mpdu then the probability of this event is calculated as follows.

$$
\begin{aligned}
P[T_{(u',v')}/T_{(u,v)}] = & p^{u'} \frac{1}{(N-u)-(v-1)} \left[\sum_{i=i_{min}}^{i_{max}} \sum_{j=0}^{i} p_{RL}^{j} \left(p_{NRL}\, p_s\right)^{(i-j)} \right] \left[p_{NRL}^{(v-i)}\, p_e \right] \\
& \left[C_{v-(i+1)}^{v'-(v-1)}\, p_e^{(v'-(v-1))}\, p_s^{(u-v'_u)} \right] \left[C_u^{v'_u}\, p_e^{v'_u}\, p_s^{(u-v'_u)} \right] \qquad (10) \\
& 1 \leq u \leq N-1,\; 1 \leq v \leq N-u, \\
& 1 \leq u'' \leq N-1,\; 1 \leq u' \leq u",\; 1 \leq v' \leq min(N-u'', v+u-1).
\end{aligned}
$$

Three important points are considered in this equation. The first point, since the v retransmitted subframes are scattered over the $(N-u)$ available positions, we multiply by the term $\frac{1}{(N-u)-(v-1)}$ which gives the probability of existence of a retransmitted mpdu in a given position. The second point, we should take into consideration the fact that the number of the retransmitted mpdus older than the first corrupted one can take many values according to the number of available positions before and after the first lost subframe. Those subframes are either transmitted with success or dropped. And finally, we should take into consideration the fact that the v' retransmitted subframes are composed of the lost subframes among the v old subframes that have not yet reached the maximum retry limit in addition to subframes lost for the first time among the u subframes transmitted for the first time. In what follows, we detail the steps of calculation of the scope of variation of i; the number of retransmitted mpdus older than the first corrupted one. The computation of v'_v and v'_u which correspond to the corrupted subframes among the v retransmitted mpdus and among the u new subframes, respectively, is further detailed.

Computation of the argument i

- if $u''=1$

$$i_{min} = i_{max} = 1$$

- if $N-(u+u'') \geq v-1$

$$
\begin{aligned}
i_{min} &= 1 \\
i_{max} &= min\,(v-v'_v, u'')
\end{aligned}
$$

- else

$$
\begin{aligned}
i_{min} &= v-(N-(u+u'')) \\
i_{max} &= min\,(v-v'_v-1, u'').
\end{aligned}
$$

Computation of v'_v and v'_u

- if $u \geq v' - 1$

$$1 \leq v'_v \leq min(v', N - (u + u''))$$

- else

$$v' - u \leq v'_v \leq min(v', N - (u + u'').$$

- $v'_u = v' - v'_v$.

Based on Eqs. 1 to 10, we build the state transition matrix P associated to our model.

$$P = P_{T_{(u',v')}/T_{(u,v)}} = P[T_{(u,v)} \rightarrow T_{(u',v')}] \\ u,\ v,\ u',\ v' \in [\,0,\ N\,]. \quad (11)$$

Then we characterize the stationary state distribution:

$$\Pi^* = \lim_{m \rightarrow \inf} (\Pi^0 P^m). \quad (12)$$

Being Π^0 refers to the initial state distribution vector and is defined as follows:

$$\Pi^0_{u,v} = \begin{cases} p^u \text{ if } \ 1 \leq u \leq N,\ v = 0 \\ 0 \ \ \text{otherwise} \end{cases} \quad (13)$$

Once done, we define the expression of the probability mass function of the discrete random variable A presenting the random number of mpdus within an A-MPDU aggregated frame.

$$P_A(A_k) = \sum_{u=0}^{A_k} \Pi^*_{(u,v=A_k-u)} \ \text{for} \ \ k = 1, 2, 3, ..., N. \quad (14)$$

The average aggregation size $E[A]$ corresponds to the first moment order of $P_A(A_k)$.

$$E[A] = \sum_{A_k=1}^{N} A_k \ P_A(A_k). \quad (15)$$

3.3 Queuing Model

We proceed to the analysis of the $M/M/1/k$ queue to determine the probabilities p^k for $1 \leq k \leq N$, defining the probability that the waiting queue contains k packets. Using the pre-defined equations from the queuing theory we get:

$$p^k = \frac{(1-\rho)\,\rho^k}{(1-\rho^k)}. \quad (16)$$

Where ρ defines the queue load factor.

4 Analysis and Discussions

In this section we start by validating the correctness of our model using ns-3 simulations. We further compare the performance of our model with the model proposed in [11]. Next, we assess the behavior of the new all aggregation, all the time transmission scheme under different network conditions. The average aggregation size is measured for 2 configurations $\rho = 1.2$ and $\rho = 3$, considering a *PER* that raises from 0 to 0.5. Simulation parameters are presented in Table 1.

Table 1. Analysis parameters.

Parameter	Designation	Value
PER	The packet error rate	[0..0.5]
ρ	The traffic load	[1.2, 3]
L_{subf}	The payload size	1500 bytes
m	The retry limit	7
N	The size of the Block Ack window	128

Figure 1 draws the evolution of the average number of aggregated mpdus face to channel quality degradation. Two different network configurations are considered: (1) light network load ($\rho = 1.2$) and (2) heavy network load ($\rho = 3$). Measurements are performed with the new proposed model and the model proposed in [11]. Ns-3 simulations are further conducted to verify the correctness of our model. Analysis results show that, under heavy network load, the two models perform almost similarly. However, under light network load, the new model shows visibly better estimations. Indeed, in [11] authors suppose that A-MPDU aggregation is all the time performed at the maximum aggregation size supported by the Block Ack window. Therefore, this model is not applicable for 802.11-ac networks under light network load. Regarding the first network configuration, ($\rho = 1.2$), in absence of errors the average aggregation size estimated by [11] is up to 128 mpdus., whereas the values estimated by ns-3 simulations and the novel model do not exceed 50 and 52, respectively. This gap is explained by the impact of the network load on the aggregation size. As the *PER* increases, this gap is reduced inspite that the channel load is constant. Indeed, from one hand, the new model shows higher average aggregation size, about 73 when $PER = 0.2$. This is explained by the raise of the number of retransmitted mpdus. From another hand, the average aggregation size calculated by [11] decreases due to the limit imposed by the Block Ack window in presence of channel errors. In very lossy environments ($PER = 0.5$), the two models show almost equal estimated values. Thus, given the significant number of erroneous subframes, the aggregation size becomes very limited which reduces the influence of network load intensity on the aggregation size. Based on these results, we can conclude that our model provides accurate estimations of the average aggregation size for the "All aggregated all the time" aggregation scheme under different network load conditions. It further reflects correctly the impact of the Block Ack window limit

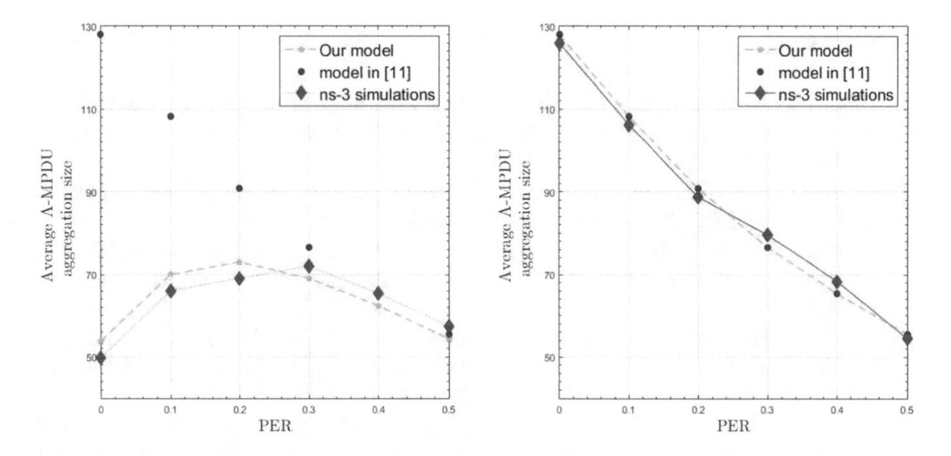

Fig. 1. The average aggregation size $E[A]$ versus the packet error rate (PER) for light and heavy network load.

on the aggregation size. We note also that existing analytical models are not applicable for the A-MPDU aggregation scheme defined by the 802.11ac standard, especially under light network load. Indeed, precise analysis of the transmission buffer behavior is indispensable to correctly model 802.11ac MAC layers.

5 Conclusion

This paper presented a new mathematical model for the all aggregated, all te time transmission scheme specified by the 802.11 ac amendment. Using markovian techniques and queuing theory, the proposed model performs accurate estimation of the average A-MPDU aggregation size under different network conditions. Indeed, it considers individual subframes losses due to channel errors as well as individual subframes discard. The proposed model captures also the impact of the Block Ack window limit on the maximum aggregation size. The accuracy of the proposal is proved through ns-3 simulations. Comparison with the previous model proposed in [11] showed that, in heavy network load, the two models show almost similar performance. However, in light network load, the novel model gives visibly better results. Indeed, the model proposed in [11] supposes that A-MPDU aggregation should be performed at the maximum allowed aggregation level, whereas our new model incorporates an M/M/1/k queue to provide precise estimation of the aggregation size for the new 802.11 all aggregated transmission scheme.

Currently we are working on the extension of this model to provide an analytic model that accounts for all the features of 802.11ac MAC layer and permits accurate analysis of different QoS metrics.

References

1. Charfi, E., Chaari, L., Kamoun, L.: Fairness of the IEEE 802.11 n aggregation scheme for real time application in unsaturated condition. In: Joint IFIP Wireless and Mobile Networking Conference, 4th edn, pp. 1–8. IEEE (2011)
2. Hajlaoui, N., Jabri, I., Jemaa, M.B.: An accurate two dimensional Markov chain model for IEEE 802.11 n DCF. Wireless Netw. **24**(4), 1019–1031 (2018)
3. Hajlaoui, N., Jabri, I., Taieb, M., Benjemaa, M.: A frame aggregation scheduler for QoS sensitive applications in IEEE 802.11n WLANs. In: Proceedings of the ICCIT, Tunisia, pp. 221-226 (2012)
4. Riley, G.F., Henderson, T.R.: The ns-3 network simulator. In: Wehrle, K., Gunes, M., Gross, J. (eds.) Modeling and Tools for Network Simulation, pp. 15–34, Springer, Heidelberg (2010). https://doi.org/10.1007/978-3-642-12331-3_2
5. Inamullah, M., Raman, B.: 11 ac frame aggregation is bottlenecked: revisiting the block ACK. In: International ACM Conference on Modeling, Analysis and Simulation of Wireless and Mobile Systems, 22nd edn, pp. 45–49 (2019)
6. Jabri, I., Mansour, K., AlOqily, I., et al.: Enhanced characterization and modeling of A-MPDU aggregation for IEEE 802.11 n WLANs. Trans. Emerg. Telecommun. Technolog. **33**, e4384 (2021)
7. Karmakar, R., Swain, P., Chattopadhyay, S., et al.: Performance modeling and analysis of high throughput wireless media access with QoS in noisy channel for different traffic conditions. In: International Conference on Communication Systems and Networks (COMSNETS), 8th edn., pp. 1–8. IEEE (2016)
8. Kim, B.S., Hwang, H.Y., Sung, D.K.: Effect of frame aggregation on the throughput performance of IEEE 802.11 n. In: IEEE Wireless Communications and Networking Conference, pp. 1740–1744. IEEE (2008)
9. Lin, Y., Wong, V.W.: Frame aggregation and optimal frame size adaptation for IEEE 802.11 n WLANs. globecom, pp. 1–6. IEEE (2006)
10. Lu, C., Wu, B., Ye, T.: A novel QoS-aware a-MPDU aggregation scheduler for unsaturated IEEE802. 11n/ac WLANs. Electronics **9**(8), 1203 (2020)
11. Mansour, K., Jabri, I., Ezzedine, T.: Revisiting the IEEE 802.11 n A-MPDU retransmission scheme. IEEE Commun. Lett. **23**(6), 1097–1100 (2019)
12. Seytnazarov, S., Choi, J.G., Kim, Y.T.: Enhanced mathematical modeling of aggregation-enabled WLANs with compressed blockACK. IEEE Trans. Mobile Comput. **18**(6), 1260–1273 (2018)
13. IEEE 802.11-2020 - IEEE Approved Draft Standard for Information Technology - Telecommunications and Information Exchange Between Systems Local and Metropolitan Area Networks - Specific Requirements - Part 11: Wireless LAN Medium Access Control (MAC) and Physical Layer (PHY) Specifications
14. IEEE Standard for Information technology technology-Telecommunications and information exchange between systems Local and metr metro- politan area networks networks-Specific requirements requirements, Part 11: Wireless LAN Medium Access Control (MAC) and Physical Layer (PHY) Specifications, 20 (2012)

A Fuzzy-Based Scheme for Slice Priority Assessment in 5G Wireless Networks

Phudit Ampririt[1](✉), Ermioni Qafzezi[1], Kevin Bylykbashi[2], Makoto Ikeda[2], Keita Matsuo[2], and Leonard Barolli[2]

[1] Graduate School of Engineering, Fukuoka Institute of Technology, 3-30-1, Wajiro-Higashi, Higashi-Ku, Fukuoka 811-0295, Japan
{bd21201,bd20101}@bene.fit.ac.jp

[2] Department of Information and Communication Engineering, Fukuoka Institute of Technology, 3-30-1,Wajiro-Higashi, Higashi-Ku, Fukuoka 811-0295, Japan
kevin@bene.fit.ac.jp, makoto.ikd@acm.org,
{kt-matsuo,barolli}@fit.ac.jp

Abstract. The resources in the Fifth Generation (5G) wireless network are restricted, and the number of devices is growing considerably faster than the system can handle. As a result, the overloading issue and priority of using resources will be major concern. For satisfying user requirements and resolve that problem, the Software-Defined Network (SDN) with Network Slicing will be a good approach for admission control and managing the resources. In this paper, we propose a fuzzy-based scheme for evaluating Slice Priority (SP) considering three parameters: Slice Traffic Volume (STV), Slice Interference from Other Slices (SIOS) and Slice Connectivity (SC). From simulation results, we conclude that the considered parameters have different effects on the SP. When STV and SC are increasing, the SP parameter is increased but when SIOS is increasing, the SP parameter is decreased.

1 Introduction

In 5G wireless networks will be billions of new devices with unpredictable traffic patterns that deliver high data rates. With the introduction of the Internet of Things (IoT), these devices will create large amounts of data on the Internet, causing congestion and QoS deterioration [1].

The 5G wireless network will provide customers with new experiences such as Ultra High Definition Television (UHDTV) and support a large number of IoT devices with long battery life and high data rates in high-density hotspot locations. Because 5G uses high frequency to deal with larger device volume and high user density, routing and switching technologies are no longer required, and coverage area is shorter than 4G [2–4].

As a consequence, several research projects are attempting to develop systems that are suited for the 5G era. One of them is the SDN [5]. Also, the mobile handover method with SDN is used to reduce changeover processing delays. In addition, introducing Fuzzy Logic (FL) to SDN controllers enhance the QoS [6–8].

L. Barolli et al. (Eds.): AINA 2022, LNNS 449, pp. 651–661, 2022.
https://doi.org/10.1007/978-3-030-99584-3_56

In our previous work [9–12], we presented a Fuzzy-based system for admission decision considering four input parameters: Quality of service (QoS), Slice Priority (SP), Service Level Agreement (SLA) and Slice Overloading Cost (SOC). The output parameter was Admission Decision (AD). In this paper, we propose a Fuzzy-based scheme for evaluating Slice Priority (SP) considering three parameters: Slice Traffic Volume (STV), Slice Interference from Other Slices (SIOS) and Slice Connectivity (SC).

The rest of the paper is organized as follows. In Sect. 2 is presented an overview of SDN. In Sect. 3, we present application of Fuzzy Logic for admission control. In Sect. 4, we describe the proposed Fuzzy-based system and its implementation. In Sect. 5, we discuss the simulation results. Finally, conclusions and future work are presented in Sect. 6.

2 Software-Defined Networks (SDNs)

The SDN is a new networking paradigm that decouples the data plane from control plane in the network. By SDN is easy to manage and provide network software based services from a centralised control plane. The SDN control plane is managed by SDN controller or cooperating group of SDN controllers. The SDN structure is shown in Fig. 1 [13,14].

- **Application Layer** builds an abstracted view of the network by collecting information from the controller for decision-making purposes. The types of applications are related to network configuration and management, network monitoring, network troubleshooting, network policies and security.
- **Northbound Interfaces** allow communication between the control layer and the application layer and can provide a lot of possibilities for networking programming. Based on the needs of the application, it will pass commands and information to the control layer and make the controller creates the best possible software network with suitable qualities of service and acceptable security.
- **Control Layer** receives instructions or requirements from the Application Layer. It contains the controllers that control the data plane and forward the different types of rules and policies to the infrastructure layer through the Southbound Interfaces.
- **Southbound Interfaces** allow connection and interaction between the control plane and the data plane. The southbound interface is defined as protocols that allow the controller to create policies for the forwarding plane.
- **Infrastructure Layer** receives orders from SDN controller and sends data among them. This layer represents the forwarding devices on the network such as routers, switches and load balancers.

The SDN can manage network systems while enabling new services. In congestion traffic situation, the SDN can control and adapt resources appropriately throughout the control plane. Mobility management is easier and quicker in forwarding across different wireless technologies (e.g. 5G, 4G, Wifi and Wimax). Also, the handover procedure is simple and the delay can be decreased.

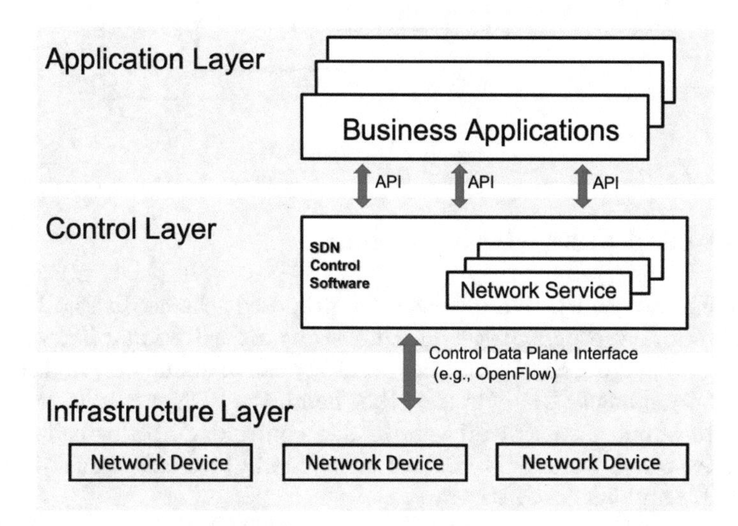

Fig. 1. Structure of SDN.

3 Outline of Fuzzy Logic

A FL system can do a nonlinear mapping of an input data vector into a scalar output and is able to simultaneously handle numerical data and linguistic knowledge. The FL can deal with statements which may be true, false or intermediate truth-value. These statements are impossible to quantify using traditional mathematics. The FL system is used in many controlling applications such as aircraft control (Rockwell Corp.), Sendai subway operation (Hitachi), and TV picture adjustment (Sony) [15–17].

In Fig. 2 is shown Fuzzy Logic Controller (FLC) structure, which contains four components: fuzzifier, inference engine, fuzzy rule base and defuzzifier. A fuzzifier is used to combine crisp values with linguistic variables that have fuzzy sets associated with them. In the first phase, by Fuzzifier are fuzzified the actual inputs to generate fuzzy inputs. The fuzzy rules can be given by an expert or can be inferred from numerical data. The fuzzy rules are expressed as a set of IF-THEN statements. By using fuzzified input values and fuzzy rules, the Inference Engine infers fuzzy output. Finally, the Defuzzification gives a crisp control output from fuzzified inputs.

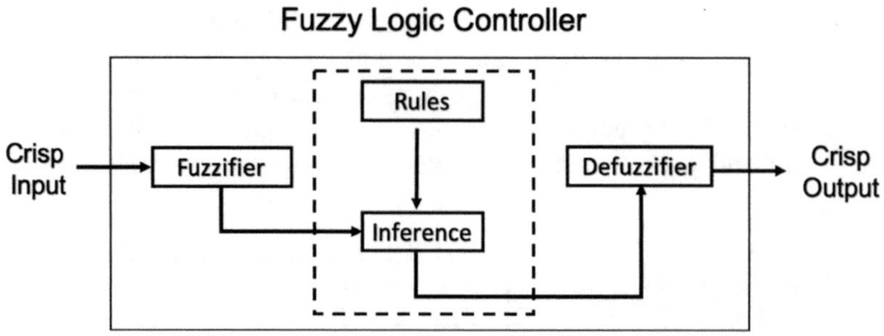

Fig. 2. FLC structure.

4 Proposed Fuzzy-Based Scheme

In this work, we use FL to implement the proposed scheme. In Fig. 3, we show the overview of our proposed system. Each evolve Base Station (eBS) will receive controlling order from SDN controller and they can communicate and send data with User Equipment (UE). On the other hand, the SDN controller will collect all the data about network traffic status and controlling eBS by using the proposed Fuzzy-based system. The SDN controller will be a communicating bridge between eBS and 5G core network.

The proposed system is called Integrated Fuzzy-based Admission Control System (IFACS) in 5G wireless networks. The structure of IFACS is shown in Fig. 4. For the implementation of our system, we consider four input parameters: Quality of Service (QoS), Slice Priority (SP), Slice Overloading Cost (SOC), Service Level Agreement (SLA) and the output parameter is Admission Decision (AD). We applied FL to evaluate QoS, SP, SOC and SLA. The QoS is considering four parameters: Slice Throughput (ST), Slice Delay (SD), Slice Loss (SL) and Slice Reliability (SR). The SOC is considering three parameters: Virtual Machine Overloading Cost (VMOC), Link Overloading Cost (LOC) and Switches Overloading Cost (SWOC).

In this paper, we apply FL to evaluate the SP by considering three parameters: Slice Traffic Volume (STV), Slice Interference from Other Slices (SIOS) and Slice Connectivity (SC). The output parameter is SP.

Slice Traffic Volume (STV): When traffic volume of a slice is higher than others, that slice also has higher priority than other slices.

Slice Interference from Other Slices (SIOS): The slice with lower interference from other slices will have higher priority.

Slice Connectivity (SC): When SC is higher, the priority of slices will be higher.

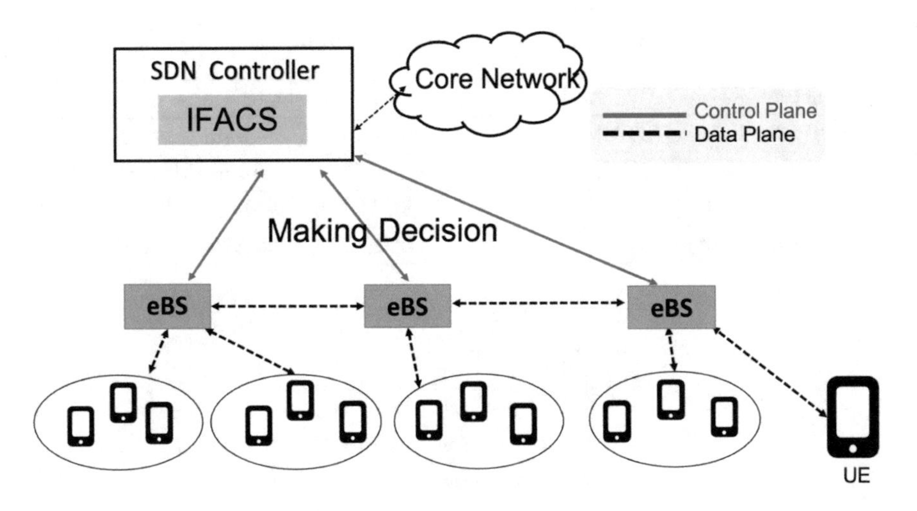

Fig. 3. Proposed system overview.

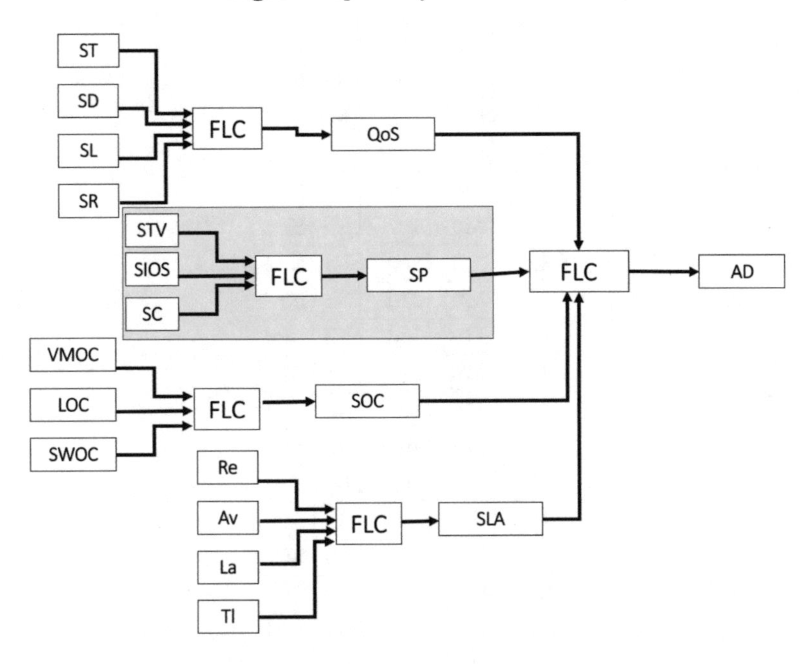

Fig. 4. Proposed system structure.

Slice Priority (SP): In the case when there are different connection requests from different mobile devices, the system will accept with higher possibility the requests from high priority slices.

The membership functions are shown in Fig. 5. We use triangular and trapezoidal membership functions because they are more suitable for real-time operations [18–21]. We show parameters and their term sets in Table 1. The Fuzzy

Table 1. Parameter and their term sets.

Parameters	Term set
Slice Traffic Volume (STV)	Small (Sa), Medium (Mu), Big (Bg)
Slice Interference from Other Slices (SIOS)	Low (Lw), Medium (Md), High (Hg)
Slice Connectivity (SC)	Short (Sh), Medium (Mi), Long (Lo)
Slice Priority (SP)	SP1, SP2, SP3, SP4, SP5, SP6, SP7

Table 2. FRB.

Rule	STV	SIOS	SC	SP
1	Sa	Lw	Sh	SP3
2	Sa	Lw	Mi	SP4
3	Sa	Lw	Lo	SP5
4	Sa	Md	Sh	SP1
5	Sa	Md	Mi	SP3
6	Sa	Md	Lo	SP4
7	Sa	Hg	Sh	SP1
8	Sa	Hg	Mi	SP1
9	Sa	Hg	Lo	SP2
10	Mu	Lw	Sh	SP4
11	Mu	Lw	Mi	SP5
12	Mu	Lw	Lo	SP6
13	Mu	Md	Sh	SP2
14	Mu	Md	Mi	SP4
15	Mu	Md	Lo	SP5
16	Mu	Hg	Sh	SP1
17	Mu	Hg	Mi	SP2
18	Mu	Hg	Lo	SP3
19	Bg	Lw	Sh	SP6
20	Bg	Lw	Mi	SP7
21	Bg	Lw	Lo	SP7
22	Bg	Md	Sh	SP4
23	Bg	Md	Mi	SP6
24	Bg	Md	Lo	SP7
25	Bg	Hg	Sh	SP2
26	Bg	Hg	Mi	SP3
27	Bg	Hg	Lo	SP5

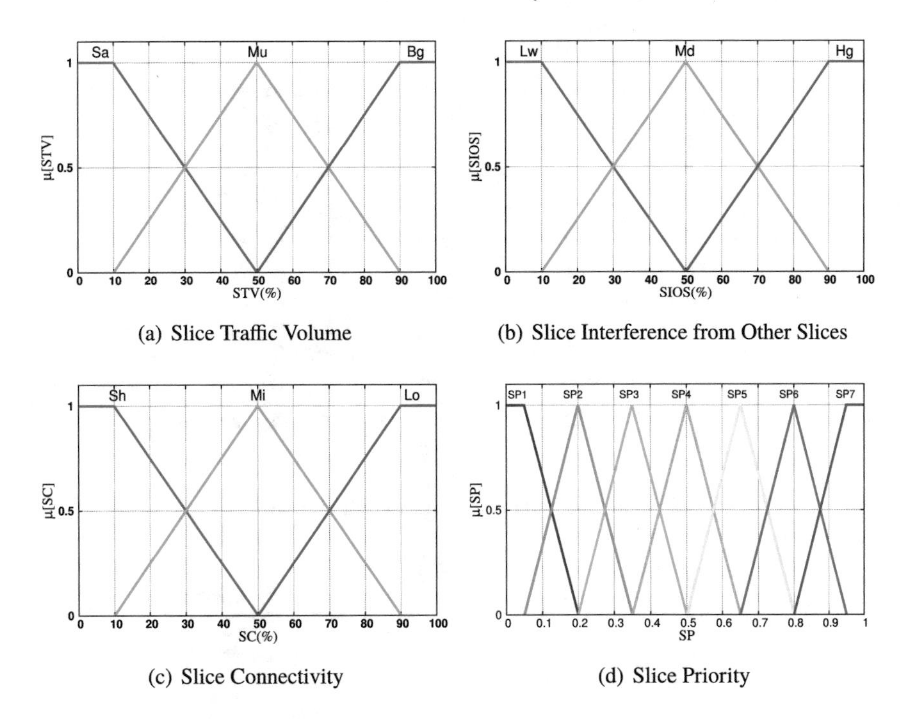

(a) Slice Traffic Volume

(b) Slice Interference from Other Slices

(c) Slice Connectivity

(d) Slice Priority

Fig. 5. Membership functions.

Rule Base (FRB) is shown in Table 2 and has 27 rules. The control rules have the form: IF "condition" THEN "control action". For example, for Rule 1: "IF STV is Sa, SIOS is Lw and SC is Sh THEN SP is SP3".

5 Simulation Results

In this section, we present the simulation result of our proposed scheme. The simulation results are shown in Fig. 6, Fig. 7 and Fig. 8. They show the relation of SP with SC for different SIOS values considering STV as a constant parameter.

In Fig. 6, we consider the STV value 10%. When SC is increased, we see that SP is increased. That mean a slice with higher SC will have higher priority. For SC 50%, when SIOS is increased from 10% to 50% and 50% to 90%, the SP is decreased by 15% and 28%, respectively.

We compare Fig. 6 with Fig. 7 to see how STV has affected SP. We change the STV value from 10% to 50%. The SP is increasing by 15% when the SIOS value is 10% and the SC is 60%. This is because the slice has a lot of traffic volume and can be considered as a higher priority slice. When the SIOS is 10%, all SP values are higher than 0.5. This indicates that the slice will have low interference from other slices and a better service is provided to the user.

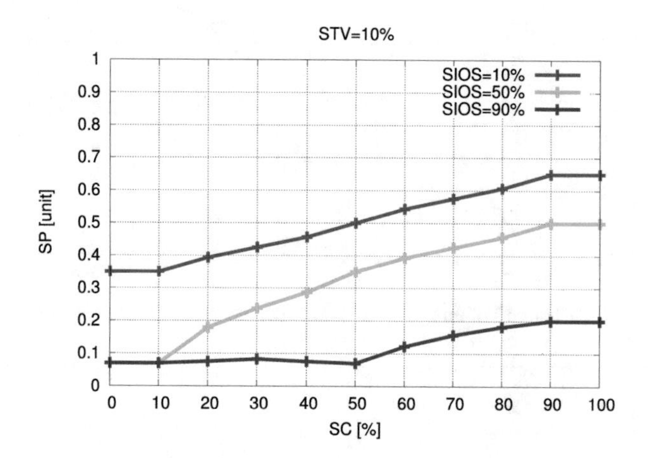

Fig. 6. Simulation results for STV = 10%.

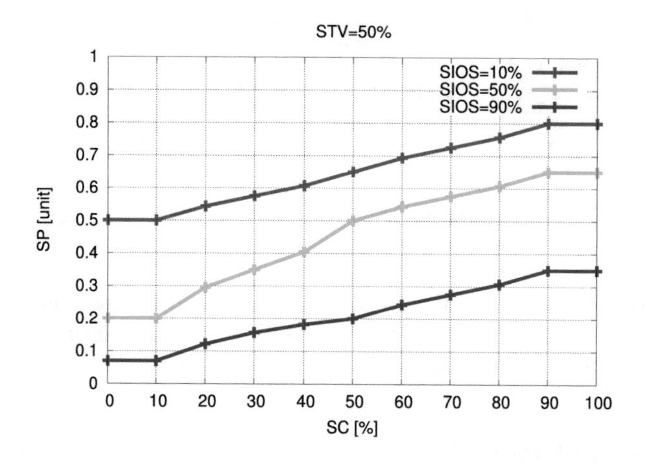

Fig. 7. Simulation results for STV = 50%.

In Fig. 8, we increase the value of STV to 90%. We see that the SLA values are increased much more compared with the results of Fig. 6 and Fig. 7. For SIOS value 50% and 10%, all SP values are higher than 0.5. Thus, the SP is good enough to satisfy the user requirements.

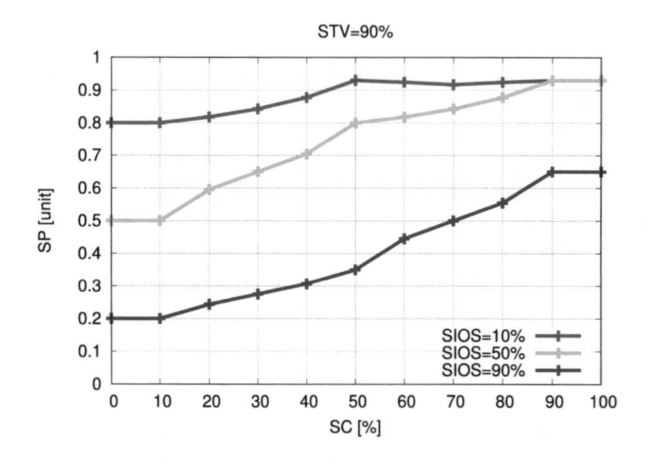

Fig. 8. Simulation results for STVC = 90%.

6 Conclusions and Future Work

In this paper, we proposed and implemented a Fuzzy-based scheme for assessing SP. The admission control system will have higher acceptance probability, when slice has high priority. In 5G wireless networks, the SP parameter will be considered as an input parameter for admission control. We evaluated the proposed scheme by simulations. From the simulation results, we found that three parameters have different effects on the SP. When STV and SC are increasing, the SP parameter is increased but when SIOS is increasing, the SP parameter is decreased.

In the future, we will consider other parameters and make extensive simulations to evaluate the proposed system.

References

1. Navarro-Ortiz, J., Romero-Diaz, P., Sendra, S., Ameigeiras, P., Ramos-Munoz, J.J., Lopez-Soler, J.M.: A survey on 5g usage scenarios and traffic models. IEEE Commun. Surv. Tutorials **22**(2), 905–929 (2020)
2. Hossain, S.: 5g wireless communication systems. Am. J. Eng. Res. (AJER) **2**(10), 344–353 (2013)
3. Giordani, M., Mezzavilla, M., Zorzi, M.: Initial access in 5g mmwave cellular networks. IEEE Commun. Mag. **54**(11), 40–47 (2016)
4. Kamil, I.A., Ogundoyin, S.O.: Lightweight privacy-preserving power injection and communication over vehicular networks and 5g smart grid slice with provable security. Internet Things **8**(100116), 100–116 (2019)
5. Hossain, E., Hasan, M.: 5g cellular: key enabling technologies and research challenges. IEEE Instrum. Measure. Mag. **18**(3), 11–21 (2015)
6. Yao, D., Su, X., Liu, B., Zeng, J.: A mobile handover mechanism based on fuzzy logic and mptcp protocol under sdn architecture*. In: 18th International Symposium on Communications and Information Technologies (ISCIT-2018), pp. 141–146, September 2018

7. Lee, J., Yoo, Y.: Handover cell selection using user mobility information in a 5g sdn-based network. In: 2017 Ninth International Conference on Ubiquitous and Future Networks (ICUFN-2017), pp. 697–702, July 2017
8. Moravejosharieh, A., Ahmadi, K., Ahmad, S.: A fuzzy logic approach to increase quality of service in software defined networking. In: 2018 International Conference on Advances in Computing,Communication Control and Networking (ICACCCN-2018), pp. 68–73, October 2018
9. Ampririt, P., Ohara, S., Qafzezi, E., Ikeda, M., Barolli, L., Takizawa, M.: Integration of software-defined network and fuzzy logic approaches for admission control in 5g wireless networks: a fuzzy-based scheme for qos evaluation. In: Barolli, L., Takizawa, M., Enokido, T., Chen, H.-C., Matsuo, K., (eds.) Advances on Broad-Band Wireless Computing, Communication and Applications, pp. 386–396, Springer, Cham (2021)
10. Ampririt, P., Ohara, S., Qafzezi, E., Ikeda, M., Barolli, L., Takizawa, M.: Effect of slice overloading cost on admission control for 5G wireless networks: a fuzzy-based system and its performance evaluation. In: Barolli, L., Natwichai, J., Enokido, T. (eds.) EIDWT 2021. Lecture Notes on Data Engineering and Communications Technologies, vol 65, pp. 24–35. Springer, Cham (2021). https://doi.org/10.1007/978-3-030-70639-5_3
11. Ampririt, P., Qafzezi, E., Bylykbashi, K., Ikeda, M., Matsuo, K., Barolli, L.: A fuzzy-based system for slice service level agreement in 5G wireless networks: effect of traffic load parameter. In: Barolli, L., Kulla, E., Ikeda, M. (eds.) EIDWT 2022. Lecture Notes on Data Engineering and Communications Technologies, vol 118, pp. 96–106. Springer, Cham (2022). https://doi.org/10.1007/978-3-030-95903-6_29
12. Ampririt, P., Qafzezi, E., Bylykbashi, K., Ikeda, M., Matsuo, K., Barolli, L.: An intelligent system for admission control in 5G wireless networks considering fuzzy logic and SDNs: effects of service level agreement on acceptance decision. In: Barolli, L. (eds) Advances on P2P, Parallel, Grid, Cloud and Internet Computing. 3PGCIC 2021. Lecture Notes in Networks and Systems, vol. 343, pp. 185–196. Springer, Cham (2022). https://doi.org/10.1007/978-3-030-89899-1_19
13. Li, L.E., Mao, Z.M., Rexford, J.: Toward software-defined cellular networks. In: 2012 European Workshop on Software Defined Networking, pp. 7–12, October 2012
14. Mousa, M., Bahaa-Eldin, A.M., Sobh, M.: Software defined networking concepts and challenges. In: 2016 11th International Conference on Computer Engineering & Systems (ICCES-2016), pp. 79–90, IEEE (2016)
15. Jantzen, J.: Tutorial on fuzzy logic. Technical University of Denmark, Dept. of Automation, Technical Report (1998)
16. Mendel, J.M.: Fuzzy logic systems for engineering: a tutorial. Proc. IEEE **83**(3), 345–377 (1995)
17. Zadeh, L.A.: Fuzzy logic. Computer **21**, 83–93 (1988)
18. Norp, T.: 5g requirements and key performance indicators. J. ICT Stand. **6**(1), 15–30 (2018)
19. Parvez, I., Rahmati, A., Guvenc, I., Sarwat, A.I., Dai, H.: A survey on low latency towards 5g: Ran, core network and caching solutions. IEEE Commun. Surv. Tutorials **20**(4), 3098–3130 (2018)

20. Kim, Y., Park, J., Kwon, D., Lim, H.: Buffer management of virtualized network slices for quality-of-service satisfaction. In: 2018 IEEE Conference on Network Function Virtualization and Software Defined Networks (NFV-SDN-2018), pp. 1–4 (2018)
21. Barolli, L., Koyama, A., Yamada, T., Yokoyama, S.: An integrated CAC and routing strategy for high-speed large-scale networks using cooperative agents. IPSJ J. **42**(2), 222–233 (2001)

Applying Machine Learning and Dynamic Resource Allocation Techniques in Fifth Generation Networks

Christos J. Bouras(✉), Evangelos Michos, and Ioannis Prokopiou

Computer Engineering and Informatics Department, University of Patras, Patras, Greece
bouras@cti.gr, {emichos,st1059554}@ceid.upatras.gr

Abstract. According to Internet of Things (IoT) Analytics, soon, the online devices in IoT networks will range from 25 up to 50 billion. Thus, it is expected that IoT will require more effective and efficient analysis methods than ever before with the use of Machine Learning (ML) powered by Fifth Generation (5G) networks. In this paper, we incorporate the K-means algorithm inside a 5G network infrastructure to better associate devices with Base Stations (BSs). We use multiple datasets consisting of user distribution in our area of focus and propose a Dynamic Resource Allocation (DRA) technique to learn their movement and predict the optimal position, RB usage and optimize their resource allocation. Users can experience significantly higher data rates and extended coverage with minimized interference and in fact, the DRA mechanism can mitigate the need for small cell infrastructure and prove a cost-effective solution, due to the resources transferred within the network.

1 Introduction

Internet of Things (IoT) refers to a system of interconnected devices that possess the ability to communicate (send/receive) data over the same shared network [1]. Based on this architecture, many advanced applications have been created, like smart houses, smart buildings and more. IoT applications are more and more utilized in today's industry and the majority of them focus on long-range communication, while at the same time, they increase the data throughput and minimize power consumption as much as possible. Big Data (BD) derives by IoT sensors and devices and is transferred to servers, which most of the times, is located in Cloud Data Centers worldwide. As a result, the demands for communication and infrastructure keep rising daily. Studies reveal that at least 25 billion devices will be online by 2022 (without including laptops, tablets and smartphones). This alarming increase also comes with an increase on the amount of data that is currently being stored. All this information undoubtedly has to be monitored and analyzed, so that we can keep learning from the available datasets and improve without any manual intervention. Using this technique, IoT devices are becoming smarter and more efficient day by day.

L. Barolli et al. (Eds.): AINA 2022, LNNS 449, pp. 662–673, 2022.
https://doi.org/10.1007/978-3-030-99584-3_57

To face these challenges, many models have been designed that focus on making everything inside an IoT network more Cloud independent. This is where 5th Generation (5G) networks come into play, offering massive connectivity and/or massive machine-type communication (mMTC). Massive access alongside with Machine Learning (ML) aims to achieve effective and secure communications for a large number of distributions to IoT devices via 5G and beyond networks. Massive access key features include low power, massive connectivity, while on the other hand, ML must cope with a massively increased complexity, reducing the number of measurements and facilitating robust decisions, promoting self-organizing networks and future predictions. Those characteristics are constantly proven to be promising for 5G networks [2]. At the same time, Heterogeneous Networks (HetNets) will also come into play by extending the existing macrocell infrastructures. This will be achieved by installing small cells in specific locations inside the macrocell (e.g. in areas near the macro cell borders) so as to provide improved coverage and throughputs for all devices near cell borders, where interference levels significantly experience spikes. At the same time, the use of ML and Artificial Intelligence (AI) is deemed highly necessary for 5G networks, as their application cellular networks is a subject that has recently gained significant interest [3].

Starting off with some of the most popular and efficient ML existing algorithms, the Decision Tree algorithm is a supervised learning algorithm that is mainly utilized in order to efficiently solve the problems of regression and classification, in contrast with other supervised learning algorithms, by classifying the information based on a certain variable [4]. The input variables and output variable correlate with each other through Linear Regression as $y = a + bx$, where y is the output and x is the input. Linear Regression strives towards finding out the coefficients a and b, based on supervised learning [5, 6]. Furthermore, the K-Nearest Neighbors algorithm (KNN) recursively loops through the existing information in order to find the K-nearest instances to the new instance, or on the other hand, the number (denoted as k) of instances that are closer to the new example. The output is either a regression problem or a common class for classification and the aim is to reduce standard deviation at each cluster's points and takes advantage of the Bayes' Theorem in order to calculate how likely is that an event will eventually occur, supposing that another event also occurred [7]. Lastly, the Random Forest algorithm involves numerous decision trees that operate together and simultaneously. Each decision tree reveals a prediction for a class and the most voted class becomes the prediction of the model [8–10]. Last but not least, the K-means algorithm is an unsupervised ML method for the processing of learning data and starts with a first group of randomly selected centroids, which are used for each cluster as starting points, and then performs iterative calculations to optimize the location of the centroid [11–13].

Regarding our motivation, the city of Patras, as well as the majority of the cities of today, has different connectional needs depending on the distribution of the users inside the network. Our goal is to use the knowledge from this user distribution in any given day and suggest the optimal positions for connectivity, as far as the small cells are concerned. When using ML, we observe that K-means often converges to clearly suboptimal local minima depending on the initial conditions possibly not giving the best results. The way we deal with this problem, using the corresponding big dataset, is shown in the following sections of this paper. Furthermore, there exists the UE-BS association issue, which relates to which is the optimal connection between a station and a network device. A data object that deviates greatly from the rest is referred to as an outlier. They signify measurement errors, poor data collection, or simply highlight variables that were not taken into account when collecting the data. They can be the result of a measurement or execution mistake. With the use of DRA also, we seek to minimize the number and effects of the outliers.

Aiming to tackle the aforementioned challenges, in this work we will incorporate the K-means unsupervised learning algorithm in a 5G geographical area, which will help towards optimizing the association problem between a device and a Base Station (BS), assuming their possible positions of the city structure. We will make use of multiple datasets that consist of device spawning (their position and datetime randomly deviate by a small margin) dividing it into 70% of training and 30% of testing dataset and we will then use K-means algorithm, with K equals the number of BS, to learn the user distribution in the network from these datasets and predict future optimal positions of the small cells based on their movement. With the use of ML, we can observe that K-means often converges to clearly suboptimal local minima depending on initial conditions and for that reason, we will be using a large dataset on the users' distribution (position with deviation of some meters) for a relatively representative sample. Additionally, we will also propose a Dynamic Resource Allocation (DRA) technique, where BSs that are low in usage can lend extra resources to neighboring stations to help tackle user congestion. Such an approach can mitigate the need for small cell infrastructure and prove a cost-effective solution, due to the resources transferred within the network. No similar work has been conducted for the specific analysis with the use of K-means.

The rest of this work is organized as follows: In the following section, we showcase our system model on which we are going to examine the association algorithm alongside with the ML mechanisms. In Sect. 3, we demonstrate our proposed mechanisms and Sect. 4 includes the evaluation and comparison and report real experiments with our findings. Section 5 discusses the conclusion and future work.

2 System Model

Starting with the energy consumption model, we consider that in this 5G network model, all BSs do operate at maximum power. This will ultimately result in the highest available throughputs for the network devices. Supposing that each macrocell holds a BS at its center, let P_i^{BS} be the power consumed by the ith BS, which is calculated as [14]:

$$P_i^{BS} = P_i^{cons} \cdot P_i^{rad} + P_i^{BS} \tag{1}$$

where P_i^{rad} corresponds to the outgoing radiated power from the BS, P_i^{cons} is the power consumed because of the feeder/amplifier losses and P_i^{BS} related to the consumed BS-related power.

Regarding the UEs, supposing that P_j^{UE} is the consumed power for the network device, when connected P_j^{UE} is calculated as [15]:

$$P_j^{UE} = P_j^{\text{loss}} \cdot \left(\sum_{a \in N_j^{ant}} \sum_{i \in N_{BS}^i} P_{j,i}^{rad,a} \right) + P_i^{\text{cons}} \tag{2}$$

where P_j^{loss} corresponds the power consumed (including system losses) for each of the antennas the device is connected to, N_j^{ant} depicts the different antennas the user is equipped with, N_{BS}^i relates to the set of antennas of a BS, $P_{j,i}^{rad,a}$ is the radiated power of the ath antenna for the jth user connected to the ith BS and lastly, P_i^{cons} corresponds to the energy required for the network user to associate with the BS.

Each UE has specific RB demands, depending on the BS it attempts to link/connect to. The device with the lowest number of RBs will attempt to associate to a BS, if the BS has enough RBs to satisfy the device itself. The device's RB demands are proportional to its data rates needs and inversely proportional to the bandwidth of the RB and the Signal-to-Interference-plus-Noise Ratio (SINR) between the UE and the BS. The equation to calculate the required amount of RBs for a device to link to a BS is computed as [16]:

$$r_{j,i} = \left\lceil \frac{th_j}{B_{RB} \cdot \log_2(1 + SINR_{j,i})} \right\rceil \tag{3}$$

where $\lceil \bullet \rceil$ corresponds to the operator for the ceiling function, th_j relates to the UE throughput demands, B_{RB} is the RB's bandwidth and $SINR_{j,i}$ denotes the signal quality between the device and the BS.

Regarding the Path Loss (PL) propagation model, in order to measure the signal losses in the simulated network, we construct the distance-dependent path loss model for the macrocell infrastructure (measured in dB) as follows [17]:

$$PL_{macro} = 128.1 + 37.6 \cdot \log_{10}(d) \tag{4}$$

where d corresponds to the distance between the transmitter and the receiver (note that this is measured in kilometers). Consequently, the channel gain can be calculated as:

$$G = 10^{-PL/10} \tag{5}$$

In our simulation, we note that we consider the fact that all BSs have an antenna height equal to 15m, as stated in the 5G NR technical specifications. Any additional wall losses are excluded from our model formulation.

Moving on to the model concerning the user throughputs, let $s_{j,i}$ be subcarrier between the jth UE and the ith BS. Regarding the overall set of subcarriers, we assume that $S_{s,j,i}$ denotes the subcarriers summation between the jth UE and the ith BS. Following the Orthogonal Frequency-Division Multiple Access (OFDMA) standard, the jth UE associated with i^{th} BS has throughput equal to:

$$R_{j,i} = \sum_{s \in S_{s,j,i}^{DL}} B_s \cdot \log_2(1 + SINR_{s,j,i}) \tag{6}$$

where B_s denotes the subcarrier bandwidth and $\boldsymbol{SINR_{s,j,i}}$ is the SINR between the BS and UE on a subcarrier s. BLER is equal to 10^{-4}. The $\boldsymbol{SINR_{s,j,i}}$ is formulated as follows (all calculations are over a subcarrier s) [14]:

$$SINR_{s,j,i} = \frac{P_{s,i}^{rad} \cdot G_{s,j,i}}{N_0 \cdot \Delta f + \sum_{i'} P_{s,i'}^{rad} \cdot G_{s,j,i'}} \tag{7}$$

where $\boldsymbol{P_{s,i}^{rad}}$ denotes the radiated power from the BS, $\boldsymbol{G_{s,j,i}}$ corresponds to the channel gain between a $\boldsymbol{j^{th}}$ UE and an $\boldsymbol{i^{th}}$ BS, $\boldsymbol{N_0}$ is the white noise power spectral density, $\boldsymbol{\Delta f}$ is the subcarrier spacing and $\sum_{i'} \boldsymbol{P_{s,i'}^{rad}} \cdot \boldsymbol{G_{s,j,i}'}$ relates to the summation of every $\boldsymbol{i'^{th}}$ BS's radiated power (which causes interference in the neighboring cells), multiplied with the channel gain between the interfering BS and the UE. Finally, in order to calculate SINR in (dB), we use the following equation:

$$SINR_{(dB)} = 10 \cdot \log_{10}\left(SINR_{s,j,i}\right) \tag{8}$$

3 Proposed Mechanisms

The K-means algorithm is an unsupervised ML technique for the processing of learning data and begins with a first group of randomly chosen centroids, which are used as the starting points for each cluster and then performs iterative calculations to optimize the centroid positions. K-means is chosen through other clustering variances because of its scalability and adaptability in large datasets as well as its guaranteed coverage. By alternating between assigning data points to clusters based on current centroids, K-means finds the best centroids selecting centroids (the center points of a cluster) based on the current assignment of data points to clusters. This attempts to make the data points of the intra-cluster as close as possible while at the same time, keeping the clusters as distinct as possible. The clustering generated is a form of vector quantization that aims to divide n observations into k clusters in which each observation belongs to non-overlapping subgroups (clusters) in which each data point belongs to only one group with the nearest mean (cluster centers or centroids), serving as the cluster prototype. This results in the data space being partitioned into Voronoi cells. K-means clustering minimizes variances within clusters using squared Euclidean distances, but not normal Euclidean distances, whereas only the geometric median minimizes Euclidean distances (the mean optimizes squared errors).

The proposed UE-BS association algorithm assumes pre-defined context information for users. Aiming at maximizing the efficiency of the proposed model using ML while respecting the pre-defined user data demands, the aforementioned problem transforms into a minimization of required RBs. The proposed low-complexity UE-BS association algorithm requires knowledge of the SINR, the system architecture, the available RBs, the throughput demands for every user and the outcome of the K-means algorithm. To achieve efficiency maximization, we begin iterating from the device with the lowest RB requirements. Repetitively, for each device, we will attempt to associate the device which has the lowest demands towards the BS to which it has maximum signal quality, or in more technical term, maximum SINR. If ML is enabled in the current simulation scenario, then we attempt to take advantage of the distribution prediction K-means produced with K equals the number of BS in the scenario using the resulting centroids as the optimal positions of BS including the corresponding users that we suggest, otherwise we continue without ML. Additionally, the best-case scenario is when both ML and DRA are enabled, offering additional RBs to BSs that are in need because of multiple reasons (device congestion, high interference, low coverage etc.). A DRA connection will always be optimal for the device, because the UE-BS association will be optimal in terms of signal coverage. Each UE-BS association is possible only if there exist remaining RBs, otherwise, we decide to select the next best candidate. As for any remaining BSs, they are discarded in this scenario.

Mechanism 1. K-means Algorithm using input datasets

```
begin
 specify number of clusters K and initialize k means
points randomly
 guess some initial cluster centers
 calculate the distance between each data point and
cluster centers
 for every µ_i = some value, i=1,...,k, do
      categorize each item to its closest mean
      update the mean's coordinates (averages of the
items categorized in that mean so far)
      assign points to nearest cluster center
c_i={j:d(xj,µi)≤d(xj,µl),l≠i,j=1,...,n}
      set the cluster centers to the mean
µ_i=1|c_i|∑jcEixj for every I
  end for
 keep repeating until there is no change to the centroids
 If no data point was reassigned then
      Stop
 end .
```

Mechanism 2. Association Algorithm using ML and DRA support

```
begin
 K = number of BSs
 for each j in N_UE do
      choose device candidate with min (T_device,BS )
      select best BS by finding max(SINR_device,BS )
      if ML is enabled then
           for each BS do
              for each device do
                   clusters[BS_l] = K-means of device on BS_l
              end for
          end for
      end if
      if the available RB_BS are enough then
         if ML is enabled then
            associate device and BS based on the clusters
        else
               associate device and BS
        end if
         if ML DRA is enabled
            for each BS do
                 if any BS needs resources then
                       if neighboring BS has enough RBs
to serve then
                       offer 15% RBs
                       update remaining RBs
                       end if
                 end if
           end for
         else
            update available RBs;
         end if
      else
           select next best BS candidate by max ( SINR_device,BS )
      end if
 end for
 end .
```

4 Performance Evaluation

In this section, we discuss the 5G network simulation scenario, where the Python programming language was used to construct the experiment analysis (datasets, K-means implementation, system model, association algorithms etc.). We consider a two-level ring topology in the geographical area of study, resulting in a total of 19 macrocells. Considering the second level of surrounding macrocells is crucial towards measuring the interference caused by neighboring cells, as it would be a mistake not to consider the negative effects of signal interference from the neighboring cells. All macro BSs are located in the center of the cell and are surrounded by small cell infrastructures that can help towards better user coverage inside the network. We consider that all BSs operate at full power, to provide the highest available throughputs to the devices inside the network.

The geographical area of interest is depicted below in Fig. 1. The map depicted below represents a larger area in the city of Patras and includes all available BSs (macrocell and smallcell infrastructures, as well as the devices distribution alongside the area of interest). To evaluate the ability of our mechanisms to efficiently predict user movement inside the network, a dataset was created based on the devices' location (with deviation of some meters) enough times for the accuracy to be objective and representative, including the (x,y) pair of the devices' positions and their timestamp, for a fixed number of hours per day. The experiment ended after gathering enough information from a whole month, which was entered in the dataset, which was then given as input to the K-means algorithm. The ultimate goal of the approach is to manage to offer better user coverage and data rates, after successfully predicting the changes in user demands each day, based on the user distribution as gathered in the dataset. This means that for the case of predicting resources that e.g. higher probability of devices being in the city center from Monday through Friday (weekdays) and them being out of the urban area on the weekends, the prediction would suggest allocating extra resources to the more crowded areas (due to the extra bandwidth available). The probability distribution is based on real-life scenarios (Fig. 2 and Table 1).

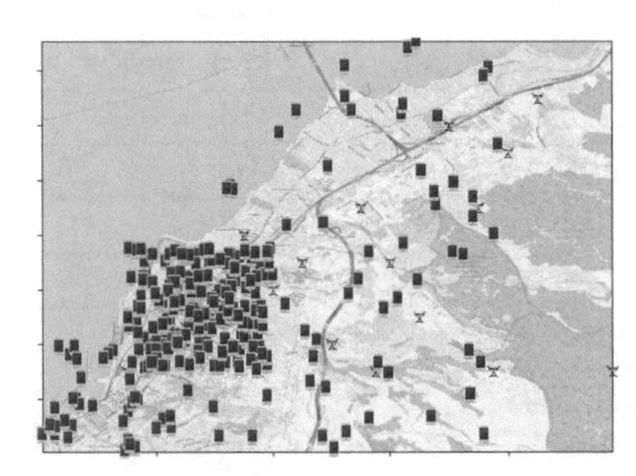

Fig. 1. Snapshot of simulation scenario in patras

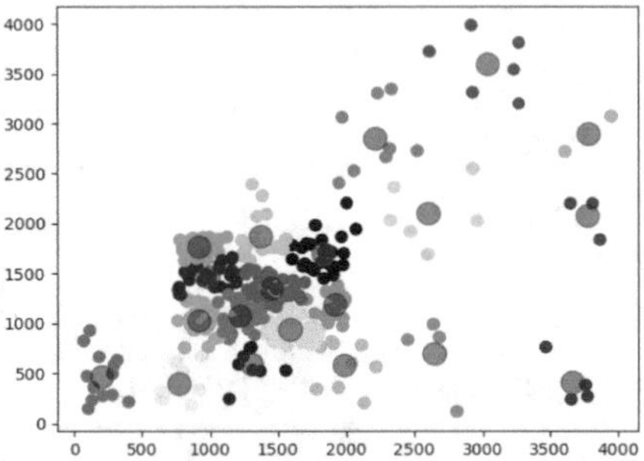

Fig. 2. Optimal smallcell positions using user movement

Figure 3 and Fig. 4 reveal the successful connections percentage, divided into macrocell and smallcell connections, for both the scenarios of weekly and weekend device congestion. The very first thing we can easily notice is that as devices increase inside the same geographical area, it is getting harder and harder for macrocells to serve all devices in the network. This is due to the fact that all macrocell BSs have a pre-defined amount of RBs devoted to them and as devices increase in the network, the more chances there are that the RBs will diminish with a higher rate. This means that such devices can attempt to connect to the additional surrounding layer of small cells, as envisioned officially in 5G networks, where due to multiple factors (e.g. no RBs remaining, outside of area coverage, high interference from neighboring BSs), it is preferable to connect to low-emission and low energy consumption smallcells to be successfully covered in the 5G network.

Additionally, we observe that the applied ML techniques, with or without DRA, can offer an increased amount of macrocells resulting in less smallcell connections because

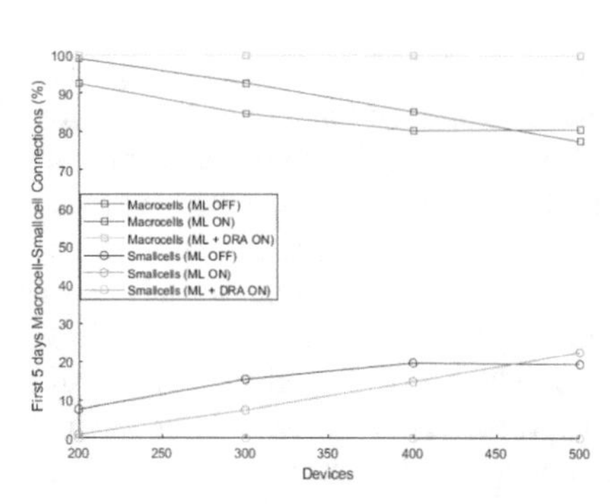

Fig. 3. Weekdays Macrocell/Smallcell Connections (%)

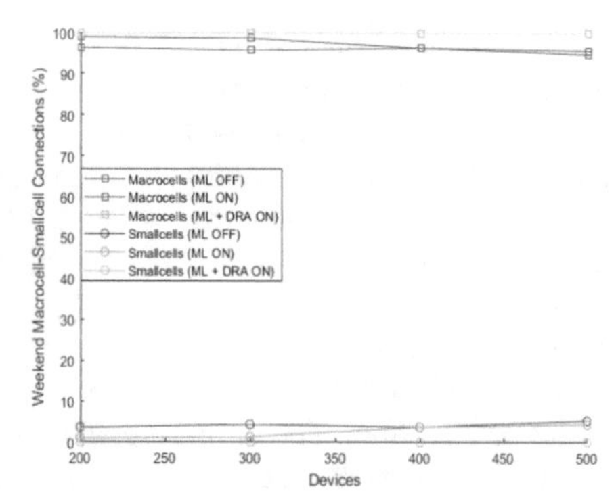

Fig. 4. Weekend Macrocell/Smallcell Connections (%)

Table 1. Experiment parameters

Parameter	Setting	Parameter	Setting
Macrocells	19	White noise density	−174 dBm/Hz
Air Protocol	5G NR	Macrocell Coverage	375 m
5G Frequency Range	FR1	Small cell Coverage	50 m
RB Bandwidth	360 kHz	BS Antenna Gain	15 dBi
Modulation Scheme	64QAM	UE Antenna Gain	0 dBi
Bandwidth	100 MHz	Macro BS $P_{i,max}^{rad}$	40 W
Carrier Frequency	3.5 GHz	Small BS $P_{i,max}^{rad}$	1 W
RBs	273	UE $P_{j,max}^{rad}$	0.2 W
Subcarrier spacing	30 kHz		

the proposed ML techniques gain knowledge from the input dataset from multiple previous instances of the geographical area gathered. Using the K-means (Mechanism 1), they suggest the optimal connection based on the existing network. When the DRA comes into play, we can see that the connections to macrocells maximize, whereas the connections to smallcells minimize. This is because the DRA mechanism takes advantage of the existing ML dataset and can accurately relocate resources from relatively empty BSs to BSs that need them the most due to device congestion (see lines 19–30 in Mechanism 2). This is an important achievement, because with this ML technique, we can mitigate the need for acquiring and installing smallcell infrastructures by relying on existing knowledge of the network's datasets.

In the figures above (Fig. 5 and Fig. 6), we observe the usage of the overall RBs available to the network for the three different simulation scenarios (without ML, with ML, and with ML combined with DRA). Studying the weekly and the weekend device

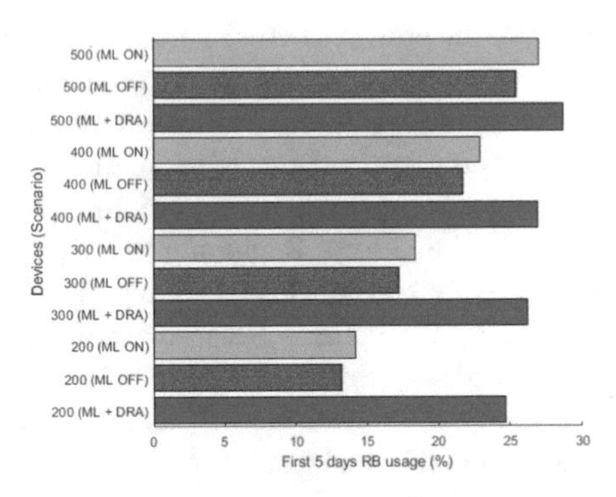

Fig. 5. Weekdays overall RB usage (%)

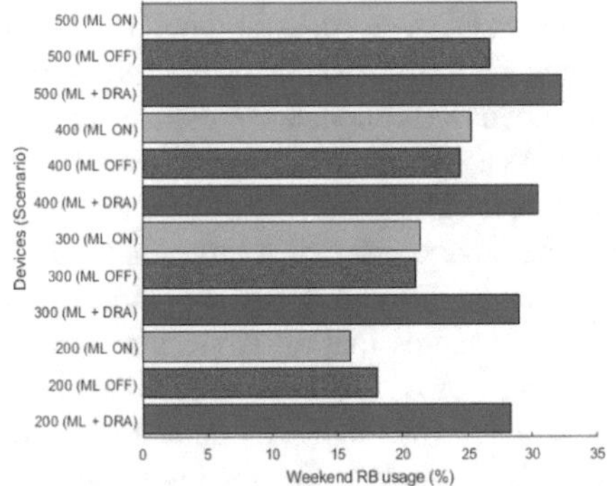

Fig. 6. Weekend overall RB usage (%)

congestion, we conclude that as the number of devices augments, more RBs are needed, since the UE-BS association algorithm relies on the device's RB demands (see (3) and lines 13–33 in Mechanism 2). The more devices in our network, the more resources are needed from the BSs. Since all BSs have a pre-defined number of RBs, the RB usage increases proportionally to the device amount and according the ML techniques applied, the RB usage augments. When DRA is applied, more RBs are being consumed, since this mechanism tries to associate the current device with the best BS available according to the ML output. If this option is not possible, resources will be relocated inside the network infrastructure for the optimal association to be completed successfully (see lines 19–30 in Mechanism 2). Thus, more and more RBs of the macrocells are needed, despite the existence of a small cell infrastructure inside a network, which leads to the ML technique with the DRA being a very cost-effective and efficient solution of the UE-BS association problem inside 5G networks.

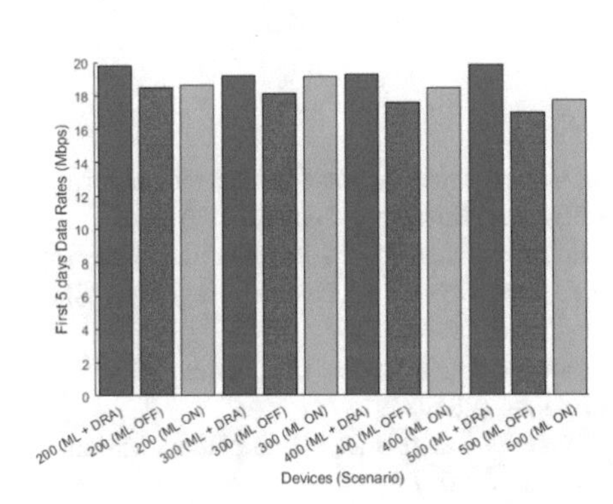

Fig. 7. Weekdays average data rates (Mbps)

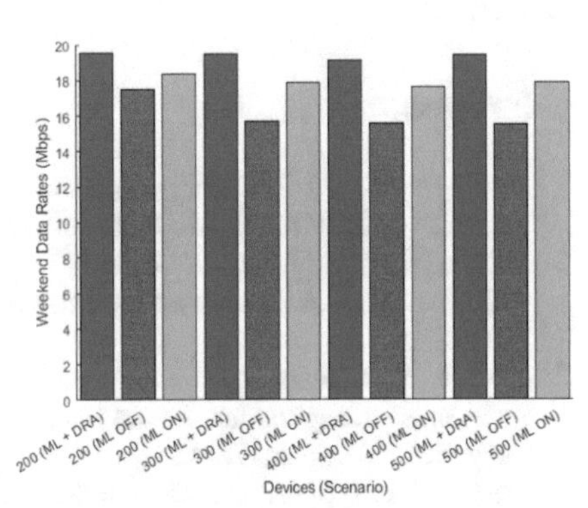

Fig. 8. Weekend average data rates (Mbps)

Figures 7 and 8 show the weekly and weekend average data rates for the connected devices in the network. We conclude that: a) more devices connected lead to lower average data rates, regardless of whether ML techniques were used or not and b) upon applying DRA, the data rates maximize, compared to the previous two scenarios. The first conclusion can be easily justified since more devices connect to the BSs with a predefined amount of RBs allocated to them causing the resources to eventually diminish at a higher rate. As a result, the number of unconnected devices is increased, due to insufficient BS resources. Thus, there must be a compromise at the throughput demands coming from all the devices so that a larger amount of them can be served. Regarding the second conclusion, through DRA, all devices can efficiently be served while scanning the ML dataset ensures that all devices are in fact served by the optimal BS in their area of coverage, providing better signal quality (so, better SINR). Since their optimal BS will be able to cover their RB demands, according to (6), the higher the RB demands and the SINR signal quality, the higher the data rates eventually will end up being (see lines 19–30 in Mechanism 2).

5 Conclusion and Future Work

Studies show that in the future, the number of devices connected in IoT networks will range from 25 up to 50 billion. As a result, IoT infrastructures will require more effective and efficient analysis methods than ever and ML techniques are envisioned to be the solution, accompanied by the coming of 5G networks. In this work, we incorporated the K-means unsupervised learning algorithm inside a 5G network infrastructure to better associate devices with BSs. We used multiple datasets consisting of user distribution (datetime included) in Patras and used K-means to learn the user movement and predict the optimal position for the connection station. Additionally, we proposed a DRA technique, where BSs that are low in usage can lend extra resources to neighboring stations to help tackle user congestion. Simulations revealed that by applying such ML mechanisms inside 5G infrastructures, users can experience significantly higher data rates and extended coverage with minimized interference. The DRA mechanism can mitigate the need for small cell infrastructure and prove a cost-effective solution, due to the resources transferred.

References

1. Rocha Neto, A., et al.: Classifying smart IoT devices for running machine learning algorithms. In: Anais do XLV Seminário Integrado de Software e Hardware, Natal (2018)
2. Polese, M., et al.: Machine Learning at the Edge: A Data-Driven Architecture with Applications to 5G Cellular Networks, Institute of Electrical and Electronics Engineers (IEEE) (2020)
3. Zantalis, F., Koulouras, G., Karabetsos, S., Kandris, D.: A review of machine learning and iot in smart transportation. Future Internet **11**(4), 94 (2019)
4. Dietterich, T.G., Bae, E.: Machine Learning Bias, Statistical Bias, and Statistical Variance of Decision Tree Algorithms, Oregon State University (2008)
5. L'Heureux, A., Grolinger, K., Elyamany, H.F., Capretz, M.A.M.: Machine learning with big data: challenges and approaches. In: Institute of Electrical and Electronics Engineers. (IEEE) Access, vol. 5, pp. 7776–7797 (2017)
6. Ray, S.: A quick review of machine learning algorithms. In: International Conference on Machine Learning, Big Data, Cloud and Parallel Computing. (COMITCon), Faridabad, India, pp. 35–39 (2019)

7. Zhang, H.: The optimality of naive bayes. In: Proceedings of the Seventeenth International Florida Artificial Intelligence Research Society Conference, FLAIRS (2004)
8. Breiman, L.: Random forests. Mach. Learn. **45**, 5–32 (2001)
9. Breiman, L.: Arcing classifiers. Ann. Stat. **26**(3), 801–849 (1998)
10. Freund, ϒ., Schapire, R.E.: Experiments with a new boosting algorithm. In: Machine Learning: Proceedings of the Thirteenth International Conference, pp. 148–156 (1996)
11. Balevi, E., Gitlin, R.D.: A clustering algorithm that maximizes throughput in 5G heterogeneous F-RAN networks. In: IEEE International Conference on Communications (ICC), Kansas City, MO, pp. 1–6 (2018)
12. Cabrera, E., Vesilo, R.: An enhanced k-means clustering algorithm with non-orthogonal multiple access (NOMA) for MMC networks. In: 28th International Telecommunication Networks and Applications Conference (ITNAC), Sydney, Australia, pp. 1–8 (2018)
13. Kumar, S., et al.: Handover forecasting in 5G using machine learning. Int. J. Eng. Technol. (UAE) **7**(31), 76–79 (2018)
14. Richter, F., Fehske, A., Fetwweis, G.: Energy efficiency aspects of base station deployment strategies for cellular networks. In: Institute of Electrical and Electronics Engineers (IEEE) 70th Vehicular Technology Conference Fall, Anchorage, AK, (2009) pp. 1–5 (2009)
15. Ghaleb, A.M., Mansoor, A.M., Ahmad, R.: An energy-efficient user-centric approach high-capacity 5G heterogeneous cellular networks. Int. J. Adv. Comput. Sci. Appl. (IJACSA) **9**(1), 405–411 (2018)
16. Holma, H., Toskala, A., (eds.): LTE for UMTS: OFDMA and SC-FDMA Based Radio Access. John Wiley & Sons, pp. 213–257 (2009)
17. 3rd Generation Partnership Project (3GPP). 3GPP TS 36.931, v13.0.0, Radio Frequency (RF) requirements for LTE Pico Node B, (Release 13) (2016)

Mesh Routers Placement by WMN-PSODGA Simulation System: Effect of Number of Mesh Routers Considering Stadium Distribution and RDVM Method

Admir Barolli[1(✉)], Kevin Bylykbashi[2], Shinji Sakamoto[3], Elis Kulla[4], and Leonard Barolli[2]

[1] Department of Information Technology, Aleksander Moisiu University of Durres, L.1, Rruga e Currilave, Durres, Albania
admir.barolli@gmail.com

[2] Department of Information and Communication Engineering, Fukuoka Institute of Technology, 3-30-1 Wajiro-Higashi, Higashi-Ku, Fukuoka 811-0295, Japan
kevin@bene.fit.ac.jp, barolli@fit.ac.jp

[3] Department of Information and Computer Science, Kanazawa Institute of Technology, 7-1 Ohgigaoka Nonoichi, Ishikawa 921-8501, Japan
shinji.sakamoto@ieee.org

[4] Department of Information and Computer Engineering, Okayama University of Science, 1-1 Ridaicho, Kita-Ku, Okayama 700-0005, Japan
kulla@ice.ous.ac.jp

Abstract. Wireless Mesh Networks (WMNs) are gaining a lot of attention from researchers due to their advantages such as easy maintenance, low upfront cost, and high robustness. However, designing a robust WMN at low cost requires the use of the least possible mesh routers but still interconnected and able to offer full coverage. Therefore, the placement of mesh routers over the area of interest is a problem that entails thorough planning. In our previous work, we implemented a simulation system that deals with this problem considering Particle Swarm Optimization (PSO) and Distributed Genetic Algorithm (DGA), called WMN-PSODGA. In this paper, we compare the results of Stadium distribution of mesh clients for 16 and 17 mesh routers with Rational Decrement of Vmax Method (RDVM) as a router replacement method. The simulation results show that 17 mesh routers enable full client coverage, better connectivity and improved load balance.

1 Introduction

The wireless networks and devices are becoming increasingly popular and they provide users access to information and communication anytime and anywhere [2,7,10,18]. Wireless Mesh Networks (WMNs) are gaining a lot of attention because of their low-cost nature that makes them attractive for providing

L. Barolli et al. (Eds.): AINA 2022, LNNS 449, pp. 674–685, 2022.
https://doi.org/10.1007/978-3-030-99584-3_58

wireless Internet connectivity. A WMN is dynamically self-organized and self-configured, with the nodes in the network automatically establishing and maintaining mesh connectivity among itself (creating, in effect, an ad hoc network). This feature brings many advantages to WMN such as low up-front cost, easy network maintenance, robustness and reliable service coverage [1]. Moreover, such infrastructure can be used to deploy community networks, metropolitan area networks, municipal and corporative networks, and to support applications for urban areas, medical, transport and surveillance systems.

Mesh node placement in WMNs can be seen as a family of problems, which is shown (through graph theoretic approaches or placement problems, e.g. [5,11]) to be computationally hard to solve for most of the formulations [22].

We consider the version of the mesh router nodes placement problem in which we are given a grid area where to deploy a number of mesh router nodes and a number of mesh client nodes of fixed positions (of an arbitrary distribution) in the grid area. The objective is to find a location assignment for the mesh routers to the cells of the grid area that maximizes the network connectivity, client coverage and consider load balancing for each router. Network connectivity is measured by Size of Giant Component (SGC) of the resulting WMN graph, while the user coverage is simply the number of mesh client nodes that fall within the radio coverage of at least one mesh router node and is measured by Number of Covered Mesh Clients (NCMC). For load balancing, we added in the fitness function a new parameter called NCMCpR (Number of Covered Mesh Clients per Router).

Node placement problems are known to be computationally hard to solve [8, 9,23]. In previous works, some intelligent algorithms have been recently investigated for node placement problem [3,6,12,14].

In [16], we implemented a Particle Swarm Optimization (PSO) based simulation system, called WMN-PSO. Also, we implemented another simulation system based on Genetic Algorithm (GA), called WMN-GA [15], for solving node placement problem in WMNs. Then, we designed and implemented a hybrid simulation system based on PSO and distributed GA (DGA). We call this system WMN-PSODGA.

In this paper, we compare the results of Stadium distribution of mesh clients for 16 and 17 mesh routers considering Rational Decrement of Vmax Method (RDVM) as a router replacement method.

The rest of the paper is organized as follows. In Sect. 2, we introduce intelligent algorithms. In Sect. 3 is presented the implemented hybrid simulation system. The simulation results are given in Sect. 4. Finally, we give conclusions and future work in Sect. 5.

2 Intelligent Algorithms for Proposed Hybrid Simulation System

2.1 Particle Swarm Optimization

In PSO a number of simple entities (the particles) are placed in the search space of some problem or function and each evaluates the objective function at its current location. The objective function is often minimized and the exploration of the search space is not through evolution [13].

Each particle then determines its movement through the search space by combining some aspect of the history of its own current and best (best-fitness) locations with those of one or more members of the swarm, with some random perturbations. The next iteration takes place after all particles have been moved. Eventually the swarm as a whole, like a flock of birds collectively foraging for food, is likely to move close to an optimum of the fitness function.

Each individual in the particle swarm is composed of three $\mathcal{D}$-dimensional vectors, where $\mathcal{D}$ is the dimensionality of the search space. These are the current position $\vec{x}_i$, the previous best position $\vec{p}_i$ and the velocity $\vec{v}_i$.

The particle swarm is more than just a collection of particles. A particle by itself has almost no power to solve any problem; progress occurs only when the particles interact. Problem solving is a population-wide phenomenon, emerging from the individual behaviors of the particles through their interactions. In any case, populations are organized according to some sort of communication structure or topology, often thought of as a social network. The topology typically consists of bidirectional edges connecting pairs of particles, so that if j is in i's neighborhood, i is also in j's. Each particle communicates with some other particles and is affected by the best point found by any member of its topological neighborhood. This is just the vector $\vec{p}_i$ for that best neighbor, which we will denote with $\vec{p}_g$. The potential kinds of population "social networks" are hugely varied, but in practice certain types have been used more frequently. We show the pseudo code of PSO in Algorithm 1.

In the PSO process, the velocity of each particle is iteratively adjusted so that the particle stochastically oscillates around $\vec{p}_i$ and $\vec{p}_g$ locations.

2.2 Distributed Genetic Algorithm

Distributed Genetic Algorithm (DGA) has been used in various fields of science. DGA has shown their usefulness for the resolution of many computationally hard combinatorial optimization problems. We show the pseudo code of DGA in Algorithm 2.

Population of individuals: Unlike local search techniques that construct a path in the solution space jumping from one solution to another one through local perturbations, DGA use a population of individuals giving thus the search a larger scope and chances to find better solutions. This feature is also known as "exploration" process in difference to "exploitation" process of local search methods.

Algorithm 1. Pseudo code of PSO.

/* Initialize all parameters for PSO */
Computation maxtime:= Tp_{max}, $t := 0$;
Number of particle-patterns:= m, $2 \leq m \in \boldsymbol{N}^1$;
Particle-patterns initial solution:= $\boldsymbol{P}_i^0$;
Particle-patterns initial position:= $\boldsymbol{x}_{ij}^0$;
Particles initial velocity:= $\boldsymbol{v}_{ij}^0$;
PSO parameter:= ω, $0 < \omega \in \boldsymbol{R}^1$;
PSO parameter:= C_1, $0 < C_1 \in \boldsymbol{R}^1$;
PSO parameter:= C_2, $0 < C_2 \in \boldsymbol{R}^1$;
/* Start PSO */
Evaluate($\boldsymbol{G}^0, \boldsymbol{P}^0$);
while $t < Tp_{max}$ **do**
  /* Update velocities and positions */
  $\boldsymbol{v}_{ij}^{t+1} = \omega \cdot \boldsymbol{v}_{ij}^t$
    $+C_1 \cdot \text{rand}() \cdot (best(P_{ij}^t) - x_{ij}^t)$
    $+C_2 \cdot \text{rand}() \cdot (best(G^t) - x_{ij}^t)$;
  $\boldsymbol{x}_{ij}^{t+1} = \boldsymbol{x}_{ij}^t + \boldsymbol{v}_{ij}^{t+1}$;
  /* if fitness value is increased, a new solution will be accepted. */
  Update_Solutions($\boldsymbol{G}^t, \boldsymbol{P}^t$);
  $t = t + 1$;
end while
Update_Solutions($\boldsymbol{G}^t, \boldsymbol{P}^t$);
return Best found pattern of particles as solution;

Fitness: The determination of an appropriate fitness function, together with the chromosome encoding are crucial to the performance of DGA. Ideally we would construct objective functions with "certain regularities", i.e. objective functions that verify that for any two individuals which are close in the search space, their respective values in the objective functions are similar.

Selection: The selection of individuals to be crossed is another important aspect in DGA as it impacts on the convergence of the algorithm. Several selection schemes have been proposed in the literature for selection operators trying to cope with premature convergence of DGA. There are many selection methods in GA. In our system, we implement 2 selection methods: Random method and Roulette wheel method.

Crossover operators: Use of crossover operators is one of the most important characteristics. Crossover operator is the means of DGA to transmit best genetic features of parents to offsprings during generations of the evolution process. Many methods for crossover operators have been proposed such as Blend Crossover (BLX-α), Unimodal Normal Distribution Crossover (UNDX), Simplex Crossover (SPX).

Mutation operators: These operators intend to improve the individuals of a population by small local perturbations. They aim to provide a component of randomness in the neighborhood of the individuals of the population. In our

Algorithm 2. Pseudo code of DGA.

/* Initialize all parameters for DGA */
Computation maxtime:= Tg_{max}, $t := 0$;
Number of islands:= n, $1 \leq n \in \boldsymbol{N}^1$;
initial solution:= $\boldsymbol{P}_i^0$;
/* Start DGA */
Evaluate($\boldsymbol{G}^0, \boldsymbol{P}^0$);
while $t < Tg_{max}$ **do**
  for all islands **do**
    Selection();
    Crossover();
    Mutation();
  end for
  $t = t + 1$;
end while
Update_Solutions($\boldsymbol{G}^t, \boldsymbol{P}^t$);
return Best found pattern of particles as solution;

system, we implemented two mutation methods: uniformly random mutation and boundary mutation.

Escaping from local optima: GA itself has the ability to avoid falling prematurely into local optima and can eventually escape from them during the search process. DGA has one more mechanism to escape from local optima by considering some islands. Each island computes GA for optimizing and they migrate its gene to provide the ability to avoid from local optima (See Fig. 1).

Convergence: The convergence of the algorithm is the mechanism of DGA to reach to good solutions. A premature convergence of the algorithm would cause that all individuals of the population be similar in their genetic features and thus the search would result ineffective and the algorithm getting stuck into local optima. Maintaining the diversity of the population is therefore very important to this family of evolutionary algorithms.

3 Proposed and Implemented WMN-PSODGA Hybrid Intelligent Simulation System

In this section, we present the proposed WMN-PSODGA hybrid intelligent simulation system. In the following, we describe the initialization, particle-pattern, gene coding, fitness function, and replacement methods. The pseudo code of our implemented system is shown in Algorithm 3. Also, our implemented simulation system uses Migration function as shown in Fig. 2. The Migration function swaps solutions among lands included in PSO part.

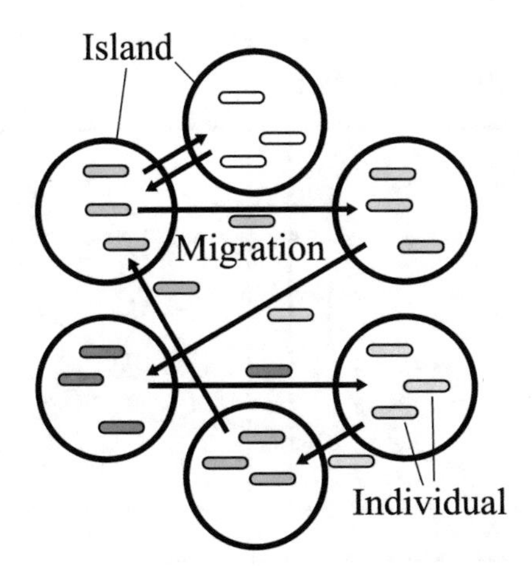

Fig. 1. Model of Migration in DGA.

Algorithm 3. Pseudo code of WMN-PSODGA system.

Computation maxtime:= T_{max}, $t := 0$;
Initial solutions: $\boldsymbol{P}$.
Initial global solutions: $\boldsymbol{G}$.
/* Start PSODGA */
while $t < T_{max}$ **do**
  Subprocess(PSO);
  Subprocess(DGA);
  WaitSubprocesses();
  Evaluate($\boldsymbol{G}^t, \boldsymbol{P}^t$)
  /* Migration() swaps solutions (see Fig. 2). */
  Migration();
  $t = t + 1$;
end while
Update_Solutions($\boldsymbol{G}^t, \boldsymbol{P}^t$);
return Best found pattern of particles as solution;

Initialization

We decide the velocity of particles by a random process considering the area size. For instance, when the area size is $W \times H$, the velocity is decided randomly from $-\sqrt{W^2 + H^2}$ to $\sqrt{W^2 + H^2}$.

Particle-Pattern

A particle is a mesh router. A fitness value of a particle-pattern is computed by combination of mesh routers and mesh clients positions. In other words, each particle-pattern is a solution as shown is Fig. 3.

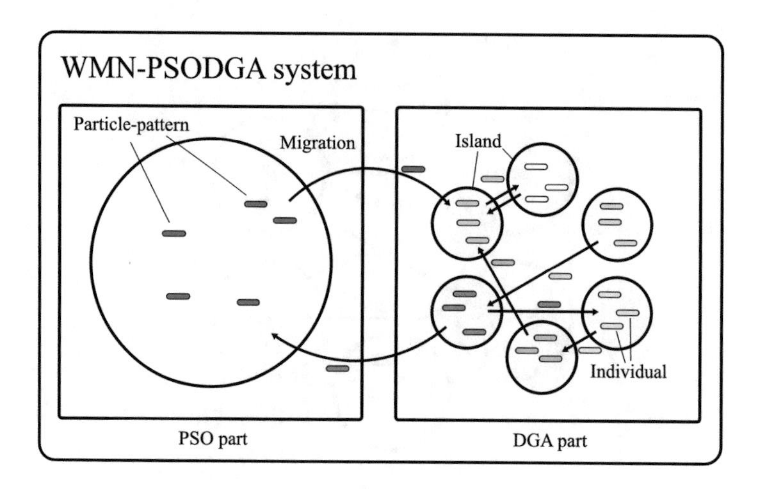

Fig. 2. Model of WMN-PSODGA migration.

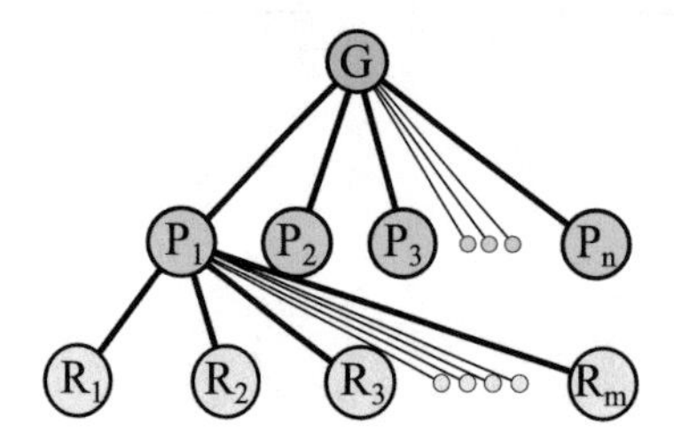

Fig. 3. Relationship among global solution, particle-patterns, and mesh routers in PSO part.

Gene Coding

A gene describes a WMN. Each individual has its own combination of mesh nodes. In other words, each individual has a fitness value. Therefore, the combination of mesh nodes is a solution.

Fitness Function

WMN-PSODGA has the fitness function to evaluate the temporary solution of the router's placements. The fitness function is defined as:

$$Fitness = \alpha \times NCMC(\boldsymbol{x}_{ij}, \boldsymbol{y}_{ij}) + \beta \times SGC(\boldsymbol{x}_{ij}, \boldsymbol{y}_{ij}) + \gamma \times NCMCpR(\boldsymbol{x}_{ij}, \boldsymbol{y}_{ij}).$$

This function uses the following indicators.

- NCMC (Number of Covered Mesh Clients)
  The NCMC is the number of the clients covered by the SGC's routers.

- SGC (Size of Giant Component)
  The SGC is the maximum number of connected routers.
- NCMCpR (Number of Covered Mesh Clients per Router)
  The NCMCpR is the number of clients covered by each router. The NCMCpR indicator is used for load balancing.

WMN-PSODGA aims to maximize the value of the fitness function in order to optimize the placements of the routers using the above three indicators. Weight-coefficients of the fitness function are α, β, and γ for NCMC, SGC, and NCMCpR, respectively. Moreover, the weight-coefficients are implemented as $\alpha + \beta + \gamma = 1$.

Router Replacement Methods

A mesh router has x, y positions, and velocity. Mesh routers are moved based on velocities. There are many router replacement methods. In this paper, we consider RDVM.

Constriction Method (CM)
CM is a method which PSO parameters are set to a week stable region ($\omega = 0.729$, $C_1 = C2 = 1.4955$) based on analysis of PSO by M. Clerc et al. [4,20].

Random Inertia Weight Method (RIWM)
In RIWM, the ω parameter is changing ramdomly from 0.5 to 1.0. The C_1 and C_2 are kept 2.0. The ω can be estimated by the week stable region. The average of ω is 0.75 [20].

Linearly Decreasing Inertia Weight Method (LDIWM)
In LDIWM, C_1 and C_2 are set to 2.0, constantly. On the other hand, the ω parameter is changed linearly from unstable region ($\omega = 0.9$) to stable region ($\omega = 0.4$) with increasing of iterations of computations [20,21].

Linearly Decreasing Vmax Method (LDVM)
In LDVM, PSO parameters are set to unstable region ($\omega = 0.9$, $C_1 = C_2 = 2.0$). A value of V_{max} which is maximum velocity of particles is considered. With increasing of iteration of computations, the V_{max} is kept decreasing linearly [19].

Rational Decrement of Vmax Method (RDVM)
In RDVM, PSO parameters are set to unstable region ($\omega = 0.9$, $C_1 = C_2 = 2.0$). The V_{max} is kept decreasing with the increasing of iterations as

$$V_{max}(x) = \sqrt{W^2 + H^2} \times \frac{T - x}{x}.$$

Where, W and H are the width and the height of the considered area, respectively. Also, T and x are the total number of iterations and a current number of iteration, respectively [17].

Table 1. The common parameters for each simulation.

Parameters	Values
Distribution of mesh clients	Stadium
Number of mesh clients	48
Number of mesh routers	16, 17
Radius of a mesh router	2.0–3.5
Number of GA Islands	16
Number of migrations	200
Evolution steps	9
Selection method	Random method
Crossover method	UNDX
Mutation method	Uniform mutation
Crossover rate	0.8
Mutation rate	0.2
Replacement method	RDVM
Area Size	32.0×32.0

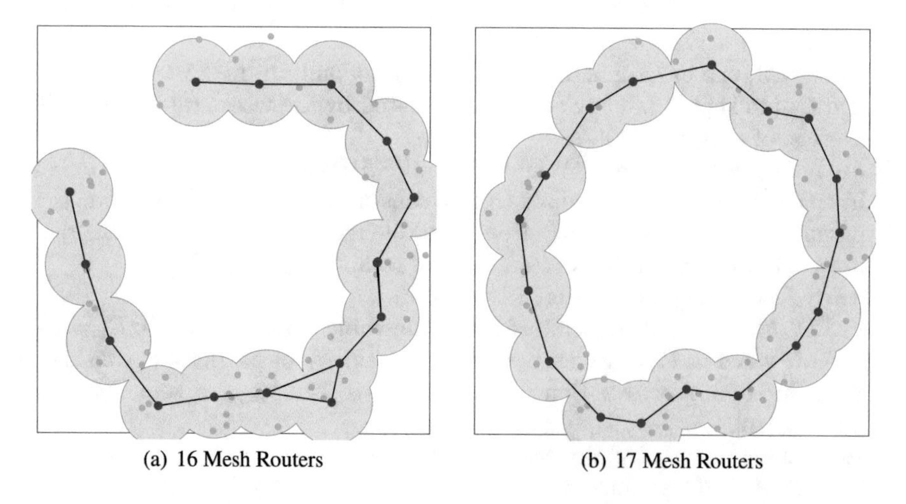

(a) 16 Mesh Routers

(b) 17 Mesh Routers

Fig. 4. Visualization results after the optimization.

4 Simulation Results

In this section we present and compare the simulation results of Stadium distribution of mesh clients for 16 and 17 mesh routers considering RDVM as a router replacement method. The weight-coefficients of fitness function were adjusted for optimization. In this paper, the weight-coefficients are $\alpha = 0.8$, $\beta = 0.1$, $\gamma = 0.1$. The number of mesh clients is 48, whereas selection, crossover and mutation methods are Random, UNDX and Uniform, respectively. Table 1 summarizes

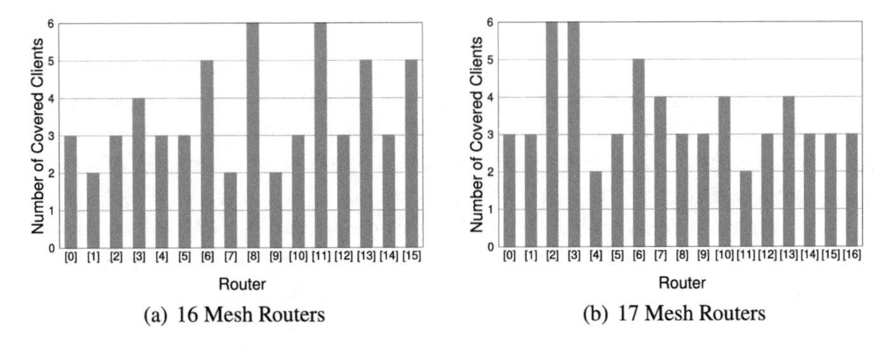

Fig. 5. Number of covered clients by each router after the optimization.

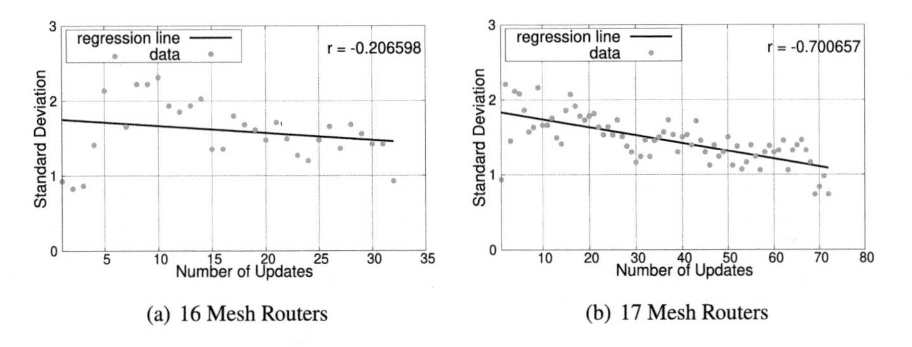

Fig. 6. Transition of the standard deviations.

the common parameters used for simulations. Figure 4 shows the visualization results after the optimization, Fig. 5 the number of covered mesh clients by each route, whereas Fig. 6 the standard deviation where r is the correlation coefficient.

As shown in Fig. 4(a), 16 mesh routers are not enough to cover all mesh clients. On the other hand, in Fig. 4(b), the simulation results show that 17 mesh routers can cover all mesh clients and enable better connectivity by forming a ring topology among mesh routers. In Fig. 5(a), Fig. 5(b), Fig. 6(a) and Fig. 6(b), we see the results in terms of load balancing. We can see which case has better results by comparing their standard deviations and their correlation coefficients. When the standard deviation is an increasing line ($r > 0$), the number of mesh clients for each router tends to be different. On the other hand, when the standard deviation is a decreasing line ($r < 0$), the number of mesh clients for each router tends to go close to each other. The standard deviation is a decreasing line in both cases, but a better load balancing is achieved when 17 mesh routers are used.

5 Conclusions

In this work, we evaluated the performance of WMNs using a hybrid simulation system based on PSO and DGA (called WMN-PSODGA). We compared the simulation results of Stadium distribution of mesh clients and RDVM router replacement method for 16 and 17 mesh routers.

The simulation results show that 17 mesh routers enable full client coverage, better connectivity and improved load balance. Good connectivity and load balancing are achieved for 16 mesh routers, too, but not all mesh clients are covered in this case.

In future work, we will consider other distributions of mesh clients and other router replacement methods.

References

1. Akyildiz, I.F., Wang, X., Wang, W.: Wireless mesh networks: a survey. Comput. Netw. **47**(4), 445–487 (2005)
2. Barolli, A., Sakamoto, S., Barolli, L., Takizawa, M.: Performance analysis of simulation system based on particle swarm optimization and distributed genetic algorithm for WMNs considering different distributions of mesh clients. In: International Conference on Innovative Mobile and Internet Services in Ubiquitous Computing, pp 32–45. Springer, Cham (2018)
3. Barolli, A., Sakamoto, S., Ozera, K., Barolli, L., Kulla, E., Takizawa, M.: Design and implementation of a hybrid intelligent system based on particle swarm optimization and distributed genetic algorithm. In: Barolli, L., Xhafa, F., Javaid, N., Spaho, E., Kolici, V. (eds.) Advances in Internet, Data & Web Technologies. EIDWT 2018. Lecture Notes on Data Engineering and Communications Technologies, vol 17, pp 79–93. Springer, Cham (2018). https://doi.org/10.1007/978-3-319-75928-9_7
4. Clerc, M., Kennedy, J.: The particle swarm-explosion, stability, and convergence in a multidimensional complex space. IEEE Trans. Evol. Comput. **6**(1), 58–73 (2002)
5. Franklin, A.A., Murthy, C.S.R.: Node placement algorithm for deployment of two-tier wireless mesh networks. In: Proceedings of Global Telecommunications Conference, pp. 4823–4827 (2007)
6. Girgis, M.R., Mahmoud, T.M., Abdullatif, B.A., Rabie, A.M.: Solving the wireless mesh network design problem using genetic algorithm and simulated annealing optimization methods. Int. J. Comput. Appl. **96**(11), 1–10 (2014)
7. Goto, K., Sasaki, Y., Hara, T., Nishio, S.: Data gathering using mobile agents for reducing traffic in dense mobile wireless sensor networks. Mob. Inf. Syst. **9**(4), 295–314 (2013)
8. Lim, A., Rodrigues, B., Wang, F., Xu, Z.: k-center problems with minimum coverage. Theoret. Comput. Sci. **332**(1–3), 1–17 (2005)
9. Maolin, T., et al.: Gateways placement in backbone wireless mesh networks. Int. J. Commun. Netw. Syst. Sci. **2**(1), 44–50 (2009)
10. Matsuo, K., Sakamoto, S., Oda, T., Barolli, A., Ikeda, M., Barolli, L.: Performance analysis of WMNs by WMN-GA simulation system for two WMN architectures and different TCP congestion-avoidance algorithms and client distributions. Int. J. Commun. Networks Distr. Syst. **20**(3), 335–351 (2018)

11. Muthaiah SN, Rosenberg CP (2008) Single Gateway Placement in Wireless Mesh Networks. Proc of 8th International IEEE Symposium on Computer Networks pp 4754–4759
12. Naka, S., Genji, T., Yura, T., Fukuyama, Y.: A hybrid particle swarm optimization for distribution state estimation. IEEE Trans. Power Syst. **18**(1), 60–68 (2003)
13. Poli, R., Kennedy, J., Blackwell, T.: Particle swarm optimization. Swarm Intell. **1**(1), 33–57 (2007)
14. Sakamoto, S., Kulla, E., Oda, T., Ikeda, M., Barolli, L., Xhafa, F.: A comparison study of simulated annealing and genetic algorithm for node placement problem in wireless mesh networks. J. Mob. Multimed. **9**(1–2), 101–110 (2013)
15. Sakamoto, S., Kulla, E., Oda, T., Ikeda, M., Barolli, L., Xhafa, F.: A comparison study of hill climbing, simulated annealing and genetic algorithm for node placement problem in WMNs. J. High Speed Networks **20**(1), 55–66 (2014)
16. Sakamoto, S., Oda, T., Ikeda, M., Barolli, L., Xhafa, F.: Implementation and evaluation of a simulation system based on particle swarm optimisation for node placement problem in wireless mesh networks. Int. J. Commun. Networks Distr. Syst. **17**(1), 1–13 (2016)
17. Sakamoto, S., Oda, T., Ikeda, M., Barolli, L., Xhafa, F.: Implementation of a new replacement method in WMN-PSO simulation system and its performance evaluation. In: The 30th IEEE International Conference on Advanced Information Networking and Applications (AINA-2016), pp 206–211 (2016)
18. Sakamoto, S., Ozera, K., Ikeda, M., Barolli, L.: Implementation of intelligent hybrid systems for node placement problem in WMNs considering particle swarm optimization, hill climbing and simulated annealing. Mobile Networks Appl. **23**(1), 27–33 (2017). https://doi.org/10.1007/s11036-017-0897-7
19. Schutte, J.F., Groenwold, A.A.: A study of global optimization using particle swarms. J. Global Optim. **31**(1), 93–108 (2005)
20. Shi, Y.: Particle swarm optimization. IEEE Connections **2**(1), 8–13 (2004)
21. Shi, Y., Eberhart, R.C.: Parameter selection in particle swarm optimization. In: Evolutionary Programming VII, pp. 591–600 (1998)
22. Vanhatupa, T., Hannikainen, M., Hamalainen, T.: Genetic algorithm to optimize node placement and configuration for WLAN planning. In: Proceedings of the 4th IEEE International Symposium on Wireless Communication Systems, pp 612–616 (2007)
23. Wang, J., Xie, B., Cai, K., Agrawal, D.P.: Efficient mesh router placement in wireless mesh networks. In: Proceedings of IEEE International Conference on Mobile Adhoc and Sensor Systems (MASS-2007), pp. 1–9 (2007)

Author Index

L. Barolli et al. (Eds.): AINA 2022, LNNS 449, pp. 687–689, 2022.
https://doi.org/10.1007/978-3-030-99584-3

Printed in the United States
by Baker & Taylor Publisher Services